INDUSTRIAL ORGANIC CHEMICALS

Starting Materials and Intermediates

VOLUME 8

Weinheim · New York · Chichester · Brisbane · Singapore · Toronto

INDUSTRIAL ORGANIC CHEMICALS

VOLUME 1
Acetaldehyde to **Aniline**

VOLUME 2
Anthracene to **Cellulose Ethers**

VOLUME 3
Chlorinated Hydrocarbons to **Dicarboxylic Acids, Aliphatic**

VOLUME 4
Dimethyl Ether to **Fatty Acids**

VOLUME 5
Fatty Alcohols to **Melamine and Guanamines**

VOLUME 6
Mercaptoacetic Acid and Derivatives to **Phosphorus Compounds, Organic**

VOLUME 7
Phthalic Acid and Derivatives to **Sulfones and Sulfoxides**

VOLUME 8
Sulfonic Acids, Aliphatic to **Xylidines**

Index

INDUSTRIAL ORGANIC CHEMICALS

Starting Materials and Intermediates

VOLUME 8

Sulfonic Acids, Aliphatic to Xylidines

Index

Weinheim · New York · Chichester · Brisbane · Singapore · Toronto

This book was carefully produced. Nevertheless, authors and publisher do not warrant the information contained therein to be free of errors. Readers are advised to keep in mind that statements, data, illustrations, procedural details or other items may inadvertently be inaccurate.

Library of Congress Card No.: Applied for.
British Library Cataloguing-in-Publication Data: A catalogue record for this book is available from the British Library.

Die Deutsche Bibliothek – CIP-Einheitsaufnahme
Industrial organic chemicals : starting materials and intermediates ;
an Ullmann's encyclopedia. – Weinheim ; New York ;
Chichester ; Brisbane ; Singapore ; Toronto : Wiley-VCH
   ISBN 3-527-29645-X
Vol. 8. Sulfonic Acids, Aliphatic to Xylidines; Index. – 1. Aufl. – 1999.

© WILEY-VCH Verlag GmbH, D-69469 Weinheim (Federal Republic of Germany), 1999
Printed on acid-free and chlorine-free paper.
All rights reserved (including those of translation in other languages). No part of this book may be reproduced in any form – by photoprinting, microfilm, or any other means – nor transmitted or translated into machine language without written permission from the publishers. Registered names, trademarks, etc. used in this book, even when not specifically marked as such, are not to be considered unprotected by law.

Composition and Printing: Rombach GmbH, Druck- und Verlagshaus, D-79115 Freiburg
Bookbinding: Wilhelm Osswald & Co., D-67433 Neustadt (Weinstraße)
Cover design: mmad, Michel Meyer, D-69469 Weinheim
Printed in the Federal Republic of Germany

Contents

1 Sulfonic Acids, Aliphatic

1. Properties and Uses 4503
2. Production 4506
3. References................. 4508

2 Tar and Pitch

1. Origin, Classification, and Industrial Importance of Tars and Pitches ... 4509
2. Properties................. 4512
3. Processing of Coke-Oven Coal Tar.. 4519
4. Processing of Low-Temperature Coal Tars 4547
5. Processing of Other Tars and Tarlike Raw Materials 4550
6. Uses of Tar Products and their Economic Importance 4552
7. Toxicology and Ecotoxicology 4553
8. References................. 4556

3 Tartaric Acid

1. Introduction 4559
2. Physical Properties........... 4560
3. Optical Isomerism: Configuration .. 4562
4. Chemical Properties 4562
5. Resources and Starting Materials .. 4563
6. Production 4563
7. Environmental Protection 4565
8. Quality Specifications 4566
9. Chemical Analysis 4566
10. Derivatives of L(+)-Tartaric Acid ... 4567
11. Uses..................... 4568
12. Economic Aspects............ 4569
13. Physiological and Toxicological Behavior.................. 4569
14. References................. 4570

4 Terephthalic Acid, Dimethyl Terephthalate, and Isophthalic Acid

1. Introduction 4573
2. Physical Properties........... 4574
3. Production 4576
4. Environmental Protection 4585
5. Quality Specifications 4586
6. Storage and Transportation 4587
7. Uses..................... 4587
8. Economic Aspects............ 4588
9. Toxicology and Occupational Health 4589
10. References................. 4590

5 Terpenes

1. General 4593
2. Acyclic Monoterpenes.......... 4598
3. Monocyclic Monoterpenes....... 4601
4. Bicyclic Monoterpenes.......... 4604
5. Acyclic Sesquiterpenes.......... 4609
6. Monocyclic Sesquiterpenes....... 4610
7. Bicyclic Sesquiterpenes 4611
8. Tricyclic Sesquiterpenes 4612
9. Acyclic Diterpenes 4613
10. Acyclic Triterpenes 4613
11. Toxicology................ 4614
12. References................. 4616

V

6 Tetrahydrofuran

1. Introduction 4621
2. Physical Properties 4622
3. Chemical Properties 4622
4. Production 4623
5. Specifications and Analysis 4626
6. Environmental Protection and Recovery 4626
7. Storage and Transport 4627
8. Uses 4627
9. Economic Aspects 4628
10. Toxicology and Industrial Hygiene .. 4628
11. References 4629

7 Thiocyanates and Isothiocyanates, Organic

1. Organic Thiocyanates 4631
2. Organic Isothiocyanates 4635
3. References 4644

8 Thiols and Organic Sulfides

1. Aliphatic Thiols, Sulfides, Disulfides, and Polysulfides 4648
2. Aromatic Thiols, Sulfides, Disulfides, and Polysulfides 4662
3. Heterocyclic Thiols, Sulfides, and Disulfides 4667
4. Toxicology 4677
5. References 4679

9 Thiophene

1. Introduction 4691
2. Physical and Chemical Properties .. 4692
3. Resources and Raw Materials 4694
4. Production 4694
5. Specifications, Transportation, Health and Safety 4695
6. Uses 4696
7. Toxicology 4705
8. References 4705

10 Thiourea and Thiourea Derivatives

1. Introduction 4707
2. Thiourea 4708
3. Thiourea Derivatives 4716
4. References 4723

11 Toluene

1. Introduction 4727
2. Properties 4728
3. Separation, Extraction, and Processing 4731
4. Integration into Refineries and Petrochemical Complexes 4734
5. Economic Aspects 4735
6. Quality Requirements 4738
7. Storage, Transport, and Safety 4738
8. Environmental Aspects and Toxicology 4739
9. References 4741

12 Toluidines

1. Introduction 4745
2. Physical and Chemical Properties ... 4746
3. Production 4747
4. Quality Specifications, Storage, and Transportation 4748
5. Uses 4748
6. Derivatives 4749
7. Toluidine Toxicology 4754
8. References 4755

13 Turpentines

1. Turpentine as a Natural Product ... 4759
2. History 4760
3. Definition and Classification 4761
4. Production 4762
5. Composition 4765
6. Physical Properties 4767
7. Chemical Properties 4769
8. Analysis 4769
9. Safety and Occupational Health ... 4770
10. Storage and Transportation 4771
11. Ecological Aspects 4771
12. Uses 4772
13. Economic Aspects 4773
14. Toxicology 4774
15. Turpentine Substitute 4776
16. Pine Oil 4776
17. References 4779

14 Urea

1. Physical Properties 4784
2. Chemical Properties 4786
3. Production 4787
4. Forms Supplied, Storage, and Transportation 4813
5. Quality Specifications and Analysis . 4815
6. Uses 4816
7. Economic Aspects 4817
8. Urea Derivatives 4818
9. References 4833

15 Vinyl Esters

1. Introduction 4839
2. Vinyl Acetate 4840
3. Vinyl Propionate 4855
4. Vinyl Esters of Higher Carboxylic Acids 4857
5. References 4861

16 Vinyl Ethers

1. Introduction 4865
2. Properties 4866
3. Production 4866
4. Environmental Protection, Waste Disposal 4872
5. Quality Specifications 4873
6. Chemical Analysis 4873
7. Storage and Transport 4873
8. Use 4874
9. Toxicology 4874
10. References 4875

VII

17 Waxes

1. Introduction 4878
2. Recent Natural Waxes 4886
3. Montan Wax 4907
4. Petroleum Waxes 4914
5. Fischer–Tropsch Paraffins 4942
6. Polyolefin Waxes 4944
7. Toxicology 4959
8. References 4963

18 Xanthates

1. Introduction 4971
2. Physical Properties 4971
3. Chemical Properties 4973
4. Production 4976
5. Quality Specifications 4978
6. Storage and Transportation 4978
7. Analysis 4978
8. Uses 4979
9. Economic Aspects 4980
10. Toxicology and Occupational Health ... 4981
11. References 4981

19 Xylenes

1. Introduction 4987
2. Properties 4990
3. Occurrence and Raw Materials .. 4994
4. Production, Separation, and Further Processing 4996
5. Integration into Refinery and Petrochemical Complexes 5005
6. Economic Aspects 5006
7. Quality Specifications and Analysis . 5010
8. Storage, Transport, and Safety 5012
9. Environmental Aspects and Toxicology 5013
10. References 5014

20 Xylidines

1. Introduction 5019
2. Physical Properties 5020
3. Chemical Properties 5020
4. Industrial Production 5021
5. Quality Specifications, Storage, and Transport 5022
6. Use 5023
7. Toxicology 5025
8. References 5026

Index

1. Author Index 5029
2. CAS Registry Number Index 5039
3. Subject Index 5077

Sulfonic Acids, Aliphatic

Kurt Kosswig, Hüls Aktiengesellschaft, Marl, Federal Republic of Germany

1. Properties and Uses 4503
2. Production 4506
3. References 4508

1. Properties and Uses

Alkanesulfonic Acids. In alkanesulfonic acids RSO_3H, a hydroxysulfonyl group is bonded via the sulfur atom to a carbon atom of an alkyl group. Alkanesulfonic acids are strong acids, and the lower members up to heptane-1-sulfonic acid are completely dissociated in dilute aqueous solution. Estimated pK values in aqueous solution of some other strong acids are compared with those of methane-, ethane-, and propanesulfonic acids below [5]:

$C_3H_7SO_3H$	−1.5
$C_2H_5SO_3H$	−1.7
CH_3SO_3	−1.9
p-$CH_3C_4H_4SO_3H$	−2.7
$C_6H_5SO_3H$	−2.8
$HOSO_3H$	−3.0
CH_3OSO_3H	−3.5
p-$O_2NC_4H_4SO_3H$	−3.8
CF_3SO_3H	−5.5
FSO_3H	−5.6

Methanedisulfonic acid (methionic acid) [503-40-2], mp 96–100 °C, has the same strength as sulfuric acid.

From heptane-1-sulfonic acid onwards the sulfonic acids increasingly associate in aqueous solution, forming micelles [6]. Pure alkanesulfonic acids attack stainless steel, but the presence of 0.5 to 1% sulfuric acid hinders corrosion.

The physical properties of some n-alkanesulfonic acids are listed in Table 1. Alkanesulfonic acids decompose on distillation at normal pressure, but can be distilled without decomposition under vacuum.

Alkanesulfonic acids darken when stored in the absence of light. They form hydrates, are hygroscopic, and dissolve readily in water. The stable alkali metal, ammonium, alkaline-earth metal, and lead salts also have high water solubility and can be recrystallized from alcohol.

Table 1. Physical data of some alkane-1-sulfonic acids

Acid	CAS no.	bp/hPa, °C	mp, °C	d_4^{25}
Methanesulfonic acid	[75-75-2]	122/1.3	20	1.4844
Ethanesulfonic acid	[594-45-6]	123/1.3	−17	1.3341
Propane-1-sulfonic acid	[5284-66-2]	136/1.3	7.5	1.2516
Butane-1-sulfonic acid	[2386-47-2]	147/0.9	−15.2	1.1906
Pentane-1-sulfonic acid	[35452-30-3]	163/1.3	15.9	1.1226
Hexane-1-sulfonic acid	[13595-73-8]	174/1.3	16.1	1.1047

Table 2. Boiling points of some perfluoroalkanesulfonic acids at normal pressure

	CAS no.	bp, °C
Perfluoromethanesulfonic acid	[1493-13-6]	162
Perfluoroethanesulfonic acid	[354-88-1]	178
Perfluorompropane-1-sulfonic acid	[423-41-6]	196
Perfluorobutane-1-sulfonic acid	[375-73-5]	210–212
Perfluoropentane-1-sulfonic acid	[2706-91-4]	224–226
Perfluorohexane-1-sulfonic acid	[355-46-4]	238–239

Alkanesulfonic acids with nonterminal sulfonic acid groups are less stable than the alkane-1-sulfonic acids, in particular 2-methylpropane-2-sulfonic acid [16794-13-1], mp 113 °C [7].

Lower molecular mass alkanesulfonic acids, especially methanesulfonic acid, being strong, nonoxidizing acids, are used as catalysts for esterification, polymerization, and alkylation, and for the production of peroxycarboxylic acids from hydrogen peroxide and carboxylic acids [8]. Their esters are used as alkylating agents. Short-chain alkanesulfonic acids such as butanesulfonic acid, have been proposed as components of industrial cleaning agents, as they can remove both organic and mineral contamination [9].

Perfluorinated Alkanesulfonic Acids [10], [11]. Perfluoroalkanesulfonic acids, which can be obtained by electrochemical fluorination of alkanesulfonic acids, followed by hydrolysis of the resulting sulfonyl fluorides, are the strongest organic acids and are even stronger than sulfuric acid. Perfluoroalkanesulfonic acids are thermally very stable and can be distilled at normal pressure (Table 2). They form crystalline hydrates.

Perfluoroalkanesulfonate anions are good leaving groups [12]. Esters of perfluoroalkanesulfonic acids are therefore used as strong alkylating agents in preparative chemistry. Trifluoromethanesulfonates are known as triflates and nonafluorobutanesulfonates as nonaflates.

Hydroxyalkanesulfonic Acids. Some hydroxyalkanesulfonic acids and their derivatives are important in industry.

The sodium salts of 1-hydroxyalkane-1-sulfonic acids are readily crystallizable compounds which are formed from aldehydes and sodium bisulfite in neutral solution. In an acidic or alkaline medium they decompose into their components. They are used for the sulfonalkylation of compounds with an active hydrogen atom, in particular sodium hydroxymethanesulfonate (from formalin and sodium bisulfite), for example:

$$RNH_2 + CH_2O + NaHSO_3 \longrightarrow RNH-CH_2SO_3Na + H_2O$$

Isethionic acid, 2-hydroxyethanesulfonic acid [107-36-8], is used as an intermediate in numerous syntheses. It is generally used in the form of its readily accessible stable sodium salt [1562-00-1], mp 192 – 194 °C, for example:

$$RCOCl + HOCH_2CH_2SO_3Na \longrightarrow RCOOCH_2CH_2SO_3Na + HCl$$

Isethionic acid is the starting material for the industrial synthesis of taurine (2-aminoethanesulfonic acid [107-35-7]), mp 317 °C dec., and N-methyltaurine (2-N-methylaminoethanesulfonic acid [107-68-6]), mp 245 °C. Taurine is a naturally occurring substance, formed in organisms by oxidation of cysteine, and is a building block of taurocholic acid. The derivatives of isethionic acid, taurine, and N-methyltaurine, especially their fatty acid derivatives, are important wetting, washing, and dispersion agents in body and household cleaning agents, and also in textile processing.

1,3-Propanesultone [1120-71-4] [13], mp 31 °C, bp 150 °C (13 hPa), d_4^{40} 1.392, the internal anhydride of 3-hydroxypropane-1-sulfonic acid [15909-83-3] is a strong alkylating agent, for example:

$$R_3N + \underset{\underset{SO_2}{O\diagdown\diagup CH_2}}{CH_2-CH_2} \longrightarrow R_3\overset{+}{N}-CH_2CH_2CH_2SO_3^-$$

Reaction products of propanesultone with alcohols, phenols, thiols, sulfides, and amides have been proposed as tensides, leather auxiliaries, antistatic agents, and bacteriostatic agents. Cellulose becomes soluble when treated with propanesultone. Polyamide fibers treated with propanesultone can be dyed more easily. However, since propanesultone has been shown to be a strong carcinogen, it is no longer important as a synthon.

Vinylsulfonic acid, ethenesulfonic acid [1184-84-5] [14], bp 95 °C (1 Pa), is derived from isethionic acid and can only be kept below 0 °C, as it polymerizes rapidly. It is preferable to use its sodium salt in aqueous solution or its esters [e.g., methyl ester, bp 54 – 60 °C (30 – 50 Pa), n_D^{20} 1.4310]. The sodium salt [3039-83-6] or the esters of vinylsulfonic acid add to compounds with activated hydrogens (sulfonethylation) for example:

$$NaHSO_3 + CH_2=CHSO_2OR \longrightarrow NaO_3SCH_2CH_2SO_2OR$$

The addition products with alcohols, phenols, thiols, and amines have been proposed as plasticizers for poly(vinyl chloride), as cross-linking agents, emulsifiers, microbicides, fungicides, and fungistatic agents. Vinylsulfonic acid and its salts are also suitable monomers for polymerization reactions [15], [16]. Esters of vinylsulfonic acid are powerful alkylating agents: phenol is converted smoothly into anisole with methyl vinylsulfonate in alkaline solution, even at room temperature.

Allylsulfonic acid, 2-propene-1-sulfonic acid [*1606-80-8*], *bp* 180 °C (0.1 Pa), is unstable and is preferably used in the form of its sodium salt [*2495-39-8*] in aqueous solution as a comonomer for polymerizations [15].

Methallylsulfonic acid, 2-methyl-1-propene-3-sulfonic acid [*3934-16-5*], is also predominantly used in the form of its aqueous sodium salt [*1561-92-8*] as a comonomer in polymerizations. It improves, for example, the dyeability of fibers, especially polyacrylonitrile fibers.

2. Production

Alkanesulfonic Acids. Alkanes do not react specifically with sulfur trioxide or oleum, alkanesulfonic acids being among a number of products. Methanesulfonic acid and sulfur trioxide give methionic acid as the only product in good yields. Starting from alkanes, alkanesulfonic acids are smoothly obtained by sulfoxidation or sulfochlorination [17]. However, mixtures of positional isomers are formed in these reactions.

Uniform, defined alkanesulfonic acids are obtained by Strecker synthesis from alkyl halides, preferably alkyl bromides, and alkali metal or ammonium sulfite, or by reaction of the corresponding Grignard reagent with sulfur dioxide and subsequent oxidation. The oxidation of thiols with nitric acid, potassium permanganate, or hydrogen peroxide/glacial acetic acid is frequently used, as is the oxidation of thiocyanates and disulfides, all of which are obtained from alkyl halides.

Starting from 1-alkenes the sodium salts of alkane-1-sulfonic acids are obtained by the addition of sodium bisulfite in a radical-initiated reaction [9], [18].

2-Hydroxyalkane-1-sulfonic acids, especially isethionic acid, are produced by the reaction of 1,2-epoxides (e.g., ethylene oxide) with aqueous sodium bisulfite. The reaction of ethylene with sulfur trioxide gives carbyl sulfate (*mp* 107–108 °C), which reacts with water to give ethionic acid [*461-42-7*] (sulfuric acid half ester of 2-hydroxyethane-1-sulfonic acid), and on heating to give isethionic acid [*107-36-8*] [19]:

$$CH_2=CH_2 + 2\,SO_3 \longrightarrow$$

$$\begin{array}{c} CH_2-CH_2 \\ |\quad\quad\; | \\ O\quad\; SO_2 \\ |\quad\quad\; | \\ SO_2-O \\ \text{Carbyl sulfate} \end{array} \left[\begin{array}{l} \xrightarrow[\text{cold}]{+H_2O} HO_3S-O-CH_2CH_2-SO_3H \\ \quad\quad\quad\quad\quad \text{Ethionic acid} \\ \\ \xrightarrow[\text{hot}]{+2\,H_2O} HO-CH_2CH_2-SO_3H + H_2SO_4 \\ \quad\quad\quad\quad\quad \text{Isethionic acid} \end{array} \right.$$

Ethionic acid is also obtained by the reaction of sulfur trioxide with ethanol.

The methyl-substituted carbyl sulfate, obtained analogously from propene and sulfur trioxide, gives 1-propene-1-sulfonic acid [*97141-05-4*] on decomposition.

1,3-Propanesultone is obtained by elimination of water from 3-hydroxypropane-1-sulfonic acid [20], whose sodium salt is formed by radical-initiated addition of sodium bisulfite to allyl alcohol [21]:

$$HOCH_2CH=CH_2 + NaHSO_3 \longrightarrow HOCH_2CH_2CH_2SO_3Na$$

3-Hydroxypropane-1-sulfonic acid is obtained on acidification.

Vinylsulfonic Acid. The alkaline hydrolysis of carbyl sulfate or ethionic acid gives the vinylsulfonate salt directly:

$$\begin{array}{c} CH_2-CH_2 \\ |\quad\quad\; | \\ O\quad\; SO_2 \\ |\quad\quad\; | \\ SO_2-O \end{array} + 3\,NaOH \longrightarrow$$

$$CH_2=CH-SO_3Na + Na_2SO_4 + 2\,H_2O$$

Sodium vinylsulfonate can also be produced by dehydrochlorination of sodium 2-chloroethane-1-sulfonate with aqueous sodium hydroxide. The chloroethanesulfonate is produced by sulfochlorination of ethyl chloride and subsequent hydrolysis of the 2-chloroethane-1-sulfonyl chloride, or by treatment of sodium bisulfite with vinyl chloride [14].

Allylsulfonic Acid. Only the sodium salt of allylsulfonic acid is used in industry. It is obtained by treatment of allyl chloride with aqueous sodium bisulfite [22].

Methallylsulfonic Acid. Dimethylcarbyl sulfate cannot be obtained from isobutene and sulfur trioxide, as the methallylsulfonic acid reacts further to give methallyldisulfonic acid:

$$CH_2=\underset{CH_3}{C}-CH_3 \xrightarrow{SO_3} CH_2=\underset{CH_3}{C}-CH_2-SO_3H$$

$$\xrightarrow{SO_3} CH_2=\underset{CH_2-SO_3H}{C}-CH_2-SO_3H$$

The formation of the latter can be suppressed by complexation of the sulfur trioxide, for example, with N,N-dialkylcarboxylic acid amides [23]. However, in industry methallylsulfonate is preferably produced from methallyl chloride and sodium sulfite [24].

3. References

General References

[1] *Beilstein,* 324–329.
[2] *Houben-Weyl,* **9,** 343–427; E **11,** 1055 ff.
[3] *Kirk-Othmer,* **22,** 45 ff.
[4] S. Patai, Z. Rapoport: *The Chemistry of Sulphonic Acids, Esters and their Derivatives,* J. Wiley & Sons, Chichester 1991.

Specific References

[5] J. F King in [4], p. 251.
[6] E. C. McBain, W. B. Dye, S. A. Jonston, *J. Am. Chem. Soc.* **61** (1939) 3210.
[7] F. Asinger, P. Laue, B. Fell, G. Gubelt, *Chem. Ber.* **100** (1967) 1696.
[8] L. S. Silbert, E. Siegel, D. Swern, *J. Org. Chem.* **27** (1962) 1336.
[9] A. Behr, N. Döring, *Fat Sci. Technol.* **92** (1990) 335.
[10] H. C. Fielding in R. E. Banks (ed.): *Organofluorine Chemicals and their Industrial Applications,* Ellis Harwood Ltd., Chichester 1979, pp. 214–234.
[11] W.-Y. Huang, Q.-Y. Chen in [4], pp. 903–946.
[12] P. J. Stang, M. Hanack, R. Subramanian, *Synthesis* **1982,** 85.
[13] R. F. Fischer, *Ind. Eng. Chem.* **56** (1964) no. 3, 41.
[14] H. Distler, *Angew. Chem.* **77** (1965) 291.
[15] D. S. Breslow, *Encycl. Polym. Sci. Eng.* **6** (1986) 564.
[16] D. A. Kanglas in R. H. Yocum, E. B. Nyquist (eds): *Funtional Monomers,* vol. **1,** Marcel Dekker, New York 1974, pp. 489–640.
[17] F. Asinger: *Die Petrolchemische Industrie,* Akademie Verlag, Berlin 1971, pp. 731–761.
[18] F. Asinger, in [17], p. 757.
[19] G. H. Weinreich, M. Jufresa, *Bull. Soc. Chim. Fr. Mem.* **5** (1965) 787.
[20] D. K. Tolopko, E. S. Pugasch, O. E. Baranovskaja, O. I. Romanyuk, *Visn. L'viv Politekh Inst.* **181** (1984) 137; *Chem. Abstr.* **101** (1984) 90801 t.
[21] *Chem. Abstr.* **104** (1986) 129455 x. O. E. Baranovskaya, A. Yu. Shuter, L. Ratkovskaya, in [20]191, (1985) 116.
[22] R. Buening et al., DD 282 819, 1988; *Chem. Abstr.* **114** (1991) 186306 e.
[23] Lion Corp., JP-Kokai 63 51 368, 1986 (O. Hizawa, M. Yoshida, M. Takahashi); *Chem. Abstr.* **109** (1988) 169858 r.
[24] Hüls AG, DE 3 337 103, 1983 (D. J. Mueller, S. Austin); *Chem. Abstr.* **103** (1985) 106612 m.

Tar and Pitch

GERD COLLIN, DECHEMA e.V., Frankfurt/Main, Federal Republic of Germany (Chaps. 1–6)
HARTMUT HÖKE, Weyl GmbH, Mannheim, Federal Republic of Germany (Chap. 7)

1.	Origin, Classification, and Industrial Importance of Tars and Pitches	4509	3.4.2. Processing of Wash Oil	4539
			3.4.3. Processing of Naphthalene Oil	4539
			3.4.4. Processing of Carbolic Oil	4539
2.	Properties	4512	3.4.5. Processing of Light Oil	4544
3.	Processing of Coke-Oven Coal Tar	4519	4. Processing of Low-Temperature Coal Tars	4547
3.1.	Survey	4519	5. Processing of Other Tars and Tarlike Raw Materials	4550
3.2.	Primary Distillation	4520		
3.3.	Processing of Coal-Tar Pitch	4525	6. Uses of Tar Products and their Economic Importance	4552
3.3.1.	Cooling	4525		
3.3.2.	Production of Electrode Pitch	4526	7. Toxicology and Ecotoxicology	4553
3.3.3.	Production of Pitch Coke	4529	7.1. Toxicology	4553
3.3.4.	Production of Special Pitches	4532	7.2. Classification and Legislation	4555
3.3.5.	Pitch–Oil Blends	4535		
3.4.	Processing of Tar Distillates	4536	7.3. Ecotoxicology	4555
3.4.1.	Processing of Anthracene Oil	4536	8. References	4556

1. Origin, Classification, and Industrial Importance of Tars and Pitches

Origin and Classification. *Tars* are liquid or semisolid products obtained by thermal decomposition of natural organic materials. *Pitches* are residues of tar distillation or meltable residues directly obtained by distillation of organic substances. Tars and pitches take their names from the source material. Tars are produced in industrial quantities by low-temperature carbonization, coking, or gasification of fossil raw materials or biomass, e.g., coal, lignite, peat, wood, and oil shale. In addition, tars result as higher boiling liquid pyrolysis products from thermal decomposition of organic secondary raw materials and waste products, e.g., pyrolysis residual oils as byproducts of steam cracking of petroleum fractions to alkenes and from pyrolysis of refuse and plastic waste.

Properties and composition of tars and pitches depend not only on the raw material but also on the temperature conditions during thermal treatment. Thus low-temperature tars are produced during low-temperature carbonization or partial gasification below 700 °C. High-temperature tars are generated by coking between 900 and 1300 °C. The differences in the properties and composition of low- and high-temperature tars are especially marked in case of coal tars. Low-temperature coal tars contain phenols, hydrogenated aromatics, aliphatic hydrocarbons, and alkenes as their main components, together with small amounts of aromatic hydrocarbons. These compounds are generated from two sources: (1) Coal bitumen, the extractable component of coal, which consists of oxygen-, sulfur-, and nitrogen-containing hydrocarbon compounds, and (2) the bitumen which arises additionally from the insoluble components by cracking of aliphatic C–C, C–O, C–N, and C–S bonds, as well as through the removal of side chains. High-temperature coking of coal generates thermally stable aromatics, which are characteristic of high-temperature coal tars. These heat-stable aromatics arise via the following reaction schemes: Paraffins are cracked pyrolytically and dehydrogenated. The resulting alkenes are converted first into hydrogenated aromatics and finally into aromatics. Alkylphenols are dealkylated and reduced to aromatic hydrocarbons.

Lignite tars differ from low-temperature coal tars mainly in an increased content of solid paraffins and a lower amount of cyclic hydrocarbons. The main source of paraffins is the extractable montan wax occurring in lignite. The tars are formed by thermal decomposition of initially present and newly formed "bitumen," similar to the case of coal. Tar obtained by high-temperature coking of lignite is similar to low-temperature lignite tar, as lignite bitumen, in contrast to coal bitumen, is already entirely decomposed and volatilized at lower temperatures.

Peat tar properties are determined by the structure of peat, which is a conglomerate of extractable bitumen, humic acid and its salts, and partially decomposed vegetation. In pyrolytic distillation "bitumen" is the source of the bulk of the tar, mainly paraffins. Ketones, methanol, monocarboxylic acids, and phenols arise from the plant components cellulose and lignin.

Wood tars are formed by thermal decomposition of the wood components cellulose, hemicelluloses, and lignin. This gives rise to acetic acid, methanol, ketones and phenols but, in contrast to tars from fossil raw materials, almost no hydrocarbons.

History. The discovery of *coal tar* goes back to J. R. Glauber, who in 1658 described the formation of "black oleum" during coal retorting. A patent for the production of pitch and tar by coking of coal with condensation of volatiles was granted in England to J. J. Becher and H. Serle in 1681. In the early 1800s coal tar was initially regarded as an annoying byproduct of town-gas and metallurgical-coke production. In 1822 the first industrial tar distilleries were built in Britain. The light oils obtained were used as solvents, the heavy oils as impregnants for building timber and railway sleepers, and the distillation residue, pitch, for the production of carbon black. In 1845 A. W. Hofmann discovered benzene in the light oil used to produce nitrobenzene as a perfume for soaps, named essence of mirbane. H. Garden (1819) had previously discovered naphthalene and J. B. A. Dumas and M. A. Laurent (1832) anthracene in coal tar. With the discovery of synthetic dyes during the second half of the 1800s, demand for these three aromatic hydrocarbons led to coal tar becoming the main source of raw materials for the flourishing organic chemical industry in central Europe. The first German tar distilleries were built in 1842 in Offenbach by E. Sell, and in 1860 in Erkner near Berlin by J. Rütgers, followed by numerous others.

Lignite tar was obtained by J. G. Krünitz in 1788 as "rock oil" when retorting central German "earth coal." In 1830 Karl Freiherr von Reichenbach isolated the main component, paraffin, by distillation. The lamp oil, obtained as a byproduct, replaced rapeseed oil, which had been used up to then for lighting. After 1858 low-temperature lignite-carbonization plants were constructed in central Germany for the production of tar-based products and town gas. During the 1930s central German lignite tar was mainly processed by hydrogenation to produce motor fuels.

Peat and wood were carbonized in early times. The Egyptians used *wood tar* for embalming corpses. Solid paraffin for candle production and liquid paraffin oil for lamp oil were obtained from *peat tar* or lignite tar during the 1800s.

Industrial Importance. Approximately 15×10^6 t/a of *coal tar* is produced worldwide, mainly in the form of high-temperature tar as a byproduct of metallurgical coke production by horizontal chamber coking. Smaller amounts are produced as byproducts of low-temperature carbonization of coal to produce solid smokeless fuel and in pressure gasification of coal by the Lurgi process. Approximately 100 coal-chemical tar refineries process coal tar to give chemical feedstocks and other aromatic products. Industrial countries with major steel production account for most of the capacity: ca. 50% in Europe and the CIS, 20% in North America, and 20% in South-East Asia, especially China and Japan. In the United Kingdom and India, low-temperature tars are obtained as byproducts of low-temperature carbonization of coal to produce smokeless fuels. Lurgi pressure gasification is a preliminary stage of Fischer–Tropsch synthesis in South Africa.

In contrast to coal tars, *lignite tars* are produced only in a few countries, e.g., as a byproduct of pressure gasification of lignite in the United States (North Dakota), the Czech Republic, Germany (Lausitz), Russia, and China. Lignite is carbonized to produce coke on a minor scale, e.g., in Germany (Rhineland and Lausitz) and India. Normally only the phenolic effluents are processed, whereas the lignite tars are gasified or burnt within the process. The total amount of lignite tar arising is ca. 10^6 t/a.

Peat tars and *wood tars* arise as byproducts in the production of activated carbon and reductive carbon by carbonization of peat and wood. The main producing countries are Russia (peat) and Brazil (wood). The total production of these tars is at most ca. 3×10^6 t/a.

Oil-shale tars are produced by low-temperature carbonization and partial gasification of oil shale, mainly in Estonia (Kiviter process). Approximately 10^6 t/a are produced in this way. Other processes involving pyrolysis of oil sand and oil shale are still at the pilot-plant stage.

Pyrolysis residual oils are tarlike byproducts of the steam cracking of naphtha and gas oil to produce alkenes. Approximately 10×10^6 t were produced in 1993. Pyrolysis oils with similar compositions are obtained by thermal decomposition of waste plastics. Industrial large-scale tests are being carried out to develop methods of waste-plastics pyrolysis.

2. Properties

In general, *tars* are dark or black viscous liquids containing small amounts of emulsified water and dispersed high-carbon solid particles (free carbon, tar sediments) entrained during formation. Organic components may crystallize from high-paraffinic tars at lower temperatures.

Pitches are generally dark or black thermoplastic materials with a broad molecular-mass spectrum. By using selective solvents it is possible to separate pitches into fractions with different properties. At temperatures below the softening range most pitches transform into glass-like solids. At higher temperatures polymerization and polycondensation reactions take place, and pitches change partly from liquid crystal intermediate stages into a nonthermoplastic state and finally into coke [5].

The chemical composition of tars and pitches varies widely and depends essentially on the type of raw materials and the temperature conditions during production. In all cases, multicomponent mixtures are produced that contain aromatic and/or aliphatic hydrocarbons as the main components, together with oxygen-, sulfur-, and nitrogen-containing heterocyclic compounds; phenols; amines; and possibly carboxylic acids, ketones, and aliphatic alcohols as minor constituents. It has been estimated that coal tar can contain more than 10 000 different constituents, most of which have boiling points above 500 °C. Multicomponent eutectics can form, especially in the case of condensed aromatics. The liquid-viscous nature of pitches and many tar oils arises from strong mutual melting-point depression of the individual components [6].

The main properties of different tars are listed in Table 1. Given approximately equal boiling range and pitch content, the density and carbon content indicate the degree of aromatization of tars. Coke-oven coal tars with average densities of ca. 1.2 g/cm^3 and carbon contents $>90\%$ contain the highest fraction of aromatic hydrocarbons, followed by coal low-temperature gasification tars with densities of ca. 1.1 g/cm^3 and carbon contents of ca. 85%. Smaller amounts of aromatics are found in tars produced by low-temperature carbonization of coal and lignite having densities of ca. 1 g/cm^3 and carbon contents of max. 85%. Peat and wood tars with carbon contents of only ca. 80% or even $< 70\%$ and varying densities, contain almost no aromatic compounds.

Table 1. Properties and composition of various tars

	Coke-oven tars	Coal tars		Lignite tars	Peat tars	Wood tars	Oil-shale tars	Pyrolysis residual oils from steam cracking of naphtha and gas oil	Pyrolysis residual oils from pyrolysis of plastic waste
		Low-temperature tars	Gasification tars*						
Density at 20 °C, g/cm³	1.14–1.25	0.96–1.05	1.05–1.14	0.95–1.05	0.94–0.98	1.08–1.20	0.90–0.97	0.96–1.10	0.93–1.05
Carbon, %	90–93	83–85	84–86	81–85	78–82	60–65	78–86	89–93,1	88–91
Hydrogen, %	5–6	8–9.5	6–8	9–11	8–10	6–8	9–12	6–9	8–10
Naphthalene, %	5–15	0–2	2–4	0–1	0–1	0–0.5		10–15	4–12
Phenols, %	0.5–5	10–45	15–25	5–30	5–25	20–40	1–30	0–trace	trace
Bases, %	0.2–2	0.5–2	0.5–4	0.1–1.5	0.5–3	trace–0.5	0.2–9	0–trace	0–trace
Solid paraffins, %	0–trace	3–15	0–5	8–20	10–30	trace	0.5–12	0–5	0–trace
Coking residue**, %	10–40	5–15	10–20	4–9	4–10	10–18	1–7	5–20	1–10
Toluene insolubles, %	2–20	0.5–10	3–10	0.2–1.5	2–8	trace–12			
Ash (800 °C), %	< 0.5	< 0.2	< 1.5	< 0.2	< 0.2	< 0.2		trace	trace
Water, %	< 5	< 5	< 5	5	1–10	3–10		< 0.1	< 0.2
Distillation analysis (DIN 1995)								0–1	0.2
Light oil up to 180 °C, %	0.2–2	0–4	0–10	0–2	0–3	10–20	10–15	0–10	40–70
Middle oil 180–230 °C, %	3–10	5–25	5–15	0–4	3–25	10–14	5–20	20–30	8–15
Heavy oil 230–270 °C, %	7–15	10–20	5–20	2–18	5–20	7–12	15–35	20–30	5–15
270–300 °C, %	3–7	8–12	5–10	5–20	10–20	5–10	0–15	10–20	3–8
300 °C up to pitch, %	12–30	15–40	10–30	20–50	20–40	25–40	0–10	2–5	2–10
Pitch, %	45–65	25–40	30–50	20–60	20–35	20–30	30–50	25–30	10–20

* From Lurgi fixed-bed pressure gasification of coal with oxygen and steam.
** Conradson method.

Oil-shale tars are hardly or only partially aromatized. However, a high content of aromatic components can be found in pyrolysis residual oils from ethylene/propene plants and from the thermal decomposition of waste plastics if the pyrolysis temperature is at least ca. 700 °C. These pyrolysis residual oils have an aromatic hydrocarbon content of ca. 90% and higher, similar to coke-oven coal tar, whereas pyrolysis oils contain almost no phenols and only small amounts of heterocyclic compounds due to the raw-material basis. The carbon content falls and the hydrogen content rises with decreasing tar aromatization. A sensitive criterion for the "aromaticity" of tars is the naphthalene content, which is ca. 10% in coke-oven coal tars and pyrolysis residual oils, ca. 3% in coal-gasification tars on average, less than 2% in low-temperature coal-carbonization tars, and max. 1% for lignite, peat, and wood tars. With increasing geological age of the raw material (i.e., wood < peat < lignite < coal), the content of phenols and solid paraffins decrease, while the "aromaticity" increases. Thus coke-oven coal tars contain on average ca. 2% phenols and only traces of solid paraffinic hydrocarbons, whereas low-temperature coal-carbonization tars, lignite tars, and peat tars contain up to ca. 30% phenols and solid paraffins. High coking values and components that are insoluble in toluene normally indicate a high degree of aromatization.

Tar properties are investigated with numerous analytical methods. Table 2 contains the analytical values of a typical coke-oven coal tar. Figure 1 shows a gas chromatogram of a coke-oven tar, and Table 3 lists the most important constituents of coke-oven coal tar. Up to now about 600 compounds have been identified in coal tars; together, they account for ca. 55% of the tar. The remaining, unidentified compounds are mainly pitch components which decompose before they vaporize.

In general, *pitches* are multicomponent mixtures of higher molecular compounds. *Coal-tar pitch* obtained from coke-oven tar is a mixture of condensed aromatic hydrocarbons and heterocyclic compounds. Figure 2 shows a high-pressure liquid chromatogram (HPLC) of coal-tar pitch. HPLC can analyze ca. 70% of coal-tar pitch [7]. Gel permeation chromatographic investigations indicate in case of higher molecular pitch fractions that they have molecular weights up to approximately 2500 [8]. The highest molecular components, which are partly insoluble, even in pitch, have higher molecular masses. The higher molecular fractions consist of condensed aromatic hydrocarbons and heterocyclics with insignificant fractions containing methyl- or hydroxyl-substituted compounds. The ^{13}C NMR spectrum shows that 97% of the carbon atoms present are aromatic [9], [10]. Higher molecular pitch fractions contain monomeric aromatics and heterocyclic aromatics, as well as oligomeric systems, in some of which aromatics containing nitrogen and other heteroelements are linked by C–C single bonds or bridges such as CH_2, O, and S [11], [12]. Very large *peri*-condensed systems arise on thermal treatment of normal pitch, whereby the average molecular mass increases and the solubility of the pitch in solvents decreases. Furthermore, anisotropic liquid crystal structures known as "mesophases" are formed [13], [14].

Table 2. Properties of coke-oven coal tars

	Average value	Range of analytical data
Density, g/cm^3	1.175	1.14 – 1.20
Water, %	2.5	0.2 – 5.0
Toluene insolubles, %	5.50	2 – 12
Quinoline insolubles, %	2.0	0.5 – 5
Coking value (platinum crucible), %	14.6	10 – 20
Carbon (waf)*, %	91.39	90 – 93
Hydrogen (waf), %	5.25	5 – 6
Nitrogen (waf), %	0.86	0.6 – 1.2
Oxygen (waf), %	1.75	1.5 – 2
Sulfur, %	0.75	0.6 – 1
Chlorine, %	0.030	trace – 0.1
Ash (800 °C), %	0.15	0.05 – 0.25
Zinc, %	0.038	trace – 0.2
Naphthalene, %	10.0	5 – 13
Distillation analysis (DIN 1995), %		
up to 180 °C		
water	2.5	0.2 – 5
light oil	0.9	0.2 – 2
180 – 230 °C	7.5	3 – 10
230 – 270 °C	9.8	7 – 15
270 – 300 °C	4.3	3 – 7
300 °C up to pitch	20.1	12 – 30
pitch (softening point 67 °C KS**)	54.5	45 – 65
distillation loss	0.5	
Crude phenols, %	1.5	0.4 – 5
Phenol, %	0.5	0.08 – 0.7
Cresols and xylenols, %	1.0	0.3 – 4.5
Low-boiling pyridine bases (in light oil), %	0.081	0.01 – 0.2
High boiling pyridine and quinoline bases (in middle and heavy oil), %	0.065	0.2 – 0.9
Unsaturated compounds (in light oil), %	0.590	0.02 – 1.0

* waf: water- and ash-free. ** KS: Krämer – Sarnow.

Properties of coke-oven tar "normal" pitch are listed below:

Density at 20 °C	1.27 – 1.28 g/cm^3
Softening point (Krämer-Sarnow)	68 ± 3 °C
Softening point (Mettler)	86 ± 3 °C
Brittle point (Fraass)	30 – 40 °C
Elemental analysis:	
Carbon	92 – 94 %
Hydrogen	4 – 5 %
Nitrogen	1.0 – 1.3 %
Sulfur	0.5 – 0.8 %
Oxygen	1 – 2 %
n-Hexane solubles	40 – 60 %
Toluene insolubles	15 – 25 %
Quinoline insolubles	3 – 7 %
Coking value (ALCAN) (DIN 51 905)	43 – 48 %

Ash (800 °C)	< 0.3%
Iron	200–300 ppm
Silicon	100–300 ppm
Vanadium	< 50 ppm
Specific heat capacity (140–210 °C)	$C_p/\text{J g}^{-1}\text{ K}^{-1} = 1.15 + 0.0037\ T/\text{°C}$
Coefficient of thermal expansion	$0.00045–0.00056\ \text{K}^{-1}$
Thermal conductivity (25–105 °C)	$0.0014\ \text{J cm}^{-1}\text{ sec}^{-1}\text{ K}^{-1}$
Surface tension (120–220 °C)	$\gamma/\text{mN cm}^{-1} = 18.4 \cdot d_t^4$ (d_t = density at measuring temperature)
Dielectric constant (25 °C)	3.3–3.4
Magnetic susceptibility	0.69×10^6

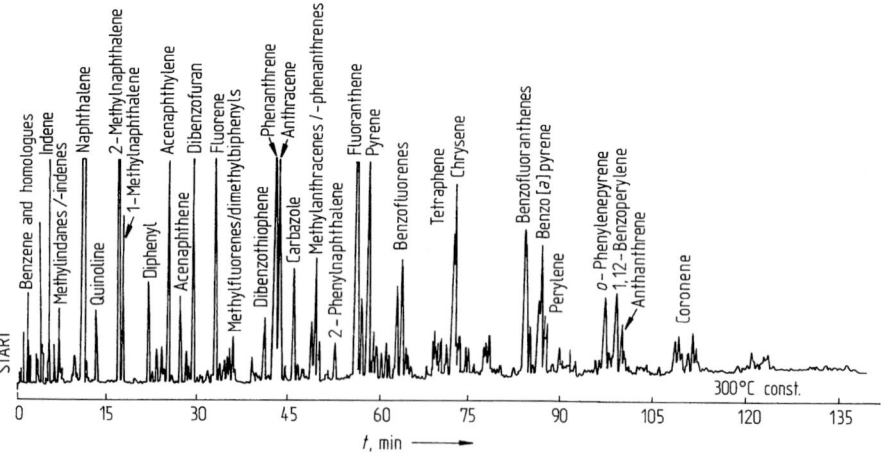

Figure 1. Gas chromatogram of a coke-oven coal tar 22 mm glass-capill. CP-sil 5 (inner diameter 0.5 mm), split, 50 °C 2 min, 2 °C/min up to 300 °C const., helium.

For reasons of clarity only selected peaks are shown with compound names. The concentrations of selected components (in wt%) are:

benzene/benzene homologues (1), indene (1), methylindanes/methylindenes (0.8), naphthalene (10), quinoline (0.3), 2-methylnaphthalene (1.5), 1-methylnaphthalene (0.5), diphenyl (0.2), acenaphthylene (2), dibenzothiophene (0.3), phenanthrene (5), anthracene (1.3), carbazole (0.7), methylanthracenes/methylphenanthrenes (1.8), 2-phenylnaphthalene (0.3), fluoranthene (3.3), pyrene (2.1), benzofluorenes (2.3).

Table 3. Important constituents of a typical coke-oven coal tar

Compound	Formula	bp at 1.013 mbar, °C	mp, °C	Average content, %
Hydrocarbons				
Naphthalene		217.95	80.29	10.0
Phenanthrene		338.4	100.5	4.5
Fluoranthene		383.5	111	3.0
Acenaphthylene		270	93	2.5
Pyrene		393.5	150	2.0
Fluorene		298	115	1.8
2-Methylnaphthalene		243.1	34.6	1.5
Anthracene		339.9	218	1.3
Chrysene		441	256	1.0
Indene		182.8	−1.8	1.0
1-Methylnaphthalene		244.4	−30.5	0.7
Benzene		80.1	5.53	0.4
Diphenyl		255.9	71	0.4
Acenaphthene		277.2	95.3	0.2
Heterocyclics				
Dibenzofuran		287	83	1.3
Carbazol		354.75	245	0.9

Table 3. continued

Compound	Formula	bp at 1.013 mbar, °C	mp, °C	Average content, %
Dibenzothiophene		331.4	97	0.4
Benzothiophene		219.9	31.3	0.3
Quinoline		273.1	−15	0.3
7,8-Benzoquinoline		340.2	52	0.2
2,3-Benzodiphenylenoxide		394.5	208	0.2
Indole		254.7	52.5	0.2
Acridine		243.9	111	0.1
Isoquinoline		243.25	26.5	0.1
Quinaldine		245.6	−2	0.1
Phenanthridine		349.5	107	0.1
Pyridine		115.26	−41.8	0.03
2-Methylpyridine		129.41	−66.7	0.02
Phenols				
Phenol		181.87	40.89	0.5
m-Cresol		292.23	12.22	0.4
o-Cresol		191.00	30.99	0.2
p-Cresol		201.94	34.69	0.2
3,5-Dimethylphenol		221.96	63.27	0.1
2,4-Dimethylphenol		210.93	24.54	0.1

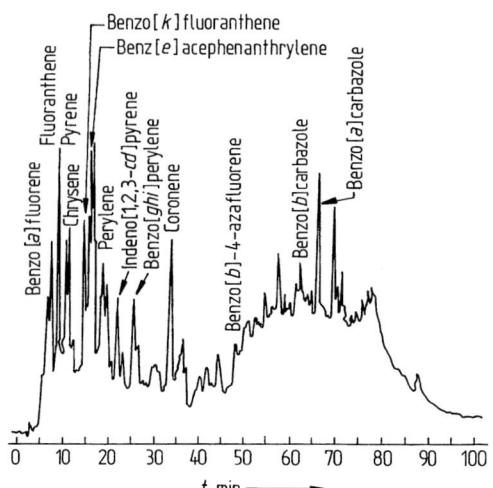

Figure 2. High-pressure liquid chromatogram of coal-tar pitch [7]
Nucleosil 5-NO$_2$, 5 µm, 2 columns of 200 mm length and 4 mm inner diameter and pre-column with 20 mm length and 4 mm inner diameter (switched one after the other, with gradient elution (chloroform – n-hexane) and UV detection at 300 nm)

3. Processing of Coke-Oven Coal Tar

3.1. Survey

The refining and processing of coke-oven coal tar preferentially isolates aromatic chemicals by a combination of six basic processes (Figure 3):

1) *Distillation* is used for primary separation of crude tar into six fractions: light oil, carbolic oil, naphthalene oil, wash oil, anthracene oil, and pitch. It is also used for further separation of primary and secondary tar fractions.
2) *Crystallization* is used to produce crystalline condensed aromatic hydrocarbons and nonpolar heterocyclics from distillate fractions (e.g., naphthalene, methylnaphthalenes, acenaphthene, dibenzofuran, fluorene, phenanthrene, anthracene, carbazole, fluoranthene, and pyrene). The aromatic oils obtained as filtrates are suitable as components for carbon-black oils, impregnating oils, fuel oils, diesel fuels, fluxing oils, and pitch – oil mixtures such as road tars, refractory pitches, anti-corrosion and protective agents for buildings.
3) *Extraction* with alkalis and acids allows the recovery of phenols and polar nitrogenous tar bases from light oil, carbolic oil, and filtered naphthalene oil.
4) *Catalytic polymerization* with Friedel – Crafts catalysts yields thermoplastic indene – coumarone resins from light oil fractions from which the phenols and bases have been extracted. The nonpolymerizable component is crude benzole, which together with coke-oven benzol is processed to benzene and benzene homologues; the higher boiling fractions are refined to solvent naphtha and heavy naphtha.

4519

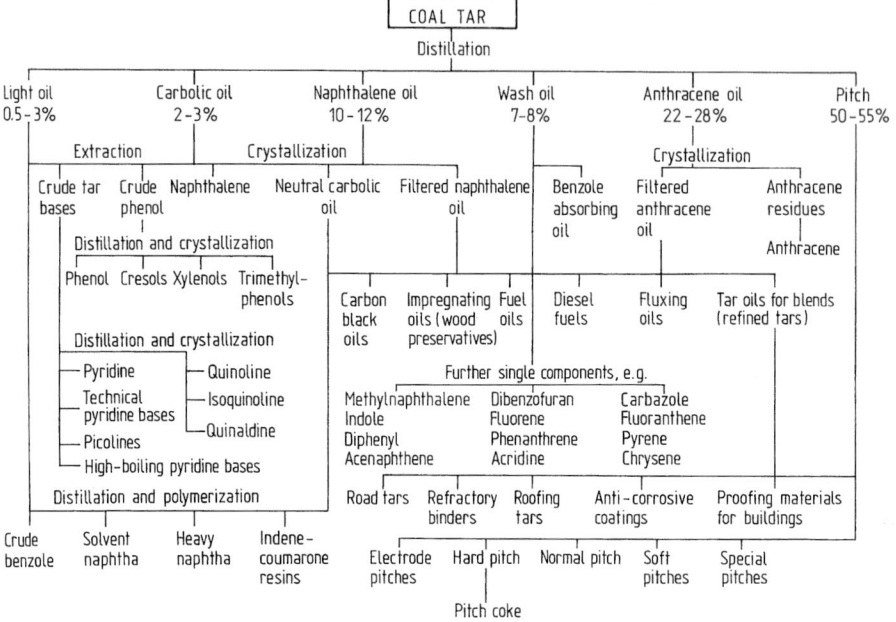

Figure 3. Processing of coal tar [1]

5) *Thermal polymerization* of coal-tar normal pitch yields higher value pitch products such as hard pitch and various special pitches.
6) Coking of pitch leads to pitch coke as technically pure carbon which is suitable for further processing into graphite, carbon electrodes, and carburizing agents.

3.2. Primary Distillation

Crude coal tar is generally pretreated prior to primary distillation. The emulsified aqueous solution (up to 5%) of ammonia, hydrogen sulfide, and ammonium chloride is removed as far as possible. Suspended fine solids (tar sediments) may also be removed.

The simplest procedure for a partial separation of water is to store crude tar at elevated temperatures. Storage for several days at 60 °C decreases the water content by 1–2%. Due to its lower density the water collects at the top of the tank, thus enabling its periodical discharge. To reduce the tar viscosity and increase the settling rate of water, crude tar can be stored in pressure tanks at 100–200 °C. This pressure dehydration can decrease the water content to ca. 1–1.5%. The settling rate can be further increased by centrifuging at 6000–7000 g. Disk centrifuges with discharge nozzles reduce the water content of crude tar down to ca. 0.5–1%.

The ammonium chloride contained in tar water dissociates into ammonia and hydrogen chloride above 225 °C, causing corrosion. The ammonium chloride content of crude tar can be greatly reduced by extraction with water prior to mechanical tar–water separation. Another method is to mix in an amount of sodium carbonate or sodium hydroxide solution equivalent to the ammonium chloride content. However, during tar distillation nonvolatile sodium chloride is formed and remains in the pitch, increasing its ash content.

Further impurities of crude tar are dispersed fine solids (tar sediments), which mainly consist of entrained fine coke particles. The partial removal of these sediments takes place together with mechanical water separation by sedimentation during storage. It is also possible to obtain a sediment concentrate from the crude tar by two- or three-phase centrifugal separators either with water as third or as second solid phase. Thus, erosion in pumps, pipes, and valves is avoided and the content of inorganic trace contaminants in the pitch primary distillation residue is decreased.

Nowadays, partly dehydrated and pretreated crude tar is distilled continuously in pipe stills and rectifying columns. Numerous distillation processes have been developed. They can be divided into flash processes, bottom-circulation processes, and processes in which these two principles are combined.

In *flash distillation* the tar is heated continuously under pressure, usually in a two-stage pipe still. It is then flashed into a system of rectification columns where the oil components are evaporated and fractionated. The first heating step dehydrates and evaporates the light oil, and in the second step the remaining oil components are evaporated, leaving pitch as the residue.

In *bottom-circulation distillation* the rectifying column bottoms are pumped continuously through pipe stills where they are provided with the thermal energy required for distillation.

These two distillation processes differ in the thermal stress on the tar fractions. In the flash process the oil fractions are subject to temperatures well above their boiling range, whereas the distillation residue pitch is not exposed to high temperatures. In the bottom-circulation process, however, the pitch is subject to high temperatures for long periods, but there is only a low thermal stress on the vaporized oil components. As several coal-tar constituents are unstable under distillation conditions, the two distillation processes differ significantly in the yield and quality of certain tar products.

Pipe stills are mostly rectangular or cylindrical, with an inner refractory lining. To maximize the heat utilization of firing gases, pipe stills are divided into radiant and convection zones and are often equipped with air preheaters. The burners of the radiant zone can be operated with gases or liquid fuels of various qualities. The tubes are made of steel; austenitic 18/8/2 chromium nickel molybdenum steels have proved to have especially good corrosion resistance.

As rectifying columns, bubble-cap columns, valve tray columns, sieve-plate columns, packed columns, and combinations thereof are used. Stoneware Raschig rings are used as packing materials. The lower regions of the dehydration columns and flash columns mostly contain simple cascade trays. In zones with high corrosion risk the columns consist of high-grade stainless steel or cast iron, otherwise of mild steel.

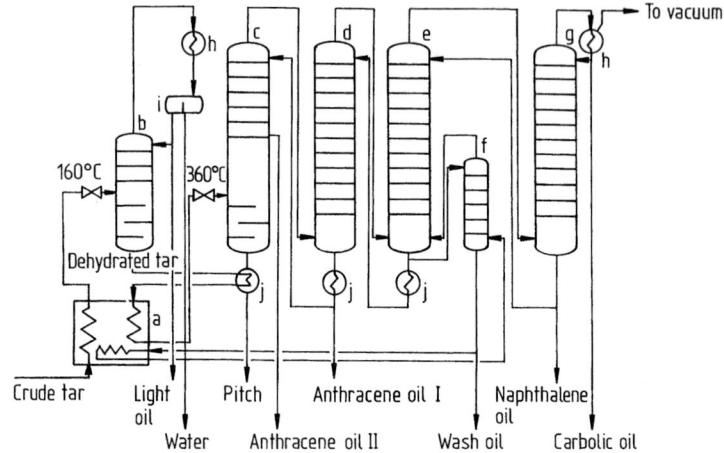

Figure 4. Continuous tar distillation with multiflash system (GfT/Koppers process)
a) Pipe still; b) Dehydration column; c) Pitch/anthracene oil II column; d) Anthracene oil I column; e) Wash oil column; f) Wash oil side-stream column; g) Naphthalene oil column; h) Condenser; i) Decanter; j) Heat exchanger

To separate high-boiling tar fractions the columns normally operate under vacuum. In some processes steam is injected to lower the evaporation temperature. Heat of condensation and the thermal energy contained in the flowing bottom products are transferred via heat exchangers to the crude tar and its intermediate products to lower the total energy demand.

The *GfT/Koppers* continuous tar distillation process is shown in Figure 4 as an example of flash distillation. In the first step the crude tar is heated to ca. 160 °C in the pipe still and then flashed through an expansion valve into the middle of the atmospheric dehydration column. The pressure before the expansion valve is up to 10 bar. Light oil and water evaporate in the dehydration column, which is equipped with bubble and overflow trays, and are separated into two phases after condensation. Part of the light oil is refluxed to the top of the dehydration column. The dehydrated (topped) tar as column bottom is preheated in a heat exchanger by the hot pitch stream, then it is heated under pressure to 360 °C in the second stage of the pipe still. At this temperature it is flashed in the pitch column at a pressure of 100–130 mbar. The pressure before the expansion nozzle is up to 5 bar. To take off the high-boiling anthracene oil II above the flash chamber, there is a column, filled with Raschig rings, which is dephlegmated at the top with the lower boiling anthracene oil I. The column tops of the vacuum system are refluxed with a split stream of the oil fraction taken from the bottom of the next column. The bottom product of the wash oil column is directed via a side column to evaporate the remaining naphthalene. This side column is additionally heated by bottom circulation, yielding an improved rectification efficiency. Concentrated ammonium chloride is extracted with water from the bottom product of the wash oil main column before the oil is refluxed to the top of the anthracene oil column. This continuous removal of ammonium chloride prevents clogging and corro-

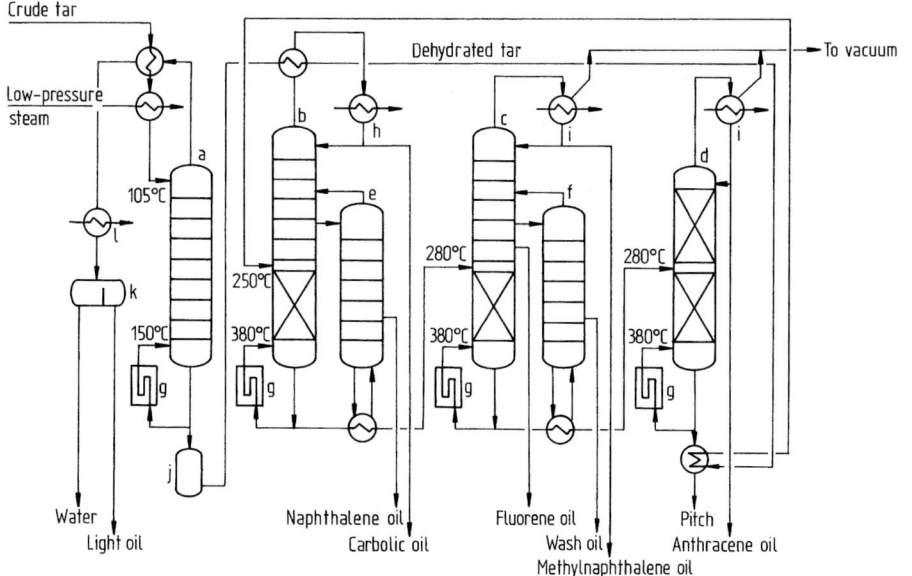

Figure 5. Continuous tar distillation (Rütgers process)
a) Dehydration column; b) Carbolic oil column; c) Methylnaphthalene oil column; d) Anthracene oil column; e) Naphthalene oil side-stream column; f) Wash oil side-stream column; g) Pipe still; h) Condenser; i) Condenser (steam generator); j) Vessel for dehydrated tar (topped tar); k) Decanter; l) Cooler

sion of the column by the salt, which condenses to form a solid in the temperature range 180–200 °C. The tar oil fractions anthracene oil I, wash oil, and naphthalene oil are separated as column bottoms. Carbolic oil is condensed in the final condenser as the top product of the naphthalene column. The uncondensed gases and vapors are extracted with oil and water before entering the vacuum equipment. Light oil and naphthalene vapors are extracted in the oil scrubber with cooled wash oil. Hydrogen sulfide is removed with water in the water scrubber. The heat consumption of the total plant is ca. 900 MJ/t of wet crude tar.

Figure 5 shows a diagram of the *Rütgers process* for continuous tar distillation using the bottom-circulation principle. Each of the main rectifying columns has its own pipe still for bottom circulation to achieve independent control of the column reflux. Condensing water, light oil vapors of the dehydration column, and waste heat vapor preheat the crude tar to 105 °C. Afterwards the crude tar is pumped into a dehydration column which contains cascade trays. The bottoms of the dehydration column are circulated through a pipe still and kept at 150 °C. After intermediate storage the dehydrated topped tar is passed through to heat exchangers where the heat of condensation of carbolic oil and the heat of pitch is used. The topped tar, at ca. 250 °C, enters the middle section of the carbolic oil column, which is operated at atmospheric pressure and is equipped with valve or Perform trays in the lower part and, bubble trays for example, in the upper part. The pipe still, with a discharge temperature of

380 °C, allows adjustment of the reflux ratio up to 16:1. A distillate fraction in the carbolic oil column is cycled to the top of a side-stream column where the naphthalene oil fraction is taken as a vapor from its base. The main-stream bottom product is led via heat exchangers to the methylnaphthalene column, which is run with a top pressure of 250 mbar and which also contains a packing or Perform trays as well as bubble trays. The reflux ratio of this column is up to 17:1. A methylnaphthalene fraction is obtained as top product. The top vapors are condensed in a low-pressure steam generator. A wash oil fraction is taken from a side column and a fluorene fraction is taken as a further side stream of the main column. The bottom product of the methylnaphthalene column is led to the anthracene-oil packed column with a top pressure of 90 mbar. Its top product is anthracene oil, and the bottom product is normal pitch. This column only requires a reflux ratio of 1.5:1. The condensation heat of the anthracene oil is used for steam generation.

Besides the product flows described, other column interconnections are possible. Due to the high reflux ratios in efficient rectifying columns the tar oils are separated into many single fractions. Thus a high-yield extraction of technically pure components can be achieved. For example, the naphthalene oil fraction contains up to 90% or more of the naphthalene occurring in crude tar at a concentration up to 90% (crystallizing point up to 75 °C). Despite the increased energy needed for high reflux ratios, the heat requirement of the plant is only ca. 880 MJ/t crude tar due to the heat recovery system.

If the production of coal-tar pitch rather than tar aromatics and oils is the main aim then the primary distillation of crude tar should be combined with a thermal treatment at increased temperature and longer time under pressure. This is the case, for example, in the production of binders or coke for carbon and graphite electrodes.

The *Cherry-T process* of Osaka Gas [15] serves as an example of such a procedure. In this process water and light oil are separated from the crude tar as a first step, usually in a dehydration column. The dehydrated tar is heated in a pipe still to almost 400 °C and afterwards thermally polycondensed in a pressure reactor. After this reactor, a flash column is used for oil/pitch separation. If necessary, the oil can also be treated thermally in a second pressure reactor. Then, after additional flashing, a secondary oil pitch and nonpolymerized oils are obtained. The total yield of primary and secondary pitch is ca. 75%, compared with 50–55% for tar distillation without intensified thermal treatment.

Whereas in the Cherry-T process, electrode pitches are the main products of primary processing, pitch coke is the main product when tar distillation is coupled with a *delayed coker*. Some Japanese plants use such a combination (see Section 3.3). Here only light, carbolic, naphthalene, and wash oils are separated by primary distillation. Soft pitch obtained as residue is pumped into the bottom of a combination tower where it is distilled together with coker oil. The bottom product of the combination tower is transferred to the pipe still of the delayed-coking plant where it is heated to almost 500 °C and pumped into the delayed-coking drum (see Section 3.3).

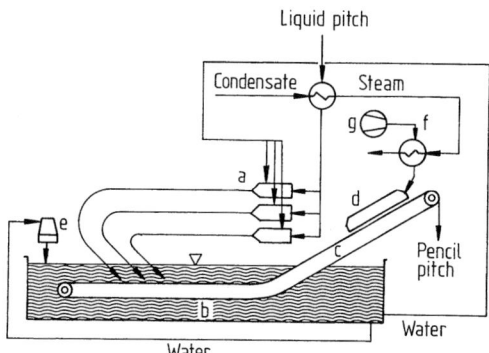

Figure 6. Direct cooling of coal-tar pitch (Rütgers pencil-pitch process)
a) Nozzles; b) Cooling water basin; c) Conveyor belt; d) Dryer; e) Cooling facility; f) Air preheating; g) Blower

Nondistillation processes for the primary processing of coke-oven coal tar, such as liquid or supercritical extraction and hydrocracking, are presently not used in large-scale production.

3.3. Processing of Coal-Tar Pitch

3.3.1. Cooling

Approximately 50–55 % of the products of a coal tar primary distillation is a coal-tar normal pitch with a softening point of 68 °C (Krämer–Sarnow). This is either processed directly in the liquid state or solidified by cooling, which converts the thermoplastic pitch into its glassy state.

Modern pitch-cooling processes operate continuously and use water as cooling medium. The pitch is obtained in granulated form.

The problem of continuous pitch cooling is to obtain a rapid heat transfer despite the relatively low thermal conductivity of pitch. If the granules do not cool completely, undesired agglomeration during storage and transport can occur. For various applications the pitch granulate must be moisture-free; therefore, cooling is followed by drying. Several continuous processes for direct cooling of molten coal-tar pitch with water have been developed.

Figure 6 shows the diagram of the Rütgers pencil-pitch process [16]. Relatively thick strings are formed by pumping pitch through heated nozzles directly into water flowing cocurrently. The water acts as cooling and transport medium. Before the liquid pitch is formed by the nozzles, it is cooled in a heat exchanger that produces low-pressure steam. The nozzles are shaped such that by adjusting the temperature and water/pitch ratio, a pitch string without internal water inclusion is obtained. The cooling water and the solidified pitch string run together through a cooling tube 20–25 m long. This tube may be bent to shorten the equipment. Each nozzle has its own cooling tube to prevent the warm, thermoplastic pitch strings merging. At the end of the cooling tube the outer

Table 4. Specifications of various electrode pitches from coal tar

	Electrode pitches for aluminum		Electrode pitches for electrographite	
	Prebaked anodes	Søderberg anodes	Binder pitches	Impregnating pitches
Softening point, °C				
Krämer–Sarnow	88–103	108–112	90–95	68–74
Mettler	106–121	126–130	108–113	87–93
Density at 20 °C, g/cm³	≥ 1.31	≥ 1.31	≥ 1.31	
Quinoline insolubles, %	5.5–13	≤ 14	7–12	≤ 5
Toluene insolubles, %	23–36	≥ 32	27–34	16–22
β-Resins, %	18–26	≥ 22	20–22	11–18
Coking value (Alcan), %	53–63	≥ 57	56–63	≥ 43
Ash (800 °C), %	0.15–0.35	0.15–0.35	0.15–0.35	≤ 0.53
Sulfur, %	≤ 0.7	≤ 0.7	≤ 0.7	≤ 0.7
Sodium, ppm	≤ 200	≤ 200	≤ 200	
Iron, ppm	≤ 200	≤ 200	≤ 200	
Silicon, ppm	≤ 200	≤ 200	≤ 200	
Distillation up to 360 °C (ASTM D 2569), %	≤ 3	≤ 3	≤ 3	*

* Flash point ≥ 260 °C (open cup).

skin of the pitch strings at least has cooled and solidified. The strings can be deposited together on a slowly moving conveyor belt running submerged in a long cooling-water basin. The cooling water is cooled and recycled. At the end of the cooling section the conveyor belt removes the cold, solidified pitch strings from the water. They pass a dryer whose airstream is heated by the low-pressure steam produced in the heat exchanger. The dried pitch strings are air-cooled and then loaded into transport and storage containers. During loading the pitch strings break into short, air-conveyable pieces. This process is also applicable to other pitches with different softening points. The throughput capacity is 1–2 t per nozzle per hour.

3.3.2. Production of Electrode Pitch

Electrode pitches are the most important products derived from coal-tar pitch. They are used as reactive binders, especially for the production of carbon anodes for aluminum production by electrolysis and furthermore as binder and impregnating pitches for the production of graphite electrodes for electric steelmaking.

Electrode pitches should have good caking and binding properties towards electrode cokes (petroleum cokes, pitch cokes), the solid components of the electrode paste. On heating, electrode pitches should give a high yield of a high-strength coke. Medium and higher molecular pitch fractions are mainly responsible for the caking and binding properties. They are analyzed by solvent fractionation (e.g., quinoline insolubles, QI; toluene insolubles, TI; β-resins = TI – QI). The volatile matter of the electrode pitch

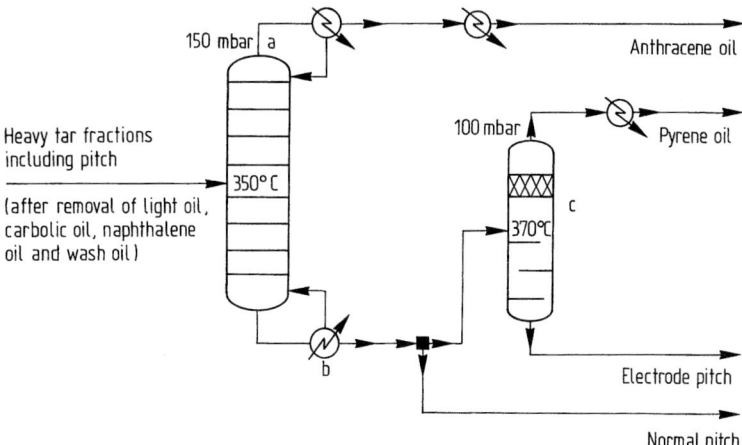

Figure 7. Flowsheet of a coal-tar primary distillation with integrated electrode-pitch fabrication (Rütgers process)
a) Pitch/anthracene oil column; b) Reboiler; c) Flash column

should degas smoothly to prevent cracking during carbonization of the electrode paste. Moreover, electrode pitches should wet electrode coke particles well and should have a desirable viscosity–temperature behavior and their properties should not change significantly on hot storage at temperatures of, e.g., 200–250 °C. The content of ash-forming constituents should be low. Table 4 lists typical characteristics of various electrode pitches. Bench-scale processes have been developed that allow the quality properties of electrode pitches to be determined more precisely. It is now possible to produce test electrodes with electrode coke on a laboratory scale. These electrodes are tested according to various methods [17], [18]. Quality aspects of the test electrodes include high mechanical strength, high electrical conductivity, and low air permeability, as well as low air reactivity at, e.g., 525 °C and reactivity to CO_2 at, e.g., 1000 °C [19], [20].

The properties required for electrode pitch are generally not fulfilled by normal pitch from conventional coal-tar primary distillation. The aim of production processes for electrode pitch is to increase the pitch coking value. This is achieved either by thermal condensation and polymerization of pitch components in secondary processes or by gentle distillation of vaporizable pitch components integrated with the tar distillation. In both processes the softening point and the content of medium-molecular β-resins are increased, and in the thermal condensation and polymerization process, the amount of quinoline insolubles rises. These "secondary" quinoline insolubles, generated during thermal retreatment, mainly consist of liquid crystalline carbonaceous mesophases. These mesophases advantageously increase the coking value of total pitch, but at the same time generally decrease its wetting ability. Therefore, in many cases electrode pitches produced without mesophase formation are preferred for electrode production [21].

Figure 7 shows the flowsheet of a thermally gentle electrode-pitch production. The process is part of the coal-tar primary distillation. In this case the coal-tar primary distillation (see Fig. 5) is extended by a flash column. By regulation of temperature and vacuum it is possible to directly control the desired increased softening points and the coking values related therewith.

The Cherry-T process of Osaka Gas (see Chap. 3.2) is a continuous thermal pitch-condensation process forming part of the coal-tar primary distillation. This process gives a high yield of electrode pitch from coal tar in two different qualities. In the first step a primary pitch yield of 55–60% from coal tar is achieved. The pitch has a softening point of 80–100 °C, 31–38% toluene insolubles, and 8–14% quinoline insolubles, thus having a β-resins content of ca. 22%. In the second step, further thermal treatment of the flash oil from the first reactor gives 40–50% of a secondary pitch with a softening point of 70–90 °C, 23–31% toluene insolubles, and a very low amount of quinoline insolubles (max. 2%) [22].

Not only the production conditions for electrode pitches but also the choice of parent coal tars is important. The desired pitch components may already have been formed to a large extent in the coal tar. Moreover, the tars have to meet the requirements concerning the content of inorganic contamination, which very often require a separation of these microparticles by centrifugation or extraction. For use as binder pitches for carbon anodes in aluminum production, for example, a low content of zinc and iron compounds is important. Under the conditions of aluminum oxide electrolysis in the melt, elemental metals are formed together with aluminum at the cathode, thus reducing the purity of the aluminum product. Furthermore, the content of alkaline earth and alkali metal compounds should be as low as possible because they catalyze the reaction of CO_2 with the carbon anode. In the case of impregnating pitches it may also be necessary to partly remove the highest molecular mass quinoline insoluble α-resins from coal tar, as the impregnating behavior depends upon the content of α-resins and upon their particle size distribution. On the other hand, α- and β-resins considerably contribute to the stability of the pitch coke skeleton formed from electrode pitch, thus giving the mechanical and thermal stability of carbon and graphite electrodes.

For large-scale production of electrodes, bituminous binding agents other than coal-tar pitch have not gained acceptance. The reason is the good coking behavior of coal-tar electrode pitches due to the high content of condensed aromatics and fixed carbon. Coal-tar pitch has the highest carbon content of comparable bituminous binding agents. In contrast to petroleum asphalt and coal extracts, the carbon content of coal-tar pitch is considerably above 90%, on average ca. 93%. The atomic C/H ratio is 1.7–1.8 compared with, for example, 1.3 for coal extract from the solvent refined coal (SRC) process and 0.8 for high-vacuum bitumen with a comparable softening point [23]. It is only possible to produce aromatized petroleum pitches from strongly aromatized residues of catalytic petroleum cracking by thermal treatment. These may serve as substitutes for coal-tar electrode pitches in some situations [24], [25].

Despite the advantages of coal-tar pitch as electrode binder, it does have environmental and toxicological drawbacks. Coal-tar pitch has a high content of cancerogenic polycyclic aromatics, especially benzo[a]pyrene. However, the uptake of harmful polycyclic aromatic hydrocarbons can be avoided if pitch fumes and dusts are not handled

Table 5. Properties of pitch cokes

	Regular coke	Needle coke
Density, g/cm^3	2.04–2.10	$\geq$ 2.12
Microstructure	coarse mosaic or fine isotropic	coarse fibrous
Coefficient of linear thermal expansion * 10^{-6} K^{-1}	2.5–5.5	< 1
Ash content, %	< 0.5	< 0.2
Elemental analysis		
C, %	$\geq$ 98	> 98
S, %	0.4–0.5	< 0.3
N, %	0.5–0.7	< 0.5
H, %	0.1–0.2	0.1–0.2
Si, %	ca. 0.03	< 0.03
Fe, %	ca. 0.03	< 0.03
V, ppm	< 5	< 5

* Measured on graphite rods, 20–200 °C.

or inhaled. Nevertheless, attempts have been made to reduce the benzo[a]pyrene content of electrode pitches. The physical removal of the cancerogenic polycyclic aromatic hydrocarbons by extraction or distillation is technically possible. The softening point of the pitch increases on removal of these low-molecular compounds. Bench-scale experiments showed that it is possible to obtain carbon electrodes with improved technical properties when using such high-melting electrode pitches [26]. In an electrode pitch with an increased melting point of 170 °C, for example, the benzo[a]pyrene content is decreased by > 60%, down to 0.3–0.4%. A further increase of the melting point up to, for instance, 260 °C by a > 99% reduction of the benzo[a]pyrene content to less than 100 or 50 ppm is possible in principle. For practical applications, however, modification of the electrode production process by using a different mixing technology for the electrode masses, or adjustment of the softening point is necessary [27].

3.3.3. Production of Pitch Coke

Like calcined petroleum coke, pitch coke is used as electrode coke, the solid component for the production of carbon and graphite electrodes. In addition, technically pure carbon pitch coke is used to produce other carbon and graphite materials, especially for high-temperature and chemically resistant equipment. It is also used as conductive filler in carbon brushes, batteries, plastics, and brake linings; as a carburizing agent in steel and cast-iron production; as adsorption coke for water purification; as an almost ash-free solid reducing agent in metallurgy and chemistry; and for the production of highly isotropic graphite for nuclear reactors. Specifications of different pitch cokes are listed in Table 5.

The most important processes for producing pitch coke are:

1) Discontinuous high-temperature coking of hard pitch in normal coke ovens with gas heating and liquid feeding, used especially in the CIS.
2) Discontinuous coking of soft pitch by the delayed-coking process in Japan and China.
3) Continuous hard-pitch coking in an indirectly heated rotary-drum reactor (still in development).

Hard pitch with a softening point > 150 °C (Krämer–Sarnow) can be obtained either from the normal pitch produced by continuous tar distillation by dehydrogenative condensation with atmospheric oxygen (air blowing) or solely by thermal treatment, similar to electrode-pitch production processes.

In the air-blowing process, air is blown through liquid normal pitch at 300 °C in a retort directly after coal-tar primary distillation [28]. There is an air distributor at the bottom of the retort, and the air is either sucked or blown through the pitch. The reaction is exothermic above 300 °C and takes place without additional heating. The temperature of the retort contents rises to ca. 400 °C. The main product is a hard pitch which can be processed to coke with a yield of up to 83%. The byproduct pitch distillate can be converted into a secondary hard pitch by distillative thermal treatment and air blowing. By using a mixture of primary and secondary hard pitch, different pitch-coke qualities can be obtained in the subsequent coking process.

The thermal condensation of normal pitch to hard pitch is usually carried out by continuous treatment of normal pitch in pipe stills under pressure followed by flash distillation to remove vaporizable components. Compared to the air-blowing process, the yield of hard pitch is lower, but the yield of high-boiling aromatic oils is higher.

In discontinuous *pitch coking in horizontal coke ovens* hard pitch is pumped at a temperature of above 300 °C into heated pitch tanks. In the Koppers process [29] several oven blocks, each with, e.g., five pitch-coke ovens, can be fed by these tanks. The hard pitch is pumped continuously via a ring main from the pitch tanks to the oven blocks. The liquid hard pitch is pumped in portions over a period of several hours into the coke ovens from this ring main. The final coking temperature can be up to 1200 °C, and the coking time 14–24 h. The pitch-coke yield is ca. 70%. The pitch-coke oil byproduct can be used for secondary-pitch production. By mixing this pitch with the primary pitch, the quality of pitch coke can be varied and the pitch-coke yield increased to 80–85%. Under these conditions ca. 250 m^3/t process gas, containing ca. 80 vol% hydrogen and ca. 12 vol% methane, are generated as a byproduct. This gas can be used for the production of hydrogen, or as low-calorific gas for heating the coke ovens and pitch tanks. With heat requirements of ca. 18 MJ per tonne of hard pitch, about 80% of the process gas is required. Analogous to metallurgical coke production, after carbonization the pitch coke is pushed mechanically out of the horizontal chamber of the coke oven and cooled in aquenching car by spraying with a calculated amount of water. It is possible to achieve a water content of 1–2% in spite of the high porosity of pitch coke. Due to the high carbonizing temperature, the pitch coke produced does not require secondary calcination.

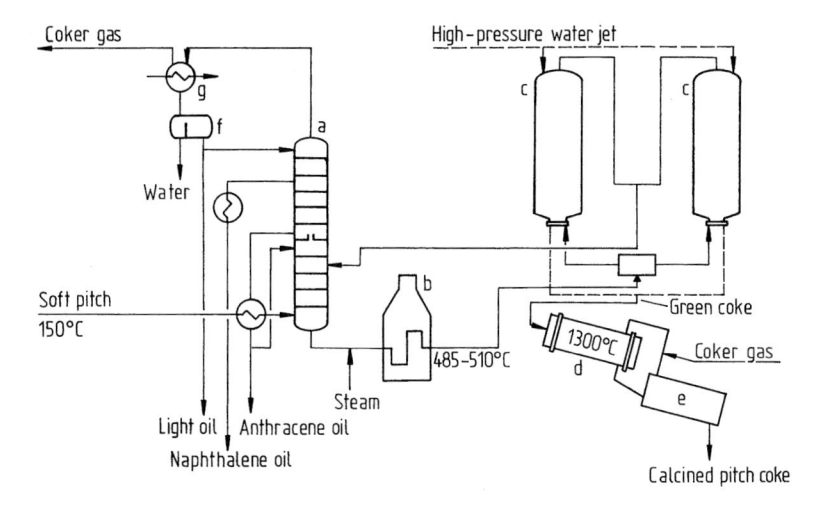

Figure 8. Production of pitch coke by delayed coking
a) Combination column; b) Pipe still; c) Coker drums; d) Rotary kiln; e) Rotary cooler; f) Decanter; g) Condenser

The second process for production of technical pitch-coke from coal tar is *delayed coking*, well known from petroleum refining. To produce a pitch coke with a low volatiles content a second calcination step is required. Soft pitch containing high-boiling aromatic oils, obtained by primary distillation of coal tar, is used as feed. The soft pitch is pumped into a combination column, the bottoms of which are fed to the pipe still of the delayed coking plant.

Figure 8 shows the process flow sheet for production of pitch coke from coal-tar soft pitch by delayed coking using the Lummus process as applied in Japan and China [30]. The soft pitch coming from the coal-tar primary distillation at 150 °C is heated in a heat exchanger by the heavy oil of the combination column and then pumped into the bottom of this column. The coker oil of the delayed-coking plant is also added to the column. Light oil, middle (naphthalene) oil, and heavy (anthracene) oil fractions are taken from the combination column. The bottoms of the column are heated in a pipe still to 485 – 510 °C by gas or fuel oil. Then they are fed from below into one of two coker drums via a switching valve. In these drums the delayed coking of pitch takes place during a 24 h cycle with a residence time of 12 h at a pressure of, for example, 5 bar. When the first coker drum is full, the feed is switched to the preheated empty coker drum. After switching, the full drum is steamed for 2 h to remove residual coker oil, which is mainly present in the upper part of the coker drum. The oil obtained is recycled to the combination column. After steaming, the coke charge is cooled directly with water. Then the coke is cut out by a high-pressure water jet, which takes 5 – 7 h. The green coke obtained is dried, broken, and calcined in a hearth furnace or a rotary drum. The pipe still and the calciner can be heated with coker gas arising as the noncondensable top product of the combination column.

Normally the green-coke yield is at least 60 % of the soft pitch fed, with approximately 3 % process gas and 35 % coker oil as byproducts. The combination column yields ca. 10% light and middle (naphthalene) oil, as well as ca. 25% heavy (anthracene) oil. The coker gas contains ca. 50% hydrogen and 50% methane. The middle oil has a boiling range of 180–310 °C and contains ca. 30% naphthalene. The heavy oil has a boiling range of 265–400 °C and a composition that differs from that of an anthracene oil from normal continuous tar distillation. Under the conditions of pressure carbonization in the delayed coker, some distillable tar aromatics are preferentially converted to coke. The green coke still has 7.5–9.5% volatiles, which are removed to a residual content of < 0.5% by calcination at up to 1300 °C.

Highly anisotropic cokes, known as *needle cokes*, are obtained via intermediate liquid-crystalline mesophases by delayed coking of pitches from which the α-resins (quinoline insolubles) have been largely removed by filtration [31] or extraction [32]. These needle cokes are used for the production of ultra-high power (UHP) graphite electrodes with high electrical conductivity and low thermal expansion coefficient.

The carbonization of hard pitch in an indirectly heated *rotary drum reactor* [33], [34] is a continuous pitch-coke production process. Hard pitch is pumped into an indirectly fired rotary drum reactor where it is carbonized with a residence time of 0.5–1.5 h at an inner-wall temperature of 500–800 °C. The gases and vapors arising are pumped in countercurrent to the carbonizing pitch, if necessary together with an inert gas. The green-coke yield is 70–75% of the hard pitch feed. The grained coke has volatile matter content of ca. 3.5 wt% and relatively high strength and density. It is calcined without pre-cooling in a second directly heated rotary drum reactor at up to 1300 °C. The yield of calcined coke is ca. 90% of the green-coke feed, and 65–70% of the hard pitch input. The advantages of this continuous process compared with the discontinuous processes of horizontal chamber coking and delayed coking are the possibility to adjust the residence times and heating periods, thus yielding a more homogeneous pitch coke.

3.3.4. Production of Special Pitches

Hard pitches, soft pitches, and extractive pitches produced by solvent extraction with considerably decreased content of α-resins (primary quinoline insolubles) are special pitches that are intermediate products of pitch carbonization. Other special pitches include plasticized pitches for protective coatings, pitches used for pitch–polymer combinations, and road-binding agents which are compatible with polymers and bitumen, as well as the newly developed and produced mesophase pitches and high-melting isotropic pitches for further processing to carbon materials.

Soft pitches with increased coking values are used as *refractory binders*. They are reactive binding agents for the production of dolomite and magnesite bricks used especially for the lining of steel converters. During brick calcination a binding coke skeleton is formed by carbonization of the soft pitch, which considerably increases the compressive strength of the preform and counteracts corrosive attack by slags. Refractory binder pitches with increased coking values are obtained by adjusting hard pitches with high-boiling aromatic oils to give the desired softening point. Also soft-pitch mixtures with phenolic resins or their precursors are used. Special formulations contain less than 50 ppm benzo[a]pyrene.

For the production of *anticorrosion protective coatings* the plasticity range (difference between softening and brittle point) of 30 K of coal-tar normal pitch is too low. To increase this plasticity range to 100 K or more, hard pitches with a high content of toluene insolubles are adjusted to the desired softening point with high-boiling tar oils, preferably by adding crystalline tar aromatics. A further increase in the plasticity range is obtained by hot-mixing these pitches with extenders such as finely ground coal, minerals, diatomaceous earth, or fly ash. Such plasticized pitches are used for long-term anticorrosion pipe coatings and as sealing compounds. The advantages compared with the bitumen coatings are the excellent resistance of coal-tar pitch to moisture, bacteria, fungi, mineral oils, and gasoline. When pitch coatings are used together with cathodic pipe protection the low water absorption of coal-tar pitch is important. The water content of a 2.5 mm tar-pitch coating is only 0.3 – 0.6 % after prolonged storage in water. Plasticized pitches without extenders are used as dip-coating agents for protecting cast-iron pipes.

To meet especially high anticorrosion requirements, coal-tar pitches are combined with polymers [35]. Especially suitable for such *pitch – polymer-combinations* are epoxy resins, polyurethanes, and some elastomers. Normally two-pack systems are produced for epoxy resin combinations. One component contains the nonhardened epoxy compound dissolved in a compatible special coal-tar pitch, and the other contains the hardener for the epoxide compound. After combination of the components the epoxy resin polymerizes and incorporates the pitch components. Usually low molecular-mass liquid epoxy resins are combined with soft coal-tar pitches. After chemical hardening, coatings with high-temperature stability and resistance to chemicals are obtained. Pitch – polyurethane two-pack systems are produced and processed similarly. One component contains the special soft coal-tar pitch and the polyol, and the other the polyisocyanate. In combination with polyurethanes, one-pack systems that cure by reaction with atmospheric moisture are also feasible. Other polymers and elastomers are mixed directly with the hot liquid pitch, whereby miscibility gaps and the decomposition temperatures of the polymers must be taken into account. It is also possible to add high-boiling or unvaporizable, preferably aromatic homogenizing agents.

By using high-boiling tar aromatics as homogenizing agents, soft coal-tar pitches with a low content of α-resins can be combined in any ratio with appropriate, preferably naphthene-based bitumens. Carbobitumen, a highly deformation resistant *road binding agent*, is a blend of soft pitch and hard bitumen, containing 20 – 30 % of a special pitch. The pitch is enriched with high-boiling crystalline aromatics [36], [37]. These crystalline aromatics cause a separation of microcrystals in the pitch – bitumen matrix when cooled to ca. 100 °C. These microcrystals crosslink the thermoplastic system and harden the binding agent at lower temperatures. Thus carbobitumen is harder at temperatures below 70 – 80 °C and, therefore, more deformation resistant than the road bitumen usually used. At higher temperatures, however, it is less viscous and, therefore, can be applied at lower temperatures than road bitumen.

Normal pitches, hard pitches, and special pitches with carbon fillers and polymers are suitable as reactive additives for *cocarbonization of weak-coking coal blends* in amounts of 2–10%. The mechanical strength of the metallurgical coke produced in horizontal chamber-coke ovens increases on addition of an amount of pitch ascertained by laboratory tests [38], [39]. With increasing temperature the pitch wets the coal grains in the coke-oven chamber. It partially dissolves the coal macerals in the softening range of the coal. During carbonization above 450 °C inert macerals and solid leaning agents such as coke dust or petroleum coke are also brought into the generated coke skeleton. Used plastics, for example, can be utilized as additives. These polymers should first be subjected to cothermolysis with pitch, to obtain a resistant reactive additive as liquid phase in the coal-softening range [40].

Further special pitches are fine-grained *foundry pitches* produced by spraying. They have a high softening point and high coking value. They serve as binding agents for the production of foundry molds. Another special pitch is *fiber-pipe pitch* which has a low content of α-resins. This is used as impregnating material for cellulose fibers in the production of pitch fiber pipes. *Clay pigeon pitch* is used as a brittle binding agent with increased softening point for clay pigeons used in sport shooting.

Mesophase pitches are pitches with a high content of carbonaceous mesophases (see Chap. 2). They are produced by heat treatment of pitches that have been filtered or extracted to remove primary α-resins [31]. By heat treatment, mesophase spherules are formed, which can be separated from the isotropic pitch matrix by settling and/or by solvent fractionation. Pitch pre-fractionation with solvents or supercritical extraction is often carried out before thermal mesophase formation. Two-step thermal pitch treatment is also utilized. In the first step the especially reactive pitch aromatics are separated by condensation to give secondary quinoline insolubles. The second thermal treatment step leads to mesophases with low to medium molecular mass and correspondingly low viscosity, which can be advantageous. Mesophase pitches can also be produced from special petroleum pitches, hydrogenated coal-tar pitches, and pyrolysis residual pitches. By using different processes special properties can be achieved [41]. Mesophase pitches also can be produced by catalytic polycondensation of naphthalene and its homologues and higher condensed polycyclic aromatic hydrocarbons in the presence of homogeneous Friedel–Crafts catalysts. Thus, after 2–17 h at 140–180 °C 80–100% mesophase pitches with a yield of 50–80% are formed from naphthalene, 1,1-binaphthylanthracene, and 1,2-benzanthracene and from a 1:1 mixture of naphthalene and phenanthrene by reaction with aluminum chloride/copper(II)chloride [42]. From naphthalene it is possible to produce mesophase pitches by using HF/BF_3 catalysts, either by one step at 260–300 °C, or by two steps, the first at 80 °C and the second at 400–480 °C under vacuum or pressure [43]. Instead of naphthalene, 1-methylnaphthalene, 2-methylnaphthalene, or mixtures thereof can also be used. Here mesophase pitches are formed by using HF/BF_3 at 260 °C under a pressure of ca. 30 bar in 4–5 h with a yield of 80–90% [44].

Mesophase pitches and high-melting isotropic pitches are intermediates in the production of new forms of technical carbon such as anisotropic and isotropic carbon fibers, mesocarbon microbeads, and sinterable carbon powders [41]. In their production the special pitches are normally first stabilized by oxidation after pre-forming by melt spinning or fine grinding and are then partly or entirely carbonized. For the production of high-modulus carbon fibers from mesophase pitches various processes have been developed. Some of these processes have been used commercially. Mesophase microbeads and sinterable carbon powders can be pressed isostatically without additional binding agents and converted by thermal treatment into highly dense isotropic carbon materials [41]. In the last few years isotropic fine-grained carbon materials have found a wide range of applications, as have carbon fiber reinforced carbon materials (CFC) [45]. Widespread areas of application for these carbon materials can be predicted for the future, especially in energy and transport technology. This is due to their low density compared to metals such as steel and aluminum, their mechanical strength over a wide temperature range, and their excellent properties at temperatures up to 4000 °C if protected against oxidation.

3.3.5. Pitch – Oil Blends

Coal-tar pitches are highly soluble in various tar oils. For manufacturing of pitch – oil blends (refined tars) the liquid pitch is mixed with tar oils.

Road tars based on pure coal tars are blends of normal pitch with middle oils (boiling range 170 – 270 °C), heavy oils (boiling range 270 – 300 °C), and anthracene oils (boiling range above 300 °C). The content of normal pitch is between 60 and 80%. These road tars are suitable for surfacing low-traffic roads. The advantages of road tars for this purpose are good adhesive strength, quick rock wetting due to their content of polar compounds, and their property of gradual surface resinification. The middle and heavy oil components evaporate during setting, leaving as the actual binder a soft pitch consisting of anthracene oil and normal pitch. The setting speed depends on the initial viscosity and the pitch/anthracene oil ratio. Normal setting road tars have a pitch/anthracene oil ratio of ca. 3.5:1. By increasing the ratio of anthracene oil II (boiling range above 350 °C) and anthracene oil I (boiling range up to 350 °C), aging-resistant road tars can be produced. Bitumen-containing road tars include, for example, 15% asphalt basic distillate bitumen with low oil content; bitumen-rich road tars contain 35 – 40% bitumen, and *pitch bitumen* contains 70 – 85% bitumen. These tar – bitumen blends with increased or predominant bitumen components combine the advantages of tar (good setting properties, resistance to mineral oil and fuel, easy processability, and good road grip) with those of bitumen (low temperature dependence of viscosity and wide plasticity range). Carbobitumen is a specially deformation resistant pitch – bitumen blend (see Section 3.3.4).

Besides bitumen, plastics such as poly(vinyl chloride), polystyrene and epoxy resins are used for the modification of road tars. Thermosetting tar – epoxy resin combinations (see Section 3.3.4) are particularly suitable for highly exposed road surfaces and airport runways due to their good wear properties and their resistance to liquid fuels.

Roofing tars used as impregnating, coating, and adhesive material for tarred felts and tarred sealing webs are usually blends of pitch and filtered anthracene oil. By using plasticized pitches or by adding extenders the plasticity and temperature stability of roofing tars is improved considerably. Tarred felts produced in this manner have the same temperature stability as bitumen felts. Moreover, they have a high resistance to water, microorganisms, and plant roots.

Pitch paints are solutions of coal-tar pitch in highly volatile aromatic solvents such as light and heavy tar-oil naphtha. Normally plasticized pitches are used or extenders are added. Like the plasticized pitches themselves, which are processed while hot, pitch paints have excellent resistance to water, bacteria, and chemicals. Anticorrosion and building-protective coatings are applied, for example, as underwater protection and as protective paints for steel, concrete, and underground brickwork [35].

Effective anticorrosion protective coatings are obtained by combining pitch with polymers. The best-known *pitch combination coatings* are two-pack systems based on soft pitches and epoxy resins (see Section 3.3.4). The coatings are resistant both to low temperatures and high temperatures up to 100 °C. They are resistant to fresh and salt water, moderately strong mineral acids and alkalis (except ammonia), salt solutions, lubricating oils, and nonaromatic motor fuels. Two- and one-pack pitch–polyurethane coatings have a similar range of applications.

3.4. Processing of Tar Distillates

This chapter deals with the further processing of primary tar distillates to fractions and the various aromatic oils. The processing to technically pure aromatic compounds is dealt under separate keywords. Only the processing to technically pure phenols and pyridine bases is described in more detail here.

3.4.1. Processing of Anthracene Oil

In the primary distillation of coal tar, ca. 20–30% of the crude tar distills as anthracene oil, which at room temperature is a semisolid, green-brown, crystalline material with a boiling range of ca. 300–450 °C. In most primary distillation plants two fractions are produced: the larger fraction being low-boiling anthracene oil I, and as the smaller, high-boiling anthracene oil II. Anthracene oil I contains phenanthrene, anthracene, and carbazole, whereas fluoranthene and pyrene are concentrated in anthracene oil II.

The crude anthracene oil or anthracene oil I is cooled in water-cooled discontinuous or continuous crystallizers and partially crystallized to give a filtered anthracene oil and anthracene residues. Filtered anthracene oil does not crystallize at normal temperature. The discontinuous cooling process is usually used. Anthracene oil at 80–100 °C is pumped from primary distillation into cylindrical vertical stirred coolers. It is cooled over 24 – 36 h to 20 – 30 °C by indirect water cooling. Early sedimentation of crystals is prevented by a slowly rotating stirrer (ca. 15 rpm). After cooling the crystal slurry is discharged via a gate through the conical lower part of the cooler and centrifuged with a peeler centrifuge. The crude anthracene oil yields 10 – 15 % anthracene residues and 85 – 90 % filtered anthracene oil. The residues are further processed to anthracene, phenanthrene, and carbazole.

Filtered anthracene oil I and high-boiling anthracene oil II are important blending components for technical *aromatic oils* and pitch – oil mixtures. Aromatic oils with anthracene oil as main component include carbon-black oils, impregnating oils (creosote), medium fuel oils, special diesel fuels, gasometer oils, fluxing oils, naphthalene wash oil and numerous others. Table 6 lists specifications for some of these aromatic tar oils.

Carbon-black oils. Due to their low sulfur content and high aromaticity, coal-tar oils are high-value raw materials for carbon black. Carbon-black oils obtained from coal tar have, due to their high content of condensed aromatics, the highest Correlation Index (Bureau of Mines). They are pyrolyzed at ca. 1500 °C by the furnace-black process to yield filler blacks, mostly for rubber, and pigment blacks, mainly for printing inks. Depending on the required quality of the carbon black the yield is 40 – 70 %.

Impregnating oils (creosote) obtained from coal tar are important for long-term wood preservation. The aromatic components of anthracene oil and other tar-oil components have, when mixed, strong antifungal properties and significant pesticidal properties for wood pests such as the house long horn beetle (*Hylotropes bajulus*), the ship worm (*Toredo navalis*), anobium beetles (*Anobium punctatum*) and termites, and furthermore inhibit dry-rot fungi. Railway sleepers, telegraph poles, pit props, timbers for use in water and other building timbers are protected against decay and insect damage over a long term by vacuum – pressure impregnation with tar oils. Coal-tar impregnating oils are not washed out by water, are noncorrosive, and have high electrical resistance. The service life of such impregnated woods is prolonged from 5 – 15 years to 30 – 50 years. The various types of impregnating oils are blends of filtered anthracene oil, wash oil, and filtered naphthalene oil.

Fuel oils based on coal tar have a short, hot flame with high radiation intensity due to their high carbon content. Because of these properties they are used in various metallurgical plants. Coal-tar fuel oils are produced in varying types which differ in boiling range, density, viscosity, and sedimentation behavior. Table 6 lists data for a light fuel oil. Heavy fuel oils for steel works may also contain pitch, which itself is used as extraheavy fuel oil. Coal-tar fuel oils have a minimum calorific value of ca. 38 MJ/kg.

Diesel fuels manufactured from coal tar are suitable for the operation of stationary, slowly running diesel motors. They are blends of filtered anthracene oil ($< 40\%$) and tar oil boiling below 300 °C ($> 60\%$). This diesel fuel contains no active sulfur (copper strip test) or corrosive chlorine ($< 0.01\%$) and has, like all coal-tar oils, a minimum calorific value of ca. 38 MJ/kg.

Table 6. Specifications of various coal-tar oils

	Impregnating oil for railway sleepers (Deutsche Bundesbahn)	Impregnating oil for telegraph poles (Deutsche Bundespost)	Impregnating oil AWPA (P 1–78) (USA)	Carbon-black oil (light quality)	Carbonblack oil (heavy quality)	Fluxing oil	Light fuel oil (DIN 51603)	Diesel fuel	Benzole absorbing oil
Density at 20 °C, g/cm³	1.04–1.15	1.04–1.15	at 15.5 °C ≥ 1.050	1.07–1.1	1.1–1.15	1.02–1.07	1.02–1.10	1.04–1.10	1.02–1.06
Clear point or free of sediment, °C	≤ 23	≤ 23		< 40	< 50	0	0	0	< 0
Water content, %	≤ 1	≤ 1	≤ 1.5	≤ 0.2		≤ 0.5	≤ 0.3 non-settleable	≤ 0.3	≤ 0.1
Toluene insolubles, %	traces	traces	≤ 0.5	≤ 0.1			≤ 1	≤ 0.1	
Coking value[a], %				≤ 1			a) ≤ 0.01[d]	≤ 0.5	
Ash content, %							b) ≤ 0.04	≤ 0.01	
Phenols, %	< 3								
Sulfur, %		< 2.3		≤ 0.8	≤ 0.8		a) ≤ 0.3[d] b) ≤ 0.8		
Correlation index[b]				≥ 140	≥ 150				
Flash point[c], °C						> 55	> 85	> 80	
Calorific value							≥ 37.7 MJ/kg	≈ 37.7 MJ/kg	
Boiling range, %	up to 235 °C: ≤ 10 up to 300 °C: 20–40 up to 355 °C: 55–75	up to 235 °C: ≤ 10 up to 300 °C: 40–60 up to 355 °C: 70–90	up to 210 °C: ≤ 2.0 up to 235 °C: ≤ 12.0 up to 270 °C: 10–35 up to 315 °C: 40–65 up to 355 °C: 60–77 residue: 23–40 density at 15.5 °C fraction 235–315 °C: ≥ 1.027 g/cm³ fraction 315–355 °C: ≥ 1.095 g/cm³	200–400 °C up to 200 °C: ≤ 20 up to 300 °C: ≥ 60 up to 350 °C: ≥ 70 up to 400 °C: ≥ 90	250–450 °C	180–360 °C		up to 250 °C: ≤ 55 up to 300 °C: ≤ 60	230–300 °C

[a] Conradson. [b] Bureau of Mines. [c] Pensky–Martens. [d] a) and b) are two types of fuel oil.

Fluxing oils for bitumen are, depending on their quality, blends of low- and high-boiling tar oils. They decrease the softening point and the viscosity of bitumen and, therefore, lower the processing temperature of road bitumen (cutback bitumen).

Other technical aromatic oils with coal-tar anthracene oil as main component are applied as solvents, flotation oils, quenching agents, and plasticizers.

3.4.2. Processing of Wash Oil

Wash oil, or Solvay oil, which makes up 7–8% of the crude tar, is obtained by coal-tar primary distillation directly as marketable benzol wash oil. It boils between 230 and 290 °C and contains, besides slight amounts of naphthalene, mainly the naphthalene homologues, indole, diphenyl, acenaphthene, dibenzofuran, fluorene, ca. 2% phenol homologues, and 4–6% quinoline bases. If suitable rectifying conditions are used, crystals will not settle out from wash oil, even at 0 °C. This serves as a good example for the mutual melting-point depression of the tar aromatics, most of which are solids.

Wash oil can be directly used as benzol wash oil. Wash oil and anthracene oil also serve as blending components for technical aromatic oils such as impregnating oil (creosote), diesel fuels, and light fuel oils. In pitch–oil blends wash oil is the heavy oil component and offers good solubility, fungicidal activity, and slow evaporation. Furthermore, wash oil is a starting material for numerous technically pure tar aromatics such as dibenzofuran; acenaphthene, indene and fluorene; indole (→ Indole) and alkylnaphthalenes (→ Naphthalene and Hydronaphthalenes).

3.4.3. Processing of Naphthalene Oil

Naphthalene oil is obtained by primary distillation of coal tar, in 10–12% yield based on crude tar. Depending on the separation efficiency of the distillation process the oil fraction, boiling range ca. 210–220 °C, has a solidification point of 65–75 °C, i.e., a naphthalene content between 73 and 90%. Besides naphthalene, the oil contains phenols, tar bases, dibenzothiophene, and small amounts of other aromatic hydrocarbons. The naphthalene oil is further processed by crystallization (→ Naphthalene and Hydronaphthalenes, → Naphthalene and Hydronaphthalenes). Phenol homologues (carbolic oil II) are generally also extracted from the filtered naphthalene oil.

3.4.4. Processing of Carbolic Oil

A yield of 2–4% carbolic oil based on crude tar, is obtained by primary distillation of coal tar as a fraction with a boiling range of 180–210 °C (carbolic oil I). It contains benzene homologues, some naphthalene, pyridine bases, and ca. 20–30% phenols. The phenols are a mixture of phenol, cresols, and dimethylphenols (xylenols).

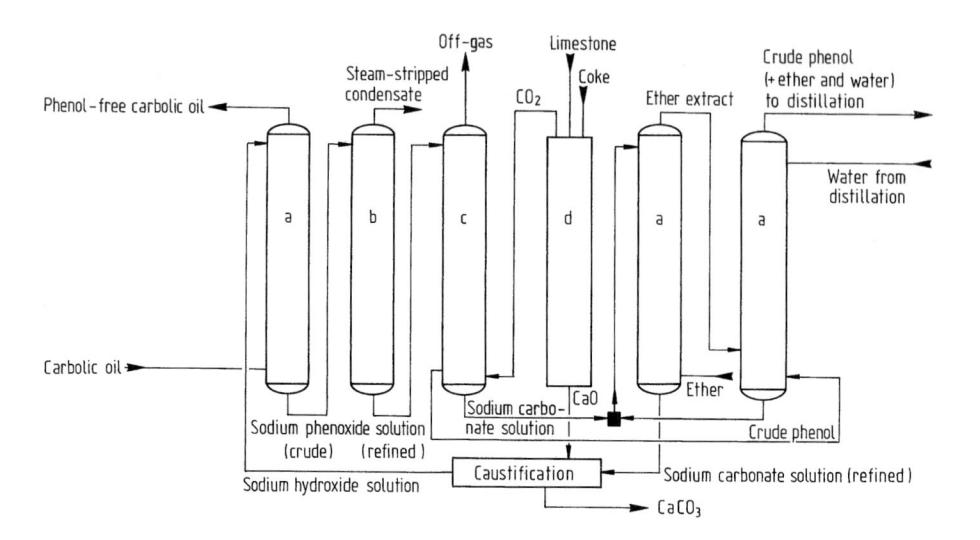

Figure 9. Isolation of crude phenol by sodium hydroxide solution extraction
a) Extraction; b) Steam stripping; c) Saturation; d) Lime kiln

Phenol and its homologues are isolated from carbolic oil I, filtered naphthalene oil (carbolic oil II) with a crude phenol content of 15–25%, light oil containing ca. 5–10% phenols, and possibly the wash oil and the fractions obtained from wash oil components with a total content of high-boiling phenol homologues of ca. 2%. Furthermore, during processing phenols extracted from the water effluents from coke-oven plants are added, either as crude phenol or sodium phenoxide solution. The coal-tar phenols are, apart from some sterically hindered dimethylphenols, readily soluble in dilute sodium hydroxide solution to form sodium phenoxides. For continuous removal of the phenols from phenol-containing oils normally a 8–12% aqueous sodium hydroxide solution is used, which dissolves only slight amounts of hydrocarbons.

Figure 9 shows the processing flow sheet for the isolation of *crude phenol* from tar fractions by extraction with sodium hydroxide solution. The process has four main steps:

1) Oil extraction with dilute sodium hydroxide solution, yielding the crude sodium phenoxide solution
2) Refining the crude sodium phenoxide solution
3) Precipitation of crude phenol from the refined sodium phenoxide solution with carbon dioxide and formation of sodium carbonate solution
4) Regeneration of the sodium hydroxide solution by caustification of the sodium carbonate solution with calcium oxide

The hot carbolic oils from the primary distillation of coal tar and naphthalene crystallization are pumped with a solution of phenols in sodium phenoxide solution, recycled fom the second step, into the mixing chamber of the first extraction step. The oil–alkali mixture is continuously recirculated at ca. 50 °C and kept at a constant level. Part of the mixture is pumped continuously into a separation column from which the heavier crude sodium phenoxide solution is drawn off at the bottom. The carbolic oil is removed at the top. About 4–6% phenol remains in the carbolic oil. In the second extraction step this carbolic oil is extracted analogously to the first step with fresh sodium hydroxide solution to yield a carbolic oil containing less than 0.5% residual phenols. The solution of phenols in sodium phenoxide solution is returned to wash the fresh carbolic oil in the first step.

The crude sodium phenoxide solution contains approximately 0.5% dissolved non-phenolic components (hydrocarbons and pyridine bases) which must be removed before the precipitation (saturation or "springing") of the crude phenols. Distillative impurity removal by steam stripping to the clear point is widely used. This takes place in continuously operating packed columns heated at the bottom with direct or indirect steam. To save energy it is favorable to connect several columns in sequence to a multistage evaporator. The vapors taken off at the top of the steam-stripping plant are condensed in a heat exchanger with the crude sodium phenoxide solution and separated into an oil and a water phase. The steam-stripped oil is fed to the light oil processing stage. The dissolved pyridine bases can be extracted from the aqueous phase with benzene.

The hot sodium phenoxide solution (ca. 100 °C) is then pumped to the top of the packed precipitation column (saturation or "springing"). Carbon dioxide is bubbled through the column. The precipitated crude phenol and the sodium carbonate solution formed flow off at the bottom and are separated.

The carbon dioxide used for precipitation of crude phenol from sodium phenoxide solution is normally produced in a lime kiln. This provides kiln gas with a carbon dioxide content of 30–35% and the calcium oxide required for the caustification of the sodium carbonate solution. The sodium carbonate solution flowing from the precipitation column still contains 3–4% physically dissolved phenols which can be isolated before caustification by ether extraction. Caustification takes place in stirred vessels at ca. 100 °C; the carbonate solution can be treated without cooling. The suspension of precipitated calcium carbonate in sodium hydroxide solution flowing out of the caustification plant is pre-separated in a sedimentation vessel. For isolation of sodium hydroxide solution the settling calcium carbonate is filtered on a vacuum rotary filter and washed. The regenerated clear sodium hydroxide solution is recycled to the carbolic oil extraction step. The filter cake of the vacuum filter consists of precipitated microcrystalline calcium carbonate with a water content of 40–50%. It can be used as neutralizing agent, for instance, in the inorganic chemical industry, for flue-gas desulfurization, or as a concrete additive.

The crude phenol from the saturation step (generally 3 parts) and the ether extract from sodium hydroxide solution refining (1 part) is subjected to extraction with water (from dehydration of the crude phenol) in a further column to demineralize the crude phenol. The crude phenol is liberated from ether and water in a continuous distillation

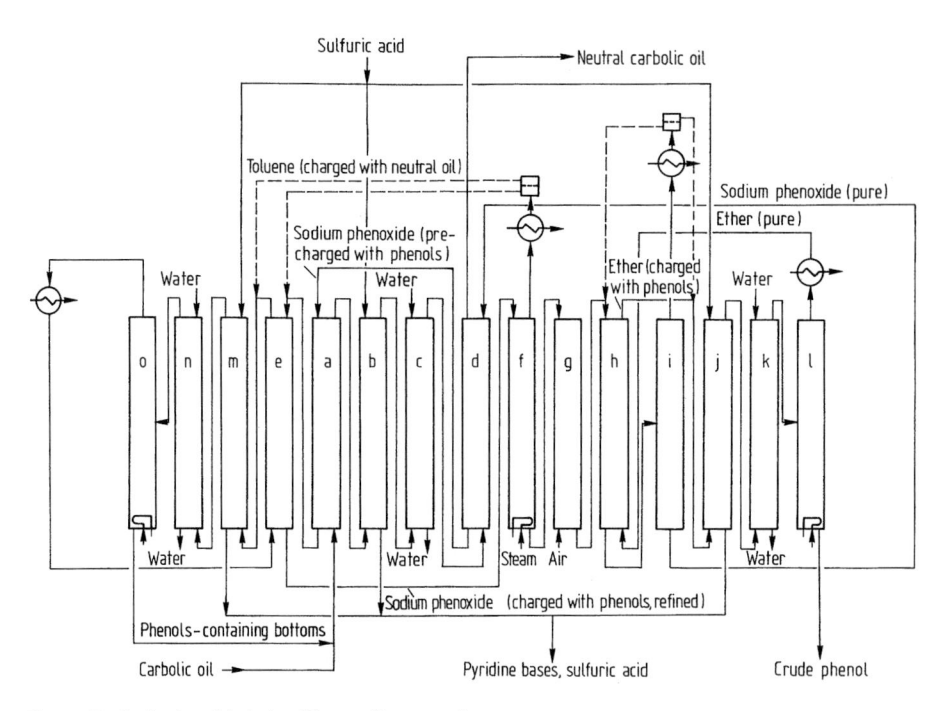

Figure 10. Crude phenol isolation (Phenoraffin process)
a) Oil extraction with sodium phenoxide (1st step); b) Oil de-basing by extraction with sulfuric acid; c) Oil extraction with water (neutralization); d) Oil extraction; e) Removal of phenols from the sodium phenoxide solution; f) Steam stripping; g) Air blowing; h) Ether extraction of the refined phenoxide solution of phenols; i) Ether stripping; j) Ether de-basing by extraction with sulfuric acid, k) Ether extraction with water (neutralization); l) Distillation column for separating ether and crude phenol; m) Toluene de-basing by extraction with sulfuric acid; n) Toluene extraction with water (neutralization); o) Toluene distillation

plant prior to distillative separation of individual phenols. This process is carried out on a large scale in a plant having a capacity of ca. 30 000 t/a [3], [46], [47].

The isolation of crude phenol by this process requires ca. 550 kg of limestone and 55 kg of coke per tonne water-free crude phenol.

The Lurgi *Phenoraffin process* is the only industrially important process in which crude phenol is extracted by selective solvents.

Figure 10 shows the processing flow sheet for crude phenol isolation by the Phenoraffin process. As selective solvent an aqueous sodium phenoxide solution is used with excellent dissolving power for phenols. The phenoxide solution can be oversaturated to 350% of the amount of stoichiometrically bound phenol. The phenols are extracted from the sodium phenoxide solution with diisopropyl ether. The plant consists of numerous extraction and distillation columns. The carbolic oil is extracted in two steps in countercurrent flow with the phenoxide solution. Between the two steps the oil is de-based with dilute sulfuric acid. Hydrocarbons and pyridine bases are extracted from the supersaturated solution of phenols in sodium phenoxide solution with, for example, toluene. After de-basing with dilute sulfuric acid and extraction with water the toluene is regenerated by distillation. The distillation bottoms, which contain phenols and hydrocarbons, are returned to the carbolic oil. The phenoxide solution after extraction with toluene and still charged with phenols is stripped with steam to remove the remaining toluene. It is then blown with air to destroy colored components and then extracted with diisopropyl ether. After the ether extraction the ether dissolved in the phenoxide solution is removed by distillation. The regenerated phenoxide solution is recirculated to the oil extraction step. The ether phase, which contains phenoxide solution, is de-based with sulfuric acid, washed with water and separated by distillation into diisopropyl ether and crude phenol. The ether is recirculated. The crude phenol is purified by distillation.

Energy and chemical requirements for production of one tonne of crude phenol by the Phenoraffin process are 5 t steam, 165 m^3 cooling water, 60 kWh electrical energy, 20 kg diisopropyl ether, and 10–200 kg sulfuric acid (depending on the amount of pyridine bases).

The crude phenol extracted from coal-tar carbolic oils by sodium hydroxide solution has the following composition: 15% water, 30% phenol, 12% o-cresol, 18% m-cresol, 12% p-cresol, 8% dimethylphenols (xylenols), 5% trimethylphenols and higher boiling phenols. The quantitative composition of the crude phenol produced in coal-tar refineries varies widely as varying amounts of sodium phenoxide solution or crude phenol from coking plant effluents are used. These mainly contain phenol and cresols but insignificant amounts of higher phenol homologues. Normally crude phenol from sodium hydroxide extraction has an ether content of ca. 2% which is decreased by water washing before distillation to ca. 0.3%.

Figure 11 shows the processing flow sheet of a continuous crude phenol distillation. The first stage is dehydration and ether removal by distillation of the crude phenol, normally under atmospheric pressure. From the dewatered, ether-free crude phenol, a residue containing higher boiling phenol homologues and unvaporizable phenolic pitch is separated in a vacuum stripping column. The stripped vapors are rectified, for instance, in four interconnected tray columns operating under a graduated vacuum of 10–140 mbar and equipped with steam-heated reboilers. The tray columns are preferably made of stainless steel. For example, in the first, 80-tray column, technically pure phenol as top product is obtained with a reflux ratio of, e.g., 12:1. From the bottom stream of this phenol column, a technically pure o-cresol is obtained as top product in the second, 65-tray column with a reflux ratio of, e.g., 25:1. From the bottom stream of this o-cresol column, a m/p-cresol mixture is obtained as top product in the third 45-tray column with a reflux ratio of, e.g., 5:1. The bottom product of this m/p-cresol column together with the bottom product of the stripping column is separated in the fourth 45-tray xylenol column into a dimethylphenol fraction and phenolic pitch. The phenolic pitch contains polymerized and oxidized phenols, which can be depolymerized to phenols by distillation with tar oils.

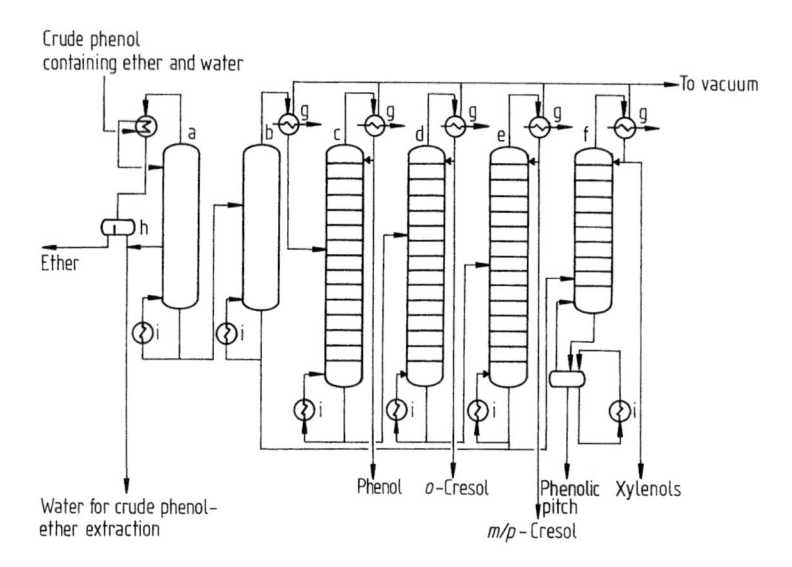

Figure 11. Distillation of crude phenol
a) Column for the separation of ether and water; b) Stripping column; c) Phenol column; d) *o*-Cresol column; e) *m/p*-Cresol column; f) Xylenol column; g) Condenser (low-pressure steam production); h) Separator; i) Reboiler

3.4.5. Processing of Light Oil

In coal-tar primary distillation, light oil is obtained azeotropically during distillative dehydration. It accounts for 0.5 – 3 % of the crude tar and has a wide boiling range of ca. 70 – 200 °C. In addition to the main components, benzene and its homologues, it contains 10 – 40 % unsaturated, polymerizable aromatic hydrocarbons, 3 – 10 % phenols and cresols, 2 – 7 % pyridine bases, and varying amounts of naphthalene. Coal-tar light oil, preferably together with carbolic oil I, is processed by distillation, extraction, and polymerization to obtain benzene and its homologues, pyridine bases, phenols, and indene – coumarone resins.

Pyridine bases are obtained from the dephenolized light oil and carbolic oil I, to which the strip oil from the sodium phenoxide solution refining can also be added, by discontinuous or continuous extraction with dilute sulfuric acid and precipitation with ammonia or sodium hydroxide solution.

Figure 12 shows the flow sheet for the recovery of pyridine bases. Extraction is carried out with 25 – 35 % sulfuric acid. After separation of acidic resins that may form the dissolved hydrocarbons are removed from the acidic solution of pyridine-base sulfates by steam stripping or solvent extraction with benzene. From this refined solution the crude pyridine is precipitated with gaseous or aqueous ammonia. The aqueous phase is separated and concentrated to yield ammonium sulfate.

The composition of crude pyridine bases from coal tar varies widely, depending mainly on the boiling range of the starting oil. The main components of a typical water-

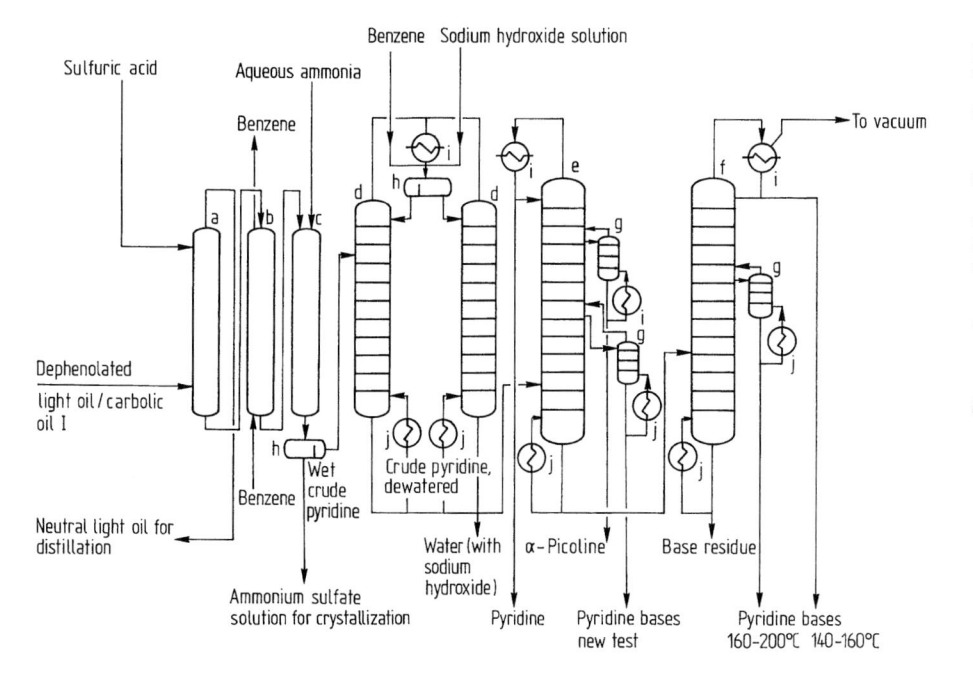

Figure 12. Recovery of pyridine bases
a) De-basing; b) Refining of the base sulfates solution; c) Base precipitation; d) Dewatering column system; e) Pyridine main column; f) Vacuum main column; g) Side column; h) Separator; i) Condenser; j) Reboiler

free tar base mixture from coal-tar light oil are listed below. The crude pyridine bases contain mainly pyridine and its methyl homologues as well as small amounts of aniline and its methyl homologues.

Pyridine	12%
α-Picoline	10%
2,6-Lutidine	6%
β-Picoline	6%
γ-Picoline	8%
2-Ethylpyridine	2%
2,5-Lutidine	4%
2,4-Lutidine	11%
2,3-Lutidine	2%
3,5-Lutidine	<1%
2,4,6- and 2,3,6-Collidine	9%
Aniline	12%
o-Toluidine	4%
m/p-Toluidine	8%
Higher boiling bases	6%

The extracted crude pyridine is distilled (Fig. 12). The crude pyridine still contains water after precipitation with ammonia. Dehydration is preferably carried out by azeotropic distillation with benzene in a continuous two-column system. In the first column the water-containing crude pyridine is separated into a water-free bottom product and an azeotropic pyridine–water top product. After condensation and cooling, sodium hydroxide solution or benzene or both are added to the top product in a separator, thus obtaining a predominantly pyridine-containing and water-containing phases. The aqueous phase is pumped to the upper tray of the second column where it is separated into a pyridine-free, aqueous sodium hydroxide solution as bottom product and an azeotropic pyridine–water mixture as top product. The top products of the first and second column are both condensed, cooled, and separated into the above-mentioned phases. The phase that contains the pyridine bases from the separation vessel is fed as reflux to the first column. The dehydrated crude pyridine is separated into pyridine and various homologue fractions by, for example, continuous rectification in a column at atmospheric pressure and a vacuum column with accompanying side columns. In the main column at atmospheric pressure, technically pure pyridine is taken off as top product. An α-picoline fraction and a base fraction ("pyridine bases new test"), with a boiling range of 130–150 °C, consisting of picolines and lutidines, are obtained in a side column. In the vacuum main column the bottom product is split into a pyridine base fraction, boiling range 160–200 °C, as top product and the base residue. A fraction containing collidine and aniline bases, boiling range 160–200 °C, is obtained in a side column. The reboilers are generally heated by pipe stills in heat exchange. By using additional side columns or by redistillation, further splitting of the base fractions is possible. The following main fractions are obtained:

1) Technically pure pyridine, *bp* 115.3 °C
2) Technically pure α-picoline, *bp* 129.4 °C
3) A *β/γ*-picoline/2,6-lutidine-fraction, boiling range 144.0–145.4 °C
4) A 2-ethylpyridine fraction, boiling range 146–157 °C
5) A lutidine fraction with 2,4- and 2,5-lutidine as main components, boiling range 157–159 °C
6) An intermediate fraction with 2,3-lutidine, 3- and 4-ethylpyridine, and 2-methyl-6-ethylpyridine as main components, boiling range 159–170 °C
7) A collidine fraction with 2,4,6- and 2,3,6-collidine as main components and small amounts of 3,5-lutidine, boiling range of 170–173 °C
8) An aniline fraction, *bp* ca. 184 °C
9) A toluidine fraction with the three toluidine isomers, boiling range 195–205 °C
10) A xylidine fraction with the six isomeric dimethylanilines, boiling range 210–225 °C

The fractions 1–7 contain pyridine and its homologues, and the fractions 8–10 contain aniline and its homologues.

Whereas pyridine and α-picoline are obtained technically pure by distillation alone, isolation of the other nitrogen bases from the distillate fractions is difficult. The majority of the pyridine bases, however, are not isolated in pure form, but are used as distillate fractions or as blends.

The commonly used technical pyridine-base blends, which are mainly employed as solvents, are classified according to boiling range. The "pyridine bases new test" are used to denaturize ethanol. The higher boiling base fractions are important anticorrosion agents. The isomeric mixture of toluidines serves as absorbant for SO_2 and as a vulcanization accelerator. The pyridine homologue fractions can be demethylized catalytically in the presence of hydrogen to give pyridine and lower pyridine homologues.

After the extraction of phenols and pyridine bases, the "neutral oil" obtained from light oil and carbolic oil I is separated by distillation into different fractions. The fraction with a boiling range of 160–185 °C is of special interest as it contains up to 60% polymerizable unsaturated hydrocarbons, with indene as main component and slight amounts of coumarone, dicyclopentadiene, and methylstyrenes. These compounds are polymerized cationically to indene–coumarone resins with Friedel–Crafts catalysts such as boron trifluoride and its complexes or aluminum chloride.

The nonpolymerizable benzene homologues which are distilled from the resins under vacuum after polymerization are combined with the remaining light-oil fractions boiling up to 185 °C, which are free of phenol and organic bases. They are processed as benzol pre-product together with the coke-oven crude benzol after refining with sulfuric acid or catalytical hydrogenation pressure refining to yield technically pure benzene, toluene, and benzene homologues. After extraction of phenols and bases, the light-oil fractions that boil above 185 °C are used as solvents for paints, varnishes, and binders.

4. Processing of Low-Temperature Coal Tars

Only in a few locations (e.g., United Kingdom, India, China, and South Africa) are coal tars from low-temperature carbonization and gasification of coal processed by distillative and extractive refining. The aim of the recovery process is mainly the isolation of the phenols (see Table 1 for composition), rather than the isolation of aromatic hydrocarbons.

Figure 13 shows the flow chart for low-temperature tar, light condensate oil, and aqueous ammonia liquor from low-temperature carbonization of highly volatile coal to produce smokeless fuels at ca. 640 °C using the Coalite process [48]–[51]. In this process one tonne of coal yields 70–75 kg of crude coal oil and about 160 L of ammonia liquor. From the crude gas, tar and light condensate oil are obtained by two-stage condensation. Analoguous to coke-oven coal tar, they are dewatered and separated into three to five fractions by continuous primary or multi-stage distillation at atmospheric pressure and under vacuum:

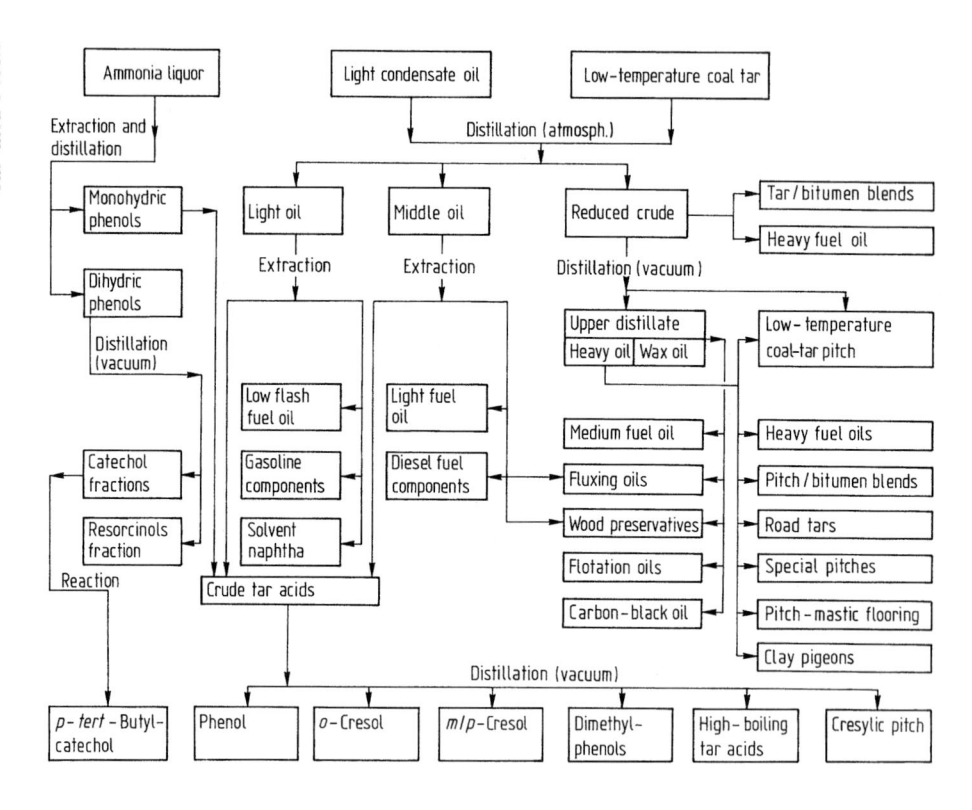

Figure 13. Processing of liquid products from low-temperature carbonization (Coalite process)

1) Light oil, boiling range ca. 90–160 °C, about 2% of crude tar, containing roughly 30% aromatic hydrocarbons, 35% alkenes, 20% alkanes + naphthenes, 10–15% phenols, and 2% tar bases.
2) Middle oil, boiling range 200–300 °C, ca. 40% of crude tar, containing roughly 45% phenols, 25% aromatic hydrocarbons, 10–15% alkenes, and 10–15% alkanes + naphthenes.
3) Upper distillate, boiling range 300–400 °C, ca. 30% of crude tar, containing ca. 65% aromatic hydrocarbons and alkanes. If required they can be separated by further vacuum distillation into a heavy oil and a wax oil with a significant content of solid paraffin hydrocarbons.
4) Low-temperature coal tar pitch as a distillation residue, ca. 25% of crude tar.

Alternatively, the residue from the base of the atmospheric fractionating column, after removal of light and middle oil (reduced crude) can be utilized, for example, as a heavy fuel oil or for making tar–bitumen blends for road construction.

Low-temperature coal-tar pitch is composed primarily of a complex mixture of hydrocarbons. It can be produced with softening points within the range of 40–180 °C (ring & ball). With a softening point of 55–60 °C it is used in production of road binders, sometimes in combination with bitumen. It is also used as an ingredient in the material prepared for coating submarine telephone cables and in the preparation of pitch–fiber pipes. The pitch that softens between 73 and 77 °C is also used for the production of pitch–fiber pipes and as a binder in briquette production. Special pitches with softening points of, for example, 82–88 °C and 100–110 °C are obtained by air blowing at elevated temperatures. These products (Florastic 85 and Florastic 100) are used for the production of colored or black pitch–mastic flooring for domestic and business premises and for the manufacture of clay pigeons. These pitch–mastic floors are resistant to oils and fats. The significantly lower content of benzo[a]pyrene in low-temperature pitch compared with pitch obtained from coke-oven coal tar is highly advantageous for these applications.

Heavy oil or upper distillate is processed to produce creosote oils with various boiling ranges. These are used as wood preservatives for railway sleepers, telegraph poles, electricity-transmission poles, and timber; fluxing oils for cutting bitumens and road tars; as carbon-black oil; and as medium industrial fuel oil.

Similar to the processing of coke-oven coal tar phenols are extracted from the *middle oil* continuously with sodium hydroxide solution to form a solution of water-soluble sodium salts known as cresylate. A neutral oil fraction then separates. Neutral oil entrained in the cresylate is removed by extraction with a lighter oil, followed by steam treatment to remove the last traces of oil. The phenolic cresylate is converted to a crude mixture of free phenols (crude tar acids) by acidification with flue gases containing carbon dioxide. Further refining of the neutral middle oil affords low-viscosity wood preservatives, fluxing oils for bitumens and tar blends, as well as components for diesel fuels and light industrial fuel oils.

Crude tar acids are also separated from *light oil* by extraction with sodium hydroxide solution. Distillation and refining of the neutral oil gives low-flash fuel oils, gasoline components, and solvent naphtha. The oil is also used as a light industrial fuel oil.

The aqueous *ammonia liquor* contains ca. 1.5 % dissolved phenols, about two-thirds of which are monohydric and one-third dihydric. These phenols are extracted from the aqueous phase with, for example, isobutyl acetate or diisopropyl ether (Lurgi Phenosolvan process. After distillation of the extractant, the crude phenol mixture is separated by continuous distillation into a lower boiling monohydric phenols fraction and a dihydric phenols fraction. The dihydric phenols fraction is separated by further distillation into catechol and resorcinols fractions. Further processing leads to catechol, 4-methylcatechol, a resorcinols fraction, and 2-methylresorcinol, and as a further downstream product, 4-*tert*-butylcatechol.

The monohydric phenol fraction isolated from ammonia liquor and the crude phenol obtained by extraction of light oil and middle oil is dehydrated by distillation and then separated by continuous distillation into the following fractions:

1) Technically pure phenol (8–10 % of the crude phenol)
2) Technically pure crude *o*-cresol (ca. 2 % of the crude phenol)
3) An *m*-/*p*-cresol fraction containing 50 % *m*-cresol (ca. 15–29 % of the crude phenol)

4) Various dimethylphenol fractions, for example a fraction with 90% 2,4- and 2,5-dimethylphenol and one with 50% *m-/p*-ethylphenol and ca. 50% 3,5-dimethylphenol. These fractions make up ca. 30–35% of the crude phenol.
5) Refined high-boiling tar acids (boiling range ca. 230–305 °C), approx. 20% of the crude phenol.
6) Cresylic pitch (R & B softening point ca. 115 °C), approx. 18% of the crude phenol.

The *high boiling tar acids* contain as main components C_3, C_4, and C_5 alkylphenols as well as 4- and 5-indanols. They are converted to black and white heavy duty disinfectants, soil sterilizers, horticultural fluids, sheep dips, foundry molding resins, agents for ore and coal flotation, engine degreasers and metal cleaners, specialized dye-markers, resins for food and beverage can linings, and germicides for impregnating toilet tissues.

The cresylic pitch is used as pulverized fuel and as a binding agent.

Because of the corrosive properties of dihydric phenols contained in low-temperature tar crude phenol, the distillation plants must be manufactured of high-quality stainless steel.

In the United Kingdom phenols and phenolic fractions from low-temperature tar crude phenols are extracted by the Coalite process. These phenols and phenolic fractions are then processed by chlorination and phosphorylation as well as by reaction with formaldehyde to give phenolic resins.

As an alternative to the chemical processing, crude tar from low-temperature carbonization of coal can be processed directly by catalytic hydrogenation to produce gasoline and diesel fuels [52].

Low-temperature coal tars obtained as byproducts of coal pressure gasification (e.g., by the Lurgi fixed-bed process) are processed similarly to the coal oil from low-temperature carbonization of coal. Attention has to be paid to the increased concentration of aromatic hydrocarbons which is due to higher generation temperatures (see Table 1). In contrast to the crude phenol resulting from lowtemperature carbonization of coal, the crude phenol, from water effluents and tar oils, contains almost no dihydric phenols but increased amounts of phenol, *o*-cresol, and *m-/p*-cresol.

Since the synthesis gas formed by coal gasification is the desired product, the crude tar formed is preferably recycled to the gasifier, thus giving an increased gas yield.

5. Processing of Other Tars and Tarlike Raw Materials

Lignite tars and crude phenols are byproducts of lignite carbonization and gasification. Large-scale plants for lignite coking and chemical processing of lignite tars in Eastern Germany [53], were shut down for economic reasons after German reunification in 1991/1992. Lignite tars are obtained in American, Czech, Indian, and German large-scale lignite gasification plants, especially in fixed-bed gasification by the Lurgi process. These tars are either recirculated to the gasifier to increase gas production (similar to coal-gasification tars) or are utilized as heavy fuel oil for steam generation. Generally only the crude phenol extracted from wastewater is chemically processed, analogous to the refining of coal-tar crude phenol to phenol, *o*-cresol and *m-/p*-cresol, xylenols, and higher boiling phenols (see Section 3.4.4and Chap. 4). The Dakota Gasification Co. in Bismarck/North Dakota runs such a plant with a phenol capacity of ca. 15 000 t/a.

Peat tars are obtained as byproducts with a yield up to 10 % during carbonization of peat to peat coke, which is used as low-ash metallurgical carbon and for production of activated carbons. Peat tars are used only as low-sulfur industrial fuel oils. Technically the distillation of peat tar to obtain oil is possible. This distillation product can be extracted to yield crude phenol, containing as its main components phenol, the three isomeric cresols, guaiacol (2-methoxyphenol), 2,4-, 2,3-, and 3,5-dimethylphenol, *p*-ethylphenol, and higher phenol homologues, as well as small amounts of catechol and its homologues. Peat-tar oils can be used to formulate wood preservatives.

Wood tars, methanol, and acetic acid are the liquid byproducts of wood carbonization to give activated carbons and wood charcoal for iron smelting. In these plants wood tar normally is used directly as a low-sulfur fuel oil. Wood tar can be separated into wood-tar pitch and various tar-oil fractions. The tar oils are suited to the formulation of wood preservatives and flotation oils. Extraction with sodium hydroxide solution gives wood-tar creosote, which contains phenol, cresols, guaiacol, and other phenol ethers as main components. Due to its disinfectant properties purified beechwood creosote, boiling range 200–220 °C, with cresols and guaiacol as main components, serves as a formulating agent for pharmaceutical preparations (e.g., cough mixtures). Furthermore, it is used as auxiliary solution in mercury maximum/minimum thermometers. Wood-tar pitch can be used as putty for the polishing of optical glasses. Crude wood tars are suitable for jute impregnation, serve as binders for briquettes and active carbon, as game deterrents, and as disinfectant formulations in veterinary medicine.

Oil-shale tars, also called shale oils, are the main liquid products of oil-shale retorting. Depending on composition (see Table 1), oil-shale tars can be refined by pressure hydrogenation and subsequent distillation to fuel oils, gasoline and diesel fuels, or can be carbonized [54]. Estonian oil-shale tars, rich in crude phenol (cucersit oil, obtained by the Kivitter process), are separated by distillation into various oil fractions, boiling up to 360 °C. They are used for the production of wood preservatives, road binders, and plasticizers [55]. The pitch residue can be carbonized to electrode coke. Phenols are extracted from water effluents and lower boiling distillate fractions and processed to phenolic resins, injection agents for soil compaction, tanning agents, and alkyl resorcins.

Pyrolysis residual oils are tarlike byproducts of steam cracking of naphtha and gas oil to produce ethylene/propene (→ Ethylene). Pyrolysis residual oils consist almost entirely of aromatic hydrocarbons (see Table 1). Due to the relatively high content of unsaturated, polymerizable aromatics they are less stable than coal tars. The main components are naphthalene and its homologues, indene and its homologues, and thermally polymerized pyrolysis residual resin (pyrolysis pitch), obtained as a dark distillation residue. By using modified coal-tar refining processes (distillation, thermal and catalytic polymerization, crystallization) the basic chemicals naphthalene, methylnaphthalenes, aromatic hydrocarbon resins and aromatic oils are obtained from pyrolysis residual oils on a large scale [56], [57]. Pyrolysis residual oils are also used as raw materials for carbon black.

Pyrolysis residual oils from scrap plastics, tyres, etc. are obtained, similarly to petrochemical pyrolysis residual oils from ethylene/propene plants, as tarlike materials. They can be processed to chemical raw materials. If the pyrolysis is carried out at ca. 700 °C in fluidized-bed or rotary-drum reactors, up to 50% pyrolysis oils, rich in aromatics, are obtained. These pyrolysis oils have a composition similar to pyrolysis residual oils from ethylene/propene plants (see Table 1). As main components they contain benzene, naphthalene and its homologues, styrene, indene and its homologues, and dark pyrolysis residual resins (pyrolysis pitch), which is obtained as distillation residue [58]–[60]. If modified processes of coal-tar refining are applied, pyrolysis oils yield benzene homologues, naphthalene and methylnaphthalenes as well as technical aromatic oils.

6. Uses of Tar Products and their Economic Importance

About 30×10^6 t/a of all kind of tars are recovered, of which ca. 15×10^6 t/a are processed to basic chemicals and other products. The main raw materials are high-temperature coal tars from coking plants. Low-temperature coal tars from coal carbonization and gasification; lignite tars, oil-shale tars, and pyrolysis residual oils are of lesser importance. Peat tars, wood tars, and pyrolysis oils from scrap are not currently processed on a large scale.

Tars are the basic raw material for:

1) Industrial carbon products: carbon electrodes for aluminum production, graphite for production of electrodes and hightemperature materials and carbon black as a rubber filler and black pigment (ca. 4×10^6 t/a)
2) Technically pure condensed aromatic hydrocarbons such as naphthalene and its homologues, anthracene, phenanthrene, acenaphthene and pyrene for synthesis of polymers, plasticizers, and dyes (ca. 1×10^6 t/a)
3) Technically pure aromatic heterocyclics such as carbazole, quinoline, pyridine, and indole for syntheses of dyes, pesticides, and pharmaceuticals (several thousand tonnes per annum)
4) Phenols for synthesis of thermosets, pesticides, disinfectants, and antioxidants (several hundred thousand tonnes per annum)

5) Thermoplastic indene–coumarone resins for the formulation of paints, varnishes, adhesives, and rubber products (ca. 100×10^3 t/a)
6) Aromatic oils as wood preservatives, solvents, and gas-washing oils (> 2×10^6 t/a)

Whereas aliphatic hydrocarbons and mononuclear aromatics presently are mainly produced from petrochemicals, polynuclear aromatics from coal tar cover a high percentage of the total requirements. In the case of naphthalene the fraction arising from coke-oven coal tar is ca. 85% worldwide. Naphthalene homologues, the polynuclear aromatics anthracene, phenanthrene, acenaphthene and pyrene, and the heterocyclics carbazole and quinoline are obtained exclusively from coal tar. The portion of phenols derived from tar is only 3% worldwide, whereas in case of phenol homologues it is ca. 40%. Approximately 20% of world aromatic hydrocarbon resin production is accounted for by indene–coumarone resins based on coal tar. In the case of binders for carbon and graphite electrodes, coal-tar electrode pitches account for over 90%. Carbon-black oils from coal tar are a basic raw material for more than a quarter of world carbon-black production (> 2 × 10^6 t/a), and 75% of oil-based wood preservatives for pressure-impregnation of timber (> 10^6 t/a) are based on coal tar.

As discussed above, tars and pitches of different types are used also as industrial fuel oils, motor fuels, binders and coating materials as well as for co-processing together with coal for gasification and coking.

7. Toxicology and Ecotoxicology

7.1. Toxicology

The acute toxicity of tars and pitches is low, but high doses of polycyclic aromatic hydrocarbons and structurally related heterocyclic systems are known to result in necrosis of the adrenal gland.

Tars or their fractions may cause irritation of the skin and eyes by direct or indirect exposure. The extent of skin irritation varies, depending on the origin of the tar, the exposure time, and involvement of light. High-temperature coal tar has the strongest, and wood tar the weakest impact on skin. Contact with unprotected human skin may lead to dermatitis, inflammation of the sebaceous glands (folliculitis), and acne. Several years of exposure can cause brownish, speckled pigmentation (melanosis) and tar or pitch warts [61, pp. 130–132].

Certain tar constituents are activated by UV light, resulting in intensified irritant skin reactions (phototoxicity) or allergic contact dermatitis (photosensitization); this may occur also on contact with minor quantities of volatiles from tar or tar fractions.

Pharmaceutical formulations of coal tars, especially in combination with exposure to UV light, have been successful in the treatment of psoriasis [62], though after excessive usage of tar ointments, melanomas or squamous-cell carcinomas developed in isolated cases [61, p. 136]. Furthermore, such combined treatment caused marked inhibition of DNA synthesis in vivo in normal and proliferative skin of hairless mice [61, p. 128].

In general, tar constituents seem to be readily available by all exposure routes and distributed throughout the body, but are also rapidly excreted [63]–[66]. On application of crude coal tar (2% solution in petrolatum) to the skin of humans for 8-h periods on two consecutive days evidence was obtained for absorption of polycyclic aromatic hydrocarbons (PAH) [65].

The potency of tumor formation can be attributed to this fraction of the mixture; PAHs are byproducts of pyrolysis of all organic materials and are generated to a major extent above 500 °C. Thus high-temperature coal tar has the highest content of highly condensed aromatic ring systems.

The experimental generation of skin carcinomas by the topical application of tars, tar fractions such as creosote, anthracene oil, and pyrolytic oil [67] of naphtha, and tar pitch is well documented [61], [68]. Case reports and epidemiological studies have provided sufficient experience about the occurrence of skin, scrotum, bladder, and other systemic cancers in exposed workers [61, pp. 121–127], [68, pp. 140–155].

Evidence has accumulated to confirm that occupational inhalation of tar or tar pitch volatiles leads to increased lung and bladder cancer mortality that is strongly associated with duration and intensity of exposure [61, pp. 134–136], [69, pp. 174–176]. There is suggestive experimental and epidemiological support for PAH causing cancer by the oral route [61], [69].

Inhalation studies with rats exposed to a condensation aerosol from tar pitch alone and in combination with carbon black [70], [71] corroborate the idea that the carcinogenic impact of PAH on the lungs is enhanced by adsorption onto respirable fine particles: After 43 weeks of exposure, the lung tumor rate was 4% with 1.1 mg/m^3 aerosol alone (= 20 µg/m^3 benzo[a]pyrene), which was less than in the control with carbon black alone (8–18%). With 2.6 mg/m^3 of aerosol (= 50 µg/m^3 benzo[a]pyrene), the tumor rate amounted to 33%, but this concentration in combination with 2 or 6 mg/m^3 of carbon black led to 89 and 72%, respectively. The majority of lung tumors were squamous-cell carcinomas. This potentiating effect can be explained by a PAH depot from which carcinogenic material is sustainably released, thus resulting in long-term exposure to the target cells of the lung.

Some 50 coal-tar compounds have been studied for their cancerogenic effect under experimental conditions, mainly on the skin of rats and mice. About 30 proved to be noncarcinogenic, and ca. 20 were positive to varying extent; for status reports see [72], [73]. The strongest carcinogens are found among systems with five or more condensed benzene rings. They are not tumorigenic by themselves, but must be metabolically converted into highly reactive alkylating dihydrodiol epoxides, the most potent of which have a so-called bay region.

Bay-region dihydrodiol
epoxide of benzo[a]pyrene

Benzo[a]pyrene, the model compound, is one of the strongest carcinogens among the PAHs, and the carcinogenic potency of tar fractions is frequently related to its concentration.

Examples of carcinogenic high-molecular N- and S-heterocycles are benzologues of acridine, carbazole, and dibenzo[bd]thiophene. Apart from PAHs, 2-naphthylamine — found in tar only in very low concentrations — is well known to specifically cause bladder cancer. This was identified primarily among workers manufacturing aniline dyes where appreciable exposure to aromatic amines occurred under former working conditions [74].

7.2. Classification and Legislation

According to IARC, there is sufficient evidence for carcinogenicity of coal tars and coal tar pitches to humans (class 1) [68, pp. 174–176]. In the United States and Germany, the classification is A1 (human carcinogen). In Germany, the occupational limit value for benzo[a]pyrene in volatiles from pitch and coke processing is set at 5 µg/m^3, from other sources at 2 µg/m^3 [75]. In the United States the amount of coal-tar/pitch volatiles is determined as benzene-soluble matter recovered from sampled dust. The TLV or MAC (maximal accepted concentration) is 0.2 mg/m^3 [76]. In the European Union, complex coal-derived materials for industrial purposes must be labelled as carcinogens unless a certain limit concentration of benzo[a]pyrene (currently 50 mg/kg) as indicator component for PAHs is met [77]. A comparative carcinogenicity study with two high-temperature coal tar oils (benzo[a]pyrene < 30 and < 300 mg/kg) is in progress [78].

7.3. Ecotoxicology

The aquatic toxicity of tar and tar-borne materials depends primarily on the physical and toxicological properties and availability of their aromatic constituents from the matrix of the mixture. However, the outcome of test procedures is very much determined by the method of preparation of test solutions and the reference basis for the computation of the effective or toxic concentrations. Therefore, the evaluation for aquatic ecotoxicological properties of mixtures which consist of sparingly soluble substances is problematic.

In particular, organic solvents or detergent-like agents used as cosolubilizers, dispersants, or emulsifiers tend to bias the measurement (see below) [79]. The choice and standardization of adequate testing methods continue to be subject of controversy and debate [80].

The highest toxicity (LC_{50} values mostly < 1 mg/L) [81], [79] was obtained with acetone solutions of high-temperature coal tar creosote. In a reliable comparative study on fish and daphnia [79], the aqueous creosote extracts yielded 96h and 48h LC_{50} values between 2 and 3.5 mg/L (referring to nominal concentration of creosote); on the other hand, about 0.7 mg/L was found when the organic solvent was used. This can be explained by the finding that the composition of the acetone solution was almost identical to that of the original material, while in the aqueous extract—reflecting more realistic conditions—less toxic components with higher water solubility were enriched.

Aqueous coal tar extracts were somewhat less toxic to fish and daphnia, with LD_{50} values ranging from about 7 to 12 mg/L (nominal). This may reflect its lower content of water soluble fractions and lower availability of the toxic constituents compared with creosote. This seems to be in agreement with findings on four tar oils in the bacteria bioluminescence assay with effective 50% doses from ca. 0.8 to 1.5 mg/L, since these data are comparable to the range of creosote toxicity mentioned above [82].

For tar pitch, the lowest LC_{50} (96 h) for fish is reported to be ca. 100 mg/L and for daphnia > 100 mg/L (nominal) [79].

8. References

General References

[1] H. G. Franck, G. Collin: *Steinkohlenteer*, Springer Verlag, Berlin 1968.
[2] G. Collin, E. Wolfrum: "Chemierohstoffe aus flüssigen Nebenprodukten der Verkokung, Verschwelung und Vergasung," in Winnacker-Küchler: *Chemische Technologie, Organische Technologie I*, 4th ed., Hanser Verlag, München 1981.
[3] G. Collin, *Erdöl, Kohle Erdgas Petrochem.* **29** (1976) 159–165; **38** (1985) 489–496.
[4] G. Collin, Proc. World Conf. Future Sources Org. Raw Mater. Chemrawn I, Toronto, July 10–13, 1978, Pergamon Press, London 1979, pp. 283–297.

Specific References

[5] M. Zander, G. Collin, *Fuel* **72** (1993) 1281–1285.
[6] H. G. Franck, *Brennst. Chem.* **36** (1955) 12–20.
[7] G. P. Blümer, R. Thoms, M. Zander, *Erdöl, Kohle Erdgas Petrochem.* **31** (1978) 197.
[8] W. Boenigk, M. W. Haenel, M. Zander, *Fuel* **69** (1990) 1226–1232.
[9] H. D. Sauerland, J. W. Stadelhofer, R. Thoms, M. Zander, *Erdöl, Kohle Erdgas Petrochem.* **30** (1977) 215–218.

[10] P. Fischer, J. Stadelhofer, M. Zander, *Fuel* **57** (1978) 345–352.
[11] M. Zander, *Erdöl, Erdgas, Kohle* **105** (1989) 373–377.
[12] M. Zander, *ACS Fuel Chem. Div. Prep.* **34** (1989) 1218–1225.
[13] J. D. Brooks, G. H. Taylor, *Nature (London)* **206** (1965) 697–699.
[14] M. Zander, *Erdöl, Kohle Erdgas Petrochem.* **38** (1985) 496–503.
[15] S. Takeshita, *CEER Chem. Econ. Eng. Rev.* **9** (1977) nos. 2/3, 25–30.
[16] Rütgerswerke, DE 976 913, 1954 (H. Krieger, E. Schweym, J. Geller).
[17] A. M. Odok, W. K. Fischer, *Light Met. (Warrendale, Pa.)* 1978, 269–286.
[18] A. Alscher, W. Gemmeke, F. Alsmeier, W. Boenigk, *Light Met. (Warrendale, Pa.)* 1987, 483–489.
[19] F. Keller, W. K. Fischer, *Light Met. (Warrendale, Pa.)* 1982, 729–740.
[20] W. K. Fischer, F. Keller, R. Perruchoud, *Light Met. (Warrendale, Pa.)* 1991, 681–686.
[21] W. Boenigk, A. Niehoff, R. Wildförster, *Light Met. (Warrendale, Pa.)* 1991, 615–619. N. I. Steward, P. H. Halley, *Light Met. (Warrendale, Pa.)* 1994, 517–524.
[22] Osaka Gas Co., DE 2 121 458, 1971 (K. Ueda, J. Kimoto, M. Moritake).
[23] G. Collin, H. Köhler, *Erdöl, Kohle Erdgas Petrochem.* **30** (1977) 257–263.
[24] C. D. Alexander, V. L. Bullough, J. W. Pendley, *Light Met. (Warrendale, Pa.)* 1971, 365–376.
[25] G. L. Ball, C. R. Gannon, J. W. Newman, *Light Met. (Warrendale, Pa.)* 1972, 127–140.
[26] W. Boenigk, A. Niehoff, R. Wildförster, *Light Met. (Warrendale, Pa.)* 1992, 581–584.
[27] W. Boenigk, J. W. Stadelhofer, Proc. Carbon '92, Essen, June 22–26, 1992, pp. 30–32.
[28] Gesellschaft für Teerverwertung, DE 642 437, 1935 (F. Meyer, O. Preiss).
[29] Koppers, DE 705 576, 1939 (H. Koppers).
[30] A. J. Gambro, D. T. Shedd, H. W. Wang, T. Yoshida, *Chem. Eng. Prog.* **65** (1969) no. 5, 75–79.
[31] A. Eckert, R. Marrett, J. W. Stadelhofer, *Erdöl, Kohle Erdgas Petrochem.* **38** (1985) 510–514.
[32] I. Moshida, Y. Q. Fei, Y. Korai, K. Fujimoto, R. Yamashita, *Carbon* **27** (1989) 375–380.
[33] Rütgerswerke, DE 3 609 348, 1986 (M. Morgenstern, C. Bertrand).
[34] G. Collin, M. Zander, Prep. 19th Biennial Conf. Carbon, University Park, PA, June 25–30, 1989, pp. 466–467.
[35] D. Stoye (ed.): *Paints, Coatings and Solvents*, VCH Verlagsgesellschaft, Weinheim 1993, pp. 91–94.
[36] Rütgerswerke, DE 2 623 574, 1976 (H. Louis, H. Köhler, R. Oberkobusch, G. Collin, V. Potschka).
[37] G. Herion, G. von Mossen, *Strasse Autobahn* **37** (1986) 3–6.
[38] H. Spengler, W. Weskamp, G. Collin, *Cokemaking Int.* **3** (1991) 25–29.
[39] B. Bujnowska, G. Collin, *Cokemaking Int.* **6** (1994) 25–31.
[40] G. Collin, B. Bujnowska, *Carbon* **32** (1994) 547–552; Prep. Carbon '94, Granada, July 5–8, 1994, pp. 80–81. J. Polaczek, G. Collin et al., DGMK-Tagungsber. 9401, Velen, April 18–20, 1994, pp. 241–245.
[41] H. Honda, *Carbon* **26** (1988) 139–156.
[42] I. C. Lewis, Prep. 20th Biennial Conf. Carbon, June 23–28, 1991, Santa Barbara, CA, pp. 156–157.
[43] I. Mochida et al., *Carbon* **28** (1990) 311–319.
[44] Y. Korai et al., *Carbon* **29** (1991) 561–567.
[45] K. J. Hüttinger, *Chemiker-Ztg.* **112** (1988) 355–366.
[46] Rütgerswerke, DE 2 919 298, 1979 (H. Spengler, D. Pachaly).
[47] Rütgerswerke, DE 3 813 634, 1988 (J. Talbiersky, K. Stolzenberg, B. Wefringhaus).
[48] W. A. Bristow, *J. Inst. Fuel* **20** (1947) 109–130.

[49] G. S. Pound, *Coke Gas* **21** (1959) 395–401, 446–453, 521–523.
[50] K. R. Payne: Yearbook Coke Oven Managers' Association 1977, pp. 154–164.
[51] A. D. Gregory: private communication.
[52] S. A. Quader, G. R. Hill, *Chem. Ing. Tech.* **43** (1971) 595–596.
[53] *Ullmann*, 4th ed., **22**, 439–441.
[54] R. Rammler, H.-J. Weiss, A. Bussmann, T. Simo, *Chem. Ing. Tech.* **53** (1981) 96–104.
[55] R. E. Joonas, W. M. Jefimow, T. A. Purre, *Chem. Ind.* **34** (1982) 739–740.
[56] Rütgerswerke, DE 1 815 568, 1968 (O. Wegener, R. Oberkobusch, G. Collin, M. Zander, H. Buffleb).
[57] J. Mostecky, M. Popl, M. Kuras, *Erdöl, Kohle Erdgas Petrochem.* **22** (1969) 388–392.
[58] H. J. Sinn, *Chem. Ing. Tech.* **46** (1974) 579–589.
[59] H. J. Sinn, W. Kaminsky, J. Janning, *Angew. Chem.* **88** (1976) 737–749.
[60] G. Collin, G. Grigoleit, G. P. Bracker, *Chem. Ing. Tech.* **50** (1978) 836–841.
[61] *IARC Monographs on the Evaluation of Carcinogenic Risks to Humans* **35** (1985).
[62] J. S. Comaish, *J. Invest. Dermatol.* **88** (Suppl.) (1987) 61s–64s.
[63] R. Modica et al., *J. Chromatogr.* **24** (1982) 352–355.
[64] C. L. Sanders et al., *J. Environ. Pathol. Toxicol. Oncol.* **7** (1986) 25–34.
[65] J. S. Storer et al., *Arch. Dermatol.* **120** (1984) 874–877.
[66] P. M. Daniel et al., *Nature (London)* **215** (1967) 1142–1146.
[67] C. S. Weil, N. I. Condra, *Am. Ind. Hyg. Assoc. J.* **38** (1977) 730.
[68] *IARC Monographs on the Evaluation of Carcinogenic Risks to Humans*, **34** (1984).
[69] *IARC Monographs on the Evaluation of Carcinogenic Risks to Humans*, updating vols. **1–42**, Suppl. 7 (1987).
[70] U. Heinrich et al. in U. Mohr (ed.): *Toxic and Carcinogenic Effects of Solid Particles in the Respiratory Tract*, 4th Int. Inhal. Symposium, Hannover, March 1–5, 1993, pp. 57–73.
[71] U. Heinrich in U. Mohr (ed.): *Assessment of Inhalation Hazards*, Springer Verlag, Berlin 1989, pp. 301–313.
[72] *IARC Monographs on the Evaluation of Carcinogenic Risks to Humans*, **32** (1983) 95–452.
[73] J. Santodonatro, P. Howard, D. Basu, "Health and Ecological Assessment of Polynuclear Aromatic Hydrocarbons," *J. Environ. Pathol. Toxicol.* **54** (1981) 1–364.
[74] S. Tola, *J. Toxicol. Environ. Health* **6** (1980) 1253–1260.
[75] List of MAK and BAT Values, VCH Verlagsgesellschaft, Weinheim 1993.
[76] Am. Conf. of Government of Industrial Hygienists (ACGIH): *Threshold Limit Values*, 1990/1991.
[77] EEC: Draft Annex Entries for Complex Coal-Derived Substances XI/51793, July 5, 1993 as Supplement of Chemical Substance Directive 67/548/EEC, Annex 1.
[78] Inter. Tar. Association, 1993 (in preparation).
[79] H. Tadokoro et al., *Ecotoxicol. Environ. Saf.* **21** (1991) 57–67.
[80] A. E. Girling et al., *Chemosphere* **24** (1992) 1469–1472.
[81] G. Sundström: "Creosote," in O. Hutzinger (ed.): *The Handbook of Environmental Chemistry*, vol. 3D, Springer Verlag, Berlin 1986, p. 158.
[82] K. G. Steinhäußer et al., *Vom Wasser* **72** (1989) 93.

Tartaric Acid

JEAN-MAURICE KASSAIAN, ETS Legré-Mante, Marseilles, France

1.	Introduction 4559	6.3.	Chemical Synthesis 4565
2.	Physical Properties 4560	7.	Environmental Protection ... 4565
3.	Optical Isomerism:	8.	Quality Specifications....... 4566
	Configuration 4562	9.	Chemical Analysis 4566
4.	Chemical Properties........ 4562	10.	Derivatives of L(+)-Tartaric
5.	Resources and Starting		Acid 4567
	Materials................ 4563	11.	Uses 4568
6.	Production 4563	12.	Economic Aspects 4569
6.1.	Production of L(+)-Tartaric	13.	Physiological and
	Acid 4563		Toxicological Behavior 4569
6.2.	Fermentation Process....... 4565	14.	References............... 4570

1. Introduction

Tartaric acid is a mixed diacid and dialcohol, empirical formula $C_4H_6O_6$.

It is also known as acidum tartaricum, 2,3-dihydroxysuccinic acid and, according to the official IUPAC nomenclature, 2,3-dihydroxybutanedioic acid. The structural formula is:

$$\text{HOOC} - \underset{\underset{\text{HO}}{|}}{\overset{\overset{\text{H}}{|}}{\text{C}}} - \underset{\underset{\text{H}}{|}}{\overset{\overset{\text{OH}}{|}}{\text{C}}} - \text{COOH}$$

The molecule contains two asymmetric carbon atoms; there are four stereoisomeric forms.

Optically Active Forms. L(+)-Dextrotartaric acid and D(−)-levotartaric acid rotate the plane of polarization of polarized light. These two forms, which are enantiomers, have identical physical and chemical properties. The optical rotatory powers are equal, but of opposite sign.

Optically Inactive Forms. DL-Tartaric acid (racemic tartaric acid), obtained by chemical synthesis, is an equimolar mixture of the two enantiomers; *meso*-tartaric acid (unresolvable tartaric acid) is inactive owing to internal compensation. These two forms have no effect on polarized light. They have the same chemical properties (identical to the active forms), but differ in certain physical properties from each other and the optically active acids.

Tartaric acid, known from antiquity as its acid potassium salt (a sparingly soluble compound which forms a deposit in wine vats), was characterized by the Swedish chemist SCHEELE, who extracted it by decomposing its calcium salt with sulfuric acid. This extraction method is still currently used. BERZELIUS, another illustrious Swedish chemist, established the structural formula in 1830. Between 1848 and 1860, PASTEUR made one of the most fundamental discoveries in organic chemistry while studying the asymmetry and optical activity of salts of tartaric acid, establishing the relationships between racemic tartaric acid and levotartaric and dextrotartaric acids.

PASTEUR showed that, if sodium ammonium tartrate is crystallized below 28 °C, a tetrahydrate is obtained in which half the crystals have supplementary hemihedral faces oriented to the left, and the other half have hemihedral faces oriented to the right. He separated these crystals by hand, using a lens, and showed that some of them yielded a new levorotatory acid, and that the others yielded the already known dextrorotatory acid [1].

L(+)-Dextrotartaric acid is the only natural form and the main form produced commercially. It is widely distributed in plants, either in its free form or as salts. It is found in many fruits, especially in grapes, where it occurs particularly as the acid potassium salt, and to a small extent as the neutral calcium salt.

During vinification, and as alcoholic fermentation proceeds, the acid potassium tartrate becomes insoluble and crystallizes on the vat walls. This deposit, called tartar, is collected and used as starting material in the production of L(+)-tartaric acid.

2. Physical Properties

The physical properties described below refer to natural L(+)-tartaric acid. They are identical to those of D(−)-tartaric acid. A comparison of the characteristic properties of the isomers are given in Table 1 [1]–[6].

Tartaric acid has molecular mass M_r 150.09. It crystallizes in the anhydrous state from solution above 5 °C, as large colorless monoclinic needles, density 1.7598 g/cm^3, melting point 169–170 °C. It has a strong acidic taste and is odorless. It is stable in air.

It decomposes on heating above 220 °C, with an odor of caramelized sugar. Tartaric acid is very soluble in water.

Table 1. Physical properties of the isomers of tartaric acid

	L(+), D(−) Tartaric acid	Racemic tartaric acid	*meso*-Tartaric acid
Crystal form	monoclinic prisms	triclinic prisms	tetragonal plates
Density, g/cm^3	1.7598	1.788	1.666
Melting point, °C (anhydrous)	169–170	206	159–160
Solubility, g/100 g H$_2$O (25 °C)	147	25	167
Solubility of potassium bitartrate, g/100 g H$_2$O (20 °C)	0.57	0.56	13.2
Solubility of calcium salt, g/100 g H$_2$O (20 °C)	0.026	0.004*	0.034
Water molecules in hydrated Ca salt	4	8	3

* 25 °C.

At various temperatures the solubility is:

t, °C	0	5	10	20
g/100 g H$_2$O	115	120	125	139
t, °C	30	40	50	60
g/100 g H$_2$O	156	176	195	218
t, °C	70	80	90	100
g/100 g H$_2$O	244	273	307	343

It is also soluble in alcohol: 100 g alcohol dissolves 20.4 g tartaric acid at 18 °C; it is less soluble in ether: 100 g diethyl ether dissolves 0.3 g tartaric acid at 18 °C.

Tartaric acid solutions have a rotatory power that varies according to concentration:

$$[A]_D^{20} = 15.050 - 0.1535\,c$$

where c is the concentration of tartaric acid, in the range 20–50% wt/vol. At 20% wt/vol, $[A]_D^{20} = +11.98$ (−11.98 for the D(−)-form).

Its enthalpy of combustion is 1149.9 kJ/mol. Its specific heat capacity (0–100 °C) is 1237 kJ kg^{-1} K^{-1}.

Tartaric acid is a diacid. Its electrolytic dissociation constants at 25 °C are $K_1 = 1.17 \times 10^{-3}$, $K_2 = 5.0 \times 10^{-5}$.

At various temperatures (°C) the relative density d_4^{15} of solutions is as follows:

1	10	20	30	40	50
1.0045	1.0469	1.0969	1.1505	1.2078	1.2696

The boiling points of solutions are: 102.2 °C for a 25% solution, 106.7 °C for a 50% solution.

The refractive index at the melting point n_D^{170} is 1.464.

3. Optical Isomerism: Configuration

The conventions using the symbols *l*, *d*, L, D to denote the enantiomers can lead to confusion, and an improved system of nomenclature for compounds containing one or more chiral centers was developed by CAHN, INGOLD, and PRELOG in 1960. It uses the symbols *R* (Latin, *rectus*, right) and *S* (*sinister*, left). This system enables the compounds to be classified without ambiguity [7], [8]:

$$
\begin{array}{ccc}
\text{COOH} & \text{COOH} & \text{COOH} \\
| & | & | \\
\text{H}-\text{C}-\text{OH} & \text{HO}-\text{C}-\text{H} & \text{H}-\text{C}-\text{OH} \\
| & | & | \\
\text{HO}-\text{C}-\text{H} & \text{H}-\text{C}-\text{OH} & \text{H}-\text{C}-\text{OH} \\
| & | & | \\
\text{COOH} & \text{COOH} & \text{COOH} \\
\mathbf{1} & \mathbf{2} & \mathbf{3}
\end{array}
$$

L(+) or (2*R*-3*R*) or [*R*-(*R**,*R**)]tartaric acid [87-69-4]

D(−) or (2*S*-3*S*) or [*S*-(*R**,*R**)]tartaric acid [147-71-7]

meso-tartaric acid [147-73-9]

DL-Racemic tartaric acid, an equimolar mixture of (**1**) and (**2**), has CAS No. [*133-37-9*].

4. Chemical Properties

Action of Heat. Tartaric acid melts between 170 and 180 °C, and isomerizes without loss of water to metatartaric acid. If heating is continued, amorphous anhydrides are obtained, which regenerate tartaric acid when boiled with water. At about 220 °C tartaric acid decomposes, swells, and ignites, leaving a carbonized residue.

Oxidation. Tartaric acid is very sensitive to oxidizing agents. The Fenton reaction involving oxidation with hydrogen peroxide in the presence of ferrous sulfate leads to the formation of dioxymaleic acid.

Reduction. Reduction with hydriodic acid leads to the formation of succinic acid.

Complexing Action. Tartaric acid acts as a complexing agent, preventing the precipitation of heavy metal salts by bases.

Boiling. Prolonged boiling of L(+)-tartaric acid solutions in the presence of alkalis (potassium and sodium hydroxides) yields the racemic and meso acids [1], [3].

5. Resources and Starting Materials

L(+)-Tartaric acid is the only one of the four isomers to be produced industrially on a large scale. The starting materials are exclusively natural residues derived from vinification. The tartaric acid in these residues exists mainly in the form of potassium bitartrate and, to a lesser extent, calcium tartrate. Depending on their tartaric acid concentration and composition, these materials are classified under four headings [9]–[12].

Tartar. This is deposited by crystallization on the walls of wine vats as the ethyl alcohol concentration increases. Tartar contains 80–90% potassium bitartrate.

Lees. These are deposited on the bottom of wine vats and contain 19–38% potassium bitartrate.

Dried Lees. These are lees that have been treated, and contain 55–70% potassium bitartrate.

Tartrate. This is produced by distilleries from wine lees or grape marc (the residue obtained after pressing grapes). After distillation to remove the alcohol, the lees and marc are precipitated with milk of lime, in the form of calcium tartrate.

6. Production

Two of the four isomers of tartaric acid are produced industrially, mainly L(+)-tartaric acid, but the racemic acid is also produced in very small amounts by chemical synthesis.

6.1. Production of L(+)-Tartaric Acid

Tartaric acid is produced mainly in Spain, France, and Italy. One of the oldest producers is Société Legré-Mante established in Marseilles in 1784. All processes for producing tartaric acid are based on the decomposition of calcium tartrate with sulfuric acid.

Two competing methods have been used to convert the starting materials to calcium tartrate: the Scheurer–Kestner acid process, and the Scheele–Lowitz neutral process, together with the Desfosses variant [9], [13]. Nowadays, only the less costly neutral process is employed.

After grinding, the dried starting material (tartar or lees) passes to a roaster where it is heated for 2 h at 160 °C, to remove the organic material that might interfere in the filtration.

After it has been roasted, the hot product is passed to a reactor where it is diluted with water and neutralized to pH 5 with milk of lime. The reaction temperature is held at 70 °C. The addition of calcium chloride or calcium sulfate in 10 % excess with respect to the stoichiometric amount gives a more complete reaction and reduces losses.

The chemical reaction that takes place is:

$$2\ KHC_4H_4O_6 + Ca(OH)_2 + CaCl_2 \longrightarrow 2\ CaC_4H_4O_6 + 2\ KCl + 2\ H_2O$$

The calcium tartrate is separated from the mother liquor by filtration on a rotary filter and washed.

The tartrate obtained from distilleries enables this first stage to be omitted.

The second stage is the decomposition of the calcium tartrate in aqueous solution with sulfuric acid, which yields tartaric acid solution and insoluble calcium sulfate

$$CaC_4H_4O_6 + H_2SO_4 \longrightarrow H_2C_4H_4O_6 + CaSO_4$$

An excess of sulfuric acid of ca. 5 % with respect to tartaric acid is necessary in order to optimize the reaction. The pulp obtained by this decomposition is filtered and washed.

The red tartaric acid solution is recovered at a concentration of ca. 200 g/L. The wash waters are recycled to the decomposition stage and the calcium sulfate is discharged as waste.

The tartaric acid solution is concentrated in vacuum evaporators at 70 °C to a concentration of 650 g/L. This new solution is then concentrated in suitable apparatus under vacuum at 70 °C until crystals appear (at ca. 1300 g/L).

This liquid–crystal mixture is then passed to crystallizers where it is slowly cooled to provide a maximum yield of crystals. After the mixture has been cooled it is dried in centrifuges, where the crystals are separated from the mother liquor. The mother liquor is subjected to repeated evaporation, granulation, and drying cycles to recover the maximum yield of granulated tartaric acid.

The granules obtained in each operation are cooled to give a tartaric acid solution of concentration 650 g/L, which is then decolorized with activated charcoal and chemically purified if excess iron and sulfuric acid are present. This operation is carried out at 70 °C.

This solution is passed through a filter press and becomes almost colorless, and is then concentrated under vacuum until crystallization starts, following which it is passed to crystallizers or cooled to provide maximum yield of crystals.

This mixture is separated by centrifugation. The mother liquor, which still contains tartaric acid, is subjected to repeated evaporation, granulation, and drying cycles before being retreated. The purified granules are dried in an oven at 140 °C and passed to a battery of sieves for separation into different grain sizes, from 2000 µm to < 100 µm (powder). Other processes are sometimes used [14], [15].

Two other production processes have been studied: one involves production of L(+)-tartaric acid by fermentation; the other involves production of racemic DL-tartaric acid by chemical synthesis.

6.2. Fermentation Process

Studies and investigations include conversion of glucose by bacteria of the species *Acetobacter suboxydans* [16]; and conversion of *cis*-epoxysuccinic acid and its sodium derivatives by bacteria of the species *Nocardia tartaricans* [17]. Publications and patents relating to other research work are listed in [18].

6.3. Chemical Synthesis

Racemic DL-tartaric acid is produced from maleic acid or its salts [19], [20]. Maleic acid in aqueous solution is oxidized by 35% hydrogen peroxide in the presence of potassium tungstate. Epoxysuccinic – tartaric acid is formed, which is hydrolyzed on boiling to racemic tartaric acid. The latter is recovered after cooling, centrifugation, washing, and drying.

Several variants of this process have been studied and published, and numerous patent applications have been filed.

Racemic DL-tartaric acid is produced commercially in a small production unit in South Africa.

7. Environmental Protection

In the production of L(+)-tartaric acid, the solid waste is discharged into bins before being sent to authorized dumps. The liquid waste is collected, treated with milk of lime, and filtered before being discharged into the sewage system, where it is retreated by municipal purification plants.

Daily analyses of the COD, suspended matter, and sulfates in the effluent are obligatory. Independent checks are carried out by authorized bodies.

Tartaric acid and its salts are biodegradable compounds.

8. Quality Specifications

L(+)-Tartaric acid, 2,3-dihydroxybutanedioic acid (2R, 3R) is used mainly in the food and pharmaceutical industries, and therefore has to satisfy extremely stringent quality criteria, defined in the pharmacopoeia of each country. The two principal publications are the European Pharmacopoeia (1986) and the United States Pharmacopoeia (1990), whose requirements are very similar.

Limiting values from the European Pharmacopoeia are:

Tartaric acid content	99.5 – 101 %
Loss on drying	$\leq 0.2\%$
Calcination residue	$\leq 0.1\%$
Rotatory power in 20% wt/vol solution	$+(12-12.8°)$
Sulfates	≤ 150 ppm
Chlorides	≤ 100 ppm
Calcium	≤ 200 ppm
Heavy metals	≤ 10 ppm
Oxalic acid	≤ 350 ppm

9. Chemical Analysis

The official analysis of tartaric acid in starting materials is the Goldenberg–Géromont method (1898–1907) as modified by the VIIth Chemistry Congress, London (1909). The accuracy of this analysis is ±0.5%. This method enables total tartaric acid to be determined without distinction between the isomers or salts derived [9], [13], [21], [22].

The first stage consists in acidifying the starting material with 5.6 mol/L hydrochloric acid to dissolve all the tartaric acid derivatives (potassium bitartrate, calcium tartrate).

These derivatives are filtered, boiled, and converted to neutral potassium tartrate with a solution of potassium carbonate at 682 g/L. After filtration followed by evaporation on a water bath, the neutral potassium tartrate is reacted with 3.5 mL pure acetic acid to potassium bitartrate. The precipitate is rendered insoluble with 100 mL ethyl alcohol and collected by vacuum filtration, washed with alcohol, taken up and solubilized with boiling water in a porcelain evaporating dish. This solution is brought to the boil and titrated against sodium hydroxide 0.2 mol/L, using neutral litmus paper as indicator.

Other, significantly less accurate methods exist. The most widely used of these is spectrophotometry, whose accuracy is ca. ±2%. The reagent used is ammonium metavanadate, which forms a colored complex with tartaric acid derivatives [23], [24]. This method is most suitable for continuous analyses during production.

10. Derivatives of L(+)-Tartaric Acid

Tartaric acid forms numerous salts and esters. The commercially best-known compounds are discussed in this Chapter [1]–[3].

Sodium potassium tartrate is commonly known as Seignette's salt or Rochelle salt. It was discovered in 1672 by Pierre Seignette, a pharmacist working in La Rochelle. Empirical formula $KNaC_4H_4O_6 \cdot 4\ H_2O$, [6381-59-5], M_r 282.23, density 1.79 g/cm^3, mp 70–80 °C, decomposes at 220 °C; solubility (g per 100 mL water) 26 at 0 °C, 66 at 26 °C.

The starting material for its production is tartar with minimum tartaric acid content 68%. This is first diluted in water or in the mother liquor of a previous batch. It is then neutralized hot with caustic soda to pH 8, decolorized with activated charcoal, and chemically purified before being filtered. The filtrate is evaporated to 42° Bé at 100 °C, and passed to granulators in which Seignette's salt crystallizes on slow cooling. The salt is separated from the mother liquor by centrifugation, accompanied by washing of the granules, and is dried in a rotary furnace and sieved before packaging.

Commercially marketed grain sizes range from 2000 μm to < 250 μm (powder).

Acid potassium tartrate (potassium bitartrate, monopotassium tartrate) is commonly called cream of tartar. $KHC_4H_4O_6$ [868-14-4], density 1.96 g/cm^3; solubility (g per 100 mL water) 0.57 at 20 °C, 6.1 at 100 °C.

This salt is mainly produced from the mother liquor of Seignette's salt. These decolorized, purified, and filtered solutions are acidified to pH 3.5 with hydrochloric or sulfuric acid. Since cream of tartar is sparingly soluble, it precipitates and is recovered by centrifugation, dried, and ground before being packaged as fine powder.

Potassium antimonyl tartrate, also known as tartar emetic, was discovered in 1631 by Adrien de Mynsicht. $KSbC_4H_4O_7 \cdot \frac{1}{2} H_2O$, [28300-74-5], M_r 333.93, density 2.61 g/cm^3; solubility (g per 100 mL water) 8.7 at 25 °C, 35.7 at 100 °C.

The compound is prepared by reacting 3 parts antimony oxide with 4 parts cream of tartar diluted in water. The reaction mixture is heated to 100 °C and held at this temperature for 4 h. After filtration, tartar emetic crystallizes in the cooled solution in a mixer. The granules are extracted by centrifugation. After drying, the granules are pulverized.

Metatartaric acid is obtained by heating tartaric acid to 170–180 °C. The conversion takes place without any loss of weight. Two or more molecules react with mutual esterification.

The compound obtained is more or less polymerized. Metatartaric acid has outstanding inhibitory properties with respect to precipitation of tartaric acid salts.

11. Uses

L(+)-Tartaric acid and its derivatives are used particularly in the food, pharmaceutical, and viniculture industries. There are also numerous applications in other fields [2], [25], [26].

L(+)-Tartaric acid (E 334) is used:

1) In the acidification of wine musts
2) As an acidifier and taste enhancer in sweets, candies, jellies, jams, fruit nectars, ice creams, gelatins, and pastes
3) In fruit, vegetable, or fish preserves, where it acts as a synergistic antioxidant, and also stabilizes the pH, color, taste, and nutritional value
4) In greases and oils, as an antioxidant
5) In the preparation of carbonated beverages
6) In the pharmaceutical industry, as an excipient or carrier for the active principal, helping to correct the basicity
7) Because of its simplicity of use, stability, and high solubility, as an acidification agent for foaming tablets and powders
8) In the cement industry, and particularly in plaster and gypsum, where its retardant action facilitates handling of these materials
9) In polishing and cleaning metals

Cream of tartar (E 336) is used:

1) To accelerate the precipitation of tartaric acid salts in wine
2) In the production of chemical yeasts
3) In the production of some drugs, by virtue of its laxative action

Seignette's salt (E 337) is used:

1) In electroplating, where it improves deposition and yield
2) In electronics and piezoelectricity
3) As a reducing agent in the silvering of mirrors
4) As a laboratory reagent; it is one of the constituents of Fehling's solution
5) In the production of cigarette paper, as a combustion regulator
6) In the pharmaceutical industry, on account of its laxative action

Potassium antimonyl tartrate is used:

1) In small amounts as an expectorant in cough syrups
2) In the treatment of some tropical diseases
3) As a mordant in the textile and leather industries

Metatartaric acid is used to inhibit tartrate crystallization in table wines.

The tartrates of fatty acid monoglycerides and diglycerides are used as emulsifiers in the bakery industry.

12. Economic Aspects

L(+)-Tartaric acid currently has a world market of ca. 30 000 t. The main producers are France, Spain, and Italy; it is also produced in small amounts in Portugal, Argentina and Japan.

The tartaric acid market has remained static over the last decade. Production is entirely dependent on vinicultural raw materials, leading to large fluctuations in the retail price of tartaric acid.

13. Physiological and Toxicological Behavior

Tartaric acid, unlike some other organic acids in fruit, is not an intermediate of the Krebs cycle, which is a major center of organic synthesis in the human body. It may be ingested in more or less large amounts with foodstuffs such as fruit and wine, where it occurs in the L(+)-form. After oral absorption, about 20% is excreted in the urine, a large proportion of the remainder undergoing bacterial breakdown in the intestines. No traces are found in the feces. The absorbed fraction is rapidly eliminated from the blood, and is either excreted by the kidneys or oxidized in various tissues.

Long-term toxicity studies in rats have been carried out with L(+) monosodium tartrate. No toxic or carcinogenic effects have been detected. Moreover, no adverse effects on renal function and pathology have been detected at a daily dose of 3 g/kg.

The toxicity of the (DL)-form has been less widely studied. One study has shown that it behaves differently from the L(+)-form. The (DL)-isomer disappears more slowly from the blood and accumulates for a longer period in the kidneys, producing a weight increase in the latter. On account of this nephrotoxic effect, this isomer has not been approved for use in foodstuffs and pharmaceuticals [27]–[30].

14. References

[1] V. Grignard: *Traité de chimie organique*, vol. **11,** Masson et Cie, Paris 1945, pp. 285–331.
[2] *Merck Index* 1989, pp. 1432–1433.
[3] A. D. Wurtz: *Dictionnaire de Chimie*, vol. **3,** Librairie Hachette et Cie, Paris 1882, pp. 198–246.
[4] A. D. Wurtz: *Dictionnaire de Chimie*, vol. **7,** Librairie Hachette et Cie, Paris 1908, pp. 663–671.
[5] H. V. Ockermann: *Source Book for Food Scientists,* Avi Publishing, Westport, Conn. 1978, p. 276.
[6] *Dictionary of Organic Compounds,* 4th ed., vol. **5,** Eyre and Spottiwoode, Oxford 1965, 2943–2945.
[7] *Vogel's Text Book of Practical Organic Chemistry,* 5th ed., J. Wiley & Sons, New York 1989, pp. 4–7.
[8] J. March: *Advanced Organic Chemistry*, 4th ed., McGraw Hill, New York 1992, pp. 94–114.
[9] J. Ventre: *Dérivés Tartriques de la Vendange,* Librairie Coulet et fils, Montpellier 1909.
[10] C. Cantarelli: "Produits secondaires de la Vinification, Rapport Général," Office International de la Vigne et du Vin, no. 501 (1972) 947–967.
[11] J. Mourgues: "Valorisation des sous produits des distilleries vinicoles," *C.R. Seances Acad. Agric. Fr.* **59** (1973) 475–480.
[12] J. Mourgues, J. Mangenet, *Ind. Aliment. Agric.* **1** (1975) january.
[13] U. Roux: *La Grande Industrie des Acides Organiques,* Librairie Dunod, Paris 1939.
[14] Montecatini, US 3 069 230, 1962 (B. Pescarolo, V. Bianchi).
[15] Orandi et Massera, US 3 114 770, 1963 (L. Dabul).
[16] T. K. Yamada, T. Kodoma, T. Obata, N. Takakashi, *J. Ferment. Technol.* **49** (1971) no. 2, 85–92. K. Yamada, Y. Minoda, T. Kodama, U. Kotera, US 3 585 109, 1971. H. K. Bhat, G. N. Quazi, S. K. Chaturvedi, C. L. Chopra, *Res. Ind.* **31** (1986) no. 2, 148–152. Skowa Chem., Mitsubishi Chem. Ind., JP Kokai 93 163 193, 1993 (N. Ootomo).
[17] Tokuyama Soda, JP Kokai 77 90 690, 1977 (K. Yutani, H. Takesue, K. Fujii, Y. Miura). Tokuyama Soda, JP Kokai 7 914 917, 1979 (H. Takesue, Y. Miura, M. Shibuya, M. Toyoshima). Tokuyama Soda., DE-OS 2 605 921, 1976 (Y. Miura et al.). Takeda Chemical Industries, DE-OS 2 619 311, 1976 (Y. Kamatani et al.).
[18] T. Huang, Y. Qian, *Gongye Weishengwu (Ind. Microbiol.)* **20** (1990) no. 6, 14–17. Z. Zhang, T. Huang, *Gongye Weishengwu (Ind. Microbiol.)* **20** (1990) no. 2, 7–12, 24. Miles Laboratory, US 2 314 831, 1943 (I. Kamlet). Institute of Microbiology, Academy of Sciences, Moldavian S.S.R., SU 514 891, 1976.
[19] J. M. Church, R. Blumberg, *Ind. Eng. Chem.* **43** (1951) no. 8, 1780–1786.
[20] Imperial Chemical Industries, GB 1 442 748, 1976 (F. Lewis, P. A. Rodriguez). Faming Zhuanli Shenqing Gongkai Shuomingshu, Peoples Republic of China (Xintong Group), 1 050 712, 1991 (X. Wu, Z. Liu et al.). Faming Zhuanli Shenqing Gongkai Shuomingshu, Peoples Republic of China, 85 100 455, 1986 (Z. Zhang et al.). Nippon Peroxide, JP Kokai 7 339 437, 1973. G. M. Grinberg, S. M. Gabrielyan, N. M. Morlyan, M. K. Madroyan, SU 322 043, 1975. Österreichische Chemische Werke, DE-OS 2 555 699, 1976 (K. Petritsch, P. Korl, F. Pogoriach). Nippon Peroxide, JP Kokai 7 785 119, 1977 (K. Kazutani et al.). S. Eisler et al., *Rev. Chim. (Bucarest)* **39** (1988) no. 8, 660–663. Kawaden Fine Chemicals, JP Kokai 6 372 647, 1988 (Nozue, Moriaki).
[21] Goldenberg, Geromont et Cie, *Z. Anal. Chem.* (1898) 312.
[22] Chem. Fabrik, former Goldenberg, Geromont et Cie, *Z. Anal. Chem.* **47** (1908) 57.
[23] H. Rebelein, *Mitteilungsbl. GDCh Fachgruppe Lebensmittelchem. Gerichtl. Chem.* **24** (1970) 14.

[24] H. Tanner, M. Sandoz, *Schweiz, Z. Obst Weinbau* **108** (1972) 10.
[25] *Pharmaceutical Codex*, vol. **11**, Pharmaceutical Press, London 1979, pp. 53, 720, 836, 916.
[26] E. F. James (ed.): *The Extra Pharmacopoeia*, 30th ed., Pharmaceutical Press, London 1993, pp. 39, 900, 904, 1408.
[27] F. P. Underhill et al., *J. Pharmacol. Exp. Ther.* **43** (1931) 359, 381.
[28] FAO/OMS reports nos. 7, 17, 21, 27.
[29] P. Dupuy, "Le Métabolisme de l'Acide Tartrique," *Ann. Technol.* **2** (1960) 139–184.
[30] T. E. Furia: *Handbook of Food Additives*. The Chemical Rubber Co., Boca Raton, Fla. 1968, pp. 261–263.

Terephthalic Acid, Dimethyl Terephthalate, and Isophthalic Acid

RICHARD J. SHEEHAN, Amoco Research Center, Amoco Chemical Company, Naperville, Illinois, United States

1.	Introduction	4573	3.7.	Alternative and Past
2.	Physical Properties	4574		Technologies 4584
3.	Production	4576	4.	Environmental Protection . . . 4585
3.1.	Amoco Oxidation	4576	5.	Quality Specifications 4586
3.2.	Amoco Purification	4578	6.	Storage and Transportation . . 4587
3.3.	Multistage Oxidation	4579	7.	Uses 4587
3.4.	Dynamit-Nobel (Witten) Process	4580	8.	Economic Aspects 4588
3.5.	Esterification of Terephthalic Acid	4582	9.	Toxicology and Occupational Health 4589
3.6.	Hydrolysis of Dimethyl Terephthalate	4583	10.	References 4590

1. Introduction

Terephthalic acid [100-21-0], and isophthalic acid [121-91-5], both $C_8H_6O_4$, have the IUPAC names 1,4- and 1,3-benzenedicarboxylic acid. Dimethyl terephthalate [120-61-6], $C_{10}H_{10}O_4$, is also known as 1,4-benzenedicarboxylic acid dimethyl ester. The acids are produced by oxidation of the methyl groups on the corresponding p-xylene [106-42-3] or m-xylene [108-38-3]. After oxidation to a carboxylic acid, reaction with methanol [67-56-1] gives the methyl ester, dimethyl terephthalate.

Terephthalic acid and dimethyl terephthalate are used to make saturated polyesters with aliphatic diols as the comonomer. Isophthalic acid is used as a feedstock for unsaturated polyesters as well as a comonomer in some saturated products. Structures are as follows:

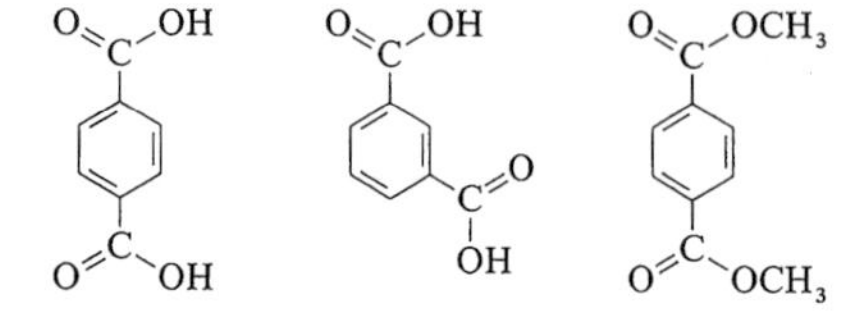

Terephthalic acid

Isophthalic acid

Dimethyl terephthalate

Terephthalic acid came to prominence through the work of WHINFIELD and DICKSON in Britain around 1940 [4]. Earlier work by CAROTHERS and coworkers in the United States established the feasibility of producing high molecular weight linear polyesters by reacting diacids with diols, but they used aliphatic diacids and diols. These made polyesters which were unsuitable to be spun into fibers. WHINFIELD and DICKSON found that symmetrical aromatic diacids yield high-melting, crystalline, and fiber-forming materials; poly(ethylene terephthalate) has since become the largest volume synthetic fiber. Worldwide, terephthalic acid plus dimethyl terephthalate ranked about 25th in tonnage of all chemicals produced in 1992, and about tenth in terms of organic chemicals.

2. Physical Properties

Terephthalic acid, M_r 166.13, is available acommercially as a free-flowing powder composed of rounded crystals. It forms needles if recrystallized slowly. Vapor pressure is low: 0.097 kPa at 250 °C, with sublimation at 402 °C and atmospheric pressure. Melting has been reported at 427 °C.

Dimethyl terephthalate, M_r 194.19, melts at 140.6 °C and has sufficient vapor pressure for vacuum distillation. The molten form is preferred commercially, but flakes and briquettes are available when long transport distances are required.

Isophthalic acid, M_r 166.13, melts at 348 °C. Vapor pressure is 0.61 kPa at 250 °C.

Both terephthalic and isophthalic acid are stable, intractable compounds with low solubilities in most solvents; isophthalic acid is 2–5 times more soluble than terephthalic acid in the same solvent. Terephthalic acid solubilities > 10 g per 100 g solvent at room temperature occur with ammonium, potassium, or sodium hydroxide, dimethylformamide, and dimethyl sulfoxide. Tetramethylurea and pyridine each dissolve ca. 7 g per 100 g.

Dimethyl terephthalate is also stable, and more soluble in common organic solvents than the acids. Its main physical properties are a melting point below the point of degradation, and a vapor pressure which allows for purification by distillation.

Other physical properties of terephthalic acid and isophthalic acid are listed in Tables 1 and 2. Some physical properties of dimethyl terephthalate are listed below:

Table 1. Physical properties of terephthalic acid and isophthalic acid

	Terephthalic	Isophthalic
Crystal density, g/cm^3, 25 °C	1.58	1.53
Specific heat, J g^{-1} K^{-1}, to 100 °C	1.20	1.22
Heat of formation, kJ/mol	−816	−803
Heat of combustion, kJ/mol, 25 °C	3189	3203
First dissociation constant	2.9×10^{-4}	2.4×10^{-4}
Second dissociation constant	3.5×10^{-5}	2.5×10^{-5}
Ignition temperature in air, °C	680	700

Table 2. Solubility of terephthalic acid and isophthalic acid (g/100 g solvent)

	25 °C	150 °C	200 °C	250 °C
Terephthalic acid solubility, g/100 g solvent				
Water	0.0017	0.24	1.7	12.6
Methanol	0.10	3.1		
Acetic acid	0.013	0.38	1.5	5.7
Isophthalic acid solubility, g/100 g solvent				
Water	0.012	2.8	25.2	
Methanol	1.06	8.1		
Acetic acid	0.23	4.3	13.8	44.5

Solubility, g/100 g solvent	
Methanol, 25 °C	1.0
Methanol, 60 °C	5.7
Ethyl acetate, 25 °C	3.5
Ethyl acetate, 60 °C	16.0
Trichloromethane, 25 °C	10.0
Trichloromethane, 60 °C	23.0
Benzene, 25 °C	2.0
Benzene, 60 °C	14.0
Toluene, 25 °C	4.3
Toluene, 60 °C	10.4
Dioxane, 25 °C	7.5
Dioxane, 60 °C	26.5
Vapor pressure, kPa	
140 °C	1.58
160 °C	2.62
200 °C	10.88
250 °C	42.60
Density at *mp*	1.07 g/cm^3
Viscosity, 180 °C	0.0071 Pa · s
Viscosity, 200 °C	0.0060 Pa · s
Specific heat, solid	1.55 J g^{-1} K^{-1}
Specific heat, liquid	1.94 J g^{-1} K^{-1}
Heat of fusion	159.1 kJ/Kg
Heat of vaporization	355.5 kJ/Kg
Heat of formation	−740.2 kJ/mol
Heat of combustion, 25 °C	4660 kJ/mol
Flash point, DIN 51 758	141 °C

3. Production

p-Xylene is the feedstock for all terephthalic acid and dimethyl terephthalate production; *m*-xylene is used for all isophthalic acid. Oxidation catalysts and conditions have been developed which give nearly quantitative oxidation of the methyl groups, leaving the benzene ring virtually untouched. These catalysts are combinations of cobalt, manganese, and bromine, or cobalt with a co-oxidant, e.g., acetaldehyde [75-07-0]. Oxygen is the oxidant in all processes. Acetic acid [64-19-7] is the reaction solvent in all but one process. Given these constant factors, there is only one industrial oxidation process, with different variations, two separate purification processes, and one process which intermixes oxidation and esterification steps.

3.1. Amoco Oxidation

About 70% of the terephthalate feedstock used worldwide is produced with a catalyst system discovered by Scientific Design [5], [6]. Almost 100% of new plants use this reaction. A separate company, Mid-Century Corporation, was established to market this technology, and subsequently purchased by Amoco Chemical. Amoco developed a commercial process, as did Mitsui Petrochemical, now Mitsui Sekka. Mitsui was an early licensee of Mid-Century. Both Amoco and Mitsui participate in joint-venture companies, and both have licensed the process. Licensees are distributed around the world, and some have relicensed the process to other companies.

A soluble cobalt – manganese – bromine catalyst system is the heart of the process. This yields nearly quantitative oxidation of the *p*-xylene methyl groups with small xylene losses [7]. Acetic acid is the solvent, and oxygen in compressed air is the oxidant. Various salts of cobalt and manganese can be used, and the bromine source can be HBr, NaBr, or tetrabromoethane [79-27-6] among others. The highly corrosive bromine – acetic acid environment requires the use of titanium-lined equipment in some parts of the process.

A feed mixture of *p*-xylene, acetic acid, and catalyst is continuously fed to the oxidation reactor (Fig. 1). The feed mixture also contains water, which is a byproduct of the reaction. The reactor is operated at 175 – 225 °C and 1500 – 3000 kPa. Compressed air is added to the reactor in excess of stoichiometric requirements to provide measurable oxygen partial pressure and to achieve high *p*-xylene conversion. The reaction is highly exothermic, releasing 2×10^8 J per kilogram *p*-xylene reacted. Water is also released. The reaction of 1 mol *p*-xylene with 3 mol dioxygen gives 1 mol terephthalic acid and 2 mol water. Only four hydrogen atoms, representing slightly over 2 wt% of the *p*-xylene molecule, are not incorporated in the terephthalic acid.

Owing to the low solubility of terephthalic acid in the solvent, most of it precipitates as it forms. This yields a three-phase system: solid terephthalic acid crystals; solvent with some dissolved terephthalic acid; and vapor consisting of nitrogen, acetic acid,

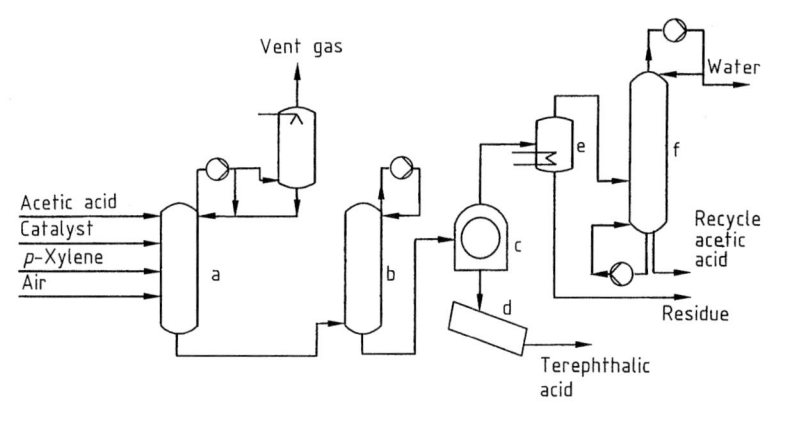

Figure 1. Catalytic, liquid-phase oxidation of *p*-xylene to terephthalic acid by the Amoco process
a) Oxidation reactor; b) Surge vessel; c) Filter; d) Dryer; e) Residue still; f) Dehydration column

water, and a small amount of oxygen. The heat of reaction is removed by solvent evaporation. A residence time up to 2 h is used. Over 98% of the *p*-xylene is reacted, and the yield to terephthalic acid is > 95 mol%. Small amounts of *p*-xylene and acetic acid are lost, owing to complete oxidation to carbon oxides, and impurities such as oxidation intermediates are present in reactor effluent. The excellent yield and low solvent loss in a single reactor pass account for the near universal selection of this technology for new plants.

The oxidation of the methyl groups occurs in steps, with two intermediates, *p*-toluic acid [99-94-5] and 4-formylbenzoic acid [619-66-9]. While 4-formylbenzoic acid is the IUPAC name of the intermediate, it is customarily referred to as 4-carboxybenzaldehyde (4-CBA).

4-Formylbenzoic acid is troublesome, owing to its structural similarity to terephthalic acid. It co-crystallizes with terephthalic acid and becomes trapped and inaccessible for completion of the oxidation. Up to 5000 ppm 4-formylbenzoic acid can be present, and this necessitates a purification step to make the terephthalic acid suitable as a feedstock for polyester production.

The slurry is passed from the reactor to one or more surge vessels where the pressure is reduced. Solid terephthalic acid is then recovered by centrifugation or filtration, and the cake is dried and stored prior to purification. This is typically referred to as crude terephthalic acid, but is > 99% pure.

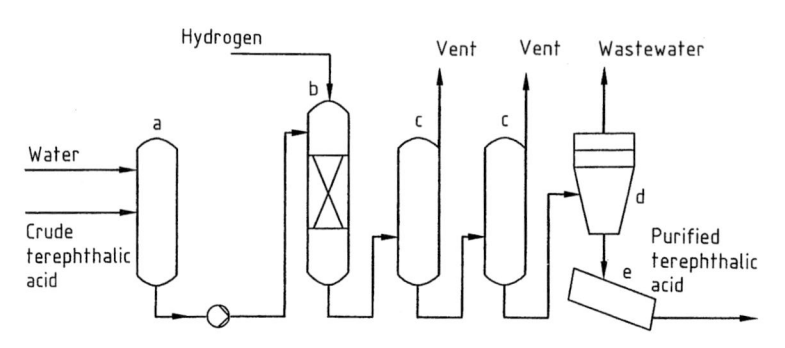

Figure 2. Purification of terephthalic acid by the Amoco process
a) Slurry drum; b) Hydrogenation reactor; c) Crystallizers; d) Centrifuge; e) Dryer

Vapor from the reactor is condensed in overhead heat exchangers, and the condensate is refluxed to the reactor. Steam is generated by the condensation and is used as a heating source in other parts of the process. Oxygen-depleted gas from the condensers is scrubbed to remove most uncondensed vapors. Similar to the reactor condensate, liquid from centrifuges or filters is sent to solvent recovery. Since the centrifugate or filtrate contains dissolved species, it is first sent to a residue still. Vapor from the still and other vents from throughout the oxidation process are sent to a solvent dehydration tower. The tower removes the water formed in the reaction as the overhead stream, and the acetic acid from the tower bottom is combined with fresh acetic acid to make up for process losses, and returned to the process.

Isophthalic acid is also produced by this process from *m*-xylene. Because isophthalic acid is several times more soluble than terephthalic acid, much less precipitates in the reactor. Consequently, isophthalic acid from this process contains much less 3-formylbenzoic acid, since it tends to stay in solution where complete oxidation can occur. Further purification was not carried out in the past, but a purified grade that is now being produced will become the standard.

3.2. Amoco Purification

The purification process developed by Amoco Chemical [8] and used on terephthalic acid from the Amoco oxidation process supplies over 60% of the terephthalate feedstock for polyester production (Fig. 2).

Crude terephthalic acid is unsuitable as a feedstock for polyester, primarily owing to the 4-formylbenzoic acid impurity concentration. There are also yellow impurities and residual amounts of catalyst metals and bromine. The Amoco purification process removes 4-formylbenzoic acid to < 25 ppm, and also gives a white powder from the slightly yellow feed.

It is necessary to make all impurities accessible to reaction, so the crude terephthalic acid is slurried with water and heated until it dissolves entirely. A solution of at least

15 wt % is obtained, and this requires a temperature ≥ 260 °C. The solution passes to a reactor where hydrogen is added and readily dissolves. The solution is contacted with a carbon-supported palladium catalyst. Reactor pressure is held above the vapor pressure of water to maintain a liquid phase.

The 4-formylbenzoic acid is converted to *p*-toluic acid in the reactor, and some colored impurities are hydrogenated to colorless compounds. The reaction is highly selective; loss of terephthalic acid by carboxylic acid reduction or ring hydrogenation is < 1%.

After reaction, the solution passes to a series of crystallizers where the pressure is sequentially decreased [9]. This results in a stepped temperature reduction, and crystallization of the terephthalic acid. The more soluble *p*-toluic acid formed in the reactor, and other impurities, remain in the mother liquor. After leaving the final crystallizer, the slurry undergoes centrifugation and/or filtration to yield a wet cake, and the cake is dried to give a free-flowing terephthalic acid powder as the product. Over 98 wt % of the incoming terephthalic acid is recovered as purified product.

As with the Amoco oxidation, this purification process is also used with isophthalic acid.

3.3. Multistage Oxidation

Several companies, mostly in Japan [10], [11], have developed processes to reduce the 4-formylbenzoic acid content to 200–300 ppm by more intensive oxidation. A separate purification step is eliminated; the concentration of 4-formylbenzoic acid is low enough for the terephthalic acid to be suitable as a feedstock for some polyester products where high feedstock purity is not critical. The product is often called medium-purity terephthalic acid, and accounts for about 11 % of the terephthalic acid produced. Most of these processes also use the catalyst system discovered by Scientific Design.

Most medium-purity terephthalic acid is produced by Mitsubishi Kasei and its licensees. They have named this product Q-PTA, and it has a typical 4-formylbenzoic acid level of ca. 290 ppm [10]. Mitsubishi has also developed a still more intensive oxidation process where the 4-formylbenzoic acid level is further reduced. The product is called S-QTA.

The oxidation of *p*-xylene in acetic acid with a cobalt–manganese–bromine catalyst is carried out as in the Amoco oxidation. The slurry is heated to 235–290 °C and oxidized further in another reactor. More catalyst can be added in addition to the temperature increase [12].

Heating gives increased terephthalic acid solubility, and as crystals dissolve, some 4-formylbenzoic acid and colored impurities are released. Although the terephthalic acid is not completely soluble at the higher temperature, the crystals can digest. Digestion is a dynamic equilibrium process wherein crystals constantly dissolve and reform. This increases the release of 4-formylbenzoic acid into solution where oxidation can be completed. While the need for a separate purification process is eliminated, another reactor is needed in the oxidation process. Also, at higher temperature, acetic acid tends to be oxidized to a larger extent to carbon oxides and water [12].

The rest of the Mitsubishi process consists of solid–liquid separation and drying to obtain the powdered product. Acetic acid must be dehydrated and recycled to the process.

Eastman Chemical has also developed a medium-purity terephthalic acid product. The process does not employ manganese, only cobalt and bromine. Two oxidation stages are used, both at 175–230 °C [13]. Instead of heating between stages to obtain increased solubility, as performed by Mitsubishi, the contents of the first reactor are sent to hydroclones where hot, fresh acetic acid displaces the mother liquor. Samarium may be added to the catalyst. A residence time up to 2 h with air addition provides sufficient digestion of the crystals to yield a 40–270 ppm level of 4-formylbenzoic acid [13]. Downstream recovery of the terephthalic acid crystals by solid–liquid separation and drying must again be performed.

3.4. Dynamit-Nobel (Witten) Process

Most dimethyl terephthalate is made by a process first developed by Chemische Werke Witten, with work also being done at a division of Standard Oil of California. Modifications were made by Hercules and Dynamit-Nobel. Hüls Troisdorf currently licenses the process with further modifications [14]. Dimethyl terephthalate is formed in four steps. First, p-xylene is passed through an oxidation reactor, where p-toluic acid is formed. It then passes to an esterification reactor, the second step, where methanol is added to form methyl p-toluate [99-75-2]. The methyl p-toluate is isolated and returned to the oxidation reactor for oxidation to monomethyl terephthalate [1679-64-7], the third step, followed by the fourth step, esterification to dimethyl terephthalate. The sequence is as follows:

The process as licensed by Hüls Troisdorf is illustrated in Figure 3. Fresh and recovered p-xylene, along with catalyst (mostly cobalt with some manganese) are combined with methyl p-toluate and fed to the liquid-phase oxidation reactor. Because bromine and acetic acid are not used, vessels lined with titanium or other expensive metals are not necessary. Oxygen supplied by compressed air is added at the bottom. Oxidation conditions are 140 – 180 °C and 500 – 800 kPa. The heat generated by oxidation is removed by vapors of unreacted p-xylene and the water of reaction. Cooling coils in the reactor are used to generate steam. The steam and reactor vapors are condensed and combined to recover p-xylene for recycle.

The oxidation effluent is then heated and sent to the esterification reactor, operated at 250 °C and 2500 kPa. Excess vaporized methanol is sparged into the esterifier, where the p-toluic acid and monomethyl terephthalate are converted noncatalytically to methyl p-toluate and dimethyl terephthalate, respectively. Overhead vapors from the esterification reactor are condensed and fed to a distillation system, where the water from the esterification is separated from methanol, which is recycled.

The remainder of the process separates the dimethyl terephthalate from methanol and methyl p-toluate, which are recycled, and residue and wastewater, which go to waste treatment.

The product from the esterifier goes to an expansion vessel. Vapor from this vessel feeds a methanol recovery column, where the methanol overhead goes to methanol recovery, and the methyl p-toluate bottoms are recycled to the oxidation reactor. Liquid from the expansion vessel feeds two vacuum distillation columns in series, which yield crude dimethyl terephthalate. The first column recovers more methyl p-toluate overhead for recycle to oxidation, and the bottoms feeds the crude dimethyl terephthalate column, where the product is taken overhead. The bottoms from the dimethyl terephthalate column, containing heavy byproducts and catalyst metals, can be mixed with water from the oxidation, which dissolves the catalyst. The resulting slurry is centrifuged; the catalyst solution is recycled, and the cake is sent to disposal.

The crude dimethyl terephthalate goes through two stages of crystallization. It is slurried with the methanol mother liquor from the second crystallization stage and dissolved by heating. Dimethyl terephthalate is crystallized from this solution on cooling, by flashing the methanol. The dimethyl terephthalate cake is separated by centrifugation, dissolved in fresh methanol, and crystallized in the same way. The wet dimethyl terephthalate cake can be melted and stored molten, or dried and flaked. There is usually a distillation column after the second centrifuge for further purification.

Mother liquor from the first centrifugation is sent to the methanol recovery system. The centrifugation yields a cleaner mother liquor which is combined with methanol recovered from the melting operation for use in dissolving the crude dimethyl terephthalate entering the crystallization section. Side-streams from throughout the process are recycled to appropriate points to maximize yield and to minimize methanol and catalyst use.

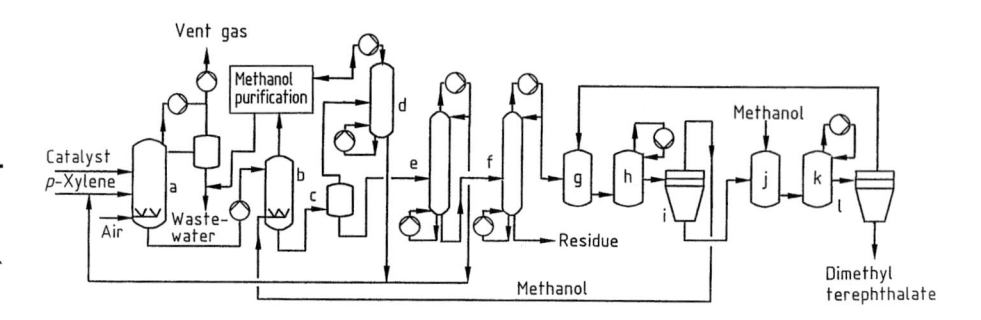

Figure 3. Production of dimethyl terephthalate by the Dynamit Nobel process
a) Oxidation reactor; b) Esterifier; c) Expansion vessel; d) Methanol recovery column; e, f) Methyl p-toluate and dimethyl terephthalate columns; g, j) Dissolvers; h, k) Crystallizers; i, l) Centrifuges

3.5. Esterification of Terephthalic Acid

Terephthalic acid can be produced and, in a separate process, esterified with methanol to dimethyl terephthalate which is then purified by distillation (Fig. 4) [15], [16]. This process can be used on highly impure terephthalic acid, because of the purification achievable by distillation.

Crude terephthalic acid and excess methanol are mixed and pumped to the esterification reactor. In this example, o-xylene [95-47-6] is used to enhance the subsequent separations. The terephthalic acid is rapidly esterified by the methanol at 250–300 °C without catalysis, although a catalyst can be used. Methanol vapor carries dimethyl terephthalate and o-xylene from the reactor to a column where o-xylene is added to scrub out monomethyl terephthalate and return it to the reactor for completion of the esterification. The vapor contains the dimethyl terephthalate product as well as methanol, o-xylene, water from the esterification, and esterified impurities in the terephthalic acid feed. Several distillation steps are needed to separate out the dimethyl terephthalate and to process the separated streams for recovery of valuable components for recycle.

The reactor overhead vapor first goes to a methanol column where methanol is removed overhead, water and some methanol form a side-draw, and the bottoms contain the dimethyl terephthalate, o-xylene, and impurities.

Next, in the o-xylene recovery column, dimethyl terephthalate purification is started; it operates at 10–20 kPa absolute so that a temperature of 200–230 °C can be used. o-Xylene is removed overhead, the methyl esters of 4-formylbenzoic acid and p-toluic acid are removed in the middle, and dimethyl terephthalate forms the bottoms. A 4-formylbenzoic acid stripping column follows, where the middle stream from the previous column is sent, so that 4-formylbenzoic ester can be removed overhead. Finally, the bottoms from both the o-xylene recovery column and the 4-formylbenzoic

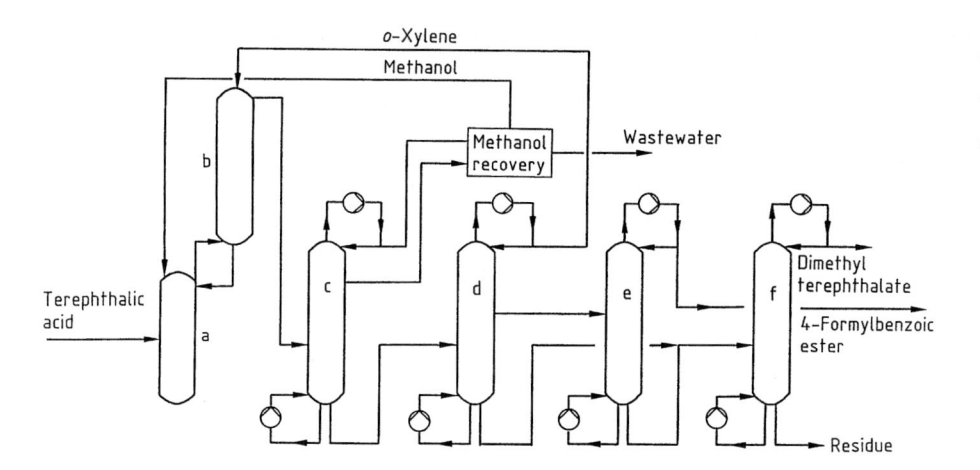

Figure 4. Esterification of terephthalic acid and separation of purified dimethyl terephthalate
a) Esterifier; b) *o*-Xylene scrubber; c) Methanol column; d) *o*-Xylene recovery column; e) 4-Formylbenzoic ester stripper; f) Purification column

acid stripper are sent to a purification column, where the dimethyl terephthalate product is taken overhead.

Methanol from the top of the methanol recovery column is sent to a purification column where the overhead contains low boilers, and the methanol from the bottoms is recycled. The methanol and water from the side-draw of the methanol recovery column are sent to a methanol dehydration column, where the water is removed and the dehydrated methanol recycled.

3.6. Hydrolysis of Dimethyl Terephthalate

High-purity terephthalic acid can be produced by hydrolysis of dimethyl terephthalate. Slightly over 2% of terephthalate feedstock is produced by this route.

Dimethyl terephthalate and recovered monomethyl terephthalate are combined with water from the methanol recovery tower of the dimethyl terephthalate process and heated to 260–280 °C at 4500–5500 kPa in a hydrolysis reactor, where the methyl esters are hydrolyzed. The overhead methanol plus water vapor is returned to the methanol recovery tower for separation. The hydrolysis reactor liquid is sent to a series of crystallizers and cyclones. After cooling to crystallize the terephthalic acid at ca. 200 °C, washing cyclones are used to remove mother liquor which contains monomethyl terephthalate. The cyclone underflow is further cooled to 100 °C, and final crystallization occurs. This slurry is centrifuged and dried to give the final product.

Monomethyl terephthalate from the cyclone overflow is recovered, again by cooling to 100 °C and centrifuging, and the cake is recycled to the hydrolysis reactor.

3.7. Alternative and Past Technologies

Oxidation of *p*- or *m*-xylene is also possible with acetic acid solvent and a cobalt catalyst with an acetaldehyde activator. Reaction temperature is 120–140 °C, and as a result the oxidation requires a residence time ≥ 2 h. Titanium-lined vessels are not required because bromine is not used. The acetaldehyde is converted to acetic acid as it promotes the reaction. After reaction, the terephthalic or isophthalic acid is recovered by centrifugation or filtration and drying, much like the Amoco process, and acetic acid is recovered for recycle by distillation.

With isophthalic acid, which has a lower concentration of 3-formylbenzoic acid, considerable purification is possible by slurrying in water to ca. 20–25 wt %, heating to 240–260 °C to dissolve, and crystallizing by cooling in batch mode. Recovery of the crystals is again by centrifugation or filtration and drying.

Eastman Chemical in the United States also used an acetaldehyde activator with a cobalt catalyst to produce terephthalic acid. Bromine is now being used in place of acetaldehyde [17].

Mobil Chemical in the United States had a commercial terephthalic acid operation which has since been abandoned [18]. As with the above processes, a cobalt catalyst was used with acetic acid solvent; the activator was 2-butanone [78-93-3]. After reaction, the crude terephthalic acid was leached by adding pure acetic acid and heating to achieve partial solubility; this removed gross impurities. Final purification was by sublimation and catalytic treatment of the vapor. Impure terephthalic acid was vaporized in a steam carrier, and catalyst and hydrogen were added to convert undesirable organic impurities. Subsequent cooling of the stream condensed the terephthalic acid, leaving impurities in the vapor. Product crystals were removed by cyclones.

A process no longer practiced is based on Henkel technology. Starting with phthalic anhydride [85-44-9], the monopotassium and dipotassium *o*-phthalate salts were formed in sequence. The dipotassium salt was isolated from solution by spray drying, and isomerized to dipotassium terephthalate under carbon dioxide at 1000–5000 kPa and 350–450 °C. This salt was dissolved in water and recycled to the start of the process, where terephthalic acid crystals formed during the production of the monopotassium salt. The crystals were recovered by filtration.

Two proposed processes which were never commercialized were by Lummus, using a dinitrile route [19] and by Eastman, using the formation of 1,4-diiodobenzene [624-38-4] with carbonylation to aromatic acids [20].

Toluene [108-88-3] is a potential feedstock for manufacture of terephthalic acid and is cheaper than *p*-xylene. Mitsubishi Gas Chemical has researched a process where a complex between toluene and hydrogen fluoride–boron trifluoride is formed, that can be carbonylated with carbon monoxide to form a *p*-tolualdehyde [*104-87-0*] complex [21]. After decomposition of the complex, *p*-tolualdehyde can be oxidized in water with a manganese–bromine catalyst system to terephthalic acid. While cheaper toluene is the feedstock and acetic acid is not required, the complexities of handling hydrogen fluoride–boron trifluoride and the need for carbon monoxide add other costs. This process has not been commercialized.

4. Environmental Protection

The three chemicals discussed here consist only of carbon, hydrogen, and oxygen, so carbon dioxide and water are the final effluents of oxidative degradation. There are traces of the cobalt–manganese–bromine catalyst in waste streams, however.

Effluents from the terephthalic acid process include water generated during oxidation and water used as the purification solvent. Gaseous emissions are mostly the oxygen-depleted air from the oxidation step. Volumes are large, but the chemicals in the streams can be effectively destroyed and removed.

Water effluents are subjected to aerobic wastewater treatment, where the dissolved species, mostly terephthalic acid, acetic acid, and impurities such as *p*-toluic acid, are oxidized to carbon dioxide and water by the action of bacteria which are acclimated to these chemicals. The bacterial growth is a sludge which can be dried and burned or spread on land. An anaerobic process has been developed to treat the wastewater [22]. Advantages include much less waste sludge production, less utility consumption, and the generation of methane, which can be burned for energy recovery.

Waste gas is scrubbed in process equipment to remove acetic acid vapors for recovery. Trace amounts of other compounds formed in the reactor can be removed by catalytic oxidation, followed by scrubbing to meet the most demanding regulations for process vents [23].

Waste streams from distillation in the dimethyl terephthalate process can be burned for energy recovery.

Another aspect of environmental protection is the recyclability of polyesters. Post-consumer containers can be ground and cleaned, and then extruded and spun into fiber to fill bedding, and quilted clothing. Textile fibers can also be made. The polyester can also be reacted to regenerate the terephthalate feedstock. High-temperature hydrolysis yields terephthalic acid. Reaction with methanol yields dimethyl terephthalate.

Table 3. Specifications and typical analyses of purified terephthalic acid

Property	Specification	Typical value
Acid number, mg KOH/g	675 ± 2	673–675
Ash, ppm	≤ 15	< 3
Metals,* ppm	≤ 9	< 2
Water, wt %	≤ 0.2	0.1
4-Formylbenzoic acid, ppm	≤ 25	15 **
p-Toluic acid, ppm	125 ± 45	125 **

* Co + Mn + Fe + Cr + Ni + Mo + Ti.
** Medium-purity terephthalic acid typically has 250 ppm 4-formylbenzoic acid and < 50 ppm p-toluic acid.

Table 4. Specifications and typical analyses of dimethyl terephthalate

Property	Specification	Typical value
Freezing point, °C	≥ 140.62	140.64
Acid number, mg KOH/g	0.03	0.01
Molten color, APHA	≤ 25	10
Color stability,* APHA	≤ 25	15
Water, wt %	≤ 0.02	0.01

* 4 h at 175 °C

5. Quality Specifications

Feedstocks for polyester production must be extremely pure; they are among the purest high-volume chemicals sold by industry. If certain impurities are present in high enough concentrations, harmful effects can be measured, e.g., monofunctional compounds can cap the polyester chain and limit molecular mass buildup. Trifunctional carboxylic acids can cause chain branching which leads to undesirable rheological and spinning properties. Colored impurities can be incorporated into the polyester. In particular, 4-formylbenzoic acid limits polyester molecular mass and causes yellowness [24]. The particle size of terephthalic acid determines how the powder flows, and the viscosity of the slurry when mixed with 1,2-ethanediol [107-21-1].

Owing to the consistent high purity of these products and the different effects of impurities, specifications put limits on specific impurities or color, rather than overall purity (Table 3). Trends in quality specifications are away from maximum or minimum values, toward a target value with an allowable range. This has been instituted to some extent by Amoco Chemical for the p-toluic acid content, for example.

In Table 3, acid number, determined by titration, is an overall purity measurement. A perfectly pure sample will have an acid number of 675.5 mg KOH/g, but the low impurity levels make the acid number meaningless as a quantitative indication of purity, and it is being phased out.

Impurities specified include p-toluic acid and 4-formylbenzoic acid, residual water, and trace metals, the ash being trace metal oxides. There are also color and particle size measurements. Polyester producers often have specific tests which they prefer for color or particle size.

Medium-purity terephthalic acid specifications address the same impurities, except that 4-formylbenzoic acid concentrations are up to 300 ppm, and residual acetic acid is present.

Typical dimethyl terephthalate specifications are shown in Table 4. Freezing point is an extremely sensitive purity indicator. Since dimethyl terephthalate is often shipped and stored in molten form, color measurements are made on the melt. A color stability test can determine the sample's resistance to degradation. This resistance is reduced by the presence of unesterified acid groups, so these residual groups are measured.

Specifications for isophthalic acid are similar to those of terephthalic acid, *m*-toluic acid and 3-formylbenzoic acid being the impurities.

6. Storage and Transportation

Terephthalic acid powder is stored in silos and transported in bulk by rail hopper car, hopper truck, or bulk container with a polyethylene liner. Hopper cars and trucks should be dedicated to this service, because of the extreme purity requirements. Shipment is also in 1 t bags, or paper bags containing ca. 25 kg. Isophthalic acid, being a much lower volume chemical, has a higher proportion shipped in bags.

Dimethyl terephthalate is stored molten in insulated heated tanks, and is preferentially shipped molten in insulated rail tank cars or tank trucks. It is also solidified into briquettes or flakes, and shipped in 1 t or 25 kg bags.

The huge production volume of terephthalic acid and dimethyl terephthalate favors the use of bulk shipment with minimized packaging costs. Rail cars are preferred where possible, and they can accommodate up to 90 t. Containers with plastic liners hold 20 t terephthalic acid, and these are the preferred ocean shipping method.

7. Uses

Terephthalic acid and dimethyl terephthalate are used almost exclusively to produce saturated polyesters. Poly(ethylene terephthalate), the alternating copolymer of terephthalic acid and 1,2-ethanediol, accounts for > 90% of this use, with a worldwide demand of 12.6×10^6 t in 1992. Textile and industrial fiber accounted for 75% of this demand. Polyester is the largest volume synthetic fiber. Food and beverage containers accounted for 13%, and constituted the fastest growing segment. Film for audio, video, and photography took 7%.

Other uses are for poly(butylene terephthalate), a high-performance molding resin made by reaction with 1,4-butanediol [*74829-49-5*], and for special industrial coatings, solvent-free coatings, electrical insulating varnishes, aramid fibers, and adhesives. A small amount of bis(2-ethylhexyl) terephthalate [*6422-86-2*] is produced as a plasticizer,

Table 5. Capacity of terephthalate feedstocks (10^3 t/a)

Feedstock *	1980	1986	1992
PTA	2800	4000	8100
MTA	420	580	1300
DMT	3300	3600	3300

* PTA purified terephthalic acid; MTA medium-purity terephthalic acid; DMT dimethyl terephthalate.

and some dimethyl terephthalate is ring-hydrogenated to produce the cyclohexane analog, 1,4-cyclohexanedicarboxylic acid [1076-97-7], for specialty polyesters and coatings.

Isophthalic acid is used as a comonomer with terephthalic acid in bottle and specialty resins. It is also a feedstock for coatings and for high-performance unsaturated polyesters, being first reacted with maleic anhydride [108-31-6] and then this resin is cross-linked with styrene [100-42-5].

8. Economic Aspects

Together, terephthalic acid and dimethyl terephthalate rank about 25th in tonnage of all chemicals, and about tenth for organic chemicals. The rapid growth of polyester use has led to an above-average growth for these feedstocks, and this is projected to continue as deeper penetration into fiber and container markets is made. The versatility of terephthalate and isophthalate polyesters is projected to drive growth in the foreseeable future.

Capacity for the three types of terephthalate feedstocks since 1980 is given in Table 5. Purified terephthalic acid has grown rapidly, and is projected to supply virtually all the growth for polyesters, about 7% annually to the year 2000. Medium-purity terephthalic acid capacity has grown by construction of a few new plants, and by conversion of some dimethyl terephthalate capacity. Dimethyl terephthalate has shown little change in capacity since 1980, and essentially none is expected in the future.

The Far East contains most of the world's population and most of the garment manufacturing industry. As a result, it is experiencing most of the growth in manufacturing capacity for polyester fibers, and most of the growth for terephthalic acid production (Table 6). Approximately 75% of the growth in terephthalate feedstocks in 1986–1992 was due to purified terephthalic acid plants built in the Far East.

Pricing is often determined in conjunction with long-term contracts, and is most influenced by the cost of p-xylene. In 1992, prices were about $ 0.60/kg for terephthalic acid and $ 0.57/kg for dimethyl terephthalate [25]. Because a given mass of terephthalic acid produces 17% more polyester than the same mass of dimethyl terephthalate, the dimethyl terephthalate price must be adjusted downward, and allowance is made for the value of the methanol generated during transesterification.

Table 6. Geographical distribution of terephthalate production (10^3 t/a)

	1980	1986	1992
North America	3300	3400	3600
Europe	1500	2100	2700
Far East	1500	2300	5800
Rest of World	220	380	600

The major producer of purified terephthalic acid is Amoco Chemical, with over 17% of the world's production in 1992. Joint ventures of Amoco, the largest being China American Petrochemical in Taiwan and Samsung Petrochemical in Korea, also produce about 17%. Other major producers are Mitsui Sekka in Japan and Imperial Chemical Industries in England and Taiwan. China is certain to become a major producer, and large plants are being planned in Southeast Asia. Larger producers of medium-purity terephthalic acid are Mitsubishi Kasei in Japan, and Eastman and DuPont in the United States. Some polyester producers have licensed technology to supply their own feedstock. All licensors are themselves producers of terephthalic acid, or have obtained the technology from a producer.

Major dimethyl terephthalate producers are DuPont, Eastman Chemical, and Hoechst Celanese in the United States, Hoechst in Europe, and Teijin Petrochemical in Japan. There are several small, older plants, mostly in Europe and China.

Isophthalic acid production capacity is 250×10^3 t/a, ca. 2% of the terephthalate total. Amoco Chemical is the major producer.

9. Toxicology and Occupational Health

Terephthalic acid, dimethyl terephthalate, and isophthalic acid all have low toxicity and cause only mild and reversible irritation to skin, eyes, and the respiratory system.

For oral ingestion of terephthalic acid, LD_{50} values for rats have been reported as 18.8 g/kg [26], and for mice as 6.4 g/kg [27], or >5 g/kg [28]. Dimethyl terephthalate LD_{50} values for rats are 6.5 g/kg [29] and 4.39 g/kg [26]. Some mortality was found at 5 g/kg oral ingestion [30], and with a 28-day oral uptake of 5% in the feed, ca. 2.5 g/kg. For isophthalic acid, LD_{50} for rats was reported as 10.4 g/kg [26].

Interperitoneal administration of terephthalic acid in rats showed LD_{50} values >1.43 g/kg [28] and 1.9 g/kg [31], with LD_{100} = 3.2 g/kg [31]. Mortality was found at 0.8 g/kg in mice and 1.6 g/kg in rats [27]. For dimethyl terephthalate, an LD_{50} of 3.9 g/kg has been reported [29], and for isophthalic acid, 4.3 g/kg [26].

Ingestion of terephthalic acid or dimethyl terephthalate results in rapid distribution and excretion in unchanged form [32]. In rats, at high ingestion rates of about 3% terephthalic acid or dimethyl terephthalate in the feed, bladder calculi are formed which are mostly calcium terephthalate; these injure the bladder wall and lead to cancer. Calculi cannot form unless the calcium terephthalate solubility is exceeded (threshold effect) [33]. Neither compound has been found to be genotoxic (Ames test) [34].

Inhalation of terephthalic acid dust appears to pose no clear hazards. Inhalation by rats for 6 h/d, 5 d/week, for 4 weeks at 25 mg/m^3 produced no fatalities [35]. Dimethyl terephthalate dust at 16.5 and 86.4 mg/m^3 for 4 h/d caused no toxicological effects after 58 days [29]. The rats intermittently reacted to the nuisance dust levels at the higher concentration.

Irritation of the skin, eyes, and mucous membranes by any of these chemicals is mild and reversible. In rabbits, the FHSA test for terephthalic acid for eyes gave scores of 14.0/110.0 and 3.5/110.0 at 1 and 24 h after exposure. The FHSA skin test score was 0.4/8.0 [35]. For isophthalic acid, eye and skin irritancy scores were 5.3/110.0 and 0.2/8.0 [36]. Irritation due to dimethyl terephthalate has also been reported to be mild [30].

Terephthalic acid, dimethyl terephthalate, and isophthalic acid can all form dust clouds. As with any flammable substance, an explosion can occur, given proper dust and oxygen concentrations. Reported limits on the explosive region for terephthalic acid dust clouds are minimum dust content of 40 g/m^3 and minimum oxygen content 12.4% at 20 °C. At 150 °C, minimum oxygen is 11.1% [37]. The maximum concentration of an explosible dust cloud has been calculated as 1400 g/m^3 [38].

Dimethyl terephthalate and isophthalic acid have a more stringent oxygen limit. In tests with carbon dioxide diluent, the minimum oxygen content was 12, 14, and 15% for dimethyl terephthalate, isophthalic acid, and terephthalic acid [39]. The oxygen content is higher for carbon dioxide diluent than for nitrogen, owing to the higher heat capacity of carbon dioxide (the 12.4% minimum oxygen content reported above for terephthalic acid was for nitrogen diluent).

For molten dimethyl terephthalate, the flash and fire points, determined by the Cleveland open-cup method, are 146 and 155 °C [1].

10. References

General References

[1] *Kirk-Othmer*, **18**, 732–777.
[2] P. Raghavendrachar, S. Ramachandran, *Ind. Eng. Chem. Prod. Res. Dev.* **31** (1992) 453–462.
[3] J. E. McIntyre, *Chem. Eng. Monogr.* **15** (1982) 400–444.

Specific References

[4] Calico Printers Assoc, GB 578 079, 1941 (J. R. Whinfield, J. T. Dickson).
[5] Mid-Century Corp., US 2 833 816, 1955 (R. S. Barker, S. A. Soffer).
[6] Mid-Century Corp., US 3 089 906, 1958 (R. S. Barker, S. A. Soffer).

[7] W. Partenheimer in D. W. Blackburn (ed.): *Catalysis of Organic Reactions*, Marcel Dekker, New York 1990, pp. 321–346.
[8] Standard Oil Company (Indiana), US 3584039, 1967 (D. H. Meyer).
[9] Standard Oil Company (Indiana), US 3931305, 1973 (J. A. Fisher).
[10] K. Matsuzawa, *Chem. Econ. Eng. Rev.* **8** (1976) no. 8, 25–30.
[11] M. Hizikata, *Chem. Econ. Eng. Rev.* **9** (1977) no. 9, 32–38.
[12] Mitsubishi Chem. Ind. Ltd., US 4877900, 1988 (A. Tamaru, Y. Izumisawa).
[13] Eastman Kodak Company, US 4447646, 1983 (G. I. Johnson, J. E. Kiefer).
[14] H. J. Korte, H. Schroeder, A. Schoengen: "The PTA Process of Hüls Troisdorf AG," AIChE Summer National Meeting, Denver, Co. 1988.
[15] E. I. DuPont de Nemours, US 2491660, 1949 (W. F. Gresham).
[16] Standard Oil Company (Indiana), US 2976030, 1957 (D. H. Meyer).
[17] *Eastman News* **43** (1988) no. 14.
[18] H. S. Bryant, C. A. Duval, L. E. McMakin, J. I. Savoca, *Chem. Eng. Prog.* **67** (1971) no. 9, 69–75.
[19] A. P. Gelbein, M. C. Sze, R. T. Whitehead, *CHEMTECH* **3** (1973) 479–483.
[20] Eastman Kodak Company, US 4705890, 1987 (G. R. Steinmetz, M. Rule).
[21] *Chem. Eng. (N.Y.)* **83** (1976) no. 17, 27–28.
[22] S. Shelley, *Chem. Eng. (N.Y.)* **98** (1991) no. 12, 90–93.
[23] T. G. Otchy, K. J. Herbert: "First Large Scale Catalytic Oxidation System for PTA Plant CO and VOC Abatement," *Ann. Air Waste Managem. Assoc. Meet. 85th*1992.
[24] W. Berger, D. Dornig, *Faserforsch. Textiltech.* **29** (1978) 256–262.
[25] *Chem. Markt. Rep.* **242** (1992) no. 4, 42.
[26] J. V. Marhold, *Sb. Vys. Toxicologickeho Vysetrien Latid A Priravku* 1972, 52.
[27] A. E. Moffitt et al., *Am. Ind. Hyg. Assoc. J.* **36** (1975) 633–641.
[28] A. Hoshi, R. Yanal, K. Kuretani, *Chem. Pharm. Bull.* **16** (1968) 1655–1660.
[29] W. J. Krasavage, *Am. Ind. Hyg. Assoc. J.* **34** (1973) 455–462.
[30] Haskel Laboratories, unpublished data, Wilmington, Delaware, USA, 1979.
[31] E. O. Grigas, R. Ruiz, D. M. Aviado, *Toxicol. Appl. Pharmacol.* **18** (1971) 469–486.
[32] A. Hoshi, K. Kuretani, *Chem. Pharm. Bull.* **16** (1968) 131–135.
[33] H. d'A. Heck, *Banbury Rep.* **25** (1987) 233–244.
[34] E. Zeiger, S. Haworth, W. Speck, K. Mortelmans, *EHP Environ. Health Perspect.* **45** (1982) 99–104.
[35] Amoco Chemical Company, Material Safety Data Sheet, Amoco TA-33, Chicago, Ill., 1993.
[36] Amoco Chemical Company, Material Safety Data Sheet, Amoco PIA, Chicago, Ill., 1993.
[37] A. D. Craven, M. G. Foster, *Combust. Flame* **11** (1967) 408–414.
[38] S. Nomur, M. Torimoto, T. Tanake, *Ind. Eng. Chem. Proc. Res. Dev.* **23** (1984) 420–423.
[39] M. Jacobson, J. Nagy, A. R. Cooper, *Rep. Invest. U.S. Bur. Mines* **5971** (1962) 25–26.

Terpenes

MANFRED EGGERSDORFER, BASF Aktiengesellschaft, Ludwigshafen, Germany

1.	General	4593	4.	Bicyclic Monoterpenes	4604
1.1.	Definition and Basic Structures	4593	4.1.	3-Carene	4604
			4.2.	Pinane	4605
1.2.	Occurrence	4596	4.3.	2-Pinane Hydroperoxide	4606
1.3.	Biosynthesis	4596	4.4.	2-Pinanol	4607
1.4.	Biodegradation	4597	4.5.	Camphene	4608
1.5.	Importance and Extraction	4597	5.	Acyclic Sesquiterpenes	4609
2.	Acyclic Monoterpenes	4598	6.	Monocyclic Sesquiterpenes	4610
2.1.	Myrcene	4598	6.1.	Bisabolene	4610
2.2.	Ocimene	4600	6.2.	Bisabolol	4611
2.3.	2,6-Dimethyl-2,4,6-octatriene	4600	7.	Bicyclic Sesquiterpenes	4611
3.	Monocyclic Monoterpenes	4601	8.	Tricyclic Sesquiterpenes	4612
3.1.	p-Menthane	4601	9.	Acyclic Diterpenes	4613
3.2.	α-Terpinene	4602	10.	Acyclic Triterpenes	4613
3.3.	Terpinolene	4602	11.	Toxicology	4614
3.4.	p-Cymene	4603	12.	References	4616
3.5.	1,8-Cineole	4604			

1. General

1.1. Definition and Basic Structures

Terpenes are natural products, whose structures are built up from isoprene units. They are classified according to the number of these units:

Monoterpenes C_{10}
Sesquiterpenes C_{15}
Diterpenes C_{20}

4593

Terpenes

Sesterpenes	C_{25}
Triterpenes	C_{30}
Tetraterpenes	C_{40}

Terpenes can be acyclic, or mono-, bi-, tri-, tetra-, and pentacyclic with 3-membered to 14-membered rings [1]–[7].

Deviations from the rule occur through rearrangement reactions or degradation of parts of the molecule during biosynthesis. The broader term terpenoids also covers natural degradation products, such as ionones, and natural and synthetic derivatives, e.g., terpene alcohols, aldehydes, ketones, acids, esters, epoxides, and hydrogenation products.

An overview of the most important basic structures demonstrates the structural variety of terpenes [8]. The compounds are normally known by their trivial names, as the systematic nomenclature is frequently less practicable for complicated structures.

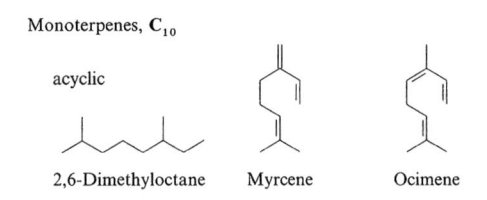

Monoterpenes, C_{10}

acyclic

2,6-Dimethyloctane Myrcene Ocimene

tetracyclic

Lanostane

Tetraterpenes, C_{40}

Perhydro-β-carotene

Polyprenes

$n = 8-12$ in plants
$n = 16-21$ in animals

monocyclic

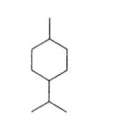

p-Menthane 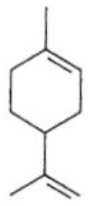   Limonene 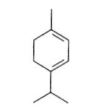   α-Terpinene

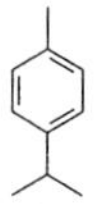

p-Cymene    Safrane

bicyclic

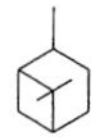

Pinane    α-Pinene 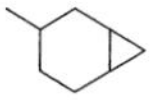   Carane

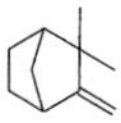

Camphor

Sesquiterpenes, C_{15}

acyclic

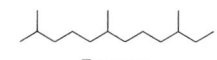

Farnesane

monocyclic

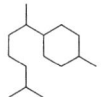

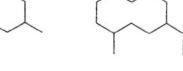

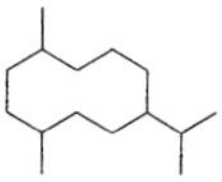

Bisabolane   Germacrane 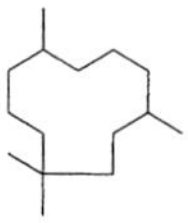   Humulane

bicyclic

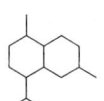

Cadinane 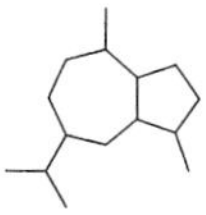   Guaiane

tricyclic

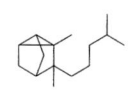

α-Santalane    Patchoulane

Diterpenes, C_{20}

acyclic

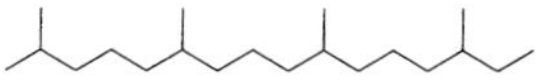

Phytane

monocyclic

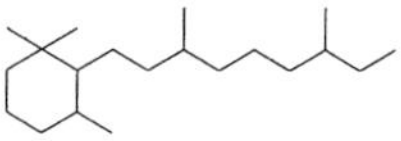

Retinane

bicyclic

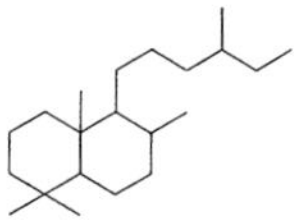

Labdane

tricyclic

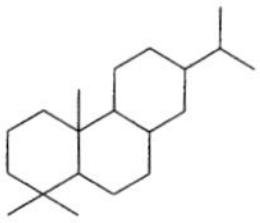

Abietane

tetracyclic

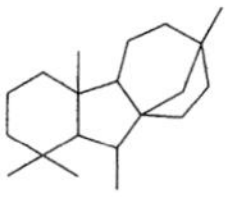

Gibbane

pentacyclic

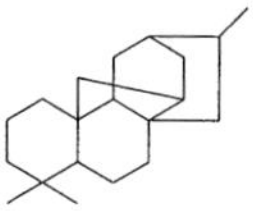

Atisane

Triterpenes, C_{30}

acyclic

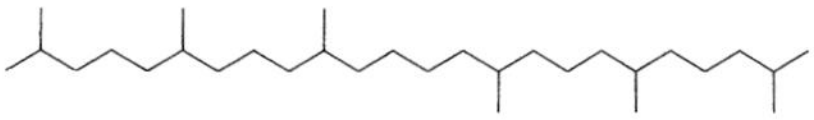

Squalane

4595

1.2. Occurrence

Terpenes occur everywhere and in all organisms, in particular in higher plants. Certain components or their combinations are often characteristic of individual types of plant, so terpenes are used for chemotaxonomy [8]. In vegetable raw materials mono- and sesquiterpenes are predominantly found in essential oils [9], [10], sesqui-, di-, and triterpenes in balsams and resins [11], tetraterpenes in pigments (carotenoids) [12], and polyterpenes in latexes [13]. Animal organisms also contain terpenes and terpenoids, which are predominantly incorporated by consumption of plants.

1.3. Biosynthesis

The biosynthesis of terpenes was explained by F. LYNEN. Starting from (3 R)-(+)-mevalonic acid (C_6), isopentenyl diphosphate (IDP) and dimethylallyl diphosphate (DMADP) [8], [14]–[16] are formed by decarboxylation and isomerization as active C_5 building blocks (Scheme 1). (3 R)-(+)-Mevalonic acid is formed from three C_2 units from acetyl-CoA [8]. Plants synthesize IDP in the chloroplasts, mitochondria, and microsomes; animals synthesize IDP in the liver.

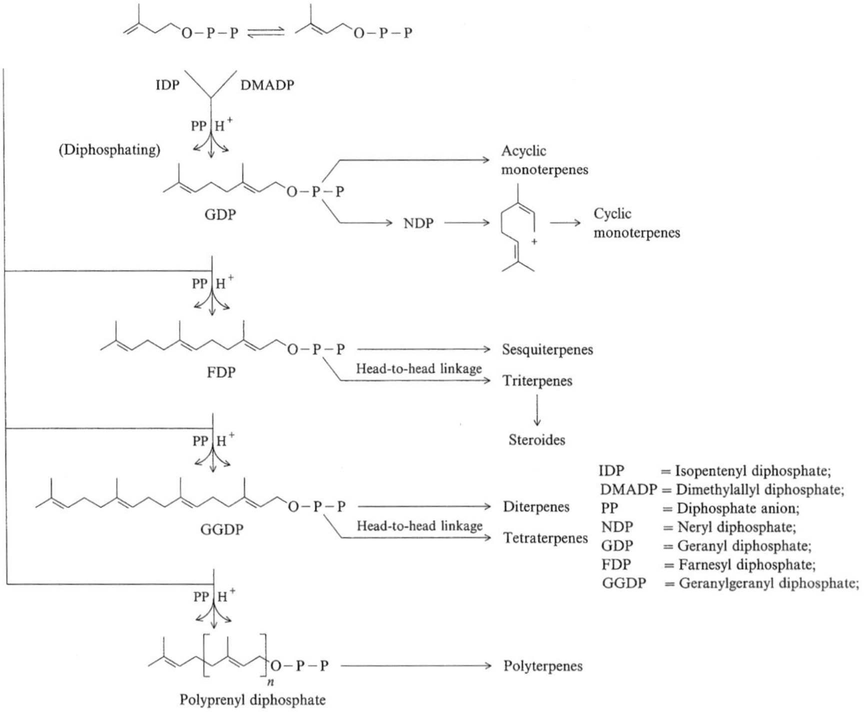

Scheme 1. Biosynthesis of terpenes.

Monoterpenes are formed by stereospecific condensation of IDP with DMADP, whereby a diphosphate anion (PP) is eliminated from DMADP. Geranyl diphosphate is thus formed in a head-to-tail condensation. Neryl diphosphate with the Z-double bond is formed when the S-proton in isopentenyl diphosphate is eliminated stereospecifically. Neryl diphosphate is also formed from geranyl diphosphate by double bond isomerization.

The hydrolysis of geranyl or neryl diphosphate by a prenolpyrophosphatase gives the monoterpene alcohols geraniol and nerol. The acyclic monoterpene hydrocarbons, such as myrcene and ocimene, are formed by dehydration and isomerization of geraniol. The monoterpene aldehydes geranial and neral are formed from geraniol, and the biosynthesis of cyclic monoterpenes involves the loss of diphosphate from neryl diphosphate. The carbonium ion formed is stabilized by ring closure, sometimes after rearrangement, elimination of a proton, or addition of an anion.

The head-to-tail condensation of IDP with geranyl diphosphate gives farnesyl diphosphate, which rearranges further to give geranylgeranyl diphosphate. A large number of sesqui-, di-, and polyterpenes are formed from farnesyl and geranyl diphosphates in the same way as the monoterpenes. Triterpenes (and hence, steroids) are formed by head-to-head condensation of two C_{15} units (farnesyl diphosphate), and tetraterpenes from two molecules of geranylgeranyl diphosphate.

1.4. Biodegradation

Terpenes are degraded by microorganisms, such as Pseudomonas and Aspergillus species [17]. Generally both acyclic and cyclic terpenes are oxidized. Plants are known to be capable of degrading as well as forming terpenes.

In animals acyclic terpenes are degraded by ω-, α- and/or β-oxidation and cyclic terpenes by hydroxylation (elimination as glucuronides).

1.5. Importance and Extraction

Since the study of organic chemistry began, it has been intensively concerned with terpenes. Many reactions, such as the Wagner–Meerwein rearrangement, theories of conformational analysis, and the Woodward–Hoffmann rules have their basis in terpene chemistry. The postulation of the isoprene rule by O. WALLACH (1887) and its confirmation and extension by L. RUZICKA (1921, 1953) gave this area of chemistry direction and unity.

Monoterpenes and a few sesquiterpenes are important economically as perfumes and fragrances. Depending on the properties, terpenes (mostly essential oils) are obtained from the corresponding plants by steam distillation, extraction, or enfleurage. The pure terpenes are mostly isolated from the extracts by fractional distillation.

As readily available, optically active products, some monoterpenes, such as (−)-α-pinene and (−)- and (+)-limonene are used for the synthesis of other optically active products and reagents for cleaving racemates [18].

Various essential oils and the terpenes isolated from them are still used in pharmaceutical preparations, e.g., turpentine, camomile oil, eucalyptus oil, and camphor. The activity of some plants used in medicine and flavoring is due to their terpene content [19]. Terpenes are also used in the production of synthetic resins [20]–[22], and as solvents or diluting agents for dyes and varnishes [23].

In the following sections only the industrially most important terpenes are dealt with, insofar as they are not part of other articles. For natural resins, gibberellic acid, vitamins A and E, triterpenes, tetraterpenes, and the other terpenes the specialist literature must be consulted.

2. Acyclic Monoterpenes

Alcohols: geraniol, nerol, linalool, myrcenol, citronellol, dihydromyrcenol, tetrahydrogeraniol, and tetrahydrolinalool.

Aldehydes and acetals: citral, citraldimethylacetal, citronellal, hydroxydihydrocitronellal, hydroxydihydrocitronellal dimethylacetal, and methoxydihydrocitronellal

Esters and nitriles: geranyl formate, acetate, isobutyrate, isovalerate, phenylacetate, and propionate; neryl acetate; linalyl acetate, butyrate, formate, isobutyrate, and propionate; lavanduyl acetate; and citronellyl acetate, formate, isobutyrate, isovalerate, propionate, and tiglate.

2.1. Myrcene

Myrcene [*123-35-3*], 2-methyl-6-methylene-2,7-octadiene, $C_{10}H_{16}$, M_r 136.23, is a colorless liquid with a pleasant odor.

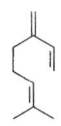

Some physical properties are listed below:

bp	166–168 °C (101.3 kPa)
d^{20}	0.7905
n_D^{20}	1.4697
mp	50 °C

Myrcene is very reactive. It polymerizes slowly at room temperature. For this reason it is stored in a cool place from which light and air are excluded. It can be stabilized by the addition of 0.1% 4-(*tert*-butyl)catechol.

Although myrcene occurs naturally in many organisms, its extraction is uneconomic; it is produced industrially by pyrolysis of β-pinene [24].

The fragmentation of linalool and linalyl acetate and the catalytic dimerization of isoprene have been described as laboratory methods [25].

Because of its functionality, myrcene is the starting material for a range of industrially important products, e.g., geraniol, nerol, linalool, and isophytol [26]. Addition of hydrogen chloride is industrially important. Depending on reaction conditions, this gives 1-chloro-3,7-dimethyl-2,6-octadiene (geranyl chloride and/or neryl chloride), 3-chloro-3,7-dimethyl-1,6-octadiene (linaloyl chloride), and 7-chloro-7-methyl-3-methylen-1-octene (myrcenyl chloride) [27]. The alcohols are produced from the chlorides by saponification.

The 1,3-diene system of myrcene reacts with a large number of dienophiles. For example, the maleic anhydride adduct is used to characterize and determine the quality of the product [28].

Isophytol is formed by transition-metal-catalyzed addition of malonic ester and subsequent acidification [29].

In the hydrogenation of myrcene 2,6-dimethyl-2,6-octadiene, 2,6-dimethyl-2,7-octadiene, and 2,6-dimethyloctane are formed, depending on the catalyst and the reaction conditions [30: 1H 260, 1EIII 1054].

Besides its main use as an intermediate for the production of terpene alcohols, myrcene is used in the production of terpene polymers [20], terpene–phenol resins [21], and terpene–maleate resins [22]. It can also be used as a solvent or diluting agent for dyes and varnishes [23].

2.2. Ocimene

Ocimene [673-84-7], (Z,E)-2,6-dimethyl-2,5,7-octatriene, $C_{10}H_{16}$, M_r 136.23, is a colorless liquid with a pleasant odor.

The Z/E mixture has the following properties:

bp	176–178 °C (101.3 kPa)
d^{20}	0.800
n_D^{20}	1.4862
mp	50 °C

Ocimene is very sensitive to oxidation and therefore can be kept for long periods only with the exclusion of air. At elevated temperatures it rearranges to alloocimene [31].

Ocimene is produced by the flash pyrolysis or photochemical isomerization of α-pinene [32].

Ocimene is mainly used as a perfume component. The derivatives ocimenol (2,6-dimethyl-5,7-octadien-2-ol), produced by acid-catalyzed hydration, and the partially and completely hydrogenated products are also important [33], [34].

2.3. 2,6-Dimethyl-2,4,6-octatriene

2,6-Dimethyl-2,4,6-octatriene exists as two stereoisomers:

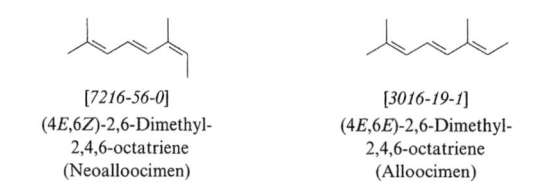

[7216-56-0]
(4E,6Z)-2,6-Dimethyl-
2,4,6-octatriene
(Neoalloocimen)

[3016-19-1]
(4E,6E)-2,6-Dimethyl-
2,4,6-octatriene
(Alloocimen)

(4E,6Z)-2,6-Dimethyl-2,4,6-octatriene, $C_{10}H_{16}$, M_r 136.23, is a colorless liquid with an intense odor, and is sensitive to oxidation.

bp	80 °C (1.9 kPa)
d^{20}	0.8161
n_D^{20}	1.5437

(4*E*,6*Z*)-2,6-Dimethyl-2,4,6-octatriene is produced as a mixture with other products by the thermolysis of α-pinene at 350 °C [35], or by heating *Z*-ocimene to 190 °C [36]. The products are separated by fractional distillation.

(4*E*,6*E*)-2,6-Dimethyl-2,4,6-octatriene, $C_{10}H_{16}$, M_r 136.23, has physical properties similar to those of the (4*E*, 6*Z*)-isomer.

bp	79 °C (1.9 kPa)
d^{20}	0.8106
n_D^{20}	1.5438

(4*E*, 6*E*)-2,6-Dimethyl-2,4,6-octatriene is formed as a mixture with the (4*E*,6*Z*)-isomer and other products in the thermolysis of pinene at 320 – 330 °C [37], by dehydration of 3,7-dimethyl-3,6-octadien-1-ol [38], and by treating *Z*-ocimene with base [39].

2,6-Dimethyl-2,4,6-octatriene is used to a small extent in the perfume industry. It is also used as a diluting agent for varnishes and dyes [23], and as a component for terpene polymers [20], [22].

3. Monocyclic Monoterpenes

3.1. *p*-Menthane

p-Menthane [*99-82-1*], 1-isopropyl-4-methylcyclohexane, $C_{10}H_{20}$, M_r 140.25, occurs widely as a *cis/trans* mixture, e.g., in essential oils in the eucalyptus fruit.

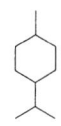

The *cis/trans* mixture of *p*-menthane is a colorless liquid with a fennel-like odor.

cis-1-Isopropyl-4-methylcyclohexane
bp	168.8 °C (99.2 kPa)
d^{20}	0.8086
n_D^{20}	1.4443

trans-1-Isopropyl-4-methylcyclohexane
mp	−86 °C
bp	168.1 °C (99.2 kPa)
d^{20}	0.7941
n_D^{20}	1.4369

p-Menthane is obtained as a *cis/trans* mixture by the catalytic hydrogenation of limonene, terpinols, and *p*-cymene [40]. Raney nickel, platinum, or copper and aluminum oxides are used as catalyst [41].

p-Menthane is predominantly used as a precursor in the production of 1-isopropyl-4-methylcyclohexane hydroperoxide, a catalyst for radical polymerization [42].

3.2. α-Terpinene

α-Terpinene [*99-85-4*], 4-isopropyl-1-methyl-1,3-cyclohexadiene, $C_{10}H_{16}$, M_r 136.23, is the main component of terpinene, the other components being the β- and γ-isomers. It occurs in various essential oils, almost always as a mixture with its isomers.

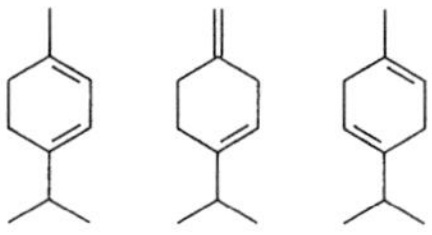

α-Terpinene is a colorless liquid with a lemon-like odor. It resinifies on prolonged storage.

bp 172.5 – 173.5 °C
n_D^{20} 1.4780

α-Terpinene is obtained by fractional distillation of pine oil [43] and by the thermolysis of α-pinene in the presence of catalysts, such as manganese oxide [44].

α-Terpinene is used in essential oils and as an intermediate for perfumes.

3.3. Terpinolene

Terpinolene [*586-62-9*], 4-isopropylidene-1-methyl-1-cyclohexane, $C_{10}H_{16}$, M_r 136.23, occurs in many essential oils and in sulfate turpentine.

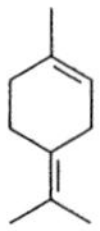

Terpinolene is a colorless liquid with an odor resembling that of turpentine.

bp 186 °C
d^{20} 0.8623
n_D^{20} 1.4861

It forms a hydroperoxide on autoxidation [45], [46].

Terpinolene used to be extracted by fractional distillation of wood turpentine [46]. It is now produced by treating α-pinene with aqueous H_3PO_4 at 75 °C [47]. It is preferable to distill terpinolene under vacuum to minimize the formation of polymeric products at elevated temperatures.

Terpinolene is mainly used to improve the odor of industrial and household products [48]–[50].

3.4. *p*-Cymene

p-Cymene [*99-87-6*], 4-(isopropyl)methylbenzene, $C_{10}H_{14}$, M_r 134.21, is the main component of sulfite turpentine and occurs in many essential oils [51].

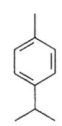

p-Cymene is a colorless liquid with an odor typical of aromatic hydrocarbons.

mp	−67.7 °C
bp	177.3 °C
d^{20}	0.8573
n_D^{20}	1.4906

The enthalpy of evaporation is 44.670 J/mol.

p-Cymene undergoes catalytic hydrogenation to 1-isopropyl-4-methylcyclohexane [40]. It must be stored in the absence of light and air.

p-Cymene is formed during the sulfite leaching of wood [52]. A range of production processes starting from mono- and bicyclic terpenes have also been described. For example, *p*-cymene is obtained in good yields from α-pinene in the presence of copper catalysts [53], [54]. In this process a significant proportion of 1-isopropyl-4-methylcyclohexane is also formed, which is dehydrogenated in the presence of palladium catalysts at 260–280 °C to *p*-cymene [55]. The Friedel–Crafts alkylation of toluene with propene is used industrially [56]. Products containing more than one isopropyl group are converted into *p*-cymene with aluminum chloride and hydrochloric acid [57].

p-Cymene is used to improve the odor of soaps and as a masking odor for industrial products. Its use as an intermediate for musk perfumes [58], the oxidation to terephthalic acid, and its addition to antiseptic preparations have also been described [59]. *p*-Cymene is also used as a solvent for dyes and varnishes.

3.5. 1,8-Cineole

1,8-Cineole [*470-82-6*], 1,8-epoxy-*p*-menthane, eucalyptol, $C_{10}H_{18}O$, M_r 154.25, occurs widely in natural essential oils, mainly eucalyptus oil.

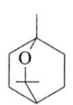

1,8-Cineole is a colorless liquid. Its odor and taste resemble those of camphor.

mp	1.55 °C
bp	176 – 176.4 °C
d^{20}	0.9232
n_D^{20}	1.4575

On warming, particularly in the presence of acids, 1,8-cineole forms other terpenes of the *p*-menthane type. It reacts with a range of substances, forming sparingly soluble products.

1,8-Cineole is extracted exclusively from eucalyptus oils with a high 1,8-cineole content. Various processes are used to separate it from the other terpenes, for example, treatment with H_2SO_4 in the cold [60], distillation in the presence of phenols, such as cresols or resorcinol [61], which form loose addition compounds, or by addition to β-naphthol [62]. 1,8-Cineole can also be enriched by rectification. The yield is increased by gasification in the presence of a chromium – nickel catalyst [63].

1,8-Cineole is used as a perfume and fragrance. In the past it was also used as an expectorant and an antiseptic.

4. Bicyclic Monoterpenes

The industrially most important bicyclic monoterpenes are subdivided into caranes, pinanes, and bicyclo[2.2.1]heptanes, depending on their basic structure.

4.1. 3-Carene

3-Carene [*498-15-7*], 3,7,7-trimethylbicyclo-[4.1.0]hept-3-ene, $C_{10}H_{16}$, M_r 136.23, occurs in a range of turpentines as the enantiomerically pure compound, but also as the racemate [64], [65].

3-Carene is a colorless compound with a sweet, penetrating odor.

bp	176–176.4 °C
d_4^{20}	0.9232
n_D^{20}	1.4575
$[\alpha]_D^{25}$	+17.6 ° (undiluted)

It readily undergoes autoxidation and resinification in air. Addition of pyrogallol or resorcinol as stabilizers is recommended [66]. On treatment with peracetic acid in glacial acetic acid and subsequent saponification, 3-carene is oxidized to 3,4-caranediol [67].

Pyrolysis at 300–580 °C in the presence of Fe^{III} oxide on carriers gives *p*-cymene as the main product [68].

3-Carene is obtained exclusively by fractional distillation of certain turpentines [69].

3-Carene is predominantly used in the perfume industry for producing essential oils [70]. It is also an intermediate in perfume synthesis and a precursor for insecticides of the pyrethroid type [71], [72].

4.2. Pinane

Pinane [*473-55-2*], 2,6,6-trimethylnorpinane, 2,6,6-trimethylbicyclo[3.1.1]heptane, $C_{10}H_{18}$, M_r 138.25.

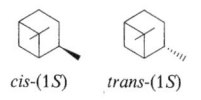

cis-(1*S*)   *trans*-(1*S*)

Pinane derivatives occur in wood, the leaves of many plants, algae, and insects. Pinane itself apparently does not occur naturally. However, it is formed on hydrogenation of pinane derivatives [73].

cis-Pinane (colorless liquid)

bp	167.2–168 °C
d_4^{20}	0.8575
n_D^{20}	1.4626
$[\alpha]_D^{25}$	−23.6 ° (undiluted)

trans-Pinane (colorless liquid)

bp	164 °C
d_4^{20}	0.854
n_D^{20}	1.4610
$[\alpha]_D^{25}$	+21.4 ° (undiluted)

The oxidation of pinane with air or oxygen gives 2-pinane hydroperoxide or 2-pinanol [74]. In industry, oxidation is carried out in the presence of catalysts [75].

The pyrolysis of (–)-(1*S*)-*cis*-pinane at 500 °C gives a mixture of (3*R*)-3,7-dimethyl-1,6-octadiene and (1*R*)-isopropyl-2,3-dimethylcyclopentane [76].

cis-Pinane is obtained, together with small quantities of *trans*-pinane, by the catalytic hydrogenation of α- and β-pinene. If (–)-α- and (–)-β-pinene are hydrogenated, the (–)-enantiomer is obtained, and from the (+)-pinenes, (+)-*cis*-pinane. In industry, α-pinene is predominantly used for the hydrogenation. Raney nickel, Pt, or Pd (and their oxides) on carriers are used as catalysts [77], [78]. The hydrogenation with Raney nickel is carried out at 2 – 10 MPa at 150 °C. In the presence of halogen compounds *cis*-pinane is obtained predominantly [79].

trans-Pinane can be obtained from β-pinene by treatment with sodium borohydride in the presence of boron trifluoride in di(ethylene glycol) and subsequent heating of the reaction mixture in propionic acid [80].

Pinane is used in industry for the production of pinane hydroperoxide [81]. It is also important as an intermediate in the production of 3,7-dimethylocta-1,6-diene, which is used to produce perfumes, such as citronellol, citronellal, and hydroxycitronellal.

4.3. 2-Pinane Hydroperoxide

2-Pinane hydroperoxide [5405-84-5], 2,6,6-trimethylbicyclo[3.1.1]heptane-2-hydroperoxide, $C_{10}H_{18}O_2$, M_r 170.25, is a colorless liquid, which is readily flammable and insoluble in water.

(–)-(1*R*)

bp	48 – 52 °C (1.3 Pa)
d_4^{20}	1.021

n_D^{20} 1.4898
$[\alpha]_D^{25}$ −27.4° (undiluted)

On heating 2-pinane hydroperoxide to > 110 °C 1-((1R)-cis-3-ethyl-2,2-dimethylcyclobutyl)ethanone is formed [82]. (1R,2S)-2-pinanol is formed in good yields by catalytic hydrogenation.

2-Pinane hydroperoxide is produced by oxidation of pinane with air or oxygen at 95 °C [83]–[85]. cis-Pinane reacts more rapidly than trans-pinane [86].

2-Pinane hydroperoxide is used as a radical initiator for polymerization reactions [87], e.g., for the polymerization of diolefins and aromatic vinyl compounds, or for hardening unsaturated polyester resins. It is also an intermediate in the production of perfumes, such as pinanol, linalool, nerol, and geraniol [88].

4.4.   2-Pinanol

2-Pinanol [473-54-1], 2,6,6-trimethylbicyclo[3.1.1]heptan-2-ol, $C_{10}H_{18}O$, M_r 154.25, is obtained as the cis or trans isomer, depending on the production process.

cis-(−)-(1R,2S)   trans-(−)-(1R,2R)

cis-2-Pinanol forms colorless crystals with a camphor-like odor.

mp   78–79.2 °C
bp   90–91 °C (1.5 kPa)
$[\alpha]_D^{20}$ −24.5° (CHCl$_3$, C = 8 g/L)

trans-2-Pinanol forms colorless needles.

mp   58–59 °C
bp   88–89 °C (0.12 kPa)
$[\alpha]_D^{17}$ −2.3° (ether, 10 g/L)

The pyrolysis of 2-pinanol at 500 °C gives linalool. (−)-Linalool is formed from (+)-cis- and (−)-trans-2-pinanol, and (+)-linalool from (−)-cis- and (+)-trans-2-pinanol. The process is used industrially [89], [90].

cis-2-Pinanol is produced industrially by the catalytic hydrogenation of 2-pinane hydroperoxide [91], [92]. Alternatively 2-pinane hydroperoxide can be treated with sodium sulfide in aqueous sodium hydroxide [93] or with sodium methoxide [94]. cis-2-Pinanol can also be obtained directly from pinane by air oxidation in the presence of alkalis, such as sodium hydroxide, at 80–100 °C [95].

2-Pinanol is used to produce linalool by pyrolysis [96].

4.5. Camphene

Camphene [5794-03-6], 2,2-dimethyl-3-methylenenorbornane, $C_{10}H_{16}$, M_r 136.23, occurs in a large number of essential oils in optically active form, both as the R and S enantiomers.

(+)-(1 R)

Camphene is a colorless, crumbly crystalline solid with a camphor-like odor. It has a tendency to sublime.

mp	52.5 °C
bp	157.8 °C (98.8 kPa)
d_4^{25}	0.84225
n_D^{54}	1.4564
$[\alpha]_D^{20}$	+106.8 ° (CHCl$_3$, 44 g/L)

Camphene is stable in air and light. At elevated temperatures, in the presence of oxygen, it undergoes autoxidation to camphenilone [97]. Peracids attack the double bond, giving camphene oxide [98]. Catalytic hydrogenation gives the saturated hydrocarbon isocamphane [99].

The extraction of camphene by distillation of turpentine is now hardly used in industry. Instead camphene is produced from α-pinene.

(+)-α-Pinene is converted into (+)-bornyl chloride by the action of dry hydrogen chloride in a Wagner–Meerwein rearrangement. Base-catalyzed dehydrohalogenation of the (+)-bornyl chloride gives racemic camphene [100], [101]. The reaction of α-pinene with borophosphoric acid in the gas phase [102], or on TiO$_2$ catalysts [103] has also been described.

Camphene is used to improve the odor of industrial products. It is an intermediate in the production of camphor, isobornyl esters, and the insecticide Toxaphen.

5. Acyclic Sesquiterpenes

Farnesenes, α-Farnesene, 2,6,10-trimethyl-2,6,9,11-dodecatetraene, $C_{15}H_{24}$, M_r 204.36.

α- and β-Farnesenes and their Z- and E-isomers occur in many essential oils, e.g., in apples [104], citrus fruits [105], and hop oil. α-Farnesene is a natural attractant for the larvae of the codlin moth [106]. It is also found in the gland secretions of certain ants [107].

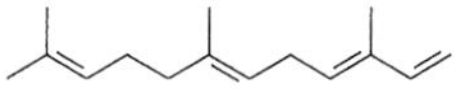

α-Farnesene
[502-61-4]

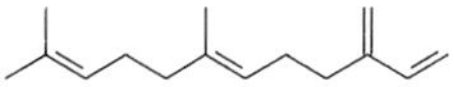

β-Farnesene
[18794-84-8]

α- and β-Farnesenes are colorless liquids with fruity odors. They are sensitive to autoxidation and can be stored for long periods only with the exclusion of light and air.

α-Farnesene
bp 98 – 102 °C (0.5 kPa)
n_D^{25} 1.4790

β-Farnesene
bp 95 – 107 °C (0.5 kPa)
d^{20} 0.8363
n_D^{20} 1.4899

Mixture
bp 129 – 132 °C (1.2 kPa)
d^{18} 0.8770
n_D^{20} 1.4995

Mixtures of α- and β-farnesenes are obtained by heating farnesol with potassium hydrogen sulfate to 160 – 170 °C [108]. Other syntheses of the mixture, involving dehydration of farnesol, nerolidol, and isoprene, have also been described [109], [110].

Farnesene is used in small quantities in the fragrance industry.

6. Monocyclic Sesquiterpenes

6.1. Bisabolene

Bisabolene, $C_{15}H_{24}$, M_r 204.33, occurs in myrrh oil and limett oil. The isomers α-, β-, and γ-bisabolene are known.

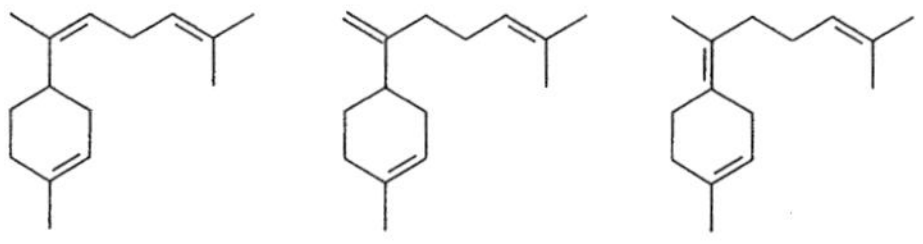

α-Bisabolene
[17627-44-0]

β-Bisabolene
[495-61-4]

γ-Bisabolene
[495-62-5]

α-Bisabolene, (E, Z)-4-(1,5-dimethyl-1,4-hexadienyl)-1-methyl-1-cyclohexane

bp 95 °C (0.02 kPa)

β-Bisabolene, 1-methyl-4-(5-methyl-1-methylene-4-hexenyl)-1-cyclohexane

bp	148 °C (2.4 kPa)
n_D^{15}	1.4893
$[\alpha]_D^{15}$	−67 ° (undiluted)

γ-Bisabolene, (E, Z)-4-(1,5-dimethyl-4-hexenylidene)-1-methyl-1-cyclohexene

bp	132 °C (1.4 – 1.6 kPa)
n_D^{25}	1.4928

The individual isomers and their mixtures are colorless liquids with pleasant, balsamic odors.

Isomer mixtures are produced by treating nerolidol and farnesol with acid [111]. Targeted syntheses of α- and β-bisabolene by the Wittig reaction have been described [112]–[114].

Bisabolene mixtures are used in the perfume and fragrance industries.

6.2. Bisabolol

Bisabolol [515-69-5], $C_{15}H_{26}O$, M_r 222.37, occurs in various plants.

(+)-α-Bisabolol, (+)-2-methyl-6-(4-methyl-3-cyclohexenyl)-6-hepten-2-ol

bp	120–122 °C (0.13 kPa)
d_4^{20}	0.9213
n_D^{20}	1.4919
$[\alpha]_D^{20}$	+51.7 ° (undiluted)

(−)-α-Bisabolol, (−)-2-methyl-6-(4-methyl-3-cyclohexenyl)-6-hepten-2-ol

bp	153 °C (1.59 kPa)
d_4^{20}	0.9211
n_D^{20}	1.4936
$[\alpha]_D$	−55.7 ° (undiluted)

rac-Bisabolol, 2-methyl-6-(4-methyl-3-cyclohexenyl)-6-hepten-2-ol

bp	157 °C (1.59 kPa)
d_4^{23}	0.9223
n_D^{23}	1.4917

The bisabolols are colorless liquids with slightly flowery odors.

rac-Bisabolol is produced by acid-catalyzed cyclization of farnesol or nerolidol [115]. The individual enantiomers are obtained by extraction from the appropriate plants.

Because of their anti-inflammatory and spasmolytic properties [117], both (−)- and rac-bisabolol are predominantly used in the cosmetics industry and to a small extent in the pharmaceutical industry.

7. Bicyclic Sesquiterpenes

Caryophyllene [87-44-5], 6,10,10-trimethyl-2-methylenebicyclo[7.2.0]-5-undecene, $C_{15}H_{24}$, M_r 204.36, occurs in many essential oils.

Caryophyllene is a colorless, oily liquid with an odor resembling that of cloves.

bp	254–256 °C
d_4^{20}	0.9019
n_D^{20}	1.4995
$[\alpha]_D^{20}$	−9.2 ° (undiluted)

Catalytic hydrogenation on platinum gives tetrahydrocaryophyllene [118], and on palladium a dihydrocaryophyllene [119]. On oxidation with potassium permanganate, a glycol (mp 120–120.5 °C) is formed [120], and with perbenzoic acid [121] or peracetic acid, a caryophyllene oxide.

Caryophyllene is best extracted from clove oil. It can also be obtained from other sources, such as certain American pine oil fractions.

Caryophyllene is used as a perfume and fragrance, for example in chewing gum (ca. 200 mg/kg). It is also used as a fixative. It can be used as an intermediate in the synthesis of other perfumes and fragrances [123].

8. Tricyclic Sesquiterpenes

Longifolene [*475-20-7*], 3,3,7-trimethyl-8-methylenetricyclo[5.4.0.9-Z9]undecane, $C_{15}H_{24}$, M_r 204.36, occurs in the *d*- and *l*-forms in a range of essential oils. Indian turpentine contains up to 20 % longifolene.

(+)-(1R, 2S, 7S, 9S)

Longifolene is a colorless, oily liquid.

bp	254–256 °C (93.9 kPa)
d_4^{18}	0.9319
n_D^{20}	1.5040
$[\alpha]_D$	+42.73 ° (undiluted)

Longifolene forms crystalline products with hydrogen halides [124]. On treatment with bromine followed by N,N-dimethylaniline, ω-bromolongifolene is formed, and with nitrogen oxides ω-nitrolongifolene [125].

Longifolene is obtained by distillation, e.g., Indian turpentine.

Longifolene is used as a solvent additive. It has been described as a starting material for the production of perfumes, e.g., by oxidation [126], or by treatment with formic acid [127].

9. Acyclic Diterpenes

Phytol [*150-86-7*], (7R, 11R)-3,7,11,15-tetramethyl-*trans*-2-hexadecen-1-ol, $C_{20}H_{40}O$, M_r 295.52, is formed by saponification of chlorophyll.

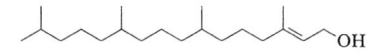

Phytol is a colorless liquid with a slightly flowery odor. The action of acids or heat effects dehydration to phytadiene.

bp	136 °C (1.3 kPa)
d_4^{21}	0.8533
n_D^{25}	1.4637

Phytol can be obtained from natural raw materials by a variety of processes [128]. The industrial synthesis of phytol starts from isophytol. The latter is first treated with formic acid to give phytyl formate, from which (Z,E)-phytol (isomer ratio 3:7) is obtained by saponification or transesterification [129].

Phytol is used in the perfume and cosmetics industries [130]. It can also be used, like isophytol, as the starting material for the synthesis of vitamins E and K_1 [131]–[133]. The hydrogenation and oxidation products are also known [134].

10. Acyclic Triterpenes

Squalene [*111-02-4*], 2,6,10,15,19,23-hexamethyltetracosa-2,*t*-6,*t*-10,*t*-14,*t*-18,*r*-22-hexaene, $C_{30}H_{50}$, M_r 410.73, occurs in liver oils of various species of shark [135], in cod-liver oil [136], in vegetable fats and oils [137], and in human skin fat.

Squalene is a colorless, air-sensitive, almost odorless liquid.

bp	223–226 °C (0.26 kPa)
d_4^{20}	0.8577
n_D^{25}	1.4961

Squalene is extracted from shark liver [138] or synthesized from hexaphenyl-1,4-butanediyldiphosphonium dibromide and 6,10-dimethyl-5,9-undecadien-2-one (geranylacetone, industrial intermediate in the vitamin E synthesis) [139].

Squalene is hydrogenated on platinum or nickel catalysts to dodecahydrosqualene (perhydrosqualene, squalane, cosbiol) [140], which is widely used in cosmetics [141], [142].

11. Toxicology

There are excellent reviews of the toxicology of terpenes [143]. The known toxicological test results (Table 1) show that most terpenes have low acute oral toxicity and, in spite of occasionally good skin resorption, low dermal toxicity. Some are irritant to the skin of rabbits in concentrated form, but a 4–16% solutions in vaseline, are generally not irritant to humans. Only 3-carene, which occurs in turpentine, has been shown to have a sensitizing potential in humans and animals.

To investigate detoxification in mammals, a metabolism study on rabbits has been carried out. On oral administration several characteristic oxidation reactions have been detected:

1) Stereoselective oxidation: formation of 3-caren-9-ol, 3-caren-9-carboxylic acid, and 3-caren-9,10-dicarboxylic acid from 3-carene
2) Regioselective oxidation: 3,10-myrcenediol, 1,2-myrcenediol, and 2-hydroxymyrcene-1-carboxylic acid from myrcene
3) Allylic oxidation: verbenol and myrtenic acid from α-pinene, and pinocarveol from β-pinene
4) Homoallylic oxidation: 6-exo-hydroxycamphene and 10-hydroxytricyclene from camphene
5) Hydrolysis of epoxides: camphenediol from camphene

With p-cymene, three different oxidation products were detected in the first step. The alcohols 2-(4-methylphenyl)-1-propanol and 2-(4-methyl-phenyl)-2-propanol are formed by the ω- and (ω-1)-oxidation of the isopropyl group, and 4-isopropylbenzyl alcohol by oxidation of the methyl group. The alcohols formed in these different oxidations are either excreted from the organism as their glucuronides or further oxidized [8], [144].

Most of the terpenes described in Table 1are licensed as food additives by the international bodies responsible (Table 2). This licensing clearly demonstrates the low toxicity of the terpenes concerned.

Table 1. Toxicity data for terpenes

Substance	Acute oral toxicity (rat), LD_{50}, g/kg	Acute oral toxicity (rabbit), LD_{50}, g/kg	Skin resorption	IRT* (rat)	Skin irritation (R = rabbit; H = human)	Irritation of rabbit eye mucous membrane	Sensitization (H = human; G = guinea pig)	Specific properties/ tests	References
Myrcene	>5	>5	+		R: moderately irritant; H: 4% not irritant		H: negative		
Ocimene	>5	>5			R: moderately irritant; H: 5% not irritant		H: negative		
α-Terpinene	1.68				H: 5% not irritant		H: negative	damage to liver and blood (forms methemoglobin)	
Terpinolene	4.39 ml/kg	>5			R: not irritant; H: 20% not irritant		H: negative		
p-Cymene	4.74	>5	+	0/6 after 6 h	H: irritant; R: not irritant; H: 4% not irritant	irritant	H: negative	affects central nervous system (narcosis)	[145]
1,8-Cineole	2.48	>5			R: not irritant; H: 16% not irritant		H: negative	cancerization study in progress	[145]
3-Carene	4.8	>5			R: irritant		H: positive G: positive		[145]
Camphene	>5	2.5	+		R: slightly irritant; H: 4% not irritant		H: negative		
Bisabolene	>5	>5		0/12 after 7 h	R: slightly irritant; H: 10% not irritant	not irritant	H: negative		[145]
Bisabolol	>5	>5		0/12 after 7 h	R: not irritant	slightly irritant			
Caryophyllene	>5	>5			R: irritant; H: 4% not irritant **		H: negative		
Phytol	>10				R: moderately irritant	not irritant			[145]

* IRT: inhalation risk test (rat, test result dependent on toxicity and volatility; 0/6 after 6 h means that after 6 h exposure in an enriched or saturated atmosphere at room temperature no animals died). ** In light petroleum jelly.

Table 2. Terpenes licensed as food additives

Substance	Licensing authority
Myrcene	FDA (GRAS classification), European Council
Ocimene	FDA, European Council
α-Terpinene	FDA
Terpinolene	FDA, European Council
p-Cymene	FDA
1,8-Cineole	FDA, European Council
Camphene	FDA, European Council
Caryophyllene	FDA, European Council

FDA = Food and Drug Administration;
GRAS = generally recognized as safe.

12. References

[1] J. C. Simonsen et al.: *The Terpenes*, 2nd ed., vols. **1–3**, Cambridge University Press, New York 1947–1951.

[2] A. A. Newmann: *Chemistry of Terpenes and Terpenoids*, Academic Press, London 1972. T. K. Devon, A. I Scott:
Handbook of Naturally Occurring Compounds, vol. **II**, Terpenes, Academic Press, London 1972.

[3] *Terpenoids and Steroids (A Spezialist Periodical Report)* vols **1–6**, The Chemical Society, Burlington House, London 1971–1976.

[4] W. Templeton: *An Introduction to the Chemistry of the Terpenoids and Steroids*, Butterworths, London 1969.

[5] W. J. Tayler, A. R. Battersby: *Cyclopentanoid Terpene Derivatives*, Marcel Dekker, New York 1969.

[6] O. Aschan: *Naphthenverbindungen, Terpene und Campherarten*, De Gruyter, Berlin 1929.

[7] J. ApSimon: *The Total Synthesis of Natural Products*, vol. **2**, J. Wiley, New York 1973.

[8] O. W. Thiele: *Lipide, Isoprenoide mit Steroiden*, Thieme Verlag, Stuttgart 1979.

[9] E. Gildemeister, F. Hoffmann: *Die ätherischen Öle*, 4th ed., vols. **1–7**, Akademie Verlag, Berlin 1960.

[10] E. Guenther: *The Essential Oils*, Van Nostrand, Princeton, N. J., 1949.

[11] W. Sandermann: *Naturharze Terpentinöl Tallöl*, Springer Verlag, Berlin 1960.

[12] O. Isler: *Carotinoide*, Birkhäuser Verlag, Basel 1971.

[13] *Beilstein*, **30** (1938) 64–76. S. Boström: *Kautschuk-Handbuch*, vols. **1–5**, Berliner Union, Stuttgart 1958–1962. L. Batteman:
The Chemistry and Physics of Rubber-Like Substances, Maclaren, London 1963.

[14] T. W. Goodwin: *Aspects of Terpenoid Chemistry and Biochemistry*, Academic Press, London 1971.

[15] D. V. Banthopre, B. V. Charlewood, M. J. O. Francis: "The Biosynthesis of Monoterpenes," *Chem. Rev.* **72** (1972) no. 2, 115.

[16] A. Cordell: "Biosynthesis of Sequisterpenes," *Chem. Rev.* **76** (1976) no. 4, 425.

[17] S. G. Cantwell, E. P. Lau, D. S. Watt, R. R. Fall, *J. Bacteriol.* **135** (1978) 324.

[18] H. Siegel, W. Himmele, *Angew. Chem.* **92** (1980) 185.

[19] W. Franke: *Nutzpflanzenkunde*, Thieme Verlag, Stuttgart 1976.

[20] C. D. Ender, *Nav. Stores Rev.* **66** (1956) 10. W. H. Wiles, *Rubber Plast. Age* **40** (1959) 249. K. Blaetuer, *Dtsch. Farben Z.* **17** (1963) 34. H. P. Preuss, *Met. Finish* **62** (1964) no. 8, 63.

[21]   *Chem. Age (London)* **63** (1950) 269. *Prakt. Chem.* **10** (1959) 383. R. Werner, *Prakt. Chem.* **11** (1960) 331.
[22]   J. Schreiber: *Chemie und chemische Technologie der Kunstharze*, Wissenschaftl. Verlagsgesellschaft, Stuttgart 1943, p. 675. E. R. Littmann, *Ind. Eng Chem.* **28** (1936) 1150.
[23]   L. Désalbres, FR 10 23 339, 1950.
[24]   J. M. Mellor, S. Munavalli, *Q. Rev. Chem. Soc.* **18** (1964) 270.
[25]   F. Sorm: "Sequisterpenes with Ten-Membered Carbon Rings, A. Review," *J. Agric. Food Chem.* **19,** (1971) 1081.
[26]   Van Amerigen-Haebler, Inc., US 2882 323, 1957 (R. Weiss).
[27]   Glidden Co., US 3031442, 1958 (R. L. Webb).
[28]   T. Sasaki, S. Eguchi, T. Jshi, *J. Org. Chem.* **34** (1969) 3749.
[29]   L. A. Goldblatt, S. Palkins, *J. Am. Chem. Soc.* **63** (1941) 3517. R. L. Burwell, Jr., *J. Am. Chem. Soc.* **73** (1951) 4421. E. T. Theimer, B. M. Mitzner, *Can. J. Chem.* **42** (1964) no. 41, 959. E. L. Patton, *Am. Perfum. Essent. Oil Rev.* **56** (1950) 118.
[30]   *Beilstein,* 1 H 260, 1 E III, 1054.
[31]   J. E. Hawkins, H. G. Hunt, *J. Am. Chem. Soc.* **73** (1951) 5379. J. E. Hawkins, W. A. Burris, *J. Org. Chem.* **24** (1959) 150.
[32]   R. L. Burwell, *J. Am. Chem. Soc.* **73** (1951) 4461.
[33]   M. C. J. Enklaar, *Recl. Trav. Chim. Pays-Bas Belg.* **26** (1907) 157, 167.
[34]   M. C. J. Enklaar, *Recl. Trav. Chim. Pays-Bas Belg.* **26** (1907) 174; *Chem. Zentrabl.* 1907 II, 56; *Ber. Dtsch. Chem. Ges.* **41** (1908) 2084.
[35]   K. Alder, A. Dreike, H. Erpenbach, U. Wicker, *Justus Liebigs Ann. Chem.* **609** (1957) 1.
[36]   T. Saskaki, S. Eguchi, H. Yamada, *Tetrahedron Lett.* 1971, 99.
[37]   H. Pines, N. E. Hoffmann, V. N. Ipatieff, *J. Am. Chem. Soc.* **76** (1954) 4412.
[38]   G. Ohloff, *Chem. Ber.* **90** (1957) 1554.
[39]   V. N. Ipatieff, H. Pines, E. E. Meisinger, *J. Am. Chem. Soc.* **71** (1949) 2934.
[40]   V. N. Ipatieff, H. Pines, E. E. Meisinger, *J. Am. Chem. Soc.* **71** (1949) 2934.
[41]   H. A. Smith, J. F. Fuzek, H. T. Meriwether, *J. Am. Chem. Soc.* **71** (1949) 3765, 3766.
[42]   USA Secr. of Agric., US 2 775 578, 1955.
[43]   W. Sandermann: *Naturharze: Terpentinöl, Tallöl* Springer Verlag, Berlin 1960, pp. 269 – 310.
[44]   BASF, DE 960 988, 1954.
[45]   M. Saito, *Kogyo Kagaku Zasshi* **61** (1958) 326 –330.
[46]   J. N. Borgein, D. A. Lister, E. J. Lorand. J. E. Reese, *J. Am. Chem. Soc.* **72** (1950) 4591 – 4596.
[47]   Haarmann u. Reimer, DE 411901, 1923 (R. Moelleken).
[48]   Monsanto Co., US 4 097 555, 1976.
[49]   Mitsubishi Chem. Ind., JP 53 121 891, 1977; *Chem. Abstr.* **90** (1979) 88066 t.
[50]   Standard Oil Co., US 4080 592, 1977.
[51]   R. G. Buttery et al., *J. Agric Food Chem.* **22** (1974) 773.
[52]   Zellstoffabrik Waldhof, DE 727 475, 1940.
[53]   BASF, DE 961 979, 1953.
[54]   Goodyear Co., EP 77289, 1981 (J. Kuczkowski, L. Wideman).
[55]   Hercules Powder Co., US 2400012, 1942.
[56]   BASF, US 3 555 103, 1967.
[57]   Mobil Oil Corp., EP 12514, 1978 (G. T. Burress).
[58]   Sumitomo Chemical KK, DE 2910493, 1978.
[59]   G. Ohloff: *Riechstoffe und Geruchssinn,* Springer Verlag, Berlin 1990.
[60]   Rheinische Kampferfabrik. DE 499 732, 1928.

[61] Newport Ind. Inc., US 2090620, 1936.
[62] Takasago Perfumery Ind. Co. JP 11 72, 1950.
[63] Union Camp Corp., EP 270023 (P. W. D. Mitchell, D. E. Sasser).
[64] B. S. Rao, J. L. Simonsens, *J. Chem. Soc.* **127** (1925) 2494–2499.
[65] N. T. Mirov, *J. Am. Pharm. Assoc. Sci. Ed.* **40** (1951) 410–413.
[66] G. Widmarks, S. G. Blohm, *Acta Chem. Scand. (1947–1973)* **11** (1957) 392–394.
[67] B. A. Arbusow, B. M. Michailow, *J. Prakt. Chem.* **127** (1930) 1–15.
[68] J. Verghese, W. K. Sondhi, B. Bushan, M. L. Joshi, *Curr. Sci.* **17** (1948) 359.
[69] B. S. Rao, J. L. Simonsens, *J. Chem. Soc.* **127** (1935) 2494.
[70] S. Arctander: *Perfume and Flavor Chemicals,* Det Hoffensbergske, Etablissement, Kopenhagen 1969.
[71] Shell Oil Co., US 4156692, 1977.
[72] Shell Int. Res., EP 2850, 1977.
[73] D. V. Banthorpe, D. Whittaker, *Chem. Rev.* **66** (1966) 643.
[74] Monsanto US 2097 744, 1934.
[75] SCM Corp., US 4254291, 1981.
[76] J. Tanaka, T. Katagiri, K. Ozawa, *Bull. Chem. Soc. Jpn.* **44** (1971) 130–132.
[77] G. S. Fisher, J. S. Stinson, R. N. Moore, L. A. Goldblatt, *Ind. Eng. Chem.* **47** (1955) 1368–1373.
[78] G. S. Fisher, L. A. Goldblatt, J. Kniel, A. D. Snyder, *Ind. Eng. Chem.* **43** (1951) 671–674.
[79] SCM-Corp., US 4018 842, 1977.
[80] G. Zweifel, H. C. Brown, *J. Am. Chem. Soc.* **86** (1964) 393–397.
[81] *Chem. Eng. News* **60** (1982) no. 11, 5.
[82] G. A. Schmidt, G. S. Fisher, *J. Am. Chem. Soc.* **76** (1954) 5426–5430.
[83] Tagasako, JP 61 200999, 1986.
[84] G. Ohloff, E. Klelin, *Tetrahedron* **18** (1962) 37.
[85] USA Secr. of Agric, US 2735870, 1950.
[86] G. S. Fisher, J. S. Stinson, L. A. Goldblatt, *J. Am. Chem. Soc.* **75** (1953) 3675–3678.
[87] Amer. Cynamid, US 306 1554, 1962.
[88] *Beilstein,* **27,** 5 E IV, 277.
[89] Studiengesellschaft Kohle mbH, DE 1150974, 1961.
[90] G. Ohlof, E. Klein, *Tetrahedron* **18** (1962) 37–42.
[91] Shell, NL 7900588, 1979.
[92] L. A. Shutikova, V. G. Charakew, M. S. Erzhanova, A. K. Alifanova, *Maslo. Zhir. Promst.* 1973, 23–25.
[93] Cocker Chemical Co., GB 1619649, 1963.
[94] G. A. Schmidt, G. S. Fisher, *J. Am. Chem. Soc.* **81** (1959) 445–448.
[95] Stephan Chemical Co., DE 2 305 363, 1973.
[96] Studiengesellschaft Kohle mbH, FR 1328113, 1963.
[97] Du Pont, *Ind. Chim. Belge Ser. 2* **11** (1940) 3.
[98] W. Hückel, *Ber. Dtsch. Chem. Ges.* **80** (1947) 41–47.
[99] P. Lipp, *Justus Liebigs Ann. Chem.* **382** (1911) 265–305.
[100] A. Meyer, DE 272562 1911.
[101] V. Hilcken, DE 439 695 1924.
[102] Schering-Kahlbaum, DE 578 569, 1931.
[103] Du Pont, US 2551 795, 1949.
[104] E. F. L. J. Anet, *Aust. J. Chem.* **23** (1970) 2101.

[105] M. G. Moshonas, P. E. Shaw, *J. Agric. Food Chem.* **28** (1980) 680.
[106] O. R. W. Sutherland, R. F. N. Hutchins, *Nature (London)* **239** (1972) 170.
[107] G. W. K. Cavill, P. J. Williams. F. B. Withfield, *Tetrahedron Lett.* 1967, 2201.
[108] C. Harries, R. Haarmann, *Chem. Ber.* **46** (1913) 1741.
[109] G. Brieger, T. J. Nestrick, C. McKenna, *J. Org. Chem.* **34** (1969) 3789.
[110] G. Brieger, *J. Org. Chem.* **32** (1967) 3720.
[111] L. Ruzicka, E. Capato, *Helv. Chim. Acta* **8** (1925) 259.
[112] F. Delay, G. Ohloff, *Helv. Chim. Acta* **62** (1979) 369.
[113] G. Brieger, T. J. Nestirck, C. McKenna, *J. Org. Chem.* **34** (1969) 3789.
[114] O. P. Vig, S. D. Sharma, P. Kumar, M. L. Sharma, *J. Indian Chem. Soc.* **52** (1975) 614.
[115] E. V. Weber, *Justus Liebigs Ann. Chem.* **238** (1887) 101.
[116] Degussa, DE-OS 2709033, 1978.
[117] M. Holub, V. Heront, F. Sorm, *Cesk. Farm* **4** (1955) 129.
[118] F. W. Semmler, E. W. Mayer, *Ber. Dtsch. Chem. Ges.* **45** (1912) 1384–1394.
[119] Shell, EP 127 911, 1983 (A. J. Mulder).
[120] E. Deussen, *Justus Liebigs Ann. Chem.* **388** (1912) 136–165.
[121] H. N. Rydon, *J. Chem. Soc.* 1939, 537–540.
[122] Soda Shangyo Co. Ltd, JP-Kokai 76 80847, 1975.
[123] BBA, GB 1580184, 1976 (H. R. Ansari, R. Clark, H. R. Wagner).
[124] J. L. Simonsen, *J. Chem. Soc.* **117** (1920) 564–578.
[125] G. R. Dupont, R. Dulou, P. Naffa, G. Ourisson, *Bull. Soc. Chim. Fr.* 1954, 1075, 1078.
[126] Tagasako Co., JP 04 360848, 1992.
[127] Tosho Corp., JP-Kokai 01 268657, 1989.
[128] E. M. Burdick, US 3248301, 1964.
[129] BASF, DE 28187150, 1978.
[130] H. Ianistyn, DE 1467955, 1964.
[131] B. Staller-Bourdillon, *Ind. Chim. Belge* **35** (1970) no. 1, 13.
[132] Eisai KK, JP-Kokai 88 108, 1976.
[133] Nisshin Flour Milling, JP-Kokai 10 235, 1960.
[134] E. Jellum, L. Eldjarn, K. Try, *Acta Chem. Scand. (1947–1973)* **20** (1966) 2535.
[135] M. Tsujmoto, *Chem. Zentralbl.* 1918 I, 638, 1048.
[136] J. C. Drummond, H. J. Channon, K. H. Coward, *Biochem. J.* **19/23** (1919/1923)1055/279.
[137] K. Täufel, H. Thaler, H. Schreyegg, *Fettchem. Umsch.* **43** (1936) 26.
[138] M. Hamaya, JP-Kokai 00 3158, 1977.
[139] D. W. Dicker, M. L. Whiting, *J. Chem. Soc.* 1958, 1994.
[140] J. M. Heilbron, T. P. Hildrich, E. D. Kamm, *J. Chem. Soc.* 1926, 3135.
[141] H. Janistyk: *Handbuch der Kosmetik und Riechstoffe,* 2nd ed., vol. **1**, Hüthig Verlag, Heidelberg 1969, p. 753.
[142] E. S. Lower, *Spec. Chem.* **1** (1981) no. 1, 34.
[143] E. L. J. Opdyke (ed): *Monographs on Fragrance Raw Materials,* Pergamon Press, Oxford 1979.
[144] T. Ishida et al.: "Biotransformations of Terpenoids in Mammals," *Chem. Abstr.* **93** (1980) 8986003.
[145] NIOSH: *Registry of Toxic Effects of Chemical Substances 1979,* **II**, Washington, D. C., 1979, p. 33.

Tetrahydrofuran

HERBERT MÜLLER, BASF Aktiengesellschaft, Ludwigshafen, Federal Republic of Germany

1.	Introduction 4621	4.6.	Pentosan/Furfural Processes . 4626
2.	Physical Properties 4622	5.	Specifications and Analysis. . . 4626
3.	Chemical Properties. 4622	6.	Environmental Protection and Recovery 4626
4.	Production 4623		
4.1.	Acetylene/Formaldehyde 4623	7.	Storage and Transport 4627
4.2.	Butadiene Acetoxylation. 4624	8.	Uses 4627
4.3.	Propylene Oxide Process 4624	9.	Economic Aspects 4628
4.4.	Maleic Anhydride Hydrogenation. 4625	10.	Toxicology and Industrial Hygiene. 4628
4.5.	*n*-Butane – Maleic Anhydride Process 4625	11.	References. 4629

1. Introduction

Tetrahydrofuran (THF) [*109-99-9*] tetramethylene oxide, oxolane, is a five-membered cyclic ether with wide application in the chemical industry.

The largest proportion of the 1.4×10^5 t of THF produced worldwide in 1992 was used as a monomer in the manufacture of poly(tetramethylene oxide) (PTMO) also known as poly(tetramethylene ether glycol) (PTMEG) and polytetrahydrofuran (PTHF), important in the production of thermoplastic polyurethanes, elastic fibers, molded elastomers, and copolyesters or copolyamides. A smaller proportion finds use as a solvent for coatings, adhesives, and special varnishes based, for example, on poly(vinyl chloride) (PVC) or polyurethanes; as an extracting agent; and as the preferred medium for organometallic syntheses.

2. Physical Properties

The pure distilled tetrahydrofuran sold commercially is a colorless, clear, volatile, polar liquid with a characteristic acetone-like odor and a minimum purity of 99.9 wt %. THF is miscible in all proportions with water, alcohols, ethers, and many common solvents. Important physical properties include the following [1], [2]:

M_r	72.1
bp	66 °C
mp	−108.5 °C
Critical temperature	267 °C
Critical pressure	5.19 MPa
Critical density	0.322 g/cm^3
d_4^{20}	0.886
n_D^{20}	1.4073
Specific heat	1.765 J g^{-1} K^{-1}
Heat of vaporization (66 °C/101.3 kPa)	435 J/kg
Heat of combustion	−35 141 kJ/kg
Explosive limits in air (25 °C)	
lower	1.5 vol %
upper	12.0 vol %
Flash point (Abel–Pensky)	−22 °C
Dipole moment	5.84 × 10^{-30} C m
Dielectric constant (20 °C)	7.6

3. Chemical Properties

THF is a valuable starting material for a wide range of important reactions. For example, cationic polymerization accompanied by ring opening leads to high molecular mass bifunctional glycol ethers with various chain lengths [3], [4]. These have achieved great economic significance in the preparation of important plastics. Ring-cleavage reactions also constitute the basis for dehydration to butadiene, oxidation to succinic acid, and carboxylation to adipic acid or γ-valerolactone [5], [6].

THF is the preferred reaction medium for carrying out Grignard reactions or reductions with, for example, LiAlH$_4$. It is also suitable for use as a ligand in coordination complexes, such as those sometimes employed in stereospecific polymerizations [7], [8].

THF reacts readily with oxygen (on contact with air, for example), the principal product being an unstable hydroperoxide [9]. Hydroquinone or 2,6-di-*tert*-butyl-*p*-cresol (BHT) (e.g., 250 mg/kg) can be added to retard peroxide formation. Distillation of peroxide-containing THF increases the peroxide concentration, resulting in a serious risk of explosion, even on a laboratory scale.

4. Production

4.1. Acetylene/Formaldehyde

A process developed by Reppe in the 1930s was for many years the preferred synthetic route to 1,4-butanediol and THF, and it is still the most common approach in Europe and the United States [10], [11]. The Reppe process involves a reaction between acetylene and formaldehyde to give 2-butyne-1,4-diol, with subsequent hydrogenation to 1,4-butanediol. The saturated diol is very readily cyclized to THF with elimination of water by acid catalysis above 100 °C. Suitable catalysts include inorganic acids, acidic aluminum silicates, and earth or rare-earth oxides.

$$HC\equiv CH + 2CH_2O \longrightarrow HOCH_2C\equiv CCH_2OH$$

$$\xrightarrow{H_2} HOCH_2CH_2CH_2CH_2OH \xrightarrow[\Delta]{H^+} \underset{O}{\bigcirc}$$

$$\Delta H = -13.4 \text{ kJ/mol}$$

Quantitative conversion and a yield of almost 100% can be achieved in an atmospheric-pressure, continuous process with aluminum oxide as the catalyst provided the THF is continuously distilled out of the reaction mixture as it forms and pure 1,4-butanediol is fed into the reactor at a rate equal to its rate of consumption. The amount of conversion per unit amount of catalyst is very high. Crude butanediol can be converted into THF very selectively and with little added energy by a medium-pressure process recently described by BASF [12]. The THF – steam mixture obtained from cyclization is first distilled in a rectification column to give the corresponding azeotrope (5.3 wt% water, *bp* 62.3 °C). Treatment with alkali hydroxide [13] and subsequent distillation provides anhydrous THF. The azeotrope problem can also be circumvented industrially by distillation under pressure.

4.2. Butadiene Acetoxylation

The Mitsubishi-Kasei Corporation in Japan produces 1,4-butanediol and THF in parallel starting from butadiene [14]. The procedure can be described by the following set of equations:

$$H_2C=CH-CH=CH_2 \xrightarrow[O_2]{AcOH} AcOCH_2CH=CHCH_2OAc$$

$$\xrightarrow{H_2} AcOCH_2CH_2CH_2CH_2OAc$$

$$\xrightarrow{H_2O} AcOCH_2CH_2CH_2OH + AcOH$$

$$\xrightarrow{H_2O} HOCH_2CH_2CH_2CH_2OH + AcOH \quad \text{or} \quad \text{(THF)} + AcOH$$

where $Ac = CH_3\overset{O}{\underset{\|}{C}}-$

Butadiene is oxidized at 3 MPa/80 °C over a palladium–tellurium catalyst with acetic acid and a nitrogen–oxygen mixture to give 1,4-diacetoxy-2-butene. The olefin is hydrogenated to 1,4-diacetoxybutane [15], which can be hydrolyzed to butanediol or THF.

4.3. Propylene Oxide Process

Arco recently commenced industrial production of 1,4-butanediol by the following route:

$$H_2\overset{O}{\overset{\diagup\diagdown}{C}}CH-CH_3 \xrightarrow{Li_3PO_4} H_2C=CH-CH_2OH \xrightarrow[cat.]{H_2/CO}$$

$$HOCH_2CH_2CH_2CHO + HOCH_2\overset{CH_3}{\underset{|}{C}H}-CHO \xrightarrow{H_2}$$

$$HOCH_2CH_2CH_2CH_2OH + HOCH_2\overset{CH_3}{\underset{|}{C}H}-CH_2OH$$

Propylene oxide is isomerized to allyl alcohol by conventional means at 250–300 °C/1 MPa over a trilithium orthophosphate (Li_3PO_4) catalyst. The allyl alcohol is then hydroformylated to 4-hydroxybutyraldehyde and the byproduct 3-hydroxy-2-methylpropionaldehyde by a process developed by Kuraray [16]. Subsequent hydrogenation of the aldehydes gives 1,4-butanediol as the main product together with 2-methyl-1,3-propanediol.

4.4. Maleic Anhydride Hydrogenation

Because of its structure, maleic anhydride is an attractive precursor for the preparation of butanediol, THF, and γ-butyrolactone:

$$\text{maleic anhydride} + H_2 \longrightarrow \text{succinic anhydride}$$

$$\xrightarrow[-H_2O]{H_2} \text{THF} + HOCH_2CH_2CH_2CH_2OH + \gamma\text{-butyrolactone}$$

This route was originally developed and adapted to industrial scale by Mitsubishi-Kasei [17], but Mitsubishi itself has since abandoned the process and replaced it by the newer butadiene acetoxylation (Section 4.2).

Recently, Davy – McKee Ltd., BP, and Sohio have introduced a similar process for the production of butanediol and THF. Shinwha Petrochemical in South Korea has commissioned a plant based on the McKee process producing 20 000 t/a of butanediol. The first step is a vapor-phase hydrogenation of ethyl maleate [18], and the ratio of THF to butanediol in the product mix can be adjusted by altering the reaction conditions.

Increasing attention is being focused on maleic anhydride as a starting material for THF, and it appears that this approach will become more important than the Reppe process in the medium term [19].

4.5. *n*-Butane – Maleic Anhydride Process

Du Pont has developed a new two-step process for producing THF from *n*-butane. The process involves oxidation to maleic anhydride with a transported abrasion-resistant oxidation catalyst, which guarantees a high yield (70 – 75%), with subsequent hydrogenation of aqueous maleic acid solution over a special rhenium-doped palladium catalyst [20]. This method also leads to either butanediol or THF as desired; it has been perfected on a pilot scale and is due to be implemented in 1994 in the Spanish province of Asturia with an annual production of 45 000 t. The process can be summarized as follows:

$$CH_3CH_2CH_2CH_3 \xrightarrow{O_2} \text{maleic anhydride} \xrightarrow{H_2} \text{THF}$$

Table 1. Quality specifications for commercial tetrahydrofuran

Characteristics	Value	Test method
THF content	min. 99.9%	GC
Water content	max. 0.03%	DIN 51777, section 21
Hazen color index	max. 10	DIN/ISO 6271

4.6. Pentosan/Furfural Processes

Because it takes advantage of a renewable raw material, THF production from furfural is currently attracting interest. Because of their widespread distribution in agricultural waste products, pentosans are especially promising as starting materials [21]. The production of THF from furfural involves catalytic decarbonylation to furan and hydrogenation of the latter to THF [22].

5. Specifications and Analysis

Table 1 illustrates typical quality specifications quoted by a leading European producer of THF. Gas chromatographic (GC) analysis can be conducted with a 30-m WCOT (fused silica) column coated with DB-1 (FID detector, He as carrier gas) using n-octane as an internal standard [2].

6. Environmental Protection and Recovery

Based on the considerations presented in the section on toxicology (Chapter 10), it can be concluded that with proper handling THF constitutes no significant risk with respect to the environment.

THF is also quite easily recovered. Thus, THF vapors from exhaust gases can be adsorbed on activated charcoal and subsequently desorbed with steam. The efficiency of the process is said to be 98–99%. In the case of high concentrations of THF in air, absorption in water is also a possibility. The THF–water azeotrope (5.3 wt% water, bp 62.3 °C) obtained on distillation can be converted into anhydrous THF ($< 0.05\%$ water) by distillation at reduced or (preferably) elevated pressure [23].

Extraction of THF from an aqueous mixture with a solvent immiscible with water (e.g., pentane), followed by phase separation and fractional distillation of the organic phase, is another drying technique practiced on an industrial scale. This results in THF containing $< 0.1\%$ water [24]. For the separation of THF from water and such organic solvents as toluene by fractional distillation, see [24]. Various methods have been developed for drying on a smaller scale [25], [26]. Regardless which method is selected, great care is required because of the danger of peroxide formation [27].

7. Storage and Transport

The handling of THF entails only the usual precautionary measures applicable to other volatile organic compounds, and THF can be stored in the usual steel containers. For transportation through piping, welded pipes are preferable; special attention should be directed toward the seals in pumps and valves.

Like other ethers (e.g., diethyl or diisopropyl ether) THF forms peroxides in air if it has not been stabilized. At a peroxide concentration > 0.1 % THF is subject to explosive decomposition on distillation. It is stabilized with 2,6-di-*tert*-butyl-*p*-cresol (BHT). In the absence of a stabilizer THF must be stored in an inert gas atmosphere.

A safety data sheet in accordance with DIN 52 900 is available for THF.

For THF the following transport regulations are in force:

GGVE/RID	Class 3, no. 3 b
	RN 301
GGVS/ADR	Class 3, no. 3 b
	RN 2301
ADNR	Class 3, no. 3 b
	RN 6301
IMDG Code	Class 3.1
UN no.	2056
S phrases	16 – 29 – 33
R phrases	11 – 19 – 36/37

8. Uses

The most important area of application for THF from the standpoint of quantity and turnover is polymerization with simultaneous ring opening to give poly(tetramethylene oxide). Available in various molecular masses, this is a key component in the production of important elastic construction materials, thermoplastics and molded elastomers based on polyurethanes, polyesters and polyamides, elastic Spandex fibers, and polyurethane coatings. Trade names for commercial PTMO include Terathane (DuPont), Polytetrahydrofuran (BASF), and Polymeg (Quaker Oats).

THF is also a versatile solvent for natural and synthetic resins as well as PVC. It is used extensively in the varnish and film industries and as a cosolvent for printing inks and adhesives.

Table 2. Major THF producers in 1992

Company	Region	Process	Capacity, 10^3 t/a
Du Pont	USA	Reppe	70
BASF	USA, Europe, Japan	Reppe	40
ARCO	USA	PO/AA*	8
GAF	USA, Europe	Reppe	18
Quaker Oats	USA	furfural	10
Mitsubishi-Kasei	Japan	butadiene	16
Toso Corp.	Japan	Reppe	3
Shinwha	Korea	butane	10
	CIS/China	furfural/Reppe	10

* PO = propylene oxide; AA = allyl alcohol.

Table 3. Worldwide production capacity and consumption data by geographical area (10^3 t/a)

	USA 1990/1995	Western Europe 1990/1995	Southeast Asia 1990/1995	China, CIS 1990
Capacity	95/133	30/50	20/40	5–10
Consumption	63/80	50/70	20/30	5–10

9. Economic Aspects

The projected annual growth rate for tetrahydrofuran is 4–7% based on anticipated increases in the consumption of polytetrahydrofuran. Major producers of THF are listed in Table 2, and regional production capacity and consumption estimates are provided in Table 3.

10. Toxicology and Industrial Hygiene

THF shows weak narcotic activity. According to very intensive investigations conducted at the Industrial Hygiene and Pharmacology Institute at BASF it is one of the least toxic solvents. This is in sharp contrast to previous reports, which were based on impure samples. Its true toxicity is evidently not greater than that of acetone.

In the investigations at BASF, THF exposure levels of 1000 ppm for 6 h a day, 5 days a week, over a period of 1 year were tolerated by cats, rabbits, and rats with no associated symptoms of poisoning. Exposure to 3000 ppm daily for 8 h over a course of almost 2 years produced as few symptoms of poisoning in rats as did inhalation over the same length of time of 3000 ppm acetone. Concentrations of 1000 ppm or 3000 ppm are completely intolerable for humans because of severe mucous membrane irritation, and would not therefore be experienced in practice. Even higher concentrations of THF vapor produce a narcotic effect similar to that of ether or acetone. Like alcohol, THF is rapidly metabolized in the living organism.

As a consequence of its universal solvent power, THF dissolves the upper keratin-containing layers of skin and mucous membranes, penetrating very rapidly into deeper parts of the skin. Thus it gives rise to very severe skin and mucous membrane irritation, but it does not display a sensitizing effect. This irritant character is apparently reinforced by peroxides, since THF that contains peroxides due to prolonged storage is a more severe skin irritant. THF that is allowed to penetrate a wound or get under the fingernails causes very unpleasant pains.

Like methanol, THF is resorbed very rapidly by the skin. For this reason, prolonged wetting of the skin with THF must be strictly avoided, and hand protection should be provided against repeated wetting.

The maximum workplace concentration for THF has been fixed at 200 mL/m^3 (ppm). This limiting concentration is a function only of local mucous membrane irritation caused by THF vapors; much higher concentrations are required to produce a narcotic effect in humans [28].

11. References

[1] A. P. Kudschadker et al.: *Key Chemicals Data Books Furan, Dihydrofuran, Tetrahydrofuran,* Thermodynamics Res. Center, Texas A & M University, College Station, Texas 1978.
[2] BASF, technical report "Tetrahydrofuran," Ludwigshafen 1991.
[3] H. Meerwein et al., *Angew. Chem.* **72** (1960) 927.
[4] P. Dreyfuss: "Polytetrahydrofuran," in *Polymer Monographs,* vol. 1.8, Gordon and Breach Sci. Publ., New York 1982.
[5] W. Reppe: *Neue Entwicklungen auf dem Gebiet der Chemie des Acetylens und Kohlenmonoxyds,* Springer Verlag, Heidelberg 1949.
[6] "Furfural" 660.5020 in: *Chemical Economics Handbook,* Stanford Res. Inst., Menlo Park, Calif. 1968.
[7] A. Wong et al., *J. Am. Chem. Soc.* **102** (1980) no. 13, 4529.
[8] R. G. Gastinger et al., *J. Am. Chem. Soc.* **102** (1980) no. 15, 4959.
[9] R. Criegee et al., *Angew. Chem.* **62** (1950) 120.
[10] J. W. Copenhaver, M. H. Bigelow: *Acetylene and Carbonmonoxide Chemistry,* Reinhold Publ. Co., New York 1949; University Microfilms, Ann Arbor, Mich. 1949.
[11] BASF, DE 1 043 342, 1957 (J. Schneiders); DE 2 930 144, 1979 (H. Müller et al.).
[12] BASF, EP 1 536 680, 1979 (H. Müller, Ch. Palm).
[13] IG Farben, DE 713 565, 1939 (W. Reppe, H.-G. Trieschmann).
[14] Mitsubishi Chem., US 3 922 300, 1975 (T. Onoda et al.).T. Onoda: "1,4-Butylenglycol and Tetrahydrofuran Production from Butadien," *AIChE Summer National Meeting,*Minneapolis, Minn., Aug. 16 – 17, 1987. Kurary, US 3 872 163, 1975 (T. Shimizu, T. Yasui).
[15] Mitsubishi Chem., US 4 010 197, 1977 (J. Toriya, K. Shirago).
[16] Kurary, US 4 567 305, 1986 (T. Shimizu, T. Yasui). Kurary & Daicel, EP-A 129 802, 1984 (M. Milsuo, M. Shini).

[17] J. Kanetaka et al., *Ind. Eng. Chem.* **62** (1970) 24.Mitsubishi Chem., GB 1 230 276, 1971. GB 1 344 557, 1974.

[18] Davy McKee Ltd., EP 143 634, 1984 (M. Sherif, K. Turner). Davy McKee Ltd., GB 2 207 914, 1987 (M. Sherif, K. Turner).

[19] BASF, DE 3 423 447, 1986 (R. Schnabel, H. Weitz); DE 3 726 509, 1989 (R. Fischer, W. Harder, K. Malsch). Huels, 3 539 151, 1986 (R. Nehring, W. Otte). UCB, EP 339 012, 1989 (J. L. Dalons, J. Markusel). General Electric, US 4 652 685, 1987 (W. James, E. Norman). General Aniline, US 4 797 382, 1989 (D. Waldo, D. Taylor). Standard Oil, US 4 301 077, 1980 (F. A. Pesa, A. M. Graham); US 4 810 807, 1988 (J. Budge, E. Pederson); US 4 827 001, 1988 (Th. Attig, A. Graham). Tonen Corp., EP 373 946, 1989 (S. Sadakatsu, I. Hiroyuki); EP 431 923, 1991 (S. Sadukatsu, I. Tatsumi).

[20] Du Pont, EP-A 189 261, 1986 (R. M. Contractor); US 4 442 226, 1982 (T. A. Bither); US 4 371 702, 1982 (T. A. Bither); EP 147 219, 1989 (A. Marby, W. Pichard).

[21] C. Godawa et al., *Inf. Chim.* **260** (1985) 171–177.

[22] N. N. Machalaba et al., *Gidroliz. Lesokhim. Prom.-st.* **1988**, no. 5, 19–20; *Chem. Abstr.* 109 151883k. Quaker Oats, US 3 233 714, 1963. Du Pont, US 2 846 449, 1956 (W. H. Banford, H. M. Manes); US 2 374 149, 1943 (G. M. Whitman). Quaker Oats, US 3 021 342, 1958 (H. W. Hemker); GB 855 255, 1959.

[23] Du Pont, CA 546 591, 1957.

[24] Du Pont, US 2 790 813, 1953 (P. F. Bente).

[25] Du Pont, US 3 074 968, 1960 (J. W. Meinherz).

[26] Du Pont, bulletin A-25077, recovery of THF.

[27] B. G. White et al., *Org. Synth.* **46** (1966) 105.

[28] BG Chemie, Merkblatt M038, 588, Tetrahydrofuran, Jedermann Verlag Dr. Otto Pfeffer, Heidelberg 1988. Kühn-Birett: *Merkblätter Gefährlicher Arbeitsstoffe*, 26. Erg.Lfg., 3/85-T08-1, Verlag Moderne Industrie, München 1977. G. Hommel: *Handbuch der gefährlichen Güter*, 5th ed., Merkblatt 192, Springer Verlag, Berlin 1993. *Ullmann*, 4th ed., **12**, 21.

Thiocyanates and Isothiocyanates, Organic

FRANK ROMANOWSKI, Degussa AG, Frankfurt, Federal Republic of Germany
HERBERT KLENK, Degussa AG, Frankfurt, Federal Republic of Germany

1.	Organic Thiocyanates	4631	2.5.	Quality Specifications and Analysis 4641
1.1.	Introduction	4631		
1.2.	Physical Properties	4632	2.6.	Storage and Transport 4641
1.3.	Chemical Properties.......	4632	2.7.	Commercial Names 4641
1.4.	Production	4633	2.8.	Uses 4641
1.5.	Use and Commercial Names ..	4634	2.8.1.	Methyl Isothiocyanate........ 4641
			2.8.2.	Other Organic Isothiocyanates .. 4642
1.6.	Toxicology...............	4635	2.9.	Toxicology and Industrial Hygiene.................. 4643
2.	Organic Isothiocyanates	4635		
2.1.	Introduction	4635	2.10.	Safety Measures and Processing Advice 4644
2.2.	Physical Properties	4636		
2.3.	Chemical Properties.......	4637	2.11.	Ecological Aspects 4644
2.4.	Production	4638	3.	References............... 4644

1. Organic Thiocyanates

1.1. Introduction

Organic thiocyanates are esters of thiocyanic acid. In the general formula $R-S-C\equiv N$, the group R can be aliphatic, alicyclic, aromatic, or heterocyclic. So far, no organic thiocyanates have been detected in plants or animals.

Special applications have been discovered and developed for only a few synthetic thiocyanates. Certain organic thiocyanates are used as vulcanization accelerators, others have been used as preservatives because of their biocidal and fungicidal properties. Small quantities are produced as fine chemicals.

Table 1. Physical properties of some thiocyanates

Name, formula	M_r	CAS	Freezing point, °C	Boiling point, °C (kPa)	Density, g/cm^3
Methyl thiocyanate, CH$_3$SCN	73.11	[556-64-9]	−51	132 (101.3)	1.074
Ethyl thiocyanate, C$_2$H$_5$SCN	87.14	[542-90-5]	−85.5	145.5 (101.9)	1.011
3-Thiocyanatopropyltriethoxy-silane, (C$_2$H$_5$O)$_3$Si(CH$_2$)$_3$SCN	263.42	[34708-08-2]	−80	95 (0.013)	1.000
2-(2-Butoxyethoxy)ethyl thio-cyanate, C$_4$H$_9$OC$_2$H$_5$OC$_2$H$_5$SCN	205.28	[112-56-1]		120 – 122 (0.266)	1.016
Methylenebis(thiocyanate), NCSCH$_2$SCN	130.18	[6317-18-6]	103 – 104		

1.2. Physical Properties [1]

The low molecular mass alkyl thiocyanates are colorless oils with an odor reminiscent of leeks. They are insoluble in water, but soluble in ethanol and ether, for example. Some physical properties are listed in Table 1.

1.3. Chemical Properties

Thiocyanates isomerize on warming to isothiocyanates. Allylic and tertiary thiocyanates isomerize rapidly, so some thiocyanates can be converted quantitatively to isothiocyanates [2]. In the case of methyl thiocyanate, it is possible to produce methyl isothiocyanate on an industrial scale by isomerization (Section 2.4).

Reducing agents cleave thiocyanates to thiols and hydrogen cyanide.

$$\text{RSCN} + [\text{H}] \longrightarrow \text{RSH} + \text{HCN}$$

The reduction can be carried out with Sn/HCl, LiAlH$_4$, or sodium in liquid ammonia.

Organic thiocyanates are converted to sulfonic acids by strong oxidizing agents:

$$\text{R}-\text{SCN} + [\text{O}] \xrightarrow{\text{H}_2\text{O}} \text{RSO}_3\text{H}$$

The most commonly used oxidizing agent is nitric acid, but alkaline hypochlorite or hydrogen peroxide can also be used. Acid hydrolysis of organic thiocyanates leads to thiols via thiocarbamates:

$$\text{R}-\text{SCN} \xrightarrow[(\text{H}^+)]{\text{H}_2\text{O}} \text{R}-\text{S}-\overset{\text{O}}{\underset{\text{NH}_2}{\text{C}}} \xrightarrow[(\text{H}^+)]{\text{H}_2\text{O}} \text{RSH} + \text{CO}_2 + \text{NH}_3$$

Alkaline hydrolysis gives the corresponding disulfides in good yields [3]:

$$2\ RSCN + 2\ OH^- \longrightarrow RSSR + OCN^- + CN^- + H_2O$$

Because of its polarity, the C≡N bond can undergo addition reactions. Intramolecular reactions for the synthesis of heterocycles are particularly important [4], [5].

1.4. Production

An important method for the production of alkyl thiocyanates is substitution of a halogen in an alkyl halide by the thiocyanate anion. The most frequently used solvents are ethanol, water, and DMF [4], [6]. As the thiocyanate anion is ambidentate, the reaction can give mixtures of thiocyanates and isothiocyanates, or can proceed selectively to either isomer [7]:

$$R-X \begin{array}{c} \xrightarrow{SCN^-} R-SCN \\ \xrightarrow{SCN^-} R-NCS \end{array}$$
$$X = \text{halogen}$$

The selectivity of the reaction depends on the solvent, temperature, and thiocyanate concentration. Methylenebis(thiocyanate) [8]–[10] and methyl thiocyanate [11], [12], for example, can be produced by this method.

Organic thiocyanates isomerize partially or completely to the corresponding isothiocyanates during the production process, depending on the type of compound. For example, only isothiocyanates can be produced from β,γ-unsaturated alkyl halides.

Aromatic thiocyanates can be produced by reaction of arenes with dicyanodisulfane. Addition of a Lewis acid increases the electrophilicity of the dicyanodisulfane so much that even unreactive aromatics react:

$$R\text{-}C_6H_5 + NC-S-S-CN \longrightarrow R\text{-}C_6H_4\text{-}SCN + HSCN$$

A further route for the production of aromatic thiocyanates involves reaction of diazonium compounds with salts of thiocyanic acid in the presence of copper salts:

$$ArN_2^+X^- + MSCN \xrightarrow{Cu\ salt} ArSCN + N_2 + MX$$
$$Ar = aryl$$

A valuable extension of the possibilities for producing organic thiocyanates is the reaction of thiols with cyanogen chloride [13], particularly as no isomeric isothiocyanate can be formed.

$$RSH + ClCN \longrightarrow RSCN + HCl$$

Other methods exist for the production of organic thiocyanates, but these are not of industrial interest.

Thiocyanic acid adds to epoxides, forming 2-hydroxyalkyl thiocyanates [14], [15]:

$$\text{epoxide} + \text{HSCN} \longrightarrow \text{HO}\diagup\diagdown\text{SCN}$$

Addition of dicyanodisulfane to olefins results in quantitative formation of dithiocyanatoalkanes; the reaction is used to detect double bonds analytically in fat chemistry [16]. Radical initiators or sunlight accelerate the reaction, so a radical addition mechanism is probable.

Reaction of alkylating agents, e.g., dialkyl sulfates or alkyl sulfonates, with salts of thiocyanic acid gives organic thiocyanates [17], [18]:

$$R^1-SO_2OR^2 + MSCN \longrightarrow R^2SCN + R^1-SO_2OM$$

An overview of production methods for thiocyanates can be found in reference [6].

1.5. Use and Commercial Names

Industrial applications of organic thiocyanates have been developed for only a few compounds.

Methyl thiocyanate is isomerized to methyl isothiocyanate on an industrial scale (Section 2.4). A small number of compounds are used as insecticides. The compounds known as Lethanes [19] are used as contact insecticides for flies, mosquitoes, and household pests. Lethane 384 [*112-56-1*] has the formula $C_4H_9OC_2H_4OC_2H_4SCN$; Lethane 60 [*301-11-1*] has the formula $C_{11}H_{23}COOC_2H_4SCN$.

Lethane 60 (2-thiocyanatoethyl dodecanoate) is tolerated better by plants than Lethane 384 (2-(2-butoxyethoxy)ethyl thiocyanate) and can thus be used for combatting greenfly.

Methylenebis(thiocyanate) (MBT) and 2-(thiocyanatomethylthio)benzothiazole [*21564-17-0*] [20] exhibit biocidal properties. Both compounds are supplied ready for use in widely differing formulations under the name Busan. The quantities required for use as biocides are predicted to rise [21].

MBT hinders the propagation of slime-forming bacteria and fungi. As Busan 110, it is preferably used in refrigerator circulation systems and in papermills as a 10% solution in inert solvents [22]. To lower the COD in the circulation water in papermills MTB is used as a stable aqueous dispersion [23]. MTB is employed as a fumigant for counteracting fungal growth [24]. Other applications of methylenebis(thiocyanate) are described in the literature [25]–[28].

2-(Thiocyanatomethylthio)benzothiazole has been known as a pesticide for a long time [29]. It is sold as a solution in an inert solvent under the name Busan. It is used as a broad-spectrum bactericide und fungicide in various areas. It is thus possible to

Table 2. Toxicological data for important organic thiocyanates

Name	RTECS	Administration route	Species	LD_{50}, mg/kg
Methyl thiocyanate	XL 1575000	oral	rat	60
		intravenous	mouse	18
Methylenebis(thiocyanate)	XL 1560000	oral	rat	161
		intravenous	mouse	3.6
2-(2-Butoxyethoxy)ethyl thiocyanate	XK 8400000	oral	rat	90
		intravenous	mouse	56
2-(Thiocyanatomethylthio)benzothiazole	XK 8150000	oral	rat	1590
		dermal	rabbit	642

preserve water-based or organic-solvent-based dyes with Busan 30 [30]. All types of wood are protected from fungi and insects by a coating of the substance. Attack of leather by mold during processing can be prevented by treatment with 2-(thiocyanatomethylthio)benzothiazole [31], [32]. An advantage of the compound over pentachlorophenol is its better environmental compatibility.

3-Thiocyanatopropyltriethoxysilane (Si 264), $(C_2H_5O)_3SiCH_2CH_2SCN$, is used as a reinforcing additive in molding compounds which can be vulcanized by heating with peroxides. The molding compounds are more readily processed, and the polymer obtained after vulcanization has markedly improved properties [33].

1.6. Toxicology

In the case of alkyl thiocyanates, the toxicity in rodents after oral and subcutaneous administration decreases with increasing molecular mass. The higher homologs have an irritant, sometimes corrosive effect on the skin and mucous membranes. They cause irreversible damage to the outer epithelium, death of tissue (necrosis), and defect formation (ulceration), with subsequent hyperplastic reconstruction of the outer epithelium. The systemic toxicity is high (Table 2). Acute poisoning leads to a high degree of circulatory disturbance in the inner organs, with increases in circulation, bleeding, and possibly death.

2. Organic Isothiocyanates

2.1. Introduction

Organic isothiocyanates are esters of isothiocyanic acid. In the general formula R–N=C=S, the group R can be aliphatic, alicyclic, aromatic, acyl, or heterocyclic.

Organic isothiocyanates do not occur naturally in the free form. They are liberated from glucosinolates in plant tissues by enzymatic degradation. Their pungent odor and

Table 3. Physical data for some isothiocyanates

Name	M_r	CAS	Freezing point, °C	Boiling point, °C (kPa)	Density, g/cm^3
Methyl isothiocyanate, CH$_3$NCS	73.11	[556-61-6]	35.9	118.7 (100.9)	1.0691 (37°C)
Allyl isothiocyanate, CH$_2$=CHCH$_2$NCS	99.15	[57-06-7]	−102.5	151.8 (100.7)	1.015 (20°C)
Phenyl isothiocyanate, C$_6$H$_5$NCS	135.18	[103-72-0]	−21	221 (100.5)	1.138 (15.5°C)
Vinyl isothiocyanate, CH$_2$=CHNCS	85.12	[1520-22-5]		46 (10.0)	1.018 (15°C)
Methoxycarbonyl isothiocyanate, CH$_3$OCONCS	117.10	[35266-49-0]		30 (0.12)	1.152 (15°C)

sharp, mustard-like taste gave these compounds the name mustard oils, by analogy with the ethereal oils of black mustard [34]. Allyl isothiocyanates are liberated from the glucosinolate sinigrin of black mustard by myrosinase [35], [36]. Methyl isothiocyanate is formed by the enzymatic degradation of glucocapparin, which is contained in the caper plant.

Glucocapparin

Many other organic isothiocyanates with, e.g., sulfur-containing or aromatic groups have become known as natural products through the degradation of the corresponding glucosinolates.

Aliphatic and aromatic isothiocyanates are important starting materials for the production of thiourea derivatives, substituted thiosemicarbazides, heterocycles, and other organic intermediates. Some isothiocyanates have nematocidal, bactericidal, and fungicidal properties. Allyl, methoxycarbonyl, and methyl isothiocyanate are produced on an industrial scale.

2.2. Physical Properties

Most aliphatic and aromatic isothiocyanates are colorless oils with a pungent odor. Methyl isothiocyanate is a solid at room temperature. Many isothiocyanates are not only relatively stable to water in the cold, but can also be steam distilled. Aliphatic isothiocyanates with the general formula R–N=C=S, in which R is an unbranched or branched alkyl group with 2–6 carbon atoms, distil azeotropically with water at 80–120°C. The boiling points of the pure compounds lie in the range 120–200 °C.

Organic isothiocyanates are readily soluble in most common organic solvents.

Table 3 shows the most important physical properties of some isothiocyanates (for further data on isothiocyanates see reference [1]).

2.3. Chemical Properties

The chemical properties of organic isothiocyanates are similar to those of organic isocyanates, but the latter are generally more reactive [37]. Most reactions of isothiocyanates involve addition of a nucleophile to the –N=C=S group. Some examples of nucleophilic addition are given below [1], [38]:

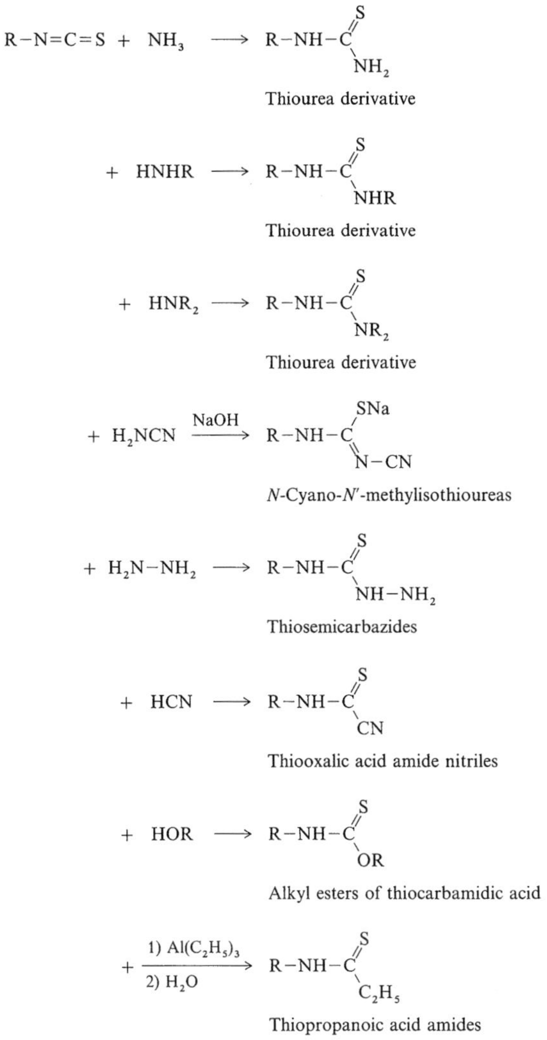

$$R-N=C=S + NH_3 \longrightarrow R-NH-\underset{NH_2}{\overset{S}{\overset{\|}{C}}}$$

Thiourea derivative

$$+ HNHR \longrightarrow R-NH-\underset{NHR}{\overset{S}{\overset{\|}{C}}}$$

Thiourea derivative

$$+ HNR_2 \longrightarrow R-NH-\underset{NR_2}{\overset{S}{\overset{\|}{C}}}$$

Thiourea derivative

$$+ H_2NCN \xrightarrow{NaOH} R-NH-\underset{N-CN}{\overset{SNa}{C}}$$

N-Cyano-*N'*-methylisothioureas

$$+ H_2N-NH_2 \longrightarrow R-NH-\underset{NH-NH_2}{\overset{S}{\overset{\|}{C}}}$$

Thiosemicarbazides

$$+ HCN \longrightarrow R-NH-\underset{CN}{\overset{S}{\overset{\|}{C}}}$$

Thiooxalic acid amide nitriles

$$+ HOR \longrightarrow R-NH-\underset{OR}{\overset{S}{\overset{\|}{C}}}$$

Alkyl esters of thiocarbamidic acid

$$+ \xrightarrow[2) H_2O]{1) Al(C_2H_5)_3} R-NH-\underset{C_2H_5}{\overset{S}{\overset{\|}{C}}}$$

Thiopropanoic acid amides

Isothiocyanates have a very important function in the synthesis of heterocycles. The large number of reactions which have been discovered is notable, and with them access to new compounds has become possible [39]–[42].

Interest in the development of corresponding synthetic processes was directed towards five- and six-ring heterocycles, but four- and seven-ring systems can also be synthesized, such as benzimidazoles, benzothiazoles, and other bicyclic heterocycles.

Amino acids can be converted to thiohydantoin derivatives [43]:

$$(HO)_2\overset{O}{\overset{\|}{P}}CH_2NHCH_2CO_2H + CH_3NCS \longrightarrow$$

$$(HO)_2\overset{O}{\overset{\|}{P}}H_2C-N\underset{O}{\overset{S}{\underset{\|}{\bigcirc}}}CH_3$$

Acyl isothiocyanates react with 1,3-diketones to 5-acyl-4H-1,3-oxazine-4-thiones [44]:

$$R\overset{O}{\overset{\|}{C}}NCS + CH_2(COR)_2 \longrightarrow \text{(5-acyl-4H-1,3-oxazine-4-thione)}$$

Anthranilic acid reacts with methyl isothiocyanate to 3-N-methyl-2-thioxotetrahydroquinoxalin-4-one [45]:

$$\text{o-H}_2N\text{-C}_6H_4\text{-CO}_2H + CH_3NCS \longrightarrow \text{(3-N-methyl-2-thioxotetrahydroquinoxalin-4-one)}$$

The isothiocyanate group –N=C=S can undergo cycloadditions with suitable reaction partners. Methyl isothiocyanate reacts with sodium azide to 5-mercapto-4-methyl-1,2,3,4-tetrazole [46]:

$$NaN_3 + CH_3N=C=S \longrightarrow \text{(5-mercapto-4-methyl-1,2,3,4-tetrazole)}$$

2.4. Production

A large number of processes are known for the production of isothiocyanates [39], [41], [47], [48]. The structure desired and sometimes the other functional groups on the organic skeleton determine the choice of production process.

The most important production methods are:

1) Reaction of alkyl and acyl halides with salts of thiocyanic acid [49], [50]
2) Isomerization of esters of thiocyanic acid [51], [52]
3) Reaction of dithiocarbamates with: hydrogen peroxide [53], [54], cyanogen chloride [55], cyanuric chloride [56], chloroformates [57], alkali hypochlorite or alkali chlorite [58], phosgene [59], heavy metal salts [60], atmospheric oxygen [61], [62], phosphorus oxychloride [63]

4) Reaction of primary amines with thiophosgene [64]
5) Reaction of thiourea derivatives with mineral acids, or acetic anhydride, or by heating [39]
6) Addition of thiocyanic acid to unsaturated compounds [65]

For vinyl isothiocyanates, special production methods are usually necessary [48].

Methyl isothiocyanate, which is industrially important, is produced by two different processes:

1) Rearrangement of methyl thiocyanate
2) Oxidation of sodium N-methyldithiocarbamate with hydrogen peroxide

The isomerization of methyl thiocyanate begins above 100 °C. The reaction is catalyzed by salts, e.g., $ZnCl_2$, alkali-metal thiocyanate [2].

$CH_3SCN \longrightarrow CH_3 N=C=S$

A patent specification by Morton Chemical [52] describes a multistep continuous process for the isomerization. The temperature of the reaction, which requires a long induction period, is 130–160 °C. In the course of reaction, a mixture is formed, which promotes the isomerization autocatalytically and favors the shift in the equilibrium. The methyl isothiocyanate formed is continuously removed from the vapor phase together with methyl thiocyanate. This mixture is subjected to fractional distillation and the methyl thiocyanate is recycled into the isomerization. The methyl isothiocyanate removed is replaced by fresh methyl thiocyanate. The side products formed (sulfides, disulfides) are removed periodically from the liquid phase. It is possible to convert up to 90% of the methyl thiocyanate into isothiocyanate by this process.

A patent specification by Stauffer [51], describes the isomerization of methyl thiocyanate in the gas phase at 300–400°C on a fixed-bed catalyst. The carrier (silica gel, activated charcoal) is impregnated with catalytically active salts, e.g., alkali thiocyanate, $ZnCl_2$. The vapor phase at the head of the reactor contains up to 55% methyl isothiocyanate. It is purified by fractional distillation and the recovered methyl thiocyanate is recycled.

The production of methyl isothiocyanate from sodium methyldithiocarbamate and hydrogen peroxide has been developed on a large industrial scale [53], [54]. The formation of methyl isothiocyanate takes place via several steps:

$$2\,CH_3NH-\overset{\underset{\|}{S}}{C}-SNa + H_2O_2 \longrightarrow$$

$$\left[CH_3NH\overset{\underset{\|}{S}}{C}-S-S-\overset{\underset{\|}{S}}{C}NHCH_3\right] + 2\,NaOH$$

$$CH_3HN-\overset{\underset{\|}{S}}{C}-S-S-\overset{\underset{\|}{S}}{C}-NHCH_3$$
$$\longrightarrow 2\,CH_3-N=C=S + S + H_2S$$

$$H_2S + 4\,H_2O_2 \longrightarrow H_2SO_4 + 4\,H_2O$$

The reaction can be represented by the overall equation:

$$2\,CH_3NH\overset{\underset{\|}{S}}{C}-SNa + 5\,H_2O_2$$
$$\longrightarrow 2\,CH_3-N=C=S + S + Na_2SO_4 + 6\,H_2O$$

With smaller quantities of hydrogen peroxide, sodium thiosulfate is predominantly formed as the side-product:

$$2\,CH_3NH\overset{\underset{\|}{S}}{C}-SNa + 4\,H_2O_2$$
$$\longrightarrow 2\,CH_3-N=C=S + Na_2S_2O_3 + 5\,H_2O$$

Industrial-scale production is as follows: ca. 40% aqueous sodium methyldithiocarbamate solution (pH 10–12) is reacted with 30–40% hydrogen peroxide at ca. 100°C. The pH during the reaction is 3–4. The reaction mixture is heated so that the methyl isothiocyanate immediately distils as an azeotrope with water. Methyl isothiocyanate (ca. 5–7%) is dissolved in the water that distils over. This can easily be separated by a short distillation. A very pure product is formed in ca. 85% yield. The sulfur precipitated during the reaction can be separated readily from the wastewater, which is readily biodegradable. The small quantity of waste-gas formed is incinerated.

Allyl isothiocyanate is produced by reaction of allyl chloride with alkali rhodanides in the two-phase system water/1,2-dichloroethane [66].

All commercial methoxycarbonyl isothiocyanate is produced in situ from methyl chloroformate and potassium thiocyanate in acetone. It is then converted immediately to thiophanate methyl (Section 2.8.2) [67].

Vinyl isothiocyanate can be produced in 63% yield in a two-step reaction. 2-Aminobromoethane is reacted with thiophosgene in ether. The 2-bromoethyl isothiocyanate obtained after work-up and distillation is converted into vinyl isothiocyanate by elimination of hydrogen bromide [68].

2.5. Quality Specifications and Analysis

The processing of industrially produced organic isothiocyanates to higher-value products and their use in pharmaceuticals requires high purity. The quality of the product is indicated by the –N=C=S content. To determine this, the product is treated with excess dibutylamine in dioxan or chlorobenzene/methanol, and the excess amine is titrated [69], [70]. Gas–liquid chromatography gives a detailed overview of the side-product spectrum.

2.6. Storage and Transport

Methyl isothiocyanate is generally stored in tanks or in 200-kg steel drums. The product is stable to storage, provided it is free of water. The containers should be stored and handled tightly closed at normal temperature in a dry, well-ventilated area, because the product has a severely irritant effect on the skin and mucous membranes and is toxic [71].

Methyl thiocyanate: UN no. 2477, EINECS 209-132-5, IMDG 6.1 PG II, RID/ADR 6.1 no. 20b.

2.7. Commercial Names

Methyl isothiocyanate is sold worldwide by Degussa (Germany) and Morton Chemical (United States).

In formulations, it is sold under the names Trapex and Di-Trapex by Schering. Di-Trapex is sold by Nor-Am under the name Vorlex in the United States. Osmose sells methyl isothiocyanate as MITC-FUME in glass and aluminum ampules. Allyl isothiocyanate is sold in Germany by Haarman & Reimer and in the United Kingdom by Rhône-Poulenc.

2.8. Uses

2.8.1. Methyl Isothiocyanate

Consumption of methyl isothiocyanate is currently < 2000 t/a; 80% is used in agrochemicals, ca. 15% in pharmaceuticals, and ca. 5% in other applications.

Agrochemicals. Methyl isothiocyanate is used directly as a 20% formulation in an inert solvent (Vorlex, Trapex, Di-Trapex) for protecting soils against fungi and nematodes.

1,3,4-Thiadiazole derivatives are known for their herbicidal properties. Methyl isothiocyanate is used for their synthesis. The most important examples are Spike (Eli Lilly, United States) [72], Ustilan (Bayer, Germany) [73], and Erbotan (Ciba-Geigy, Switzerland) [74].

$$R\underset{S}{\overset{N-N}{\diagup\!\!\!\diagdown}}N-\underset{CH_3}{\overset{O}{\overset{\|}{C}}}-NHCH_3$$

Spike   R = *t*-butyl
Ustilan   R = ethylsulfonyl
Erbotan   R = trifluormethyl

Pharmaceuticals. Methyl isothiocyanate is an important basic chemical for the synthesis of the H_2-blockers ranitidine (Zantac, Glaxo) [75], [76] and cimetidine (Tagamet, Smith Kline & French) [77].

5-Mercapto-1-methyltetrazole is obtained by treating methyl isothiocyanate with sodium azide. It is a side-group in the cefalosporin antibiotics Latamoxef [78] and Cefamandol [79].

Other Uses. Methyl isothiocyanate is very effective in the protection of wood against rotting [80] and premature decay.

A special application of decay prevention for telegraph poles in the United States has been developed by Osmose Wood Preserving. Four holes, on average, are bored at an angle of 45° in the poles. Ampules of methyl isothiocyanate (MITC-FUME) are inserted into the holes and the latter are closed with wood plugs. The methyl isothiocyanate diffuses slowly into the wood. After ca. 30 d the substance can be detected at 60 – 80 cm from the opening of the ampule in dry Douglas pine [81]. The moisture content of the wood determines the rate of migration, the effectiveness, and also the time at which further treatment is required. The decision as to whether a secondary impregnation with MITC-FUME is necessary is made on inspection of the poles after 6 years [82]. Effectiveness is estimated to last 12 years, so that the second inspection cycle is within the expected period of activity. The lifetime of the telegraph poles is expected to be doubled by subsequent treatment with MITC-FUME [83].

2.8.2. Other Organic Isothiocyanates

Allyl isothiocyanate is mainly used in the sugar industry because of its fungicidal properties. It protects sugar beet from fungi during storage [84]. Allyl isothiocyanate is also known as an insecticide and an antimicrobial agent, and is used in perfumes. It is used for combatting the cabbage maggot fly *H. brassicae*.

Table 4. Toxicological data for organic isothiocyanates

Name	RTECS	Administration route	Species	LD_{50}, mg/kg
Methyl isothiocyanate	PA 9625000	oral	rat	67
		dermal	rabbit	174
		inhalation	rat	29.6
Allyl isothiocyanate	NX 8225000	oral	rat	112
		dermal	rabbit	88
Phenyl isothiocyanate	NX 9275000	oral	mouse	87

2-(Hydroxyethyl)allylthiourea, produced from allyl isothiocyanate, has been used for producing light-sensitive paper. Methoxycarbonyl isothiocyanate [67], [85] is treated in situ with o-phenylenediamine to give thiophanate methyl. Thiophanate methyl [23564-05-8] is a systemic fungicide, sold under the names Topsin M (Nippon Soda, Japan), Mildothane (May & Baker, United Kingdom), and Cycosin (American Cyanamid, United States). In 1980, United States consumption of thiophanate methyl was 150 t.

2.9. Toxicology and Industrial Hygiene

As vapors, organic isothiocyanates have a severely irritant effect on the skin, eyes, and respiratory tract. Direct contact with the body leads to severe damage to the areas affected. Skin resorption is possible. Oral administration leads to general disturbance of function, cramp, and unconsciousness. Acute toxicity after oral administration in animals is generally high (Table 4).

Methyl Isothiocyanate [86] – [88]. The vapor has a severely irritant effect on the skin and mucous membranes of the eyes and respiratory tract, and gives rise to a burning pain. On dermal contact, first- or second-degree acid burns are possible. Without immediate treatment, corneal opacity of the eyes is also possible. After swallowing, burning, pain, vomiting, cramp, unconsciousness, complete loss of reflexes, reduction in blood pressure, and death can occur. Methyl isothiocyanate can damage the liver and kidneys and cause lung edema.

Allyl Isothiocyanate [89]. Local and systemic toxic effects are very similar to those of methyl isothiocyanate.

2.10. Safety Measures and Processing Advice

In all work with organic isothiocyanates a protective mask with combination filter B, thick protective clothing, and safety glasses must be worn. Work areas must be well ventilated and, because sight can be impaired in the event of an accident, straight-line, free escape routes must be provided. Apparatus and opened vessels must be extracted while filling and emptying. Vapors form explosive mixtures with air, so ignition sources must be excluded.

Methyl isothiocyanate which has leaked must be covered with cold water, and the crusts formed added in portions to 10% aqueous ammonia, and disposed of separately. Spilled allyl isothiocyanate should be covered with activated charcoal and the latter disposed of carefully.

2.11. Ecological Aspects

Organic isothiocyanates, particularly methyl and allyl isothiocyanates, represent a hazard to bodies of water. Measures should be taken to avoid contamination of lakes, streams, and groundwater, under all circumstances.

3. References

[1] E. E. Reid: *Organic Chemistry of Bivalent Sulfur*, vol. **VI**, Chem. Publ., New York 1966.
[2] A. Fava in N. Kharasch, Cal. Y. Meyers (eds.): *The Chemistry of Organic Sulfur Compounds*, vol. **2**, Pergamon Press, New York 1966, pp. 73–91.
[3] E. Hoggart, W. A. Sexton, *J. Chem. Soc.* 1947, 815.
[4] R. G. Guy in S. Patai (ed.): *The Chemistry of Cyanates and their Thio Derivatives*, part 2, J. Wiley & Sons, Chichester 1977, pp. 819–878.
[5] B. Schulze, M. Mühlstädt, *Z. Chem.* **19** (1979) 41–48.
[6] *Houben/Weyl*, **E4**, 941–969.
[7] D. Knoke, K. Kottke, R. Pohloudek-Fabini, *Pharmazie* **28** (1973) 574, 617; **32** (1977) 195–215.
[8] Sumitomo Chem., DE 1 232 574, 1967 (T. Kawanami, G. Suzukamo).
[9] Nalco Chem., DE 1 668 494, 1967 (J. Matt, E. W. Hunter, L. A. Goretta).
[10] Kinkai Kagaku, JP-Kokai 6 007 2858, 1985 (K. Shinya, H. Seuda, S. Kojima).
[11] Morton, DE 1 295 549, 1966 (J. T. Venerable, J. Miyashiro, A. W. Seiling).
[12] Bayer, DE 1 183 903, 1965 (K. Goliasch, E. Grigat, R. Pütter).
[13] Degussa, DE 1 270 553, 1966 (Ch. Kosel).
[14] H. D. Vogelsang, T. Wagner-Janregg, R. Rebling, *Justus Liebigs Ann. Chem.* **569** (1950) 183.
[15] F. G. Weber, H. Ließert, R. Radeglia, *Pharmazie* **35** (1980) 330.
[16] H. P. Kaufmann: *Analyse der Fette und Fettprodukte*, vol. **I**, Springer Verlag, Berlin 1958, p. 570.

[17] Dow Chemical, US 2 939 875, 1958 (U. D. Floria).
[18] E. J. Corey, R. B. Mitra, *J. Am. Chem. Soc.* **84** (1962) 2938.
[19] K. H. Büchel in R. Wegler (ed.): *Chemie der Pflanzenschutz- und Schädlingsbekämpfungsmittel*, vol. **1**, Springer Verlag, Berlin 1970, pp. 457–459.
[20] Buckman Laboratories, US 1 129 575, 1966.
[21] Kline & Company, Biocides-Industry Report, Investext, March 26, 1990, pp. 1–4.
[22] Buckman Laboratories, Product Data Sheet, Memphis, Tenn., July 1981.
[23] W. D. Henkels, A. Schenker, J. F. Schuetz, *Wochenbl. Papierfabr.* **112** (1984) no. 23/24, 869.
[24] Yamato Chem. Ind., JP-Kokai 59 010 530, 1984 (A. W. Chester, Chu Yang Feng).
[25] Ichikawa Gosei Kagaku, JP-Kokai 03 007 205, 1991 (Y. Igarashi, T. Tsunoda, K. Keisake, R. Imai).
[26] Lester Technologies Corp., US 4 975 109, 1990 (L. A. Friedman, R. F. McFarlin).
[27] T. C. Presuell, D. D. Nicholas, *For. Prod. J.* **40** (1990) 57–61.
[28] W. R. Grace and Co., EP 282 203, 1988 (D. Clarkson, R. P. Cliffort).
[29] T. Egli, E. Sturm in R. Wegler (ed.): *Chemie der Pflanzenschutz- und Schädlingsbekämpfungsmittel*, vol. **6**, Springer Verlag, Berlin 1981, p. 380.
[30] Buckman Laboratories, product information "BUSAN 30L".
[31] H. Gattner, W. Lindner, H.-U. Neuber, *Leder* **39** (1988) 66.
[32] W. M. Fowler, A. E. Russel, C. G. Whiteley, *J. Am. Leather Chem. Assoc.* **85** (1990) 243.
[33] Degussa, DE 4 100 217, 1992 (U. Görl, S. Wolff).
[34] A. W. Hofmann, *Ber. Dtsch. Chem. Ges.* **1** (1868) 25.
[35] F. Hoffmann, *Chem. Unserer Zeit* **12** (1978) 182.
[36] A. Kjaer in N. Kharasch (ed.): *Organic Sulfur Compounds*, vol. **1**, Pergamon Press, New York 1961, p. 409.
[37] *Houben/Weyl*, **IX**, 768.
[38] S. J. Assony in N. Kharasch (ed.): *Organic Sulfur Compounds*, vol. **1**, Pergamon Press, New York 1961, p. 326.
[39] S. Sharma, *Sulfur Rep.* **8** (1989) 327.
[40] A. K. Mukerjee, R. Ashare, *Chem. Rev.* **91** (1991) 1.
[41] M. Avalos et al., *Heterocycles* **33** (1992) 973–1010.
[42] G. Giesselmann, K. Huthmacher, H. Klenk, F. Romanowski, *Chem. Ztg.* **114** (1990) 215–224.
[43] Schering, DE 3 427 794, 1986 (F. Blume et al.).
[44] H. Dehne, P. Krey, *J. prakt. Chem.* **324** (1982) 915.
[45] Ch. H. Chan, *Heterocycles* **26** (1987) 3193.
[46] H. W. Altland, *J. Org. Chem.* **41** (1976) 3395.
[47] *Houben/Weyl*, **E4**, 834–883.
[48] K. Schulze, B. Schulze, Ch. Richter, *Z. Chem.* **29** (1989) 41.
[49] Ciba-Geigy, DE 3 504 016, 1985 (M. Böger, J. Drabek).
[50] B. S. Drach, E. P. Sviridov, A. V. Kirsanov, *Zh. Org. Khim.* **8** (1972) 1872.
[51] Stauffer Chem., US 2 954 393, 1960 (J. N. Haimsohn, G. E. Lukes).
[52] Morton Chem., DE 1 253 265, 1962 (J. T. Venerable, J. Miyashiro, P. L. Weyna).
[53] Degussa, DE 2 105 473, 1971 (G. Giesselmann, K. Günther).
[54] Degussa, DE 4 001 020, 1990 (G. Giesselmann, W. Schwarze).
[55] Degussa, DE 2 603 508, 1976 (G. Giesselmann, R. Vanheertum, G. Schreyer).
[56] Degussa, DE 1 935 302, 1969 (W. Schwarze, W. Weigert).
[57] Bayer, DE 1 178 423, 1963 (R. Heusch).
[58] Bayer, DE 952 084, 1956 (E. Schmidt, R. Schnegg, F. Zaller, F. Moosmüller).

[59] Schering, DE 1 068 250, 1957 (H. Werres).
[60] G. Losse, H. Weddige, *Justus Liebigs Ann. Chem.* **636** (1960) 144.
[61] Stauffer Chem., US 4 713 467, 1987 (J. E. Telschow, D. A. Bright).
[62] Story Chemical, US 3 923 852, 1975 (A. G. Zeiler, H. Babad).
[63] DD 43 996, 1965 (D. Martin, E. Beyer, H. Gross).
[64] Bayer, DE 1 172 257, 1963 (E. Mühlbauer, E. Meisert).
[65] M. S. Kharasch, E. M. May, F. R. Mayo, *J. Am. Chem. Soc.* **59** (1937) 1580.
[66] IG Farbenindustrie, DE 705 561, 1941 (W. Flemming, H. Buchholz).
[67] May & Baker, DE 2 129 960, 1982 (D. A. Eichler et al.).
[68] Dow Chemical, US 2 757 190, 1956 (G. D. Jones, R. L. Zimmerman).
[69] A. G. Williamson, *Analyst (London)* **77** (1952) 372.
[70] M. Ottnad, N. A. Jenny, C.-H. Röder, *Anal. Methods Pestic. Plant Growth Regul.* **10** (1976) 563–573.
[71] Degussa, DIN Sicherheitsdatenblatt "Methylisothiocyanat," Hanau, 1991.
[72] Air Products and Chemicals, DE 2 100 057, 1971 (T. Cebalo, J. F. Alderman).
[73] Bayer, DE 1 816 568, 1968 (C. Metzger, H. Hack).
[74] Bayer, DE 1 770 467, 1968 (D. Rücker, C. Metzger, L. Eue, H. Hack).
[75] Glaxo Group, GB 2 160 204, 1985 (J. F. Seager, R. Dansey).
[76] Union Quimico Farmazeutika, ES 2 003 781, 1988 (I. Lopez Molina, A. D. Coto).
[77] Smith Kline & French, US 3 894 151, 1975 (J. W. Black, M. E. Parsons).
[78] Eli Lilly, US 4 452 778, 1984 (L. G. Brier).
[79] Eli Lilly, US 3 641 021, 1972 (Ch. W. Ryan).
[80] J. J. Morrell, M. E. Corden, *For. Prod. J.* **36** (1986) 26.
[81] J. N. R. Ruddick: "Fumigant Movement in Canadian Wood Species," The International Research Group on Wood Preservation, *15th Annual Meeting,* Stockholm, Sweden 1984.
[82] J. J. Morrell: "The Use of Fumigants for Controlling Decay of Wood: A Review of Their Efficacy and Safety," International Research Group on Wood Preservation, *20th Annual Meeting,* Lappeenranta, Finland, 22–26 May 1989.
[83] J. J. Morrell: *Market Potential for Safer Fumigants for Wood Preservation,* Oregon State University, Corvallis 1986.
[84] I. Leifertova, M. Lisa, J. Stanek, N. Hejtmankova, *Folia Pharm. (Prague)* **5** (1983) 43.
[85] Nippon Soda, DE 1 930 540, 1969 (T. Noguchi et al.).
[86] Industrieverband Agrar: *Wirkstoffe in Pflanzenschutz- und Schädlingsbekämpfungsmittel: physikalisch-chemische und toxikologische Daten,* 2nd ed., BLV Verlag, München 1990, p. 284.
[87] Kühn/Birett: Methylisothiocyanat M 45, *Merkblätter gefährliche Arbeitsstoffe,* ecomed Verlag, Landsberg 1992.
[88] Neue Datenblätter für gefährliche Arbeitsstoffe nach der Gefahrstoffverordnung, Methylisothiocyanat Datenblatt 1032, Weka Verlag, Augsburg, 1992.
[89] Kühn-Birett: Allylisothiocyanat A 106, *Merkblätter gefährliche Arbeitsstoffe,* ecomed Verlag, Landsberg, 1992.

Thiols and Organic Sulfides

KATHRIN-MARIA ROY, Langenfeld, Federal Republic of Germany

The author would like to thank Bayer AG for undertaking the literature research for this article

1.	**Aliphatic Thiols, Sulfides, Disulfides, and Polysulfides**	4648	2.2.1.	By Reduction	4663
			2.2.2.	By Thiolation	4664
1.1.	Introduction	4648	2.3.	**Uses of Aromatic Thiols**	4665
1.2.	**Properties of Aliphatic Thiols**	4648	2.4.	**Production of Aryl Sulfides, Disulfides, and Polysulfides**	4666
1.2.1.	Physical Properties	4648			
1.2.2.	Chemical Properties	4650	2.5.	**Uses of Aryl Sulfides, Disulfides, and Polysulfides**	4666
1.3.	**Production of Aliphatic Thiols**	4651			
1.3.1.	From Alcohols	4651	3.	**Heterocyclic Thiols, Sulfides, and Disulfides**	4667
1.3.2.	From Alkenes	4652			
1.3.3.	Other Production Methods	4654	3.1.	Introduction	4667
1.4.	**Construction Materials, Storage, and Transport**	4655	3.2.	**Physical and Chemical Properties**	4667
1.5.	**Uses of Aliphatic Thiols**	4656	3.3.	**Production**	4670
1.6.	**Alkyl Sulfides, Disulfides, and Polysulfides**	4657	3.3.1.	Heterocyclic Thiols by Substitution	4670
1.6.1.	Physical and Chemical Properties	4657	3.3.2.	Heterocyclic Thiols by Cyclization	4671
1.6.2.	Production of Alkyl Sulfides	4659	3.3.3.	Heterocyclic Sulfides and Disulfides	4673
1.6.3.	Production of Alkyl Disulfides	4660			
1.6.4.	Production of Alkyl Polysulfides	4661	3.4.	**Uses of Heterocyclic Thiols, Sulfides, and Disulfides**	4674
1.6.5.	Uses of Alkyl Sulfides, Disulfides, and Polysulfides	4661	3.4.1.	Pharmaceuticals	4675
2.	**Aromatic Thiols, Sulfides, Disulfides, and Polysulfides**	4662	3.4.2.	Agricultural Chemicals	4675
			3.4.3.	Other Uses	4676
2.1.	**Physical and Chemical Properties**	4662	4.	**Toxicology**	4677
2.2.	**Production of Aromatic Thiols**	4663	5.	**References**	4679

1. Aliphatic Thiols, Sulfides, Disulfides, and Polysulfides

1.1. Introduction

Aliphatic thiols, R–SH, were formerly known as mercaptans because of their affinity for mercury compounds (Latin: *corpus mercurium aptans*). They can be regarded as organic derivatives of hydrogen sulfide or thio analogs of alcohols.

According to IUPAC nomenclature the names of aliphatic thiols are constructed by adding the ending "thiol" to the name of the corresponding alkane (e.g., ethanethiol). If substituents with higher priorities are present, the prefix "mercapto" is used instead (e.g., mercaptoacetic acid).

Thiols are found in a wide range of natural materials. Low molecular mass thiols are responsible for the aromas of various foods, such as cheese, milk, coffee, cabbage, and bread. They are probably liberated during processing through enzymatic cleavage of sulfides, disulfides, and other sulfur-containing substances [12]. Thiols can also be detected in a number of plants [13]. Methanethiol occurs in radishes and various alliums, such as onions, leeks, and garlic. The characteristic odor of fresh grapefruit juice is due to the presence of 1-*p*-menthene-8-thiol in high dilution [14]. Asparagus contains 3-mercapto- and 3,3′-dimercaptoisobutyric acids [15]. Butanethiol and 2-methylbutanethiol have been found in the defense secretions of skunks [16]. Low molecular mass alkanethiols are formed in the degradation of biological material during putrefaction [17], and consequently are found in varying quantities in most crude petroleum, depending on its origin [18].

L-Cysteine, one of the natural amino acids found in proteins and a substance that participates in an important reversible redox system with the disulfide cystine, is of considerable biochemical interest, as is coenzyme A (CoA–SH), whose acetyl derivative (CoA–S–Ac) is a key participant in the metabolism of carbohydrates, fats, and proteins.

In 1834 W. C. Zeise achieved the first thiol synthesis by preparing ethanethiol from barium hydrogensulfide and calcium ethyl sulfate [19].

1.2. Properties of Aliphatic Thiols

1.2.1. Physical Properties

With the exception of methanethiol, which is gaseous at room temperature, aliphatic thiols with up to 16 carbon atoms are colorless or yellowish liquids. They are characterized by extremely offensive odors, perceptible even at very high dilution. The odor threshold for ethanethiol is less than 0.001 ppm, and that for the *S*-enantiomer of 1-*p*-

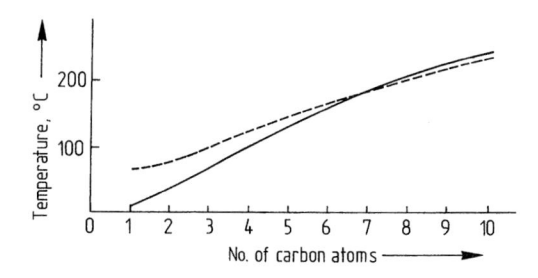

Figure 1. Comparison of boiling points for n-alcohols and n-thiols containing 1–10 carbon atoms
—— RSH   ---- ROH

Table 1. Physical data for selected aliphatic thiols

Thiol	CAS registry no.	M_r	bp, °C/kPa	d_4^{20}	n_D^{20}	Flash point, °C
CH_3-SH	[74-93-1]	48.11	6.2	0.8665		
CH_3CH_2-SH	[75-08-1]	62.13	35.0	0.8391	1.43105	−17
$CH_3-(CH_2)_2-SH$	[107-03-9]	76.16	67.8	0.8415	1.43832	−20
$(CH_3)_3C-SH$	[75-66-1]	90.19	64.2	0.8002	1.4232	−24
$CH_3-(CH_2)_3-SH$	[109-79-5]	90.19	98.4	0.8416	1.4430	12
$CH_3-(CH_2)_5-SH$	[111-31-9]	118.24	152.6	0.838	1.4497	20
⬡-SH	[1569-69-3]	116.23	158	0.950	1.4921	43
$CH_3-(CH_2)_7-SH$	[111-88-6]	146.30	199.1	0.8433	1.4540	68
$(CH_3)_3C-CH_2-C(CH_3)_2-SH$	[141-59-3]	146.30	155.5	0.845	1.456	
$CH_3-(CH_2)_{11}-SH$	[112-55-0]	202.41	142−5/2	0.8450	1.4589	
tert.-$C_{12}H_{25}$-SH (isomeric mixture)	[25103-58-6]	202.41	ca. 233	0.86	1.467	

menthene-8-thiol is only 2×10^{-8} ppm [13]. The odor sensitivity is highly concentration dependent, however, particularly for the lower thiols. At very low concentrations the C_1 to C_4 thiols have an odor similar to that of coal gas. That of other low molecular mass thiols has been described as pleasantly fresh or fruity, but at higher concentrations these compounds have extremely nauseous odors.

The bond energy of the S–H bond (339 kJ/mol) is lower than that of the O–H bond (462 kJ/mol), so hydrogen bonding to sulfur is also much weaker than to oxygen. This in turn means that thiols are less highly associated than the corresponding alcohols, as manifested in higher volatilities and lower boiling points for simple aliphatic thiols compared with their oxygen counterparts. This boiling point difference diminishes as the chain length increases, however, and is eventually reversed. Thus, from 1-octanethiol onward the straight-chain thiols have higher boiling points than the corresponding alcohols (Fig. 1).

The physical properties of important alkanethiols are summarized in Table 1. The high flammability of thiols containing up to six carbon atoms is particularly noteworthy. All the corresponding flash points lie below room temperature.

Unlike their oxygen analogs, aliphatic thiols are only slightly soluble in water, but they are readily soluble in hydrocarbons, ethers, and alcohols. Ethane-, propane-, and butanethiols form azeotropes with aliphatic hydrocarbons, but not with benzene or toluene.

Aliphatic thiols are only weakly acidic, but they are significantly more acidic than the corresponding alcohols, which explains the ready solubility of low molecular mass alkanethiols in dilute base. However, the solubility decreases sharply with increasing molecular weight.

$$R-SH + Na^+OH^- \longrightarrow R-S^-Na^+ + H_2O$$

The resulting thiolate salts, RS^-Na^+, are less basic but more strongly nucleophilic than the equivalent alkoxide salts.

1.2.2. Chemical Properties

Salt Formation. Aliphatic thiols form thiolates (formerly known as mercaptides) with numerous metal cations. Many of these salts are sparingly soluble, so their formation can be used for the qualitative and quantitative determination of thiols. Many mercury thiolates, $Hg(SR)_2$, have been used for the identification of thiols because of their defined melting points. Silver thiolates have also been used for the qualitative and quantitative detection of thiols. The formation of lead thiolate constitutes the basis for the "doctor test," a method of determining the thiol content of crude petroleum. Heavy metals, in particular mercury and cadmium, can be removed from wastewater with SH-group-containing resins or ion exchangers [20].

Thiolate formation is also utilized in the treatment of heavy-metal poisoning (copper, lead, mercury, zinc, cobalt, and gold). The thiol groups of essential enzymes that have been blocked by such metal ions are liberated again with the aid of the more strongly chelating thiols. The water-soluble substances 2,3-dimercaptopropane-1-sulfonic acid [4076-02-2] and dimercaptosuccinic acid [304-55-2] are being increasingly used for this complexation reaction, alongside 2,3-dimercapto-1-propanol [59-52-9], also known as dimercaprol or British Anti-Lewisite (BAL), and penicillamine [52-67-5], which have been known for a long time [21]. Mercaptoacrylic acids are also said to be suitable antidotes for lead and nickel poisoning [22]. Zinc and cadmium salts of cylcohexanethiol can be used as catalysts, crop-protection agents, and polymer stabilizers [23]. Certain gold thiolates are used in the treatment of arthritis. Organic antimony thiolates can be used as passivators for catalysts employed in the cracking of hydrocarbons [24]. Thiolate-coordinated transition metals play an important role in many physiological processes [25].

Oxidation. Thiols are readily converted into the corresponding disulfides by mild oxidizing agents.

$$2\,R-SH \underset{\text{Red.}}{\overset{\text{Ox.}}{\rightleftarrows}} R-S-S-R$$

This facile oxidizability is the basis of quantitative thiol determination, and it is used industrially for sweetening crude-petroleum distillates. In this process the thiol-containing acidic fractions are oxidized with atmospheric oxygen under basic conditions in the presence of such catalysts as cobalt phthalocyanine monosulfonate [26] or EDTA complexes of copper(II) [27] or cobalt(II) [28]. Many other reagents can also be used for oxidizing thiols to disulfides (see Section 1.6.3). The thiol–disulfide redox system also plays an important role in a range of biological processes (glutathione, cysteine–cystine) [29].

Thiols and disulfides are oxidized to sulfonic acids by stronger oxidizing agents via a series of isolable intermediates. These oxidizing agents include nitric acid, potassium permanganate, chlorine gas [30], a mixture of hydrogen peroxide and aqueous hydrochloric acid [31], and atmospheric oxygen in the presence of catalytic amounts of dimethylsulfoxide and hydrogen halide [32].

Addition Reactions. Being strong nucleophiles, aliphatic thiols add to multiple bonds to give thioethers. This reaction, which is also used for the synthesis of unsymmetrical sulfides (see Section 1.6.2), produces a normal Markovnikov adduct (→ Alcohols, Aliphatic) under ionic reaction conditions and an anti-Markovnikov adduct in the presence of radical-forming agents.

$$R^1-CH=CH_2 + R^2-SH \begin{array}{c} \xrightarrow{H^+} R^1-CH(S-R^2)-CH_3 \\ \xrightarrow{h\cdot v \text{ or peroxide}} R^1-CH_2-CH_2-S-R^2 \end{array}$$

Thiols add to carbonyl compounds in the same way as alcohols. With aldehydes or ketones the products are thioacetals or thioketals [33], and with carboxylic acids and their derivatives they are S-esters of thiocarboxylic acids [34]. Other transformations of thiols are described in [35].

1.3. Production of Aliphatic Thiols

1.3.1. From Alcohols

The reaction of hydrogen sulfide with alcohols is used mainly for industrial syntheses of low molecular mass thiols.

$$ROH + H_2S \longrightarrow R\text{-}SH + H_2O$$

Such reactions are carried out in the gas phase > 300 °C, and generally require a catalyst. The thorium oxide [36] once used for this purpose has now largely been replaced by basic catalysts on a carrier of basic aluminum oxide. These catalysts favor thiol formation, whereas disulfides are the main products in acidic medium [37]. Promoters for thiol formation include a range of metal oxides, potassium and sodium tungstates, and heteropolyacids and their salts [37]–[40]. Thioethers and ethers may form as byproducts, along with olefins (except in the case of methanol).

The method described is used mainly for the industrial production of *methanethiol* [38], [41], although ethanethiol is also produced by this route [42]. The method is not suitable for primary thiols with more than three carbon atoms because of the small boiling-point differences between the products and the starting alcohols (see Fig. 1).

Tertiary thiols can be produced in the liquid phase by reacting alcohols with hydrogen sulfide in the presence of sulfuric acid. This method leads to *triphenylmethanethiol* [3695-77-0] in 75% yield [43], and is used industrially to produce *2-methyl-2-propanethiol* (*tert*-butanethiol) [44].

The reaction of alcohols with phosphorus pentasulfide is frequently employed for this purpose as well, and has been patented for alkanethiols [45]. C_4 to C_{16} alcohols are converted in this way into dialkyl dithiophosphates, which give the corresponding thiols on acid hydrolysis.

A more recent method uses Lawesson's reagent [*19172-47-5*] for the conversion of alcohols into thiols [46].

Lawesson's Reagent

1.3.2. From Alkenes

A second industrially important process for aliphatic thiol production is the addition of hydrogen sulfide to alkenes. In principle, both Markovnikov (**I**) and anti-Markovnikov (**II**) adducts can result from this reaction, but formation of one isomer or the other can be favored by the use of suitable catalysts.

The Markovnikov product (**I**) is formed preferentially under ionic conditions in the presence of oxides, sulfides, acids, or elemental sulfur. Branched alkenes react more rapidly than straight-chain or terminal alkenes. Secondary thiols are prepared by the addition of hydrogen sulfide to linear alkenes in the presence of acid catalysts [47]. The analogous reaction of branched alkenes to give tertiary thiols is catalyzed by aluminum chloride, Al/Si oxides [48], zeolites [49], or cation exchange resins [50].

This method is exploited for the industrial production of a variety of thiols. Reaction is carried out in either the gas or liquid phase depending on the catalyst used. The yield of thiol increases with excess hydrogen sulfide (3–10 mol per mole of alkene), high pressure (≤ 7 MPa), and relatively low reaction temperature. Dialkyl sulfides are formed as byproducts. *Ethanethiol* is prepared from ethylene and hydrogen sulfide in the gas phase over aluminum oxide catalysts containing potassium tungstate [51] or nickel and tungsten oxides [52]. *Cyclohexanethiol* is obtained with cyclohexene and hydrogen sulfide in the gas phase over nickel sulfide on silica gel [53]. The reaction of cyclohexene with hydrogen sulfide over cobalt molybdate has also been described [54]. The industrial production of "*tert*-dodecyl mercaptan," a mixture of isomeric thiols, starts from oligomers of isobutylene or propylene. In the presence of hydrogen sulfide and a boron trifluoride catalyst, "*tert*-dodecyl mercaptan" can be prepared from triisobutylene at $-40\,°C$ [41] and from propylene tetramer at $60\,°C$ [55]. Aluminum chloride [41], ethyl aluminum dichloride [56], a mixture of boron trifluoride with phosphoric acid and a low molecular weight alcohol [57], and sodium oxide on zeolite [58] have also been described as catalysts. The industrial production of *2,4,4-trimethyl-2-pentanethiol* is similar to that of *tert*-dodecyl mercaptan. Terpenes are also used as industrial starting materials in this reaction. Thus, 1-*p*-menthene-8-thiol [71159-90-5], a component of the grapefruit aroma, is obtained selectively and in high yield by the addition of hydrogen sulfide to limonene in the presence of inorganic Lewis acid catalysts [59].

Electron-withdrawing groups, such as the carbonyl, cyano, nitro, carboxyl, or ester functions, activate alkenes toward nucleophilic addition of hydrogen sulfide. Basic catalysts are particularly suitable for these reactions, including amines [60], anion exchange resins [61], and magnesium oxide [61].

$$CH_2=CH-X + H_2S \xrightarrow{\text{base}} HS-CH_2-CH_2-X$$

$$X = -CO-CH_3, -CN, -NO_2, -COOH, -COOR$$

Formation of the anti-Markovnikov product (**II**) results from a radical addition of hydrogen sulfide to alkenes. In industry this reaction is usually initiated with UV light ($\lambda = 253.7$ nm) [62], [63]. Longer wavelength radiation ($\lambda = 300-400$ nm) can also be used in the presence of photosensitizers [62]–[64]. Among the chemical radical initiators, azobisisobutyronitrile is particularly suitable [65]. This can also be used in the presence of accelerators, such as organic phosphines or phosphites [66]. Peroxides, on the other hand, can oxidize the hydrogen sulfide to elemental sulfur, which then acts as a catalyst for the ionic reaction pathway (see above) [67].

The method described is used mainly for the industrial production of primary thiols. The formation of dialkyl sulfides as byproducts is minimized by ensuring that excess hydrogen sulfide is present. Thus, *ethanethiol* is obtained from ethylene and hydrogen sulfide in the liquid phase in the presence of azobisisobutyronitrile and tributylphosphine [68]. *1-Dodecanethiol* is prepared by photochemically catalzyed addition of hydrogen sulfide to 1-dodecene. The olefin is reacted with a five-fold excess of hydrogen sulfide in the presence of α,α-diethoxyacetophenone as initiator at 45 °C and 3 MPa [63]. The industrial preparation of *1-octanethiol* is similar to that of 1-dodecanethiol.

1.3.3. Other Production Methods

Other methods for the industrial-scale preparation of thiols include the reductive thiolation of carbonyl compounds, epoxide ring-opening with hydrogen sulfide, and nucleophilic substitution of alkyl halides or tosylates.

Aldehydes and ketones are reduced to thiols by hydrogen sulfide and hydrogen in the presence of metal sulfides [69]. This process is used for the industrial production of cyclohexanethiol [70].

$$\text{\textbackslash}C=O + H_2 + H_2S \xrightarrow{\text{cat.}} \text{\textbackslash}CH-SH + H_2O$$

Base-catalyzed cleavage of epoxides with hydrogen sulfide gives mercaptoalcohols. Symmetrical bis(1-hydroxyalkyl) sulfides are formed as byproducts. Inorganic bases, amines and anion exchange resins, zeolites [71], or guanidine [72] can be used as catalysts. In industry, preparation of *mercaptoethanol* [60-24-2] by addition of H_2S to ethylene oxide is catalyzed by either a cation exchange resin [73] or the byproduct thiodiglycol [74].

$$H_2C-CH-R + H_2S \xrightarrow{\text{base}} H_2C-CH-R$$
$$\quad\backslash\!/\qquad\qquad\qquad\quad |\quad\ |$$
$$\quad O\qquad\qquad\qquad\qquad HS\ \ OH$$

The reaction of alkyl halides or tosylates with metal hydrogensulfides is frequently used to prepare thiols on a laboratory scale [75].

$$R-X + M-SH \xrightarrow[-MX]{} R-SH$$
$$X = Cl, Br, I, -O\text{-tosyl}$$

Alcohols are the usual solvents. The competing formation of sulfide can be suppressed by the presence of hydrogen sulfide or sulfuric acid. Sulfide formation is also strongly inhibited in the reaction of alkyl halides with a mixture of hydrogen sulfide and ammonia or alkylamines [76]. C_2 to C_{12} alkanethiols form in essentially quantitative yield from the corresponding bromides and sodium hydrogensulfide in dimethylsulfoxide [77]. The nucleophilic substitution reaction is used industrially to produce *1-dodecanethiol* from 1-chloro- or 1-bromododecane [78].

Substitution of –SH for chlorine occurs particularly readily if the process is activated by the presence of electron-withdrawing groups (e.g., C=O, C≡N) or heteroatoms (O, S, Si) in the α-, β-, or γ-positions [79]. This method is used to prepare mercaptocarboxylic acids and thiols containing additional functional groups [79], [80].

Table 2. Transport restrictions applicable to aliphatic thiols

Thiol	ADR/RID		IDMG Code	
	Class	Number		
CH_3-SH	2	3bt	2	UN 1064
C_2H_5-SH	3	18b	3.1/II	UN 2363
$n\text{-}C_4H_9-SH$	3	3b	3.2/II	UN 2347
$t\text{-}C_4H_9-SH$	3	3b	3.2/II	UN 1228
$cyclo\text{-}C_6H_{11}-SH$	3	31c	3.3/III	UN 3054
$n\text{-}C_8H_{17}-SH$	3	32c		
$n\text{-}C_{12}H_{25}-SH$	3	32c		
	6.1	21c		
$t\text{-}C_{12}H_{25}-SH$	6.1	21c		

$$Cl-(CH_2)_n-COOH \xrightarrow[2)\ HCl]{1)\ NaSH} HS-(CH_2)_n-COOH$$

$$Cl-CH_2CH_2-O-R \xrightarrow{NaSH} HS-CH_2CH_2-O-R$$

Besides the processes commonly used in industry for thiol production, a number of generally applicable laboratory methods are also available, such as the reaction of an alkyl halide with thiourea, thiosulfate, or dithio- or trithiocarbonate. Details of these methods can be found in the general literature.

1.4. Construction Materials, Storage, and Transport

Manufacturing plant and transport facilities for thiols are preferentially made from (coated) normal steel, stainless steel, aluminum, or copper-free alloys. Thiols in containers made of carbon steel should be handled dry and under a blanket of protective gas to prevent the formation of pyrophoric iron–sulfur complexes. Copper and copper-containing alloys are not suitable for use because they are rapidly attacked by thiols.

Stringent demands apply to the impermeability of thiol piping systems. Seals are normally based on asbestos or fluoroplastics.

Regulations in force with respect to the transport of industrially important thiols are summarized in Table 2.

1.5. Uses of Aliphatic Thiols

Many aliphatic thiols are important starting materials for the synthesis of crop-protection agents (especially thiophosphoric acid esters) [81], pharmaceuticals, and polysulfides (see Section 1.6.4). They are also widely used as polymerization regulators in rubber and plastics manufacture. *Methanethiol* is important as a starting material for the industrial production of the amino acid DL-methionine, used as an animal-feed additive. Methanethiol is also used as a gas odorizer, as are other low molecular mass thiols (in particular, ethanethiol). For applications of thiolates and chelate-forming thiols see Section 1.2.2.

Examples, drawn mainly from the recent extensive patent literature, are provided below.

Pharmaceuticals. Mercaptoamino acids and their derivatives display a broad spectrum of activity. *N*-Isobutyryl- and *N*-pivaloylcysteine are said to exhibit improved bioavailability compared with *N*-acetylcysteine [*616-91-1*], which is used as a mucolytic [82].

Vanadyl–cysteine complexes can be used for oral treatment of insulin-dependent diabetes [83], and substituted cysteinamides for the localized treatment of inflammatory skin diseases [84]. Certain mercaptoamino acids, such as the 2S-enantiomer of 1-(3-mercapto-2-methylpropionyl)-L-proline (Captopril [*62571-86-2*]), act as antihypertensives [85].

Other mercaptoamino acid derivatives have been described as analgetics [86] and antiarthritic agents [87].

Diaminodithiols are chelating agents, and thus useful in radioactive diagnosis [88]. Thiol-substituted dioxolanes [89] and cyclohexenes [90] are mucolytics. Mercapto-substituted propanamides exhibit psychotropic [91] and analgesic activity [91], [92].

Agricultural Chemicals. Aminodithiols have been described as effective low-toxicity insecticides [93]. Bismuth dithiolates can be used to combat coccidiosis in poultry and as insecticides [94].

Thiol-substituted 2-propanols are fungicides with low toxicity [95]. Bis(dithiobenzyl)nickel complexes absorb IR radiation and can be used as plant-growth regulators [96].

Other Uses. *N*-Substituted1-amino-2-propylmercaptans are used as antioxidants in lubricating oils [97]. 4-Mercaptobutyramides can act as reducing agents in the permanent waving of hair. They are said to be less toxic and accompanied by less noxious odors than mercaptoacetic acid [98]. Mercaptoalkyl esters of acrylic and methacrylic acids are used as binders for metallic fillings for teeth [99]. Many thiol-substituted terpenes [100], esters [101], ketones [102], and mercaptoalkyloxathiolanes and

-oxathianes [103] have been patented as fragrances and aroma substances (e.g., for foods, cosmetics, or pharmaceuticals).

1.6. Alkyl Sulfides, Disulfides, and Polysulfides

1.6.1. Physical and Chemical Properties

Sulfides (R^1–S–R^2) are the sulfur analogs of ethers, and are thus derivatives of hydrogen sulfide. They were referred to previously as thioethers or sulfanes. Alkyl *disulfides* (R^1–S–S–R^2) and *polysulfides* (R^1–S_x–R^2) are derived from di- and polysulfanes. Symmetrical sulfides contain R groups that are identical ($R^1 = R^2$), whereas unsymmetrical ones have different R groups ($R^1 \neq R^2$). Sulfides are named in the same way as ethers (e.g., diethyl sulfide, methyl phenyl sulfide). Di- and polysulfides are named either as sulfides or as sulfane derivatives (e.g., dimethyl disulfide, diethyltrisulfane). Most low molecular mass alkyl sulfides have intense, unpleasant odors. Organic sulfides and disulfides are widely distributed in natural materials. For example, they occur (together with thiols) in onions and garlic [104], and have been found to be the source of various well-known aromas, including those of asparagus and coffee. In polecats, mink, and skunks such compounds act as pheromones [105]. The cyclic polysulfide lenthionine confers a characteristic odor on cooked mutton and the shiitake mushroom [106]. Many disulfides are components of biologically active substances (e.g., lipoic acid, cystine, holomycin) and physiologically important peptides (e.g., insulin, oxytocin). Here the activity is associated directly with the presence of the disulfide group. Aliphatic sulfides are also present in crude petroleum, accounting for up to 45% of the total sulfur, but disulfides are virtually absent [107].

Aliphatic sulfides, disulfides, and polysulfides are colorless to yellow or orange liquids. Most are insoluble in water, but readily soluble in ether, acetone, chloroform, and ethanol. Physical data for a few representative compounds are provided in Table 3.

Alkyl sulfides, like their oxygen analogs, are very weak bases. Because of nucleophilicity attributable to the free electron pair on sulfur, they form stable addition compounds with halogens and salts of heavy metals. Aliphatic sulfides also form crystalline trialkylsulfonium salts with alkyl halides or dialkyl sulfates.

$$\begin{array}{c} R^1 \\ \diagdown \\ R^2 \end{array}\!\!S + R^3\!-\!X \longrightarrow \begin{array}{c} R^1 \\ \diagdown \!+ \\ R^2 \end{array}\!\!S\!-\!R^3 \; X^-$$

X = halogen, O–SO_2–O–R^3

Depending on the reaction conditions, oxidation of sulfides leads to sulfones or sulfoxides [107]. Sulfonic acids are formed under forcing conditions, and these are then further oxidized to sulfuric acid [108].

Table 3. Physical data for selected alkyl sulfides, disulfides, and polysulfides

Sulfide	CAS registry no.	M_r	bp, °C/kPa	d_4^{20}	n_D^{20}	Flash point, °C
$(CH_3)_2S$	[75-18-3]	62.13	37.3	0.846	1.4355	−36
$(CH_3CH_2)_2S$	[352-93-2]	90.19	92.1	0.8362	1.4430	−9
$(CH_2=CH–CH_2)_2S$	[592-88-1]	114.21	138	0.887	1.4889	46
$[CH_3(CH_2)_2]_2S$	[111-47-7]	118.24	142	0.838	1.4487	28
$[CH_3(CH_2)_3]_2S$	[544-40-1]	146.30	188	0.838	1.4530	76
$[CH_3CH_2CH(CH_3)]_2S$	[626-26-6]	146.30	165	0.839	1.4500	39
$[(CH_3)_3C]_2S$	[107-47-1]	146.30	149	0.815	1.4506	48
$[CH_3(CH_2)_5]_2S$	[6294-31-1]	202.40	230	0.849	1.4587	>110
$(CH_3)_2S_2$	[624-92-0]	94.20	109	1.046	1.5253	24
$(CH_3CH_2)_2S_2$	[110-81-6]	122.25	152	0.993	1.5073	40
$(CH_2=CH–CH_2)_2S_2$	[2179-57-9]	146.28	187	1.008	1.5410	62
$[CH_3(CH_2)_3]_2S_2$	[629-45-8]	178.36	226	0.9382	1.4926	93
$(CH_3)_2S_3$	[3658-80-8]	126.27	42.5/0.4	1.2013	1.6012	
$(CH_3)_2S_4$	[5756-24-1]	158.33	59/0.1	1.3065	1.6612	
$(CH_3)_2S_5$	[7330-31-6]	190.39	70/0.01			

$$R-S-R \xrightarrow{[O]} R-\overset{O}{\underset{}{S}}-R \xrightarrow{[O]} R-\overset{O}{\underset{O}{S}}-R \xrightarrow{[O]}$$

$$R-SO_3H \xrightarrow{[O]} H_2SO_4$$

Sulfides can be reduced to hydrocarbons or, depending on the reducing agent and the substrate, thiols or hydrogen sulfide.

$$R-S-R \xrightarrow{Red.} R-H + R-SH \xrightarrow{Red.} 2R-H + H_2S$$

The sulfide contraction reaction, a desulfurization with simultaneous C–C bond formation, is important in organic synthesis [109].

Reaction conditions determine whether disulfides are oxidized to thiosulfinic or thiosulfonic acid esters or to sulfonic acids.

$$R^1-S-S-R^2 \begin{cases} \xrightarrow{R-C(O)OOH} R^1-\overset{O}{\underset{}{S}}-S-R^2 \text{ thiosulfinic acid ester} \\ \xrightarrow{Ph-C(O)OOH} R^1-\overset{O}{\underset{O}{S}}-S-R^2 \text{ thiosulfonic acid ester} \\ \xrightarrow{O_2/H_2O} R^1-SO_3H + R^2-SO_3H \end{cases}$$

The reduction of disulfides to thiols is a reversible reaction of great physiological importance. Hydrogen, zinc in the presence of acid, complex hydrides, or sodium are generally selected as the reducing agents [110]. Disulfides can be reduced selectively with sodium hydrogentelluride in good yield under very mild conditions in the presence of other functional groups [111]. Organic polysulfides can also be reduced to thiols.

1.6.2. Production of Alkyl Sulfides

Many of the procedures for the synthesis of dialkyl sulfides correspond to those for aliphatic thiols, including the alkylation of hydrogen sulfide or alkali metal sulfides with alcohols, alkenes, or alkyl halides to give symmetrical sulfides (Section 1.3). *Dimethyl sulfide*, the industrially most important sulfide, is produced by treating hydrogen sulfide with excess methanol over an aluminum oxide catalyst. Unsymmetrical sulfides are formed in a similar way from thiols and their salts [112], [113], but they can also be obtained by alkylation of thiols with acidic C–H compounds [114]. The alkylation of thiols with alkyl halides in the presence of K_2CO_3 in DMF occurs particularly rapidly, and in almost quantitative yield [115]. Unsymmetrical sulfides can also be made by exchange reactions involving symmetrical sulfides and thiols, catalyzed by phosphotungstic or phosphomolybdic acids [116].

$$R^1-S-R^1 + R^2-SH \xrightarrow{cat.} R^1-S-R^2 + R^1-SH$$

Cleavage of an epoxide ring with a thiol proceeds analogously to the cleavage with hydrogen sulfide (Section 1.3.3). Bases (such as sodium hydroxide or triethylamine [117]), metal salts (e.g., lithium perchlorate [118]), and acids [119] have all been used as catalysts. The resulting β-hydroxythioethers are important intermediates for a large number of other products.

Alkylthio-substituted carboxylic acids and their thioesters can be prepared from lactones by treatment with thiols in the presence of an acid catalyst at 100–350 °C [120].

$$\text{(lactone)} + R-SH \longrightarrow R-S-(CH_2)_3-COOH + R-S-(CH_2)_3-C(=O)S-R$$

The reductive thiolation of carbonyl compounds (Section 1.3.3) can also be carried out with thiols instead of hydrogen sulfide. The borane–pyridine complex in trifluoracetic acid [121] or treatment with boron trifluoride monohydrate followed by triethylsilane [122] can be used to accomplish the reduction.

$$\begin{matrix} R^1 \\ \diagdown \\ R^2 \end{matrix} C=O + R^3-SH \xrightarrow{cat.} \begin{matrix} R^1 \\ \diagdown \\ R^2 \end{matrix} CH-S-R^3$$

α-Substituted dialkyl sulfides can be prepared from sulfoxides by means of the Pummerer rearrangement. The deoxygenation of sulfoxides to sulfides is a particularly important approach with respect to asymmetric syntheses. Several methods are known for achieving this reduction [123]. The systems sodium iodide–titanium(IV) chloride [124] and sodium iodide–sulfonic acid [125] have recently been described as especially efficient and mild reagents. The selective reduction of various functionalized sulfoxides can be effected using Lawesson's reagent [126] (see Section 1.3.1).

1.6.3. Production of Alkyl Disulfides

Dialkyl disulfides are often formed as byproducts in the synthesis of sulfides. However, the most important synthetic method is the mild oxidation of thiols. In industry this oxidation is carried out in aqueous alkaline solution using air or oxygen. The reaction is catalyzed by radical initiators (copper, cobalt, nickel, or iron salts, or UV light) [127]. Even high molecular mass disulfides can be prepared in this way [128]. Since in this case water-insolubility of the products can be a disadvantage, the reaction is sometimes carried out in a two-phase mixture in the presence of phase-transfer catalysts [129]. For the oxidation of tertiary thiols a catalyst system consisting of alkali-metal and/or alkaline-earth hydroxides and cobalt molybdate is particularly suitable [130]. The oxidation of thiols with atmospheric oxygen is also exploited industrially for sweetening crude petroleum distillates (see Section 1.2.2).

Other potential oxidizing agents include hydrogen peroxide, halogens (in particular, iodine) [131], and potassium hexacyanoferrate(III) [132]. The fire hazard is lowered and the yield of dialkyl disulfide increased using carbon tetrachloride in aqueous alkali as the oxidizing agent together with a phase-transfer catalyst [133].

Interest has recently been shown in the oxidation of thiols with nitrogen oxides. Whereas symmetrical sulfides are obtained in good yields with nitric oxide and nitrogen dioxide [134], dinitrogen tetroxide leads first to thionitrites [135], which may then react with an additional equivalent of thiol to give symmetrical or unsymmetrical disulfides [136].

$$R^1-SH \xrightarrow{N_2O_4} R^1-SNO \xrightarrow{R^2-SH} R^1-S-S-R^2$$

The analogous coupling of thiols with iron(III) nitrate on a bentonite carrier to give symmetrical disulfides takes place under very mild conditions [137].

An important method for the production of unsymmetrical disulfides is the thioalkylation of thiols with sulfenyl halides [138].

$$R^1-SH + R^2-S-Cl \longrightarrow R^1-S-S-R^2$$

1.6.4. Production of Alkyl Polysulfides

Alkyl polysulfides are produced industrially by treating thiols with sulfur in the presence of bases (alkali-metal hydroxides or tertiary amines) [139]. Mixtures of various polysulfides are frequently formed in this reaction, so the isolation, purification, and characterization of defined compounds can be difficult. Higher molecular mass dialkyl trisulfides can be obtained in very good yields from the corresponding thiols and sulfur with magnesium oxide as the base at 30–100 °C [140]. The reaction of thiols with electrolytically formed S_x^{2+} gives trisulfides in 53–74% yields [141]. Alkyl polysulfides containing an average of five sulfur atoms can be prepared at room temperature in the presence of thioxanthates as catalysts [142].

The reaction of dihaloalkanes with alkali polysulfides gives polymeric polysulfides with the general formula $-[R-S_x-]_n$. These polysulfide rubbers (thioplastics) are widely used in industry because of their resistance toward chemicals and light and their impermeability toward gases.

1.6.5. Uses of Alkyl Sulfides, Disulfides, and Polysulfides

The most important industrial application of aliphatic sulfides and disulfides is as intermediates in the production of pharmaceuticals, crop-protection agents, and a wide range of other substances [143]. Alkyl sulfide groups themselves are also present in the active components of many pharmaceuticals and crop-protection agents, especially insecticides [144]. *Dimethyl sulfide*, the industrially most important sulfide, is widely used in the form of the borane complex $(CH_3)_2S \cdot BH_3$ [*13292-87-0*] for the reduction of a variety of functional groups [145]. Dimethyl sulfide is also the starting material for production of the solvent dimethyl sulfoxide (DMSO).

Organic disulfides have a variety of applications as well. L-Cystine [*56-89-3*] and lipoic acid [*1077-28-7*] are therapeutic agents for the liver. Holomycin and the pharmaceutically active component in garlic, di(1-propenyl) disulfide [146], exhibit antibacterial activity. Dialkyl disulfides with 2–8 carbon atoms per alkyl residue can be used as insecticides and acaricides, or to combat algae [133]. Higher molecular mass disulfides are used as additives in the plastics industry for chain transfer during polymerization, and as antioxidants. Di- and polysulfides are used as high-pressure additives for lubricating oils.

Table 4. Physical data for several thiophenols

R	CAS registry no.	M_r	bp, °C/kPa	mp, °C	Flash point, °C
H	[108-98-5]	110.18	169	−15	50
4-CH$_3$	[106-45-6]	124.21	195	43	68
4-Cl	[106-54-7]	144.62	206	50	> 110
2,5-Cl$_2$	[5858-18-4]	179.07	132/3.9	28	> 110
2,6-Cl$_2$	[24966-39-0]	179.07		49	> 110
2-OH	[637-89-8]	126.18	150/3.3	34	> 110
2-O-Me	[7217-59-6]	140.20	99/1.1		95
2-NH$_2$	[137-07-5]	125.19	71/0.03	20	79
2-COOH	[147-93-3]	154.19		167	

2. Aromatic Thiols, Sulfides, Disulfides, and Polysulfides

2.1. Physical and Chemical Properties

Aromatic thiols, Ar–SH, are sulfur analogs of phenols. This relationship is also manifested in many names, where the prefix thio is added to the trivial name of the corresponding phenol (e.g., thiosalicylic acid, thiocresol). According to IUPAC rules, aromatic thiols are supposed to be referred to as mercaptoarenes (e.g., mercaptoaniline) or arenethiols (e.g., 2-naphthalenethiol).

Aromatic thiols differ only slightly from alkanethiols with respect to acidity, salt formation, odor, and solubility. Many thiophenols are volatile in steam. Table 4 provides an overview of physical properties for several important arenethiols.

Unlike aliphatic thiols, arenethiols occur only to a very limited extent in natural materials. One reason for this may be the more facile oxidation of aromatic thiols, which are converted into disulfides simply on standing in air. The powerful nucleophilicity of arenethiols is exemplified primarily by substitution and addition reactions, most of which lead to thioethers. Such reactions correspond largely to those of aliphatic thiols (see Section 1.2.2). Unlike phenol derivatives, derivatives of aromatic thiols undergo a number of rearrangements. A review of the reactions of aromatic thiols can be found in [11].

2.2. Production of Aromatic Thiols

2.2.1. By Reduction

The reduction of sulfonyl chlorides, which are readily accessible from the corresponding aromatic systems by sulfochlorination, is one of the most important methods for producing arenethiols.

$$Ar-H + HO-SO_2Cl \xrightarrow{-H_2O} Ar-SO_2Cl \xrightarrow{Red.} Ar-SH$$

Many different reducing agents can be employed. The use of zinc or tin in dilute acid is still important in industry [147]. For example, *thiophenol* [148], *halothiophenols* [149], or *aromatic dithiols* [150] continue to be produced by this method. The resulting levels of ecologically harmful metal salts can be considerably reduced by the use of such catalysts as lead [151] or various harmless metal salts [152].

Sulfonyl chlorides can be hydrogenated directly in the presence of metal sulfide catalysts to the corresponding thiophenols at 150–275 °C and a pressure of 3–17 MPa [153]. This process is used to produce *p-thiocresol*. The preparation of *2,5-dichlorothiophenol* by pressurized hydrogenation of the corresponding sulfonyl chloride over a platinum catalyst is similar [154].

Another commonly used reducing agent is red phosphorus in the presence of hydrogen iodide in aqueous [155] or glacial acetic acid solution [156]. Two-step reduction of a sulfonyl chloride is also practiced on an industrial scale, as in the preparation of thiophenol. Depending on the reaction conditions, sodium arenesulfinates or disulfides are formed initially by sulfite reduction, and these are subsequently converted into thiophenols by pressurized hydrogenation in the presence of metal sulfides [157].

$$Ar-SO_2Cl \xrightarrow[NaOH]{Na_2SO_3} Ar-SO_2^- Na^+ \longrightarrow Ar-S-S-R$$
$$\searrow_{H_2, cat.} \quad \swarrow_{H_2, cat.}$$
$$Ar-SH$$

The reduction of aromatic sulfonic acids with triphenylphosphine in the presence of a reaction accelerator has also been described [158].

The tendency of aromatic thiols to undergo oxidation means that the actual products are often disulfides, so reduction of the latter and the corresponding polysulfides is frequently important. This indirect route to arenethiols often leads to better yields than the direct synthesis despite the additional reaction step. Many reducing agents are suitable for the reductive cleavage of disulfides. Pressurized hydrogenation in the presence of a catalyst is used in industry [159]. For halosubstituted compounds the preferred catalysts are Raney cobalt or palladium on charcoal [160]. Other reducing agents include metallic aluminum in liquid ammonia with inorganic halogen as the catalyst [161], zinc and acid in the presence of another metal [162], lithium aluminum hydride [163], hydrogen sulfide [164], and alkali-metal sulfides [165]. Functionalized aromatic disulfides are reduced selectively and under mild conditions using lithium (tri-*tert*-butoxy)aluminum hydride, where halogen, nitro, and carboxyl groups are spared from attack [166]. High yields of arenethiols are also obtained in the presence of a wide range of functional groups with a phosphine reagent bonded to a polymeric carrier [167]. For other reducing agents, particularly those used in the laboratory, see [2].

2.2.2. By Thiolation

A nucleophilic exchange between aromatically bonded halogen and the thiol group occurs under mild conditions only in the presence of other strongly electron withdrawing substituents, such as nitro or sulfonyl groups [168]. Either sodium hydrogensulfide or sodium disulfide is used as the thiolating agent. Poor yields and selectivities are often observed with sodium sulfide, particularly with weakly activated haloaromatics. Chloronitrobenzenes can be converted into the corresponding thiophenols either with retention [165], [169] or simultaneous reduction of the nitro group [170], depending on the reaction conditions.

$$Cl-C_6H_4-NO_2 \xrightarrow{Na_2S} HS-C_6H_4-NH_2$$
$$Cl-C_6H_4-NO_2 \xrightarrow{Na_2S_2} HS-C_6H_4-NO_2$$

Polychlorinated benzenes also react with sodium hydrogensulfide to give the corresponding thiophenols [171].

The thiolation of haloaromatics is often used in the industrial preparation of arenethiols. *Pentachlorothiophenol* is obtained from hexachlorobenzene by treatment with sodium sulfide and sulfur in methanol, or with sodium hydrogensulfide [172], [173]. Various halothiophenols can be produced by a similar method in such polar solvents as DMF, sulfolane, dimethylsulfoxide (DMSO), and *N*-methylpyrrolidone [174]. Mercaptoanilines are formed from halonitrobenzenes with simultaneous reduction of the nitro group [175]. Aromatic dithiols can be obtained in a particularly pure state and high yield from polyhalogenated aromatics by reaction with hydrogen sulfide and sulfur in the presence of iron [176]. Sodium alkanethiolates can be used for the thiolation of nonactivated aryl halides. The aryl alkyl sulfide formed initially is converted into the desired aromatic thiol by acid hydrolysis [177]. Nucleophilic substitution of an aryl iodide with thiourea in the presence of transition metal complexes leads to *S*-arylisothiuronium salts, which can by hydrolyzed to arenethiols [178].

Another thiolating agent, one that is also suitable for the production of alkoxythiophenols, is disulfur dichloride. Disulfides are formed initially, and these are then reduced as described above (see Section 2.2.1). Various metal salts can serve as catalysts for this thiolation [179]. Similarly, monohalogenated benzenes can be converted into halothiols via polysulfides [180].

Other preparative methods applicable to arenethiols include the catalytic reaction of phenols with hydrogen sulfide in the gas phase [181] and reductive cleavage of aryl alkyl sulfides [182].

2.3. Uses of Aromatic Thiols

The most important industrial applications of arenethiols are as intermediates in the production of pharmaceuticals, agrochemicals [183], dyes, and pigments, and as chemicals for the electronics industry. Thiophenols are used in the rubber industry as polymerization regulators, plasticizers, and stabilizers [184], as well as for rubber reclamation. Aromatic dithiols and mercaptoanilines are used in the plastics industry as monomers and modifiers [185].

Thiophenolamides act as collagenase inhibitors, and can be used for treating arthritic conditions [186]. Thiophenolate-containing preparations can dissolve seals and coatings based on polysulfides [187].

Methyl and ethyl 2-mercaptobenzoates have been patented as scents [188]. Substituted mercaptoanilines are flotation agents for lead and zinc ores [189]. Polythiols can be used as optical materials in plastics [190], and for the production of lightweight reversible electrodes with high theoretical energy density [191]. The use of mercaptophenyl-substituted pyrrolinones as herbicides and growth regulators for crop-protection purposes has been described [192].

2.4. Production of Aryl Sulfides, Disulfides, and Polysulfides

The production processes for aromatic sulfides, disulfides, and polysulfides correspond largely to those for the related aliphatic compounds. Symmetrical aryl sulfides can be obtained by arylation of hydrogen sulfide or its salts with alkyl halides [193]. Similarly, unsymmetrical or mixed (aliphatic–aromatic) sulfides are formed from thiols or thiolates on treatment with alkyl or aryl halides [194]. The addition of thiols to multiple bonds [195] and the reaction of sulfenyl chlorides with arenes or phenols [196] also lead to unsymmetrical sulfides. Sulfides are formed as well in the cleavage of disulfides with phenols [197] or aromatic amines [198].

2.5. Uses of Aryl Sulfides, Disulfides, and Polysulfides

Aryl sulfides are used mainly as intermediates in the production of biologically active substances and dyes, and as monomers in the plastics industry. The applications mentioned below are derived in most cases from the recent extensive patent literature.

Tetrasul and Mercaptodimethur are examples of commercial insecticides. Other aryl sulfides have been introduced as fungicides [199], herbicides [200], microbial growth inhibitors [201], acaricides and nematocides [202], miticides [203], and gametocides [204]. Aryl thioethers are components of color developers in recording materials [205] and of the photosensitive elements in photocopiers and printers [206]. Certain monomeric aryl sulfides can be polymerized to materials with interesting optical properties, such as transparency or a high refractive index [207]. In the plastics industry arylthio compounds can be used as hardeners [208], antioxidants [209], or heat and light stabilizers [210]. Other aryl sulfides function as lubricating-agent additives [211], hair dyes [212], and flotation agents [213]. Aromatic sulfides are used in the pharmaceutcial sector for treating immunological diseases [214], skin diseases [215], and asthma [216], and also as muscle relaxants [217].

Aromatic disulfides have been patented as stabilizers for polycarbonate plastics [218], corrosion inhibitors in transport systems for crude petroleum and natural gas [219] and antioxidants for use in lubricating agents [220].

Aryl polysulfides in rubber act simultaneously as antioxidants and cross-linking agents [221].

Poly(p-arylenesulfides) are industrially important as thermoplastic polycondensates with good mechanical and thermal characteristics [222].

3. Heterocyclic Thiols, Sulfides, and Disulfides

3.1. Introduction

Heterocyclic thiols and their derivatives are widely distributed in nature. For example, the eggs of such marine invertebrates as sea urchins, starfish, and squid contain relatively high concentrations of mercaptohistidine and the corresponding methyl derivatives, known as "ovothiols" [223]. These compounds function as biological antioxidants in the fertilized eggs [224].

	R^1	R^2	Ovothiol
	H	H	A
	CH_3	H	B
	CH_3	CH_3	C

2-Methyl-3-furanthiol and its disulfide and S-methyl derivatives have all been identified as aroma components of cooked meat [225], and they have also been incorporated into synthetic flavorings [226]. 2-Methyl-3-furanthiol has been shown to be present in the aromas of roasted coffee [227] and canned tuna [228]. S-Substituted furans, thiophenes, thiazoles, and 3-thiazolines contribute to the odor of yeast extracts [229]. Heterocyclic sulfides occur in cytokinins of the bacteria *Escherichia coli* [230] and in the phallotoxins, toxic substances from the "death cup" mushroom *Amanita phalloides* [231].

3.2. Physical and Chemical Properties

Heterocyclic thiols are generally colorless or pale yellow solids with relatively faint characteristic odors. Table 5 provides an overview of the physical properties of several heterocyclic thiols.

Just like the corresponding aliphatic or aromatic compounds, heterocyclic thiols form salts with many metals, some of which have therapeutic applications (see Section 3.4).

Nitrogen-containing heterocyclic thiols are subject to the same tautomeric equilibrium between thiol and thione forms exhibited by the analogous oxygen species.

Spectroscopic studies [232] have shown that this equilibrium is displaced in the direction of the thione form. Reactions of heterocyclic thiols can take place from either the thiol or the thione form.

Table 5. Physical data for selected heterocyclic thiols

Formula, name	CAS registry no.	M_r	mp, °C
Imidazole-2-thiol	[872-35-5]	100.14	228–231
2-Thiazoline-2-thiol	[96-53-7]	119.21	105–107
[1H]-1,2,4-Triazole-3-thiol	[3179-31-5]	101.13	221–224
5-Methyl-1,3,4-thiadiazole-2-thiol	[29490-19-5]	132.21	188–189
1-Methyl-[1H]-tetrazole-5-thiol	[13183-79-4]	116.15	125–128
Pyridine-2-thiol	[2637-34-5]	111.17	128–130
2-Mercaptonicotinic acid	[38521-46-9]	155.18	270 (decomp.)
Pyrimidine-2-thiol	[1450-85-7]	112.15	230 (decomp.)
Benzothiazole-2-thiol	[149-30-4]	167.25	182
6-Purinethiol hydrate	[6112-76-1]	170.19	>300
2-Quinolinethiol	[2637-37-8]	161.23	174–176
2-Mercapto-4[3H]-quinazolinone	[13906-09-7]	178.21	>300

The mild oxidation of heterocyclic thiols leads to disulfides, just as with the corresponding aliphatic and aromatic systems. Catalytic oxidation of 2-mercapto-1,3-benzothiazol with atmospheric air is not practical, so the corresponding disulfide is prepared industrially by oxidation with sodium chlorate [233].

$$2 \text{ (benzothiazole-SH)} \xrightarrow[\text{H}_2\text{O, 30°C}]{\text{NaClO}_3, \text{HCl}} \text{benzothiazole-S-S-benzothiazole}$$

Stronger oxidizing agents like potassium permanganate or potassium peroxide transform heterocyclic thiols into sulfonic acids [234]. Treatment of 3-mercapto-1,2,4-triazole or 2-mercaptopyridinoimidazole with nitric acid leads to oxidative desulfurization [4].

$$\text{(X=N, Y, N-H, SH)} \xrightarrow{\text{HNO}_3} \text{(X=N, Y, N-H)}$$

$$-X=Y-\ :\ -N=CH-\ \text{or}\ \text{(pyridine)}$$

However, a more widespread desulfurization method for heterocyclic thiols is the reductive approach, in which Raney nickel [235] or sodium borohydride [236] is usually the reagent of choice.

Oxidation of heterocyclic thiols with hypochlorite in the presence of excess amine results in the corresponding sulfenamides [237], and disulfides are subject to similar treatment [238].

$$\text{Het-SH} + \text{HN}(R)(R) \xrightarrow[\text{NaOH}]{\text{NaOCl}} \text{Het-S-N}(R)(R)$$

Het: (thiazole with R' substituents) or (2-methylpyrimidine)

Heterocyclic thiols or sulfides are easily transformed into amino or hydrazo derivatives by substitution [239].

$$\text{Het-S-R} + \text{H}_2\text{N-R'} \xrightarrow{-\text{HS-R}} \text{Het-NH-R'}$$

Moreover, heterocyclic thiols are subject just like their aliphatic and aromatic counterparts to such transformations as etherification, esterification, or rearrangement.

3.3. Production

A wide variety of methods is available for the preparation of heterocyclic thiols. Thus, a thiol function can either be introduced by substitution into an existing heterocyclic system, or else the heterocycle itself may be constructed by cyclization of a sulfur-containing precursor.

3.3.1. Heterocyclic Thiols by Substitution

Generally speaking, heterocycles undergo substitution to thiols under substitution conditions similar to those applicable to activated benzene derivatives. One frequently applied method is the thiolation of a halo-substituted heterocycle, taking advantage of a wide range of thiolation reagents. Particularly suitable in this context are sodium or potassium hydrogensulfide [240] and thiourea [241]. Other possibilities include elemental sulfur [242], thiosulfate [243], alkali thiocyanates, and thiocarboxylic acid derivatives.

Heteroaromatic organomagnesium or organolithium compounds react with sulfur to give thiolates, hydrolysis of which leads to the corresponding thiols. This method permits the preparation of mercaptothiophenes in yields up to 70% [244].

Similarly, the mercapto function can be introduced into N-protected 2-aminopyridines with dibenzylsulfide [245].

Just like aryl amines, heteroaromatic amines are subject to conversion by way of diazonium salts into the corresponding thiols [246].

Heterocycles bearing oxygen substituents can also be transformed into thiols. Suitable reagents include especially phosphorus pentasulfide [247], hydrogen sulfide [248], and the Lawesson reagent [249], just as in the case of alkanethiols (see Section 1.3.1).

Another reaction that has been used to prepare heteroaromatic thiols is the reduction of a sulfonic acid with iodine and triphenylphosphine [250].

3.3.2. Heterocyclic Thiols by Cyclization

Cyclocondensations with such sulfur-containing species as carbon disulfide, thiourea, or thiocyanates are especially important reactions for the synthesis of heterocyclic thiols.

Carbon Disulfide. 2-Mercapto-1,3-benzothiazoles are prepared on an industrial scale from anilines by cyclocondensation with carbon disulfide and sulfur. One reaction of this type exploited industrially, that of o-aminothiophenols, occurs even at low temperature [251].

o-Phenylenediamines react analogously to 2-mercaptobenzimidazoles, whereby potassium xanthogenate can be utilized in place of carbon disulfide if so desired [252].

1,2-Diaminoalkanes are transformed by carbon disulfide into 2-mercaptoimidazolines [253].

The cyclization of 2-aminobenzonitriles with carbon disulfide or potassium xanthogenate leads to mercapto-substituted quinazolines [254].

Cycloaddition of carbon disulfide to sodium azide produces the sodium salt of 5-mercapto-1,2,3,4-thiatriazole [255].

$$\text{CS}_2 + \text{NaN}_3 \longrightarrow \text{Na}^+ {}^-\text{S}-\underset{\underset{N-N}{}}{\overset{\overset{S}{\diagup \diagdown}}{\underset{}{C}-N}}$$

Thiourea. Condensation of carbonyl compounds with thiourea derivatives represents another important approach to the synthesis of heterocyclic thiols. Thus, cyclocondensation of α-hydroxyketones with thiourea results in 2-mercaptoimidazoles [256].

The analogous conversion of β-dicarbonyl compounds to 2-mercaptopyrimidines has been subject to especially widespread utilization [257].

The usual starting materials are β-diketones or β-ketoaldehydes along with the corresponding acetals and acetoacetic or malonic ester derivatives. The latter lead to thiobarbiturates, which are used as short-duration narcotics.

Thiourea reacts with two equivalents of an aldehyde and one equivalent of amine to provide mercapto-substituted 1,3,5-triazines [258].

Thiosemicarbazide undergoes a reaction similar to that of thiourea, and can be transformed with aliphatic or aromatic carboxylic acids into 3-mercapto-1,2,4-triazoles [259].

$$H_2N-C(=S)-NH-NH_2 + R-COOH \longrightarrow R\text{-[triazole]-SH}$$

R: H, alkyl, aryl

Thiocyanates. Potassium thiocyanate is a frequently employed reagent that lends itself to cyclization with aminocarbonyl compounds to 2-mercaptoimidazoles [260].

$$R-CO-CH(R)-NH_2 \xrightarrow{KSCN} \text{2-mercaptoimidazole}$$

Isothiocyanates react with acylhydrazides in the presence of sodium alcoholate to give 3-mercapto-1,2,4-triazoles [261].

$$R^1-N=C=S + H_2N-NH-CO-R^2 \xrightarrow{NaOC_2H_5} \text{3-mercapto-1,2,4-triazole}$$

Other cyclization processes are described in [262].

3.3.3. Heterocyclic Sulfides and Disulfides

The synthetic methods applicable to heterocyclic sulfides and disulfides correspond largely to those for aliphatic and aromatic compounds (see Sections 1.6 and 2.4). Thus, heteroaromatic sulfides can be prepared just like their aromatic analogues by heteroarylation of such inorganic compounds as sodium sulfide or dichlorosulfane [263].

$$2\ \text{pyrrole-R} \xrightarrow{Na_2S \text{ or } SCl_2} \text{bis(pyrrolyl)sulfide}$$

Unsymmetrical sulfides can be obtained from heteroaryl halides by treatment with sodium alkanethiolates or thiols [264]. Good yields are also observed in the palladium-catalyzed reaction of heteroaromatic halogen compounds with organotin sulfides [265].

Sulfenyl chlorides can be transformed into unsymmetrical sulfides with alkenes [266] or arenes [267].

$$\text{dihydropyran} + Cl-S-CH_3 \longrightarrow \text{2-Cl-3-(SCH}_3\text{)-tetrahydropyran}$$

$$O_2N\text{-}C_6H_3(NO_2)\text{-}S\text{-}Cl + \text{thiophene} \xrightarrow{SnCl_4} O_2N\text{-}C_6H_3(NO_2)\text{-}S\text{-thiophene}$$

4,5-Dinitroimidazoles are converted in high yield to thioethers through substitution of a nitro group [268].

Table 6. Heterocyclic thiols and sulfides used as pharmaceuticals

Thiol/Sulfide	Application	Ref.
2-Thiobarbituric acid	anaesthetic, narcotic, hypnotic, sedative	[269]
6-Thiopurine	cytostatic agent, antileukemic	[270]
2-Thiouracil	thyreostatic	[271]
Cephalosporin	antibiotic	[272]
2-Pyridinethiol-N-oxide Na salt, Zn salt, disulfide	antimycotic, antiseborrhoeic, fungicide, bactericide	[273]

For the preparation of heterocyclic disulfides see the methods described in Section 1.6.3.

3.4. Uses of Heterocyclic Thiols, Sulfides, and Disulfides

Numerous heterocyclic thiols and their sulfur derivatives have found application as intermediates in the synthesis of pharmaceuticals, crop-protection agents, and dyes. Additional applications are described in the sections that follow.

3.4.1. Pharmaceuticals

Many heterocyclic thiols and sulfides are among the active ingredients in commercially available drugs. Table 6 presents a few examples.

A wide range of such substances with antibiotic activity has been developed in the last few years, including systems with skeletons that differ from those of penicillin and cephalosporin. A review of the recent patent literature is incorporated into [274]. The description that follows provides a few examples of the broad spectrum of pharmaceutical activity associated with heterocyclic thiols and their derivatives.

Mercapto- (or alkylthio-) triazoles [275], imidazoles [276], imidazolines [277], and pyridines [278] inhibit the activity of dopamine-β-hydroxylases. 7-Thiosubstituted oxoquinolinecarboxylic acids [279], 2-amino-3-mercaptopyridines [280], and thiopyridine derivatives of 5-nitro-2-vinylfuran [281] display antibacterial activity, and the latter are also fungicides. Analgesic, antipyretic, and anti-inflammatory activity has been reported for certain thiopyridine derivatives [282], 2-mercaptoimidazoles [283], 3-amino-4-mercapto-6-methylpyridazines [284], and 3-mercaptotriazoles [285]. Mercapto-substituted imidazoles [286] and triazoles [287] have been patented as medicinal fungicides. 2-Mercaptobenzoxazoles can be used for both prophylaxis and treatment in the case of liver diseases [288]. Various heterocyclic thiols and sulfides inhibit the secretion of stomach acid, and can be utilized in ulcer treatment [289]. Mercaptopyrrolidine derivatives have been described as cough-prevention agents [290]. Mercaptopyrazolo-pyrimidines and -triazines represent potential schistosomicides [291]. Furanyl and pyranyl sulfides display immune-controlling characteristics [292]. Thio-substituted pyridazinones [293], imidazoles [294], and other heterocycles [295] display antiallergic and antiasthmatic activity. The gold salt of thioglucose shows in vitro antiviral activity toward the immune-deficiency virus HIV-1 [296].

3.4.2. Agricultural Chemicals

Heterocyclic thiols and sulfides are included as crop-protection agents in numerous commercial products. For example, several herbicides contain methylthio-1,3,5-triazines [297].

R	Commercial product
– i-Pr	Semeron
– $(CH_2)_3$–O–CH_3	Gesaran
– CH_2–CH_3	Ingran

Other effective herbicides of considerable economic significance are the 3-methylthio-1,2,4-triazine-5(4H)-ones, which inhibit photosynthesis [298]–[300].

	R	Commercial product
	—NH$_2$	Sencor
	—N=CH—CH(CH$_3$)$_2$	Tantizon

(Structure: t-Bu-substituted triazinone with S-CH$_3$ and N-R groups)

Mercapto-substituted pyrazoles [301] and imidazole-5-carboxylic acids [302] as well as 5-mercaptopyridine-3-carboxylic acid derivatives [303] have also been reported to show herbicidal activity.

A broad spectrum of activity has been reported in the extensive patent literature with respect to 1,2-pyridazinones containing thio substituents. For example, they can be used not only as pesticides in veterinary-medicine and crop-protection applications [304], but also as plant fungicides [305]. Other compounds effective as pesticides are derivatives of 1,2,5-thiadiazol-3-thiol [306] and heteroaryl vinyl sulfides [307]. 1-Oxopyridyl disulfides can be utilized as feed supplements in the raising of poultry [308].

3.4.3. Other Uses

Heterocyclic thiols and their salts are employed in the photographic sector as stabilizers and as protection against fogging. Patents have been issued in this context covering pyrazoles and triazoles [309], thiadiazoles [310], pyrrolidines [311], and benzimidazoles [312] with free mercapto groups, but also heterocyclic sulfides [313]. Other photographic auxiliaries contain mercaptopyridines [314].

2-Mercaptobenzothiazole and its derivatives have been used to protect copper and copper alloys against corrosion. Corrosion-inhibiting properties have also been reported for 2-mercaptopyridines [315], 4,5-bismercapto-1,2-dithiol-3-one and its salts [316], and certain pyridine thioethers [317].

A few heterocyclic thiols are effective as stabilizers for microbicidal 3-isothiazolones [318].

Mercaptobenzimidazole and its derivatives have proven valuable in preventing the aging of rubber [319]. 2-Mercaptobenzothiazole is an important vulcanization catalyst in the rubber industry. Other compounds suggested for this purpose include mercapto-substituted triazine [320] and pyrimidine [321] derivatives. Derivatives of 2,5-dimercaptothiadiazole [322] as well as aminobis- and aminotris(alkylthio)triazines have been suggested as stabilizers for natural and synthetic rubber [323].

Heterocyclic anthraquinone thioethers have been described as dyes for liquid crystals [324]. Thioethers [325] and disulfides [326] of furans and thiophenes are cited in the recent patent literature as flavoring food additives.

2-Mercaptobenzothiazole plays a role in analysis as a reagent for cadmium as well as for the determination of copper, lead, bismuth, silver, mercury, thallium, gold, platinum, and iridium. 4-Amino-3-hydrazino-5-mercapto-1,2,4-triazole (Purpald, Aldrich) is a reagent specific for aldehydes [327]. Other analytical applications of heterocyclic thiols and sulfides are discussed in [328].

Cosmetic applications include the use of 2-mercaptopyridine to prevent the formation of melanin in the skin [329], and of certain heterocyclic thiols as active ingredients in deodorants [330].

4. Toxicology [331]–[335]

Most thiols are classed as moderately toxic compounds. Since many thiols and sulfides are readily detectable on the basis of their odors at levels far below the danger point, these odors serve to provide a warning. Exceptions include 2,4,4-trimethyl-2-pentanethiol (*tert*-octanethiol), cyclohexanethiol, and thiophenol, which are significantly more toxic than short-chain thiols. Toxicity decreases with longer-chain thiols.

The SH group plays a significant role in biological metabolism (e.g., in metabolite transfer), and for this reason thiols may exhibit either inhibitory or accelerating effects on metabolic processes. A comprehensive treatment of the metabolic, biochemical, and toxicological aspects of xenobiotics containing thiol or sulfide functions is presented in [333].

The discussion that follows constitutes a brief description of the toxic properties of a few representative thiols, sulfides, and disulfides.

Methanethiol (Methyl Mercaptan). The greatest risk with respect to this compound involves inhalation, and the corresponding toxic properties are comparable to those of hydrogen sulfide. Low concentrations lead to irritation of the eyes and mucous membranes, headache, nausea, and vomiting. Irritation of the respiratory tract may lead to edema in the lungs. Higher concentrations result in increased blood pressure and pulse rate, coma, severe anemia, and death. Inflammation of the skin (dermatitis) is observed to accompany chronic exposure to low concentrations. MAK: 0.5 ppm (1 mg/m^3); TLV: 0.5 ppm.

Ethanethiol (ethyl mercaptan) has shown moderately toxic effects in animal experiments in conjunction with inhalation, oral ingestion, or intraperitoneal application. Low concentrations (4 ppm) lead in the case of humans to irritation of the skin and eyes, headache, and nausea, and respiratory rate also decreases at an exposure level of 50 ppm. Extended exposure to 4 ppm of ethanethiol results in a decrease in odor sensitivity. MAK: 0.5 ppm (1 mg/m^3); TLV: 0.5 ppm.

n-Propanethiol (propyl mercaptan) has been shown in animal studies to display little toxicity upon oral administration, and it is essentially nontoxic with respect to inhalation. Its effect as an irritant to mammalian skin and eyes is also low. On the other hand, human subjects residing adjacent to a field treated with the pesticide Ethioprop (Mocap) displayed health impairments that persisted for as long as six weeks as a result of the release of propanethiol. The symptoms described included headache, sore throat, diarrhea, and irritation of the eyes, as well as hay fever and asthma attacks [336].

Butanethiols. The inhalation of 50–500 ppm of *n-butanethiol* over a period of one hour led in the case of seven human subjects to modest to severe signs of poisoning, including such symptoms as weakness, increased respiratory rate, pain in the neck, nausea, and depression of the central nervous system (diminished reaction capability). MAK: 0.5 ppm (1.5 mg/m^3); TLV: 0.5 ppm. *tert-Butanethiol* has been shown in animal experiments to be somewhat less toxic than the straight-chain isomer. Both butyl thiols are eye irritants. MAK and TLV values for these compounds have not yet been established.

Cyclohexanethiol causes eye irritation and severe skin damage. Animal experiments (rats) involving oral ingestion have revealed a low LD_{50} of 558 mg/kg.

tert-Octanethiol has been found in animal studies to be toxic by oral ingestion, inhalation, or intraperitoneal application, and moderately toxic upon resorption through the skin. The lowest lethal dose (intraperitoneal) is reported for rats at 11 mg/kg.

Dodecanethiols. The toxicity of *n-dodecanethiol* is low. Chronic exposure has been observed to produce skin irritation and sensitization. In the case of *tert-dodecanethiol* (isomeric mixture) animal studies based on oral administration produced a very low LD_{50} of 309 mg/kg. Skin and eye irritation was also noted.

Thiophenol (phenyl mercaptan) has a severe irritating effect on the skin and eyes, and can lead to acute dermatitis. Headache and dizziness may accompany working with the compound. Animal studies have shown that chronic inhalation of high concentrations may cause damage to the liver, lungs, and kidneys. No MAK value has yet been established; TLV: 0.5 ppm. Thiophenol as well as *o-* and *p-aminothiophenols* cause the transformation of oxyhemoglobin into methemoglobin in human erythrocytes [337]. *p-Chlorothiophenol* and *pentachlorothiophenol* cause severe eye irritation, and the former may be carcinogenic.

Alkyl sulfides are generally described as moderately toxic compounds that may lead to hemolytic anemia and allergic dermatitis. *Dimethyl sulfide* is responsible for skin irritation and severe eye irritation. Certain diaryl sulfides [e.g., 2,2'-thiobis(4,6-dichlorophenol) and 4,4'-thiodianiline] are suspected of being carcinogenic.

2-Mercaptobenzothiazole has a sensitizing effect, and with chronic exposure (e.g., through the use of rubber gloves) it may induce skin reactions [338]. *2-Mercapto-1-methylimidazole* displays teratogenic effects in humans, and it is suspected of being a carcinogen. *6-Mercaptopurine derivatives* influence cell division, and have been employed as cytostatic agents. They have been found to be teratogenic in animal studies. *Methylthio-substituted triazines*, which are used as herbicides, are only slightly toxic. The corresponding LD_{50} or LC_{50} values derived from animal studies are above the level of 1000 mg/kg [339]. Similar values have been established for a methylthiotriazinone that is the active ingredient in the herbicide Sencor [340].

The professional literature should be consulted regarding the diverse toxic side effects of the many pharmacologically active heterocyclic thiols and sulfides.

5. References

General References

[1] *Ullmann*, 4th ed., **23**, 175.
[2] *Houben-Weyl*, **E11**, 32, 129; **9**, 55.
[3] N. Kharasch (ed.): *Organic Sulfur Compounds*, Pergamon Press, New York 1961–1966.
[4] G. C. Barrett in: *Comprehensive Organic Chemistry*, vol. 3, Pergamon Press, Oxford 1979, p. 3.
[5] J. L. Wardell in S. Patai (ed.): *The Chemistry of the Thiol Group*, Wiley-Interscience, London 1974.
[6] *Kirk-Othmer*, **22**, 946.
[7] H. Goldwhite in: *Rodd's Chemistry of Carbon Compounds*, 2nd ed., vol. **IB**, Elsevier Publ., Amsterdam 1965, p. 73.
[8] D. W. Brown in: *Rodd's Chemistry of Carbon Compounds*, 2nd ed., vol. **IB**, 2nd Suppl., Elsevier Publ., Amsterdam 1991, p. 344.
[9] A. Ohno in S. Oae (ed.): *Organic Chemistry of Sulfur*, Plenum Press, New York 1977, p. 119.
[10] S. R. Sandler, W. Karo in H. H. Wasserman (ed.): *Organic Functional Group Preparations*, 2nd ed., vol. **1**, Academic Press, New York 1983.
[11] *Rodd's Chemistry of Carbon Compounds*, Elsevier Publ., Amsterdam, 2nd ed., vol. **3A**, 1971, p. 421; Suppl. 3A, 1983, p. 241.

Specific References

[12] A. J. Virtanen, *Angew. Chem.* **74** (1962) 374; *Angew. Chem. Int. Ed. Engl.* **1** (1962) 199.
[13] A. E. Johnson, H. E. Nursten, A. A. Williams, *Chem. Ind. (London)* 1971, 556.
[14] E. Demole, P. Enggist, G. Ohloff, *Helv. Chim. Acta* **65** (1982) 1785.
[15] H. Yanagawa et al., *Tetrahedron Lett.* **13** (1972) 2549–2552.
[16] E. E. Beach, A. White, *J. Biol. Chem.* **127** (1939) 87.
[17] T. Janowski, J. Gawlik, S. Zimnal, *VDI Ber.* **226** (1975) 123.
[18] E. H. Phelps, *Proc. World Pet. Congr. 7th.*, **9** (1967) 201.
[19] W. C. Zeise, *J. Prakt. Chem.* **1** (1834) 257.
[20] C.-D. S. Lee, W. H. Daly, *Adv. Polym. Sci.* **15** (1974) 123.
[21] H. V. Aposhion, *Annu. Rev. Pharmacol. Toxicol.* **23** (1983) 193. L. Magos, *Br. J. Pharmacol.* **56** (1976) 479.
[22] B. L. Sharma et al., *Jpn. J. Pharmacol.* **45** (1987) 295–302.
[23] Pädagogische Hochschule, N. K. Krupskaja, DD 276 481, 1990 (K. Andras, H. R. Hoppe, R. Sziburies, H. Selzer).
[24] Sakai Chem. Ind., US 4 873 351, 1989 (K. Fujita, T. Wachi, Y. Ikeda).
[25] B. Krebs, G. Henkel, *Angew. Chem.* **103** (1991) 785–804. J. H. Dawson, M. Sono, *Chem. Rev.* **87** (1987) 1255–1276.
[26] UOP, US 4 490 246, 1983 (T. A. Verachtert); US 4 498 977, 1983 (R. R. Frame).
[27] Co. Française de Raffinage, EP 91 845, 1983 (C. Marty, P. Engelhard).
[28] Inst. Française du Petrole, FR 2 560 889, 1984 (P. H. Bigeard, J. P. Franck, P. Ganne, H. Mimoun).
[29] Y. M. Torchensky: *Sulfur in Proteins*, Pergamon Press, Oxford 1981. H. Sies, *Naturwissenschaften* **76** (1989) 57.
[30] Monsanto, US 4 966 731, 1990 (Y. Chou).

[31] Pennwalt, EP 313 939, 1988 (A. Husain, G. A. Wheaton).
[32] Atochem North America, EP 424 616, 1990 (A. Husain, G. A. Wheaton).
[33] E. Campaigne in N. Kharasch (ed.): *Organic Sulfur Compounds*, vol. **1,** Pergamon Press, New York 1961, chap. 14.
[34] S. Ahmad, J. Iqbal, *Tetrahedron Lett.* **27** (1986) 3791.T. Imamoto, M. Kodera, M. Yokoyama, *Synthesis* 1982, 134.T. Kunieda, Y. Abe, M. Hirobe, *Chem. Lett.* 1981, 1427.
[35] *Kirk-Othmer*, **22,** 953. D. J. R. Massy, *Synthesis* 1987, 589.
[36] P. Sabatier, A. Mailhe, *C. R. Acad. Sci.* **150** (1910) 1217.R. C. Kramer, E. E. Reid, *J. Am. Chem. Soc.* **43** (1921) 880.
[37] H. O. Folkins, E. L. Miller, *Proc. Am. Pet. Inst. Sect. 3* **42** (1962) 188.
[38] H. O. Folkins, E. L. Miller, *Ind. Eng. Chem. Prod. Res. Dev.* **1** (1962) 271. A. V. Mahkina et al., *React. Kinet. Catal. Lett.* **36** (1988) 159.
[39] Pennwalt, US 3 035 097, 1962 (T. E. Deger, B. Buchholz, R. H. Goshorn).
[40] O. Weisser, S. Landa: *Sulphide Catalysts, their Properties and Applications,* Pergamon Press, Oxford 1973, p. 268.
[41] P. Bapsères, *Chim. Ind. (Paris)* **90** (1963) 358.
[42] Toyo Chem. JP 50 040 511, 1973 (S. Ohno, R. Orita, S. Akita).
[43] M. P. Balfe, J. Kenyon, C. E. Scarle, *J. Chem. Soc.* 1950, 3309.
[44] Mitsui Toatsu, JA 51 048 606, 1974 (T. Hayakawa, T. Ichikawa).
[45] Albright & Wilson, GB 917 921, 1963 (H. Coates, P. A. T. Hoye).
[46] T. Nishio, *J. Chem. Soc., Chem. Commun.* 1989, 205.
[47] Societé Nationale Elf Aquitaine (Production), Demande brevet français 88 018 79, 1988 (E. Arretz).
[48] Pennwalt, US 2 951 875, 1960 (B. Loev, R. H. Goshorn).
[49] Pennwalt, US 4 102 931, 1978 (B. Buchholz).
[50] Societé Nationale Elf Aquitaine (Production), EP 101 356, 1982 (E. Arretz, A. Mirassou, C. Landoussy, P. Augé).
[51] Pennsalt Chem., NL 142 406, 1963.
[52] Societé Nationale Elf Aquitaine (Production), FR 2 330 674, 1975 (E. Arretz, A. Pfister).
[53] Seitetsu, JP 79 132 551, 1978 (M. Kawamura et al.).
[54] Phillips Petroleum, US 3 963 785, 1975 (D. H. Kubicek).
[55] J. F. Frantz, K. I. Glass, *Chem. Eng. Prog.* **59** (1963) 68.
[56] M. A. Korshunov, SU 518 489, 1974;*Chem. Abstr.* **85** (1977) 159410.
[57] Phillips Petroleum, US 3 214 386, 1965 (R. D. Franz, P. F. Warner).
[58] Pennwalt, BE 866 426, 1977 (B. Buchholz).
[59] Ogawa, JP-Kokai 63 201 162, 1987 (H. Masuda, H. Kikuiri, S. Mihara).
[60] R. Dahlbom, *Acta Chem. Scand.* **5** (1951) 690. Soda Aromatic Co., JP 1 052 711, 1987 (T. Shirakawa, K. Kinoshita, K. Suzuki, T. Eto).
[61] Phillips, EP 208 323, 1985 (J. S. Roberts).
[62] Phillips, US 3 050 452, 1962 (R. P. Louthan).
[63] Pennwalt, US 4 140 604, 1978 (D. A. Dimmig).
[64] Shell, US 2 411 983, 1946 (W. E. Vaughan, F. F. Rust). Societé Nationale Elf Aquitaine (Production), EP 5 400, 1979 (J. Ollivier, G. Souloumiac, J. Suberlucq).
[65] Du Pont, US 2 551 813, 1949 (P. S. Pinkney).
[66] Stauffer, US 3 689 568, 1969 (R. J. Eletto).
[67] F. W. Stacey, J. F. Harris, *Org. React. (N.Y.)* **13** (1963) 150.

[68] Nippon Oils & Fats, US 3 459 809, 1969 (S. Ishida, T. Yoshida). Stauffer, BE 750 555, 1969 (D. J. Martin, E. D. Weil).
[69] M. W. Farlow, W. A. Lazier, F. K. Signaigo, *Ind. Eng. Chem.* **42** (1950) 2547. Phillips Petroleum, US 3 994 980, 1976 (D. H. Kubicek).
[70] Seitetsu, JP 79 135 747, 1979 (M. Kawamura et al.).
[71] Pennwalt, US 4 281 202, 1980 (B. Buchholz, C. B. Welsh, H. C. Henry).
[72] Societé Nationale Elf Aquitaine (Production), EP 274 934, 1988 (E. Arretz, P. Augé, A. Mirassou, C. Landoussy).
[73] Uniroyal, CA 882 159, 1971 (G. S. Pande).
[74] Societé Nationale Elf Aquitaine (Production), US 4 507 505, 1985 (E. Arretz).
[75] L. M. Ellis, E. E. Reid, *J. Am. Chem. Soc.* **54** (1932) 1674.
[76] J. E. Bittell, J. L. Speier, *J. Org. Chem.* **43** (1978) 1687.
[77] A. M. Vasil'tov, B. A. Trofimov, S. V. Amosova, *Zh. Org. Khim.* **19** (1983) 1339; *Chem. Abstr.* **99** (1983) 70162x.
[78] Dow Corning, BE 834 487, 1974 (J. L. Speier). A. G. Koscova et al., *Zh. Prikl. Khim. (Leningrad)* **38** (1965) 170.
[79] Akzo, US 3 927 085, 1974 (A. G. Zenkel, A. Toth, H. Nagerlein, G. Meyer).
[80] Stauffer, US 4 740 623, 1988 (J. B. Heather). Societé Nationale Elf Aquitaine (Production). Demande brevet français 8 804 693, 1988 (Y. Labat).
[81] Crown Zellerbach, US 3 326 980, 1964 (D. W. Goheen).
[82] Draco, EP 317 540, 1987 (A. R. Hallberg, P. A. S. Tunek).
[83] Panmedica, EP 305 264, 1987 (R. Lazaro, G. Cros, J. H. McNeil, J. J. Serrano).
[84] L. R. Morgan, US 4 827 016, 1985.
[85] Schering, EP 355 784, 1988 (M. F. Czarniecki, L. S. Lehmann). Societé Civile Bioprojet, EP 419 327, 1989 (J. C. Plaquevent et al.). Squibb and Sons, EP 361 365, 1988 (N. G. Delaney).
[86] Ono, EP 341 081, 1988 (M. Kawamura, Y. Arai, H. Aishita).
[87] Santen, EP 326 326, 1988 (T. K. Morita, S. Mita, T. Iso, Y. Kawashima). Beecham, EP 322 184, 1987 (R. E. Markwell, S. A. Smith, L. M. Gaster).
[88] Nihon Medi-Physics, EP 322 876, 1987 (H. Yamauchi et al.). NeoRx, EP 344 724, 1988 (A. R. Fritzberg, A. Srinivasan, D. S. Jones, D. W. Wilkening). Du Pont, EP 279 417, 1987 (P. L. Bergstein, E. H. Cheeseman, A. D. Watson).
[89] Proter, EP 288 871, 1987 (P. Braga).
[90] Camillo Corvi, EP 343 367, 1988 (P. Braga).
[91] Roussel-Uclaf, EP 280 627, 1987 (J. P. Vevert, J. C. Gasc, F. Delevallee, F. Petit).
[92] Dainippon, JP 1 149 763, 1987 (T. Mimura et al.).
[93] Takeda, JP 64 000 063, 1986 (H. Uneme, H. Mitsuders, T. Uekado).
[94] Nitto, JP 63 225 391, 1986 (H. Imazaki et al.).
[95] Takeda, EP 421 210, 1989 (K. Itoh, K. Okonogi).
[96] Midori, JP 1 061 492, 1987 (H. Nakasumi, T. Kitao, M. Oizumi).
[97] M. A. Allakhverdiev, *Neftekhimiya* **30** (1990) 570; *Chem. Abstr.* **114** (1990) 26893r.
[98] L'Oreal, EP 368 763, 1988 (J. Maignan, M. Colin, G. Lang).
[99] Kurakay, JP 3 246 360, 1987 (M. Kawashima, I. Komura, J. Yamauchi).
[100] Firmenich, EP 54 847, 1982 (E. P. Demole, P. Enggist). Givaudan, US 3 896 175, 1972 (D. Helmlinger et al.). Fritzsche Dodge, GB 1 546 283, 1979 (B. J. Willis, F. Fischetti, D. I. Lerner). International Flavors and Fragrances, EP 369 668, 1988 (R. M. Boden, J. A. McGhie). BASF K & F, EP 418 680, 1989 (P. A. Christenson, R. G. Eilerman, P. J. Riker, B. J. Drake).

[101] Soda Aromatic, JP 1 052 711, 1987 (T. Shirakawa, K. Kinoshita, K. Suzuki, T. Eto).
[102] Ogawa, JP 1 102 019, 1987 (H. Masuda, H. Kikuiri, Y. Shishido, S. Mihara).
[103] General Foods, US 4 496 600, 1985 (T. H. Parliment).
[104] E. Block, *Spektrum Wissensch.* 1985, 66.
[105] H. Schildknecht et al., *Angew. Chem.* **88** (1976) 228.
[106] *Römpp Chemie Lexikon*, 9th ed., Thieme Verlag, Stuttgart 1992, p. 2484.
[107] E. B. Krein in S. Patai, Z. Rappoport (eds.): *The Chemistry of Sulphur-Containing Functional Groups*, Suppl. S, Wiley & Sons, New York 1993, p. 984.
[108] E. Block in S. Patai (ed.): *The Chemistry of Ethers, Crown Ethers, Hydroxyl Groups and Their Sulphur Analogues*, part 1, Suppl. E, Wiley & Sons, New York 1980, p. 539.
[109] S. Blechert, *Nachr. Chem. Tech. Lab.* **28** (1980) 577.
[110] S. Oae, H. Togo, *Kagaku (Kyoto)* **38** (1983) 48; *Chem. Abstr.* **98** (1983) 197543g.
[111] F. Kong, X. Zhou, *Synth. Commun.* **19** (1989) 3143.
[112] Kuraray, JP 61 155 365, 1986 (T. Onishi, S. Suzuki, M. Shiono, Y. Fujita).
[113] AS Urals Bashkir, SU 1 574 595, 1988 (U. M. Dzhemilev, A. G. Ibragimov, A. B. Morozov, E. G. Galkin); *Chem. Abstr.* **114** (1991) 5808b.
[114] Humboldt-Universität Berlin, DD 266 350, 1987 (E. Wenschuh et al.).
[115] J. M. Khurana, P. K. Sahoo, *Synth. Commun.* **22** (1992) 1691.
[116] Phillips Petroleum, US 4 837 192, 1989 (J. S. Roberts).
[117] M. E. Peach in S. Patai (ed.): *The Chemistry of the Thiol Group*, Wiley, New York 1974, p. 771.E. J. Corey et al., *J. Am. Chem. Soc.* **102** (1980) 1436, 3663.
[118] P. Crotti et al., *Synlett* 1992, 303.
[119] Nippon Mining, JP 4 352 764, 1992 (T. Katagiri, K. Furuhashi).
[120] Atochem, EP 446 661, 1991 (B. Buchholz, R. B. Hager, M. J. Lindstrom).
[121] Y. Kikugawa, *Chem. Lett.* **1981**, 1157. Y. Kikugawa, JP 5 824 557, 1983; *Chem. Abstr.* **99** (1983) 22095.
[122] G. A. Olah, Q. Wang, N. J. Trivedi, G. K. S. Prakash, *Synthesis* 1992, 465.
[123] Reviews: J. Drabowicz, T. Numata, S. Oae, *Org. Prep. Proced. Int.* **9** (1977) 63. M. Madesclaire, *Tetrahedron* **44** (1988) 6537.
[124] R. Balicki, *Synthesis* 1991, 155.
[125] J. Drabowicz, B. Dudzinski, M. Mikolajczyk, *Synlett* 1992, 252.
[126] H. Bartsch, T. Erker, *Tetrahedron Lett.* **33** (1992) 199.
[127] T. J. Wallace et al., *Ind. Eng. Chem. Process Des. Dev.* **3** (1964) 237.
[128] Phillips Petroleum, EP 316 942, 1989 (S. H. Presnall).
[129] Shell, EP 288 104, 1988 (E. M. Van Kruchten, C. M. J. Leenaars).
[130] Phillips Petroleum, US 4 277 623, 1981 (D. H. Kubicek).
[131] M. Goodrow et al., *Tetrahedron Lett.* **23** (1982) 3231. J. Nakayama et al. *Sulfur Lett.* **1** (1982) 25.J. Drabowicz, M. Mikolajczyk, *Synthesis* 1980, 32.
[132] F. S. Jargensen, J. P. Snyder, *J. Org. Chem.* **45** (1980) 1015.
[133] Organic Chem. Tech. Inst., SU 1 712 358, 1992 (O. E. Orlov, L. V. Kaabak, B. A. Kusnetsov, Y. I. Baranov).
[134] A. Pryor, D. F. Church, C. K. Govindan, G. Crank, *J. Org. Chem.* **47** (1982) 156.
[135] Waku Pure Chem. Ind., JP 59 172 456, 1984 (S. Daikiyo, Y. Kin, D. Fukushima).
[136] Waku Pure Chem. Ind., JP 59 172 457, 1984 (S. Daikiyo, Y. Kin, D. Fukushima). S. Oae et al., *J. Chem. Soc., Perkin Trans. 1* 1978, 913.
[137] A. Cornélis, N. Depaye, A. Gerstmans, P. Laszlo, *Tetrahedron Lett.* **24** (1983) 3103.

[138] L. Field, W. S. Hanley, I. McVeigh, *J. Org. Chem.* **36** (1971) 2735. Bayer, DE 3 014 717, 1981 (R. Schubart, U. Eholzer, T. Kempermann, E. Roos).
[139] Rhein Chemie, EP 25 944, 1979 (K. Morche, K. Nuetzel, M. Sauerbier, K. Schilling).
[140] Nippon Soda, WO 83/2 773, 1983 (Y. Koyama, H. Nakazawa, T. Yamashita, H. Moriyama).
[141] Electricité de France, FR 2 669 042, 1992 (G. Le Guillanton, T. Do Quang, J. Simonet).
[142] Societé Nationale Elf Aquitaine (Production), EP 356 318, 1990 (Y. Vallee, Y. Labat).
[143] Nippon Soda, WO 90/543, 1990 (S. Tazaki et al.).
[144] Squibb, US 4 888 424, 1989 (J. E. Sundeen, D. Floyd, V. G. Lee).Troponwerke, DE 3 012 000, 1981 (H. Jacobi, W. Opitz).
[145] H. C. Brown, S. U. Kulkarni, *J. Org. Chem.* **44** (1979) 2417, 2422.
[146] Nippon Miktron, JP 1 117 856, 1989 (T. Hiramitsu).
[147] R. Adams, C. S. Marvel, *Org. Synth. Coll.* **1** (1964) 504.
[148] Nippon Soda, JA 3 075 165, 1961.
[149] Sumitomo Seika, JP 5 186 418, 1991 (M. Suzuki, N. Kitagishi, A. Fujisawa, J. Ookawa).
[150] Yaroslav Polytechnic Inst., SU 1 421 736, 1987 (E. M. Alov et al.); *Chem. Abstr.* **110** (1988) 192428.
[151] Jenapharm, DD 251 343, 1986 (G. Langbein, H. J. Siemann, S. Berbig, A. Bocker). Sumitomo Seika, JP 3 153 664, 1989 (T. Morishita, M. Sato, T. Yagi, K. Kaneda).
[152] Sumitomo Seika, JP 3 170 456, 1989 (T. Morishita, M. Sato, N. Kitagishi, A. Onoe).
[153] K. Itabashi, *Yuki Gosei Kagaku Kyokaishi* **19** (1961) 601; *Chem. Abstr.* **55** (1961) 23412.
[154] Ciba-Geigy, EP 2 755, 1978 (H. Thies, F. von Kaenel).
[155] F. Vögtle, R. G. Lichtenthaler, M. Zuber, *Chem. Ber.* **106** (1973) 719. Bayer, DE 1 643 385, 1967 (H. Leuchs, H. W. Hammen).
[156] A. W. Wagner, *Chem. Ber.* **99** (1951) 375. J. Morgenstern, R. Bayer, *Z. Chem.* **8** (1968) 106.
[157] Uniroyal, NL 6 611 430, 1967.
[158] Waku Pure Chem. Ind., JP 63 054 355, 1989 (S. Daikyo, K. Fujimori, H. Togo).
[159] O. Weisser, S. Landa: *Sulphide Catalysts, their Properties and Applications*, Pergamon Press, Oxford 1973, p. 193.Seitetsu, EP 474 259, 1991 (M. Kawamura et al.).
[160] Bayer, FR 2 008 330, 1970.
[161] Nippon Kasei, JP 62 209 057, 1986 (M. Saito, R. Sato).
[162] Konica, JP 62 226 957, 1986 (S. Sugita, K. Masuda, S. Nakagawa).
[163] W. Rundel, *Chem. Ber.* **101** (1968) 2956.
[164] Phillips Petroleum, US 4 005 149, 1977 (D. H. Kubicek).
[165] C. C. Price, G. W. Stacy, *J. Am. Chem. Soc.* **68** (1946) 498.
[166] S. Krishnamurthy, D. Aimino, *J. Org. Chem.* **54** (1989) 4458.
[167] R. A. Amos, S. M. Fawcett, *J. Org. Chem.* **49** (1984) 2637.
[168] Bayer, DE 2 903 505, 1979 (H. Hagemann, E. Klauke, G. M. Petruck). F. G. Bordwell, H. M. Anderson, *J. Am. Chem. Soc.* **75** (1953) 6019.
[169] Y. Takikawa, S. Takizawa, *Nippon Kagaku Kaishi* 1972,756; *Chem. Abstr.* **77** (1972) 5081.
[170] S. K. Jain, D. Chandra, R. L. Mital, *Chem. Ind. (London)* 1969, 989.
[171] K. R. Langgille, M. E. Peach, *J. Fluorine Chem.* **1** (1972) 407.
[172] Bayer, US 3 560 573, 1971 (M. Blazejak, J. Haydn).
[173] ICI, DE 2 347 462, 1972 (D. M. Dawson, D. J. Harper).
[174] Sumitomo, JP 4 182 463, 1990 (M. Suzuki, H. Hata, M. Yoshikawa, S. Oe).
[175] Cassella, EP 409 009, 1990 (W. Bauer, W. Steckelber). Hoechst, DE 2 127 898, 1971 (S. Planker, K. Baessler). Bayer, DE 1 670 760, 1966.

[176] Mitsui Toatsu, JP 5 117 225, 1991 (T. Kobayashi et al.).Ricoh, JP 4 074 160, 1990 (R. Momiyama, K. Nagai).
[177] Phillips Petroleum, US 4 978 796, 1990 (J. E. Shaw).J. E. Shaw, *J. Org. Chem.* **56** (1991) 3728.
[178] K. Takagi, *Chem. Lett.* **1985,** 1307. Nippon Chem. Ind., JP 62 201 862, 1986 (K. Takagi).
[179] Konishiroku Photo, JP 62 036 354, 1985 (S. Sugita, K. Masadu, S. Nakagawa).Konishiroku Photo, JP 62 067 063, 1985 (S. Sugita, K. Masuda, S. Nakagawa).
[180] Sumitomo, JP 5 140 086, 1992 (M. Suzuki, N. Kitagishi, S. Kimura, A. Fujisawa).
[181] Monsanto, US 4 088 698, 1978 (N. A. Fishel, D. A. Gross).Mitsubishi Petrochem., JP 8 036 409, 1980 (A. Sakurada, N. Hirowatari).
[182] W. E. Truce, J. J. Breiter, *J. Am. Chem. Soc.* **84** (1962) 1621. A. Ferretti, *Org. Synth. Coll.* **5** (1973) 419.
[183] Sumitomo, AU 8 814 024, 1988 (T. Haga, S. Hashimoto, E. Nagano, R. Yoshida).
[184] Soc. France Hoechst, EP 368 697, 1989 (Y. Christidis).
[185] Mitsubishi Gas Chem., JP 2 075 656, 1988 (K. Okabe, S. Ametani, S. Murabayashi).
[186] Beecham, EP 358 305, 1990 (I. Hughes).
[187] Commonwealth of Australia, WO 87/01724, 1986 (W. Mazurek).
[188] Naarden, EP 219 901, 1986 (P. J. Devalois, H. J. Wille, H. L. Vandenheuv).
[189] Consiglio Nat. Ricerche, DE 3 601 286, 1986 (G. Bornengo, F. Carlini, A. Marabini, V. Alesse).
[190] Showa Denko, JP 3 212 430, 1990 (T. Arakawa, N. Minorikawa, S. Maruyama, H. Takoshi).-Mitsui Toatsu, JP 63 165 392, 1986 (K. Sasagawa, M. Imai).
[191] Matsushita, EP 569 037, 1993 (Y. Sato, T. Sotomura, K. Takeyama, H. Uemachi).
[192] Bayer, EP 408 891, 1989 (H. L. Elbe et al.).
[193] Bayer, EP 302 321, 1988 (H. Hagemann, K. Sasse, R. Fischer).
[194] Ciba-Geigy, EP 258 190, 1987 (C. Frey, B. Dill). BASF, DE 1 593 769, 1967 (H. Eilingsfeld, H. Scheuermann).
[195] Hoechst, DE 4 131 355, 1991 (J. Dannheim). Chisso, JP 3 287 573, 1990 (T. Onishi).
[196] Nisso Petrochem Ind., JP 62 077 361, 1985 (M. Hoyko, K. Kokoma, T. Hida). Fuji Photo Film, JP 2 286 655, 1989 (T. Kawagishi, Y. Shimada, H. Tamai).
[197] DuPont, US 4 792 633, 1986 (P. W. Wojtkowski). Societé Nationale Elf Aquitaine, EP 318 394, 1988 (C. Vottero, Y. Labat, J. M. Poirier).
[198] Ethyl Corp., US 4 670 597, 1985 (P. F. Ranken, R. L. Favis). Ethyl Corp., WO 92/15560, 1992 (D. E. Ballhoff, J. F. Ballhoff).
[199] Shell, EP 411 720, 1990 (A. C. G. Gray, T. W. Naisby, S. J. Turner, T. M. Machin).Sumitomo, JP 5 039 257, 1991 (S. Tsuyunashi, K. Kumagai, N. Nabeshima, M. Kato). BASF, EP 379 098, 1990 (H. Wingert et al.).
[200] Stauffer, ICI,EP 249 150, 1987 (C. G. Knudsen). Nihon Bayer, JP 5 117 109, 1991 (T. Goshima et al.). BASF, EP 402 751, 1990 (U. J. Vogelbacher et al.).ICI, GB 2 225 014, 1989 (D. Cartwright). Rhône-Poulenc, EP 351 332, 1989 (P. Desbordes, M. Euvrard).
[201] Dow Chem., WO 91/11503, 1991 (C. J. Deford et al.).
[202] Hoechst, EP 336 199, 1989 (M. Kern et al.). Kureha, EP 255 935, 1987 (S. Kato, T. Hayaoka).
[203] Sumitomo, JP 5 229 905, 1992 (T. Takagaki, H. Kishida, K. Umeda, T. Ishiwatari).
[204] Plant Science Res. Inst., SU 1 632 393, 1980 (M. A. Fedin, T. K. Alsing, T. A. Kuznetsova).
[205] Ricoh, JP 62 011 676, 1985 (K. Taniguchi, H. Furuya). Fuji Photo Film, DE 3 704 627, 1987 (M. Takashima, M. Satomura, K. Ichikawa).
[206] ICI, EP 427 404, 1990 (P. Mistry, P. Patel).
[207] Mitsubishi, EP 424 144, 1990 (Y. Tamura, F. Watari). Toray, Green Cross, JP 2 289 622 (K. Kato, C. Ohama).

[208] Bayer, EP 570 798, 1993 (K. Hentschel, U. Walter). Ethyl Corp., US 4 757 119, 1987 (P. L. Wiggins, G. G. Knapp).
[209] Sumitomo, JP 4 031 489, 1990 (H. Hayami). Bayer, DE 3 938 422, 1989 (B. Kohler, K. Stahlke, A. Sommer, K. Sommer).
[210] Ciba-Geigy, EP 307 358, 1988 (H. O. Wirth, H. H. Friedrich).
[211] Ciba-Geigy, EP 224 442, 1986 (H. R. Meier, G. Knobloch). Ethyl Petroleum, US 4 946 610, 1989 (W. Y. Lam, G. P. Liesen). Lubrizol, WO 88/09804, 1988 (M. F. Salomon).
[212] Clairol, Bristol Myers, US 4 799 934, 1987 (M. I. Lim, J. S. Anderson).
[213] Phillips Petroleum, US 5 132 008, 1991 (M. D. Bishop, J. E. Shaw).
[214] SmithKline Beecham, EP 327 306, 1989 (J. G. Gleason et al.).Searle, EP 293 895, 1988 (R. A. Mueller, R. A. Partis). Ciba-Geigy, EP 228 045, 1986 (A. Vonspreche et al.). Lilly, GB 2 184 121, 1986 (S. R. Baker, A. Todd). Kanegafuchi, EP 211 363, 1986 (T. Shiraishi et al.).
[215] BASF, EP 386 452, 1990 (B. Janssen, H. H. Wuest).
[216] Green Cross, EP 120 426, 1988 (Y. Naito et al.).
[217] Taisho, JP 4 145 062, 1990 (K. Tomizawa et al.).
[218] Bayer, Miles, US 5 214 078, 1992 (G. Fennhoft et al.).
[219] Kennelley, CA 2 058 266, 1991 (K. J. Kennelley et al.).
[220] Lubrizol, US 4 894 174, 1988 (M. F. Salomon).
[221] Hüls, DE 3 941 002, 1989 (G. Horpel, K. H. Nordsiek).
[222] Bayer, EP 23 313, 1979 (K. Idel, J. Merten). Hay, US 5 239 052, 1990 (A. S. Hay, Z. Y. Wang).
[223] F. Rossi, G. Nardi, A. Palumbo, G. Prota, *Comp. Biochem. Physiol. B Comp. Biochem.* **80B** (1985) 843.E. Turner, R. Klevit, L. J. Hager, B. M. Shapiro, *Biochemistry* **26** (1987) 4028.
[224] E. Turner, L. J. Hager, J. Lisa, B. M. Shapiro, *Science (Washington D.C.)* **242** (1988) 939. T. P. Holler, P. B. Hopkins, *J. Am. Chem. Soc.* **110** (1988) 4837.
[225] U. Gasser, W. Grosch, *Z. Lebensm. Unters. Forsch.* **190** (1990) 3. G. MacLeod, J. Ames, *Chem. Ind. (London)* 1986, 175.
[226] U. Gasser, W. Grosch, *Z. Lebensm. Unters. Forsch.* **190** (1990) 511.
[227] J. Blank, A. Sen, W. Grosch, *Z. Lebensm. Unters. Forsch.* **195** (1992) 239.
[228] D. A. Withycombe, C. J. Mussinan, *J. Food Sci.* **53** (1988) 658, 660.
[229] P. Werkhoff et al., *Chem. Mikrobiol. Technol. Lebensm.* **13** (1991) 30.
[230] W. J. Burrows et al., *J. Biochem.* **9** (1970) 1867.
[231] T. Wieland et al., *Liebigs Ann. Chem.* **12** (1981) 2318.
[232] A. R. Katritzky, J. M. Lagowski in A. R. Katritzky (ed.): *Advances in Heterocyclic Chemistry*, vol. **1**, Academic Press, New York 1963, p. 396. A. R. Katritzky, *Chimia* **24** (1970) 134.
[233] Bayer, DE 1 108 692, 1958 (E. Kracht, F. Eichler, J. Farana).
[234] Y. H. Kim, G. H. You, *Bull. Korean Chem. Soc.* **2** (1981) 122.
[235] G. Pettit, E. E. van Tamelen, *Org. React. (N.Y.)* **12** (1962) 356. Y. Nakayama et al., *J. Heterocycl. Chem.* **18** (1981) 631.
[236] Y. Sanemitsu et al., *J. Heterocycl. Chem.* **18** (1981) 1053. S. D. Aster, S. S. Yang, G. D. Berger, *Org. Prep. Proced. Int.* **23** (1991) 658.
[237] Monsanto, FR 1 549 002, 1969 (J. J. d'Amico). D. I. Banks, P. Wiseman, *J. Appl. Chem.* **18** (1968) 202. E. Kühle: *The Chemistry of Sulfenic Acids*, G. Thieme Verlag, Stuttgart 1972. Monsanto, DE 2 201 989, 1972 (J. J. d'Amico).
[238] M. D. Bentley et al., *J. Chem. Soc., Chem. Commun.* 1971, 1625.
[239] E. C. Taylor, W. A. Ehrhart, *J. Am. Chem. Soc.* **82** (1960) 3138. H. Hartmann, S. Scheithauer, *J. Prakt. Chem.* **311** (1969) 827.

[240] Nissan Chem. Ind., JP 62 004 272, 1987 (T. Sato, K. Morimoto, S. Yamamoto). Nissan Chem. Ind., JP 61 109 777, 1986 (T. Ogura, Y. Kawamura, H. Suzuki).Ricoh, JP 3 048 661, 1989 (K. Nagai). V. Konecny, S. Kovac, *Chem. Pap.* **41** (1987) 827.

[241] Ishihara Sangyo Kaisha, JP 63 203 632, 1987 (Y. Tsujii, H. Uenishi, T. Kimura).Ishihara Sangyo Kaisha, JP 1 275 562, 1988 (R. Nasu, S. Kimura, A. Mori, R. Koto).

[242] T. Kamiyama, S. Enomoto, M. Inoue, *Chem. Pharm. Bull.* **33** (1985) 5184.

[243] Ricoh, JP 4 049 282, 1990 (K. Nagai, R. Momyama).

[244] E. Jones, I. M. Moodie, *Org. Synth.* **50** (1970) 104. J. Cymerman-Craig, J. W. Loder, *Org. Synth. coll. vol.* **V** (1973) 667.

[245] Bayer, DE 3 545 124, 1985 (A. Klausener, G. Fengler).

[246] C. Corral, J. Lissavetzky, A. S. Alvarez-Insua, A. M. Valdeolmillos, *Org. Prep. Proced. Int.* **17** (1985) 163. Takeda Chem. Ind., EP 220 695, 1987 (I. Aoki, T. Kuragano, N. Okajima, Y. Okada).

[247] R. Gompper, W. Elser, *Org. Synth. coll. vol.* **V** (1973) 780. R. F. Langler et al., *Can. J. Chem.* **49** (1971) 481. SmithKline Beckman, EP 308 159, 1987 (L. I. Kruse, S. T. Ross).

[248] S. D. Aster, S. S. Yang, G. D. Berger, *Org. Prep. Proced. Int.* **23** (1991) 658.

[249] S.-O. Lawesson, *Nouv. J. Chim.* **4** (1980) 47. K. Clausen, S.-O. Lawesson, *Bull. Soc. Chim. Belg.* **88** (1979) 305 Poli Ind. Chim., EP 515 995, 1992 (S. Poli, G. Coppi, G. Signorelli).

[250] Waku Pure Chem. Ind., JP 1 230 557, 1988 (S. Daikyo, K. Fujimori, H. Togo).

[251] C. Ferri: *Reaktionen der organischen Synthese*, G. Thieme Verlag, Stuttgart 1978.

[252] J. A. van Allan, B. D. Deacon, *Org. Synth. coll. vol.* **IV** (1963) 569.

[253] C. F. H. Allen, C. O. Edens, J. A. van Allan, *Org. Synth. coll. vol.* **III** (1955) 394. SmithKline Beckman, WO 89/6126, 1989 (L. I. Kruse, T. B. Leonard, S. T. Ross).

[254] E. C. Taylor, R. N. Warrener, A. McKillop, *Angew. Chem.* **78** (1966) 333. H. J. Kabbe, *Synthesis* **1972**, 268.

[255] E. Lieber, E. Ofterdahl, C. N. R. Rao, *J. Org. Chem.* **28** (1963) 194. C. Cristophersen, *Acta Chem. Scand.* **25** (1971) 1162.

[256] Taisho Pharmaceut. Co., JP 2 083 372, 1988 (Y. Yoshikawa et al.).

[257] H.-C. Chiang, J.-R. Ko, *Heterocycles* **19** (1982) 529. Dynamit Nobel, EP 46 856, 1980 (H. Peeters, W. Vogt). H. Petersen, *Synthesis* **1973**, 258. D. G. Crosby, R. B. Berthold, H. E. Johnson, *Org. Synth. coll. vol.* **V** (1973) 703. H. M. Foster, H. R. Snyder, *Org. Synth. coll. vol.* **IV** (1963) 638.

[258] H. Petersen, *Synthesis* **1973**, 251.

[259] C. Ainsworth, *Org. Synth. coll. vol.* **V** (1973) 1070.

[260] SmithKline Beckman, US 4 798 843, 1987 (L. I. Kruse). Ciba-Geigy, Janssen Pharmaceutica, EP 277 387, 1986 (G. R. E. Van Lommen et al.). Janssen Pharmaceutica, EP 275 603, 1986 (W. R. Lutz, W. G. Verschueren, H. Fischer, G. R. E. Van Lommen).

[261] SmithKline Beckman, EP 359 505, 1988 (L. I. Kruse, J. A. Finkelstein). SmithKline Beckman, EP 323 737, 1978 (L. I. Kruse, J. A. Finkelstein).

[262] Akad. Wissenschaft DDR, DD 252 372, 1986 (A. Rumler et al.). N. Okajima, Y. Okada, *Synthesis* **1989**, 398. Technische Universität Dresden, DD 224 320, 1984 (L. Jakisch, H. Viola, R. Mayer). Karl-Marx-Universität Leipzig, DD 276 283, 1988 (D. Briel et al.).

[263] M. J. Broadhurst, R. Grigg, A. W. Johnson, *J. Chem. Soc., Perkin Trans. 1* 1972, 1124.

[264] R. A. Rossi, S. M. Palacios, *J. Org. Chem.* **46** (1981) 5300. Taiho Yakuhin, JP 56 100 777, 1980. G. M. Kulkarni, V. D. Patil, *J. Indian Chem. Soc.* **67** (1990) 693. Ube Ind., EP 283 271, 1988 (H. Yoshioka et al.). J. Matsumoto et al., *Chem. Pharm. Bull.* **38** (1990) 2190.

[265] C. Jixiang, G. T. Crisp, *Synth. Commun.* **22** (1992) 683.

[266] S. Z. Ivin, V. K. Promonenkov, *Zh. Obshch. Khim.* **37** (1967) 489; *Chem. Abstr.* **68** (1968) 114701.

[267] R. D. Schütz, W. L. Fredericks, *J. Org. Chem.* **27** (1962) 1301.

[268] BASF, EP 293 742, 1988 (H. Köhler, T. Dockner).

[269] Abbott, US 2 876 225, 1956 (G. H. Donnison). E. Miller et al., *J. Am. Chem. Soc.* **58** (1936) 1090.G. Renzoni, *Fac. Farm.* **36** (1965) 42. M. Steinmann et al., *J. Med. Chem.* **16** (1973) 1354. Lilly, US 2 561 689, 1949.

[270] Burroughs Wellcome, US 3 056 785, 1960 (G. B. Elion, G. H. Hitching). P. Nuhn et al., *Pharmazie* **34** (1979) 285.

[271] A. Kleemann, J. Engel: *Pharmazeutische Wirkstoffe*, vol. **5,** G. Thieme Verlag, Stuttgart 1982. D. A. McGimty, W. G. Bywater, *J. Pharmacol.* **84** (1945) 342. Parke & Davis, US 2 153 711, 1934. Ciba, US 2 666 764, 1951 (A. E. Lanzilotti, A. C. Shabica, J. B. Ziegler). G. W. Anderson et al., *J. Am. Chem. Soc.* **67** (1945) 2197.

[272] P. Actor et al., *J. Antibiot.* **28** (1975) 471. Eli Lilly, US 3 928 592, 1974 (J. M. Greene, J. M. Indelicato). Bristol-Myers, US 3 796 717, 1972 (R. Smith, L. Cheney). Takeda Chem. Ind., DE 2 715 385, 1977 (M. Ochiai, A. Morimoto, Y. H. Matsushita). Warner-Lambert, EP 15 771, 1979 (T. H. Haskell, D. Schweiss, T. F. Mich, T. P. Culbertson). Sankyo, DE 2 455 884, 1974 (H. Nakao et al.). H. Nakao, *J. Antibiot.* **32** (1979) 320. Eli Lilly, US 3 943 126, 1973 (C. W. Ryan, W. B. Blanchard).

[273] Olin Mathieson, US 2 745 826, 1953 (M. A. Dolliver, S. Semnoff). Olin Mathieson, GB 761 171, 1956.

[274] Fujisawa, EP 272 456, 1987 (M. Murata et al.). Sanraku-Ocean, JP 60 163 882, 1984 (T. Yoshioka et al.). Rhône-Poulenc, EP 248 703, 1987 (J. C. Barriere, C. Cotrel, J. M. Paris).Schering, EP 238 285, 1987 (M. P. Kirkup, A. S. Boland). Fujisawa, EP 111 281, 1983 (T. Takao et al.). Meiji Seika Kaisha, EP 209 751, 1986 (T. Tsuruoka et al.). Banyu, JP 58 154 588, 1982. Bristol-Myers, DE 3 312 533, 1983 (C. U. Kim, P. F. Misco, Jr.). Hoffmann-LaRoche, US 4 348 518, 1978 (M. Montavon, R. Reiner). Asahi, EP 68 403, 1982 (J. Nishikido, E. Kodama, M. Shibukawa).

[275] SmithKline Beckman, EP 323 737, 1987 (L. I. Kruse, J. A. Finkelstein). SmithKline Beckman, EP 359 505, 1988 (L. I. Kruse, J. A. Finkelstein).

[276] SmithKline Beckman, US 4 798 843, 1987 (L. I. Kruse). SmithKline Beckman, EP 371 730, 1988 (L. I. Kruse, S. T. Ross). SmithKline Beckman, US 4 749 717, 1987 (L. I. Kruse).

[277] SmithKline Beckman, WO 89/6126, 1987 (L. I. Kruse, T. B. Leonard, S. T. Ross).

[278] SmithKline Beckman, EP 308 159, 1987 (L. I. Kruse, S. T. Ross).

[279] J. Matsumoto et al, *Chem. Pharm. Bull.* **38** (1990) 2190.

[280] Bayer, DE 3 545 124, 1985 (A. Klausener, G. Fengler).

[281] D. Vegh, J. Kovac, M. Kriz, CS 223 450, 1982; *Chem. Abstr.* **105** (1986) 226324g.

[282] Degussa, EP 149 088, 1984 (G. Scheffler et al.).

[283] Taisho, JP 2 083 372, 1988 (Y. Yoshikawa et al.).

[284] R. Bluth, *Pharmazie* **37** (1982) 285, 441. R. Bluth, DD 150 150, 1980.

[285] Warner-Lambert, US 4 962 119, 1989 (D. H. Boschelli et al.).

[286] Sumitomo, EP 54 974, 1981 (I. Saji).

[287] Pfizer, EP 91 309, 1983 (K. Richardson, P. J. Whittle).

[288] Yamanouchi, JP 62 039 583, 1985 (H. Iwamoto, A. Tanaka, K. Nishikiori, H. Ikadai).

[289] Wyeth, John & Brothers, US 4 374 992 (R. Crossley). Sankyo, JP 3 014 566, 1989 (A. Yoshida, K. Oda, K. Tabada). Ohta, JP 60 218 375, 1984 (Y. Hasegawa et al.).

[290] Proter, EP 184 861, 1985 (L. Dall'Asta, G. Coppi, M. E. Scevola). Poli Ind. Chim., EP 515 995, 1992 (S. Poli, G. Coppi, G. Signorelli).
[291] M. R. H. Elmoghayar, *Arch. Pharm. (Weinheim, Ger.)* **316** (1983) 697.
[292] Taiho, JP 56 100 777, 1981.
[293] Nissan Chem. Ind., JP 61 260 018, 1985 (M. Mutsuto, K. Tanigawa, K. Shikada, R. Sakota).
[294] SmithKline Beckman, EP 168 950, 1986 (W. E. Bondinell, D. T. Hill, B. M. Weichman).
[295] Yamanouchi, EP 181 779, 1985 (M. Kiyoshi, M. Toshiyasu, H. Hiromu, T. Kenichi).
[296] T. Okada, B. K. Patterson, S. Q. Ye, M. E. Gurney, *Virology* **192** (1993) 631.
[297] Ciba-Geigy, CH 337 019, 1955. Geigy, DE-OS 1 186 070, 1963. H. Gysin, J. R. Geigy, *Chem. Ind. (London)* 1962, 1393. Geigy, US 3 207 756, 1963 (E. Knusli, W. Stammbach).
[298] W. Draber, K. Dichoré, K. H. Büchel, *Naturwissenschaften* **55** (1968) 446.
[299] Bayer, US 3 671 523, 1970 (K. Westphal, W. Meiser, L. Eue, H. Hack). Bayer, US 4 058 526, 1976 (W. Merz, G. Schümmer).
[300] Bayer, DE-OS 2 407 144, 1974 (R. R. Schmidt, L. Eue, C. Metzger, K. Dichoré).
[301] Sankyo, AU 515 639, 1981 (T. Konotsune, K. Kawakubo).
[302] Janssen Pharmaceutica, Ciba-Geigy, EP 273 531, 1986 (W. R. Lutz, G. R. E. Van Lommen, V. Sipido, W. G. Verschueren). Janssen Pharmaceutica, Ciba-Geigy, EP 289 066, 1987 (M. F. L. De Bruyn, G. R. E. Van Lommen, W. R. Lutz).
[303] Monsanto, JP 62 273 955, 1987 (M. G. Dolson, K. F. Lee, K. L. Spear).
[304] Sumitomo, JP 1 180 882, 1988 (H. Tomioka, H. Fujimoto, I. Fujimoto). Ube, EP 283 271, 1988 (H. Yoshioka et al.). Nissan Chem. Ind., JP 1 211 569, 1988 (Y. Nakajima, K. Hirata, M. Kudo). Nissan Chem. Ind., EP 183 212, 1985 (T. Ogura et al.). Nissan Chem. Ind., EP 199 281, 1986 (T. Ogura et al.). Nissan Chem. Ind., EP 232 825, 1987 (Y. Nakajima et al.). Hoechst, DE 3 733 220, 1987 (G. Salbeck et al.). BASF, DE 3 742 266, 1987 (J. Leyendecker et al.). Sumitomo, JP 1 197 479, 1988 (H. Tomioka, H. Fujimoto).
[305] Sankyo, EP 89 650, 1983 (H. Takeshiba, T. Kinoto, T. Jojima). Nissan Chem. Ind., WO 9208715, 1991 (Y. Nakajima, J. Watanabe, Y. Hirohara, T. Mita).
[306] Schering, DE 3 822 371, 1988 (D. Huebl, E. A. Pieroh, H. Joppien, D. Baumert).
[307] Bayer, DE 3 216 416, 1982 (U. Heinemann, P. E. Frohberger, W. Brandes).
[308] SmithKline Beckman, EP 104 836, 1983 (G. K. Menon, W. J. Sanders).
[309] Konica, Nissan Chem. Ind., JP 63 223 635, 1987 (H. Sakamoto, M. Fujiwara, I. Sakata, S. Yamamoto).
[310] Olin, US 4 252 962, 1980 (E. F. Rothgery).
[311] Chugai Shashin Yakuhin, JP 62 178 247, 1986 (M. Ito, N. Watanabe, H. Mukai).
[312] Fuji Photo Film, GB 2 083 243, 1980 (K. Shiba, S. Nakao, T. Toyama).
[313] Agfa-Gevaert, EP 40 771, 1980 (E. Ranz, H.-D. Schütz, J. W. Lohmann). Fuji Photo Film, JP 57 026 848, 1980 (I. Itoh, T. Toyoda, M. Yamada). Eastman Kodak, EP 54 414, 1980 (H. W. Altland, D. D. F. Shiao).
[314] Fuji Photo Film, JP 1 292 339, 1988 (M. Goto, K. Morimoto). Mitsubishi Paper Mills, JP 4 012 349, 1990 (T. Miura, H. Seiyama, Y. Tsubakii).
[315] F. Zucchi et al., *Werkst. Korros.* **44** (1993) 264.
[316] Technische Universität Dresden, DD 246 322, 1986 (G. Reinhard et al.).
[317] Ciba-Geigy, EP 330 613, 1989 (E. Phillips, R. C. Wasson).
[318] Rohm & Haas, US 5 210 094, 1991 (P. F. D. Reeve).
[319] IG Farben, DE 557 138, 1931 (M. Bögemann). DuPont, US 3 895 025, 1973 (A. L. Goodman). Bayer, DE 2 703 313, 1977 (R. Schubart, K. Wagner, K. F. Zenner). Bayer, DE 3 008 159, 1980 (R. Schubart, G. P. Langner).

[320] Sankyo, JP 53 027 641, 1976 (Y. Nakamura, K. Mori, A. Umehara). Degussa, DE 2 850 338, 1978 (K. Hentschel, F. Bittner). Bayer, Degussa, DE 2 109 244, 1971 (U. Eholzer, T. Kempermann). H. Ahne et al., *Kautsch. Gummi Kunstst.* **28** (1975) 135.

[321] T. Kempermann, *Bayer-Mitt. Gummi-Ind.* **50** (1978) 29; **51** (1979) 25.

[322] Toyo Linoleum, JP 56 026 943, 1979 (K. Miyahara, T. Matsukura, I. Niwa). Vanderbilt, US 4 704 426, 1986 (L. A. Doe, Jr.).

[323] Geigy, DE 1 470 832, 1961; *Chem. Abstr.* **62** (1965) 11834. Mitsubishi Chem. Ind., JP 56 129 249, 1980 (A. Sugio et al.).

[324] Bayer, DE 3 314 467, 1983 (M. Blunck, U. Claussen, F. W. Kroeck, R. Neeff).

[325] Haarman und Reimer, DE 3 831 980, 1988 (R. Emberger et al.).Haarman und Reimer, DE 4 016 536, 1990 (R. Emberger et al.).

[326] Givaudan-Roure, WO 93/07134, 1992 (U. Huber).

[327] R. G. Dickinson, N. W. Jacobson, *J. Chem. Soc., Chem. Commun.* 1970, 1719. H. Brandl, *Prax. Naturwiss. Chem.* **40** (1991) 25.

[328] U. Schmidt, G. Pfleiderer, F. Bartkowiak, *Anal. Biochem.* **138** (1984) 217. B. Narayana, M. R. Gajendragad, *Asian J. Chem.* **5** (1993) 121.

[329] Sansei Pharmaceut. Co., JP 1 228 908, 1988 (Y. Oyama).

[330] Procter & Gamble, EP 483 426, 1990 (R. Chatterjee).

[331] R. J. Lewis, Sr. (ed.): *Sax's Dangerous Properties of Industrial Materials*, 8th ed., vol. III, Van Nostrand Reinhold, New York 1992.

[332] E. E. Sandmeyer in G. D. Clayton, F. E. Clayton (eds.): *Patty's Industrial Hygiene and Toxicology*, 3rd ed., vol. **2A,** Wiley-Interscience, New York 1981, p. 2061.

[333] L. A. Damani (ed.): *Sulphur-Containing Drugs and Related Organic Compounds*, Ellis Horwood Limited, Chichester 1989.

[334] *Threshold Limit Values for Chemical Substances and Physical Agents in the Workroom Environment*, American Conference of Governmental Industrial Hygienists (ACGIH), 1978.

[335] E. J. Fairchild, H. F. Stokinger, *Am. Ind. Hyg. Assoc. J.* **19** (1958) 171.

[336] R. G. Ames, J. W. Stratton, *Arch. Environ. Health* **46** (1991) 213.

[337] P. Amrolia, S. G. Sullivan, A. Stern, R. Munday, *J. Appl. Toxicol.* **9** (1989) 113.

[338] H. T. H. Wilson, *Br. J. Dermatol.* **81** (1969) 175.A. A. Fischer, *J. Dermatol. Surg.* **1** (1975) 63.

[339] Industrieverband Pflanzenschutz e. V. (ed.): *Wirkstoffe in Pflanzenschutz- und Schädlingsbekämpfungsmitteln*, Pressehaus Bintz-Verlag, Offenbach 1982.

[340] E. Löser, G. Kimmerle, *Pflanzenschutz-Nachr. Bayer* **25** (1972) 186.

Thiophene

JOHN WESLEY PRATT, Synthetic Chemicals Limited, Wolverhampton, West Midlands, United Kingdom

1.	Introduction 4691	6.	Uses 4696
2.	Physical and Chemical Properties 4692	6.1.	Oxygenated Derivatives 4697
		6.2.	Bromothiophenes 4700
3.	Resources and Raw Materials . 4694	6.3.	Thiophene 4702
4.	Production 4694	6.4.	Polysubstituted Derivatives .. 4703
5.	Specifications, Transportation, Health and Safety 4695	7.	Toxicology 4705
		8.	References 4705

1. Introduction

The extensive literature on thiophenes is indicative of the commercial and research interest in these five-membered heterocyclics. Some hundreds of patents appear each year, many considering thiophene derivatives as alternative products to, e.g., a parent benzenoid molecule.

The thiophene-containing molecule must possess cost-effective advantages or uniqueness. As well as a means of producing an alternative "me too" effect chemical, the inherently greater chemical sensitivity of the thiophene ring compared to that of benzene may also improve biodegradability or metabolic characteristics.

The process to manufacture thiophene (and its homologs), is capable of unlimited expansion, with consequential lower unit costs, but it remains a fine-chemical business (ca. 1000 t/a).

This article concentrates on the products of current commerce. Chapter 6 is estimated to cover over 90% of the present volume; the remainder consists of 25 or more minor, but highly efficacious products, mainly pharmaceuticals, together with several which are still in the very early stages of development.

2. Physical and Chemical Properties

The physical characteristics of thiophene (thiofurfuran) are closer to benzene than any other five-membered heterocyclic. This enables it to replace a benzenoid moiety in the whole spectrum of aromatic chemicals [1] and, as illustrated in Chapter 6, sometimes to introduce worthwhile advantages over the analogous benzenoid compounds.

Some physical and physicochemical properties of thiophene are listed below:

Boiling point, 100 kPa	84.16 °C
Melting point	−38.30 °C
Density	
d_4^4	1.0873
d_4^{20}	1.06485
d_4^{25}	1.05887
Refractive index	
n_D^{20}	1.52890
n_D^{25}	1.52572
Viscosity	
at 20 °C	0.662 mPa · s
at 25 °C	0.621 mPa · s
Vapor pressure	
at 20 °C	8.4 kPa
at 25 °C	10.61 kPa
at 60 °C	45.32 kPa
at 110 °C	143.24 kPa
at 130 °C	270.07 kPa
Enthalpy of vaporization	32.483 kJ/mol
Critical temperature	317 °C
Critical pressure	4.53 MPa
Dipole moment	0.53–0.63 D
Dielectric constant	2.76
Flash point	−6.7 °C
Enthalpy of combustion	−2807 kJ/mol
Enthalpy of formation	82.13 kJ/mol

The numbering of the thiophene ring is orthodox:

Nomenclature can be summarized as

2-Thenyl   2-Thienyl   2-Theonyl

The usefulness of thiophene and its derivatives lies in its chemical sensitivity. It is a π-electron-rich aromatic (Fig. 1); the literature illustrates the ease of electrophilic substitution in the 2- and 5-positions (Fig. 2). As in other five-membered heterocyclics,

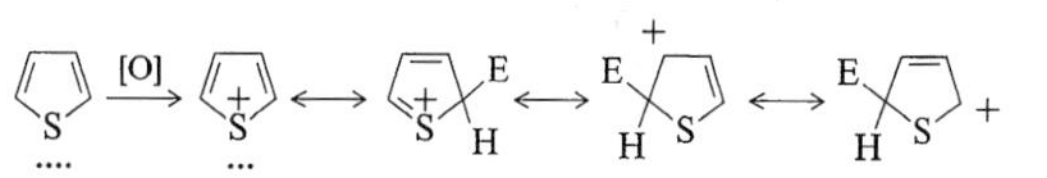

Figure 1. Delocalization of π-electrons in the thiophene ring

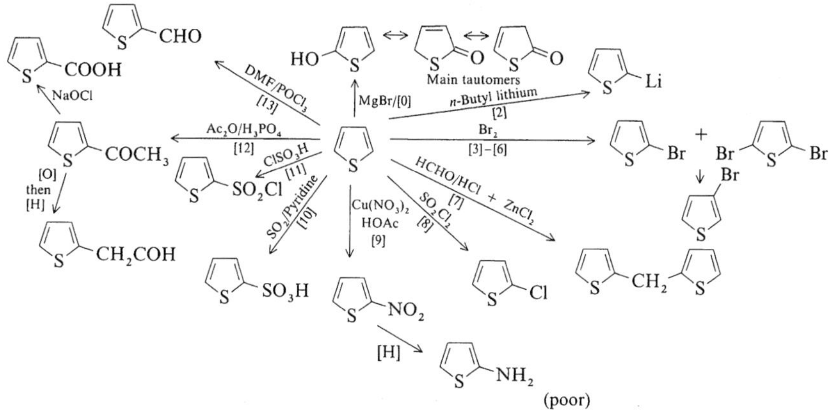

Figure 2. Electrophilic substitution reactions of thiophene

the S atom of the thiophene ring plays its part in the formation of cationic reaction intermediates. Thus, the reactivity of thiophene is more akin to that of phenol and aniline than to the comparatively unreactive benzene ring.

With the 2-position occupied by an *o/p* or *m*-directing group, a second group is substituted in the 5-position of the thiophene ring. With such a directing group in the 3-position, mixed isomers are formed.

As can be seen from Figure 2, many 2-substituted derivatives can be produced by orthodox and comparatively mild organic chemical reactions. However, decomposition sometimes interferes with yield expectations, e.g., reduction of an $-NO_2$ group to an $-NH_2$ group ruptures the thiophene ring under normal benzenoid conditions.

While 2-substitution is practically 100%, single-percentage amounts of the 3-isomer may be formed. Routes and reaction conditions to limit this effect need to be optimized.

The order of reactivity toward electrophilic substitution is:

Pyrrole >> thiophene > furan >> benzene   e.g., bromination of thiophene is 1.25×10^7 faster than that of benzene; chlorination 1.7×10^7 faster.

Irradiation causes thiophene to react with H_2O [14]:

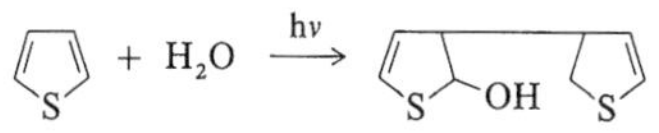

Heating thiophene does not yield C_2H_2, but bithenyl [14]:

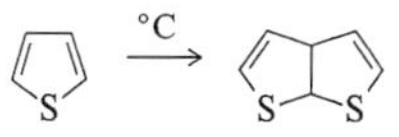

3. Resources and Raw Materials

Thiophene occurs in the light hydrocarbon fraction distilled from coal tar. The proportion of thiophene depends on the type of coal, the temperature of carbonization, and the type of retort. Typical quantities in a tar-derived benzol stream are 0.4 – 1.4 wt % [15].

Traditionally, thiophene is removed from coal tar benzol by washing with H_2SO_4, more rarely by solvent extraction, to produce a sweeter, marketable product.

The very large volumes of light hydrocarbons produced from coal carbonization plants represent a considerable quantity of associated thiophene, but the quality demands of the principal markets for thiophene require a purity which cannot be obtained economically from coal tar.

Commercialization of thiophene for the fine chemicals intermediate market requires a synthetic high-purity source. This was initiated in the 1960s in the United States by Pennsalt/Pennwalt, now Atochem. Midland-Yorkshire Tar Distillers/Synthetic Chemicals first produced synthetic thiophene on a commercial scale in Europe in 1970.

A key factor in the commercial production of pure thiophene and its homologs is the specificity and cleanliness of the processes used. Vapor phase (VP) cyclization methods are the processes of choice.

The original VP process, using furan and H_2S as the sulfur source [16], is no longer preferred. Synthetic Chemicals operates its own VP process [17], using CS_2 plus a selected C_4 skeleton. A similar VP process is patented by Elf Aquitaine [18].

Other non-VP processes may be described as laboratory methods and include: (1) syntheses from C_4 unsaturates, using NHS, Li_2S, SO_2, and S as sulfur source; (2) succinic salts and γ-diketones with, e.g., P_4S_7 or P_2S_5. Such processes result in very complex reaction products unsuitable for large-scale production.

4. Production

Since the current international market for thiophene is < 1000 t/a, the type of VP installation required for its manufacture is much smaller than a typical petrochemical unit. However, the process is capable of expansion to match any required demand.

Tight operating temperatures of 500 – 530 °C, catalyst performance, and regeneration plus overall control, particularly with respect to environmental requirements, are key factors in the successful operation of such a unit (Fig. 3).

Typical feedstock portfolio is shown in Figure 4.

The advent of a range of commercially available alkylthiophenes, including those substituted in the difficult 3-, (4-) position, resulted in an alternative raw material to synthesize a range of 3-, (4-) side-chain thiophenes, other than from, e.g., 3-bromothiophene [19].

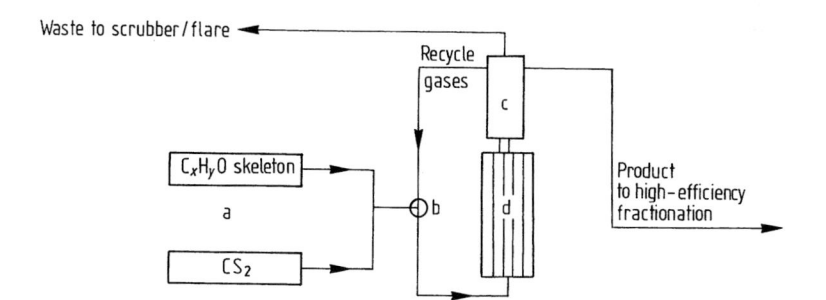

Figure 3. Thiophene production
a) Raw materials tank farm; b) Mixer/flow control; c) Condenser; d) Multitube, fixed-bed VP (vapor phase) reactor

Figure 4. Feedstock for thiophene and its derivatives

5. Specifications, Transportation, Health and Safety

Data for thiophene, together with its 2-methyl, 3-methyl, 2-acetyl, and 2-bromo derivatives are given in Table 1.

Table 1. Specifications, transportation, health and safety of thiophene and derivatives

	Thiophene	2-Methyl-thiophene	3-Methyl-thiophene	2-Acetyl-thiophene	2-Bromo-thiophene
Specification					
CAS No.	[110-02-1]	[554-14-3]	[616-44-4]	[88-15-3]	[1003-09-4]
Appearance	clear liquid	clear liquid	clear liquid	yellow liquid ($\geq 9°$)	clear/darkening
Odor	aromatic	aromatic	aromatic		mild pungent
Purity (GLC), wt%	99	98	98	99	98
Boiling point, °C	83–85 (95%)	113	115	213–214 (95%)	153–154
Flash point, °C	−7	15.5	15.5	96	
Miscibility with water	almost immiscible	almost immiscible	almost immiscible	slightly miscible	immiscible
Transportation					
Tarrif code	29340 100	293490 900	293490 900	293490 900	293490 900
Classification	class 3.2 highly flammable	class 3.2 highly flammable/irritant	class 3.2 highly flammable/irritant	class 6.1 toxic	class 6.1 toxic
Packaging group	II	II	II	II	II
IMDG page No.	3284	3230	3230	6231	6231
UN (identification)	2414	1993 NOS	1993 NOS	2810 NOS	2810 NOS
ADR	item. 3°(b)	item. 3°(b)	item. 3°(b)	item. 21°	item. 20°(b)
Health and Safety*					
R phrases	R11 R21/22/20 R38/45 R52/53	R11 R36/37/38	R11 R36/37/38	R23 R24	R23/24/25 R36/37/38
S phrases	S9/16/33/43/53/61	S9/16/33/43	S9/16/26/33/43	S18/24/30/41/44	S28/36/37/39/44

* Risk and Safety phrases as legislated by the European Commission and stated fully in United Kingdom. Statutory Instruments 1982 No. 1496, Health and Safety Executive, Sudbury, Suffolk, United Kingdom.

6. Uses

Successful commercial derivatization is governed by the physical and chemical characteristics of the basic thiophene nucleus, founded on an economic source of guaranteed quality.

By far the greatest volume of simple thiophenes are produced by means of VP cyclizations (see Chapter 4). Derivatives of these homologs account for practically all the growth of this business in the last three decades, in areas as diverse as pharmaceuticals, veterinary drugs, agrochemicals, to dyes, photography, and polymers.

6.1. Oxygenated Derivatives

The oxygenated derivatives of thiophene form the largest group of the intermediates market, 70–75% of the world total.

Thiophene 2-aldehyde (T2A), C_5H_4OS M_r 112.15, bp 198 °C, is prepared as shown in Figure 2 [13]. It was the first oxygenated derivative to reach a market of several hundred t/a, about two decades ago.

It is used to produce a range of antihistamines, including methapyrilene (**1**) [20], methaphenilene (**2**) [21], and thenalidine (**3**) [22]. However, this usage has practically disappeared.

The anthelmintic pyrantel (**4**) [23] is a major market for T2A, enhanced by new formulations and the development of the medicinal use beyond the original veterinary market.

Other pharmaceuticals containing T2A are azosemide (**5**) [24] a diuretic, and teniposide (**6**) [25] an antineoplastic.

Thiophene 2-carboxylic acid, T2CA, $C_5H_4O_2S$, M_r 128.15, bp 260 °C, mp –130 °C, is prepared via oxidation of 2-acetylthiophene (see Fig. 2).

The sodium salt of T2CA is used to treat influenza (trophires) [26]. Other salts, e.g., Li, have minor medicinal uses. A US Patent [27] describes the use of the Na salt as a lubricating oil thickener.

Suprofen (**7**) [28] a nonsteroidal anti-inflammatory (NSAI), and tienilic acid (ticrynafen) (**8**) [29], a potent diuretic, were both very large prospective markets for T2CA or its acid chloride, but were almost completely withdrawn in the 1980s. It is interesting to compare the success of Roussel Hoechst's NSAI tiaprofenic acid (**9**) in this market.

The 5-nitro derivative of T2CA is used to produce nifurzide, an intestinal antibacterial.

Pfizer's Tenidap anti-inflammatory (**10**) [30] is a developing drug which uses the acid chloride of T2CA as the starting intermediate.

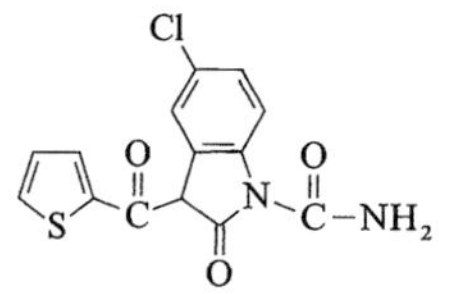

10

Thiophene 2-acetic acid, T2AA, $C_6H_6O_2S$, M_r 142.18. The use of 7-aminocephalosporanic acid (7ACA), following the classical work by the National Research and Development Council in the 1950s [31] to produce a family of semisynthetic cephalosporins, resulted in important first-generation antibiotics. Cephaloridine (**11**) [32] and cephalothin [33] were the major products licensed from this work. The second-generation cefoxitin (**12**) [34] came a few years later.

T2AA, or its acid chloride, is used to form the 7-side-chain of these antibiotics. Many processes have been designed to synthesize T2AA (Cl), including chlormethylation of thiophene until the late 1980s in the United States, and starting from T2A in the FAMSO process in Japan.

However, these processes are now superseded, and the reduction of 2-acetylthiophene followed by oxidation is the process of choice.

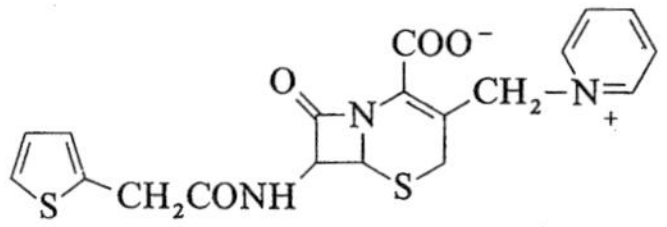

11

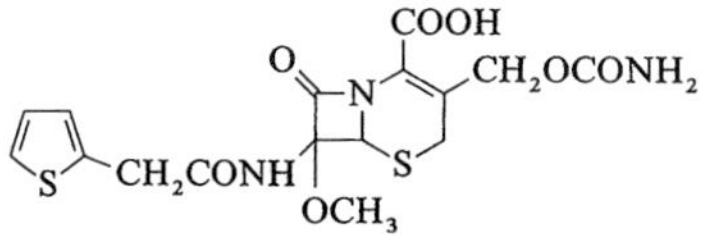

12

Ketotifen fumarate (**13**) [35] is an antiasthma drug which is thought to possess prophylactic properties, and this market is growing.

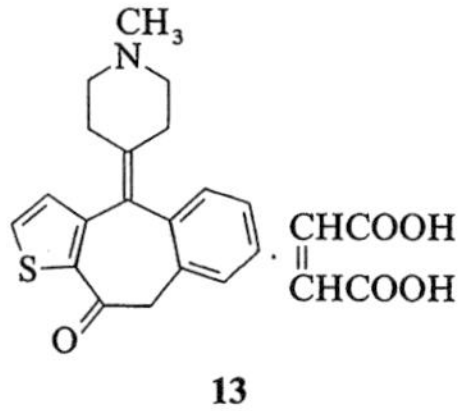

13

Pizotyline (**14**) [36] an antimigraine agent, is synthesized from T2AA (Cl).

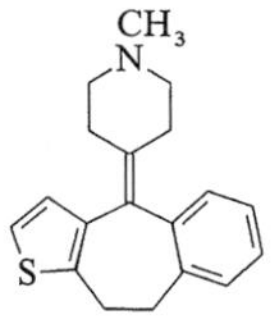

14

2-Acetylthiophene (2AT) C_6H_6OS, bp 214 °C, mp 9 °C, is prepared as shown in Figure 2. As described above, 2AT is used as the precursor to T2AA (Cl) and T2CA (Cl). It is important that the process used to synthesize 2AT yields the lowest amount of 3-acetylthiophene isomer.

Thiophene Dicarboxylic Acids. Adipic acid, or its dichlorinated derivative, provides a means of forming thiophene 2,5-dicarboxylic acid [37]. Thiophene aldehydes and other substituents exhibit pronounced fluorescence [38]:

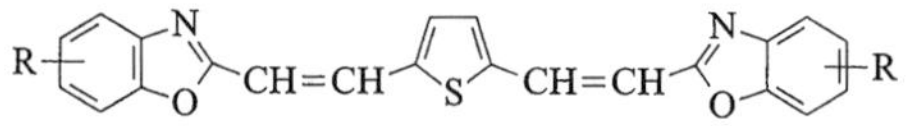

More complicated thiophene polybenzoxazoles are used as optical brighteners, and may be synthesized via the 2,5 carboxylic acid groups.

6.2. Bromothiophenes

The classical use of bromine as a leaving group has played an important part in the laboratory and commercial synthesis of many thiophene derivatives and products.

2-Bromothiophene (2BT), C_4H_3Br, M_r 78.17, bp 153.5 °C, is easily produced to a good purity (see Fig. 2) [3].
Tipepidine (**15**) [39], an antitussive, is produced from 2BT.

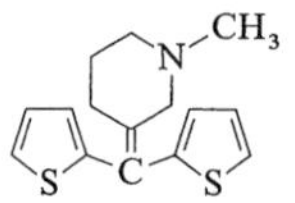

15

3-Bromothiophene (3BT), C_4H_3Br, bp 160 °C. In order to obtain 3- (4-) substituted thiophenes, more elaborate chemistry is required than straightforward electrophilic substitution. Successive bromination to 2,3,5-tribromothiophene can be followed by removal of the 2- and 5-Br atoms (see Fig. 2), but, a newer, simpler process is now patented [6] which is essentially free of 2-Br isomer.
3-Bromothiophene may be used to produce thiophene 3-malonic acid (TMA), the side-chain intermediate for the production of ticarcillin and temocillin from 6-aminopenicillinic acid (6APA). Timentin is a powerful antibiotic, produced from ticarcillin

plus clavulanic acid [40]. An alternative route to TMA is via 3-methylthiophene (see below).

The growing interest in conducting polymers based on polythiophenes is extensively reported [41], [42]. In order to produce a workable polythiophene, a fairly long side-chain may be introduced at the 3-position of the thiophene ring. Choice of the structure of this side-chain is a compromise between the workability and the maintenance of required conductivity. It generally lies between C_4 and C_{12} *n*-alkyl. Polymerization is via the 2- and 5-positions.

Removal of the 3-Br atom and replacement by an alkyl group is a means of obtaining the desired monomer.

$R = C_5 - C_{10}$ alkyl
Conducting polythiophene

Tetrahydrothiophene (THT), C_4H_8S, M_r 88.18, *bp* 121.1 °C, *mp* −96 °C, is easily confused in import/export statistics with thiophene itself, particularly since it may also be called thiophane; THT was certainly produced in commercial quantities before thiophene, and it remains a commercial product.

Manufacture is not from thiophene, but by VP replacement of the oxygen atom in tetrahydrofuran by sulfur from H_2S. The plant and catalyst is very similar to a thiophene unit. However, THT is used for its very strong odor, either on its own or, more commonly, within a cocktail, e.g. alkylsulfides, as a powerful gas odorant. The saturation of comparatively sweet-smelling thiophene thus produces vastly different odor characteristics.

2-Methylthiophene (2MT), C_5H_6S, *bp* 112.5 °C, *mp* 63.4 °C, is commercially produced to 98 wt% purity by VP cyclization (see Fig. 4).

It is a feasible raw material for the production of several of the important 2-substituted thiophenes. It may be used to manufacture 2-chloromethylthiophene (2-thenyl chloride), but this and further syntheses must be completed in situ, because of its instability.

3-Methylthiophene (3MT), *bp* 115 °C, is commercially produced by an analogous process to 2MT, and to the same purity.

Thenyldiamine (**16**) [43], an antihistamine, has been produced since the late 1940s. The synthesis requires a source of 3MT to prepare the 3-thenylbromide precursor.

16

Morantel (**17**) [44] is a potent antihelmintic, differing from pyrantel, only in the additional 3-methyl group on the thiophene ring. It is an important market for 3MT.

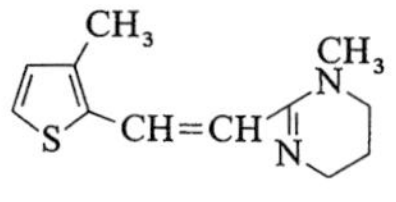

17

As stated above, the semisynthetic penicillins ticarcillin (**18**) [42] and timentin possess a 3-thienyl side-chain; 3MT, via free radical bromination of the methyl side-chain, is an alternative precursor to TMA.

The antifungal tioconazole (**19**) [45] is synthesized from 2-chloro-3-methylthiophene.

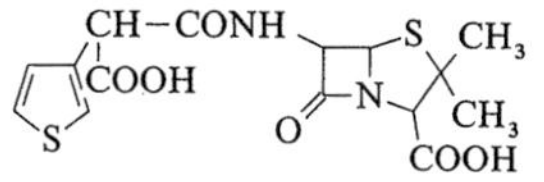

18

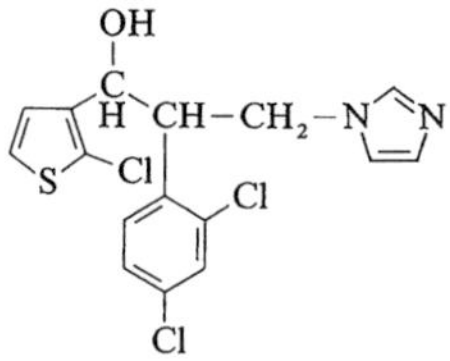

19

Cetiedil (**20**) [46] is a peripheral vasodilator which incorporates a thiophene ring in the 3-position; 3MT provides a means of synthesis.

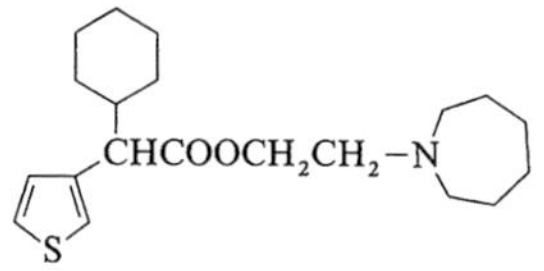

20

6.3. Thiophene

The use of thiophene mixed with other constituents to produce a denaturant has recently been developed [47]. Since thiophene is unknown in fermented alcohol, the addition of a marker containing thiophene facilitates the identification of alcohol denatured in this way. The thiophene is very difficult to remove.

Uses of thiophene itself have otherwise been restricted to analytical techniques, where a standard mass of sulfur is required for measurement control. Thiophene may also be used for quantitative deactivation of cracker catalysts.

6.4. Polysubstituted Derivatives

The commercial availability of a range of thiophenes substituted with an amino group, together with other ring substituents, is based on a modification of the Gewald reaction [48], for 2-NH_2 derivatives. Substitution of an amino group in the 3-position requires custom-designed chemistry [49], [50].

Methyl-3-amino-4-methylthiophene 2-carboxylate $C_7H_{10}NO_2S$. The 3-amino group is used to attach a side-chain to the thiophene ring for manufacture of the local anesthetic carticaine (**21**) [51]. This pharmaceutical is also produced with a 5-chloro group on the thiophene ring.

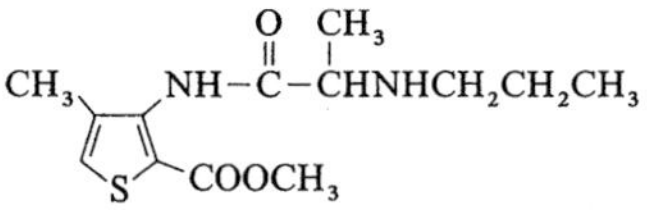

21

Methyl-3-aminothiophene 2-carboxylate $C_6H_7NO_2S$. This derivative is used to produce a sulfonylurea post-emergent herbicide, Harmony (M6316 DuPont) (**22**), which has extremely rapid biodegradeability [52]. This characteristic, together with the very low dose requirements, are important environmental safeguards.

The NSAI tenoxicam (**23**) [53] is one of a range of oxicams which happens to have a thiophene moiety and may be synthesized from this aminothiophene carboxylate.

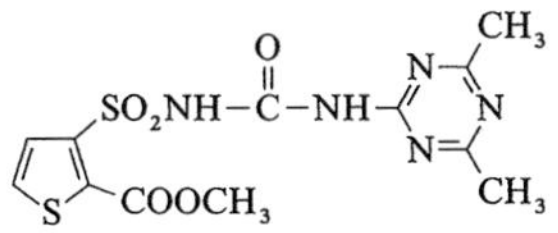

22

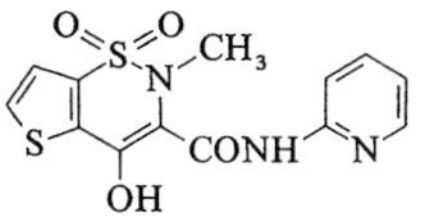

23

The amino group, when in the 2-position of a substituted thiophene derivative, may be diazotized to a range of monoazo disperse dyes [54]. Coupling is with a pyrazolone or, e.g., phenol, and colors toward the yellow/green part of the spectrum are produced.

Substituted thiophene molecules are found in various components of animal and vegetable flavorings. The combination of H_2S with unsaturated aldehydes may be the source of such thiophene derivatives in, e.g., cooked beef, pork liver, onion, and coffee [55], [56].

As well as the fused ring structures illustrated above, which are valuable commercial products, others are worth mentioning, because of their potential markets.

Table 2. Toxicological data for thiophene and derivatives

	Thiophene	2-Methyl-thiophene	3-Methyl-thiophene	2-Acetyl-thiophene	2-Bromothiophene
Acute Toxicity					
Oral LD_{50}	1400 mg/kg (rat)	3200 mg/kg (mouse)	2300 mg/kg (rat)	50 mg/kg (mouse)	200–250 mg/kg (rat)
Dermal LD_{50}	830 mg/kg (rabbit)	no data	2000 mg/kg (guinea pig)	370 mg/kg (rat)	134 mg/kg (rat)
Inhalation LC_{50}	9500 mg/kg (mouse)	no data	10 g/m^3, 2 h (rat)	1460 mg/m^3, 1 h (rat)	1.04 mg/L, 4 h (rat)
Irritation					
Skin	yes	expected	yes	not reported	yes
Eye	yes	expected	yes	predicted	yes
Skin sensitization	no data	no data	not found in tests	no data	no data
Occupational exposure limits	none specified	none specified	none specified	2 ppm (provisional)	1 ppm (provisional)
Ecotoxicity	possibly only slight harm to aquatic organisms	possibly only slight harm to aquatic organisms	possible only slight harm to aquatic organisms	harmful to aquatic organisms	expected only slight harm to aquatic organisms
Persistence	expect ultimate biodegradability	expect ultimate biodegradability	expect ultimate biodegradability	expect ultimate biodegradability	expect very slow biodegradability
Stability					
Materials to avoid	strong oxidizers	strong oxidizers	strong oxidizers	strong oxidizers	strong oxidizers
Conditions	stable under normal conditions	stable under normal conditions	stable under normal conditions	stable under normal conditions	slow decomposition unless cool/dark
Combustion products	toxic oxides of S+CO, CO_2	toxic oxides of S+CO, CO_2	toxic oxides of S+CO, CO_2	toxic oxides of S+CO, CO_2	HBr during storage (prolonged), CO, CO_2

Ticlopidine (**24**) [57] is a potent pharmaceutical, used to prevent blood platelet aggregation in the treatment of strokes. One of three possible isomeric structures of the thiophene analog of saccharin, is 10^3 times sweeter than sucrose, and possesses all the characteristics of saccharin itself.

24

Many methods are available to synthesize fused thieno rings to a nitrogen heterocyclic [58], [59].

7. Toxicology

Toxicological data for thiophene and its derivatives are given in Table 2.

8. References

[1] L. S. Fuller, J. W. Pratt, F. S. Yates, *Manuf. Chem. Aerosol News* (1978) no. 5; *Inf. Chim.* (1979) no. 10, 19.
[2] A. Gilman, D. A. Shirley, *J. Am. Chem. Soc.* **71** (1949) 1870.
[3] I. Hirao, *Yakugaku Zasshi* **73** (1953) 1023; *Chem. Abstr.* **48** (1954) 10723.
[4] C. Troyanowsky, *Bull. Soc. Chim. Fr.* 1955, 424.
[5] S. Gronowitz, *Ark. Kemi* **7** (1954) 267; *Acta Chem. Scand.* **13** (1959) 1054.
[6] Shell Int. Res., EP 029 586, 1989 (P. Grosvenor, L. Fuller).
[7] Y. Golfarb, Y. L. Danyusherskii, *Bull. Acad. Sci. USSR Div. Chem. Sci. (Engl. Transl.)* 1956, 1385.
[8] E. Campaigne, W. M. Le Suer, *J. Am. Chem. Soc.* **70** (1948) 1555.
[9] N. I. Putskin, A. N. Sopokin, *Chem. Abstr.* **51** (1957) 16419.
[10] A. P. Terentev, G. M. Kaltski, *Chem. Abstr.* **46** (1952) 11178.
[11] A. Buzas, J. Teste, *Bull. Soc. Chim. Fr.* 1960, 793.
[12] H. Hartough, I. Kosak, *J. Am. Chem. Soc.* **69** (1947) 3093.
[13] E. Campaigne, W. L. Archer, *J. Am. Chem. Soc.* **75** (1953) 989.
[14] Private Communication.
[15] *Ullmann*, 4th ed., **22,** 416.
[16] Pennsalt Chem. Corp., GB 963 385, 1964 (T. E. Deger, B. Bucholtz, R. H. Goshorn).
[17] Midland York Tar Dist., GB 1 345 203, 1974 (N. R. Clark, W. E. Webster).
[18] Soc. Nat. Elf Aquitaine, GB 1 585 647, 1981.
[19] J. A. Clarke, O. Meth-Cohn, *Tetrahedron Lett.* **52** (1975) 4705–4708.
[20] Monsanto, US 2 581 868, 1952.
[21] Lenard, Solmssen, *J. Am. Chem. Soc.* **70** (1948) 2066.
[22] Sandoz, US 2 717 251, 1955; US 2 757 175, 1956.
[23] Pfizer, US 3 502 661, 1968/70.
[24] Boehringer Mannheim, DE 1 815 922.
[25] Sandoz, US 3 524 824, 1970.
[26] Syde, Malleray, *Bull. Soc. Chim. Fr.* 1963, 1276.
[27] Standard Oil, US 2 576 031, 1951.
[28] Janssen, GB 1 446 239, 1975.
[29] CERPHA, DE 2 048 372, 1971.
[30] Pfizer, US 4 556 672, 1971.
[31] Nat. Res. Dev. Corp., GB 810 196, 1959; US 3 082 155, 1963.
[32] Glaxo, FR 1 384 197, 1965.
[33] Lilly, BE 618 663, 1962.
[34] Merck, DE 2 129 675, 191; DE 2 203 653, 1971.
[35] Sandoz, DE 2 111 071, 1971.
[36] Sandoz, US 3 272 826, 1966; BE 636 717, 1964.
[37] *Tetrahedron Lett.* **42** (1970) 3719–3722.

[38] R. E. Atkinson, F. E. Hardy, *J. Chem. Soc. B* 1971, 357–361.
[39] Tanabe, JP 17 988, 1962.
[40] Beecham, US 3 282 926, 1966.
[41] *Eur. Chem. News,* April 9, 1991. R. D. McCullough, R. D. Lowe, *J. Chem. Soc., Chem. Commun.* 1992, no. 1, 70–72.
[42] M. L. Blohm, P. C. Van Dort, J. E. Pickett, *CHEMTEC* 1992, no. 2, 105–107. K.-H. Kocher, *Hoechst High Chem. Mag.* 1992, no. 12, 39–42.
[43] E. Campaigne, W. M. Le Suer, *J. Am. Chem. Soc.* **71** (1949) 333.
[44] Pfizer, GB 1 120 587, 1968.
[45] Pfizer, US 4 062 966, 1977 (Gymer).
[46] Innothera, FR 1 460 571, 1966 (Pons, Robba).
[47] ACNA, EP-Appl. 424 886, 1991.
[48] K. Gewald, *Chem. Ber.* **98** (1965) 3571. K. Gewald, *Chem. Ber.* **99** (1966) 94.
[49] H. Fiesselmann, *Chem. Abstr.* **55** (1961) 6497c/17651d.
[50] Shell Int. Res., EP 298 542, 1989.
[51] Hoechst, US 3 855 243, 1974.
[52] Meeting of Weed Soc. of America, Seattle **25** (1980) 10.
[53] Hoffmann-LaRoche, DE 2 537 070, 1976.
[54] ICI, GB 1 394 365; GB 1 394 367; GB 1 394 368.
[55] M. Shankaranarayana, *CRC Crit. Rev. Food Technol.* **4** (1974) 395.
[56] R. Wilson, *J. Agric. Food. Chem.* **21** (1973) 873.
[57] Parcor (Sanofi), US 4 127 580, 1978 (E. Braye).
[58] Schneller, *Int. J. Sulfur Chem.* (1974) 309–316. J. M. Barker, *Adv. Heterocycl. Chem.* **21** (1977) 65.
[59] Croda Synthetic Chem., EP 0 003 645, 1982 (O. Meth-Cohn).

Thiourea and Thiourea Derivatives

BERND MERTSCHENK, SKW Trostberg AG, Trostberg, Federal Republic of Germany (Chap. 2)

FERDINAND BECK, SKW Trostberg AG, Trostberg, Federal Republic of Germany (Chap. 2)

WOLFGANG BAUER, Cassella AG, Frankfurt/Main, Federal Republic of Germany (Chap. 3)

1.	Introduction............ 4707	2.9.	Legal Aspects........... 4712
2.	Thiourea............. 4708	2.10.	Economic Aspects....... 4713
2.1.	Physical Properties...... 4708	2.11.	Toxicology........... 4713
2.2.	Chemical Properties...... 4708	3.	Thiourea Derivatives...... 4716
2.3.	Production............ 4710	3.1.	Physical Properties...... 4716
2.4.	Environmental Protection.. 4711	3.2.	Chemical Properties...... 4716
2.5.	Quality Requirements..... 4711	3.3.	Production............ 4716
2.6.	Chemical Analysis....... 4711	3.4.	Uses................ 4718
2.7.	Storage and Transportation. 4712	3.5.	Toxicology........... 4719
2.8.	Uses................ 4712	4.	References............ 4723

1. Introduction

Thiourea is a chemically interesting compound, which has three functional groups: amino, imino, and thiol. This results from tautomerism between thiourea and isothiourea:

$$\underset{\text{Thiourea}}{\overset{H_2N}{\underset{H_2N}{>}}C=S} \; \rightleftharpoons \; \underset{\text{Isothiourea}}{\overset{H_2N}{\underset{HN}{>}}C-SH}$$

Because of this polyfunctionality and also because of its complex-forming properties, thiourea has been widely used for more than 30 years [1]–[5]. Its most important uses are as a starting material for nitrogen- and sulfur-containing heterocycles and formamidinesulfinic acid (thiourea dioxide), as a reaction partner for aldehydes, and as a component of addition compounds and complexes.

Thiourea derivatives are thermally stable, mostly crystalline compounds. They can be produced in good yields from readily available starting materials by simple syntheses. This class of compounds, described in detail in several review articles [6]–[11], has gained importance in the pharmaceutical sector, in plant protection, in various technical applications, and in the synthesis of heterocycles.

Table 1. Solubility of thiourea

In H$_2$O						
Temperature, °C	0	20	40	60	80	100
Solubility, g/100 g	4.9	13.7	30.7	70.9	138	233
In MeOH						
Temperature, °C			25	40.7	53.7	61.9
Solubility, g/100 g			11.9	16.4	22	24.6
In EtOH						
Temperature, °C		20	31.9	45	58	64.7
Solubility, g/100 g		3.6	4.7	6.3	8.5	9.8

2. Thiourea

2.1. Physical Properties

Thiourea [62-56-6], 2-thiourea, CH$_4$N$_2$S, M_r 76.12, d_4^{20} 1.405, is a white, odorless solid, and crystallizes in a rhombic bipyramidal structure with four molecules per unit cell. It is the only known molecular crystal which is ferroelectric within a certain temperature range. Space group $D_{2h}^{16} - P_{nma}$.

Thiourea is soluble in protic and aprotic polar solvents (Table 1) and insoluble in nonpolar ones.

Thiourea has no sharp melting point, as it rearranges relatively rapidly to ammonium rhodanide (NH$_4$SCN) from 135 °C. On rapid heating, a melting point of 180 °C has been determined. The heat of combustion at constant volume is 1482.2 kJ/mol (19.47 kJ/g), and at constant pressure 1485.6 kJ/mol (19.52 kJ/g).

2.2. Chemical Properties

Although it is known in only one form, thiourea can also react as the tautomeric iminothiol (see Section 3.2). Salts such as (NH$_2$)$_2$CS · HNO$_3$ or (NH$_2$)$_2$CS · H$_2$SO$_4$ are formed with strong acids [3]. Addition compounds and complexes can be synthesized with metal oxides or inorganic salts in acid solution, e.g., [Cr{(NH$_2$)$_2$CS}$_3$]Cl$_3$.

In alkaline solution, a large number of heavy metal ions can be precipitated quantitatively as sulfides with thiourea. These precipitates are readily filtered, so that the method can be used for analytical determination of metal salts [12]. The production of light-sensitive PbS layers [13] and the purification of mercury-containing wastewater [14] are based on this reaction.

Thiourea forms clathrates (inclusion compounds) with branched aliphatic and cyclic hydrocarbons [15]. This property can be used, e.g., for separating branched and straight-chain aliphatic hydrocarbons [16].

Thiourea can be alkylated and acylated, and reacts with aldehydes and ketones. With formaldehyde, for example, it forms N-(hydroxymethyl)thiourea ($H_2N–CS–NH–CH_2–OH$), which condenses to resin-like products on heating [17]. By condensation of thiourea with chloroketones, diketones, hydroxyketones, oxocarboxylic acids, and similar ketone derivatives, a large number of heterocycles can be produced. From 2-haloketones or 2-haloaldehydes (in the enol form) 2-aminothiazole derivatives are formed:

$$\begin{matrix} R-C-OH \\ \| \\ R_1-C-Cl \end{matrix} + \begin{matrix} HN \\ \diagdown \\ C-NH_2 \\ \diagup \\ HS \end{matrix} \longrightarrow \begin{matrix} R-C-N \\ \| \quad \diagdown \\ R_1-C \quad C-NH_2 \\ \diagdown \diagup \\ S \end{matrix}$$

1,1,-Dihaloalkanes condense with thiourea to 1,3,5-dithiazines:

$$R-CH\begin{matrix}Cl\\ \\Cl\end{matrix} + 2HS-C\begin{matrix}NH_2\\ \\ \| \\NH\end{matrix} \longrightarrow \begin{matrix}H\\ \diagdown \\R\end{matrix}C\begin{matrix}S-C\diagup^{NH}_{\diagdown NH}\\ \\S-C\diagdown_{NH}\end{matrix}$$

2-Hydroxyketones give 4-imidazoline-2-thiones:

$$\begin{matrix}R-CH-OH\\ \| \\R_1-C=O\end{matrix} + \begin{matrix}H_2N\\ \diagdown \\ C=S\\ \diagup \\ H_2N\end{matrix} \longrightarrow \begin{matrix}R\diagdown C-N\diagup^H\\ \| \quad \diagdown \\ C \quad C=S\\ \diagup \diagdown \diagup \\ R_1 \quad N\\ \quad |\\ \quad H\end{matrix}$$

With 2-oxocarboxylic acids (e.g., pyruvic acid) 2,3-dihydro-2-thioxo-4H-imidazole-4-one derivatives are formed:

$$\begin{matrix}R-C=O\\ | \\O=C-OH\end{matrix} + \begin{matrix}H_2N\\ \diagdown \\ C=S\\ \diagup \\ H_2N\end{matrix} \longrightarrow \begin{matrix}R\diagdown C=N\\ \| \quad \diagdown \\ C \quad C=S\\ \diagup \diagdown \diagup\\ O \quad N\\ \quad |\\ \quad H\end{matrix}$$

Malonic esters react with thiourea to thiobarbituric acid:

$$\begin{matrix}O\\ \|\\ H\diagdown C-OR\\ C\\ \diagup \diagdown \\ H \quad C-OR\\ \|\\ O\end{matrix} + \begin{matrix}H_2N\\ \diagdown \\ C=S\\ \diagup \\ H_2N\end{matrix} \longrightarrow \begin{matrix}O \quad H\\ \|\quad |\\ H\diagdown C-N\\ C \quad\diagdown\\ \diagup\diagdown \quad C=S\\ H\quad C-N\\ \|\quad |\\ O\quad H\end{matrix}$$

2-Oxocarboxylic acid esters condense with thiourea to thiouracils, and 1,3-diketones to mercaptopyrimidines:

4709

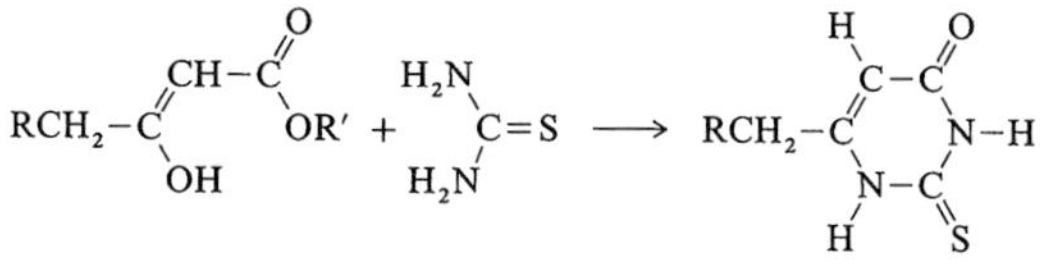

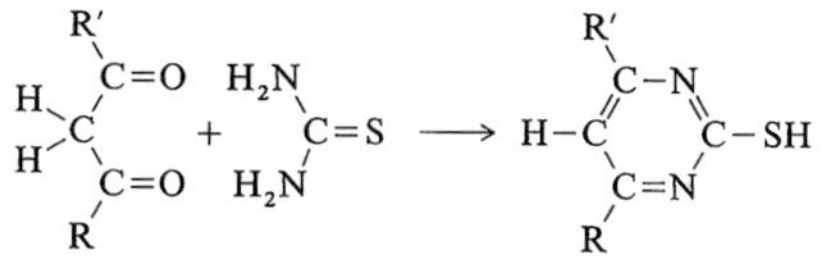

2.3. Production

Thiourea was first prepared by REYNOLDS by thermal rearrangement of ammonium rhodanide at ca. 150 °C:

$$NH_4SCN \rightleftharpoons (NH_2)_2CS$$

As this is an equilibrium reaction and separation of the two substances is quite difficult, better methods of preparation have been investigated.

The reaction between carbon disulfide and ammonia or ammonium carbonate under pressure at ca. 140 °C has not achieved industrial importance [18].

Production of thiourea is now carried out by treating technical-grade calcium cyanamide ($CaCN_2$) with hydrogen sulfide or one of its precursors, e.g., ammonium sulfide or calcium hydrogen sulfide [19]. The calcium cyanamide must contain at least 23 % nitrogen and must be free from calcium carbide (CaC_2), as otherwise explosive acetylene is liberated by water or hydrogen sulfide.

In Germany, thiourea is produced in a closed system by reaction of calcium cyanamide with hydrogen sulfide:

$$CaCN_2 + 3H_2S \longrightarrow Ca(SH)_2 + (NH_2)_2CS$$

$$2CaCN_2 + Ca(SH)_2 + 6H_2O \longrightarrow 2(NH_2)_2CS + 3Ca(OH)_2$$

$$Ca(OH)_2 + CO_2 \longrightarrow CaCO_3 + H_2O$$

A reaction vessel is charged continuously with an aqueous suspension of calcium cyanamide and a mixture of hydrogen sulfide and carbon dioxide, such that the above reactions go to completion. The converted reaction mixture is removed continuously from the reactor and excess H_2S is removed by stripping. Waste-gas from the reactor contains hydrogen sulfide, which is purified in a scrubber connected to the reactor.

The stripped thiourea solution is filtered to remove the precipitated calcium carbonate formed as one of the reaction products. The filtrate is evaporated in vacuum and thiourea is crystallized by cooling. It is filtered, washed, and dried in an air stream.

2.4. Environmental Protection

In Germany, the scrubbed waste gases resulting from the production and filtration of thiourea are treated in an incinerator. All installations where thiourea dust is handled are equipped with air extraction. The extracted air is collected and, together with the waste air from the product drier, is purified in a scrubber.

Mother liquors containing thiourea are removed batchwise from the production process and are used in high-temperature incineration processes to remove NO_x. Side products of low solubility separated from the mother liquor are incinerated externally in a licensed plant.

The precipitated calcium carbonate arising in the production of thiourea is reprocessed (e.g., in cement factories or brick works). This is a contribution to the conservation of natural lime resources.

2.5. Quality Requirements

The German producer gives the following specification for the standard product:

Thiourea	$\geq 99.0\%$
Water	$\leq 0.30\%$
Ash	$\leq 0.10\%$
Rhodanide (SCN⁻)	$\leq 0.15\%$
Other N-containing compounds	$\leq 0.5\%$
Iron	≤ 10 mg/kg

Special qualities with higher purity are also produced.

2.6. Chemical Analysis

Tests of the content (purity) of thiourea can be carried out titrimetrically with Ce(IV) and potassium iodate solutions [20]. Since the commercial product is of relatively high quality, it is more efficient to determine the impurities as a means of quality control. The main impurity, dicyandiamide, is best determined by TLC. The incineration residue (ash) and the water content are also measured. A quantitative determination of sulfate, heavy metals, and rhodanide can also be carried out.

Traces of thiourea (e.g., in water) can be determined by HPLC (C_{18} reverse-phase with water as the mobile phase and subsequent UV detection at 245 nm) [21].

Reaction of thiourea with sodium nitrite in acetic acid solution to give thiocyanate, which gives a red coloration with Fe^{3+} ions, is also suitable for the determination of trace amounts. The red coloration is used to determine thiourea photometrically at 450 nm [22].

2.7. Storage and Transportation

Thiourea is packed in polyethylene bags (≤ 50 kg) or in folding containers (≤ 1500 kg). Bags and containers must be kept tightly closed, and in a cool, dry place. The contents must be protected from the action of light. Thiourea is not regarded as dangerous goods in transport regulations (RID/ADR, etc.).

2.8. Uses

Thiourea has a wide range of uses, e.g., for producing and modifying textile and dyeing auxiliaries [23], [24], in the production and modification of synthetic resins [25], in repro technology [26], in the production of pharmaceuticals (sulfathiazoles, tetramisole [27], and cephalosporins [28], in the production of industrial cleaning agents (e.g., for photographic tanks [29] and metal surfaces in general [30], [31]), for engraving metal surfaces [32], as an isomerization catalyst in the conversion of maleic to fumaric acid [33], in copper refining electrolysis [34], in electroplating (e.g., of copper) [35], and as an antioxidant (e.g., in biochemistry) [36].

Other uses are as an additive for slurry explosives [37], as a viscosity stabilizer for polymer solutions (e.g., in drilling muds [38]), and as a mobility buffer in petroleum extraction [39]. The removal of mercury from wastewater from chlorine–alkali electrolysis [14] and gold and silver extraction from minerals [40], [41] are also of economic importance.

2.9. Legal Aspects

Thiourea is included in international inventories of existing chemicals, e.g., EINECS (European Union), TSCA Inventory (United States), DSL (Canada), ACOIN and AICS (Australia), and MITI (Japan). In Germany and in the European Union, it is classified as a suspected carcinogen (MAK III B [42], [43]; EU cancer category 3). Similar evaluations have been made by the International Agency for Research on Cancer (Lyon, France) [44], and within the framework of the National Toxicology Program in the

United States [45]. Within the European Union, thiourea must be labeled as a hazardous substance with hazard symbol Xn (St Andrew's cross – harmful), and R 22 (harmful if swallowed), R 40 (possible risk of irreversible effects), S 22 (do not breathe dust), S 24 (avoid contact with skin), and S 36/37 (wear suitable protective clothing and gloves). In the European Union Council Regulation on Existing Chemicals, thiourea is named in Annex I [46]. In Germany, the substance is classified in water hazard class 2 (hazardous for waters).

2.10. Economic Aspects

The quantity of thiourea produced and sold worldwide is ca. 10 000 t/a. The only producer in Western Europe is SKW Trostberg, Germany. Thiourea is also produced by Sakai Chemicals in Japan and at least seven Chinese companies.

According to estimations of SKW Trostberg, the largest producer, the quantity of thiourea produced worldwide is divided into the following main areas of use:

Intermediate in the production of thiourea dioxide	ca. 30 %
Ore leaching	25 %
Diazo papers	ca. 15 %
Catalyst in fumaric acid synthesis	ca. 10 %

The remainder is divided between smaller areas of use (Section 2.8).

The United States EPA recently carried out a survey of the uses of thiourea in the United States. It is used, for example, in animal glue liquefiers, in silver and metal cleaners, in the production of flame-retardant resins, and as a vulcanization accelerator [45].

2.11. Toxicology [42]–[44], [47]

Acute Toxicity and Local Tolerance. The oral LD_{50} for mice has been reported to be 1000 mg/kg [48], for rabbits 10 000 mg/kg [49], and for rats 1750 mg/kg [50] and 6200 – 6300 mg/kg [51]. After intraperitoneal administration LD_{50} values varying between 4 and 1340 mg/kg were found for rats [52], [53]. Lung edema and pleural effusion have been described as symptoms. The dermal LD_{50} is > 2800 mg/kg for rabbits [50] and > 6810 mg/kg for rats [51]. After inhalation of dust for 4 h (particle size 0.8 – 4.7 µm) LC_{50} for rats was > 170 mg/m^3 [50], after aerosol inhalation (10 % solution) > 195 mg/m^3 [50]. An alcoholic solution (droplet size < 10 µm) gave LC_{50} > 900 mg/m^3 [51].

In skin tolerance tests (normal and abraded rabbit skin, 24 h application), slight to distinct erythema and slight edema were observed [50]. A 4 h application of 0.5 g gave no reaction [51]. Application of a 10 % solution into the conjunctival sac of a rabbit's eye gave no reaction [50], but 100 mg of the solid substance gave a slight reddening and swelling [51]. Thiourea does not have a sensitizing effect in the guinea pig [51].

Subacute and Chronic Toxicity. In long-term application of thiourea, the effect on the thyroid gland stands out [54]. The substance disturbs incorporation of iodine in the synthesis of thyroxine and thus leads to a release of thyrotropic hormone from the pituitary gland, followed by severe stimulation of the thyroid gland, resulting in hyperplasia, adenomas, and, under certain circumstances, formation of thyroid carcinomas. Administration of thyroxine extensively prevents these effects.

Nonspecific effects of thiourea include hemosiderosis in the spleen, lymph nodes, and intestinal villi [55], weight loss of the liver, kidneys and spleen [54], anemia, leukopenia, increased erythropoiesis, and reduced bone growth.

Some examples of long-term tests are given below: rats received 0.625 % thiourea in the diet for 14 days. This resulted in 52 % growth inhibition [56]; 2.5 g per kilogram food (over 13 weeks) did not affect the body weight of mice [57]. Administration of 2.5 ppm in the drinking water of rats over 13 weeks gave no reaction [58]; after 27.5 mg kg^{-1} d^{-1} for 53 weeks in the drinking water normal growth was observed in rats [59]. Subcutaneous application in the mouse of 500 mg kg^{-1} d^{-1} for 10 d led to a slight decrease in the colloid content of the thyroid gland [60].

Mutagenicity [42], [43], [47]. In the Ames test with *Salmonella typhimurium* (TA 98, 100, 1535, 1537, 1538; 10 – 5000 μg per plate) thiourea was shown to be nonmutagenic, with and without activation (S9 mix) [51], [61] – [63]. A DNA repair test with isolated rat hepatocytes (0.3 – 30 mmol thiourea) also gave negative results [51]. A further test of this type gave a slightly positive result, as did a test with Chinese hamster cells V79 [64]. A CHO/HGPRT cell mutation test [65], a test for mitotic recombination (*Saccharomyces cerevisiae* D3) [66], and a micronucleus test (rat, 2350 mg/kg, oral) [50] were also negative. A host-mediated assay (mouse) with TA 1530, TA 1538 (125 mg/kg thiourea, i.p.) showed a slight increase in the mutation rate, but the same test with TA 1535 and *Saccharomyces cerevisiae* (1000 mg/kg thiourea, i.p.) did not [48].

Carcinogenicity. A cell transformation test with Syrian golden hamster embryo cells (0 – 100 g/mL) was negative [67].

Experiments on mice with different test periods (4 – 10 months) and dose levels (0.1 – 0.5 % in drinking water) showed a slight enlargement of the thyroid gland and slight thyroid hyperplasia [68]. Application of 2 % thiourea in the diet over 1.5 years led to hyperplasia and cystic changes in the thyroid gland, but not to carcinomas [69]. In rats receiving 0.25 % in drinking water for 5 – 23.5 months, adenomas and carcinomas in the thyroid gland were observed [70]; in other tests (0.2 % in drinking water) tumors in the orbita, nose, and ear regions [71] were observed as well as in the liver (0.01 – 0.25 % thiourea in the diet) [72].

Tumor formation is associated with excessive stimulation of the thyroid gland by thyrotropic hormone from the pituitary gland. The fact that thiourea has only a weak mutagenic potential also indicates an epigenetic mechanism for tumor formation. Since tumors are also found occasionally in organs other than the thyroid gland, thiourea is classified as a suspected carcinogen [42], [43].

Embryotoxicity. In mice and rats, maternally toxic oral doses of 1000 mg/kg body weight were found to be embryotoxic [73]. In rat fetuses, whose dams had received 2000 ppm in drinking water, maturation defects were apparent in the central nervous system, skeleton, and eyes [74]. These effects are attributed to the thyreostatic activity of thiourea. Therefore, at dosages which are not sufficient to inhibit thyroid function effects are not expected [42]. A single oral dose of 480 mg/kg was found to be neither maternally toxic nor teratogenic in rats [75].

Absorption and Metabolism. In humans, thiourea is rapidly and completely absorbed from the gastrointestinal tract following oral administration. Most is excreted unchanged in the urine [76]. Small amounts appear as sulfate and ether sulfate, as shown by tests with radioactive material in rats [77]. A whole-body autoradiography with ^{14}C-thiourea showed enrichment and longer residence time in the thyroid gland compared with other tissues [78]. Following cutaneous application of 2 g/kg as an aqueous solution in rabbits, ca. 4% of the dose was detected in the urine, but only 0.1% after application of the solid [50].

Effects on Humans. Thiourea has been used temporarily in therapy because of its thyroid-depressant activity. After initial doses of 2 – 3 g/d, dosages of 100 – 200 mg/d were administered, and finally 25 – 70 mg was recommended as the daily dose. Doses of 10 – 15 mg/d were ineffective in 5 of 8 people [79]. Its use in therapy has been frequently associated with side effects, e.g., gastrointestinal disorders, pain in the muscles and joints, leukopenia, fever, and reddening of the skin. There are only a few reports on observations in the workplace [80], [81]. Apart from hypothyroidism, symptoms appeared which were similar to side effects known from therapeutic use.

In the use and processing of thiourea, contact dermatitis has occasionally been reported, e.g., on use in silver polish and diazo (blueprint) paper, sometimes after UV irradiation [82] – [86].

A Dutch expert committee recommended an occupational exposure limit of 0.5 mg/m^3 [47]. This value is also cited in the documentation of the German Research Association with reference to the thyroid gland [42], [43].

3. Thiourea Derivatives

3.1. Physical Properties

Melting points and solubilities of thiourea derivatives are given in [7], [9], [11].

The free energy of activation for rotation about the C–N bond determined by dynamic NMR spectroscopy is in the range $\Delta G^+ = 29 - 50$ kJ/mol [87].

Various physical methods, e.g., neutron diffraction [88] and IR spectroscopy [89], confirm a strong contribution of the polar, planar structure I to the ground state:

$$\begin{array}{c} H \\ \diagdown \\ N-H \\ / \\ ^-S-C \\ \diagdown \\ N^+-H \\ / \\ H \end{array}$$

I

Monoalkyl-substituted thioureas exist as mixtures of E and Z isomers. In monoaryl substitution products, the E isomer is preferred [90].

$$\begin{array}{cc}
\begin{array}{c} H \\ \diagdown \\ N-H \\ / \\ S=C \\ \diagdown \\ N-R \\ / \\ H \end{array} &
\begin{array}{c} H \\ \diagdown \\ N-H \\ / \\ S=C \\ \diagdown \\ N-H \\ / \\ R \end{array} \\
(E) & (Z)
\end{array}$$

3.2. Chemical Properties

The chemistry of thiourea and its derivatives is dominated by two structural features:

1) The nucleophilicity of sulfur gives isothiuronium salts in good yields on protonation [91], alkylation, and arylation [9].
2) The polyfunctional character of the thiourea system can be used to synthesize 4-, 5-, 6-, and 7-membered heterocycles [9], [92].

Halogenation and oxidation reactions of monoaryl-substituted thioureas give 2-aminobenzothiazoles [93], important in the production of, e.g., polyester dyes [94].

3.3. Production

Methods for the production of thioureas can be divided into seven groups [11]:

1) Reaction of alkyl, aryl, and heteroaryl isothiocyanates with ammonia, or primary or secondary amines can be used universally, is simple to carry out on a preparative scale, and gives mono-, di-, and trisubstituted thioureas in good yields:

$$R^1-N=C=S + H-N\begin{matrix}R^2\\R^3\end{matrix} \longrightarrow S=C\begin{matrix}N(R^2)-R^3\\N(R^1)-H\end{matrix}$$

2) Reaction of primary amines with carbon disulfide is of fundamental importance for the synthesis of *N,N*-diaryl-substituted thioureas and results in quantitative yield with aromatic amines. Dithiocarbamates are formed as isolable intermediates (→ Dithiocarbamic Acid and Derivatives), which are converted to thioureas with loss of hydrogen sulfide:

$$CS_2 + 2\,ArNH_2 \longrightarrow \left[S=C\begin{matrix}S^-\\N(H)-Ar\end{matrix}\right] H_3N^+-Ar$$

$$\xrightarrow{-H_2S} S=C\begin{matrix}N(H)-Ar\\N(H)-Ar\end{matrix}$$

Nitroarenes can also be used as starting materials if the reaction is carried out in an alkaline medium [95].

3) Primary and secondary amines react with thiophosgene, forming thiocarbamoyl chlorides as intermediates, which are then converted to di- and tetrasubstituted thioureas:

$$S=C\begin{matrix}Cl\\Cl\end{matrix} + H-N\begin{matrix}R^1\\R^2\end{matrix} \longrightarrow S=C\begin{matrix}Cl\\N(R^1)-R^2\end{matrix} \xrightarrow{HN(R^3)R^4} S=C\begin{matrix}N(R^3)-R^4\\N(R^1)-R^2\end{matrix}$$

However, other thiocarbonic acid derivatives, e.g., thiocarbonylbisazoles or diesters of trithiocarbonic acid, react only with primary amines.

4) By analogy with the thiourea synthesis from ammonium thiocyanate, primary alkyland arylammonium thiocyanates are converted by thermolysis into monosubstituted thioureas in good yields:

$$R-\overset{+}{N}H_3 \; SCN^- \xrightarrow{\Delta H} S=C\begin{matrix}NH_2\\N(H)-R\end{matrix}$$

5) Addition of hydrogen sulfide to alkyl- and aryl-substituted cyanamides offers an effective route to N-mono- and N,N-disubstituted thiourea derivatives by analogy with the industrial thiourea synthesis from calcium cyanamide:

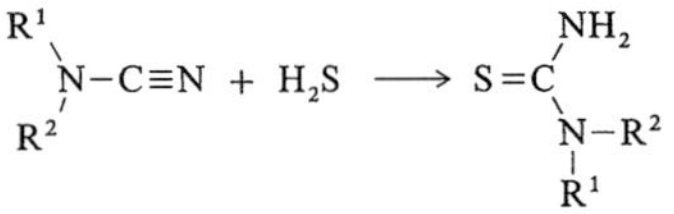

$$\begin{array}{c} R^1 \\ \diagup \\ N-C\equiv N \\ \diagdown \\ R^2 \end{array} + H_2S \longrightarrow S=C \begin{array}{c} NH_2 \\ \diagdown \\ N-R^2 \\ | \\ R^1 \end{array}$$

6) For the production of sterically hindered, tetrasubstituted thioureas, sulfurization of the corresponding urea derivatives with phosphorus pentasulfide is suitable:

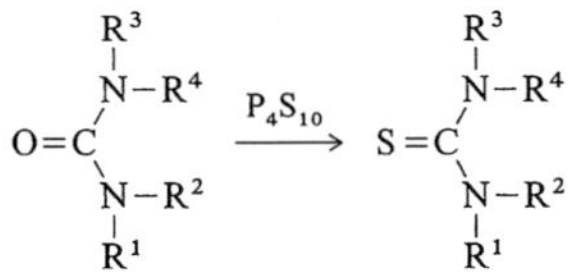

$$O=C\begin{array}{c} R^3 \\ | \\ N-R^4 \\ \diagdown \\ N-R^2 \\ | \\ R^1 \end{array} \xrightarrow{P_4S_{10}} S=C\begin{array}{c} R^3 \\ | \\ N-R^4 \\ \diagdown \\ N-R^2 \\ | \\ R^1 \end{array}$$

7) Cyclic thioureas can be obtained by the methods known for open-chain thioureas, and by the reaction of diamines with thiourea [96]:

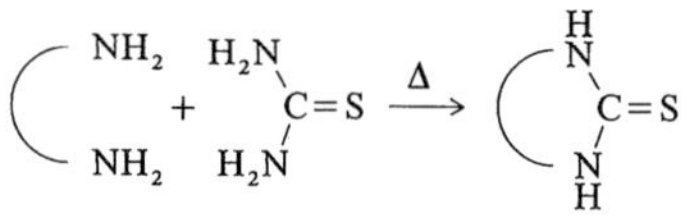

3.4. Uses

Analysis. For the quantitative determination of thiourea compounds, titration with phenyl iodosoacetate is suitable [97]. Chromatographic methods have been developed for separation [98].

Because of their complex-forming properties [99], thiourea derivatives can be used as analytical reagents [100] and for the extraction of noble metals [101].

Industrial Applications. The thiourea derivatives listed in Table 2 are used as accelerators for the vulcanization of polychloroprene and ethylene–propylene–diene terpolymers (EPDM) [102].

N-Substituted thiourea derivatives are used as corrosion inhibitors, as antioxidants in the rubber industry, in electroplating processes [8], and in the production of light-sensitive materials [103].

Biological Activity. Some thioureas are used as active substances in human and veterinary medicine [104]–[106], as insecticides [107]–[109], and as rodenticides [110]. The structures of thiourea derivatives which are important as pharmaceuticals and as plant protection agents and pesticides are given in Tables 3 and 4.

Table 2. Uses of thiourea derivatives as vulcanization accelerators

Compound	CAS No.	Polymer	Trade name	Producer
N,N,N'-trimethylthiourea	[2489-77-2]	polychloroprene	Thiate EF-2	Vanderbilt
N,N'-diethylthiourea	[105-55-5]	polychloroprene, EPDM*	Pennzone, Robac DETU, Sanceller EUR, Thiate H	Pennwalt, Robinson, Sanshin, Vanderbilt
N,N'-dibutylthiourea	[109-46-6]	polychloroprene, EPDM*	Robac DBTU, Thiate U	Robinson, Vanderbilt
N,N'-diphenylthiourea	[102-08-9]	polychloroprene, EPDM*	A-1 Thiocarbanilid, Stabilizer C	Monsanto, Mobay
ethylenethiourea (imidazolidine-2-thione)	[96-45-7]	polychloroprene	Akrochem ETU-22, Robac 22, Sanceller 22, Valcazit NPV/C	Akron Chemical, Robinson, Sanshin, Mobay
bis-piperazinyl thione derivative	[14764-02-4]	polychloroprene	Vulafor 322	Vulnax

* EPDM = ethylene–propylene–diene terpolymers.

3.5. Toxicology

The toxicological properties of a selection of alkyl- and aryl-substituted thiourea derivatives are described in [111].

More recent studies have dealt with metabolism and the effect of steric factors on the toxicity of N-phenylthiourea derivatives [112], and with the mutagenic, teratogenic, and carcinogenic properties of imidazolidine-2-thione (ethylenethiourea) [96-45-7], a metabolite of ethylene-bisthiocarbamate fungicides [113].

Thiourea and Thiourea Derivatives

Table 3. Pharmacologically active thiourea derivatives

Compound	CAS No.	Use	Generic name	Trade name	Producer
(structure)	[22232-54-8]	thyrotherapeutic agent (thyreostatic)	carbimazole	Atirozidina Basolest Neo-Carbimazole Neo-Mercazole Neo-Morphazole Neo-Thyreostat Neo-Tomizol Neral Tyrazol	Confas Pharmachemie Landerlan Schering Nicholas Herbrand Novag Conradson Neofarma
(structure)	[790-69-2]	fungicide, dermatological agent	loflucarban	Fluonilid	Continental Pharma
(structure)	[56-04-2]	thyrotherapeutic agent (thyreostatic)	methylthiouracil	Methiacil Methiocil MTU Thimecil Thyreostat Thyrostabil Tiouracil	Schwarz Helvepharm Philopharm Physicians' Drug Herbrand Streuli Lefa
(structure)	[15599-39-0]	antiseptic	noxytiolin	Gynaflex Noxyflex-S	Geistlich Geistlich
(structure)	[51-52-5]	thyrotherapeutic agent (thyreostatic)	propylthiouracil	Propycil Propyl-Thiocil Propyl-Thyracil Prothiucil Thyreostat II Tioil Tirostaz	Kali-Chemie Teva Merck-Frosst Donau-Pharmazie Herbrand Pharmacia Yurtoglu
(structure)	[515-49-1]	antiinfective agent	sulfathiourea	Badional Fontamide Salvoseptyl	Bayer Specia Alkaloida

Table 3. continued

Compound	CAS No.	Use	Generic name	Trade name	Producer
(structure)	[77-27-0] [337-47-3] Na-salt	ultrashort narcotic, anesthetic	thiamylal	Bio-Tal Bio-Tel Surital Thioseconal	Philips Roxane Philips Roxane Parke Davis Lilly
(structure)	[6028-35-9]	thyreostatic	thiabenzazolin	Thyreocordon	Donau-Pharmazie
(structure)	[92-97-7]	tuberculostatic	thiocarbanidin	Thioban	Parke Davis
(structure)	[76-75-5] [71-73-8] Na-salt	narcotic	thiopental	Farmotal Intraval Leopental Nesdonal Pentothal Thio-Barbityral Thionembutal Tiobarbital Trapanal	Farmitalia Carlo Erba May & Baker Leo Specia Abbott Amino Abbott Miro Byk Gulden
(structure)	[910-86-1]	tuberculostatic	thiocarlide	Amixyl Disoxyl Isoxyl	Inibsa Ferrosan Continental Pharma, Inibsa
(structure)	[52406-01-6]	anthelmintic (veterinary)	uredofos	Sansalid	Röhm & Haas

Thiourea and Thiourea Derivatives

Table 4. Thiourea derivatives as plant protection agents and pesticides

Thiourea derivative	CAS No.	Use	Generic name	Trade name	Producer
(4-Cl, 2-CH₃ phenyl)-N'-methyl-N,N-dimethylthiourea structure	[28217-97-2]	acaricide, anti-tick agent	chloromethiuron	Dipofene	Ciba-Geigy
diafenthiuron structure	[80060-09-9]	insecticide, acaricide	diafenthiuron	Pegasus Polo	Ciba-Geigy Ciba-Geigy
thiophanate methyl structure	[23564-05-8]	fungicide	thiophanate methyl	Topsin M Cercobin M Mildothane Cycosin	Nippon Soda Nippon Soda Rhone-Poulenc American Cyanamid
thiophanate structure	[23564-06-9]	fungicide, anthelmintic	thiophanate	Topsin Nemafax Verdamax	Nippon Soda May & Baker May & Baker

4. References

General References

[1] *Beilstein*, 4th ed., **E IV 3**, 342 ff. (1977).
[2] *Houben-Weyl*, **9**, 799–803, 884–899.
[3] *Gmelin*, 8th ed., **(14)**, part D6, Kohlenstoff, 1978.
[4] SKW Trostberg AG, Produktstudie Thioharnstoff, Trostberg 1980.
[5] BUA, Stoffdossier Thioharnstoff, in press.
[6] D. C. Schröder, *Chem. Rev.* **55** (1955) 181–221.
[7] E. E. Reid: *Organic Chemistry of Bivalent Sulfur*, Chem. Publishing Co., New York 1963, pp. 11–65.
[8] V. V. Mozolis, S. P. Iokubaitite, *Russ. Chem. Rev. (Engl. Transl.)* **42** (1973) 587–595.
[9] F. Duus in D. Barton, W. D. Ollis (eds.): *Comprehensive Organic Chemistry*, vol. **3**, Pergamon Press, Oxford 1979, pp. 452–486.
[10] D. H. Reid: *Organic Compounds of Sulphur, Selenium, and Tellurium*, The Chemical Society, London, vol. **1**, 1970, pp. 209–216; vol. **2**, 1973, pp. 236–243; vol. **3**, 1975, pp. 267–279; vol. **4**, 1977, pp. 142–153; vol. **5**, 1979, pp. 139–151; vol. **6**, 1981, pp. 165–174.
[11] *Houben-Weyl*, **E 4**, 484–521.

Specific References

[12] R. Bauer, I. Wehling, *Fresenius' Z. Anal. Chem.* 199 (1964) 171–183.
[13] F. Kicinsky, *Chem. Ind. (London)* **1** (1948) 54–57.
[14] D. Napoli, G. Ratti, G. Tubiello, *Quad. Ist. Ric. Acque* **45** (1979) 347–361.
[15] K. Gawalek: *Einschlußverbindungen*, VEB Deutscher Vlg. der Wissenschaften, Berlin 1969, pp. 226–236.
[16] W. Schlenk, *Justus Liebigs Ann. Chem.* **573** (1952) 142–162.
[17] BASF, DE-OS 1 241 612, 1962 (H. Scheuermann et al.).
[18] G. Inghilleri, *Gazz. Chim. Ital.* **39** (1909) 634–639.
[19] S. A. Miller, B. Bann, *J. Appl. Chem.* **6** (1956) 89–93.
[20] R. P. Agarwal, S. K. Ghosh, *Chim. Anal. (Paris)* **47** (1965) 407–488.
[21] SKW Trostberg AG, unpublished.
[22] K. Hayashi et al., *Bunseki Kagaku* **26** (1977) 514–515.
[23] Farbenfabriken Bayer, FR 1 576 094, 1969.
[24] Ciba Geigy, DE-OS 2 059 373, 1971 (J. Wegmann).
[25] SKW Trostberg, DE-OS 2 027 085, 1971 (H. Michaud et al.).
[26] Ricoh Co., Ltd., GB 1 218 933, 1971 (T. Iwaoka et al.).
[27] S. A. Janssen, US 3 274 209, 1960 (A. Raeymaekers).
[28] Lonza, EP 0 045 005, 1984 (A. Huwiler et al.).
[29] Degussa, DE-OS 2 511 433, 1975 (H. Knorre et al.).
[30] Trio Chemical Works, US 2 907 716, 1959 (R. J. Wassermann).
[31] Geigy, FR 1 577 582, 1968 (J. Casanova).
[32] IBM, US 4 588 471, 1986 (J. Griffith et al.).
[33] W. Schliesser, *Angew. Chem.* **74** (1962) 429–430.
[34] F. J. Kaltenböck, H. Wöbking, *Metall (Berlin)* **40** (1986) 1144–1146.
[35] A. Hung, *J. Electrochem. Soc.* **132** (1985) 1047–1049.

[36] E. Van Driesche et al., *Anal. Biochem.* **141** (1984) 184–188.
[37] ICI, DE-OS 2 553 055, 1976 (G. H. McCallum et al.).
[38] Philips Petroleum, US 4 124 073, 1978 (D. R. Wier).
[39] Shell Int. Res. Maatschappij B.V., DE-OS 2 715 026, 1977 (S. L. Wellington).
[40] T. Groenewald, *J. South Afr. Inst. Min. Metall.* 1977, 217–223.
[41] SKW Trostberg AG, company brochure, Laugung und Gewinnung von Edelmetallen mit Thioharnstoff, Trostberg 1984.
[42] D. Henschler: *Gesundheitsschädliche Arbeitsstoffe. Toxikologisch-arbeitsmedizinische Begründung von MAK-Werten, Thioharnstoff,* Verlag Chemie, Weinheim 1988.
[43] DFG: *Occupational Toxicants. Critical Data Evaluation for MAK Values and Classification of Carcinogens,* vol. **1**, Verlag Chemie, Weinheim 1991, p. 301.
[44] *IARC Monogr. Eval. Carcinog. Risk Chem. Man* **7** (1974) 95; Suppl. 4 (1982) 270.
[45] U.S. Department of Health and Human Services: *National Toxicology Program. Sixth Annual Report on Carcinogens,* National Institute of Environmental Health Services, Research Triangle Park, NC, 1991.
[46] Council of the European Communities, Council Regulation (EU) of March 23, 1993 on the Evaluation and Control of the Environmental Risks of Existing Substances, Official Journal of the European Communities, L 84, vol. **36**, 5th April 1993.
[47] Dutch Expert Committee for Occupational Standards: *Health-Based Recommended Occupational Exposure Limits for Thiourea,* Directorate-General of Labour of Ministery of Social Affairs and Employment, Monograph RA II/90, 1990.
[48] V. F. Simmon, H. S. Rosenkranz, E. Zeiger, L. A. Poirier, *JNCI J. Natl. Cancer Inst.* **62** (1979) 911–918.
[49] F. B. Flinn, J. M. Geary, *Contr. Boyce Thompson Inst.* **11** (1940) 241–247.
[50] A. P. de Groot et al., Berichte von CIVO (Central Institute for Nutrition and Food Research), TNO, in Zeist (Holland) an SKW Trostberg AG 1975–1983.
[51] F. Korte, H. Greim: berprüfung der Durchführbarkeit von Prüfungsvorschriften und der Aussagekraft der Grundprüfung des E. Chem. G. UFOPLAN-Nr. 10704006/01, 1981.
[52] S. H. Dieke, C. P. Richter, *J. Pharmacol. Exp. Ther.* **83** (1945) 195–202.
[53] G. S. Wiberg, H. C. Grice, *Food. Cosmet. Toxicol.* **3** (1965) 597–603.
[54] E. Huf, F. Auffarth, *Naunyn-Schmiedebergs Arch. Exp. Pathol. Pharmakol.* **206** (1949) 394–415.
[55] L. Arvy, M. Gabe, *Compt. Rend. Soc. Biol.* **144** (1950) 487–488.
[56] C. C. Smith, *J. Pharmacol. Exp. Ther.* **100** (1950) 408–420.
[57] H. P. Morris, C. S. Dubnik, A. J. Dalton, *J. Natl. Cancer Inst.* **7** (1946) 159–169.
[58] M. R. Osheroff, Hazleton Laboratories America Inc., 13-Week Drinking Water Study in Rats with Thiourea, HLA Study No. 2319-119, Bericht an SKW Trostberg AG 1987.
[59] A. Hartzell, *Contr. Boyce Thompson Inst.* **12** (1942) 471–480.
[60] R. P. Jones, *J. Pathol. Bacteriol.* **58** (1946) 483–493.
[61] J. McCann, E. Choi, E. Yamasaki, B. N. Ames, *Proc. Natl. Acad. Sci. USA* **72** (1975) 5135–5139.
[62] H. S. Rosenkranz, L. A. Poirier, *JNCI J. Natl. Cancer Inst.* **62** (1979) 873–891.
[63] V. F. Simmon, *JNCI J. Natl. Cancer Inst.* **62** (1979) 893–899.
[64] K. Ziegler-Skylakakis, S. Rossberger, U. Andrae, *Arch. Toxicol.* **58** (1985) 5–9.
[65] M. L. Augustine, N. K. Poulsen, J. E. Heinze, *Environ. Mutagen.* **4** (1982) 389–390.
[66] V. F. Simmon, *JNCI J. Natl. Cancer Inst.* **62** (1979) 901–909.
[67] R. J. Pienta, J. A. Poiley, W. B. Lebberz, *Int. J. Cancer* **19** (1977) 642–655.
[68] E. Vazquez-Lopez, *Br. J. Cancer Res.* **3** (1949) 401–414.
[69] A. Gorbman, *Cancer Res.* **7** (1947) 746–758.

[70] H. D. Purves, W. E. Griesbach, *Br. J. Exp. Pathol.* **28** (1947) 46–53.
[71] A. Rosin, H. Ungar, *Cancer Res.* **17** (1957) 302–305.
[72] O. G. Fitzhugh, A. A. Nelson, *Science (Washington D.C., 1883)* **108** (1948) 626–628.
[73] S. Teramoto, M. Kaneda, H. Aoyama, Y. Shirasu, *Teratology* **23** (1981) 335–342.
[74] M. Kern, Z. Tatàr-Kiss, P. Kertai, I. Földes, *Acta Morphol. Acad. Sci. Hung.* **28** (1980) 259–267.
[75] J. A. Ruddick, W. H. Newsome, L. Nash, *Teratology* **13** (1976) 263–266.
[76] R. H. Williams, G. A. Kay, *Am. J. Physiol.* **143** (1945) 715–722.
[77] J. Schulman, R. P. Keating, *J. Biol. Chem.* **183** (1950) 215–221.
[78] P. Slanina, S. Ullberg, L. Hammarström, *Acta Pharmacol. Toxicol.* **32** (1973) 358–368.
[79] A. W. Winkler, E. B. Man, T. S. Danowski, *J. Clin. Invest.* **26** (1947) 446–452.
[80] A. G. Zaslavskaja, *Klin. Med. (Moscow)* **42** (1964) 129–132.
[81] Y. N. Talakin et al., *Gig. Tr. Prof. Zabol.* **11** (1990) 18–20.
[82] A. Dooms-Goossens et al., *Br. J. Dermatol.* **116** (1987) 573–579.
[83] A. Dooms-Goossens et al., *Contact Dermatitis* **19** (1988) 133–135.
[84] D. S. Nurse, *Contact Dermatitis* **6** (1980) 153–154.
[85] J. C. Van der Leun, *Arch. Dermatol.* **113** (1977) 1611.
[86] J. K. Kellett, M. H. Beck, G. Auckland, *Contact Dermatitis* **11** (1984) 124.
[87] D. R. Eaton, K. Zaw, *J. Inorg. Nucl. Chem.* **38** (1976) 1007.
[88] D. Mullen, G. Heger, W. Treutmann, *Z. Kristallogr.* **148** (1978) 95.
[89] Y. Mido, T. Yamanaka, R. Awata, *Bull. Chem. Soc. Jpn.* **50** (1977) 27.
[90] W. Walter, K. P. Rueß, *Justus Liebigs Ann. Chem.* **743** (1971) 167.
[91] J. T. Edward, I. Lantos, G. D. Derdall, S. C. Wong, *Can. J. Chem.* **55** (1977) 812, 2331.
[92] T. S. Griffin, T. S. Woods, D. L. Klayman in A. R. Kartritzky, A. J. Boulton (eds.): *Advances in Heterocyclic Chemistry*, vol. **18**, Academic Press, New York 1975, pp. 100–145.
[93] Hoechst, EP 207 416, 1986 (H. Rentel, T. Papenfuhs); EP 529 600, 1992 (S. Dapperheld, H. Volk, K. Peter).
[94] M. A. Weaver, L. Shuttleworth, *Dyes Pigm.* **3** (1982) 81.
[95] Olin Mathieson Chem., US 3 636 104, 1972 (E. H. Kober, G. F. Ottmann).
[96] A. V. Bogatskii, N. G. Luk'yanenko, T. I. Kirichenko, *Chem. Heterocycl. Compd. (Engl. Transl.)* **19** (1983) 577.
[97] K. K. Verma, *Fresenius' Z. Anal. Chem.* **275** (1979) 287.
[98] H. Nakamura, Z. Tamura, *J. Chromatogr.* **96** (1974) 195.
[99] L. Beyer, J. Hartung, R. Köhler, *J. Prakt. Chem.* **333** (1991) 373.
[100] S. Singh, S. P. Mathur, R. S. Thakur, M. Katyal, *Acta Cien. Indica Chem.* **13** (1987) 40.
[101] S. Singh, S. P. Mathur, R. S. Thakur, L. Keemti, *Orient. J. Chem.* **3** (1987) 203.
[102] *Kirk-Othmer*, **20**, 337.
[103] Fuji Photo Film, JP 0 504 449, 1991 (H. Kawakami, K. Iwakura); *Chem. Abstr.* **118** (1993) 263906u.
[104] S. N. Pandeya, P. Ram, V. Shankar, *J. Sci. Ind. Res.* **40** (1981) 458.
[105] M. Negwer: *Organic-Chemical Drugs and their Synonyms*, 6th ed., Akademie Verlag, Berlin 1987.
[106] Swiss Pharmaceutical Society (ed.): *Index Nominum, International Drug Directory 1990/1991*, Medpharm Scientific Publ., Stuttgart 1990.
[107] C. R. Worthing, R. J. Hance (eds.): *The Pesticide Manual*, 9th ed., The British Crop Protection Council, Farnham 1991.
[108] K. H. Büchel: *Chemistry of Pesticides*, J. Wiley & Sons, New York 1983, p. 315.
[109] J. Drabek et al., *Spec. Publ. R. Soc. Chem.* **79** (1990) 170.

[110] R. Wegler: *Chemie der Pflanzenschutz- und Schädlingsbekämpfungsmittel,* Springer Verlag, Berlin 1970, pp. 605–606.
[111] *Ullmann,* 4th ed., **23,** pp. 172–173.
[112] A. E. Miller, J. J. Bischoff, K. Pae, *Chem. Res. Toxicol.* 1988, 169.
[113] R. A. Frakes, *Regul. Toxicol. Pharmacol.* **8** (1988) 207.

Toluene

JÖRG FABRI, VEBA AG, Düsseldorf, Federal Republic of Germany

ULRICH GRAESER, VEBA Öl AG, Gelsenkirchen, Federal Republic of Germany

THOMAS A. SIMO, Lurgi Öl-Gas-Chemie GmbH, Frankfurt/Main, Federal Republic of Germany

1.	Introduction	4727
2.	Properties	4728
2.1.	Physical Properties	4728
2.2.	Chemical Properties	4729
2.3.	Properties as a Motor Fuel Component	4731
3.	Separation, Extraction, and Processing	4731
3.1.	Separation	4731
3.2.	Processing	4733
4.	Integration into Refineries and Petrochemical Complexes	4734
5.	Economic Aspects	4735
6.	Quality Requirements	4738
7.	Storage, Transport, and Safety	4738
8.	Environmental Aspects and Toxicology	4739
8.1.	Environmental Aspects	4739
8.2.	Toxicology	4739
9.	References	4741

1. Introduction

In the early industrial era, toluene was produced from aromatic fractions, mostly obtained by the coking of hard coal. The importance of toluene as a raw material increased dramatically during World War I, when it was used extensively for the production of the explosive trinitrotoluene (TNT). Since the quantity of toluene produced from coking plants was not sufficient, it was also obtained by direct distillation from aromatic crude oils from Java and Borneo, and by thermal cracking of other naphtha fractions with a narrow boiling range [1].

Sulfuric acid was initially used for refining crude toluene, leading to serious disposal problems. This process began to be replaced by hydrogen refining from ca. 1950.

During World War II, the demand for toluene rose again as a result of increased demand for high-octane aircraft fuels. Production of sufficient quantities of toluene was achieved by development of the catalytic reforming process, in which naphtha fractions from crude petroleum can be converted to reformate with high aromatic content. To isolate pure toluene from reformate, extraction and extractive distillation processes were developed. Pyrolysis gasoline, another source of toluene, became available as a side product of olefin production by steam cracking from about the mid-1960s.

Since then the development of the chemical uses of toluene has lagged behind the increase in its availability. Its relatively low status in chemicals production is mainly because of its lower economic attractiveness compared to the alternative raw material benzene, e.g., for producing phenol or caprolactam. Consequently, considerable quantities of toluene are converted to benzene by hydrodealkylation. Comparatively small quantities of toluene are used for the production of explosives, pharmaceuticals, dyes, and plastics.

A significant proportion of toluene is not isolated in pure form, but is added to motor fuels as a mixture with other aromatics (reformate or pyrolysis gasoline). Because of its high resistance to knocking, the use of toluene as a motor fuel component is frequently more economic than extraction and subsequent use in chemical processes.

History [2]. Toluene (methylbenzene) was discovered by P. S. PELLETIER and P. WALTER in 1837, during the preparation of coal gas from pine resin. In 1841, H. SAINTE-CLAIRE-DEVILLE described the isolation of toluene by the dry distillation of tolu balsam, a solid resinous material with a vanilla-like odor, extracted from the trunk of *Myroxylon balsamum* Harms var. *genium*, a tall tree found in the high plateaus and mountain ranges of South America. In the extraction process toluene is formed from benzyl cinnamate (3-phenyl-2-propenoic acid phenylmethyl ester).

C. B. MANSFIELD identified toluene in coal tar in 1849. It had been given the name toluin by J. J. BERZELIUS in 1843; later, J. S. MUSPRATT and A. W. HOFMANN changed the name to toluol. In English, toluene refers to the pure substance, whereas toluol refers to a commercial product, of variable purity.

2. Properties

2.1. Physical Properties

Toluene [*108-88-3*], methylbenzene, C_7H_8, is a colorless, flammable liquid of low viscosity with a benzene-like odor. It is a good solvent for fats, oils, tars, resins, sulfur, phosphorus, iodine, etc. It is completely miscible with most organic solvents, e.g., alcohols, ethers, ketones, phenols, esters, and chlorohydrocarbons. Toluene is only slightly soluble in water, 0.047 g/100 mL at 16 °C. At 15 °C the solubility of water in toluene is also very low: 0.4 g/100 mL [2].

Some physical data are listed in the following [2]:

Molecular mass	92.13
mp	−94.991 °C
bp (100 kPa)	110.625 °C
Critical temperature	320.8 °C
Critical pressure	4.133 MPa
Critical compressibility	0.260
Critical molecular volume	0.32 L/mol
Density at 100 kPa and 25 °C	0.8631 g/cm^3
Surface tension at 100 kPa and 25 °C	26.75 mN/m
Viscosity at 100 kPa and 20 °C	0.5864 mPa · s

Thermal conductivity of liquid at 100 kPa and 0 °C		0.1438 W m^{-1} K^{-1}
Enthalpy of evaporation at 100 kPa and *bp*		32.786 kJ/mol
Lower calorific value		
gas		40.97 kJ/g
liquid		40.52 kJ/g
Flash point		4 °C
Ignition temperature		552 °C
Explosion limits in air		
upper		1.27 vol%
lower		6.75 vol%
Refractive index n_D^{20}		1.49693

Table 1. Data for binary azeotropes of toluene

Component	*bp* at 0.1 MPa, °C		Toluene
	Component	Azeotrope	content of azeotrope, wt%
Water	100	84.1	79.7
Methanol	64.7	63.8	31
Acetic acid	118.5	105.0	66
Ethanol	78.3	76.6	33
Glycol	197.4	110.2	93.5
Ethylenediamine	116.9	103	70
Propanol	97.3	92.5	49
1,2-Propanediol	187.8	110.5	98.5
1,2-Propanediamine	120.9	105	68
Dioxan	101.8	101.4	20
Butanol	117.8	105.7	73
Isobutanol	108.0	100.9	55.5
sec-Butanol	99.5	95.3	45
Pyridine	115.5	110.1	78.8
tert-Amyl alcohol	101.7	100.0	44
Isoamyl alcohol	131.3	110.0	88

Binary and ternary azeotropes of toluene are listed in Tables 1 and 2, respectively. Figure 1 shows the vapor pressure as a function of temperature.

2.2. Chemical Properties

Toluene is similar to benzene in its chemical properties, but the methyl group provides additional reactivity. The aromatic nucleus can be hydrogenated to methylcyclohexane. Oxidizing agents preferably attack the methyl group, giving benzaldehyde and then benzoic acid. The latter can be decarboxylated to phenol (Dow and California Research) or hydrogenated to cyclohexanecarboxylic acid (cf. the Snia-Viscosa ϵ-caprolactam process [6], → Caprolactam).

Table 2. Data for ternary azeotropes of toluene

Components		bp, °C		Content in the azeotrope, wt%		
		Component	Azeotrope	Toluene	A	B
A	Water	100	76.3	48.7	13.1	38.2
B	Isopropanol	82.3				
A	Butanol	117.8	108.7	67.4	11.9	20.7
B	Pyridine	115.5				
A	Pyridine	115.5	110.2	87.3	8.6	4.1
B	Isoamyl alcohol	131.3				

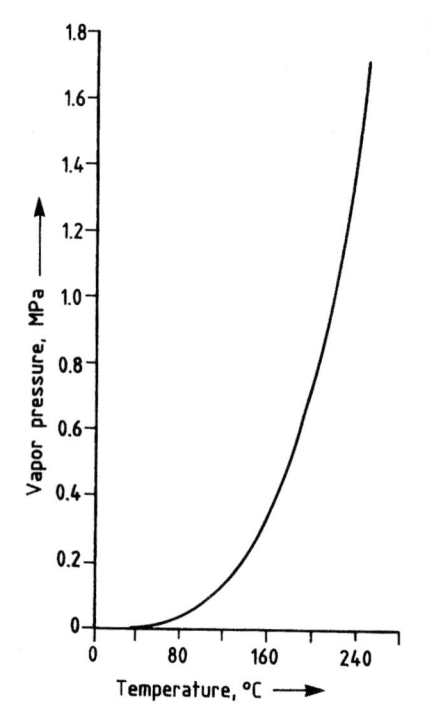

Figure 1. Vapor pressure of toluene as a function of temperature [3], [4]

Both the nucleus and the side chain can be chlorinated. Nitration leads to o- and p-nitrotoluenes, which are separated by freezing, solid–liquid separation, and distillation. Further nitration gives di- and trinitrotoluenes; TNT is used as an explosive.

Some nitrotoluenes are important organic intermediates, as are the toluidines obtainable from them by reduction (starting materials for dye syntheses and vulcanization accelerators), and the dinitriles derived from dinitrotoluenes, which give toluene diisocyanates. The latter are starting materials for polyurethanes.

Alkylation of toluene with propylene gives methylcumene isomers (cymenes). Saponification of the cumene hydroperoxides results in the isomeric cresols and acetone (Sumitomo Chemical [7]).

Table 3. Octane numbers and vapor pressures of toluene and motor fuel

	Toluene	Unleaded super (DIN 51 607/EN 228)
Research octane number (RON), pure	110	
Motor octane number (MON), pure	98	
Research octane number (RON), typical blend	124	95
Motor octane number (MON), typical blend	112	85
Reid vapor pressure, kPa	0.8	9 (winter)
		7 (summer)

Toluenesulfonic acids and acid chlorides are, like chlorotoluene, starting materials for dyes, pharmaceuticals, and saccharin.

2.3. Properties as a Motor Fuel Component

The usefulness of toluene as a motor fuel component depends on how its properties can be combined with those of the other components. Because of the high resistance to knocking and the comparatively low vapor pressure, toluene is highly valued as a motor fuel component [8], [9]. In Table 3 the octane number and the vapor pressure values (Reid vapor pressure) of toluene are compared with the specification requirements for unleaded super gasoline (DIN 51 607/EN 228).

3. Separation, Extraction, and Processing

3.1. Separation

It is possible to synthesize toluene industrially by alkylation of benzene with methanol [10], and by cyclization of *n*-heptane with subsequent aromatization [11], [12]. However, for economic reasons toluene is extracted from the raw materials described below. They contain toluene, its homologs, other hydrocarbons, and heteroatomic organic compounds. These raw materials originate from:

Reformates from crude petroleum distillates
Liquid products from the pyrolysis of hydrocarbons (steam cracking)
Liquid products from the gasification or coking (pyrolysis) of coal, lignite, etc.

These product streams with their characteristic compositions and treatments have been described in detail [13]–[16] (→ Xylenes). An important aspect of the extraction

Table 4. Aromatic extraction processes

Process	Licenser	Solvent
Sulfolan	Shell-UOP	tetrahydrothiophene dioxide
Arosolvan	Lurgi	N-methylpyrrolidone
Morphylex	Krupp-Koppers	N-formylmorpholine
Formex	Snamprogetti	N-formylmorpholine
IFP	IFP	dimethyl sulfoxide
Mofex	Leunawerke	methylformamide

of toluene is the fact that the pyrolysis products (from steam cracking, coking, etc.) must be hydrogenated before pure toluene can be extracted. The unsaturated components are converted to saturated ones and the heteroatoms such as sulfur, nitrogen, and oxygen are removed. In the case of reformates such pretreatment is usually unnecessary.

For the separation of toluene from other components within the same boiling range several methods are available, depending on quality requirements. Fine fractionation is now suitable only for the production of toluene with lower purity, and involves significant losses in fores and tails. Azeotropic distillation uses entrainers, such as methanol [13], to separate toluene from nonaromatics; a nonaromatics–methanol fraction with a lower *bp* than the methanol–toluene azeotrope distills at the column head, while pure toluene is removed from its base. Methanol is recovered from the distillate by washing with water.

For economic reasons, extractive distillation is now used only for the separation of toluene from nonaromatics. Technical grade solvents with higher *bp* than toluene have proved to be suitable extraction agents, e.g., N-methylpyrrolidone (Distapex process, Lurgi Öl-Gas-Chemie [14], [17]), and morpholine (Morphylane process, Krupp-Koppers [18]). Extractive distillation essentially involves two distillation columns between which the extraction agent is circulated. The toluene-containing material is charged to the extraction column; the extraction agent is charged to the column head. The extraction agent – toluene mixture leaves the column at the bottom, and is separated into pure toluene and extraction agent in a second recovery column. The product obtained at the distillation head of the extraction column contains the nonaromatic components of the starting material and the extraction agent. This fraction is separated in the recovery column into raffinate (nonaromatics) and extraction agent. The latter is then combined with the main portion of extraction agent from the recovery column, and fed back into the extraction column.

If a toluene fraction is a complex aromatic mixture (e.g., benzene–toluene–xylene = BTX), as most raw materials are, then liquid–liquid extraction is necessary. The solvents which have proved suitable industrially are shown in Table 4.

A range of other solvents, ethylene glycol, triethylene glycol, tetraethylene glycol, propylene carbonate, etc., have been proposed, but are no longer economically important.

3.2. Processing

Most of the pure toluene produced is converted to benzene by dealkylation. The catalytic conversion takes place in a hydrogen atmosphere at 550–650 °C, 3–10 MPa. This process, used by UOP, Air Products (Houdry-Lummus), and BASF, is described in more detail elsewhere (→ Benzene).

Another route is thermal dealkylation, using technologies developed by Atlantic Richfield (HDA process), Gulf (THD process), Mitsubishi Petrochemical (MHC process), and UOP. A somewhat higher temperature is used (600–800 °C), with a hydrogen atmosphere. The other features of these process (yield, purity, hydrogen consumption, etc.) are similar to those of the catalytic process (→ Benzene).

Toluene is sometimes used for production of xylenes by transalkylation and disproportionation. In transalkylation, one molecule of benzene and one molecule of xylene are formed from two molecules of toluene. The ratio of o- to m- to p-xylene corresponds to that present at thermal equilibrium. Ethylbenzene is formed in small quantities.

Disproportionation is the typcial reaction when only toluene is used. Transalkylation is used for additional extraction of xylenes starting from toluene and C_9 aromatics (A_9). Methyl groups are transferred from higher alkylated aromatics to less alkylated ones.

The following technologies have achieved industrial importance:

1) Tatoray (UOP)
2) Xylenes-Plus (Atlantic Richfield)
3) LTD, low-temperature disproportionation (Mobil)

The Tatoray process gives ca. 38 % conversion at each pass. Consequently, after removal of the benzene from the reaction product, a toluene fraction is obtained and recycled, then the xylenes and possibly heavier aromatics are distilled from the Tatoray product (Fig. 2) [2], [19]. Publications from UOP [20] on the MSTDP (Mobil selective toluene disproportionation process) [21], [22] describe the characteristics of the Tatoray process and the integration of the disproportionation–transalkylation into an aromatics complex. The Tatoray process was commissioned in nine industrial plants between 1985 and 1993, and is under construction in a further three units. The MSTDP process was successfully commissioned in three industrial plants between 1989 and 1992.

Development work for the production of xylenes from toluene by alkylation with methanol has continued [23], although the economic environment and the availability of toluene do not justify immediate industrial applications.

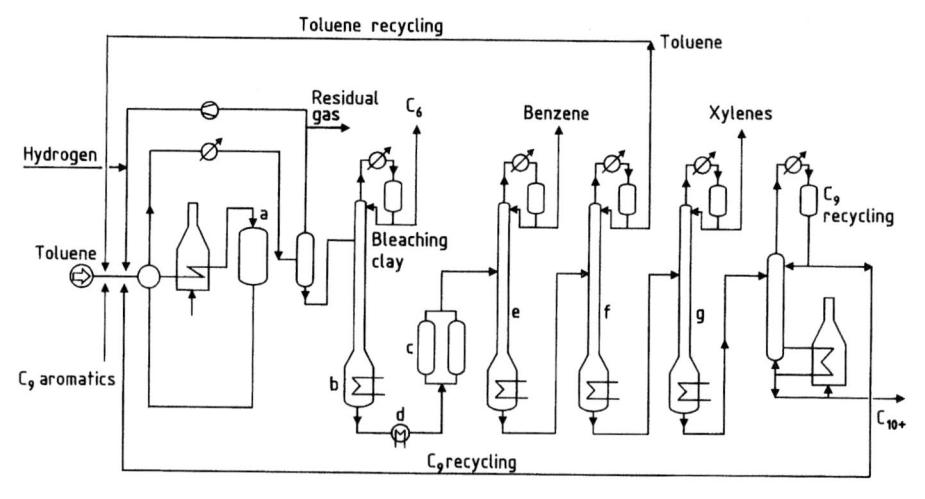

Figure 2. The Tatoray process [2], [19]
a) Reactor; b) Stripper; c) Bleaching clay columns; d) Heater; e) Benzene column; f) Toluene column; g) Xylene column; h) C_9 column

4. Integration into Refineries and Petrochemical Complexes

Backward Integration into Refineries. Production of toluene from crude petroleum occurs almost exclusively in major refineries and petrochemical complexes [24]. Because of its great importance for the fuel sector, toluene is either added to the motor fuel pool or is extracted and further processed, the proportions varying according to its current economic value [25]. In addition to the capacity for isolation of toluene, a corresponding hydrodealkylation or disproportionation capacity is often retained, so that there can be a flexible response to changes in demand and price between benzene and toluene. The principle of incorporation of toluene production into the structure of a petrochemical refinery is shown in Figure 3.

Removal of toluene from reformates or pyrolysis gasoline reduces the octane number in the motor fuel pool, and can lead to a significant gap in the boiling range of the fuel, as is also the case in benzene and xylene extraction. This is known as a gap fuel characteristic, and must be compensated for by adding alternative components [26]. Addition of pure toluene to the motor fuel pool reduces the benzene content by dilution, which is often welcome in terms of environmental and specification aspects. The octane number is raised. However, in view of a possible future limitation on the total aromatic content in motor fuel in the United States and possibly also western Europe [27]–[30], removal of toluene from reformate and pyrolysis gasoline could help meet the new specification requirements for gasoline [31]–[33].

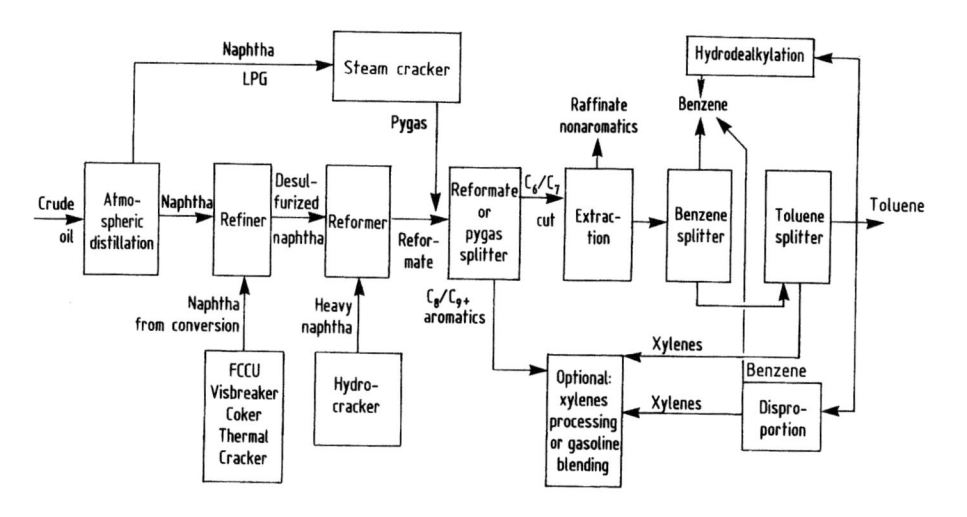

Figure 3. Incorporation of toluene production into a petrochemical refinery

Forward Integration into Chemical Processes. Toluene can be used as a solvent without further processing. In the area of petrochemistry, conversion processes (hydroalkylation or disproportionation) giving benzene and xylenes are the most important. The large number of further processing routes in the chemical industry involve a far greater extent of conversion. Toluene diisocyanate (TDI), phenol, caprolactam, nitrotoluene, and phthalates are produced in the largest quantities (Fig. 4).

5. Economic Aspects

As in the case of benzene and xylenes, toluene production and extraction is affected by the pricedifferential compared with naphtha (Fig. 5) and by the relationship, in terms of net product, between separation and further processing as a raw material for chemicals production compared with direct use as a motor fuel component, in the form of reformate or pyrolysis gasoline [25], [34]. The first aspect is essentially fixed by the economics of reforming naphtha, the second by the economic situations in the mineral oil and chemical industries.

In the 1980s, in western European refineries with appropriate extraction capacities, about one-third of the toluene produced was left in pyrolysis gasoline and reformate, in spite of the possibilities for extraction, and was added to motor fuel directly in this form. In the near future, no significant change in the production and consumption of toluene is expected (Table 5).

Table 6 lists toluene producers, production capacities, and raw material sources in western Europe. Other important production capacities for toluene are in the United States, the CIS, Japan, and Canada [36]–[38].

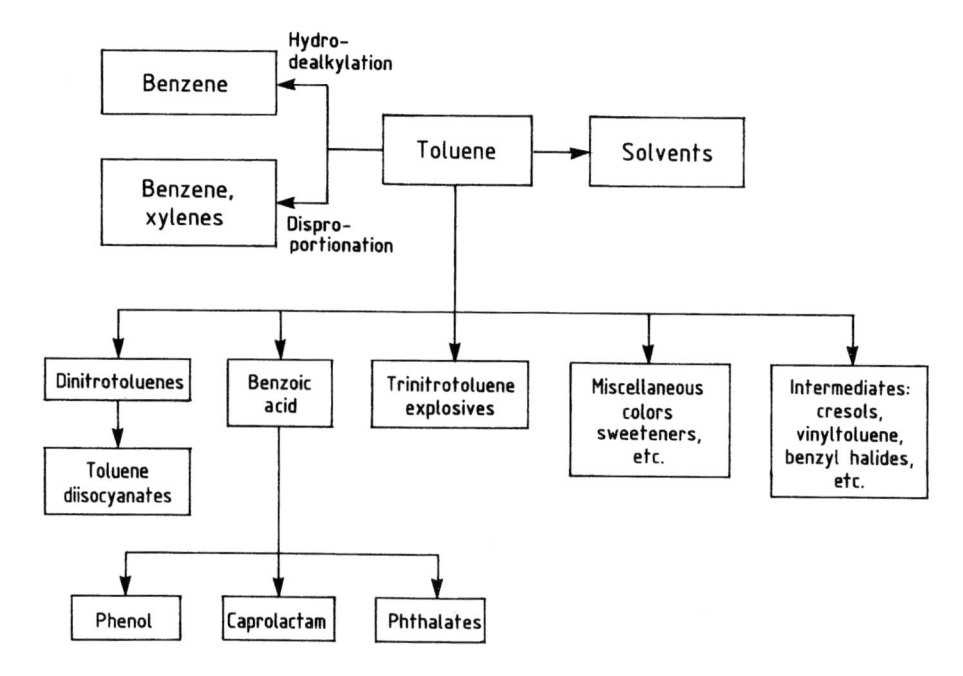

Figure 4. Further processing of toluene

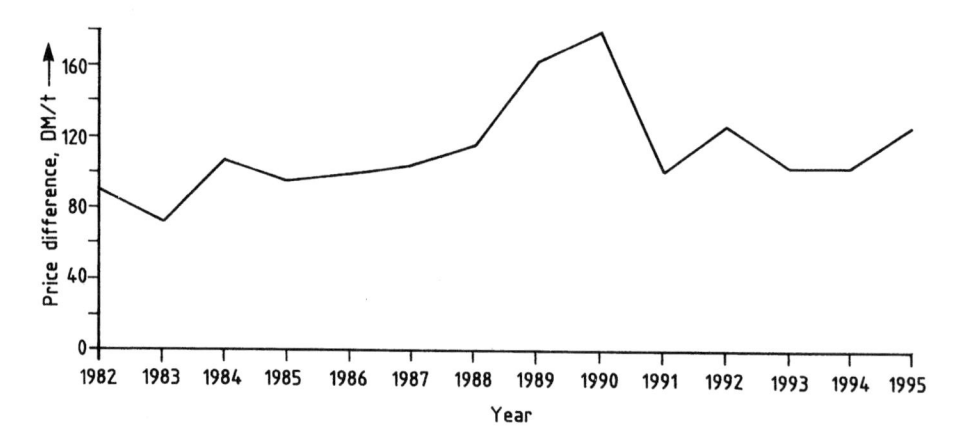

Figure 5. Price difference between toluene and naphtha in western Europe [34]

Table 5. Production and consumption of toluene in western Europe (in 10^3 t)

	1983	1986	1990	1992	1995	2000[a]
Production						
From pyrolysis gasoline[b]	1460	1519	1733	1600	1667	1978
From reformate	1114	1653	1675	1378	1325	1325
From coal	23	67	66	64	62	62
Total production	2597	3239	3474	3042	3054	3365
Net imports[a]	100	120	150	150	200	200
Total supply	2697	3359	3624	3192	3254	3565
Consumption						
Dealkylation/disproportionation	993	994	1468	1028	1106	1106
Solvent	423	510	469	465	459	446
TDI	184	160	242	237	250	273
Phenol	115	100	152	116	130	159
Caprolactam	75	95	10	17	17	17
Nitrotoluenes	17		20	20	20	20
Phthalates	8		10	10	10	10
Other chemical products[c]	228	210	228	234	239	246
Total chemical consumption	2043	2069	2599	2127	2231	2277
Use as a fuel component[d]	654	1290	1025	1065	1023	1288

[a] Estimate.
[b] Estimate; only ca. 45% of the toluene in pyrolysis gasoline is extracted.
[c] For example, benzoic acid, benzaldehyde, benzyl chloride, cresols.
[d] Difference.

Table 6. Producers, capacities, and raw material sources for toluene in western Europe in 1994 (in 10^3 t)

Country	Producers	From reformate	From pyrolysis gasoline	From coal	Total
Austria	Voest	0	0	4	4
Belgium	Finaneste; Sopar	0	65	8	73
France	Atochem; Shell; Total	30	40	0	65
Germany	SOW; EC; Shell; DEA; PCK; Ruhr Öl; Wintershall	475	650	60	1185
U.K.	BP; ICI; Conoco; Shell	465	70	18	555
Italy	Enichem; Praoil; Carboquimice	290	205	0	495
Netherlands	Exxon	255	0	0	255
Portugal	Petrogal	140	0	0	140
Spain	Cepsa	280	0	0	280
Total		1805	845	92	2727

Table 7. Quality requirements for toluene

	Industrial grade	Nitration grade	Method of determination
According to ASTM	D 362	D 841	
Density at 20 °C, g/mL	0.860 – 0.874	0.8690 – 0.8730	D 891
Color (Hazen)	≤20	≤20	D 1209
Boiling range, °C	±2	±1	D 850, D 1078
Acidity	≤0.005 %		D 847
Color after acid treatment	≤no. 4	≤no. 2	D 848
Sulfur content	no H_2S/SO_2	no H_2S/SO_2	D 853
According to DIN 51 633			
Color (Hazen)	20		DIN 53 409
Density at 15 °C, g/mL	0.869 – 0.872		DIN 51 757
Refractive index at 20 °C	≥1.4963		DIN 51 423, part 2
bp, °C	110.6		
Reaction with sulfuric acid	≤0.2		DIN 51 762
Bromine consumption	≤0.1 g/100 mL		DIN 51 774, part 3
Total sulfur	≤0.0002 wt %		DIN-EN 41
Active sulfur	below detection limit		DIN 51 764
Hydrogen sulfide	none		DIN 51 766
Doctor test	negative		DIN 51 765

6. Quality Requirements

Quality requirements for toluene depend on the type of further processing. Apart from individual specification agreements between producers and customers, there are also generally accepted quality specifications for toluene, e.g., ASTM, EN, ISO, DIN, BS [39], [40]. In the ASTM specifications there is a standard quality (Industrial Grade Toluene, ASTM D 362) and a higher purity quality (Nitration Grade Toluene, ASTM D 841) (Table 7). The German standard is DIN 51 633, comparable to ASTM D 841 (Table 7).

7. Storage, Transport, and Safety

To guarantee safe storage, transfer, and transport of flammable liquids Technical Rules for Flammable Liquids (TRbF) have been drawn up in Germany [41], [42], and are continually updated.

For toluene, the same TRbF numbers apply as for xylenes (→ Xylenes).

Because the flash point lies below 21 °C, toluene is assigned to hazard class A I. On storage and transport, toluene must be labeled as highly flammable and slightly toxic [43]. For transport in Germany, it is assigned to class 3, cipher 3b of the GGVS/GGVE regulations governing the transport of hazardous goods by road and rail [44]. The explosion limits of toluene are 1.27 and 6.75 %.

8. Environmental Aspects and Toxicology

8.1. Environmental Aspects

Leakage of toluene into soil or bodies of water constitutes a serious pollution problem. In Germany, toluene is assigned to WGK 2 [42]. The solubility of toluene in water is very low (0.1 g/L at 20 °C). With a vapor pressure of 2.9 kPa (at 20 °C) toluene evaporates relatively rapidly in air; besides the direct toxicological effects (see Section 8.2) secondary reactions with other air pollutants are also important. The reaction with nitrogen oxides in the presence of sunlight leads to the formation of ozone and smog, and can cause severe ecotoxicological pollution [27], [45]. Even though toluene undergoes these types of reaction comparatively slowly, extensive measures are increasingly being taken to reduce vapor emissions, both in solvents (where toluene is being replaced to some extent as a result of other developments, e.g., solvent-free paints), and in the fuel sector (in which the emission of toluene into the atmosphere is being significantly lowered by the installation of corresponding vapor retention systems —activated charcoal canisters in passenger cars or gas displacement devices at gasoline stations, storage depots, and refineries) [46], [47]. When toluene is used a a fuel component, it affects the composition of exhaust gases [27], [48].

8.2. Toxicology

From 1 September 1993, the MAK value for toluene was reduced to 50 ppm (190 mg/m^3) [49], [50]. This corresponds to a halving of the 1992 value. With regard to the effect on pregnancy, a parallel reassignment was made to group C (previously group B). Extensive medical data on the effects of toluene were used for the toxicological evaluation. Results of animal experiments supplement the data.

General Activity Profile. Inhalation of toluene vapor affects the central nervous system, giving rise to symptoms such as headaches, dizziness, or coordination difficulties [51]–[55]. Loss of consciousness can occur at higher concentration [56]. On acute skin contact with liquid toluene, removal of oils and drying of the skin can occur, with development of dermatitis as a possible consequence. The direct irritant effect of toluene is relatively low [56], [57].

Acute and Subacute Toxicity. In animal experiments the following toxicity values were found [58]:

rat, oral, $LD_{50} = 5000$ mg/kg

rabbit, percutaneous, LD$_{50}$ = 12 124 mg/kg
mouse, 8 h inhalation, LC$_{50}$ = 5320 mg/m^3

On exposure for several hours to < 1000 mL/m^3 toluene in inhaled air, no significant effects were established in rats [59]. In the concentration range ca. 2000 mL/m^3, in rats, pulse rate was increased many times and a state of exhaustion was recorded [55]; at higher concentrations, a narcotic effect predominated [60] – [62].

In humans, concentrations up to ca. 800 mL/m^3 affect the central nervous system, causing confusion, headache, nausea, or coordination difficulties as the most obvious effects [51] – [55]. Weakness and tiredness are also observed. The effect of toluene on the central nervous system can lead to hallucination and addiction (solvent sniffing). In three cases of death following frequent deliberate inhalation of large quantities of toluene-containing solvents, the lethal dose for humans was estimated as 2000 mL/m^3 within 30 min [63]. At high concentrations, disruption of kidney function is observed [64].

Chronic Toxicity. On prolonged exposure, toluene accumulates in the brain [65]. As with acute intoxication, this leads to damage to the central nervous system. However, in animal experiments this is clearly visible only at concentrations of > 500 mL/m^3 [66]. Only at much higher doses were damage to the liver [67] and cardiotoxic effects [68] established in rats and mice. However, in animal experiments, after initial suspicions, there was no confirmation of a leukemogenic effect similar to that found with benzene.

Many toxic effects which have been recorded for humans with high exposure to solvents cannot be attributed to toluene according to current knowledge, but mainly to small quantities of benzene present as an impurity [69], [70]. The effects of acute intoxication on the central nervous system with the symptoms already mentioned are, however, characteristic. Damage to internal organs through chronic inhalation of comparatively low toluene concentrations (up to ca. 200 – 400 mg/m^3) is not unambiguously detectable [71]. However, the chronic inhalation of high doses of toluene (solvent abuse) leads to damage to the cerebellum, disruption of the metabolism, and muscle weakness [72], [73].

Pharmacokinetics and Metabolism. In humans, the main route for taking in toluene is by inhalation of vapor. Toluene retention in the lungs on prolonged exposure is ca. 50%, and can be raised by bodily exertion [74], [75]. Half-lives between ca. 2 min and ca. 3.5 s have been reported [76], [77]. According to more recent discoveries, the percutaneous resorption rate on skin contact with liquid toluene is 0.17 mg cm^{-2} h^{-1}, considerably lower than assumed previously [78].

The metabolic conversion of toluene mainly occurs in the liver. Degradation takes place over many steps and involves the formation of benzaldehyde and benzoic acid, and their conjugation with glycine to give hippuric acid, which is excreted in the urine. Phenolic metabolites (*o*- and *p*-cresol) have also been detected in the urine in small quantities (up to ca. 0.2%) [79], [80].

Mutagenicity, Teratogenicity, and Carcinogenicity. In the Ames test using *Salmonella typhimurium* strains, no mutagenic effects could be detected with toluene, with or without metabolic activation [81], [82].

On testing the effect of toluene on mice, no increase in chromosomal aberrations was established [83]. In a more recent study, in humans exposed to toluene vapor in the concentration range 200–300 mL/m^3 for ca. 16 a, significant increase in structural changes was observed (chromatid cleavage and exchange), which could be detected up to two years after exposure had terminated [84]. In a large number of animal experiments, there was no certain evidence for teratogenic effects [85]. No indication of a carcinogenic or cocarcinogenic potential has been found for toluene [86].

9. References

[1] H.-G. Franck, J. W. Stadelhofer in: *Industrielle Aromatenchemie*, Springer Verlag, Berlin 1987, pp. 102–103.
[2] *Ullmann*, 4th ed., **23**, 301.
[3] L. H. Horsley, *Anal. Chem. Ser.* **21** (1949) no. 7, 831–873.
[4] B. Riediger: *Verarbeitung des Erdöls*, Springer Verlag, Berlin 1971, p. 89.
[5] Dechema Chemistry Data Series, Vapor-Liquid Data Collection, Frankfurt 1993. C. L. Yaws, *Chem. Eng. (N.Y.)*, Sept. 1975, 73–81.
Beilstein, Suppl. 3, (V) part 2, 663–775. F. D. Rossini et al.:
Selected Values of Physical and Thermodynamic Properites of Hydrocarbons and Related Compounds, Carnegie Press, Pittsburgh 1953.
[6] M. B. Sherwin, *Chem. Eng. Prog. (Nov.)* **26** (1979) 31. K. Weissermel, H.-J. Arpe: *Industrielle Organische Chemie*, 3rd ed., Verlag Chemie, Weinheim – New York 1990, pp. 273–274. F. Asinger:
Die Petrolchemische Industrie, Akademie Verlag, Berlin 1971, pp. 505–640, 1439–1441.
[7] K. Weissermel, H.-J. Arpe: *Industrielle Organische Chemie*, 3rd ed., Verlag Chemie, Weinheim – New York 1990, p. 374, 377.
[8] K. Owen, T. Coley: *Automotive Fuels Handbook*, Society of Automotive Engineers Inc. (SAE), Warrendale, PA 1990, pp. 144–145.
[9] W. E. A. Dabelstein, A. A. Reglitzky, K. Reders, N. Lucht: *100 Jahre Kraftstoffe für den Straßenverkehr*, Shell Technischer Dienst, Hamburg 1989, pp. 7, 8, 16–25.
[10] *Winnacker-Küchler*, 3rd ed., vol. **3**, p. 817.
[11] *Ullmann*, 3rd ed., **17**, 446.
[12] *Beilstein*, **4 (5)**, 652.
[13] K. H. Eisenlohr et al., *Erdoel Kohle Erdgas Petrochem.* **16** (1963) no. 6, 523.
[14] K.-H. Eisenlohr, *Erdoel Kohle Erdgas Petrochem.* **20** (1967) 82–89.
[15] J. J. Wise, A. J. Silvestri, *Oil Gas J.* **74** (1976) 140. C. D. Chang, A. J. Silvestri, *J. Catal.* **47** (1977) 24a.
[16] A. Chauvel, G. Lefebre: *Petrochemical Processes*, vol. **1**, Editions Technip IFP, Paris 1989, pp. 154–159, 165–178.
[17] Lurgi Öl-Gas-Chemie GmbH, company brochure, Lurgi Distapex Process – Recovery of Pure Aromatics, no. 1640e/9.91/10.

[18] K. Lackner, *Erdoel Kohle Erdgas Petrochem.* **34** (1981) no. 1, 26–30.
[19] Toray Industries, Inc.: Transalkylation of Aromatics, Tokyo, Japan, March 1982.K. Weissermel, H. J. Arpe: *Industrielle Organische Chemie*, Verlag Chemie, Weinheim–New York 1978, pp. 313–314. *Hydrocarbon Process* **60** (1981) no. 11, 139.
[20] P. J. Kuchar, D. Y. Lin, V. Zukauskas, D. Brkie: "Isomar and Tatoray Production Flexibility," *Proc. 1988 UOP Technology Conference*, Des Plaines, Ill. 1989. E. C. Haun, M. W. Golem, S. Sapuntzakies, P. P. Piotrowski:
"The Modern Aromatics Complex," *Proc. 1988 UOP Technology Conference*, Des Plaines, Ill. 1989.
[21] F. Gerra, L. L. Beckenridge, W. M. Guy, R. A. Sailor, *Oil Gas J.* **90** (1992) 60 ff.
[22] *Hydrocarbon Process.* **70** (1991) no. 3, 141.
[23] M. W. Anderson, J. Klimowski, *Nature (London)* **339** (1989) 200–203.
[24] W. Keim, A. Behr, G. Schmitt: *Grundlagen der Industriellen Chemie*, Verlag Salle-Sauerländer, Frankfurt/Main 1986, p. 165.
[25] Chem Systems Ltd.: *Petroleum and Petrochemicals Economics,* Ann. Rep. 1 and 2, London, Aug. 1991.
[26] K. Owen, T. Coley: *Automotive Fuels Handbook,* Society of Automotive Engineers Inc., Warrendale, PA 1990, pp. 564–565.
[27] *Octane Week* 1993, July 19, 6.*Octane Week* 1993, May 24, 1, 4, 11.
[28] A. M. Hochhauser et al., *SAE-Paper* **912322** (1991) 6–9.
[29] W. J. Koehl: *SAE-Paper* **912321** (1991)14–16, 28–29.
[30] J. J. Wise, *Fuel Reformulation* 1992, May/June, 64–69.
[31] D. B. Anthony, R. Ragsdale, *Fuel Reformulation* 1992, July/August, 13–18.
[32] UOP: *The Clean Air Act And The Refining Industry,* Des Plaines, Ill. 1991, p. 17.UOP: *Process Solutions for Reformulated Gasoline,* Des Plaines, Ill., 1991.
[33] W. P. Barry: *Gasoline Regulations in USA and Influence To Aromatics Market,* Paper presented to JAIA Annual General Meeting, Tokuyama, Japan, 14 Nov. 1990.
[34] Chem Systems: *Benzene in the 1990's,* London 1991.
[35] Parpinelli Technon: *West European Petrochemical Industry,* 1993 Aromatics Report and Tables, Mailand 1993, p. 202.
[36] H.-G. Franck, J. W. Stadelhofer in: *Industrielle Aromatenchemie,* Springer Verlag, Berlin 1987, p. 134.
[37] A. Chauvel, G. Lefebre: *Petrochemical Processes,* vol. **1,** Editions Technip IFP, Paris 1989, p. 299.
[38] F. Asinger, *Erdöl Erdgas* **106** (1990) no. 6, 263 ff.
[39] H.-G. Franck, J. W. Stadelhofer in: *Industrielle Aromatenchemie,* Springer Verlag, Berlin 1987, p. 133.
[40] A. Chauvel, G. Lefebre: *Petrochemical Processes,* vol. **1,** Editions Technip IFP, Paris 1989, p. 298.
[41] Degener, Krause, VbF/TRbF, vol. **5** (Vorschriften), 16. Lfg., June 1989, pp. 31 ff.
[42] BArbBL, no. 9 (1990) pp. 65 ff.
[43] Kühn, Birett, *Merkblätter gefährliche Arbeitsstoffe,* 58. Erg.-Lfg. 3/92-T 013-1 (1992).
[44] Berufsgenossenschaft Chemie (ed.): *Unfallmerkblatt für den Straßentransport,* MED-Verlagsgesellschaft, Landsberg 1987, Ausgabe 5/87.
[45] J. Fabri, A. Reglitzky, M. Voisey, *Erdöl Erdgas* **107** (1991) no. 1, 28.
[46] K. Owen, T. Coley: *Automotive Fuels Handbook,* Society of Automotive Engineers Inc., Warrendale, PA 1990, pp. 158–160, 490.
[47] DGMK-Tagungsbericht Nr. 8902, Hamburg 1989, pp. 1–71.
[48] TÜV Rheinland: *Aromaten im Abgas von Ottomotoren,* Verlag TÜV-Rheinland, Köln 1988.

[49] *Gefahrstoffe 1993*, Universum Verlagsanstalt, Wiesbaden 1993.
[50] Berufsgenossenschaft der chemischen Industrie, Anlage 4 zu den Unfallverhütungsvorschriften, Jedermann-Verlag Dr. Otto Pfeffer, Heidelberg, Sep. 1993, pp. 3 ff.
[51] E. Browning: *Toxicity and Metabolism of Industrial Solvents*, Elsevier, Amsterdam – London – New York 1965, p. 66.
[52] H. W. Gerarde in G. D. Clayton, F. E. Clayton (eds.): *Industrial Hygiene and Toxicology*, 2nd ed., vol. 2, Wiley-Interscience, New York – London – Sidney 1963, pp. 1222, 1226.
[53] G. G. Fodor: *Schädliche Dämpfe*, VDI-Verlag, Düsseldorf 1972, p. 97.
[54] B. J. Dean, *Mutat. Res.* **47** (1978) 75.
[55] V. A. Benignus, *Neurobehav. Toxicol. Teratol.* **3** (1981) 407.
[56] E. O. Longley et al., *Arch. Environ. Health* **14** (1967) 481.
[57] H. W. Gerarde: *Toxicology and Biochemistry of Aromatic Hydrocarbons, Toluene*, Elsevier, Amsterdam – London – New York – Princeton 1960, p. 140.
[58] MERCK Sicherheitsdatenbank MS-Safe, state as of Jan. 1, 1991.
[59] M. Ikeda et al., *Toxicol. Lett.* **9** (1981) 255.
[60] C. P. Carpenter et al., *Toxicol. Appl. Pharmacol.* **36** (1976) 473.
[61] N. W. Lazarew, *Naunyn Schmiedebergs Arch. Exp. Path. Pharmakol.* **141** (1929) 223.
[62] W. Estler, *Arch. Hyg. (Berlin)* **114** (1935) 261.
[63] K. Nomiyama, H. Nomiyama, *Int. Arch. Occup. Environ. Health* **41** (1978) 55.
[64] E. Reisin, A. Teicher, R. Jaffe, H. E. Eliahou, *Br. J. Ind. Med.* **32** (1975) 163.
[65] A. Fabre, *Bull. Soc. Chim. Biol.* **28** (1946) 764.
[66] K. Bättig, E. Grandjean, *Arch. Environ. Health* **9** (1964) 745.
[67] J. V. Bruckner, R. G. Peterson, *Toxicol. Appl. Pharmacol.* **61** (1981) 27.
[68] V. Morvai, A. Hudak, G. Ungvary, B. Varga, *Acta Med. Acad. Sci. Hung.* **33** (1976) 275.
[69] L. Greenberg, M. R. Mayers, H. Heimann, S. Moskowitz, *JAMA J. Am. Med. Assoc.* **118** (1942) 573.
[70] T. Matsushita et al., *Ind. Health* **13** (1975) 115.
[71] W. Bänfer, *Zentralbl. Arbeitsmed. Arbeitsschutz* **11** (1961) 35.
[72] J. W. Boor, H. I. Hurtig, *Ann. Neurol.* **2** (1977) 440.
[73] C. M. Fischmann, J. R. Oster, *JAMA J. Am. Med. Assoc.* **241** (1979) 1713.
[74] J. Piotrowski, *Med. Pr.* **18** (1967) 129.
[75] K. H. Cohr, J. Stockholm, *Scand. J. Work Environ. Health* **5** (1979) 71.
[76] K. Nomiyama, H. Nomiyama, *Int. Arch. Arbeitsmed.* **32** (1974) 85.
[77] A. Sato et al., *Int. Arch. Arbeitsmed.* **33** (1974) 169.
[78] A. Sato, T. Nakayima, *Br. J. Ind. Med.* **35** (1978) 43.
[79] O. M. Bakke, R. R. Scheline, *Toxicol. Appl. Pharmacol.* **16** (1970) 691.
[80] P. Pfäffli et al., *Scand. J. Work Environ. Health* **5** (1979) 286.
[81] R. P. Bos et al., *Mutat. Res.* **88** (1981) 273.
[82] Litton Bionetics Inc.: *Mutagenicity Evaluation of Toluene*, API Med. Res. Publ. 26-60020, American Petroleum Institute (API), Washington, D.C., Jan. 1978.
[83] P. Gerner-Smidt, U. Friedrich, *Mutat. Res.* **58** (1978) 313.
[84] M. Bauchinger et al., *Mutat. Res.* **102** (1982) 439.
[85] Litton Bionetics Inc.: *Teratology Study in Rats: Toluene*, API Med. Res. Publ. 26-60019, American Petroleum Institute (API), Washington, D.C., Jan. 1978.
[86] J. E. Gibson, J. F. Hardisty, *Fundam. Appl. Toxicol.* **3** (1983) 315.

Toluidines

JOSEPH S. BOWERS, JR., First Chemical Corporation, Pascagoula, MS 39568, United States

1.	Introduction 4745	6.	Derivatives 4749
2.	Physical and Chemical Properties 4746	6.1.	N-Alkyltoluidines 4749
		6.2.	Chlorotoluidines 4749
3.	Production 4747	6.3.	Toluidinesulfonic Acids 4751
4.	Quality Specifications, Storage, and Transportation 4748	6.4.	Chlorotoluidinesulfonic Acids . 4752
		7.	Toluidine Toxicology 4754
5.	Uses 4748	8.	References 4755

1. Introduction

Early commercial production of toluidines involved the nitration of toluene followed by a dissolving metal reduction, usually with iron and hydrochloric acid, to give a mixture of toluidine isomers. These early mixed toluidines were commercially important as dye intermediates. Today the isomers are separated by fractional distillation at the nitrotoluene stage, and the individual compounds typically reduced by catalytic hydrogenation. Current commercial applications for mixed toluidines, and more importantly for the individual toluidine isomers, include their use as: intermediates for dyes and pigments, pesticides, and herbicides; vulcanizing accelerators for the rubber industry; polymer curing agents; corrosion inhibitors; pharmaceutical intermediates; and intermediates for numerous other fine and speciality chemical areas.

Table 1. Physical properties of the toluidines

	o-Toluidine	m-Toluidine	p-Toluidine
CAS registry no.	[95-53-4]	[108-44-1]	[106-49-0]
fp, °C	−14.7	−30.4	44.0
bp, °C (101.3 kPa)	200.2	203.3	200.5
ϱ (20 °C)	0.9984	0.989	0.9619
Vapor pressure, kPa	0.20 (50 °C)	0.13 (41 °C)	0.13 (42 °C)
	5.3 (110 °C)	3.3 (100 °C)	8.0 (140 °C)
Heat of combustion, kJ/mol (20 °C)	4038	4039	4013
Heat of vaporization, J/g (101.3 kPa)	398	534.94	524.77
n_D^{20}	1.5728	1.5670	1.5534
Solubility in water, wt% (25 °C)	1.5	1.6	sl. sol.
Flash point, °C (TOC)	85.0	86.0	87.0
Autoignition temperature, °C	482	482	482

2. Physical and Chemical Properties

Table 1 provides a list of selected physical properties of the three isomers of toluidine (aminotoluene), C_7H_9N, M_r 107.2.

The chemical properties of the toluidines are greatly influenced by delocalization over the aromatic ring of the two unshared electrons on nitrogen, as well as, to a lesser extent, by the electron-donating inductive effect of the methyl group.

The toluidines are relatively weak bases ($pK_b \approx 9$) when compared to aliphatic amines ($pK_b \approx 2-3$) [1] (→ Amines, Aliphatic). However, toluidines do still display many properties similar to those of aliphatic amines, such as alkylation, acylation, and salt formation with strong acids. The toluidines differ from aliphatic amines in their reaction with nitrous acid, in that they form relatively stable diazonium salts which are in turn very important in the textile-dye industry.

Reactions involving the aromatic ring (e.g., substitution reactions; → Amines, Aromatic) are influenced by the strong ring-activating effect of the amine group, and to a lesser extent that of the methyl group. The amine group is a strong *ortho–para* director, and only if the amine group is deactivated by extensive protonation or quaternization does the methyl group exert any significant directing effect.

3. Production

The manufacture of toluidines begins with a mixed-acid (nitric/sulfuric acid) mononitration of toluene, which produces all three isomers (*ortho*, *meta*, and *para*), usually in the ratio 15:1:9, respectively. Since the isomeric toluidines themselves cannot be effectively separated by distillation, distillative separation of the isomers is achieved at the nitrotoluene stage.

The literature describes a multitude of methods for reducing nitrotoluenes to toluidines. (→ Amines, Aromatic ; → Aniline). These range from dissolving-metal reductions, such as the Béchamp process [2], [3] of years ago, to catalytic hydrogenation. Currently, high-volume manufacture of toluidines utilizes the same type of continuous vapor-phase hydrogenation process used for aniline manufacture [4] (→ Aniline). The catalysts include various supported metals [5], [6], Raney nickel [7], [8], copper [4], molybdenum, tungsten, vanadium, and noble metals [9]–[12]. The catalyst is suspended in a column or bed, and a gaseous mixture of the nitrotoluene and excess hydrogen is passed over it at about 250 °C. The gaseous products are condensed, the aqueous and organic layers are separated, excess hydrogen is recycled, and the final crude product is dried and distilled.

Batch liquid-phase catalytic hydrogenation is a common process for the production of lowervolume specialty toluidines or amines too high in molecular mass or too sensitive for vapor-phase reactions. In these cases, liquid nitro compounds can either be reduced neat or else dissolved in appropriate aromatic solvents, aliphatic alcohols, or esters. Catalysts include Raney nickel and supported noble metals. Reduction is conducted between room temperature and 150 °C under hydrogen pressures in the range 350–3500 kPa.

There are several alternative possibilities for reducing nitrotoluenes to toluidines, utilizing such reducing agents as hydrazine [13], alkaline sulfide [14], and sodium hydrosulfite [15]. These methods rarely find commercial application unless the substrate nitro compound has a functional group that might be adversely affected by catalytic hydrogenation.

Additional manufacturing techniques involve the amination of corresponding phenol derivatives [16], [17]. This method has been used commercially to produce *m*-toluidine from *m*-cresol.

4. Quality Specifications, Storage, and Transportation

Typical purity of each of the commercially available toluidine isomers is ca. 99.5% minimum, with a maximum of 0.5% total for all other isomers. Moisture is usually in the range 0.1–0.2% maximum [18]. Additional specifications important for specific isomer applications are sometimes stated, such as a 1 °C distillation range for 5% to 95% of a distilled *o*-toluidine, or a freezing point minimum of 43.0 °C for *p*-toluidine. Assays and moisture content are usually established by GC analysis.

Toluidines are typically stored in steel tanks or drums. All the toluidine isomers exhibit solvent-like properties, so plastics, wax coatings, or other nonmetallic materials are rarely permitted in the context of storage or handling. Teflon TFE or FEP fluorocarbon resins are the preferred materials for gaskets and packaging. When stored in the presence of air all the toluidine isomers tend to discolor, particularly over an extended period of time.

Toluidines are typically shipped in steel drums, tank trucks, rail tank cars, or marine vessels. For transportation by truck, rail, or ship the applicable hazard classification is UN 1708, RQ.

5. Uses

Since *o*-toluidine is the isomer produced in greatest volume, it has found the most commercial applications. By far the single largest use for *o*-toluidine is in the preparation of 6-ethyl-*o*-toluidine, an intermediate in the manufacture of the two very large-volume herbicides Metolachlor [19] and Acetochlor [20]. Another important use area for *o*-toluidine is rubber chemicals. Thus, *o*-toluidine is used in the manufacture of a rubber antioxidant developed by Goodyear called Wingstay as well as di-*o*-tolylguanidine, a nonstaining rubber accelerator (→ Amines, Aromatic). Acetoacet-*o*-toluidine, 3-hydroxy-2-naphthoyl-*o*-toluidine, 2-toluidine-5-sulfonic acid, and *o*-aminoazotoluene are four of the more important dye and pigment intermediates manufactured from *o*-toluidine. In addition, *o*-toluidine is used to manufacture epoxy resin hardeners such as methylene-bis-2-methylcyclohexylamine [21], fungicide intermediates such as 2-amino-4-methylbenzothiazole [22], and *o*-fluorobenzoyl chloride [23]. Certain pharmaceutical intermediates are also prepared starting with *o*-toluidine [24].

Both *m*- and *p*-toluidines still find considerable application in the dye and pigment industries, as well as in the manufacture of triarylmethane dyes. Two compounds derived from *m*- and *p*-toludine, *N,N*-dimethyl-*m*-toluidine and *N,N*-dimethyl-*p*-toluidine, have found use as accelerators for peroxide-cured epoxy and unsaturated polyester resins [25]–[27]. *m*-Toluidine is used in the manufacture of three important photographic color developers: color developer 2 uses *N,N*-diethyl-*m*-toluidine as an intermediate, and color developers 3 and 4 are based on *N*-hydroxyethyl-*N*-ethyl-*m*-toluidine. Just as with *o*-toluidine, the *m*- and *p*-toluidines find many applications as starting materials and intermediates in the manufacture of various fine and specialty chemicals [28]–[30].

6. Derivatives

6.1. *N*-Alkyltoluidines

N-Alkyltoluidines can be prepared from unalkylated toluidines with such common alkylating agents as alkyl halides, sulfates, and carbonates. However, large-scale manufacture of *N*-alkyltoluidines usually involves one of two alternative methods: acid-catalyzed alkylation with lower unbranched alcohols or ethers [31], or reductive alkylation with lower aldehydes or ketones over active-metal catalysts (e.g., nickel, platinum, or palladium) and under hydrogen pressure [32]. Toluidines can also be alkylated with reactive three-membered-ring compounds such as ethylene oxide and aziridine.

Important examples include:

N-Methyl-*o*-toluidine, [611-21-2], M_r 121.18, *bp* 207–208 °C, ϱ 0.97
N-Ethyl-*o*-toluidine, [94-68-8], M_r 135.21, *bp* 218–219 °C, ϱ 0.94
N,N-Dimethyl-*o*-toluidine, [609-72-3], M_r 135.21, *bp* 221–222 °C, ϱ 0.94
N-Methyl-*m*-toluidine, [696-44-6], M_r 121.18, *bp* 206–207 °C, ϱ 0.95
N-Ethyl-*m*-toluidine, [102-27-2], M_r 135.21, *bp* 221 °C, ϱ 0.96
N,N-Dimethyl-*m*-toluidine, [121-72-2], M_r 135.21, *bp* 215–216 °C, ϱ 0.93
N-Ethyl-*N*-(2-hydroxyethyl)-*m*-toluidine, [91-88-3], M_r 179.26, *bp* 114–115 °C (0.1 kPa), ϱ 1.02
N-Methyl-*p*-toluidine, [623-08-5], M_r 121.18, *bp* 208–209 °C, ϱ 0.94
N-Ethyl-*p*-toluidine, [622-57-1], M_r 135.21, *bp* 217–218 °C, ϱ 0.95
N,N-Dimethyl-*p*-toluidine, [99-97-8], M_r 135.21, *bp* 211–212 °C, ϱ 0.94

6.2. Chlorotoluidines

There exist four monochlorinated isomers each for *o*- and *m*-toluidine, and two for *p*-toluidine. Some are rather difficult to prepare, and thus do not find commercial application. Chlorotoludines are usually prepared by reducing the nitro group of the corresponding chloronitrotoluene. These reductions have traditionally been accomplished by dissolving-metal reductions, such as the Béchamp process. Today, catalytic hydrogenation is the method of choice, especially since catalysts with improved selectivity almost eliminate dehydrochlorination side reactions [33], [34]. The intermediate chloronitrotoluenes are obtained by either chlorination of nitrotoluenes or nitration of chlorotoluenes. Direct chlorination of toluidines can also be accomplished by first protecting the amine function. Chlorotoluidines find commercial applications in such areas as azo dyes, agricultural chemicals, and pharmaceuticals.

3-Chloro-o-toluidine, 3-chloro-2-methylaniline [87-60-5] (**1**), M_r 141.6, mp 2 °C, bp 241–242 °C, ϱ 1.19, is produced by the catalytic hydrogenation of 3-chloro-6-nitrotoluene [35], which is in turn obtained via the acidic chlorination of o-nitrotoluene followed by separation of isomers by distillation. An alternate method of preparing the nitro intermediate involves nitration of o-chlorotoluene, which also produces a mixture of isomers that can be separated by distillation. The compound is used in the manufacture of azo dyes, and as an intermediate in the manufacture of 2,6-dichlorotoluene.

4-Chloro-o-toluidine, 4-chloro-2-methylaniline [95-69-2] (**2**), M_r 141.6, mp 29–30 °C, bp 240–241 °C, is produced by the direct chlorination of acetyl-protected o-toluidine. After removal of the protecting group with base, the 4-chloro-o-toluidine is separated by distillation from the 6-chloro-o-toluidine isomer, described below. This compound is employed in the manufacture of the pesticide Chlordimeform [36], and for making acid azo dyes. Use of the pesticide is declining because of concern over its carcinogenicity [37], [38], and Chlordimeform has in fact been banned in several countries.

5-Chloro-o-toluidine, 5-chloro-2-methylaniline [95-79-4] (**3**), M_r 141.6, mp 20–22 °C, bp 237 °C, is produced by reduction of the nitro group in 4-chloro-2-nitrotoluene, which is itself prepared by nitration of p-chlorotoluene followed by distillatory separation of 4-chloro-2-nitrotoluene from the accompanying 4-chloro-3-nitrotoluene isomer (see below 6-chloro-m-toluidine).

6-Chloro-o-toluidine, 6-chloro-2-methylaniline [87-63-8] (**4**), M_r 141.6, mp 8–9 °C, bp 220 °C, ϱ 1.15, is co-produced along with 4-chloro-o-toluidine, as described above, and is used as an intermediate in the manufacture of azo and xanthene dyes [39] as well as pharmaceuticals [40], [41].

6-Chloro-m-toluidine, 6-chloro-3-methylaniline [95-81-8] (**5**), M_r 141.6, mp 29–30 °C, bp 228–230 °C, is produced by the nitro reduction of 4-chloro-3-nitrotoluene, where the nitro intermediate is co-produced in the nitration of p-chlorotoluene described above under 5-chloro-o-toluidine. It is used in the manufacture of azo dyes.

3-Chloro-*p*-toluidine, 3-chloro-4-methylaniline [*95-74-9*] (**6**), M_r 141.6, mp 24–25 °C, bp 237–238 °C, is produced by the catalytic hydrogenation [8] of 2-chloro-4-nitrotoluene, obtained in the chlorination of *p*-nitrotoluene. It serves as an intermediate in the manufacture of the herbicide Chlortoluron [42], [43] and the red pigment intermediate "2B-Acid."

6.3. Toluidinesulfonic Acids

There are ten distinct toluidinesulfonic acid isomers. Several of the isomers are difficult to prepare, and only five have found commercial use. Toluidinesulfonic acids find application in the manufacture of direct dyes, acid dyes, pigments, and triarylmethane dyes.

The toluidinesulfonic acids are produced either by sulfonation of the appropriate nitrotoluene, or by nitration of a toluenesulfonic acid followed by catalytic hydrogenation or chemical reduction of the nitro group. Other methods involve direct sulfonation of the appropriate toluidine isomer.

Commercial toluidinesulfonic acids are light yellow to red. Often they are hygroscopic solids without defined melting points. Just as with other amino acids, the toluidinesulfonic acids form zwitterionic inner salts that are sparingly soluble in water and dilute mineral acids, but very soluble in aqueous caustic (→ Amino Acids).

3-Amino-4-methylbenzenesulfonic acid, 3-amino-*p*-toluenesulfonic acid [*618-03-1*] (**7**), M_r 187.2, is produced by the low-temperature sulfonation of *o*-nitrotoluene with oleum followed by catalytic hydrogenation or chemical reduction. It is used in the manufacture of azo dyes and nonlinear optical materials [44].

4-Amino-3-methylbenzenesulfonic acid, 4-amino-*m*-toluenesulfonic acid [*98-33-9*] (**8**), M_r 187.2, is produced by the sulfonation of *o*-toluidine. The conditions entail heating at 150–200 °C in concentrated sulfuric acid, or heating the *o*-toluidine sulfate salt either neat or in an inert solvent. The compound is used as an intermediate in the manufacture of azo and diazo acid dyes [45], pigments [46], and fluorescent brighteners [47].

4-Amino-2-methylbenzenesulfonic acid, 4-amino-*o*-toluenesulfonic acid [*133-78-8*] (**9**), M_r 187.2, is produced by sulfonation of *m*-nitrotoluene to give 4-nitro-2-methylbenzenesulfonic acid, followed by catalytic hydrogenation or chemical reduction. Much older technology involves baking *m*-toluidine sulfate or heating *m*-toluidine with sulfamic acid. The compound is used in the manufacture of azo dyes.

5-Amino-2-methylbenzenesulfonic acid, 5-amino-*o*-toluenesulfonic acid [*118-88-7*] (**10**), M_r 187.2, is produced by the sulfonation of *p*-nitrotoluene with oleum or SO_3, followed by catalytic hydrogenation or chemical reduction. It is used as an intermediate in the manufacture of azo and anthraquinone acid dyes.

2-Amino-5-methylbenzenesulfonic acid, 6-amino-*m*-toluenesulfonic acid [*88-44-8*] (**11**), M_r 187.2, is produced by direct sulfonation of *p*-toluidine with either concentrated sulfuric acid at high temperature or various concentrations of oleum [48], or by heating *p*-toluidine sulfate salt neat [49] or in an inert solvent [50]. The compound is better known as 4B Acid, and is used as an intermediate in the manufacture of several azo acid dyes, plastics, and printing-ink pigments (e.g., Pigment Red 57, Lithol Rubine), as well as fluorescent brighteners [51].

6.4. Chlorotoluidinesulfonic Acids

There are thirty different chlorotoluidinesulfonic acid isomers, only a few of which have commercial applications. The commercial potential of these compounds is determined by the availability of appropriate starting materials and the selectivity of the manufacturing process. Many of the possible isomers are prepared as mixtures containing other isomers, and are unusable as such for most commercial applications. Thus, only reactions that produce essentially one isomer are of significance. Commercial applications center around the manufacture of dyes and pigments for the textile, paint, printing ink, and plastics industries.

Chlorotoluidinesulfonic acids can be produced by a variety of methods depending on the choice of starting materials. Chlorination of nitrotoluenesulfonic acids or sulfonation of chloronitrotoluenes both result in chloronitrotoluenesulfonic acids, which can then be catalytically hydrogenated or chemically reduced to the desired product. Care must be taken in selecting the hydrogenation conditions, because catalytic dehalogenation can become the major reaction path with certain catalysts (e.g., very active platinum or palladium catalysts). Some nickel and poisoned platinum and palladium catalysts minimize the extent of dehalogenation [33]. Additional manufacturing methods for chlorotoluidinesulfonic acids include the chlorination of toluidinesulfonic acids and the sulfonation of chlorotoluidines.

Chlorotoluidinesulfonic acids also form zwitterionic inner salts, with solubility properties much like those of the toluidinesulfonic acids.

2-Amino-3-chlorotoluene-5-sulfonic acid, 4-amino-3-chloro-5-methylbenzenesulfonic acid [6387-14-0] (**12**), M_r 221.7, can be produced by the chlorination of acyl-protected 2-aminotoluene-5-sulfonic acid [52], by baking the sulfate salt of 6-chloro-2-methylaniline, or by sulfonation of 6-chloro-3-methylaniline with oleum. It serves as an intermediate in the manufacture of such azo acid dyes as Direct Red 108, C.I. 28165.

2-Amino-6-chlorotoluene-3-sulfonic acid, 2-amino-4-chloro-3-methylbenzenesulfonic acid [79380-22-6] (**13**), M_r 221.7, is produced either by sulfonation of 3-chloro-2-methylaniline with oleum at 70–90 °C or by heating the corresponding sulfate salt at 150–200 °C, the "baking method" [53]. It is an intermediate in the manufacture of azo dyes.

2-Amino-6-chlorotoluene-4-sulfonic acid, 3-amino-5-chloro-4-methylbenzenesulfonic acid [6387-27-5] (**14**), M_r 221.7, is produced by sulfonating 6-chloro-2-nitrotoluene with 20% oleum to give 6-chloro-2-nitrotoluene-4-sulfonic acid, followed by catalytic hydrogenation [54] or chemical reduction. It is used as the diazo component for a series of acid azo dyes intended primarily for leather dying (e.g., Acid Brown 105, C.I. 33530).

3-Amino-6-chlorotoluene-4-sulfonic acid, 2-amino-5-chloro-4-methylbenzenesulfonic acid [88-53-9] (**15**), M_r 221.7, was traditionally prepared by a route that proceeds from *p*-toluenesulfonic acid via sequential chlorination, nitration, isomer separation, and reduction. More recent modifications begin with *m*-toluidine and involve acylation, chlorination, de-acylation, and sulfonation [55]. Yet another route uses *o*-chlorotoluene as starting material and proceeds via nitration, catalytic hydrogenation, and sulfonation of the amine [56]. Thiopolyols have been shown to be effective dehalogenation inhibitors during hydrogenation [57]. The compound is known commercially by several common names (e.g., Red Lake C Amine, Lake Red C Amine, C-Acid), and constitutes the diazo component for various azo dyes, metallized dyes, and pigments intended primarily for printing inks and plastics (e.g., Pigment Red 52, C.I. 15860, and Pigment Red 53, C.I. 15585).

4-Amino-6-chlorotoluene-3-sulfonic acid, 2-amino-4-chloro-5-methylbenzenesulfonic acid [88-51-7] (**16**), M_r 221.7, is made from *p*-nitrotoluene by first chlorinating to obtain 2-chloro-4-nitrotoluene, which is catalytically hydrogenated to the amine and finally sulfonated by heating the sulfate salt in *o*-dichlorobenzene to 100–120 °C [58], in sulfolane to 160 °C [59], or in 1,2-dichloroethane in an autoclave to 140 °C [60]. The compound is commonly known as Brilliant Toning Red Amine or Red 2B Acid, and is used as an intermediate in the manufacture of acid azo dyes. The azo dyes and their metallized salts are in turn used as pigments (e.g., Pigment Red 48, C.I. 15865) for paints, printing inks, and plastics [61], [62].

7. Toluidine Toxicology

The *o*-, *m*-, and *p*-toluidines as well as the chlorotoluidines belong to a group of chemicals known as potential "blue lip" compounds. This refers to the serious condition that results from overexposure to toluidines via inhalation, skin absorption, or even ingestion. All the toluidines are easily absorbed through the skin and mucous membranes.

Toluidines in the blood stream cause the conversion of hemoglobin to methemoglobin. The latter does not combine with oxygen, and in high concentrations renders the blood stream incapable of carrying oxygen from the lungs to the rest of the body. The earliest manifestation of poisoning is a bluish discoloration (*cyanosis*) of the lips, mucous membranes, fingernail beds, earlobes, and tongue. Minor levels of oxygen deficiency lead to a temporary sense of well-being and exhilaration. As the lack of oxygen increases, such symptoms as headache, drowsiness, and occasional nausea and vomiting appear. More profound decreases in blood oxygen lead to stupor, unconsciousness, and even death if treatment is not obtained immediately. All the toluidines are considered skin and eye irritants, and may cause an allergic skin reaction in some people. Pre-existing diseases of the central nervous system, cardiovascular system, liver, or bone marrow may be aggravated by excessive exposure to the toluidines. Toxicity characteristics are summarized in Table 2.

The December 1990 NIOSH Alert associated worker exposure to a combination of aniline and *o*-toluidine with an increased risk of bladder cancer; the effects of aniline and *o*-toluidine could not be separated. NIOSH concluded that *o*-toluidine was the more likely agent to be responsible for the observed bladder cancers [64]. No studies currently described in the literature directly link worker exposure to *m*- and *p*-toluidine to a carcinogenic effect in humans.

Table 2. Toxicity data for the isomeric toluidines

	o-Toluidine	m-Toluidine	p-Toluidine
Inhalation (LC_{50})	ND*	ND	ND
Humans (TCL_0)**	25 mg/m^3	ND	ND
Ingestion, mg/kg (LC_{50})			
Rats	670	450	656
Mice	520	470	330
Rabbits	840	ND	270
Skin Contact:			
Rabbits	LD_{50} (3250 mg/kg)	mildly irritating (500 mg/24 h)	irritating (500 mg/24 h)
Eye Contact:			
Rabbits	corneal damage	mildly irritating (20 mg/24 h)	severe irritation (100 mg/24 h)

*ND = not determined.
**TLC_0 = lowest published toxic concentration.

o-Toluidine, as the hydrochloride, has been tested for carcinogenicity in mice and rats by oral administration, and it was found to produce neoplasms at various sites in both species [65], [66]. No neoplasms were observed following subcutaneous injection in hamsters [67]. p-Toluidine showed cancerous potential in mice but not in rats, and m-toluidine did not produce definitive cancerous activity in either animal species. All three toluidines have been assigned a TLV of 2 ppm.

Chlorinated toluidines have received considerable attention as potential carcinogens. Animal tests have shown both 4-chloro-o-toluidine and 5-chloro-o-toluidine to be carcinogenic in mice, but not in rats [68], [69]. No convincing evidence was found for carcinogenicity in mice or rats with 3-chloro-p-toluidine [70]. A recent study established a strong correlation between worker exposure to 4-chloro-o-toluidine and increased incidence of bladder cancer [71], [72]. On the basis of this study a commission of the Deutsche Forschungsgemeinschaft categorized the compound as a human carcinogen, making it the first example of a monocyclic aromatic amine to be so designated.

All toxicity data indicate that care should be taken to minimize exposure when working with any toluidine, and extreme caution should be exercised when working with o-toluidine and its derivatives.

8. References

[1] R. T. Morrison, R. N. Boyd: *Organic Chemistry*, 2nd ed., Allyn and Bacon, Inc., Boston 1966, pp. 720–721.
[2] I. G. Farbenind., GB 274 562, 1926.
[3] I. G. Farbenind., GB 279 283, 1926.
[4] Lonza Ltd., DE 1 809 711, 1969 (L. Szigeth).
[5] Ruhrchemie, DE 1 543 836, 1972 (W. Rottig).
[6] Texaco Development Corp., DE 2 210 564, 1972 (J. Knifton, R. Suggitt).

[7] Nippon Kayaku Co., JP 49 042 617, 1974 (K. Akiyoshi, S. Shimozawa, H. Danjo, S. Shizuhiko, S. Furuhashi).
[8] Bayer, DE 2 743 610, 1979 (J. Zander).
[9] GAF Corp., US 3 093 685, 1963 (H. Freyermuth, D. Graham, E. Hort).
[10] Bayer, DE 2 135 154, 1973 (H. Thelen, W. Schwerdtel, K. Halcour, W. Swodenk).
[11] Clayton Aniline Co., EP 2 308, 1979 (J. Hildreth, D. Haslam, D. Allen).
[12] Du Pont de Nemours, US 4 212 824, 1980 (R. Seagroves).
[13] P. M. G. Bavin, *Can. J. Chem.* **36** (1958) 238.
[14] W. W. Hartman, H. L. Silloway, *Org. Synth. Coll.* **3** (1955) 83.
[15] L. F. Fieser, M. Fieser, *J. Am. Chem. Soc.* **56** (1934) 1565.
[16] ICI, DE 564 436, 1936.
[17] Ashi Chemical Ind., JP 7 773 829, 1977 (S. Enomoto, T. Kamiyama, I. Tsutomu).
[18] First Chemical Corp., Toluidine Sales Specifications, Pascagoula, Miss., 1993.
[19] Ciba-Geigy, DE 2 328 340, 1973 (C. Vogel, R. Aebi).
[20] Monsanto, US 4 258 196, 1981 (J. Chupp, R. Goodin).
[21] BASF, FR 1 514 526, 1968.
[22] Eli Lilly & Co., DE 2 250 077, 1973 (C. Paget).
[23] ICI, EP 15 756, 1980 (K. Parry, P. Worthington, W. Rathmell).
[24] Bayer, DE 2 262 830, 1974.
[25] R. F. Storey, D. L. Smith: "Improved Room-Temperature Cure of Unsaturated Polyester Resins," *Mod. Plast.* **61** (1984) 58.
[26] First Chemical Corp. and The University of Southern Mississippi, US 4 666 978, 1987 (R. Storey, M. Hogue, A. Bayer).
[27] Bayer, DE 3 202 090, 1983 (E. Eimers, D. Margotte, R. Dhein, K. Kraft, W. Kloler).
[28] Nippon Kayaku, JP 92 020 950, 1992.
[29] Kawaguchi Chemical Ind., JP 82 139 542, 1981.
[30] Ciba-Geigy, BE 800 471, 1973.
[31] Biller, Michaelis & Co., US 2 991 311, 1958 (M. Thoma).
[32] A. P. Bonds, H. Greenfield, "The Reductive Alkylation of Aromatic Amines With Formaldehyde," *Chem. Ind. (London)* **47** (1992) 65.
[33] H. Greenfield, F. S. Dovell, "Metal Sulfide Catalysts for Hydrogenation of Halonitrobenzenes to Haloanilines," *J. Org. Chem.* **32** (1967) 3670.
[34] Bayer, US 4 230 637, 1980 (J. Zander).
[35] DuPont de Nemours, DE 2 615 079, 1975 (J. Kosak).
[36] Ciba Corp., US 3 496 270, 1970 (C. Counselman).
[37] K. Hooper, J. LaDou, J. Rosenbaum, S. Book: "Regulation of Priority Carcinogens and Reproductivity or Developmental Toxicants," *Am. J. Ind. Med.* **22** (1992) 793.
[38] S. Xue et al.: "Assessment on the Carcinogenic Risk of Chlordimeform," *Huanjing Huaxue* **10** (1991) 54.
[39] Hoechst, DE 2 504 925, 1976 (M. Haehnke, R. Neeb, T. Papenfuhs).
[40] Beecham Group PLC, EP 61 907, 1982 (D. Smith).
[41] Otsuka Pharmaceutical Co., FR 2 477 542, 1981 (K. Banno, T. Fujiok, M. Osaki, K. Nakagawa).
[42] E. Gy, T. Gyogyszervegyeszeti Gyar, HU 19 950, 1981 (I. Bitter, A. Kis-Tamos, F. Jurak, L. Kocsis).
[43] Valles Industria Organica, ES 2 027 074, 1992 (M. Camps-Anaya, M. Riba-Garcia).
[44] Sony Corp., JP 05 210 130, 1993 (Y. To).
[45] Nippon Kayaku Co., EP 190 947, 1986 (R. Ueno, S. Itoh, S. Fujimoto, Y. Komatsu, H. Tsuchiya).

[46] Toyo Ink Mfg. Co., FR 2 564 099, 1985 (T. Funatsu, M. Hayashi, Y. Ohtomo).
[47] Showa Chemical Ind., JP 53 102 329, 1978 (K. Suzuki).
[48] Bayer, DE 3 401 572, 1985 (H. Blank, H. Emde, R. Schimp).
[49] Ciba-Geigy Corp., US 4 681 710, 1987 (R. Miller, J. Billini, R. Schneider).
[50] Bayer, DE 3 114 830, 1982 (H. Emde, H. Blank, P. Schnegg).
[51] Ciba-Geigy, DE 2 233 429, 1973 (F. Guenter, W. Fringeli, P. Liechti).
[52] BASF, DE 218 370, 1911.
[53] Hoechst, DE 398 049, 1927.
[54] Dinippon Ink and Chemical, Inc., JP 57 014 571, 1982.
[55] Dainichiseika Color and Chemical Mfg. Co., JP 52 062 245, 1977 (K. Noguchi, T. Fukuda, S. Hitotsuyanagi).
[56] Hoechst, DE 3 141 286, 1983 (T. Papenfuhs, R. Berthold).
[57] Dainippon Ink and Chemicals, Inc., JP 57 021 362, 1982.
[58] Combinatul Chimic, RO 76 842, 1981 (L. Vladutiu, C. Atanasiu-Meves, A. Spiliadis, D. Dobrescu, V. Aldea).
[59] Bayer, DE 3 144 830, 1982 (H. Emde, U. Blank, P. Schnegg).
[60] Bayer, DE 3 114 829, 1982 (H. Emde, U. Blank, P. Schnegg).
[61] Toyo Ink Mfg. Co., JP 03 072 574, 1991 (Y. Otomo, H. Sugamo).
[62] Hoechst, DE 3 833 226, 1990 (W. Deucker).
[63] First Chemical Corp., Toxicity Data from Respective Material Safety Data Sheets, Pascagoula, Miss., 1992.
[64] E. Ward et al.: "Excess Number of Bladder Cancers in Workers Exposed to Ortho-toluidine and Aniline," *JNCI J. Natl. Cancer Inst.* **83** (1991) 501.
[65] *IARC Monogr.* **27** (1982) 155.
[66] S. Hecht, K. El-Bayoumy, A. Rinenson, E. Fiala: "Comparative Carcinogenicity of o-Toluidine Hydrochloride and o-Nitrosotoluene in F-344 Rats," *Cancer Lett. (Shannon, Irel.)* **16** (1982) 103.
[67] S. Hecht, K. El-Bayoumy, A. Rivenson, E. Fiala: "Bioassay for Carcinogenicity of 3,2′-Dimethyl-4-nitrosobiphenyl, o-Nitrosotoluene, Nitrosobenzene and the Corresponding Amines in Syrian Golden Hamsters," *Cancer Lett. (Shannon, Irel.)* **20** (1983) 349.
[68] National Cancer Institute: "Bioassay of 4-Chloro-*o*-toluidine for Possible Carcinogenicity," DHEW Publ. no. NIH-79-1721, Washington, D.C., 1978, US Govt. Printing Off.
[69] National Cancer Institute: "Bioassay of 5-Chloro-*o*-Toluidine for Possible Carcinogenicity," DHEW Publ. no. NIH-79-1743, Washington, D.C., 1978, US Govt. Printing Off.
[70] National Cancer Institute: "Bioassay of 3-Chloro-*p*-Toluidine for Possible Carcinogenicity," DHEW Publ. no. NIH-78-1701, Washington, D.C., 1978, US Govt. Printing Off.
[71] M. Stasik: "Carcinomas of the Urinary Bladder in 4-Chloro-o-toluidine Cohort," *Int. Arch. Occup. Environ. Health* **60** (1988) 21.
[72] M. Stasik: "Urinary Bladder Cancer from 4-Chloro-o-toluidine," *Dtsch. Med. Wochenschr.* **116** (1991) 1444.

Turpentines

Individual keyword: → *Terpenes.*

MANFRED GSCHEIDMEIER, Hoechst Aktiengesellschaft, Werk Gersthofen, Augsburg, Federal Republic of Germany (Chaps. 1–13, 15–16)

HELMUT FLEIG, BASF Aktiengesellschaft, Ludwigshafen, Federal Republic of Germany (Chap. 14)

1.	Turpentine as a Natural Product 4759	9.	Safety and Occupational Health 4770
2.	History 4760	10.	Storage and Transportation . . 4771
3.	Definition and Classification . 4761	11.	Ecological Aspects 4771
4.	Production 4762	12.	Uses 4772
4.1.	Gum Turpentine 4762	13.	Economic Aspects 4773
4.2.	Wood Turpentine 4763	14.	Toxicology 4774
4.3.	Sulfate Turpentine 4763	15.	Turpentine Substitute 4776
4.4.	Sulfite Turpentine 4764	16.	Pine Oil 4776
4.5.	Destructively Distilled Wood Turpentine 4765	16.1.	Production 4776
		16.2.	Properties 4776
5.	Composition 4765	16.3.	Storage and Transportation . . 4777
6.	Physical Properties 4767	16.4.	Uses 4778
7.	Chemical Properties 4769	16.5.	Economic Aspects 4778
8.	Analysis 4769	17.	References 4779

1. Turpentine as a Natural Product

Turpentine, which occurs in many natural forms, is a component of conifer balsam and plays an important role in the ecosystem of the pine forest [5]. It is part of the plant metabolism, and its individual components are important for the protective system of the conifers: for rapid closure of wounds to the bark by resin formation, for repelling insects (e.g., myrcene repels the pine beetle), and for protection against attack by bacteria (effect of, e.g., 3-carene).

The volatile terpenes (and polymers derived from them) emitted to the atmosphere are partially absorbed by the forest soil and contribute to humus formation [6]. This emission occurs to a great extent, particularly in summer. In Russian pine forests, terpene concentrations of 3.5–35 µg per cubic meter of air have been measured [7], [8]. In similar investigations in the state of Pennsylvania (total area 117 350 km^2, of which 62 000 km^2 is forest) an emission of 3400 t of terpene hydrocarbons was found for one day in the month of August [9]. The annual worldwide biogenic production of terpenes is estimated to be 10^9 t [10]. Since this quantity is degraded by natural routes, the environmentally relevant terpenes can be considered problem-free with regard to air pollution. Only in combination with nitrogen oxides and with UV radiation may ozone be formed.

2. History

In ancient times turpentine obtained from conifer balsam was very important as a fuel for lamps and torches; as a solvent; as a fragrance; and as an embalming oil. Cuneiform script texts and archaeological finds indicate that turpentine was known by around 3000 BC in Mesopotamia [11] and Pakistan [12]. The preparation and use of distilled cedarwood oil ("turpentine oil" from *Pistacia terebinthus*) were described in early Egyptian culture (e.g., for embalming [13]). Turpentine was also known to the Greeks, Celts, and Romans. In alchemists' books of the 13th century, turpentine oil was described as *aqua ardens*, and in the distillery book of the Strasbourg doctor H. BRUNSCHWYGH (1450–1543) as *spiritus terpenthinae*.

In China, turpentine was used early on as a solvent for rosin (colophony) for impregnating sail canvas and rigging and sealing ships' hulls. This kind of use was very important among the maritime powers in the Middle Ages.

In Europe, *gum spirit of turpentine* (also known as *gum turpentine*) was produced in the 16th century in Poland, Germany, Scandinavia, and Russia; in the 19th century in France, Spain, Portugal, and Finland; and since 1902 in Greece. In North America, turpentine has presumably been extracted since 1606 in Nova Scotia. Since 1710 it has been produced to a greater extent in North Carolina, and later in other states. For a long time a mixture of turpentine and spirit—the lamp fuel camphine—was an important commercial product.

In ca. 1870, steam distillation of turpentine was introduced, followed in 1910 by extraction from tree stumps left after deforestation. In this way, the cheaper *wood turpentine* and *pine oil* came on the market.

Sulfate turpentine, a byproduct of pulp production, was obtained in small quantities in Norway and Sweden already before 1910, in Finland since 1911, and in the United States since 1918. Only since 1930 has it been available in large quantities as a low-priced raw material [14]. From 1954 onward its production in the United States overtook that of all turpentines of other origins.

The first elemental analysis of turpentine was carried out in 1818 by J. J. HOUTON DE LABILLARDIERE. A little later the essential oils of gum turpentine were investigated scientifically by J. B. DUMAS (1833), J. J. BERZELIUS, J. VON LIEBIG, and F. WÖHLER (1837). In 1862, M. BERTHELOT began his studies on terpene hydrocarbons (e.g., pinene hydrochloride), and 1877 W. TILDEN began to characterize the hydrocarbon components of spirit of turpentine by means of their crystalline nitrosyl chlorides. In 1884, O. WALLACH succeeded in structurally characterizing the components of turpentine oils, and in 1893 A. VON BAEYER began exploring the field of cyclic terpenes [15]. More recently, the biosynthesis of monoterpenes has been investigated by many scientists [16]–[21].

3. Definition and Classification

Turpentine [8006-64-2] (oil of turpentine, turpentine oil, spirit of turpentine) is the collective name for the volatile components of the crude resins (balsam or oleoresin, Latin: *resina terebinthi*) from conifers, particularly the pine family. The turpentines represent a subgroup of plant and tree saps (essential oils).

Depending on the starting material and production method, the following types of turpentine are defined according to DIN 53 248 (April 1977): gum spirit of turpentine, steam-distilled wood turpentine, refined sulfate wood turpentine, and destructively distilled wood turpentine.

The use of the term turpentine without further specification is not permitted. Usually, the name of the country of origin is added (e.g., Greek gum spirit of turpentine). The addition of products that are not typical components of turpentine must be declared.

Gum spirit of turpentine (Latin: *oleum terebinthinae*, French: *essence de térébinthine*, German: *Balsamterpentinöl*) according to DIN 53 248 denotes only the pure essential plant oil obtained by distillation of the resin exudate (balsam or turpentine) from living trees of the genus *Pinus* at up to 180 °C (under normal pressure). The name must not be used for material obtained after the extraction of several components (e.g., pinene).

Steam-distilled wood turpentine (French: *essence de bois distillée à la vapeur*, German: *Wasserdampfdestilliertes Wurzelterpentinöl*) is a turpentine obtained either directly from the steam distillation of finely chopped wood from roots and stumps of trees of the *Pinus* family or after extraction with solvents, which is otherwise unchanged.

Refined sulfate wood turpentine (French: *essence de bois dite au sulfate* or *essence de papeterie de sulfaté desodorisée*, German: *Gereinigtes Sulfatterpentinöl*) is a byproduct of pulp production using the soda or sulfate pulping process, that is subsequently purified.

Crude sulfate turpentine (French: *essence de papeterie de sulfate brute*, German: *Rohsulfatterpentinöl*), which is not purified, is now rarely marketed, because the sulfur compounds it contains produced a nauseating odor.

Destructively distilled wood turpentine (French: *essence de bois par destillation destructive*, German: *Gereinigtes Kienöl*) is the name given to the oil obtained by dry distillation of resin-rich pinewood followed by purification. It must not be called turpentine or oil of turpentine.

Sulfite turpentine and pine oil are not defined according to DIN 53 248, but are generally related to the other turpentines (see Section 4.4, Chap. 16). Sulfite turpentine is a byproduct of pulp production by the sulfite pulping process. Its composition (main component p-cymene) differs significantly from that of the other turpentines [22].

The term *naval stores* relates to all rosin and oil products extracted from conifer wood, such as turpentine, rosin, pine oil, or tall oil. It originates from the time when the United States was still an English colony and when these types of product were used widely in shipbuilding (e.g., in paint, to seal ships' hulls, or to treat rigging) and were therefore stored and sold in naval stores.

4. Production

4.1. Gum Turpentine

Gum turpentine is extracted in many countries with warm climates (more frequently in the Northern Hemisphere) from the wood resin or crude balsam of various conifers, particularly those of the pine family [23]. Incisions made in the trunks of fully grown trees are deep enough to allow crude balsam to flow out of the resin channels. It can then be collected in removable vessels. These mostly V-shaped cuts ("lachte") may be applied only to a small section of the circumference to protect the tree and must be extended upward or downward several times a month. A further development involves inserting poly(ethylene terephthalate) bottles or small polyethylene bags into holes that have been bored in the bark and injected with an irritant. In this way, not only is the tree protected, but oxidation and contamination of the balsam and evaporation of the terpenes are avoided [24].

The crude balsam collected is first freed of foreign material, such as sand, wood, etc. To this end, either in molten form or dissolved in turpentine, the balsam is filtered through a finely meshed sieve. To remove irritants, etc., the product is washed with hot water and distilled. Normal distillation has been replaced extensively by more gentle steam distillation [3]. Rosin remains as the turpentine-free residue. It is removed as the molten resin and transferred to metal drums.

In certain operations, solid high-boilers (pitch) can also be formed. These compounds are found in wood extracts in quantities of 2 to 8% and consist mainly of triglycerides, fatty acids, resin acids, and waxes [25].

The annual yield of crude balsam per tree is ca. 1.5 kg for Scots pine, 2.5 – 4 kg for black pine, and ca. 0.6 kg for spruce. With Central American species, more than 5 kg can be obtained. The average proportion of turpentine is 13 – 25%.

By means of resinification using irritants — i.e., treating cuts in the tree with mineral acid, particularly sulfuric acid; with the herbicides diquat and paraquat; or with 2-chloroethylphosphonic acid (CEPA) — the yield can be increased two- to fivefold [24], [26], [27].

4.2. Wood Turpentine

Wood turpentine is now produced in only a few places, such as the United States. It is extracted from finely chopped stump wood either by steam distillation or with solvents such as naphtha or chlorohydrocarbons (under pressure if required). The extract is subjected to either fractional or steam distillation. Wood turpentine is obtained as the lower-boiling fraction and pine oil as the higher-boiling one. Limonene can sometimes be isolated as an intermediate fraction [28], [3]. Ca. 2.2 wt% turpentine, 1.5 wt% pine oil, and 0.4 wt% limonene can be extracted from stumps of yellow pine that are at least eight years old. From stumps of Scots pine, only ca. 0.75 wt% turpentine is obtained. The yield increase with the age of the stump because the hollows between the dead and dried-out cells fill with turpentine balsam. This enrichment can be accelerated by boring holes in the tree just above the roots two years before felling and filling the holes with paraquat [29].

4.3. Sulfate Turpentine

Sulfate turpentine is isolated as a byproduct in pulp production and is most important in the United States, Scandinavia, and Russia.

In the kraft (sulfate) pulping process, debarked and finely chipped wood is heated to 150 – 180 °C in an aqueous digestion liquor containing NaOH, Na_2S, Na_2CO_3, and small quantities of Na_2SO_4, Na_2SO_3, and $Na_2S_2O_3$ in large pressure vessels at 7 – 13 bar for 1 – 5 h. Crude sulfate turpentine is condensed from the waste gas from the digester and is separated from the water. The process can also be carried out continuously in horizontal or vertical tube digesters [30] – [34], enabling more than 74% of the turpentine originally present to be extracted.

Crude sulfate turpentine still contains 5–15 wt% sulfur compounds with very nauseating odors, mainly methanethiol, dimethyl sulfide, and dimethyl disulfide [35]. These can be oxidized with sodium hypochlorite solution (bleaching liquor) at ca. 60 °C to the less volatile sulfonic acids, sulfoxides, or sulfones [36], which can be removed by washing with water or by distillation. Additional methods for removing sulfur compounds are treatment with chloride of lime [37] or sodium peroxide [38], careful distillation [39], and blowing air or inert gas through the product. These measures may be used individually or in combination. Even extensively purified sulfate turpentine contains traces of sulfur and chlorine compounds, which give a characteristic musty odor.

The yield of sulfate turpentine is 6–16 kg per tonne of pulp for pine, ca. 2 kg per tonne for spruce, and 2–3 kg per tonne for fir trees. The location and date of the wood harvest and the length of the storage period before processing affect the yield. In winter, up to 30% more turpentine can be obtained than in summer. Wood chips can lose up to 80% of their turpentine content in ten weeks [40].

About 55% of the pulp factories worldwide use the sulfate digestion process.

4.4. Sulfite Turpentine

Sulfite turpentine is a byproduct of pulp production by the sulfite process. Very small wood chips are heated in pressure vessels (pulp digesters) with pulping liquor. The latter consists mainly of $Ca(HSO_3)_2$ and SO_2 in acid bisulfite pulping (pH 1.5–2); $Mg(HSO_3)_2$ in bisulfite pulping (pH 4); and Na_2SO_3, $NaHCO_3$, $(NH_4)_2SO_3$, and NH_3 in the neutral sulfite process (pH 7–9). Wood chips are first impregnated at temperatures that increase to 110 °C under alternating reduced and overpressure. Then they are digested at 125–160 °C and 7–13 bar. The overall process takes 6–10 h [34], [41]–[44]. After the waste gas has been driven off, the crude oil separates on top of the pulping liquor as a yellow-brown, weakly green-fluorescing liquid. From this the almost colorless sulfite turpentine, *bp* 176–180 °C, is obtained by neutralization with sodium hydroxide or lime and distillation (possibly using steam) in a yield of 0.3–1 kg per tonne of pulp [37]. Sulfite turpentine consists of 70–90% *p*-cymene, which is formed from the terpene components originally present in a reaction catalyzed by hydrogensulfite. It also contains dipentene, borneol, and other mono- and sesquiterpenes [22] and, when mild digestion conditions are used, significant quantities of α-pinene.

Table 1. Species of pine that are important for turpentine extraction

Habitat	Systematic name	Common name
Western Europe	*P. pinaster* Sol.	French maritime pine
Southeastern Europe	*P. nigra* Arnold	black pine, Austrian pine
Greece	*P. halepensis* Mill.	Aleppo pine
Mid- and Eastern Europe and Russia	*P. sylvestris* L.	Scots pine
Siberia	*P. pumila*	
China	*P. massoniana*	Chinese red pine
	P. langbianensis	
	P. elliottii Engelm.	slash pine
India	*P. longifolia* Roxb.	
Southeast Asia	*P. armandi*	
Indonesia	*P. mercussii* Junk	Mindoro pine
United States	*P. palustris* Mill.	longleaf pine
	P. ponderosa Laws	yellow pine
Southwest USA	*P. edulis* Engelm.	
Southeast USA	*P. taeda* L.	loblolly pine
	P. elliottii Engelm.	slash pine
South Brazil	*P. elliottii* Engelm.	slash pine
	P. taeda	loblolly pine
Argentina	*P. taeda* L.	loblolly pine
	P. elliottii Engelm.	slash pine
Mexico	*P. oocarpa* Schiede	
Honduras	*P. montezuma* and others	
Northern Brazil	*P. caribaea* M.	
Venezuela	*P. caribaea* M.	

4.5. Destructively Distilled Wood Turpentine

The production of destructively distilled wood turpentine by the dry distillation of resin-rich pinewood (predominantly stumps) is no longer important. For more information, see [45]. This product contains little α-pinene, but high quantities of limonene (dipentene), *p*-cymene, camphene, α-terpinene, and sylvestrene.

5. Composition

The main components of turpentines are mono- and bicyclic monoterpenes ($C_{10}H_{16}$; M_r 136.2). They also contain smaller quantities of terpene–oxygen compounds (e.g., terpene alcohols), sesquiterpenes ($C_{15}H_{24}$), and other compounds (see Fig. 1).

The composition of different turpentines depends on the species of pine from which they are extracted [46]. About 105 species and varieties have now been described which can be distinguished to only a limited extent on the basis of turpentine composition (their "terpene pattern"). Species of pine that are important for turpentine extraction are listed in Table 1.

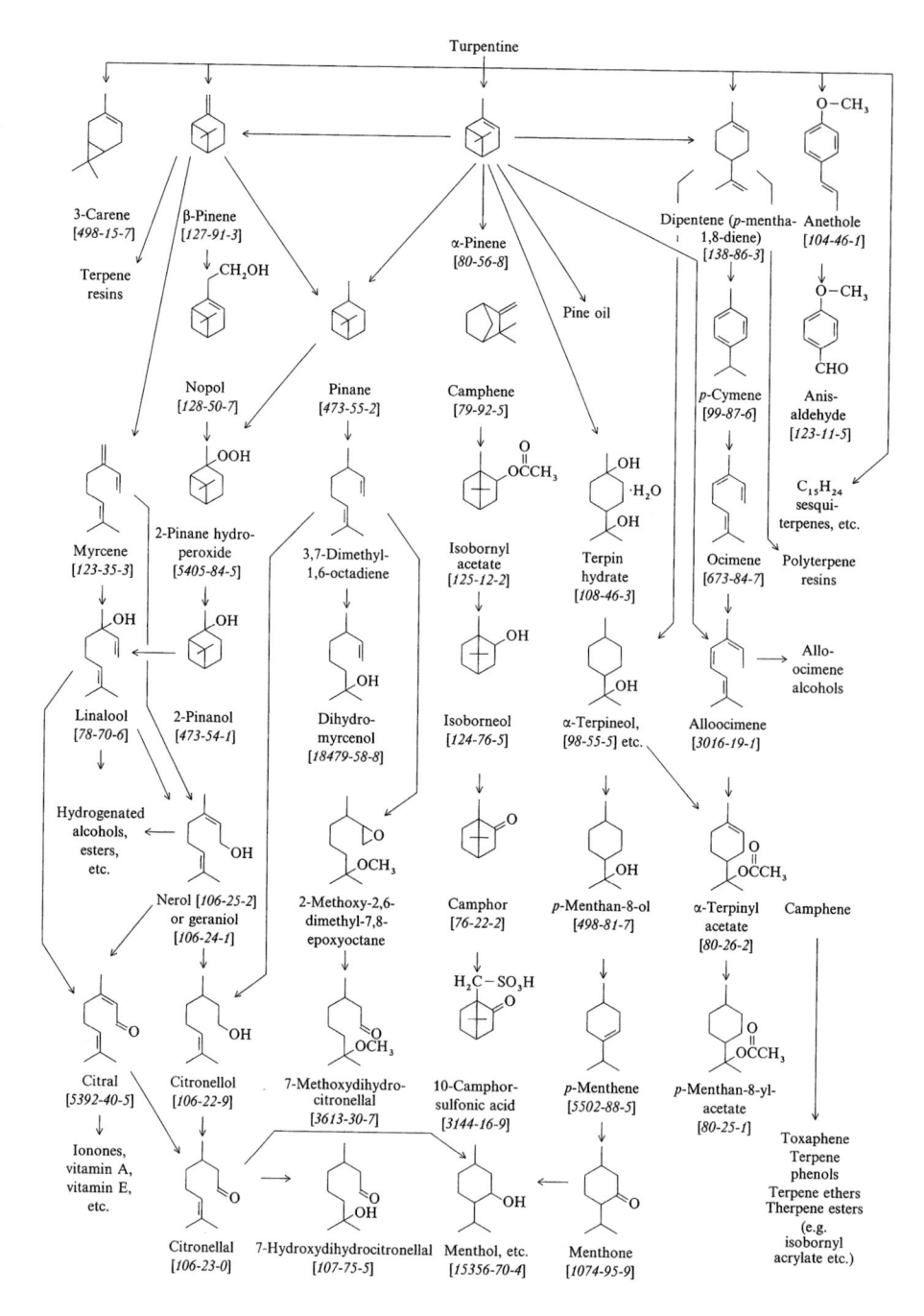

Figure 1. Turpentine constituents and downstream products

The habitat of the trees generally has little effect on the ratio of components in the turpentine oil extracted. However, for some varieties of pine (e.g., *Pinus ponderosa* Laws) the turpentine varies significantly depending on the climate, age, and habitat of the trees [47], [48].

α-Pinene is the main component of most turpentine oils. Many turpentines also contain large quantities of β-pinene, *d*- or *l*-limonene (or its racemate dipentene), 3-carene, or α-terpinene (sometimes up to 15%). Table 2 gives a simplified overview of the average composition and physical properties of various turpentines. The turpentine oils on the market are not specified according to species of pine but according to producer country, so that rather pronounced deviations in composition are possible, depending on starting material. More details can be found in [1]–[3], [49], [50] (with many additional references).

Besides the components listed in Table 2, turpentines can also contain α- and γ-terpinene, α- and β-phellandrene, terpinolene, myrcene, *p*-cymene, bornyl acetate, anethole, sesquiterpenes (such as longifolene; up to 20% in *Pinus longifolia* Roxb.), heptane, nonane, and many other compounds [51]. Some turpentines do not contain α-pinene as the main constituent but, for example, β-pinene (75% in *Pinus clausa* Engelm.), *l*-limonene (80% in *Pinus pinceana* Gordon), β-phellandrene (96% in *Pinus contorta* Douglas), or heptane (95% in *Pinus sabiniana* Douglas).

Turpentines usually contain 0.3–2.5 wt% of nonvolatile components in the evaporation residue, or considerably more if resinification occurs due to improper storage or handling.

Turpentines from spruce or fir contain significantly less α-pinene but higher quantities of camphene, α-terpinene, phellandrene, myrcene, cyclofenchene, and bornyl acetate, for example.

6. Physical Properties

Turpentines are colorless to pale-yellow liquids with low viscosity and a characteristic odor. In gum turpentine this odor is pleasantly aromatic. The physical data can vary within certain limits, depending on composition (see Table 2):

Boiling range (101.3 kPa)	150–177 °C
Solidification point	< −40 °C
ϱ (20 °C)	0.855–0.872 g/mL
$[\alpha]_D^{20}$	−40 to +48°
n_D^{20}	1.465–1.478
Relative evaporation rate (diethyl ether: 1)	ca. 170
Calorific value	ca. 40 000 kJ/kg

Turpentines are miscible with most organic solvents, but immiscible with water. They are also good solvents for hydrocarbon resins, waxes, paints and coatings, fats, and oils. Ca. 0.06 wt% of water is absorbed at 20 °C.

Table 2. Simplified overview of the composition of turpentines of different origins

Type, origin	Proportion in wt% of					Density at 20 °C, g/mL	Rotation, $[\alpha]_D^{20}$	Distillation range at 101.3 kPa, °C
	α-Pinene	β-Pinene	Camphene	3-Carene	Limonene (dipentene)			
Gum turpentine								
Greece	92–97	1–3	ca. 1	0–1	0–2	0.860–0.865	+34° to +48°	151–157
Mexico	70–95	2–15	2–15	1–2	0–4	0.862–0.868	+13° to +35°	151–161
China	70–95	4–15	1–2	0–5	1–3	0.860–0.865	−20° to +30°	151–160
Portugal	70–85	10–20	ca. 1		1–4	0.860–0.870	−24° to −38°	151–173
South America	45–85	5–45	1–3	0–2	2–4	0.860–0.870	−20° to	
Indonesia	65–85	1–3	ca. 1	10–18	1–3	0.865–0.870	+20° to +35°	152–162
France	65–75	20–26	ca. 1		1–5	0.865–0.871	−28° to −33°	152–173
Russia	40–75	4–15	1–5	0–20	0–5	0.855–0.865		152–172
Poland	40–70	2–15	ca. 1	0–25	1–5	0.855–0.865	+15° to +20°	152–172
USA–Canada	40–65	20–35	ca. 1	0–4	2–20	0.860–0.870	−10° to −30°	151–174
New Zealand	30–50	40–60						
India	20–40	5–20	1–5	45–70		0.850–0.865	−5° to −15°	153–174
Sulfate turpentine								
Finland	55–70	2–6	ca. 1	7–30	ca. 4	0.860–0.870	+2° to +21°	151–177
Sweden	50–70	4–10	ca. 1	15–40	1–3	0.860–0.870	+2° to +18°	150–175
USA	40–70	15–35	1–2	2–10	5–10	0.864–0.870	+2° to +20°	152–172
Russia	55–70	1–5	1–8	10–25	3–8	0.858–0.868	+2° to +20°	152–172
Wood turpentine								
USA	75–85	0–3	4–15		5–15	0.860–0.875	−26° to +10°	152–175

Turpentine is not listed in either DAB 10 or the European Pharmacopoeia.

7. Chemical Properties

The chemical properties of turpentine are determined by its major components, particularly the reactive pinenes. Typical reactions include isomerization, disproportionation, and polymerization catalyzed by acids, acid salts, surfactants, or Friedel–Crafts catalysts. Hydrates can be formed in the presence of water. Addition reactions take place with halogens, hydrogen halides, and nitrosyl chloride. Turpentine reacts with phosphorus pentasulfide and hydrogen sulfide to form complex sulfur compounds and thiols. Carboxylic acids add to give esters, and alcohols to give ethers. Pyrolysis yields mixtures of acyclic and monocyclic terpenes, pyronenes, hydroterpenes, and *p*-cymene. Moist air leads to autoxidation of α-pinene to sobrerol (*trans*-*p*-menth-6-ene-2,8-diol) and resins. These reactions can be inhibited by the addition of 0.005–1.0 % of certain amines or phenols acting as antioxidants (e.g., diphenylamine, thymol, catechol, resorcinol, pyrogallol, naphthols, or 2,6-di-*tert*-butyl-*p*-cresol) [52], [53].

8. Analysis

German Standards. The usual tests that should be referred to in test reports are listed in DIN 53 248 (solvents for paints, turpentine, and destructively distilled wood turpentine, April 1977). Besides appearance, color, and odor, the following characteristics should be determined:

Density	(DIN 51 757, ISO 279)
Refractive index	(DIN 53 169, ISO 280)
Flash point	(DIN 53 213; ISO 3679)
Acid number	(DIN 53 402, ISO 709)
Distillation range	(DIN 53 171 with limitations according to DIN 53 248, ISO 3405)
Residue on evaporation and material insoluble in sulfuric acid	(DIN 53 248, ISO 4715)

Determination of the residue on evaporation should be carried out under a weak stream of nitrogen to avoid oxidation and, thus, erroneous results. Determination of the bromine number of turpentines has not been standardized internationally [54].

U.S. Standards. In the United States, specifications according to ASTM D 13-92 for appearance, color, odor, and physical data of various turpentines are used. Sampling and testing must be carried out according to ASTM D 233-92. The sulfur present in sulfate turpentine, mostly in the form of volatile compounds, can be determined according to ASTM D 1266-91 [55].

Steam-distilled wood turpentine always contains traces of benzaldehyde. Detection of the latter is characteristic of this turpentine [56].

Methods of Analysis. Gas chromatography is used most frequently for the determination of individual components (DIN 51 405, ASTM E 260-91) [57] – [60]. Capillary columns with surface coatings [e.g., 50-m steel capillary columns with poly(propylene glycol) or poly(phenyl ether) coating or 25- or 50-m glass capillary columns with Carbowax 20 M or dimethylsilicone coating] are particularly effective. Accurate evaluation of gas chromatograms is occasionally difficult because of the possibility of complex reactions, such as isomerization or dehydration, in the injection unit or the column [61].

Head space gas chromatography (HSGC) has proved useful for the investigation of natural products [62].

9. Safety and Occupational Health

Turpentines are volatile, flammable liquids whose vapors are 4.7 times heavier than air; they can form explosive vapor – air mixtures on being heated above the flash point.

Flash point	32 – 35 °C
Ignition temperature	220 – 240 °C
Explosion limits (101.3 kPa)	ca. 0.8 – 6 vol% (45 – 340 g per cubic meter of air)
Ignition group (according to VDE 0165)	G 3

Handling of turpentine above the flash point must be carried out in explosion-proof apparatus [63]. Furthermore, the fact that turpentine can self-ignite in contact with air on distribution over large surface areas (e.g., on cleaning cloths or insulating materials) must be taken into account. Turpentine also ignites on contact with concentrated nitric acid.

MAK value	100 ppm (560 mg/m^3) [64]
Short-term exposure limit	5 min 2×MAK
	8 times per day
Sensitization	hypersensitive reactions of skin and respiratory tract are possible
Skin absorption	poisoning through skin absorption is possible

When handling turpentine, inhalation of the vapor, swallowing, and contact with the skin or eyes should be avoided (see Chap. 14). According to GefStoffV the following container labeling for turpentine is stipulated: St. Andrew's Cross and symbol for flammable liquids. In Germany further security information is given in VBG 1 (UVV 1 of BG Chemie), Kühn-Birett Merkblatt T 02, and Hommel-Merkblatt 190.

Both "spirit of turpentine" [*8006-64-2*], EINECS no. 232-350-7 and "turpentine, its extracts and physically modified derivatives" [*9005-90-7*], EINECS no. 232-688-5 are listed in the register of existing substances (EINECS) according to ChemG (September 1981).

10. Storage and Transportation

Turpentines should be stored in locations that are as cool as possible. Light, air, and acidic substances should be excluded to minimize changes in composition. The regulations governing equipment for storage, filling, and transport of substances hazardous to water and the regulations governing flammable liquids (VbF) must be adhered to.

The following classifications are used for the storage and transportation of turpentine:

VbF	group A class II
GGVE/GGVS	class 3, item 31 c
RID/ADR/ADNR	class 3, item 31 c
IATA/DGR	2319
IMDG code	class 3.3—2319
UN no.	2319

According to the GGVS for road transportation of turpentine, labeling with orange hazard notices and the labeling no. 30/2319 is stipulated. The same applies to α-pinene (labeling no. 30/2368).

11. Ecological Aspects

Being truly natural products, turpentines and their individual components are completely biodegradable usually in a few days, depending on the degree of dilution, temperature, availability of air, and presence of bacteria. Nevertheless, according to current estimation of the potential hazard to water, and in view of the great variety of turpentines they are assigned to the German water hazard class II (WGK II). Below their solubility limit, they do not represent any hazard to biological wastewater-treatment plants.

Turpentines are not classified in TA Luft, but by analogy to pinenes, they may be assigned to TA Luft hazard class III.

The ozone depletion potential (ODP) and the global warming potential (GWP) of turpentines are given as zero. Even the Clean Air Act of 1990 does not list terpenes among the 190 substances that cause air pollution [65].

12. Uses

The direct use of turpentine is now limited mainly to that as a solvent or dilution agent, e.g., for natural or modified binders (mainly alkyd resins), for oils, coatings, resins, paints (also artists' paints), and floor and shoe polishes. It accelerates, for example, the drying of coatings and paints by peroxide formation, in particular of those based on drying oils or other film-forming agents. Higher-boiling turpentine fractions (containing large quantities of 3-carene) are suitable as pesticides for use in forestry [66].

Although the importance of unfractionated turpentine as a solvent has clearly decreased as a result of the use of cheaper petrochemical products, turpentine has attracted increasing interest as a raw material for the chemical industry. By fractionating distillation under reduced pressure, it can be separated into fractions or individual components. Separation by process-scale chromatography has also been described [67]. The main product is usually α-pinene, whose purity can reach 99%. β-Pinene [68], 3-carene, and monocyclic terpenes [α-terpinene, limonene (dipentene), and phellandrene] are obtained in smaller quantities. Higher-boiling fractions are obtained from the residue by steam distillation. Terpene alcohols, sesquiterpenes, and diterpenes are thus obtained. The resin remaining has a composition similar to rosin and can be used for the same applications.

From the residue of sulfate turpentine distillation, estragole [*140-67-0*] (1-allyl-4-methoxybenzene) and anethole (1-propenyl-4-methoxybenzene), mostly contaminated with caryophyllene, can also be obtained.

From the different components of turpentine, particularly α- and β-pinene, a large number of substances are synthesized, some on an industrial scale. This is shown in simplified form in Figure 1(see also [69]–[74] with many additional references).

The following possible uses of turpentine and its downstream products are known (→ Terpenes):

1) Flavors and fragrances
2) Pharmaceuticals (e.g., camphor, menthol)
3) Disinfectants (e.g., terpineol, *p*-cymene)
4) Pesticides (e.g., toxaphene, 3-carene, terpene halides and thiocyanates)
5) Cleaning agents (e.g., dipentene, *d*- or *l*-limonene)
6) Solvents (e.g., turpentine and its individual components *p*-cymene, *p*-menthane, and pine oil)
7) Resins (e.g., β-pinene polymers)
8) Varnish resins (e.g., terpene phenols, isobornyl acrylate, and terpene–maleate resins)
9) Melt adhesives (e.g., terpene polymers, terpene phenols)
10) Polymerization accelerators for rubber (e.g., pinane or *p*-menthane hydroperoxide)
11) Antioxidants (e.g., isobornyl phenols)

12) Plasticizers (e.g., camphor, esters of pinic acid)
13) Lubricating oil additives (e.g., sulfur-containing terpenes)
14) Traction fluids (e.g., terpene ethers)
15) Textile auxiliaries (e.g., terpineol, pine oil)
16) Flotation aids (e.g., pine oil)

In case of overproduction sulfate turpentine may occasionally be used as a fuel (e.g., in Canada).

13. Economic Aspects

The worldwide production of turpentines in 1990 was ca. 315 000 t, of which ca. 190 000 t was sulfate turpentine. A breakdown of production among the most important producer countries is shown in Table 3 [75], with the proportion of sulfate turpentine given in each case [73]. According to this table the most important producer is still the United States. Although turpentine production has stagnated or decreased in many countries, its importance has increased markedly since ca. 1980 in the People's Republic of China and South America, where significant afforestation is being carried out. In these countries, predominantly gum turpentine is produced, whereas gum and wood turpentine production has become insignificant in the United States [75] (Table 4), the former because of the high extraction costs and the latter because of the dwindling supply of tree stumps.

The price of turpentine (and downstream products) is subject to considerable variation, as can be seen for *American sulfate turpentine* between 1977 and 1993 (mid-year prices in dollars per gallon, prices in DM per 100 kilogram in parentheses).

1977	0.51	(31.20)
1979	0.72	(40.30)
1980	1.16	(64.30)
1981	1.74	(120.00)
1982	0.95	(70.40)
1983	0.69	(53.80)
1984	0.64	(55.60)
1985	0.75	(67.30)
1986	0.97	(64.20)
1987	0.83	(45.50)
1988	1.09	(58.50)
1989	2.10	(120.50)
1990	1.90	(93.70)
1991	1.49	(75.50)
1992	1.10	(52.30)
1993	0.70	(35.30)

The price of *gum turpentine* can differ according to origin and quality. It is also subject to large variations that do not, however, inevitably parallel those of sulfate turpentine.

Table 3. Turpentine production (1000 t/a) in the most important producer countries with percentage of sulfate turpentine; 1990

Country	1960	1970	1980	1990	Sulfate turpentine, %
United States	99	94	86	105	97
Russia	28	42	30	54	85
China	10	30	65	48	2
Portugal	15	20	22	13	5
Finland		11	12	12	100
Sweden	2	9	9	10	100
Brazil				10	
Argentina			1	7	30
Mexico	9	8	11	6	10
Indonesia				6	0
New Zealand				4	0
India	5	7	8	4	0
Spain	12	9	5	2	90

Table 4. Proportions of different types of turpentine (given in %) in the United States

Type of turpentine	1950	1960	1970	1980	1990
Gum turpentine	38.4	19.7	4.5	1.0	0.1
Wood turpentine	33.5	26.9	16.4	8.0	2.9
Sulfate turpentine	27.4	53.4	79.1	91.0	97.0
Destructively distilled wood turpentine	0.7				

The average prices for Chinese gum turpentine (cost insurance freight, cif) may be cited as an example (source: Stat. Bundesamt Wiesbaden):

1989	$ 70.30 per 100 kg
1990	$ 68.40 per 100 kg
1991	$ 76.60 per 100 kg
1992	$ 68.00 per 100 kg
1993	$ 75.90 per 100 kg

Turpentines and their downstream products are expected to be very important in the future since they represent a continually renewable source of raw materials. Their yields can be increased considerably by cultivation of pine species that are richer in turpentine and less susceptible to pests or unfavorable weather conditions; by extending the area under cultivation; and by intensification of resinification using irritants.

14. Toxicology

Oil of turpentine (in vapor form also) causes characteristic and pronounced irritation of the skin and of the mucous membranes of the respiratory organs and eyes. It also leads to sensitization [76]. The LD_{50} values of 1.4 – 5.2 g/kg for rat [77] and the human lethal dose of 60 – 100 g per person [78], which was derived from cases of poisoning, show that the acute oral toxicity is relatively low. In the rat, the *acute inhalation toxicity* (LC_{50}) after 1–6 h of exposure is 12–20 mg/L [77].

Apart from biological scattering, variations can be explained by the varying compositions of different types of turpentines. The LD_{50} values (oral, rat) or turpentine are in the same range as those of its main components [79], [80]:

α-Pinene	2.1–3.7 g/kg
β-Pinene	4.7–5 g/kg
Camphene, *d*- and *l*-limonene	>5 g/kg
3-Carene	4.8 g/kg

The potential *skin irritation and sensitization* is greater in the case of old resinified turpentine because of the oxidation products formed. Of the main components, 3-carene clearly causes sensitization in humans and animals. Camphene and *l*-limonene do not have a sensitizing effect in humans, and α-pinene shows this effect only after oxidation – it has little or no effect as a purified product. The results for β-pinene and *d*-limonene are contradictory. Here, too, purity plays a part in the sensitizing potency [80]. The irritating effect on the mucous membranes of humans is pronounced: liquid turpentine and its vapors irritate the eyes and the respiratory tract at concentrations of >100 ppm (560 mg/m^3).

The *oral intake* of turpentine can cause irritation of the stomach, intestines, and urogenital tract, as well as vomiting, spasms, nausea, and kidney lesions.

Turpentine oil is easily absorbed through the skin. Although this route of application can cause damage, turpentine was used in the past in human and veterinary medicine to improve blood circulation in the skin.

In industry, however, uptake of turpentine is taken in primarily via the respiratory organs. Inhalation of higher concentrations (>100 ppm) causes not only irritation of the respiratory organs, but also headache, dizziness, increased pulse rate, and in rare cases, reversible impairment of kidney functions. Hardly any information exists on the chronic damaging effects caused by inhalation of low concentrations. Part of the turpentine absorbed parenterally is excreted unchanged by respiration. The largest part is oxidatively metabolized and excreted in urine as glucuronide [81].

The odor of turpentine is perceptible at concentrations as low as ca. 200 ppm (1.12 g/m^3). Since even this concentration causes irritation of the eyes and respiratory tract, it acts as a warning signal.

In the case of the inadvertent ingestion of turpentine, vomiting must be prevented and a doctor called immediately.

15. Turpentine Substitute

Turpentine substitute is frequently used instead of turpentine, mainly for economic reasons. The name is a collective term for colorless, flammable liquids that are similar to turpentine in their volatility and solubilizing capacity for resins, coatings, waxes, etc. The term includes high-boiling naphthas (140–170 °C), particularly those with high aromatics contents, as well as hydrogenated hydrocarbons, decalin, and tetralin. Essential oils can be added to improve the odor. *Hydroterpine* (hydrogenated destructively distilled pine oil) and distillation and rearrangement products of turpentine, consisting mainly of monocyclic monoterpenes, are more closely related to turpentine.

Trade names include Depanol I, Dipentene, Solvenol, Terposol, Unitene, and Yarmor [82].

The orange-peel terpenes may also be considered to belong to this group. They consist mainly of *d*-limonene. Those produced industrially are being used increasingly as cleaning agents, e.g., for printed circuit boards in microelectronics (as a replacement for chlorofluorocarbons).

16. Pine Oil

16.1. Production

Originally the name pine oil was given only to the higher-boiling distillation fractions (boiling range > 180 °C) obtained in the production of wood turpentine (steam-distilled pine oil, see Section 4.5) or of American pine oil (destructively distilled pine oil) [2], [83].

As a result of the decrease in production of these natural pine oils and the simultaneous increase in demand, *synthetic pine oils* have become more important. They currently account for more than 90 % of sales. Synthetic pine oils are mostly produced by hydration of α-pinene or pinene-rich turpentine fractions by use of dilute mineral acids, usually sulfuric acid, at elevated temperature [84]. After separation of the aqueous phase, they can be purified by vacuum or steam distillation. Higher-grade products are obtained by carrying out the hydration at moderate temperature to give the solid terpin hydrate. The latter is dehydrated in a weakly acidic medium, after separation of the liquid phase, and then distilled.

16.2. Properties

Pine oils are flammable, colorless to yellow, oily liquids with pleasant odors. Their main components are terpene alcohols, particularly α-terpineol.

Natural pine oils can contain the following:

50–70%	α-terpineol
5–10%	borneol (and isoborneol)
5–10%	fenchol
5–15%	other terpene alcohols (e.g., *p*-menthan-8-ol, β-terpineol, terpin)
0–10%	terpinyl and bornyl acetate
0–3%	monoterpenes (e.g., pinenes, dipentene, terpinolene)
5–10%	other components (e.g., cineols, camphor, anethole, estragole, sesquiterpenes)

Synthetic pine oils generally do not contain borneol, camphor, esters, anethole, and sesquiterpenes. The proportions of the components can vary within wide limits:

50–95%	α- and γ-terpineol
3–20%	other terpene alcohols (e.g., β-terpineol, terpinen-1-ol, and fenchol)
0–30%	terpenes (e.g., 3-carene, terpinene, dipentene, and terpinolene)
1–10%	1,4- and 1,8-cineol

Pine oils are characterized according to their terpene alcohol content (e.g., pine oil 70). The terpene alcohol content is normally ca. 60–80% but can be 20–95% after appropriate fractionation.

Pine oils are insoluble in water, but can be emulsified by use of dispersion agents such as soaps. Their *physical properties* lie mostly within the following limits:

ϱ (20 °C)	0.90–0.94 g/mL
n_D^{20}	1.475–1.488
Distillation range (101.3 kPa)	180–225 °C
Flash point	45–90 °C

The *chemical properties* of pine oils are determined mainly by the α-terpineol content.

16.3. Storage and Transportation

Pine oils should be stored in closed containers with the exclusion of heat sources. The relevant regulations governing equipment for storage, filling, and transportation of substances hazardous to water and regulations governing flammable liquids (VbF) must be adhered to. The classifications listed in Table 5 are in use, depending on the flash point.

Pine oil is not cited in the register of existing substanes (EINECS) according to the ChemG. However, the CAS registry no. and the EINECS no. of its main component terpineol [*8000-41-7*], EINECS no. 232-268-1, may be assigned to it.

Table 5. Regulations governing storage and transportation of pine oils

Regulation	Flash point	
	Below 55 °C	Above 55 °C
VbF	group A, class II	group A, class III
GGVE/GGVS	class 3, item 31c	class 3, item 32c
RID/ADR/ADNR	class 3, item 31c	class 3, item 32c
IATA – DGR	1987, at flash point > 60.5 °C no limitations	
IMDG-Code	class 3.3 – 1987	class 3.4 – 1987
UN-No.	1987	1987
Transport labeling according to regulations governing hazardous substances	30/1987	1987

Table 6. Annual production and sales of pine oil in the southern United States

Year	Production, t	Export, t
1960	34 200	8 000
1970	48 400	13 000
1980	40 400	16 000
1990	28 500	6 600

16.4. Uses

Pine oils are used mainly as foaming agents in the flotation of copper, lead, and zinc ores [85], [86] and sylvinite [87]. They have also been employed for the separation of poly(vinyl chloride) from mixtures of plastics [88]. Pine oils are good solvents for many paints and can be used as deodorants and disinfectants, as antifoams, and as additives for insecticides and wood preservatives [89]. Furthermore, they are used for the production of lubricants, emulsions, shoe and floor polishes, metal cleaners, and textile auxiliaries [90].

In the chemical industry, pine oil is used for the isolation of α-terpineol, which crystallizes from pine oil on cooling to less than ca. −2 °C (depending on concentration) or is obtained from pine oil distillation.

16.5. Economic Aspects

Besides a few South American and European (French and Austrian) pine oil-producing companies, the main producers are the southern states of the United States with annual sales listed in Table 6 [75]. The decrease in sales can be explained mainly by its reduced use in household disinfectants [65]. The most important pine oil-importing countries are Canada, Brazil, the United Kingdom, and Germany.

Trade names (besides the common term pine oil): Arizole, Dertol, Flotation Terpene-S, and Unipine.

17. References

General References

[1] E. Guenther: *The Essential Oils*, vol. VI, D. van Nostrand, New York 1952.
[2] E. Gildemeister, F. Hoffmann: *Die ätherischen Öle*, 4th ed., vol. **IV**, Akademie Verlag, Berlin 1956.
[3] W. Sandermann: *Naturharze, Terpentinöl, Tallöl*. Springer Verlag, Berlin 1960.
[4] *Kirk-Othmer*, 2nd ed., vol. **16**, pp. 680 ff; vol. **20**, pp. 748 ff.

Specific References

[5] G. Ourisson, *Pure Appl. Chem.* **62** (1990) no. 7, 1401.
[6] R. A. Stepen, T. M. Korsunova, *Sev. Biol. Nauk* **2** (1988) 43.
[7] V. A. Isidorov, I. G. Zenkevich, B. V. Joffe, *Atmos. Environ.* **19** (1985) no. 1, 1.
[8] R. C. Evans et al., *Bot. Gaz. (Chicago)* **143** (1982) no. 3, 304.
[9] K. B. Lamb, H. H. Westberg, T. Quarles, D. L. Flyckt, Washington State Univ. Report 1984, EPA-600/3-84-001, order no. PB84-124981.
[10] *Kosmos* **4** (1986) April, 87.
[11] M. Levey, *Scientia Assoc.* **91** (1956) 145.
[12] P. Rovesti, *Dragoco Rep.* **24** (1977) no. 3, 59.
[13] K. Volke, *Chem. Unserer Zeit* **27** (1993) no. 1, 42.
[14] E. v. Romaine, *Chem. Ind. (N.Y.)* **45** (1939) 259, 402.
[15] E. Gildemeister, F. Hoffmann: *Die ätherischen Öle*, 4th ed., vol. **I**, Akademie Verlag, Berlin 1956, p. 5.
[16] W. Sandermann, *Holzforschung* **16** (1962) 65.
[17] S. Juvonen, *Acta Bot. Fenn.* **71** (1966) 92.
[18] Y. Poltavchenko, G. Rudakov, *Biol. Nauki (Moscow)* **15** (1972) 95. *Chem. Abstr.* **77** (1972) 123 879.
[19] R. Croteau, *SPC Soap Perfum. Cosmet.* **53** (1980) no. 8, 428.
[20] M. H. Beale, *Nat. Prod. Rep.* **8** (1991) no. 5, 441.
[21] R. Croteau, Washington State Univ. Report 1991, DOE/ER/12027-6 order no. DE 92004298.
[22] M. Bukala et al., *Przegl. Papier.* **13** (1957) 360.
[23] W. B. Critchfield, G. L. Little: *Geographic Distribution of the Pines of the World*, United States Department of Agriculture, Mish. Publ., Washington 1971.
[24] A. W. Hodges, *Naval Stores Rev.* **103/4** July/Aug. 1993.
[25] G. C. Hoffmann, T. S. Brush, R. L. Farrell, *Naval Stores Rev.* 102/3, May/June 1992.
[26] J. M. Conley et al., *Tappi* **60** (1977) 114.
[27] O. J. Schwarz, F. J. Ryan, *Proc. Plant Growth Regul. Work Group* **7** (1980) 199.
[28] G. C. Harris, *Tappi Monogr. Ser.* **6** (1948) 167.
[29] R. G. Fajans: "Hercules Pinex Program," 7th Int. Naval Stores Meeting, Washington 1980.
[30] W. P. Lawrence, *Pap. Trade J.* **124** (1947) 128.
[31] S. Witek, *Przegl. Papier.* **14** (1958) 105.
[32] H. F. J. Wenzl: *Kraft Pulping*, Lockwood Publ. Co., New York 1967.
[33] G. S. Landry, C. D. Stillwell, *Naval Stores Rev.* **94** (1984) no. 5, 16.
[34] G. A. Smook: *Handbook for Pulp und Paper Technologists*, 2nd ed., Angus Wilde Publ., Vancouver 1992.

[35] T. T. Collins, M. G. Schmitt, *Pap. Ind. Pap. World* **26** (1944) 1137.
[36] L. E. Larrabec et al., EP 194 286, 1923.
[37] Chem. Fabrik vorm. Schering, SE 56 056, 1922.
[38] C. Schmidt, DE 340 126, 1919.
[39] J. J. Jefischew et al., *Bum. Promst.* **29** (1954) 23.
[40] D. Zinkel, J. Russel: *Naval Stores – Production, Utilization,* Pulp. Chem. Assoc. Inc., New York 1989.
[41] T. T. Collins, *Pap. Ind. Pap. World* **27** (1945) 537, 719.
[42] N. N. Neppenin: *Chemie und Technologie der Zellstoffherstellung,* vol. **1,** Akademie-Verlag, Berlin 1960.
[43] H. F. J. Wenzl: *Sulphite Pulping Technology,* Lockwood Trade Journal Co., New York 1965.
[44] O. V. Ingruber, M. Kocurek, A. Wong: *Pulp and Paper Manufacture,* 3rd ed., Joint Textbook Committee of the Paper and Pulp Industry, Montreal – Atlanta 1989.
[45] *Ullmann,* 4th ed., **22,** 555.
[46] N. T. Mirov: *The Genus Pinus,* The Ronald Press Co., New York 1967.
[47] N. T. Mirov, *Physiol. For. Trees Symp.* 1957, 251 (publ. 1958).
[48] J. Weck, *Forstwiss. Centralbl.* **77** (1958) 193.
[49] *Miltitzer Berichte über Ätherische Öle, Riechstoffe usw.,* collection of brief reports on essential oils, flavors, and fragrances, published annually until 1986, VEB Chemisches Werk Miltitz, German Democratic Republic.
[50] *Ullmann,* 4th ed., **22,** 557.
[51] J. P. Vitè et al., *Dragoco Rep.* **27** (1980) no. 11/12, 247.
[52] W. Lützkendorf, *Seifen Öle Fette Wachse* **77** (1951) 221.
[53] S. A. Ivanov et al., *Seifen Öle Fette Wachse* **97** (1971) 171.
[54] E. Hezel, *Dtsch. Farben Z.* **5** (1951) 379.
[55] W. C. Smith, *Ind. Eng. Chem. Anal. Ed.* **3** (1931) 345.
[56] S. R. Snider, *Ind. Eng. Chem. Anal. Ed.* **17** (1945) 108.
[57] K. Miltenberger, G. Keicher, *Farbe + Lack* **69** (1963) 677.
[58] R. F. Taylor, B. H. Davies, *J. Chromatogr.* **103** (1975) 327.
[59] R. Hiltunen, S. Raisanen, *Planta Med.* **41** (1981) 174.
[60] W. Bruhn, *Dragoco Rep.* **34** (1987) no. 6, 167.
[61] W. Bruhn, *Dragoco Rep.* **21** (1974) no. 6, 115; **25** (1978) no. 9, 171.
[62] J. Pohjala, R. Hiltunen, M. v. Schantz, *Flavour Fragrance J.* **4** (1989) 121.
[63] Explosionsschutzrichtlinien der BG Chemie (ZH 1/10; Sept. 1990) und Merkblatt T 033 (4/92) der BG Chemie (Guidelines on explosion protection and memorandum T 033 of BG Chemie).
[64] Anlage 4 "Grenzwerte" (1993) UVV (Appendix 4 "limit values" of UVV).
[65] S. G. Traynor, Pulp Chemicals Association, Int. Naval Stores Conference, Wien 1992.
[66] A. K. Oshkaev, G. V. Stadnitskii, *Izv. Vyssh. Uchebn. Zaved. Lesn. Zh.* 1980, no. 6, 19.
[67] M. Pointet et al., *Chem. Ind. (Düsseldorf)* **32** (1980) 690.
[68] Int. Flavors & Fragrances Inc., US 3 987 121, 1976.
[69] J. M. Derfer, *Parfum. Flavor.* **3** (1978) 45.
[70] V. Krishnasamy et al., *Chem. Petro-Chem. J.* **10** (1979) 35.
[71] R. E. Close, *Manuf. Chem. Aerosol News* **12** (1980) 45.
[72] J. Verghese, *Parfum. Flavor.* **6** (1981) 23.
[73] F. A. Dawson, Pulp Chemicals Association, Intern. Naval Stores Conference, Orlando, Fl. 1993.
[74] K. Bruns, *Parfüm. Kosmet.* **61** (1980) no. 12, 457.
[75] Naval Stores Rev., Int. Yearbooks, published until 1992.

[76] U.S. Dept. of Health and Human Services (ed.): Occupational Health Guideline for Turpentine, Washington D.C., Sept. 1978.
[77] Hercules Inc., Hercules Terpene Hydrocarbons and Solvents – Summary of Toxicological Investigations, Bull. T. 108, Wilmington, Del., 1962.
[78] S. Moeschlin: Klinik und Therapie der Vergiftungen, Thieme, Stuttgart 1980.
[79] O. A. Pickett, J. M. Schantz, *Ind. Eng. Chem.* **26** (1934) 709.
[80] E. L. J. Opdyke (ed.): Monographs of Fragrance Raw Materials, Pergamon Press, Oxford 1979.
[81] G. D. Clayton, F. E. Clayton (eds.): *Patty,* 4th ed., vol. **2,** 1994.
[82] Th. Kunzmann, *Seifen Öle Fette Wachse* **103** (1977) 470.
[83] R. W. Clements, D. N. Collins, *Naval Stores Rev.* **59** (1949) no. 40, 17.
[84] *Kirk-Othmer,* 2nd ed., vol. **19,** p. 819.
[85] L. Desalbres, *Rev. Ind. Miner.* **38** (1956) 379.
[86] A. W. Troizki: *Die Flotation,* Fachbuchverlag, Leipzig 1956, p. 115.
[87] Kali-Chemie, DE 917 361, 1954.
[88] Mitsui Mining and Smelting Co., Japan Kokai JA 7 532 274, 1973.
[89] A. Davidsohn, *Seifen Öle, Fette Wachse* **83** (1957) 15, 43.
[90] *Seifen Öle Fette Wachse* **94** (1968) 175.

Urea

Jozef H. Meessen, DSM Stamicarbon, Geleen, The Netherlands (Chaps. 1–7)
Harro Petersen, BASF Aktiengesellschaft, Ludwigshafen, Federal Republic of Germany (Chap. 8)

1.	Physical Properties	4784	5.	Quality Specifications and
2.	Chemical Properties	4786		Analysis ... 4815
3.	Production	4787	6.	Uses ... 4816
3.1.	Principles	4787	7.	Economic Aspects ... 4817
3.1.1.	Chemical Equilibrium	4787	8.	Urea Derivatives ... 4818
3.1.2.	Physical Phase Equilibria	4792	8.1.	Thermal Condensation
3.2.	Challenges in Urea Production Process Design	4794		Products of Urea ... 4818
			8.2.	Alkyl- and Arylureas ... 4819
3.2.1.	Recycle of Nonconverted Ammonia and Carbon Dioxide	4794	8.2.1.	Transamidation of Urea with Amines ... 4819
3.2.2.	Corrosion	4797	8.2.2.	Alkylation of Urea with Tertiary Alcohols ... 4820
3.2.3.	Side Reactions	4799	8.2.3.	Phosgenation of Amines ... 4821
3.3.	Description of Processes	4800	8.2.4.	Reaction of Amines with Cyanates (Salts) ... 4821
3.3.1.	Conventional Processes	4800		
3.3.2.	Stripping Processes	4802	8.2.5.	Reaction with Isocyanates ... 4822
3.3.2.1.	Stamicarbon CO_2-Stripping Process	4802	8.2.6.	Acylation of Ammonia or Amines with Carbamoyl Chlorides ... 4822
3.3.2.2.	Snamprogetti Ammonia- and Self-Stripping Processes	4804	8.2.7.	Aminolysis of Esters of Carbonic and Carbamic Acids ... 4822
3.3.2.3.	ACES Process	4806	8.3.	Reaction of Urea and Its Derivatives with Aldehydes ... 4823
3.3.2.4.	Isobaric Double-Recycle Process	4808		
3.3.3.	Other Processes	4809	8.3.1.	α-Hydroxyalkylureas ... 4823
3.4.	Effluents and Effluent Reduction	4809	8.3.2.	α-Alkoxyalkylureas ... 4825
			8.3.3.	α,α′-Alkyleneureas ... 4827
3.5.	Product-Shaping Technology	4811	8.3.4.	Cyclic Urea–Aldehyde Condensation Products ... 4828
4.	Forms Supplied, Storage, and Transportation	4813	9.	References ... 4833

Abbreviations:
CRH, %   critical relative humidity
ΔH_S, kJ/mol   integral heat of solution
m, mol/kg   urea molality, moles of urea per kilogram of water
$P_{(s)H_2O}$, Pa   water vapor pressure of a saturated urea solution
P_v, Pa   vapor pressure

Urea [57-13-6], $CO(NH_2)_2$, M_r 60.056, plays an important role in many biological processes, among others in decomposition of proteins. The human body produces 20–30 g of urea per day.

In 1828, WÖHLER discovered [1] that urea can be produced from ammonia and cyanic acid in aqueous solution. Since then, research on the preparation of urea has continuously progressed. The starting point for the present industrial production of urea is the synthesis of BASAROFF [2], in which urea is obtained by dehydration of ammonium carbamate at increased temperature and pressure:

$$NH_2COONH_4 \rightleftarrows CO(NH_2)_2 + H_2O$$

In the beginning of this century, urea was produced on an industrial scale by hydration of cyanamide, which was obtained from calcium cyanamide:

$$CaCN_2 + H_2O + CO_2 \rightarrow CaCO_3 + CNNH_2$$
$$CNNH_2 + H_2O \rightarrow CO(NH_2)_2$$

After development of the NH_3 process (HABER and BOSCH, 1913, the production of urea from NH_3 and CO_2, which are both formed in the NH_3 synthesis, developed rapidly:

$$2\,NH_3 + CO_2 \rightleftarrows NH_2COONH_4$$
$$NH_2COONH_4 \rightleftarrows CO(NH_2)_2 + H_2O$$

At present, urea is prepared on an industrial scale exclusively by reactions based on this reaction mechanism.

1. Physical Properties [3], [4]

Pure urea forms white, odorless, long, thin needles, but it can also appear in the form of rhomboid prisms. The crystal lattice is tetragonal–scalenohedral; the axis ratio $a:c = 1:0.833$. The urea crystal is anisotropic (noncubic) and thus shows birefringence. At 20 °C the refractive indices are 1.484 and 1.602. Urea has an *mp* of 132.6 °C; its heat of fusion is 13.61 kJ/mol.

Physical properties of the melt at 135 °C follow:

ϱ	1247 kg/m^3
Molecular volume	48.16 m^3/kmol
η	3.018 mPa · s
Kinematic viscosity	2.42×10^{-6} m^2/s
Molar heat capacity, C_p	135.2 J mol^{-1} K^{-1}
Specific heat capacity, c_p	2.25 kJ kg^{-1} K^{-1}
Surface tension	66.3 × 10^{-3} N/m

In the temperature range 133–150 °C, density and dynamic viscosity of a urea melt can be calculated as follows:

$$\rho = 1638.5 - 0.96\,T$$

$$\ln \eta = 6700/T - 15.311$$

The density of the solid phase at 20 °C is 1335 kg/m^3; the temperature dependence of the density is given by 0.208 kg m^{-3} K^{-1}.

At 240–400 K, the molar heat capacity of the solid phase is [5]

$$C_p = 38.43 + 4.98 \times 10^{-2}\,T + 7.05 \times 10^{-4}\,T^2 - 8.61 \times 10^{-7}\,T^3$$

The vapor pressure of the solid phase between 56 and 130 °C [6] can be calculated from

$$\ln P_v = 32.472 - 11\,755/T$$

Hygroscopicity. The water vapor pressure of a saturated solution of urea in water $P_{(s)H_2O}$ in the temperature range 10–80 °C is given by the relation [7]

$$\ln P_{(s)H_2O} = 175.766 - 11\,552/T - 22.679 \ln T$$

By starting from the vapor pressure of pure water P_{H_2O}, the critical relative humidity (CRH) then can be calculated as

$$\mathrm{CRH} = \left(P_{(s)H_2O}/P_{H_2O}\right) 100$$

The CRH is a threshold value, above which urea starts absorbing moisture from ambient air. It shows the following dependence on temperature:

25 °C	76.5 %
30 °C	74.3 %
40 °C	69.2 %

At 25 °C, in the range of 0–20 mol of urea per kilogram of water, the integral heat of solution of urea crystals in water ΔH_s as a function of molality m is given by [8]:

$$\Delta H_s = 15.351 - 0.3523\,m + 2.327 \cdot 10^{-2}\,m^2 - 1.0106 \cdot 10^{-3}\,m^3 + 1.8853 \cdot 10^{-5}\,m^4$$

Table 1. Solubility of urea in various solvents (solubility in wt% of urea)

Solvent	Temperature, °C					
	0	20	40	60	80	100
Water	39.5	51.8	62.3	71.7	80.2	88.1
Ammonia	34.9	48.6	67.2	78.7	84.5	90.4
Methanol	13.0	18.0	26.1	38.6		
Ethanol	2.5	5.1	8.5	13.1		

Urea forms a eutectic mixture with 67.5 wt% of water with a eutectic point at −11.5 °C.

The solubility of urea in a number of solvents, as a function of temperature is summarized in Table 1 [9], [10].

2. Chemical Properties

Upon *heating*, urea decomposes primarily to ammonia and isocyanic acid. As a result, the gas phase above a urea solution contains a considerable amount of HNCO, if the isomerization reaction in the liquid phase

$$CO(NH_2)_2 \rightleftharpoons NH_4NCO \rightleftharpoons NH_3 + HNCO$$

has come to equilibrium [11]. In dilute aqueous solution, the HNCO formed hydrolyzes mainly to NH_3 and CO_2. In a more concentrated solution or in a urea melt, the isocyanic acid reacts further with urea, at relatively low temperature, to form biuret ($NH_2-CO-NH-CO-NH_2$), triuret ($NH_2-CO-NH-CO-NH-CO-NH_2$), and cyanuric acid $(HNCO)_3$ [12]. At higher temperature, guanidine [$CNH(NH_2)_2$], ammelide [$C_3N_3(OH)_2NH_2$], ammeline [$C_3N_3OH(NH_2)_2$], and melamine [$C_3N_3(NH_2)_3$] are also formed [13], [14].

Melamine can also be produced from urea by a catalytic reaction in the gas phase. To this end, urea is decomposed into NH_3 and HNCO at low pressure, and subsequently transformed catalytically to melamine.

Urea reacts with NO_x, both in the gas phase at 800–1150 °C and in the liquid phase at lower temperature, to form N_2, CO_2, and H_2O. This reaction is used industrially for the removal of NO_x from combustion gases [15], [16].

Reactions with Formaldehyde. Under *acid conditions*, urea reacts with formaldehyde to form among others, methyleneurea, as well as dimethylene-, trimethylene-, tetramethylene-, and polymethyleneureas. These products are used as slow-release fertilizer under the generic name ureaform [17]. The reaction scheme for the formation of methyleneurea is given below:

$$\underset{\text{Urea}}{\begin{array}{c}NH_2\\|\\C=O\\|\\NH_2\end{array}} + \underset{\text{Formaldehyde}}{\begin{array}{c}O\\||\\H-C-H\end{array}} \longrightarrow \underset{\text{Methyleneurea}}{\begin{array}{c}NH_2\\|\\C=O\\|\\N=CH_2\end{array}} + H_2O$$

Methyleneurea reacts with additional molecules of formaldehyde to yield dimethyleneurea and other homologous products.

$$\underset{\text{Dimethyleneurea}}{\begin{array}{c}N=CH_2\\|\\C=O\\|\\N=CH_2\end{array}}$$

The reactions of urea with formaldehyde under *basic conditions* are used widely for the production of synthetic resins. As a first step, methylol urea instead of methyleneurea is formed:

$$\underset{}{\begin{array}{c}NH_2\\|\\C=O\\|\\NH_2\end{array}} + \begin{array}{c}O\\||\\H-C-H\end{array} \longrightarrow \underset{\text{Monomethylolurea}}{\begin{array}{c}CH_2OH\\|\\NH\\|\\C=O\\|\\NH_2\end{array}}$$

This product subsequently reacts with formaldehyde to dimethylol urea, $CO(NHCH_2OH)_2$, and further polymerization products. Since urea is also the raw material for the production of melamine, from which melamine–formaldehyde resins are produced, it is the most important building block in the production of amino resins.

When urea is applied as fertilizer to soil, it *hydrolyzes* in the presence of the enzyme urease to NH_3 and CO_2, after which NH_3 is bacteriologically converted into nitrate and, as such, absorbed by crops [17].

3. Production

3.1. Principles

3.1.1. Chemical Equilibrium

In all commercial processes, urea is produced by reacting ammonia and carbon dioxide at elevated temperature and pressure according to the Basaroff reactions:

$$2\,NH_3\,(l) + CO_2\,(l) \rightleftharpoons NH_2COONH_4 \qquad \Delta H = -117\text{ kJ/mol} \qquad (1)$$
$$NH_2COONH_4 \rightleftharpoons NH_2CONH_2 + H_2O \qquad \Delta H = +15.5\text{ kJ/mol} \qquad (2)$$

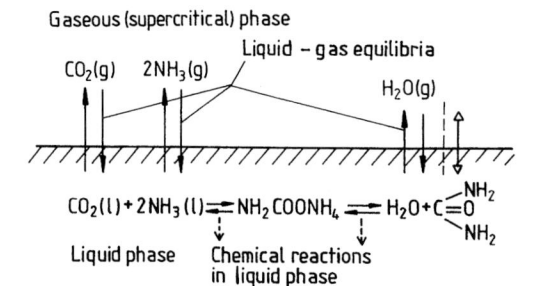

Figure 1. Physical and chemical equilibria in urea production

Gaseous (supercritical) phase
Liquid – gas equilibria
$CO_2(g)$ $2NH_3(g)$ $H_2O(g)$

$CO_2(l) + 2NH_3(l) \rightleftharpoons NH_2COONH_4 \rightleftharpoons H_2O + C(=O)(NH_2)(NH_2)$

Liquid phase   Chemical reactions in liquid phase

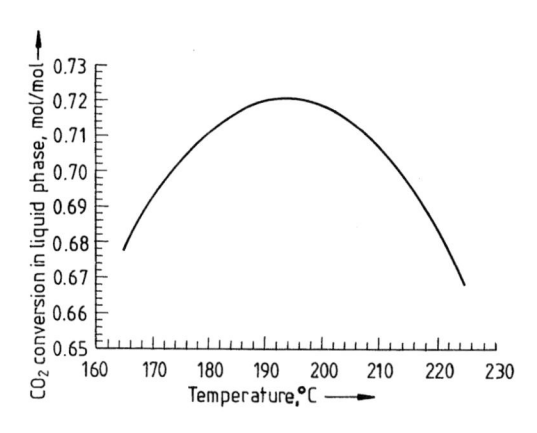

Figure 2. Carbon dioxide conversion at chemical equilibrium as a function of temperature $NH_3:CO_2$ ratio = 3.5 mol/mol (initial mixture); $H_2O:CO_2$ ratio = 0.25 mol/mol (initial mixture)

A schematic of the overall process and the physical and chemical equilibria involved is shown in Figure 1. In the first reaction, carbon dioxide and ammonia are converted to ammonium carbamate; the reaction is fast and exothermic. In the second rection, which is slow and endothermic, ammonium carbamate dehydrates to produce urea and water. Since more heat is produced in the first reaction than consumed in the second, the overall reaction is exothermic.

Processes differ mainly in the conditions (composition, temperature, and pressure) at which these reactions are carried out. Traditionally, the composition of the liquid phase in the reaction zone is expressed by two molar ratios: usually, the molar $NH_3:CO_2$ and the molar $H_2O:CO_2$ ratios. Both reflect the composition of the so-called initial mixture [i.e., the hypothetical mixture consisting only of NH_3, CO_2, and H_2O if both Reactions (1) and (2) are shifted completely to the left].

First attempts to describe the chemical equilibrium of Reactions (1) and (2) were made by FREJACQUES [19]. Later descriptions of the chemical equilibria can be divided into regression analyses of measurements [20], [21] and thermodynamically consistent analyses of the equilibria [20], [22]. As far as the most important consequences of these equilibria on urea process design are concerned, the methods correspond closely to each other: The achievable conversion per pass, dictated by the chemical equilibrium as a function of temperature, goes through a maximum (Figs. 2 and 3). This effect is

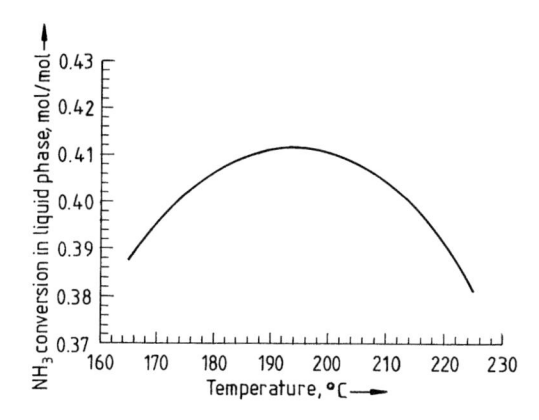

Figure 3. Ammonia conversion at chemical equilibrium as a function of temperature
$NH_3:CO_2$ ratio = 3.5 mol/mol (initial mixture); $H_2O:CO_2$ ratio = 0.25 mol/mol (initial mixture)

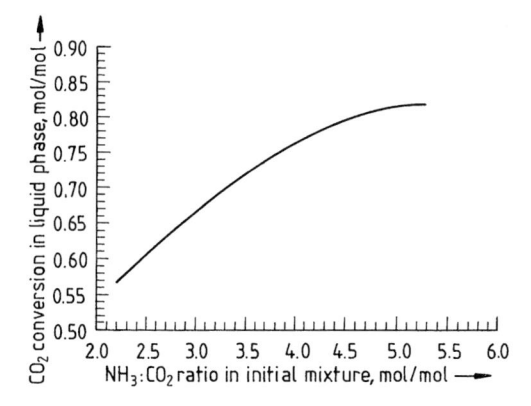

Figure 4. Carbon dioxide conversion at chemical equilibrium as a function of $NH_3:CO_2$ ratio
T = 190 °C; $H_2O:CO_2$ ratio = 0.25 mol/mol (initial mixture)

usually attributed to the fact that the ammonium carbamate concentration as a function of temperature goes through a maximum. This maximum in the ammonium carbamate concentration can be explained, at least qualitatively, by the respective heat effects of Reactions (1) and (2). However, this mechanism cannot explain the observed conversion maximum fully and quantitatively; other contributing mechanisms have been suggested [23].

The influence of the composition of the initial mixture on the chemical equilibrium can be explained qualitatively by Reactions (1) and (2) and the law of mass action:

1) Increasing the $NH_3:CO_2$ ratio (increasing the NH_3 concentration) increases CO_2 conversion, but reduces NH_3 conversion (Figs. 4 and 5).
2) Increasing the amount of water in the initial mixture (increasing the $H_2O:CO_2$ ratio) results in a decrease in both CO_2 and NH_3 conversion (Figs. 6 and 7).

In these cases, too, a full quantitative description cannot be derived simply from the law of mass action and Reactions (1) and (2). Other, not yet fully understood reaction mechanisms probably contribute to the chemical equilibria to a minor extent.

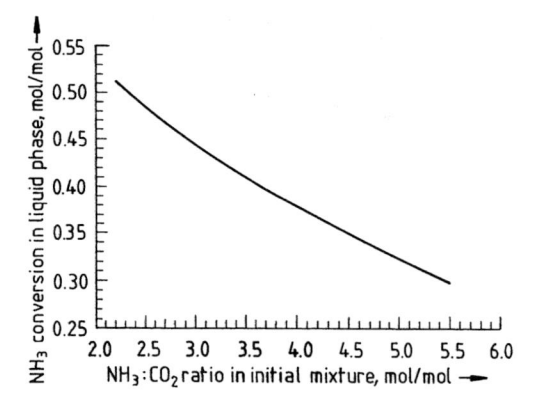

Figure 5. Ammonia conversion at chemical equilibrium as a function of $NH_3:CO_2$ ratio
$T = 190\,°C$; $H_2O:CO_2$ ratio = 0.25 mol/mol (initial mixture)

Figure 6. Carbon dioxide conversion at chemical equilibrium as a function of $H_2O:CO_2$ ratio
$T = 190\,°C$, $NH_3:CO_2$ ratio = 3.5 mol/mol (initial mixture)

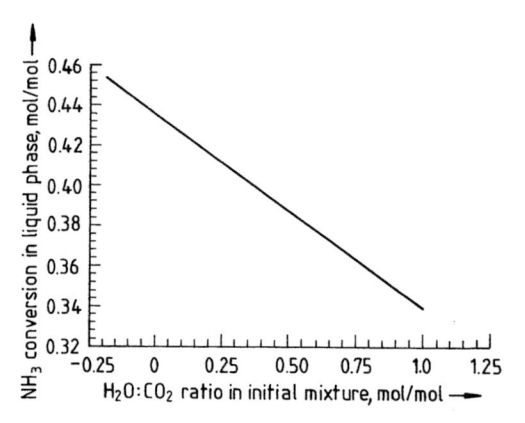

Figure 7. Ammonia conversion at chemical equilibrium as a function of $H_2O:CO_2$ ratio
$T = 190\,°C$; $NH_3:CO_2$ ratio = 3.5 mol/mol (initial mixture)

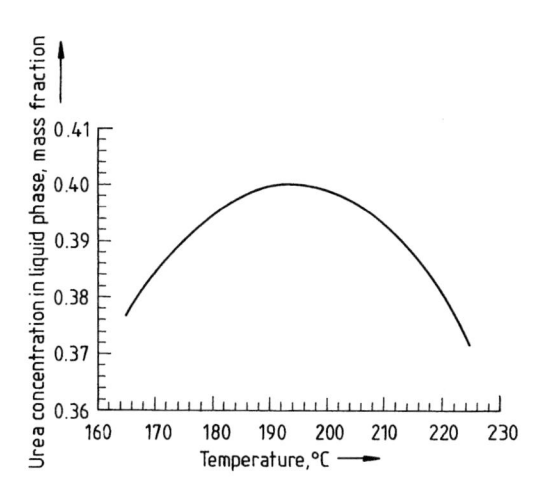

Figure 8. Urea yield in the liquid phase at chemical equilibrium as a function of temperature

$NH_3:CO_2$ ratio = 3.5 mol/mol (initial mixture);
$H_2O:CO_2$ ratio = 0.25 mol/mol (initial mixture)

In Figures 2, 4, and 6, the conversion at chemical equilibrium is expressed as CO_2 conversion, that is, the amount of CO_2 in the initial mixture converted into urea (plus biuret), if no changes occur in overall NH_3, CO_2, and H_2O concentrations in the liquid phase. This way of representing the chemical equilibrium is consistent with the presentation usually found in the traditional urea literature. However, it is based on the arbitrary choice of CO_2 as the key component. Historically, this may be justified by the fact that in early urea processes, CO_2 conversion was more important than NH_3 conversion. For the present generation of stripping processes, however, giving a higher weight to CO_2 conversion is not justified. Comparing, e.g., Figs. 4 and 5, shows that an arbitrary choice of one of the two feedstock components as yardstick to evaluate optimum reaction conversion can easily lead to faulty conclusions.

Ultimately, project economics (investment and consumptions) will dictate the choice of process parameters in the reaction section. Without going into such time- and place-dependent economic considerations, one can argue that the urea yield (i.e., the concentration of urea in the liquid phase) is a better tool for judging optimum process parameters than CO_2 or NH_3 conversion. Figure 8 illustrates that urea yield as a function of temperature also goes through a maximum; the location of this maximum is of course composition dependent. Figure 9 again shows the detrimental effect of excess water on urea yield; thus, one of the targets in designing a recycle system must be to minimize water recycle.

Figure 10 shows that the urea yield as a function of $NH_3:CO_2$ ratio reaches a maximum somewhat above the stoichiometric ratio (2:1). This is one of the reasons that all commercial processes operate at $NH_3:CO_2$ ratios above the stoichiometric ratio. Another important reason for this can be found from the physical phase equilibria in the $NH_3-CO_2-H_2O-$urea system.

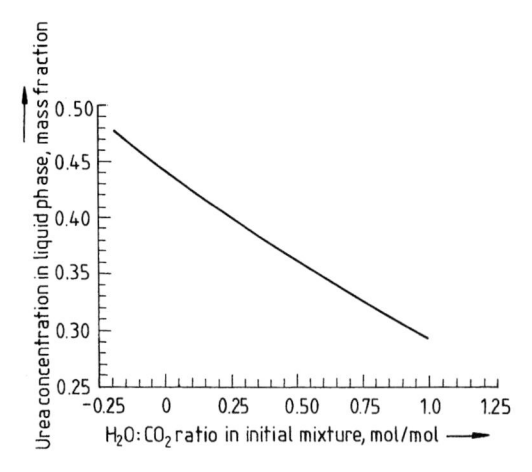

Figure 9. Urea yield in the liquid phase at chemical equilibrium as a function of $H_2O:CO_2$ ratio

$T = 190\,°C$; $NH_3:CO_2$ ratio = 3.5 mol/mol (initial mixture)

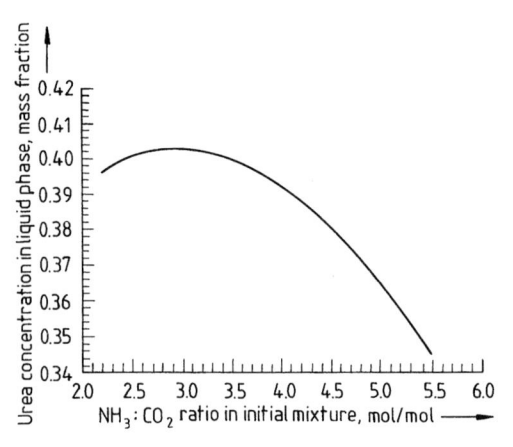

Figure 10. Urea yield in the liquid phase at chemical equilibrium as a function of $NH_3:CO_2$ ratio

$T = 190\,°C$; $H_2O:CO_2$ ratio = 0.25 mol/mol (initial mixture)

3.1.2. Physical Phase Equilibria

In urea production, the phase behavior of the components under synthesis conditions is important. In all commercial processes, conditions are such that pressure and temperature are well above the critical conditions of the feedstocks ammonia and carbon dioxide; i.e., both components are in the supercritical state. The chemical interaction between NH_3 and CO_2 (mainly the formation of ammonium carbamate) results in a strongly azeotropic behavior of the "binary" system NH_3-CO_2. An approach to the description of the phase equilibria if urea and water are added to the NH_3-CO_2 system was given by KAASENBROOD and CHERMIN [24]. If a less volatile solvent C (water) is added to an azeotropic system A–B (NH_3-CO_2) at a pressure where both components A and B are supercritical, then the $T-X$ liquid and gas planes for the ternary system thus formed assume a special shape owing to the peculiar path described by the boiling points of the changing solutions (Fig. 11). Sections through the liquid plane for constant solvent content are analogous to the liquid line for the binary

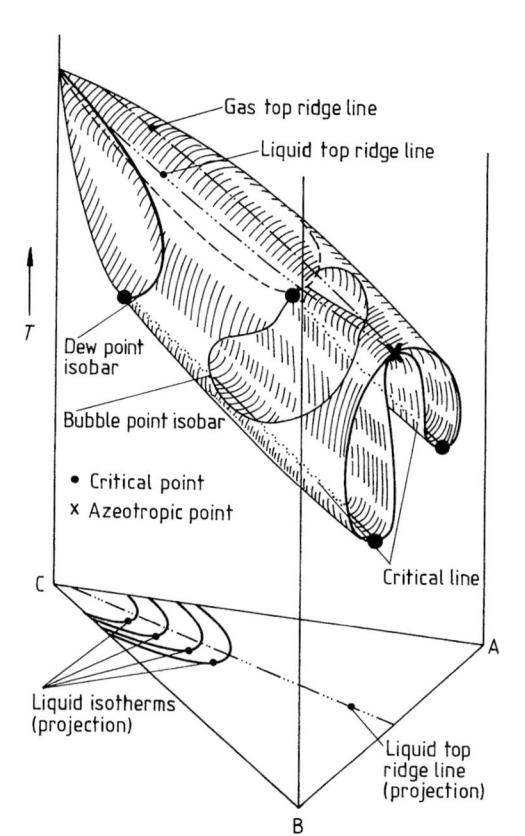

Figure 11. Liquid–gas equilibrium in a ternary system with binary azeotrope at constant pressure
The system A–B forms a binary azeotrope; C is a solvent for both A and B. The pressure is such that both A and B are supercritical, whereas the pressure is below the critical pressure of C.

system. The liquid plane for the ternary systems appears as a ridge in the $T-X$ space. If the peak points of this ridge are linked up, the top ridge line is obtained. The points on this line do not have the same A:B ratio as the maximum for the binary azeotrope, because A and B are not soluble in solvent C to the same extent. The A:B ratio changes and the boiling point increases as the percentage of C increases.

Analogous to the description of Figure 11, the equilibria in the $NH_3-CO_2-H_2O$–urea system under urea synthesis conditions show a maximum in temperature at a given pressure as a function of $NH_3:CO_2$ ratio. A full description of the phase equilibria in this system is even more complex than the aforementioned hypothetical A–B–C system, since the solid–liquid (S–L) and solid–gas (S–G) equilibria interfere with the liquid–gas (L–G) equilibria.

The strongly azeotropic behavior of the NH_3-CO_2 system, and the associated temperature maximum (or pressure minimum) in the ternary and quaternary systems with water and urea, are of practical importance in the realization of commercial urea processes. Carbon dioxide is less soluble than ammonia in water and urea melts. As a result, the pressure gradient at constant temperature is much steeper on the CO_2-rich side of the top ridge line. Moreover, this difference in solubility also causes the pressure minimum (or temperature maximum) to shift toward higher $NH_3:CO_2$ ratios as the amount of solvent (water and urea) increases. Practically, this means that in order to achieve relatively low pressures at a given temperature, the $NH_3:CO_2$ ratio in all commercial processes is chosen well above the stoichiometric ratio (2:1). In some processes, this ratio is chosen on the pressure minimum (on the top ridge line, i.e., at a ratio of ca. 3:1), whereas in other processes an even greater excess of ammonia is used.

3.2. Challenges in Urea Production Process Design

Like any process design, a urea plant design has to fulfil a number of criteria. Most important items are product quality, feedstocks and utilities consumptions, environmental aspects, safety, reliability of operation and a low initial investment. Since the urea process already has half a century of commercial scale history, it will be clear that compromises between the aforementioned, partly conflicting, criteria are well established. Also resulting from the age of urea process design is the observation that a process can only be successful if acceptable and competing solutions to all of these criteria can be combined into one process design. Apart from applying straightforward normal engineering approaches, the challenge of finding an optimum synergy between partly conflicting criteria, focuses in urea plant design essentially on a few peculiarities:

1) The thermodynamic limit on the conversion per pass through the urea reactor, combined with the azeotropic behavior of the NH_3–CO_2 system, necessitates a cunning recycle system design.
2) The intermediate product ammonium carbamate is extremely corrosive. A proper combination of process conditions, construction materials, and equipment design is therefore essential.
3) The occurrence of two side reactions—hydrolysis of urea and biuret formation—must be considered.

3.2.1. Recycle of Nonconverted Ammonia and Carbon Dioxide

The description of the chemical equilibria in Section 3.1.1 indicates that the conversion of the feedstocks NH_3 and CO_2 to urea is limited. An important differentiator between processes is the way these nonconverted materials are handled.

Once–Through Processes. In the very first processes, nonconverted NH_3 was neutralized with acids (e.g., nitric acid) to produce ammonium salts (such as ammonium nitrate) as coproducts of urea production. In this way, a relatively simple urea process scheme was realized. The main disadvantages of the once-through processes are

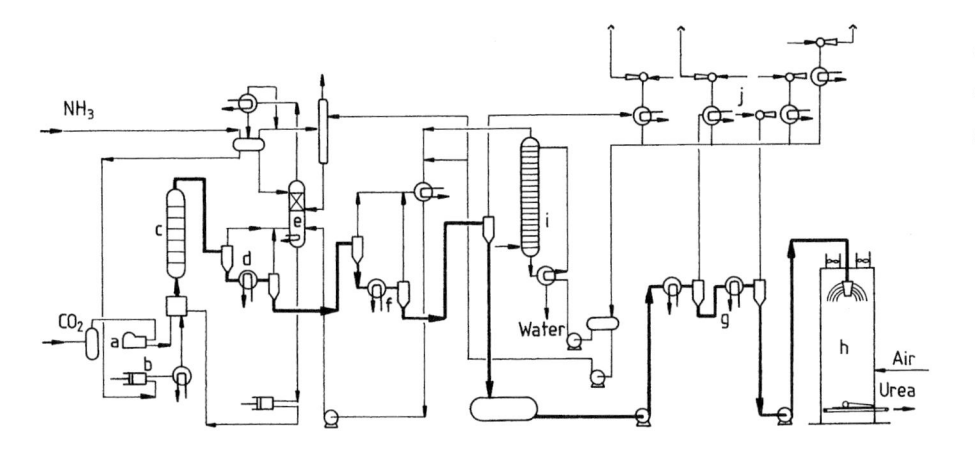

Figure 12. Typical flow sheet of a conventional urea plant
a) CO_2 compressor; b) High-pressure ammonia pump; c) Urea reactor; d) Medium-pressure decomposer; e) Ammonia–carbamate separation column; f) Low-pressure decomposer; g) Evaporator; h) Prilling; i) Desorber (wastewater stripper); j) Vacuum condensation section

the large quantity of ammonium salt formed as coproduct and the limited amount of overall carbon dioxide conversion that can be achieved. A peculiar aspect of this historic development is a recent partial "revival" of these combined urea–ammonium nitrate production facilities (UAN plants, see Section 3.3.3).

Conventional Recycle Processes. Once-through processes were soon replaced by total-recycle processes, where essentially all of the nonconverted ammonia and carbon dioxide were recycled to the urea reactor. In the first generation of total-recycle processes, several licensors developed schemes in which the recirculation of nonconverted NH_3 and CO_2 was performed in two stages. Figure 12 is a typical flow sheet of these, now called conventional, processes. The first recirculation stage was operated at medium pressure (18–25 bar); the second, at low pressure (2–5 bar). The first recirculation stage comprises at least a decomposition heater (d), in which carbamate decomposes into gaseous NH_3 and CO_2, while excess NH_3 evaporates simultaneously. The off-gas from this first decomposition step was subjected to rectification (e), from which relatively pure NH_3 (at the top) and a bottom product consisting of an aqueous ammonium carbamate solution were obtained. Both products are recyled separately to the urea reactor (c). In these processes, all nonconverted CO_2 was recycled as an aqueous solution, whereas the main portion of nonconverted NH_3 was recycled without an associated water recycle. Because of the detrimental effect of water on reaction conversion (see Figs. 6, 7, and 9), achieving a minimum CO_2 recycle (and thus maximum CO_2 conversion per reaction pass) was much more important than achieving a low NH_3 recycle. All conventional processes therefore typically operate at high $NH_3:CO_2$ ratios (4–5 mol/mol) to maximize CO_2 conversion per pass. Although some of these conventional processes, partly equipped with ingenious heat-exchanging net-

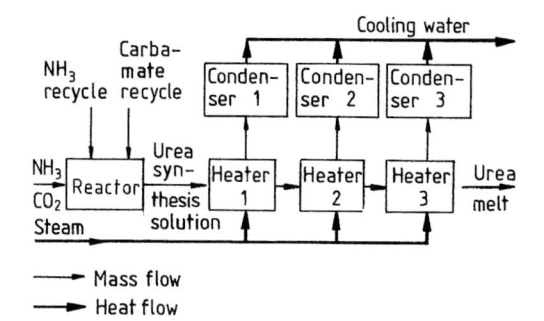

Figure 13. Conceptual diagram of the heat balance of a conventional urea process
Heat to each subsequent heater is supplied in the form of steam; the heat is used only once ($N=1$).

works, have survived until now (see Section 3.3.1), their importance decreased rapidly as the so-called stripping processes were developed.

Stripping Processes. In the 1960s, the Stamicarbon CO_2-stripping process was developed, followed later by other processes (see Section 3.3.2). Characteristic of these processes is that the major part of the recycle of both nonconverted NH_3 and nonconverted CO_2 occurs via the gas phase, such that none of these recycles is associated with a large water recycle to the synthesis zone. Another characteristic difference between conventional and stripping processes in terms of the recycle scheme, can be found in the way heat is supplied to the recirculation zones. The energy balance of the conventional processes is shown in Figure 13. In this first-generation urea process, the heat supplied to the urea synthesis solution was used only once; therefore, this type of process can be referred to as an $N=1$ process. Such a process required about 1.8 t of steam per tonne of urea.

The energy balance of a stripping plant is shown in Figure 14. As in conventional plants, heat must be supplied to the urea synthesis solution to decompose unconverted carbamate and to evaporate excess ammonia and water. However, a distinct difference in the heat balance with respect to the conventional process is that only the heat in the first heater (the high-pressure stripper) is imported. This heat is recovered in a high-pressure carbamate condenser (unconverted ammonia and carbon dioxide are condensed to form ammonium carbamate) and reused in the low-pressure heaters. The heat supplied is effectively used twice; thus, the term $N=2$ process is used. The average energy consumption of the stripping process is 0.8 – 1.0 t of steam per tonne of urea.

In the 1980s, some processes were described that aim at a greater reduction of energy consumption by a further application of this multiple effect to $N=3$ (Fig. 15) [25]–[29]. As can be seen from Figures 14 and 15, the steam requirement for process heating is reduced in these types of processes. However, whether the total energy consumption for the process is also reduced is doubtful, if the full capabilities of a $N=2$ type of process are exploited and if the total energy supply scheme, including the energy supply to the carbon dioxide compressor drive, are taken into consideration [30]. Moreover, it seems that the emphasis in urea technology now is shifting from low energy consumption toward other factors, such as more durable construction materials, more modern process control systems, and simple process design [31].

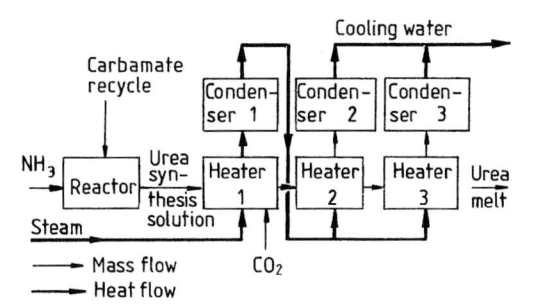

Figure 14. Conceptual diagram of the heat balance of a stripping plant

Heat supplied to the first heater (the stripper) is recovered in the first condenser (high-pressure carbamate condenser) and subsequently used again in the low-pressure heaters (decomposers and water evaporators); the heat is effectively used twice ($N=2$).

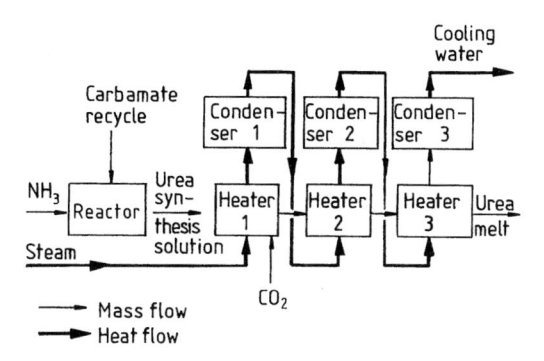

Figure 15. Further heat integration of a stripping plant in conceptual form

Heat supplied to the first heater (the stripper) is effectively used three times ($N=3$).

3.2.2. Corrosion [32]

Urea synthesis solutions are very corrosive. Generally, ammonium carbamate is considered the aggressive component. This follows from the observation that carbamate-containing product streams are corrosive whereas pure urea solutions are not. The corrosiveness of the synthesis solution has forced urea manufacturers to set very strict demands on the quality and composition of construction materials. Awareness of the important factors in material selection, equipment manufacture and inspection, technological design and proper operations of the plant, together with periodic inspections and nondestructive testing are the key factors for safe operation for many years.

Role of Oxygen Content. Since the liquid phase in urea synthesis behaves as an electrolyte, the corrosion it causes is of an electrochemical nature. Stainless steel in a corrosive medium owes its corrosion resistance to the presence of a protective oxide layer on the metal. As long as this layer is intact, the metal corrodes at a very low rate. Passive corrosion rates of austenitic urea-grade stainless steels are generally between < 0.01 and (max.) 0.10 mm/a. Upon removal of the oxide layer, activation and, consequently, corrosion set in unless the medium contains sufficient oxygen or oxidation agent to build a new layer. Active corrosion rates can reach values of 50 mm/a. Stainless steel exposed to carbamate-containing solutions involved in urea synthesis can be kept in a passivated (noncorroding) state by a given quantity of oxygen. If the oxygen content drops below this limit, corrosion starts after some time — its onset depending on process conditions and the quality of the passive layer. Hence, introduction of oxygen and maintenance of a sufficiently high oxygen content in the various process streams are prerequisites to preventing corrosion of the equipment. Although some alternatives have been mentioned [33], [34], the use of oxygen to influence the redox potential has become common practice in urea manufacture ever since it was initially suggested [35], [36].

From the point of view of corrosion prevention, the condensation of $NH_3 - CO_2 - H_2O$ gas mixtures to carbamate solutions deserves great attention. This is necessary because — notwithstanding the presence of oxygen in the gas phase — an oxygen-deficient corrosive condensate is initially formed on condensation. In this condensate the oxygen is absorbed only slowly. This accounts for the severe corrosion sometimes observed in cold spots inside gas lines. The trouble can be remedied by adequate isolation and tracing of the lines.

When condensation constitutes an essential process step — for example, in high-pressure and low-pressure carbamate condensers — special technological measures must be taken. These measures can involve ensuring that an oxygen-rich liquid phase is introduced into the condenser, while appropriate liquid – gas distribution devices ensure that no dry spots exist on condensing surfaces.

Not only condensing but also stagnant conditions are dangerous, especially where narrow crevices are present, into which hardly any oxygen can penetrate and oxygen depletion may occur.

Role of Temperature and Other Process Parameters in Corrosion. *Temperature* is the most important technological factor in the behavior of the steels employed in urea synthesis. An increase in temperature increases active corrosion, but more important, above a critical temperature it causes spontaneous activation of passive steel. The higher-alloyed austenitic stainless steels (e.g., containing 25 wt% chromium, 22 wt% nickel, and 2 wt% molybdenum) appear to be much less sensitive to this critical temperature than 316 L types of steel.

Sometimes, the $NH_3:CO_2$ *ratio* in synthesis solutions is also claimed to have an influence on the corrosion rate of steels under urea synthesis conditions. Experiments have showed that under practical conditions this influence is not measurable because the steel retains passivity. Spontaneous activation did not occur. Only with electrochemical activation could 316 L types of steel be activated at intermediate $NH_3:CO_2$ ratios. At low and high ratios, 316 L stainless steel could not be activated. The higher-alloyed steel type 25 Cr 22 Ni 2 Mo showed stable passivity, irrespective of the $NH_3:CO_2$ ratio, even when activated electrochemically. Of course, these results depend on the specific temperature and oxygen content during the experiments.

Material Selection. Corrosion resistance is not the only factor determining the choice of construction materials. Other factors such as mechanical properties, workability, and weldability, as well as economic considerations such as price, availability, and delivery time, also deserve attention.

Stainless steels that have found wide use are the austenitic grades AISI 316 L and 317 L. Like all Cr-containing stainless steels, AISI 316 L and 317 L are not resistant to the action of sulfides. Hence it is imperative in plants using the 316 L and 317 L grades in combination with CO_2 derived from sulfur-containing gas, to purify this gas or the CO_2 thoroughly.

In stripping processes, the process conditions in the high-pressure stripper are most severe with respect to corrosion.

In the *Stamicarbon CO_2-stripping process*, a higher-alloyed, but still fully austenitic stainless steel (25 Cr 22 Ni 2 Mo) was chosen as construction material for the stripper tubes. This choice ensures better corrosion resistance than 316 L or 317 L types of material but still maintains the advantages of workability, weldability, reparability, and the cheaper price of stainless steel-type materials.

In the *Snamprogetti stripping processes*, titanium usually is chosen for this critical application, although mechanically bonded bimetallic 25 Cr 22 Ni 2Mo – zirconium tubes have also been suggested to improve corrosion resistance [37], [38].

In the *ACES process*, duplex alloys (ferritic – austenitic) are used as construction material for the stripper tubes.

3.2.3. Side Reactions [39]

Three side reactions are of special importance in the design of urea production processes:

Hydrolysis of urea:

$$CO(NH_2)_2 + H_2O \rightarrow NH_2COONH_4 \rightarrow 2\, NH_3 + CO_2 \quad (3)$$

Biuret formation from urea:

$$2\, CO(NH_2)_2 \rightarrow NH_2CONHCONH_2 + NH_3 \quad (4)$$

Formation of isocyanic acid from urea:

$$CO(NH_2)_2 \rightarrow NH_4NCO \rightarrow NH_3 + HNCO \quad (5)$$

All three side reactions have in common the decomposition of urea; thus, the extent to which they occur must be minimized.

The *hydrolysis reaction* (3) is nothing but the reverse of urea formation. Whereas this reaction approaches equilibrium in the reactor, in all downstream sections of the plant the NH_3 and CO_2 concentrations in urea-containing solutions are such that Reaction (3) is shifted to the right. The extent to which the reaction occurs is determined by temperature (high temperatures favor hydrolysis) and reaction kinetics; in practice, this means that retention times of urea-containing solutions at high temperatures must be minimized.

The *biuret reaction* (4) also approaches equilibrium in the urea reactor [20], [22]. The high NH_3 concentration in the reactor shifts Reaction (4) to the left, such that only a small amount of biuret is formed in the reactor. In downstream sections of the plant, NH_3 is removed from the urea solutions, thereby creating a driving force for biuret formation. The extent to which biuret is formed is determined by reaction kinetics; therefore, the practical measures to minimize biuret formation are the same as described above for the hydrolysis reaction.

Reaction (5) shows that formation of isocyanic acid from urea is also favored by low NH_3 concentrations. This reaction is especially relevant in the evaporation section of the plant. Here, low pressures are applied, resulting in a transfer of NH_3 and HNCO into the gas phase and, consequently, low concentrations of these constituents in the liquid phase. Together with the relatively high temperatures in the evaporators, this shifts Reaction (5) to the right. The extent to which this reaction occurs is again determined by kinetics. The HNCO removed via the gas in the evaporators collects in the process condensate from the vacuum condensers, where low temperatures shift Reaction (5) to the left, again forming urea. As a result of this mechanism of chemical entrainment, attempts to minimize entrainment from evaporators with physical (liquid–gas) separation devices are destined to be unsuccessful.

3.3. Description of Processes

3.3.1. Conventional Processes

As explained in Section 3.2, conventional processes have generally been replaced by stripping processes. Only two conventional processes may still have some importance in the near future.

Urea Technologies Inc. (UTI) Heat Recycle Process (see Fig. 16). Ammonia (containing passivating air), recycled carbamate, and about 60% of the feed CO_2 are charged to the top of an open-ended coil reactor (c) operating at 210 bar. Ammonium carbamate is formed within the coil, exits the coil at the bottom, and then flows up and around it—the exothermic heat of carbamate formation in the coil driving the endothermic dehydration of carbamate to urea outside the coil. The reactor is claimed to achieve a uniform temperature profile in this way. In the reactor, a relative high $NH_3:CO_2$ ratio (4.2:1) is applied. The reactor effluent is depressurized and subcooled, and the flashed gases are released before the first decomposer (f). Gases leaving the first decomposer separator (g) are mixed with about 40% of the feed carbon dioxide and partially condensed in the heat recovery section (i–k). The two combined gas streams are then further condensed and form the carbamate recycle flow. Urea as an

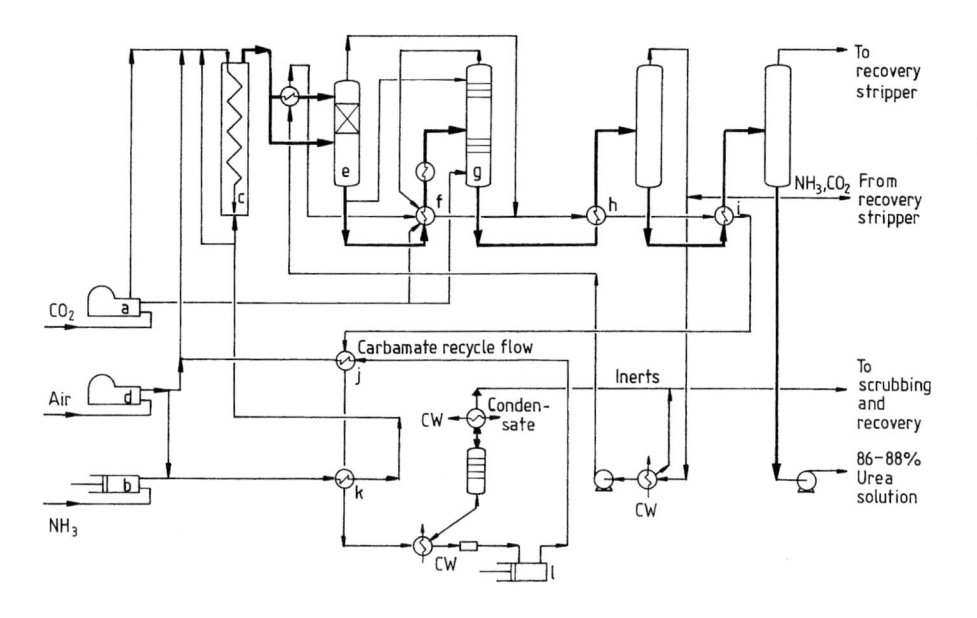

Figure 16. UTI heat recycle urea process

a) CO_2 compressor; b) High-pressure ammonia feed pump; c) Urea reactor with internal coil; d) Air compressor; e) Liquid distributor (used as first flash vessel); f) First decomposer; g) First separator; h) Second decomposer; i) Urea concentrator; j) Carbamate heater; k) Ammonia heater; l) High-pressure carbamate recycle pump

CW = Cooling water

86–88% solution is concentrated by evaporation (i) before granulation or prilling. The process is applied in eight small-scale and two medium-scale plants.

MTC (Mitsui Toatsu Corporation) Conventional Processes of Toyo Engineering Corporation. The conventional processes developed by Toyo Engineering Corporation (TEC) were successfully commercialized until the mid-1980s. The continuous evolution of these processes is reflected in their sequential nomenclature:

TR – A	Total-Recycle A Process
TR – B	Total-Recycle B Process
TR – C	Total-Recycle C Process
TR – CI	Total-Recycle C Improved Process
TR – D	Total-Recycle D Process

Partial-recycle versions of these processes were also realized. These TEC MTC conventional processes were applied in more then 70 plants. Recently, the licensor of these processes has announced a stripping process (the ACES process; see Section 3.3.2.3); this probably means the end of the TEC MTC conventional process line.

3.3.2. Stripping Processes

3.3.2.1. Stamicarbon CO_2-Stripping Process (Figs. 17 and 18)

The synthesis stage of the Stamicarbon process consists of a urea reactor (c), a stripper for unconverted reactants (d), a high-pressure carbamate condenser (e), and a high-pressure reactor off-gas scrubber (f). To realize maximum urea yield per pass through the reactor at the stipulated optimum pressure of 140 bar, an $NH_3:CO_2$ molar ratio of 3:1 is applied. The greater part of the unconverted carbamate is decomposed in the stripper, where ammonia and carbon dioxide are stripped off. This stripping action

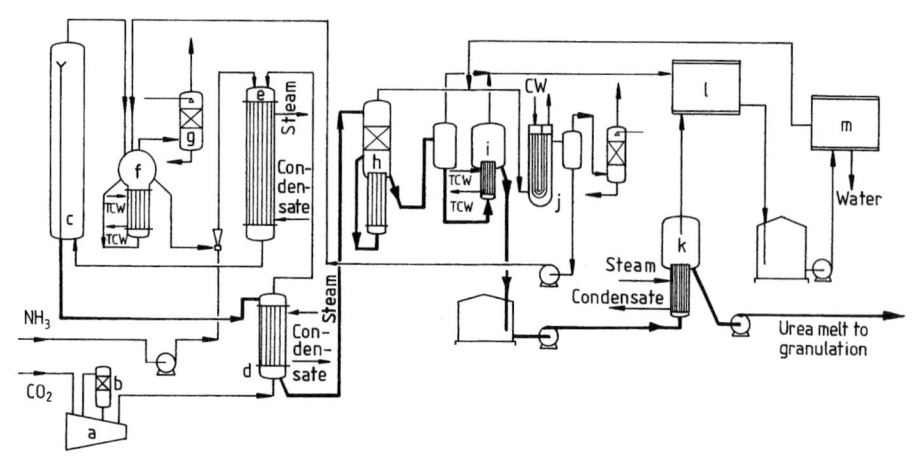

Figure 17. Stamicarbon CO_2-stripping urea process (The process suitable for combination with a granulation plant is shown here; combination with prilling is also possible.)
a) CO_2 compressor; b) Hydrogen removal reactor; c) Urea reactor; d) High-pressure stripper; e) High-pressure carbamate condenser; f) High-pressure scrubber; g) Low-pressure absorber; h) Low-pressure decomposer and rectifier; i) Pre-evaporator; j) Low-pressure carbamate condenser; k) Evaporator; l) Vacuum condensation section; m) Process condensate treatment
CW = Cooling water; TCW = Tempered cooling water

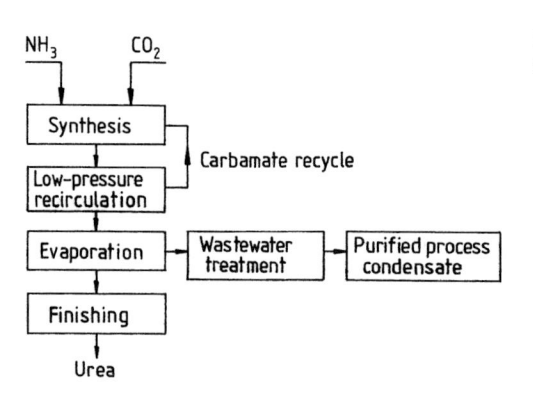

Figure 18. Functional block diagram of the Stamicarbon CO_2-stripping urea process

is effected by countercurrent contact between the urea solution and fresh carbon dioxide at synthesis pressure. Low ammonia and carbon dioxide concentrations in the stripped urea solution are obtained, such that the recycle from the low-pressure recirculation stage (h, j) is minimized. These low concentrations of both ammonia and carbon dioxide in the stripper effluent can be obtained at relatively low temperatures of the urea solution because carbon dioxide is only sparingly soluble under such conditions.

Condensation of ammonia and carbon dioxide gases, leaving the stripper, occurs in the high-pressure carbamate condenser (e) at synthesis pressure. As a result, the heat liberated from ammonium carbamate formation is at a high temperature. This heat is used for the production of 4.5-bar steam for use in the urea plant itself. The condensation in the high-pressure carbamate condenser is not effected completely. Remaining gases are condensed in the reactor and provide the heat required for the dehydration of carbamate, as well as for heating the mixture to its equilibrium temperature. In a recent improvement to this process, the condensation of off-gas from the stripper is carried out in a prereactor, where sufficient residence time for the liquid phase is provided. As a result of urea and water formation in the condensing zone, the condensation temperature is increased, thus enabling the production of steam at a higher pressure level [40].

The feed carbon dioxide, invariably originating from an associated ammonia plant, always contains hydrogen. To avoid the formation of explosive hydrogen–oxygen mixtures in the tail gas of the plant, hydrogen is catalytically removed from the carbon dioxide feed (b). Apart from the air required for this purpose, additional air is supplied to the fresh carbon dioxide input stream. This extra portion of oxygen is needed to maintain a corrosion-resistant layer on the stainless steel in the synthesis section. Before the inert gases, mainly oxygen and nitrogen, are purged from the synthesis section, they are washed with carbamate solution from the low-pressure recirculation stage in the high-pressure scrubber (f) to obtain a low ammonia concentration in the subsequently purged gas. Further washing of the off-gas is performed in a low-pressure absorber (g) to obtain a purge gas that is practically ammonia free. Only one low-pressure recirculation stage is required due to the low ammonia and carbon dioxide concentrations in the stripped urea solution. Because of the ideal ratio between ammonia and carbon dioxide in the recovered gases in this section, water dilution of the resultant ammonium carbamate is at a minimum despite the low pressure (about 4 bar). As a result of the efficiency of the stripper, the quantities of ammonium carbamate for recycle to the synthesis section are also minimized, and no separate ammonia recycle is required.

The urea solution coming from the recirculation stage contains about 75 wt% urea. This solution is concentrated in the evaporation section (k). If the process is combined with a prilling tower for final product shaping, the final moisture content of urea from the evaporation section is ca. 0.25 wt%. If the process is combined with a granulation unit, the final moisture content may vary from 1 to 5 wt%, depending on granulation requirements. Higher moisture contents can be realized in a single-stage evaporator, whereas low moisture contents are economically achieved in a two-stage evaporation section.

When urea with an extremely low biuret content is required (at a maximum of 0.3 wt%), pure urea crystals are produced in a crystallization section. These crystals are separated from the mother liquor by a combination of sieve bends and centrifuges and are melted prior to final shaping in a prilling tower or granulation unit.

The process condensate emanating from water evaporation from the evaporation or crystallization sections contains ammonia and urea. Before this process condensate is purged, urea is hydrolyzed into ammonia and carbon dioxide (l), which are stripped off with steam and returned to urea synthesis via the recirculation section. This process condensate treatment section can produce water with high purity, thus transforming this "wastewater" treatment into the production unit of a valuable process condensate, suitable for, e.g., cooling tower or boiler feedwater makeup. Since the introduction of the Stamicarbon CO_2-stripping process, some 125 units have been built according to this process all over the world.

3.3.2.2. Snamprogetti Ammonia- and Self-Stripping Processes [41]–[47]

In the first generation of NH_3- and self-stripping processes, ammonia was used as stripping agent. Because of the extreme solubility of ammonia in the urea-containing synthesis fluid, the stripper effluent contained rather large amounts of dissolved ammonia, causing an ammonia overload in downstream sections of the plant. Later versions of the process abandoned the idea of using ammonia as stripping agent; stripping was achieved only by supply of heat ("thermal" or "self"-stripping). Even without using ammonia as a stripping agent, the $NH_3:CO_2$ ratio in the stripper effluent is relatively high, so the recirculation section of the plant requires an ammonia–carbamate separation section, as in conventional processes (see Fig. 19).

The process uses a vertical layout in the synthesis section. Recycle within the synthesis section, from the stripper (h) via the high-pressure carbamate condenser (f), through the carbamate separator (e) back to the reactor (b), is maintained by using an ammonia-driven liquid–liquid ejector (c) [43], [45] (see Fig. 20). In the reactor, which is operated at 150 bar, an $NH_3:CO_2$ molar feed ratio of ca. 3.5 is applied. The stripper is of the falling film type [46]. Since stripping is achieved thermally, relatively high temperatures (200–210 °C) are required to obtain a reasonable stripping efficiency. Because of this high temperature, stainless steel is not suitable as construction material for the stripper from a corrosion point of view; titanium and bimetallic zirconium–stainless steel tubes have been used [37], [38].

Off-gas from the stripper is condensed in a kettle-type boiler (f) [44]. At the tube side of this condenser the off-gas is absorbed in recycled liquid carbamate from the medium-pressure recovery section. The heat of absorption is removed through the tubes, which are cooled by the production of low-pressure steam at the shell side. The steam produced is used effectively in the back end of the process.

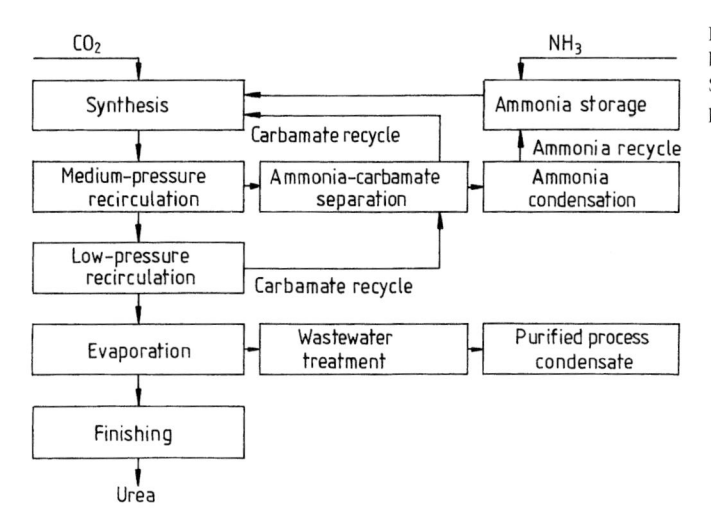

Figure 19. Functional block diagram of the Snamprogetti self-stripping process

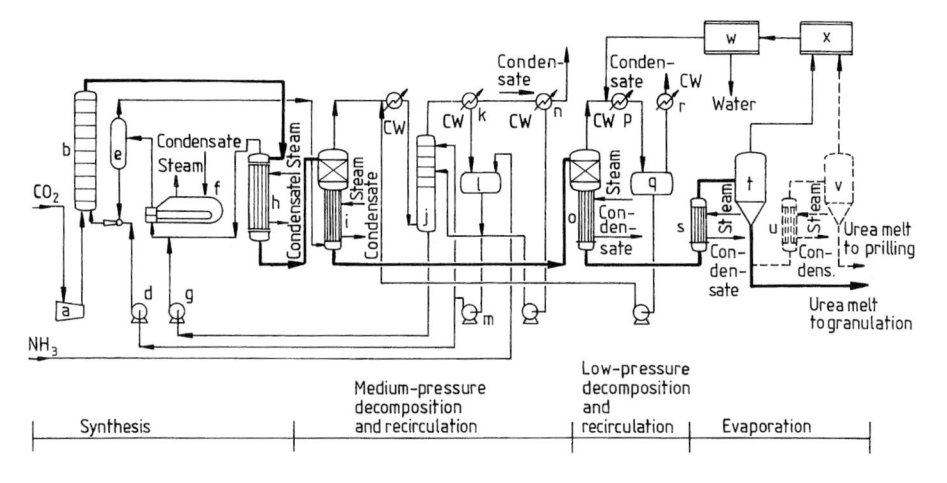

Figure 20. Schematic of the Snamprogetti self-stripping process
a) CO$_2$ compressor; b) Urea reactor; c) Ejector; d) High-pressure ammonia pump; e) Carbamate separator; f) High-pressure carbamate condenser; g) High-pressure carbamate pump; h) High-pressure stripper; i) Medium-pressure decomposer and rectifier; j) Ammonia–carbamate separation column; k) Ammonia condenser; l) Ammonia receiver; m) Low-pressure ammonia pump; n) Ammonia scrubber; o) Low-pressure decomposer and rectifier; p) Low-pressure carbamate condenser; q) Low-pressure carbamate receiver; r) Low-pressure off-gas scrubber; s) First evaporation heater; t) First evaporation separator; u) Second evaporation heater; v) Second evaporation separator; w) Wastewater treatment; x) Vacuum condensation section
CW = Cooling water

In the medium-pressure decomposition and recirculation section, typically operated at 18 bar, the urea solution from the high-pressure stripper is subjected to the decomposition of carbamate and evaporation of ammonia (i). The off-gas from this medium-pressure decomposer is rectified. Liquid ammonia reflux is applied to the top of this rectifier (j); in this way a top product consisting of pure gaseous ammonia, and a bottom product of liquid ammonium carbamate are obtained. The pure ammonia off-gas is condensed (k) and recycled to the urea synthesis section. To prevent solidification of ammonium carbamate in the rectifier, some water is added to the bottom section of the column to dilute the ammonium carbamate below its crystallization point. The liquid ammonium carbamate–water mixture obtained in this way is also recycled to the synthesis section. The purge gas of the ammonia condensers is treated in a scrubber (n) prior to being purged to the atmosphere.

The urea solution from the medium-pressure decomposer is subjected to a second lowpressure decomposition step (o). Here, further decomposition of ammonium carbamate is achieved, so that a substantially carbamate-free aqueous urea solution is obtained. Off-gas from this low-pressure decomposer is condensed (p) and recycled as an aqueous ammonium carbamate solution to the synthesis section via the medium-pressure recovery section.

Concentrating the urea–water mixture obtained from the low-pressure decomposer is performed in a single or double evaporator (s–v), depending on the requirements of the finishing section. Typically, if prilling is chosen as the final shaping procedure, a two-stage evaporator is required, whereas in the case of a fluidized-bed granulator a single evaporation step is sufficient to achieve the required final moisture content of the urea melt. In some versions of the process, heat exchange is applied between the off-gas from the medium-pressure decomposer and the aqueous urea solution to the evaporation section. In this way, the consumption of low-pressure steam by the process is reduced.

The process condensate obtained from the evaporation section is subjected to a desorption–hydrolysis operation to recover the urea and ammonia contained in the process condensate.

Up to now, about 70 plants have been designed according to the Snamprogetti ammonia- and self-stripping processes.

3.3.2.3. ACES Process [29], [48], [49]

The ACES (i.e., Advanced Process for Cost and Energy Saving) process has been developed by Toyo Engineering Corporation. Its synthesis section consists of the reactor (a), stripper (d), two parallel carbamate condensers (e), and a scrubber (f)—all operated at 175 bar (see Figs. 21 and 22).

The reactor is operated at 190 °C and an $NH_3:CO_2$ molar feed ratio of 4:1. Liquid ammonia is fed directly to the reactor, whereas gaseous carbon dioxide after compression is introduced into the bottom of the stripper as a stripping aid. The synthesis mixture from the reactor, consisting of urea, unconverted ammonium carbamate, excess ammonia, and water, is fed to the top of the stripper. The stripper has two functions. Its upper part is equipped with trays where excess ammonia is partly separated from the stripper feed by direct countercurrent contact of the feed solution with the gas coming from the lower part of the stripper. This prestripping in the top is

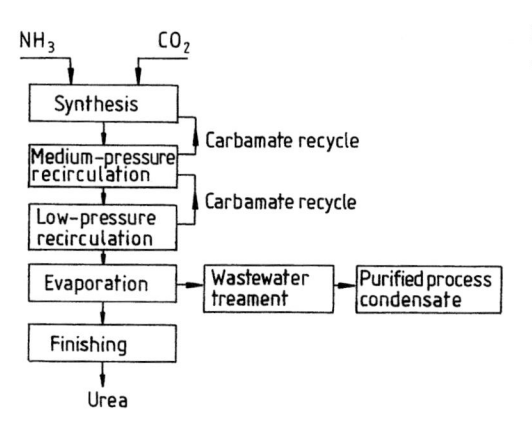

Figure 21. Functional block diagram of the ACES urea process

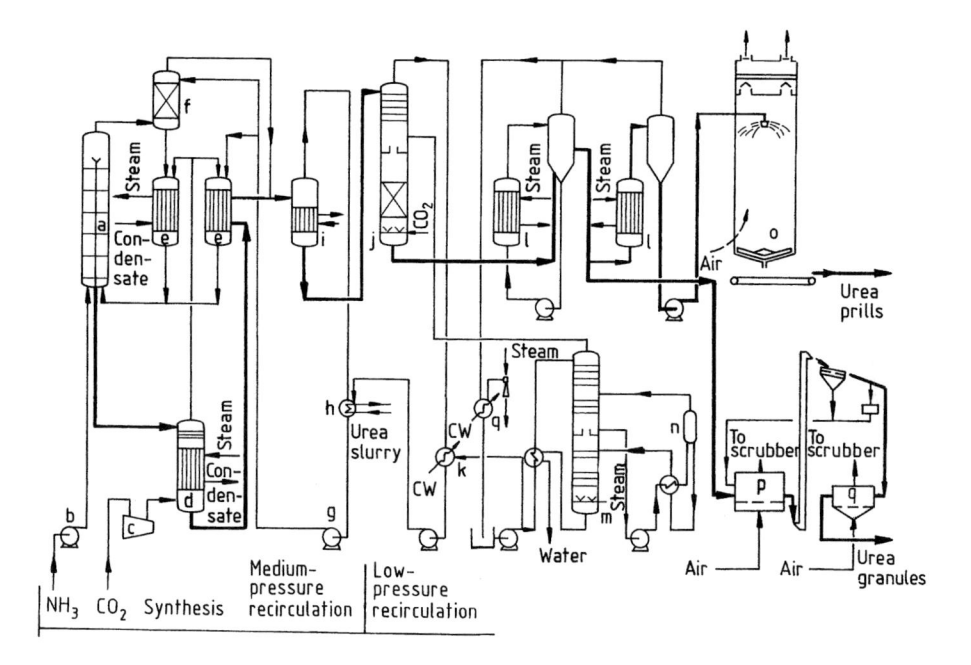

Figure 22. Schematic of the ACES process
a) Urea reactor; b) High-pressure ammonia pump; c) CO$_2$ compressor; d) Stripper; e) High-pressure carbamate condensers; f) High-pressure scrubber; g) High-pressure carbamate pump; h) Medium-pressure absorber; i) Medium-pressure decomposer; j) Low-pressure decomposer; k) Low-pressure absorber; l) Evaporators; m) Process condensate stripper; n) Hydrolyzer; o) Prilling tower; p) Granulation section; q) Surface condensers
CW = Cooling water

said to be required to achieve effective CO$_2$ stripping in the lower part. In the lower part of the stripper (a falling film heater), ammonium carbamate is decomposed and the resulting CO$_2$ and NH$_3$ as well as the excess NH$_3$ are evaporated by CO$_2$ stripping and steam heating. The overhead gaseous mixture from the top of the stripper is introduced into the carbamate condensers (e). Here, two units in parallel are installed, where the gaseous mixture is condensed and absorbed by the carbamate solution coming from the medium-pressure recovery stage. Heat liberated in the high-pressure carbamate condensers is used to generate low-pressure steam in one of the condensers and to heat the urea solution from the stripper after the pressure is reduced to about 19 bar in the shellside of the second carbamate condenser. The gas and liquid from the carbamate condensers are recycled to the reactor by gravity flow. The urea solution from the stripper, with a typical NH_3 content of 12 wt%, is purified further in the subsequent medium- and low-pressure decomposers (i, j), operating at 19 and 3 bar, respectively. Ammonia and carbon dioxide separated from the urea solution here are recovered through stepwise absorption in the low- and medium-pressure absorbers (h, k). Condensation heat in the medium-pressure absorber is transferred directly to the aqueous urea solution feed in the final concentration section. In this final concentration section (l), the purified urea solution is concentrated further either by a two-stage evaporation up to 99.7% for urea prill production or by a single evaporation up to 98.5% for urea granule production. Water vapor formed in the final concentrating section is condensed in surface condensers (q) to form process condensate. Part of this condensate is used as an absorbent in the recovery sections, whereas the remainder is purified in the process condensate treatment section by hydrolysis and steam stripping, before being discharged from the urea plant.

The highly concentrated urea solution is finally processed either through the prilling tower (o) or via the urea granulator (p). Instead of concentration via evaporation, the ACES process can also be combined with a crystallization section to produce urea with low biuret content.

Until now, the ACES process has been used in seven urea plants.

3.3.2.4. Isobaric Double-Recycle Process [26], [28], [50]

The isobaric double-recycle (IDR) stripping process, developed by Montedison, is characterized by recycle of most of the unreacted ammonia and ammonium carbamate in two decomposers in series, both operating at the synthesis pressure. A high molar $NH_3:CO_2$ ratio (4:1 to 5:1) in the reactor is applied. As a result of this choice of ratio, the reactor effluent contains a relatively high amount of nonconverted ammonia. In the first, steam-heated, high-pressure decomposer, this large quantity of free ammonia is mainly removed from the urea solution. Most of the residual ammonia, as well as some ammonium carbamate, is removed in the second high-pressure decomposer where steam heating and CO_2 stripping are applied. The high-pressure synthesis section is followed by two lower-pressure decomposition stages of traditional design, where heat exchange between the condensing off-gas of the medium-pressure decomposition stage and the aqueous urea solution to the final concentration section improves the overall energy consumption of the process. Probably because of the complexity of this process, it has not achieved great popularity so far. The IDR process or parts of the process are used in four revamps of older conventional plants.

3.3.3. Other Processes

Urea – Ammonium Nitrate (UAN) Production. Mixtures of urea (mp 133 °C) and ammonium nitrate (mp 169 °C) with water have a eutectic point at −26.5 °C [51]. As a result solutions with high nitrogen content can be made with solidification temperatures below ambient temperature. These mixtures, called UAN solutions, are used as liquid nitrogen fertilizers.

UAN solutions can be made by mixing the appropriate amounts of solid urea and solid ammonium nitrate with water or, alternatively, in a production facility specially designed to produce UAN solutions. In this latter category the Stamicarbon CO_2-stripping technology is especially suitable. In a partial-recycle version of this process, unconverted ammonia emanating from the stripped urea solution and from the reactor off-gas is neutralized with nitric acid. The ammonium nitrate solution thus formed and the urea solution from the synthesis section are mixed to yield a product solution with the desired nitrogen content (32 – 35 wt%) directly. Such a plant designated for the production of UAN solutions is cheaper than the separate production of urea and ammonium nitrate in investment and in operating costs, because evaporation, final product shaping for both urea and amonium nitrate, and wastewater treatment sections are not required.

Integrated Ammonia – Urea Production. Both feedstocks required for urea production, ammonia and carbon dioxide, are usually obtained from an ammonia plant. Since an ammonia plant is a net heat (steam) producer and a urea plant is a net heat (steam) consumer, it is normal practice to integrate the steam systems of both plants. Since both processes usually contain a process condensate treatment section where volatile components are removed by steam stripping, the advantages of combining these sections have been explored [52] – [54]. Several attempts for further integration of mass streams of both processes have been published [55] – [59]. Despite the claimed reduction in both capital and raw material cost, these highly integrated process schemes have not gained acceptance mainly because of their increased complexity.

3.4. Effluents and Effluent Reduction

Gaseous Effluents. There are potentially two sources for air pollution from a urea plant: (1) gaseous ammonia emission from continuous process vents, and (2) urea dust and ammonia emissions from the finishing section (prilling or granulation).

Gaseous Emissions from Process Vents. Noncondensable gases enter the urea process as contaminants in the raw materials, as process air introduced for corrosion protection, and as air leaking into the vacuum sections of the process. At places where these noncondensable gases are vented, proper measures should be taken to minimize ammonia losses. The present state of the art allows reduction of these losses to

< 0.2 kg of ammonia per tonne of urea produced. This is realized by using conventional absorption techniques. Special attention is required to avoid explosive gas mixtures originating from combustibles (hydrogen, methane) present in the carbon dioxide and ammonia feedstocks, and the air introduced for anticorrosion purposes. Catalytic combustion of the hydrogen present in carbon dioxide, and dosing of nitrogen or excessive amounts of combustibles have been suggested to avoid the risks of formation of explosive gas mixtures [60].

Gaseous Emissions from the Finishing Section (*Prilling/Granulation*). In the prilling processes, urea dust is produced, mainly from evaporation and subsequent sublimation of urea, but also partly via a chemical mechanism: formation of ammonia and isocyanic acid in the urea melt, evaporation of these components, followed by sublimation of this ammonia – isocyanic acid mixture to form urea dust in the colder air. Urea dust formed in this way is typically very fine (0.5 – 2 μm). Removal of this fine urea dust from the prilling tower exhaust gas is a technical challenge. Because of the small particle size, dry cyclones cannot be used. Instead, wet impingement type devices have proved successful in removing a major part of the dust from the air.

Many new urea plants use granulation instead of prilling as the finishing technique. The main purpose is to improve the size and strength of the product, but less difficulty in controlling dust emissions is also an advantage. The dust produced in these granulation devices (fluidized-bed granulation or drum granulation) is typically much coarser than the fume-like dust produced in a prilling tower.

This results basically from the much shorter contact time of liquid urea with air. Because of the coarser dust and the smaller air quantities to be handled, wet scrubbers provide adequate and much simpler dust emission control for granulation units compared to the apparatus required for emission control in prilling.

Liquid Effluents. The process condensate produced from the evaporation or crystallization sections of the plant contains 3 – 8 wt% ammonia and 0.2 – 2 wt% urea. Two techniques are known to remove these pollutants:

1) Biological treatment [61] – [63]
2) Chemical hydrolysis and steam stripping to remove ammonia from the condensate

Biological treatment seems to have gained only slight acceptance. The method involving chemical hydrolysis and steam stripping recycles urea (in the form of ammonia) and ammonia to the synthesis section for the production of urea, whereas with biological treatment the urea and ammonia present in the feed to wastewater treatment are lost to urea production. Several chemical hydrolysis and steam-stripping systems are described below.

Stamicarbon System. In the Stamicarbon system [64] first the bulk of ammonia is removed by pre-desorption of the process condensate, followed by hydrolysis of the urea with steam at 170 – 230 °C to form ammonia and carbon dioxide via ammonium carbamate. The hydrolyzer is a vertical bubble-washer column, operated in countercurrent with respect to gas and liquid flow. Ammonia remaining in the process condensate is then removed by further steam stripping (desorption). Since both the pre-desorber and the desorber operate at low pressure (1–5 bar), low-pressure steam as produced in the urea synthesis section can be used as stripping agent. The combination of pre-desorption and countercurrent operation of the hydrolyzer ensures that the chemical equilibrium of the hydrolysis reaction does not limit the minimum achievable urea content in wastewater to concentrations <1 ppm. Also, the remaining NH_3 concentration is <1 ppm.

Snamprogetti System. The Snamprogetti system [65] also includes a system of pre-desorption, hydrolysis, and final desorption with steam stripping. In this process, however, the hydrolyzer is built as a horizontal column with cross-flow operation with respect to gas and liquid flow. Hydrolysis is carried out at somewhat higher temperature (230–236 °C) and pressure (33–37 bar) than in the Stamicarbon process. Also, this process claims minimum achievable NH_3 and urea concentrations <1 ppm in the liquid effluent.

Toyo Engineering Corporation also offers a system of pre-desorption and hydrolysis, followed by final desorption, as part of its ACES and conventional urea processes. Urea and ammonia concentrations <5 ppm in the effluent are claimed.

UTI System. In the UTI system [66], desorption and hydrolysis take place in a single column, in which the hydrolysis of urea back to ammonium carbamate is carried out concurrently with the stripping of ammonia from the liquid phase. Because of this concept, both the steam required for steam stripping and the steam required to bring the process condensate to a sufficiently high temperature for hydrolysis must be at medium pressure. A small amount of carbon dioxide is fed to the bottom of the column. The system is claimed to achieve residual NH_3 concentrations <10 ppm.

Other Systems. Some systems have been proposed [52]–[54] in which the wastewater treatment sections of the urea and ammonia plants are combined. Like the systems described above, they also remove urea by hydrolysis to ammonia and carbon dioxide, which are subsequently removed by transferring them into a steam-containing gas phase. The principal difference between these systems and the aforementioned methods is that the ammonia and carbon dioxide produced are recycled to the ammonia plant reforming system, rather than to the urea synthesis section.

3.5. Product-Shaping Technology

Prilling. For a long time, prilling has been used widely as the final shaping technology for urea. In prilling processes, the urea melt is distributed in the form of droplets in a prilling tower. This distribution is performed either by showerheads or by using a rotating prilling bucket equipped with holes. Urea droplets solidify as they fall down the tower, being cooled countercurrently with up-flowing air. The prilling process has several drawbacks:

1) The size of the product is limited to a maximum average diameter of about 2.1 mm. Larger-size product would require uneconomically high prilling towers; moreover, larger droplets tend to be unstable.
2) Very fine dust is formed in the prilling process (see Section 3.4). Removal of this dust is a technically difficult and expensive operation.
3) The crushing strength and shock resistance of prills are limited, making prilled product less suitable for bulk transport over long distances. This problem can be largely overcome with appropriate techniques to improve the physical properties, such as seeding [67] to improve shock resistance or addition of formaldehyde to improve crushing strength, and to suppress the caking tendency.

Granulation [29], [68] – [70]. The drawbacks of prilling have initiated the development of several granulation techniques. These techniques deviate from the prilling technique in that the urea melt is sprayed on granules, which gradually increase in size as the process continues. The heat of solidification is removed by cooling air or, for some granulation techniques, evaporation of water. Since the contact time between liquid urea and air in these processes is much lower than in prilling, the dust formed in granulation processes is much coarser; therefore, it can be removed much more easily from the cooling air.

Drum granulation systems have been developed, for example, by C & I Girdler, Kaltenbach – Thuring, and Montedison. *Pan granulation processes* have also been developed, for example, by Norsk Hydro and the Tennessee Valley Authority (TVA). Recently *spouted-bed* and *fluidized-bed granulation techniques* were introduced. A spouted-bed technique was developed by Toyo Engineering Corporation; however, the fluidized-bed technology from Hydro Agri seems to be most successful at this time.

All granulation processes require the addition of formaldehyde or formaldehyde-containing components. Also common to all granulation techniques is that they yield products with larger diameters compared to prilling. However, their capabilities in this respect differ from each other to quite an extent.

Although the improvements brought about by granulation are beyond doubt, prilling techniques still have a place because of the lower investment and lower variable costs associated with prilling compared to granulation.

4. Forms Supplied, Storage, and Transportation

Forms Supplied. Urea may be supplied either in solid form or as a liquid. For *liquid compound fertilizers*, urea is a favorite ingredient. It is generally used in combination with ammonium nitrate as an aqueous solution to obtain liquids containing 32 – 35 wt% nitrogen. These solutions are designated as UAN-32 to UAN-35.

The *solid forms* are generally classified as granular or prilled products, because of the differences in handling properties. Prilled product is considered less suitable for bulk transportation because prills have lower crushing strength, a lower shock resistance, and a higher caking tendency than granules. Because of this, prilled products are usually marginally cheaper than granulated product. Granulated product usually also has a larger diameter (2.0 – 2.5 mm) than prills (1.5 – 2.0 mm), making granules more suitable for bulk blending to produce compound fertilizers.

Special Grades. The majority of urea is designated as "fertilizer grade"; however, some special forms have found limited application:

Technical Grade. Technical-grade urea should be without additions; color, ash-, and metal content are sometimes also specified. For urea used to produce urea – formaldehyde resins, its content of pH-controlling trace components is important. Because of this, technical-grade urea at present is mostly traded as a performance product, rather than being bound to narrow specification limits. The fitness of the product for use is judged by application-specific tests.

Low-Biuret Grade. A maximum biuret content up to 1.2 wt% is considered acceptable for nearly all fertilizer applications of urea. Only for the relative small market segment of foliar spray to citrus crops is a lower biuret content (max. 0.3%) desirable.

Feed Grade. Some urea is also used directly as a feed component for cattle. Urea used for this purpose should be free of additions. Feed-grade urea is supplied in the form of microprills with a mean diameter of about 0.5 mm.

Slow-Release Grades. Studies show that only 50 – 60% of fertilizer nitrogen applied to soil is usually recovered by crop plants. Several attempts have been made to increase this percentage by slowing the release of fertilizer to the ground via coating or additions [71].

Urea Supergranules. Granulated product with a very large diameter (up to 15 mm) has found limited application for deep placement in wetland rice [72] and forest fertilization.

Storage. The shift from bagged to bulk transport and storage of prilled and granulated urea has called for warehouse designs in which large quantities of urea can be stored in bulk. These warehouses should be designed in such a way that the product suffers little degradation. Degradation may result from: (1) segregation of fines; (2) disintegration; and (3) absorption, loss, or migration of water.

Segregation of fines can be avoided through uniform product spreading during pouring. *Disintegration* can be minimized by

1) Providing the product pouring system with a pouring height adjuster
2) Design of "product-friendly" reclaiming systems, because reclaiming the product by means of payloaders and tractor shovels invariably leads to product disintegration

Caking and subsequent product disintegration at unloading are known to result from water absorption. What is not commonly known, however, is that excessive drying of the product during storage also leads to a higher caking tendency and that migration of water from warm product in the bulk of a pile to the cold surface leads to crust formation. Thus, attempts to decrease water absorption through refrigeration or air conditioning, dehumidification, or space heating may cause the air in the warehouse to become too dry or may result in too great a temperature difference between the product and the surrounding air. Instead, the warehouse (especially the roof) should be airtight and thoroughly insulated. The caking tendency of urea can be reduced by addition of small amounts of formaldehyde (up to 0.6 wt%) to the urea melt or by addition of surfactants to the solid product.

Transportation. Urea prills and granules are transported by bulk transport in trucks, ships, rail cars, etc. To withstand numerous and rapid loading and unloading operations, product for bulk transport should have a high initial physical stability. Great demands are made, especially on the shock resistance of the product, e.g., at seaport loading and unloading facilities. In addition, a number of "good housekeeping" rules should be adhered to:

1) Do not load or unload during rain
2) Make sure that the means of transport is clean and dry
3) Close the ship's hold when rain is imminent
4) Do not replace the air above the product or ventilate the holds
5) Cover the product (e.g., by polyethylene sheeting) during prolonged transport
6) Product should be spread rather then poured solely from one point to prevent dust coning due to segregation
7) Restrict the pouring height to avoid unnecessary disintegration

Liquid Fertilizer Transport. Liquid fertilizers are transported by tank cars, railway tanks, ships, and pipelines. Although liquid fertilizer is generally accepted as the most economic form to distribute, the solid form is still the most popular by far. Distribution of large quantities of liquid fertilizer requires a complex infrastructure and is limited at present to large farm units in developed countries. Transport and storage of UAN solutions in carbon steel lines and tanks require the addition of a corrosion inhibitor to the solution.

Table 2. Typical product specifications for fertilizer-grade urea

Specification	Prilled product	Granulated product
Nitrogen content, wt%	min. 46	min. 46
Biuret content, wt%	max. 1	max. 1
Water content, wt%	max. 0.3	max. 0.25
Crushing strength, bar	20–25	30–60
Shock resistance, wt%	min. 85	100
Product size		
1.0–2.4 mm, wt%	90–95	
1.6–4.0 mm, wt%		95
Bulk density (loose), kg/m^3	730	750–790

5. Quality Specifications and Analysis

Typical quality specifications for fertilizer-grade urea are summarized in Table 2. The capabilities of a modern urea plant are better than the typical trade data given in this table.

The total *nitrogen content* is usually determined by digesting urea with sulfuric acid to yield ammonium sulfate. The ammonia content is then determined by distillation and titration. Alternatively, the total N content may be determined by the Kjeldahl method or by using a method based on hydrolyzing urea with urease followed by titration of the ammonia formed.

The *water* content is usually determined with Karl Fischer reagent.

Biuret is determined by the formation of a violet-colored complex of biuret with copper(II) sulfate in an alkaline medium and subsequent measurement of the absorbance of the colored solution at 546 nm.

Crushing strength is defined as the force required per unit cross-sectional area of a granule to crush the granule or, if it is not crushed, the force at which it is deformed by 0.1 mm. A single granule is subjected to a force that is increased at a constant rate, the force at breakage (or at 0.1-mm deformation) being recorded.

The *shock resistance* of granules is defined as the weight percentage of a sample that is not crushed when subjected to a specified shock load. To determine shock resistance, a sample of prills or granules is shot against a metal plate by means of compressed air under normalized conditions. The amount of nondamaged product that remains after the test is determined.

The *granulometry* of the product is measured by conventional sieve techniques.

6. Uses

Urea is used for soil and leaf fertilization (more than 90% of the total use); in the manufacture of urea–formaldehyde resins; in melamine production; as a nutrient for ruminants (cattle feed); and in miscellaneous applications.

Soil and Leaf Fertilization. Worldwide, urea has become the most important nitrogenous fertilizer. Urea has the highest nitrogen content of all solid nitrogenous fertilizers; therefore, its transportation costs per tonne of nitrogen nutrient are lowest. Urea is highly soluble in water and thus very suitable for use in fertilizer solutions (e.g., "foliar feed" fertilizers). Urea is also used as a raw material for the production of compound fertilizers. Compound fertilizers may be produced by mixing in urea melts or urea solutions before shaping the compound fertilizers or by mixing solid urea prills or granules with other fertilizers (bulk blending). In the latter case, the product sizes must match to prevent segregation of the products during further handling.

Urea–Formaldehyde Resins A significant proportion of urea production is used in the preparation of urea–formaldehyde resins. These synthetic resins are employed in the manufacture of adhesives, molding powders, varnishes, and foams. They are also used to impregnate paper, textiles, and leather.

Melamine Production. At present, nearly all melamine production is based on urea as a feedstock. Since ammonia is formed as a coproduct in melamine production from urea (see Chap. 1), integration of the urea and melamine production processes is beneficial.

Feed for Cattle and other Ruminants. Because of the activity of microorganisms in their cud, ruminants can metabolize certain nitrogen-containing compounds, such as urea, as protein substitutes. In the United States this capability is exploited on a large scale. In Western Europe, by contrast, not much urea is used in cattle feed.

Other Uses. On a smaller scale, urea is employed as a raw material or auxiliary in the pharmaceutical industry, the fermenting and brewing industries, and the petroleum industry. A recent development is the application of urea for the removal of NO_x from flue gases. Urea is also used as a solubilizing agent for proteins and starches, and as a deicing agent for airport runways.

7. Economic Aspects

The predicted growth in *demand* for urea until 1997 is slightly more than 3% per year, bringing the total urea demand in 1997 to some 89×10^6 t/a. Most of the growth will occur in Asia, with China and India in the lead. A little more than 7% of the worldwide demand for urea is from industry, in which Europe takes a leading role, ahead of North America and the industrialized countries of Asia.

The predicted growth in *capacity* up to 1997 will be around 1.5% per year, resulting in a total installed capacity of some 100×10^6 t by 1997. Between 1994 and 1997, new plants will account for an increase in capacity of ca. 6.5×10^6 t, whereas closure of old plants will result in a reduction of the installed capacity by some 2×10^6 t. The world's capacity utilization can thus be calculated to be ca. 89%.

Because of geopolitical changes during the early 1990s, different economic laws have become valid in the former Soviet Union and Eastern Europe, causing those countries to supply to the world market large amounts of prilled urea at prices far below the cost to many producers [e.g., sales prices, free on board (FOB) Black Sea $ 75 per tonne in 1993]. This situation is not expected to last long, although FOB prices at Black Sea ports will generally remain lower compared to other places. In recent years urea prices have shown large fluctuations. For instance, in 1994–1995 lows of $ 100 per tonne and highs of $ 250 per tonne have been reported. Predictions of future developments of urea prices are therefore highly speculative.

To cope with tough competition on the world market, producers have the tendency to build plants with very high single line capacities (2000 t/d or more), in which operating reliability is of extreme importance. Uninterrupted operating periods of more than one year are often achieved.

Furthermore, producers are increasingly shifting production facilities (both new plants and relocations) to places where natural gas is plentiful and cheap. Europe seems to have lost its competitive edge in export markets due to its expensive feedstocks.

The investment costs (in 10^6 $) for a present state-of-the-art total-recycle urea plant are estimated to be:

1000-t/d plant (single line)	43
1500-t/d plant (single line)	52
2000-t/d plant (single line)	62

8. Urea Derivatives

8.1. Thermal Condensation Products of Urea

Thermolysis of urea gives biuret, triuret, and cyanuric acid, and in a special process, melamine is produced.

Biuret [108-19-0], imidodicarbonic diamide, $H_2NCONHCONH_2$, mp 193 °C, is produced by heating urea in inert hydrocarbons at 110–125 °C or in the melt at 127 °C [73], [74].

Triuret [556-99-0], diimidotricarbonic diamide, $H_2NCONHCONHCONH_2$, mp 231 °C, is produced by decomposing urea in a thin film at 120–125 °C [75]. It is also obtained by treating 2 mol of urea with 1 mol of phosgene in toluene at 70–80 °C [76], [77].

Cyanuric acid [108-80-5], 1,3,5-triazine-2,4,6(1H,3H,5H)-trione (**1**), is formed on heating urea in the presence of zinc chloride, sulfuryl chloride, or chlorosulfuric acid [78].

$$6\,H_2NCONH_2 \xrightarrow{-3\,NH_3} 3\,H_2NCONHCONH_2 \xrightarrow{-3\,NH_3}$$

1 (1,3,5-triazine-2,4,6-triol)

Melamine [108-78-1], 1,3,5-triazine-2,4,6-triamine (**2**) is produced industrially from urea [79].

$$6\,H_2NCONH_2 \xrightarrow{NH_3} \mathbf{2} + 3\,CO_2 + 6\,NH_3$$

Table 3. Production processes for substituted ureas

Starting materials	Product
Urea and amines	ureas substituted at one or both nitrogen atoms and cyclic ureas
Urea and tertiary alcohols	ureas substituted at one nitrogen atom
Phosgene and amines	ureas substituted at one nitrogen atom
Isocyanates and NH_3 or amines	ureas substituted at one or both nitrogen atoms
Carbamoyl chloride and NH_3 or amines	ureas substituted at one or both nitrogen atoms
Esters of carbonic or carbamic acids	ureas substituted at one or both nitrogen atoms
Urea and aldehydes or ketones	ureas substituted at one or both nitrogen atoms and cyclic ureas

8.2. Alkyl- and Arylureas

Most simple substituted alkylureas are crystalline products. Tetramethyl- and tetraethylurea and some cyclic ureas are liquids. Alkylated and arylated ureas are used in the production of plant protection agents, in pharmaceutical and dye chemistry, as plasticizers, and as stabilizers. Alkylureas and polyalkyleneureas are used as additives in the production of aminoplastics.

Various processes can be used for the production of substituted ureas, the most important of which are listed in Table 3.

8.2.1. Transamidation of Urea with Amines

The transamidation of urea with amines is one of the most important industrial production processes for substituted ureas. Monosubstituted ureas are produced by condensation of urea with a sufficiently basic amine in a 1:1 molar ratio in the melt at 130–150 °C [80]. Symmetrically disubstituted ureas can be produced from 2 mol of amine and 1 mol of urea at 140–170 °C [81]. Instead of the amines, amine salts can also be reacted with urea in the melt or by prolonged boiling in aqueous solution [82].

Monomethylurea [598-50-5], $CH_3NHCONH_2$, *mp* 102 °C, is produced industrially by passing monomethylamine into a urea melt [83]. Monomethylurea is used for the synthesis of theobromine.

Phenylurea [64-10-8], $C_6H_5HNCONH_2$, *mp* 147 °C, can be produced by heating an aqueous solution of aniline hydrochloride and urea to its boiling point [84].

Symmetric N,N'-dimethylurea [96-31-1], $CH_3HNCONHCH_3$, *mp* 105 °C, is used for the synthesis of caffeine by the Traube method and for the production of formaldehyde-free easy-care finishing agents for textiles [84].

Hexamethylenediurea, [2188-09-2] $H_2NCONH(CH_2)_6NHCONH_2$, is obtained by heating a mixture of hexamethylenediamine with an excess of urea at 130–140 °C, with elimination of ammonia [85], [86].

N,N'-Diphenylurea [102-07-8], carbanilide, $C_6H_5HNCONHC_6H_5$, mp 238 °C, can be produced in high yields by heating 2 mol of aniline with 1 mol of urea in glacial acetic acid [86], [87].

Polyalkyleneureas can be obtained by treating di-, tri-, and tetraalkylamines in concentrated aqueous solution with urea at ca. 110 °C. They are used, for example, to modify melamine–formaldehyde impregnating resins and binders for derived timber products. Examples of polyalkyleneureas include the following:

1,2-Ethylenediurea, $H_2NCONH-CH_2CH_2-HNCONH_2$, mp 198 °C
1,3-Propylenediurea, $H_2NCONH-CH_2CH_2CH_2-HNCONH_2$, mp 186 °C
Diethylenetriurea, $H_2NCONH-CH_2CH_2-N(CONH_2)-CH_2CH_2-HNCONH_2$, mp 215 °C
Dipropylenetriurea, $H_2NCONH-(CH_2)_3-N(CONH_2)-(CH_2)_3-HNCONH_2$
2-Hydroxypropylene-1,3-diurea, $H_2NCONH-CH_2CH(OH)-CH_2-HNCONH_2$, mp 147 °C

Cycloalkyleneureas. *2-Imidazolidinone* [120-93-4], ethyleneurea (**3**), mp 131 °C [88], [89]; *2-oxohexahydropyrimidine* [65405-39-2], propyleneurea (**4**), mp 260–265 °C [90]; and *2-oxo-5-hydroxyhexahydropyrimidine*, 5-hydroxypropyleneurea (**5**), are produced industrially in the melt above 180 °C, preferably at 200–230 °C, by condensation of 1,2-ethylenediamine or 1,3-propylenediamine with urea and elimination of ammonia. These cycloalkyleneureas are used in the form of their N,N'-dihydroxymethyl compounds for easy-care finishes for cellulose-containing textiles [91].

8.2.2. Alkylation of Urea with Tertiary Alcohols

Urea can be alkylated with tertiary alcohols in the presence of sulfuric acid [92], [93].

tert-Butylurea, $(CH_3)_3CHNCONH_2$, mp 182 °C, is produced by treating urea with 2 mol of *tert*-butanol in the presence of ca. 2 mol of concentrated sulfuric acid at 20–25 °C with ice cooling [92]–[94].

8.2.3. Phosgenation of Amines

Symmetrically disubstituted ureas are produced in good yields by passing phosgene into solutions of amines in aromatic hydrocarbons [95]. In some cases, aqueous solutions or suspensions of amines can also be reacted with phosgenes [96]. The phosgenation of mixed aliphatic–aromatic amines is carried out industrially in the presence of sodium hydroxide at 40–60 °C [97]. For the production of substantive dyes, aminosulfone or aminocarboxylic acid groups are bonded by means of phosgenation [98]

$$2\ RNH_2 + COCl_2 \rightarrow RNH-CO-NHR + 2\ HCl$$

Tetramethylurea [632-22-4], $(CH_3)_2NCON(CH_3)_2$, *bp* 156.5 °C, is produced from dimethylamine and phosgene and is used as an aprotic solvent [99].

Asymmetric Diphenylurea, $(C_6H_5)_2NCONH_2$, *mp* 189 °C, is produced from diphenylamine, phosgene, and ammonia:

$$COCl_2 + (C_6H_5)_2NH + 3\ NH_3 \rightarrow (C_6H_5)_2NCONH_2 + 2\ NH_4Cl$$

Symmetric dimethyldiphenylurea, *mp* 121–127 °C, is obtained by treating phosgene with monomethylaniline and sodium hydroxide.

$$COCl_2 + 2\ HN\begin{matrix}CH_3\\C_6H_5\end{matrix} + 2\ NaOH \longrightarrow \begin{matrix}CH_3\\NCON\\C_6H_5\end{matrix}\begin{matrix}CH_3\\ \\C_6H_5\end{matrix} + 2\ NaCl + 2\ H_2O$$

Symmetric dialkyldiarylureas are used under the name Centralite as plasticizers and stabilizers for nitrocellulose and propellants [97].

8.2.4. Reaction of Amines with Cyanates (Salts)

The salts of aliphatic or aromatic amines react with potassium cyanate at 20–60 °C to give substituted ureas in high yields [100], [101].

$$R-NH_2 \cdot HX + KNCO + H_2O \rightarrow R-HNCONH_2 + KX + OH^-$$

4-Ethoxyphenylurea [150-69-6] (dulcin) is produced from potassium cyanate and *p*-phenetidine hydrochloride in aqueous solution at room temperature [100]. This production method can be applied to aminosulfonic and aminocarboxylic acids, whereby the betaine-like salts formed by these acids react with potassium cyanate to give ureas that are substituted at only one NH_2 group [101]. Sulfonamides react with potassium cyanate to give potassium salts of the corresponding sulfonylureas, from which the sulfonylureas are obtained by acidification [102]:

$$RSO_2NH_2 + KNCO \rightarrow RSO_2\underset{K^+}{N^-}CONH_2 \rightarrow$$
$$RSO_2HNCONH_2$$

8.2.5. Reaction with Isocyanates

Symmetrically disubstituted ureas can also be produced from isocyanates by prolonged heating in aqueous solution [103], [104]:

$$RNCO \xrightarrow{H_2O} RHNCOOH \xrightarrow{RNCO} RHNCONHR + CO_2$$

When ammonia or primary or secondary amines are reacted with isocyanates, the corresponding substituted ureas are obtained in almost quantitative yield [104]–[106]. This process is particularly suitable for the production of unsymmetrically substituted ureas [105], [106].

$$RNCO + HN\begin{subarray}{l}R^1\\R^2\end{subarray} \longrightarrow RHNCON\begin{subarray}{l}R^1\\R^2\end{subarray} \quad R^1, R^2 = H, \text{alkyl, aryl}$$

8.2.6. Acylation of Ammonia or Amines with Carbamoyl Chlorides

Ammonia and primary or secondary amines react with carbamoyl chlorides to give the corresponding urea derivatives in good yields [107].

Tetraphenylurea, $(C_6H_5)_2NCON(C_6H_5)_2$, [108] is obtained in quantitative yield by heating diphenylcarbamoyl chloride with diphenylamine.

8.2.7. Aminolysis of Esters of Carbonic and Carbamic Acids

Esters of carbonic and carbamic acids (carbonates and carbamates) react with amines at elevated temperatures to give symmetrically disubstituted ureas [109]–[112].

8.3. Reaction of Urea and Its Derivatives with Aldehydes

8.3.1. α-Hydroxyalkylureas

The industrial production of α-hydroxyalkylureas is limited to the addition of formaldehyde or glyoxal to urea, monoalkylureas, symmetrical dialkylureas, and cyclic ureas. It involves acid- or base-catalyzed additions that are generally equilibrium reactions [113]–[130].

Urea can bond with up to 4 mol of formaldehyde. However, only monohydroxymethyl- and N,N'-dihydroxymethylurea can be isolated in pure form [131], [132]. Tri- and tetrahydroxymethylureas are formed only as nonisolable intermediates, for example, in the synthesis of trimethoxymethylurea (**6**), [133], [134]; N,N'-dialkoxymethyl-4-oxomethyltetrahydro-1,3,5-oxadiazine (**7**) [135], [136]; and N,N'-dihydroxymethyl-2-oxo-5-alkyltetrahydro-1,3,5-triazines (**8**) [137]–[140].

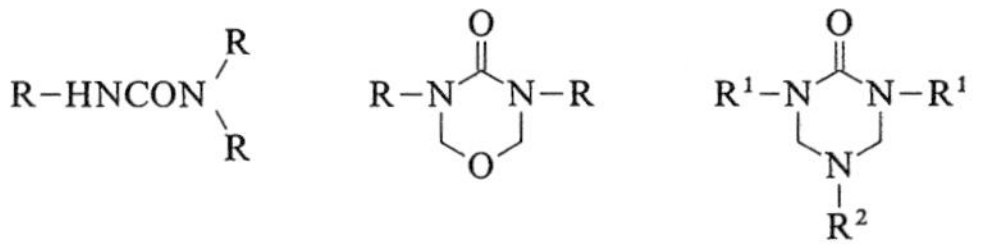

$$\text{R–HNCON} \diagup_{\diagdown R}^{R} \qquad \underset{6}{} \qquad \underset{7}{R-N \overset{O}{\underset{\diagdown O \diagup}{\diagup \diagdown}} N-R} \qquad \underset{8}{R^1-N \overset{O}{\underset{\diagdown N \diagup}{\diagup \diagdown}} N-R^1}$$

R = Alkoxymethyl
R^1 = Hydroxymethyl
R^2 = Alkyl

N,N'-Dihydroxymethylurea [140-95-4], HOCH$_2$HNCONHCH$_2$OH [131], [132] is produced industrially by charging 2 mol of formaldehyde per mole of urea to a stirred vessel. The solution is neutralized with triethanolamine. Urea is added with cooling, and the temperature must not exceed 40 °C in this slightly exothermic reaction. After a few hours the reaction mixture is cooled to room temperature and dihydroxymethylurea crystallizes out. The product is dried in a spray-drying tower.

Hydroxymethyl derivatives of cyclic ureas can be produced by reaction of these substances with formaldehyde in an alkaline medium [141].

Hydroxymethyl derivatives of urea and cyclic ureas (**9**)–(**15**) are used in easy-care finishes for textiles [91].

The addition of higher aldehydes to urea, mono- and symmetrically disubstituted ureas, and cyclic ureas generally gives unstable α-hydroxyalkyl compounds. Electron-withdrawing and electron-donating substituents next to the α-hydroxyalkyl group affect the stability of these compounds and also their ability to undergo condensations.

For example, the chloral–urea derivatives (**16**) exhibit considerable differences in reactivity compared with the *N*-hydroxymethyl (**17**) and *N*-α-hydroxyethyl compounds (**18**). These differences are exemplified by a decrease in the H-acidity of the OH groups [142]. Chloral compounds (**16**) can form alkali-metal salts, whereas the corresponding salts of *N*-hydroxymethyl (**17**) and *N*-α-hydroxyethyl compounds (**18**) are unknown. Condensation of chloral compounds with nucleophiles is possible only under extreme reaction conditions. However, *N*-α-hydroxyethyl compounds can be converted smoothly with alcohol into *N*-α-alkoxyethyl compounds in basic and sometimes even in neutral media. The reactivity of *N*-hydroxymethyl compounds lies between that of the chloral and N-α-hydroxyethyl compounds. The 4-hydroxycycloalkyleneureas (cyclic N-hemiacetals), which are obtained by treating suitable aldehydes with ureas, are stable. For example, 2-oxo-4,5-dihydroxyimidazolidines are formed by cyclization of urea, or its mono- or symmetrically disubstituted derivatives, with glyoxal [143]–[145]:

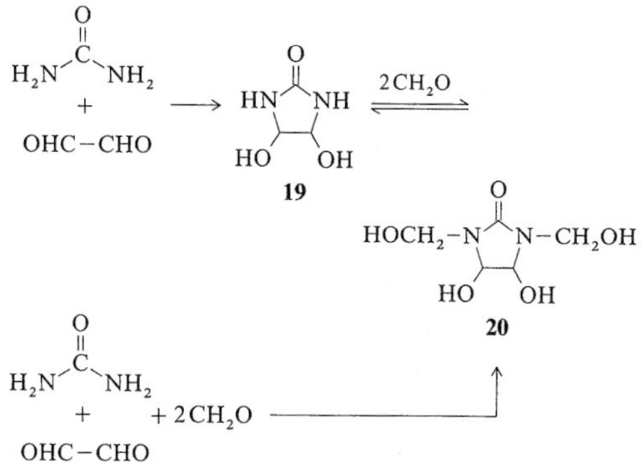

N,N'-Dihydroxymethyl-2-oxo-4,5-dihydroxyimidazolidine (**20**) is produced by hydroxymethylation of 4,5-dihydroxy-2-oxoimidazolidine (**19**) in weakly acidic to weakly alkaline aqueous solution or, more elegantly, by direct reaction of urea with glyoxal and formaldehyde in the appropriate molar ratio in weakly acidic to neutral solution at 40–80 °C, sometimes in the presence of catalytically active buffers [130].

N,N'-Dimethyl-2-oxo-4,5-dihydroxyimidazolidine and its derivative in which the OH groups are partly acetalized with methanol are used as formaldehyde-free cross-linking agents for easy-care finishes for cellulose-containing textiles.

2-Oxo-4-hydroxyhexahydropyrimidines also belong to the group of α-hydroxyalkylureas. These compounds can be produced industrially by cyclocondensation of urea with active enolizable aldehydes (see Section 8.3.4).

8.3.2. α-Alkoxyalkylureas

Condensation of N-hydroxymethylureas with alcohols to give N-alkoxymethylureas (ureidoalkylation of alcohols) is of great industrial importance. Pure hydroxymethylureas and an excess of alcohol are reacted in the presence of catalytic amounts of acid. The nature and quantity of the acid catalyst depend on the reactivity of the N-hydroxymethyl compound, its stability to hydrolysis, and the formation of byproducts and polycondensation products. At elevated temperature the reaction can be carried out under weakly acidic conditions, whereby the equilibrium position must be adjusted by variation of the concentration and the molar ratios [91], [146], [147]. Sometimes, ureidomethylation of alcohols is better at room temperature in the presence of strong acids.

The alcohol-modified urea–formaldehyde condensation products are used as resins for heat- or acid-curing coatings. Besides the water-soluble or almost solvent-free resins, which are becoming increasingly important for environmental reasons, a wide range of aminoplastic resins for coatings exist that are readily soluble in common paint solvents. To convert urea–formaldehyde resins to resins that are soluble in organic solvents, the highly polar N-hydroxymethyl groups obtained in the initial reaction between urea and formaldehyde are acetalized with alcohols, mainly butanol and isobutanol, as well as ethanol and methanol or their mixtures.

The alcohol-modified urea–formaldehyde resins are produced industrially by passing aliphatic alcohols into aqueous solutions of urea and formaldehyde or solutions of hydroxymethylated ureas in the presence of small quantities of acid at 90–100 °C so that the water formed and excess alcohol distill off. The molar ratios vary between 2 and 4 mol of formaldehyde and 2 and 5 mol of alcohol per mole of urea. The process is carried out in a reactor equipped with an adequately dimensioned heat exchanger, a vacuum pump, a distillation column, and for alcohols that are sparingly soluble in water, a water separator.

Some *4-hydroxycycloalkyleneureas* are so reactive that they can be converted into the N-α-alkoxy compounds (**21**), (**22**) even in a neutral medium by heating with alcohol [148]:

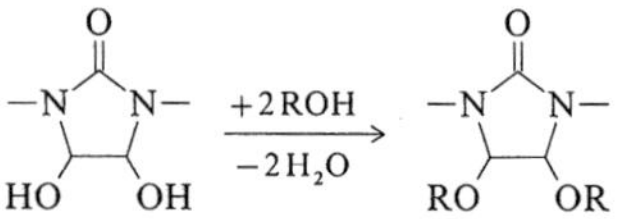

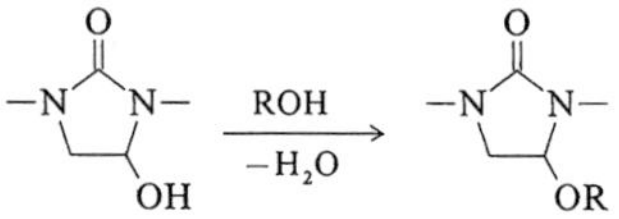

To shift the equilibrium in the N-α-ureidoalkylation of alcohols in the direction of the N-α-alkoxyalkyl compounds, the water formed during the reaction must be removed. In industrial production processes an aprotic entrainer (e.g., an aromatic) is used. In the ureidomethylation of alcohols that are immiscible or sparingly miscible with water, an excess of alcohol is used and water is removed by azeotropic distillation.

Higher-boiling alcohols can also be ureidomethylated by transacetalization of the N-methoxymethyl compounds.

Polymerizable compounds are obtained by ureidoalkylation of unsaturated alcohols (e.g., allyl alcohol [149], [150]). Tetraallyloxymethyltetrahydroimidazo[4,5-*d*]imidazole-2,5(1*H*,3*H*)-dione (**23**) has achieved importance as a polymerizable coating component [151], [152]:

$$\text{structure 23: tetrasubstituted glycoluril with } ROH_2C-, CH_2OR \text{ groups on nitrogens, two C=O} \quad R = -CH_2-CH=CH_2$$

8.3.3. α,α'-Alkyleneureas

α,α'-Alkyleneureas are obtained by condensation of urea or its derivatives with aldehydes in a weakly acidic medium. The aldehyde group first adds to the urea to form an α-hydroxyalkylurea, which then reacts with a second molecule of urea or with another α-hydroxyalkylurea to give linear or branched α,α'-alkyleneureas:

$$H_2NCONH_2 + OHC-R \rightleftharpoons H_2NCONH-CH(R)-OH$$
$$\xrightarrow{+H_2NCONH_2} H_2NCONH-CH(R)-HNCONH_2 + H_2O$$
$$\xrightarrow{+nH_2NCONH-CH(R)-OH}_{-(n-1)H_2O} H_2NCONH-[-CH(R)-HNCONH-]_n-CH(R)-HNCONH_2$$

Reaction of equimolar quantities of urea and formaldehyde in acidic solution gives polymethyleneureas as a result of stepwise ureidomethylations:

$$H_2N-\overset{O}{\overset{\|}{C}}-NH_2 + CH_2O \rightleftharpoons H_2N-\overset{O}{\overset{\|}{C}}-NH-CH_2OH$$

$$\xrightarrow{+H_2N-\overset{O}{\overset{\|}{C}}-NH_2/H^+}_{-H_2O} H_2N-\overset{O}{\overset{\|}{C}}-NH-CH_2-NH-\overset{O}{\overset{\|}{C}}-NH_2$$

$$\xrightarrow{+CH_2O} HOCH_2-HN-\overset{O}{\overset{\|}{C}}-NH-CH_2-NH-\overset{O}{\overset{\|}{C}}-NH_2$$

$$\xrightarrow{-H_2O/H^+} H_2N-\overset{O}{\overset{\|}{C}}-NH-CH_2-NH-\overset{O}{\overset{\|}{C}}-NH-CH_2-NH-\overset{O}{\overset{\|}{C}}-NH_2$$

$$\xrightarrow{+CH_2O/H^+} \text{Polymethyleneurea}$$

KADOWAKI obtained polymethyleneureas containing up to five urea groups joined by methylene bridges by means of a stepwise synthesis [135]. Because of their extreme insolubility, no higher polymethyleneureas have yet been isolated. Polymethyleneureas are used as slow-release nitrogen fertilizers, for example.

Isobutylidenediurea (**24**), which is sparingly soluble in water, is obtained by condensation of urea with isobutyraldehyde in a molar ratio of 2:1 in a weakly acidic medium:

$$2\,H_2NCONH_2 + \underset{\underset{CH_3}{|}}{\overset{\overset{CH_3}{|}}{HC}}-CHO \xrightarrow[-H_2O]{H^+}$$

$$H_2NCONH-\underset{\underset{CH\diagdown CH_3}{\diagup}}{\overset{|}{CH}}-HNCONH_2$$

24

Isobutylidenediurea is also a slow-release nitrogen fertilizer used in various special fertilizer formulations. Isobutylidenediurea is produced by a continuous process. According to a patent published by Mitsubishi Chemical Industries [153], urea is charged continuously with a screw feed via a belt weigher and is reacted with a stoichiometric quantity of isobutyraldehyde in the presence of semiconcentrated sulfuric acid in a mixer. In the last section of the mixer the reaction product is neutralized by injecting dilute aqueous potassium hydroxide solution. Theproduct is dried by using plate driers and processed by sieving, filtering, and grinding.

8.3.4. Cyclic Urea – Aldehyde Condensation Products

Almost all cyclizations of urea and its derivatives with aldehydes involve an α- or a vinylogous ureidoalkylation [113]–[115], [146]. If the urea bears a nucleophilic substituent on the second nitrogen atom, cyclocondensation occurs [154]. Saturated and unsaturated cyclic ureas with five, six, seven, or eight ring atoms, bicyclic and polycyclic heterocycles (with both uncondensed rings and rings anellated in the 1,2- or 1,3- position), and spiro compounds can be produced this way [154], [155].

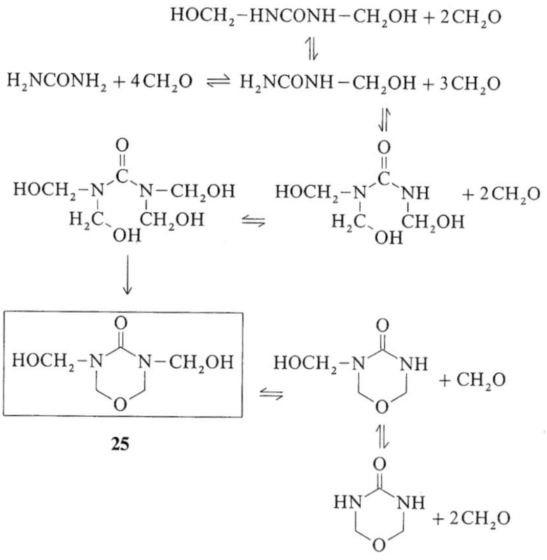

$$HOCH_2-HNCONH-CH_2OH + 2CH_2O$$
$$\Updownarrow$$
$$H_2NCONH_2 + 4CH_2O \rightleftharpoons H_2NCONH-CH_2OH + 3CH_2O$$
$$\Updownarrow$$

[Structures: HOCH₂-N(CH₂-CH(OH))-C(=O)-N(CH₂OH)-CH₂OH ⇌ HOCH₂-N(CH₂-CH(OH))-C(=O)-NH + 2CH₂O]

↓

[Boxed structure 25: 4-oxotetrahydro-1,3,5-oxadiazine with HOCH₂-N and N-CH₂OH substituents] ⇌ [same ring with HOCH₂-N and NH] + CH₂O

$$\Updownarrow$$

[Ring with HN and NH] + 2CH₂O

Industrially important reactions are those of urea with formaldehyde, acetaldehyde, isobutyraldehyde, and their mixtures. Treatment of urea with formaldehyde in a molar ratio of $1:\geq 4$ gives an equilibrium mixture of hydroxymethyl derivatives. On ureidomethylation of a hydroxymethyl group bonded to the second nitrogen atom of the urea, cyclocondensation to hydroxymethylated 4-oxotetrahydro-1,3,5-oxadiazines (**25**) occurs [135], [136].

4-Oxo-3,5-dialkoxymethyltetrahydro-1,3,5-oxadiazines are used as cross-linking agents for easy-care finishing of textiles. These compounds can be obtained directly by condensation of urea with formaldehyde and alcohols [135], [136]:

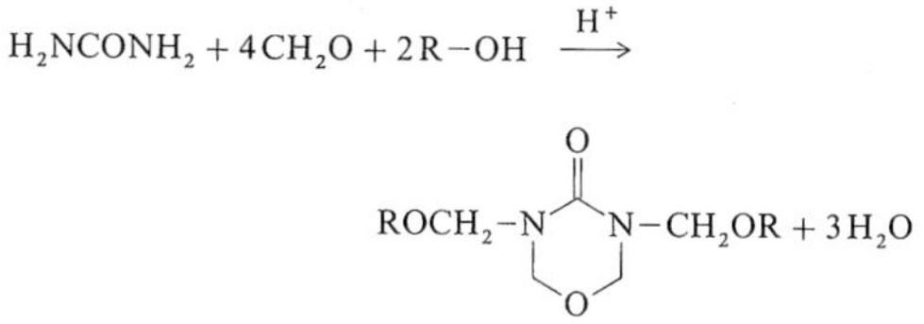

$$H_2NCONH_2 + 4CH_2O + 2R-OH \xrightarrow{H^+} ROCH_2-N\underset{O}{\overset{\overset{O}{\parallel}{C}}{\diagdown}}N-CH_2OR + 3H_2O$$

The hydroxymethyl and methoxymethyl derivatives of *2-oxo-5-alkylhexahydro-1,3,5-triazines* have achieved importance as cross-linking agents for easy-care finishing of cellulose-containing fabrics. These compounds are obtained by cyclizing ureidomethylation of urea with formaldehyde and a primary amine [137]–[140], [154]:

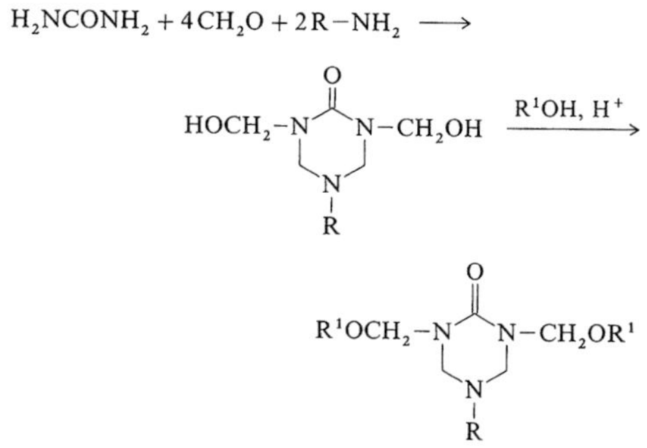

$$H_2NCONH_2 + 4CH_2O + 2R-NH_2 \longrightarrow$$

[Structure: HOCH₂-N, N-CH₂OH on a 6-membered ring with C=O and N-R] →(R¹OH, H⁺)

[Structure: R¹OCH₂-N, N-CH₂OR¹ on a 6-membered ring with C=O and N-R]

Bicyclic Ureas. 4,5-Dihydroxy- or 4,5-dialkoxyimidazolidin-2-ones can be converted into bicyclic ureas, such as *tetrahydroimidazo[4,5-d]-imidazole-2,5(1H,3H)-dione* (**26**), by means of a double α-ureidoalkylation with urea:

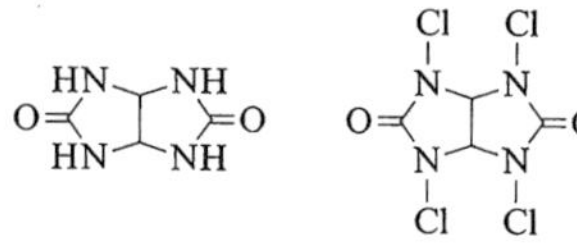

[Structures of compound 26 (tetrahydroimidazo[4,5-d]imidazole-2,5(1H,3H)-dione with HN/NH groups) and compound 27 (tetrachloro derivative with N-Cl groups)]

26   27

Tetrahydroimidazo[4,5-*d*]imidazole-2,5(1H,3H)-dione [496-46-8], acetylenediurea (**26**), can be obtained directly by condensation of glyoxal with excess urea in an acidic medium [156], [157]. A solution of urea is acidified to pH <3 with sulfuric acid, and a 40% glyoxal solution is added slowly. At the end of the reaction the molar ratio of glyoxal to urea is adjusted to at least 1:2.5. Reaction temperature should not exceed 70 °C. The precipitated product is neutralized by decantation and stirring with water.

In the form of its tetrahydroxymethyl derivative (**15**) the product has achieved importance as a cross-linking agent in easy-care textile finishes [158]. The saturated and unsaturated *N*-alkoxymethyl compounds (**23**) are used as cross-linking agents in the paint and coating industry.

The tetrachloro compound (**27**) is used as a chlorine-transfer agent for mild chlorination reactions and as a bleaching agent for textiles [159]–[161].

The synthetic possibilities for cyclizing α-ureidoalkylations with compounds containing enolizable carbonyl groups are manifold. Aldehydes and ketones are particularly important because they react not only as nucleophiles but also as carbonyl components in these cyclocondensations. Under mild conditions (i.e., in weakly acidic media), aliphatic and aromatic aldehydes generally react only as carbonyl components in reactions with urea and its derivatives, giving linear or branched condensation products. In a more strongly acidic medium and at elevated temperature, however, an α-ureidoalkylation at nucleophilic α-carbon atom in the aldehyde occurs:

$$R^1-HN-\overset{\overset{X}{\|}}{C}-NH-R^2 + 2H\overset{\overset{R^3}{|}}{\underset{\underset{R^4}{|}}{C}}-CHO \rightleftharpoons$$

R³ R¹–N–C(=X)–N–R²
 \\ | |
 CH–HC   CH–
 /   \\   /
R⁴   HO
       \\
        HC
       / \\
      R³ R⁴

28

$$\swarrow H^+/-H_2O \qquad \searrow \underset{-2H_2O}{R^5OH}$$

      X   X
      ‖   ‖
R³ R¹–N   N–R²   R⁵OH   R³ R¹–N   N–R²
 \\ | | ⇌   \\ | |
  CH—   —OH   —H₂O   CH—   —OR⁵
 /   R³ R⁴   /   R³ R⁴
R⁴   R⁴

29   **30**

In the cyclocondensation of urea or its mono- or symmetrically disubstituted derivatives with aldehydes, which have at least one activated hydrogen in α-position to the carbonyl group, 2-oxo-4-hydroxyhexahydropyrimidines (**29**) are formed in the presence of an acid [113], [154], [162], [163]. Symmetrical di-α-hydroxyalkyl compounds (**28**) and poly-α-alkyleneureas are formed as intermediates. They all contain the groups necessary for cyclocondensation involving an α-ureidoalkylation, i.e., the NH component, 1 mol of aldehyde as the carbonyl component, and the second mole of aldehyde as the nucleophilic component with activated α-hydrogen. In cyclocondensation, a bond is formed between the nucleophilic α-carbon atom of one α-hydroxyalkyl group and the carbon α to the urea nitrogen in the second α-hydroxyalkyl group to give 2-oxo-4-hydroxyhexahydropyrimidines (**29**).

Preparation of 2-Oxo-4-hydroxy-5,5-dimethyl-6-isopropylhexahydropyrimidine (*mp* 220–222 °C). One liter of a 30% aqueous urea solution (5 mol) is treated with 100 mL of 50% sulfuric acid in a stirred tank reactor with reflux cooling. Then, 10 mol isobutyraldehyde is added with stirring. After being heated at 85 °C for 2 h, the product is filtered and washed with water.

If these cyclocondensations are carried out in the presence of alcohols, 2-oxo-4-alkoxyhexahydropyrimidines (**30**) are formed [163], [164].

These cyclocondensations can also be carried out between urea and two different aldehydes, at least one of which must have an enolizable carbonyl group. For example, the cyclocondensation of 1 mol of urea with 1 mol of formaldehyde and 1 mol of isobutyraldehyde in the presence of an acid gives 2-oxo-4-hydroxy-5,5-dimethylhexahydropyrimidine (**31**) [164]

$$R^1-HN-\overset{\overset{O}{\|}}{C}-NH-R^2 + CH_2O + H\overset{\overset{CH_3}{|}}{\underset{\underset{CH_3}{|}}{C}}-CHO \xrightarrow{H^+}$$

$$\underset{\textbf{31}}{\text{R}^1-N\underset{\underset{CH_3}{}}{\overset{\overset{O}{\|}}{\diagup}}\underset{CH_3}{\diagdown}N-R^2\text{-OH}} + H_2O$$

The 2-oxo-4-hydroxy-5,5-dimethylhexahydropyrimidines are cyclic *N*-hemiacetals, whose OH groups can undergo nucleophilic substitutions similar to those undergone by *N*-hydroxymethylureas. 2-Oxo-4-hydroxy- and 2-oxo-4-alkoxy-5,5-dialkylhexahydropyrimidines are used in the form of their *N,N'*-dihydroxymethyl compounds in noncrease finishes for cellulose-containing textiles [91], [165].

Preparation of *N,N'*-Dihydroxymethyl-2-oxo-4-hydroxy(methoxy)-5,5-dimethylhexahydropyrimidine. Urea is treated with formaldehyde in the molar ratio 1 : 1 at pH > 9 and 50 – 60 °C in the presence of an excess of methanol. Isobutyraldehyde is added in the presence of a strong mineral acid to bring about cyclocondensation. The reaction mixture is rendered alkaline, and hydroxymethylation is carried out with formaldehyde.

2-Oxo-4-hydroxyhexahydropyrimidines (**32**), which are not or only mono-substituented in the 5-position, react with ureas to give 2-oxo-4-ureidohexahydropyrimidines (**32**).

<chemical structure: 32 + H₂NCONH₂ →(H, -H₂O) 33>

2-Oxo-4-ureido-6-methylhexahydropyrimidine [*1129-42-6*], crotonylenediurea (**33**), is obtained either by condensation of urea with crotonaldehyde in the presence of acid [166] or by the industrially more straightforward route involving condensation of urea with acetaldehyde in the molar ratio 1 : 1 in the presence of acid [167], [168]. 2,7-Dioxo-4,5-dimethyldecahydropyrimido[4,5-*d*]pyrimidine is formed as a byproduct in the second route [168] – [170].

$$2H_2NCONH_2 + 2CH_3CHO \longrightarrow \textbf{34}$$

$$\downarrow$$

$$2H_2NCONH_2 + CH_3CH=CH-CHO$$

2-Oxo-4-ureido-6-methylhexahydropyrimidine is produced industrially in a continuous process employing a stirred tank cascade. A 70% urea solution is treated with acetaldehyde at a molar ratio of 1:1 in the presence of a catalytic quantity of 75% sulfuric acid. The exothermic reaction is kept at 38–60 °C by controlled cooling. The pH is initially kept above 3 and in the final reactors below 2 by addition of sulfuric acid. Average residence time in the cascade is 40 min. In the last stirred tank, the reaction mixture is neutralized with aqueous potassium hydroxide solution. Drying is carried out in a spray tower.

2-Oxo-4-ureido-6-methylhexahydropyrimidine is used as a slow-release nitrogen fertilizer [171]. This fertilizer is characterized by its extreme insolubility in water and is therefore not washed out of the soil by rain or irrigation. It decomposes as a result of acid hydrolysis induced by humic acids during the growth period of plants, bringing about mineralization of the nitrogen it contains.

2-Oxo-4-hydroxy-5,5-dimethylhexahydropyrimidyl-N,N'-bisneopentals can be produced from urea, formaldehyde, and isobutyraldehyde in acid-catalyzed cyclo- and linear condensations. These bisneopentals react further according to a Claisen–Tishchenko reaction to give soft and hard resins with good light stability for the paint and coatings industry, depending on the molar ratios of the starting materials [172], [173].

9. References

[1] F. Wöhler, *Ann. Phys. Chem.* **2** (1828) no. 12, 253–256.
[2] A. I. Basaroff, *J. Prakt. Chem.* **2** (1870) no. 1, 283.
[3] J. Berliner, *Ind. Eng. Chem.* **28** (1936) no. 5, 517–522.
[4] L. Vogel, H. Schubert, *Chem. Tech. (Leipzig)* **32** (1980) no. 3, 143–144.
[5] A. A. Kozyro, S. V. Dalidovich, A. P. Krasulin, *J. Appl. Chem. (Leningrad)* **59** (1986) no. 7, 1353–1355. (*Zh. Prikl. Khim. (Leningrad)* **59** (1986) no. 7, 1456–1459.).
[6] A. P. Krasulin, A. A. Kozyro, G. Ya. Kabo, *J. Appl. Chem. (Leningrad)* **60** (1987) no. 1, 96–99. (*Zh. Prikl. Khim. (Leningrad)* (1987) no. 1, 104–108).
[7] G. Midgley, *The Chem. Eng.* (1977) Dec., 856–866.
[8] E. P. Egan, B. B. Luff, *J. Chem. Eng. Data* **11** (1966) no. 2, 192–194.
[9] A. Seidell: *Solubilities of Organic Compounds*, 3rd ed., Van Nostrand Company, New York 1941, vol. **2**.
[10] A. Seidell, W. F. Linke: *Solubilities of Inorganic and Organic Compounds*, Suppl. 3rd. ed., Van Nostrand, New York 1952.
[11] V. I. Kucheryavyi, G. N. Zinov'ev, L. K. Skotnikova, *J. Appl. Chem. (Leningrad)* **42** (1969) no. 2, 409–410 (*Zh. Prikl. Khim.* **42** (1969) no. 2, 446–447).
[12] R. I. Spasskaya, *J. Appl. Chem. (Leningrad)* **46** (1973) no. 2, 407–409. (*Zh. Prikl. Khim.* **46** (1973) no. 2, 393–396).
[13] G. Ostrogovich, R. Bacaloglu, *Rev. Roum. Chim.* **10** (1965) 1111–1123.
[14] G. Ostrogovich, R. Bacaloglu, *Rev. Roum. Chim.* **10** (1965) 1125–1135.

[15] W. R. Epperly, *Chem. Tech. (Heidelberg)* **21** (1991) no. 7, 429–431.
[16] A. Lasalle, et al., *Ind. Eng. Chem. Res.* **31** (1992) no. 3, 777–780.
[17] B. K. Banerjee, P. C. Srivastava, *Fert. Technol.* **16** (1979) nos. 3–4, 264–288.
[18] *Kirk-Othmer*, 3rd. ed., vol. **2**, pp. 440–469.
[19] M. Frejacques, *Chim. Ind. (Paris)* **60** (1948) no. 1, 22–35.
[20] S. Inoue, K. Kanai, E. Otsuka, *Bull. Chem. Soc. Jpn.* **45** (1972) no. 5, 1339–1345 (Part I); *Bull. Chem. Soc. Jpn.* **45** (1972) no. 6, 1616–1619 (Part II).
[21] D. M. Gorlovskii, V. I. Kucheryavyi, *J. Appl. Chem. (Leningrad)* **54** (1981) no. 10, 1898–1901. (*Zh. Prikl. Khim.* **53** (1980) no. 11, 2548–2551).
[22] W. Durisch, S. M. Lemkowitz, P. J. van den Berg, *Chimia* **34** (1980) no. 7, 314–322.
[23] S. M. Lemkowitz, J. Zuidam, P. J. van den Berg, *J. Appl. Chem. Biotechnol.* **22** (1972) 727–737.
[24] P. J. C. Kaasenbrood, H. A. G. Chermin, paper presented to The Fertilizer Society of London, 1st Dec., 1977.
[25] Stamicarbon, EP 0 212 744, 1987 (J. H. Meessen, R. Sipkema).
[26] Montedison, EP 0 132 194, 1985 (G. Pagani).
[27] Toyo Engineering Corporation, GB 2 109 372, 1983 (S. Inoue et al.).
[28] *Eur. Chem. News*, Aug. 9 (1982) 15–16.
[29] T. Jojima, B. Kinno, H. Uchino, A. Fukui, *Chem. Eng. Prog.* **80** (1984) no. 4, 31–35.
[30] E. Dooyeweerd, J. Meessen, *Nitrogen* **143** (1983) May–June, 32–38.
[31] *Nitrogen* **194** (1991) Nov.–Dec., 22–28.
[32] R. De Jonge, F. X. C. M. Barake, J. D. Logemann, *Chem. Age India* **26** (1975) no. 4, 249–260.
[33] Montedison, EP 0 096 151, 1983 (G. Pagani, G. Faita, U. Grassini).
[34] Urea Casale, EP 0 504 621, 1992 (V. Lagana).
[35] Stamicarbon, US 2 727 069, 1955 (J. P. M. van Waes).
[36] Stamicarbon, US 3 720 548, 1973 (F. X. C. M. Barake, C. G. M. Dijkhuis, J. D. Logemann).
[37] C. Miola, H. Richter, *Werkst. Korros.* **43** (1992) 396–401.
[38] Snamprogetti, US 4 899 813, 1990 (S. Menicatti, C. Miola, F. Granelli).
[39] E. M. Elkanzi, *Res. Ind.* **36** (1991) no. 4, 254–259 (Eng.).
[40] Unie van Kunstmestfabrieken, EP 0 155 735, 1985 (K. Jonckers).
[41] Snamprogetti, GB 1 542 371, 1979 (U. Zardi, V. Lagana).
[42] V. Lagana, G. Schmid, *Hydrocarbon Process.* **54** (1975) no. 7, 102–104 (Eng.).
[43] U. Zardi, F. Ortu, *Hydrocarbon Process.* **49** (1970) no. 4, 115–116 (Eng.).
[44] Snamprogetti, GB 1 506 129, 1978.
[45] Snamprogetti, US 3 954 861, 1976 (M. Guadalupi, U. Zardi).
[46] Snamprogetti, US 3 876 696, 1975 (M. Guadalupi, U. Zardi).
[47] *Nitrogen* **185** (1990) May–June, 22–29.
[48] Toyo Engineering Corp., US 4 301 299, 1981 (S. Inoue, H. Ono).
[49] Toyo Engineering Corp., GB 2 109 372, 1983 (S. Inoue et al.).
[50] Montedison, GB 1 581 505, 1980 (G. Pagani).
[51] A. Seidell, W. F. Linke: *Solubilities of Inorganic and Organic Compounds*, Suppl. 3rd ed., D. Van Nostrand Co., New York 1952, p. 393.
[52] Snamprogetti, US 4 327 068, 1982 (V. Lagana, U. Zardi).
[53] Unie van Kunstmestfabrieken, US 4 410 503, 1983 (P. J. M. van Nassau, A. M. Douwes).
[54] The M. W. Kellogg Co., US 5 223 238, 1993 (T. A. Czup-pon).
[55] Snamprogetti, US 4 235 816, 1980 (V. Lagana, F. Saviano).
[56] Snamprogetti, GB 1 520 561, 1978 (G. Pagani).
[57] Snamprogetti, GB 1 470 489, 1977 (A. Bonetti).

[58] Snamprogetti, GB 1 560 174, 1980 (V. Lagana, F. Saviano).
[59] V. Lagana, *Chem. Eng. (N.Y.)* **85** (1978) no. 1, 37–39.
[60] Snamprogetti, GB 2 060 614, 1981 (V. Lagana, F. Saviano, V. Cavallanti).
[61] S. K. Saxena, A. Ali, *Indian J. Environ. Prot.* **9** (1989) no. 11, 831–833 (Eng.).
[62] R. M. Krishnan, *Fert. News* (1992) May, 53–57.
[63] Ting-Chia Huang, Dong-Hwang Chen, *J. Chem. Technol. Biotechnol.* **55** (1992) no. 2, 191–199.
[64] Unie van Kunstmestfabrieken, EP 0 053 410, 1982 (J. Zuidam, P. Bruls, K. Jonckers).
[65] Snamprogetti, EP 0 417 829, 1991 (F. Granelli).
[66] I. Mavrovic: "Pollution Control in Urea Plants," *Br. Sulfphur's 12th Int. Conf. Nitrogen 88*, Geneva, March 27–29, 1988.
[67] Unie van Kunstmestfabrieken, EP 0 037 148, 1981 (M. H. Willems, J. W. Klok).
[68] *Fert. Int.* **296** (1991) April, 32–37.
[69] R. M. Reed, J. C. Reynolds, *Chem. Eng. Prog.* **69** (1973) no. 2, 62–66.
[70] *Nitrogen* **95** (1975) May/June, 31–36.
[71] R. D. Hauck, M. Koshino: *Fertilizer Technology & Use*, 2nd ed. Soil Science Society of America, Madison 1971, pp. 455–494.
[72] N. K. Savant, E. T. Craswell, R. B. Diamond, *Fert. News* **28** (1983) no. 8, 27–35.
[73] G. Wiedemann, *Justus Liebigs Ann. Chem.* **68** (1848) 325.
[74] Kurzer, *Chem. Rev.* **56** (1956) 95–197.
[75] *Zh. Prikl. Khim. (Leningrad)* **42** (1969) no. 13, 713.
[76] I. G. Farbenind., DE 689 421, 1940.
[77] G. Kränzlein, H. Keller, H. Schiff, *Justus Liebigs Ann. Chem.* **291** (1896) 374.
[78] R. C. Haworth, F. G. Mann, *J. Chem. Soc.* 1943, 603.
[79] American Cyanamid, US 2 760 961, 1956 (Mackey).
[80] A. Fleischer, *Ber. Dtsch. Chem. Ges.* **9** (1876) 995.
[81] A. v. Baeyer, *Justus Liebigs Ann. Chem.* **131** (1864) 252.
[82] T. L. Davis, K. C. Blanchard, *J. Am. Chem. Soc.* **45** (1923) 1816. US 1 785 730, 1927 (T. L. Davis).
[83] Knoll, DE 896 640, 1942.
[84] T. L. Davies, K. C. Blanchard, *Org. Synth.* **3** (1923) 95.
[85] Du Pont, US 2 145 242, 1937 (H. W. Arnold).
[86] *Houben-Weyl*, **VIII, III,** 151.
[87] A. Sonn, *Ber. Dtsch. Chem. Ges.* **47** (1914) 2440.
[88] Schweizer, *J. Org. Chem.* **15** (1950) 471.
[89] *Houben-Weyl*, **VIII, III,** 164.
[90] McKay, Coleman, *J. Am. Chem. Soc.* **72** (1950) 3205.
[91] H. Petersen: "Chemical Processing of Fibers and Fabrics, Functional Finishes," Part A, in M. Lewin, S. B. Sello (eds.): *Handbook of Fiber science and Technology*, vol. 2, Marcel Dekker, New York 1983, pp. 48–327.
[92] *Org. Synth.* **29** (1949) 18.
[93] *Houben-Weyl*, **VIII, III,** 153.
[94] L. J. Smith, O. M. Emeron, *Org. Synth. Coll.* **III** (1955) 151.
[95] G. M. Dyron, *Chem. Rev.* **4** (1927) 138.
[96] W. Hentschel, *J. Prakt. Chem.* **27** (1883) no. 2, 499.
[97] CIOS, XXVII/80, 15.
[98] Bayer, DE 116 200, 1899.*Friedländer* **6**, 200.BASF, DE 46 737, 1888. *Friedländer* **2**, 450. Bayer, DE 131 513, 1901. *Friedländer* **6**, 968.Bayer, DE 216 666, 1908. *Friedländer* **9**, 372. Bayer, DE 493 811, 1926. Ges. Chem. Ind., *Friedländer* **16**, 1059.

[99] A. Lüttringhaus, H. W. Dirksen, *Angew. Chem. Int. Ed. Engl.* **3** (1964) 260.
[100] J. Berlinerblau, *J. Prakt. Chem.* **30** (1884) no. 3, 103.
[101] Höchst, DE 205 662, 1906. *Friedländer* **9**, 395.
[102] Du Pont, US 2 390 253, 1943 (C. O. Henke); *Chem. Abstr.* **40** (1946) 1876.
[103] A. Wurtz, *Hebd. Seances Acad. Sci.* **27** (1848) 242.
[104] A. Wurtz, *Justus Liebigs Ann. Chem.* **71** (1849) 329.
[105] A. W. Hofmann, *Justus Liebigs Ann. Chem.* **74** (1850) 14.
[106] A. Wurtz, *Justus Liebigs Ann. Chem.* **80** (1851) 347.
[107] W. Michler, *Ber. Dtsch. Chem. Ges.* **9** (1876) 396, 711.
[108] W. Michler, C. Escherich, *Ber. Dtsch. Chem. Ges.* **12** (1879) 1162.
[109] *Houben-Weyl, VIII*, III, 160.
[110] H. Eckenroth, *Ber. Dtsch. Chem. Ges.* **18** (1885) 516.
[111] C. A. Bischoff, A. von Hedenström, *Ber. Dtsch. Chem. Ges.* **35** (1902) 3437.
[112] T. Wilm, G. Wischin, *Justus Liebigs Ann. Chem.* **147** (1868) 162.
[113] H. Petersen, *Angew. Chem.* **76** (1964) 909.
[114] H. Petersen, *Test. Rundsch.* **16** (1961) 646.
[115] H. Petersen, *Melliand Textilber.* **43** (1962) 380.
[116] L. E. Smythe, *J. Phys. Chem.* **51** (1947) 369.
[117] L. E. Smythe, *J. Am. Chem. Soc.* **73** (1951) 2735; **74** (1952) 2713; **75** (1953) 574.
[118] L. Bettelheim, J. Cedwall, *Sven. Kem. Tidskr.* **60** (1948) 208.
[119] G. A. Growe, C. C. Lynch, *J. Am. Chem. Soc.* **70** (1948) 3795; **72** (1950) 3622.
[120] J. I. de Jong, J. de Jonge, *Reol. Trav. Chim. Pays-Bas* **71** (1952) 643, 661, 890; **72** (1953) 88.
[121] J. Lemaitre, G. Smets, R. Hart, *Bull. Soc. Chim. Belg.* **63** (1954) 182.
[122] J. Ugelstadt, J. de Jonge, *Reol. Trav. Chim. Pays-Bas* **76** (1957) 919.
[123] M. Okano, Y. Ogata, *Bull. Soc. Chim. Belg.* **63** (1954) 182.
[124] R. Kvéton, F. Hanousek, *Chem. Listy* **48** (1954) 1205; **49** (1955) 63.
[125] N. Landquist, *Acta Chem. Scand.* **9** (1955) 1127, 1459; **10** (1956) 244; **11** (1957) 776.
[126] B. R. Glutz, H. Zollinger, *Helv. Chim. Acta* **52** (1969) 1976.
[127] P. Eugster, H. Zollinger, *Helv. Chim. Acta* **52** (1969) 1985.
[128] H. Petersen, *Textilveredlung* **2** (1967) 744.
[129] H. Petersen, *Textilveredlung* **3** (1968) 160.
[130] H. Petersen, *Chem. Ztg.* **95** (1971) 625.
[131] A. Einhorn, A. Hamburger, *Ber. Dtsch. Chem. Ges.* **41** (1908) 24.
[132] A. Einhorn, A. Hamburger, *Justus Liebigs Ann. Chem.* **361** (1908) 122.
[133] H. Petersen, *Text. Res. J.* **40** (1970) 335.
[134] BASF, GB 1 241 580, 1968 (H. Petersen).
[135] H. Kadowaki, *Bull. Chem. Soc. Jpn.* **11** (1936) 248; *Chem. Zentralbl.* (1936 II) 3535.
[136] Sumitomo Chem. Co., DE 1 123 334, 1962 (T. Oshima).
[137] A. M. Paquin, *Angew. Chem.* **60** (1948) 267.
[138] W. J. Burke, *J. Am. Chem. Soc.* **69** (1967) 2136.
[139] DuPont, US 2 304 624, 1942 (W. J. Burke).
[140] DuPont, US 2 321 989, 1942 (W. J. Burke).
[141] BASF, GB 1 077 344, 1965. BASF, CH 478 287, 1966. DuPont, US 2 436 355, 1946.
[142] H. Petersen in: *Kunststoff-Jahrbuch*, 10th. ed., Wilhelm Pansegrau Verlag, Berlin 1968, p. 46.
[143] H. Pauli, H. Sauter, *Ber. Dtsch. Chem. Ges.* **63** (1930) 2063.
[144] BASF, DE 910 475, 1951 (H. Scheuermann, B. von Reibnitz, A. Wörner); *Chem. Zentralbl.* (1955) 702.

[145] H. Petersen, *Textilveredlung* **3** (1968) 51.
[146] H. Petersen, *Chem. Ztg.* **95** (1971) 692.
[147] H. Petersen, *Text. Res. J.* **41** (1971) 239.
[148] BASF, US 4 219 494, 1980 (H. Petersen).
[149] E. R. Atkinson, A. H. Bump, *Ind. Eng. Chem.* **44** (1952) 333.
[150] Societe Nobel Française, DE 962 119, 1952 (P. A. Talet); *Chem. Zentralbl.* (1955) 1625.
[151] BASF, DE 1 049 572, 1959 (H. Willersinn, H. Scheuermann, A. Wörner).
[152] BASF, DE 1 067 210, 1960 (H. Willersinn, H. Scheuermann, A. Wörner).
[153] Mitsubishi Chem. Ind., US 3 322 528, 1961; DE 1 543 201, 1965 (M. Hamamoto, Y. Sakaki).
[154] H. Petersen, *Synthesis* 1973, 243–292.
[155] H. Petersen, *Angew. Chem.* **79** (1967) 1009.
[156] BASF, US 2 731 472, 1956 (B. von Reibnitz).
[157] H. Petersen, *Synthesis* 1973, 254.
[158] BASF, DE 859 019, 1941 (J. Lintner, H. Scheuermann); *Chem. Zentralbl.* (1953) 4779.
[159] Biltz, Behrens, *Ber. Dtsch. Chem. Ges.* **43** (1910) 1984.
[160] US 2 638 434, 1953; US 2 649 381, 1953.
[161] Chemische Werke Hüls, DE 1 020 024, 1957 (H. Steinbrink, I. Amende).
[162] BASF, DE 1 230 805, 1962 (H. Petersen, H. Brandeis, R. Fikentscher).
[163] G. Zigeuner, W. Rauter, *Monatsh. Chem.* **96** (1965) 1950.
[164] BASF, DE 1 231 247, 1962 (H. Petersen, H. Brandeis, R. Fikentscher).
[165] H. Bille, H. Petersen, *Text. Res. J.* **37** (1967) 264; *Textilveredlung* **2** (1967) 243.
[166] A. M. Paquin, *Kunststoffe* **37** (1947) 165.
[167] BASF, DE 1 223 843, 1962 (H. Petersen, H. Brandeis, R. Fikentscher).
[168] G. Zigeuner, E. A. Gardziella, G. Bach, *Monatsh. Chem.* **92** (1961) 31.
[169] G. Zigeuner, M. Wilhelmi, B. Bonath, *Monatsh. Chem.* **92** (1961) 42.
[170] G. Zigeuner, W. Nischk, *Monatsh. Chem.* **92** (1961) 79.
[171] BASF, DE 1 081 482, 1959; US 3 190 741, 1963 (J. Jung, H. Müller von Blumencron, C. Pfaff, H. Scheuermann).
[172] BASF, EP 0 002 793, 1977; US 4 220 751, 1980 (H. Petersen, K. Fischer, H. Klug, W. Trimborn).
[173] BASF, DE 2 757 220, 1977; EP 0 002 794, 1977; US 4 243 797, 1981 (H. Peterson et al.).

Vinyl Esters

Günter Roscher, Hoechst Aktiengesellschaft, Frankfurt/Main, Federal Republic of Germany

1. Introduction 4839
2. Vinyl Acetate 4840
2.1. Properties 4840
2.2. Production 4840
2.3. Quality Specifications, Analysis, Storage, Transport, and Toxicology 4852
2.4. Use, Economic Importance ... 4854
3. Vinyl Propionate 4855
3.1. Properties 4855
3.2. Production 4856
3.3. Use 4857
4. Vinyl Esters of Higher Carboxylic Acids 4857
4.1. Reaction with Acetylene in the Gas Phase 4858
4.2. Reaction with Ethylene and Oxygen 4858
4.3. Transvinylation and Other Processes 4858
4.4. Reaction with Acetylene in the Liquid Phase 4859
4.5. Individual Vinyl Esters 4860
5. References 4861

1. Introduction

The first vinyl esters, vinyl trichloroacetate and vinyl acetate, were prepared by Klatte in 1912 by treatment of acetylene with the corresponding carboxylic acid and mercury salts as catalyst. Initially, the reaction was used for the industrial production of vinyl chloroacetate. The polymer obtained from this vinyl ester was used during World War I as a varnish for airplanes. Vinyl acetate is currently the most important vinyl ester. Vinyl propionate and vinyl esters of higher carboxylic acids are now also produced industrially (see Chap. 3 and 4).

2. Vinyl Acetate

2.1. Properties

Physical Properties. Vinyl acetate [*108-05-4*] CH$_3$CO$_2$CH=CH$_2$, M_r 86.09, is a colorless, flammable liquid with a characteristic, slightly pungent odor, *bp* 72.8 °C, density at 20 °C 0.932 g/mL, *mp* −93.2 °C, viscosity 0.43 mPA · s, vapor pressure 12 kPa at 20 °C, 42.6 kPa at 50 °C, coefficient of cubic expansion 0.0014 K^{-1}, flashpoint −8 °C, ignition temperature 385 °C. Lower/upper flammability limits in air 2.3/13.4 vol%, ignition group (VDE 0165) G 2, specific heat 1.926 kJ/kg; heat of evaporation 379.3 kJ/kg at 72.7 °C, heat of combustion 2082.0 kJ/mol, refractive index n_D^{20} 1.3956, heat of polymerization 1035.8 kJ/kg, solubility of water in vinyl acetate 0.9 wt% at 20 °C, solubility of vinyl acetate in water 2.3 wt% at 20 °C, azeotrope with water *bp* 66 °C/100 kPa, water content 7.3 wt%.

Chemical Properties. The chemical property which is exploited almost exclusively is the capacity to polymerize (see Section 2.4).

Other reactions of vinyl acetate: halogens give 1,2-dihaloethyl acetates [6], hydrogen halides give 1-haloethyl acetates [7], acetic acid gives ethylidene diacetate (see Section 2.2.), hydrogen cyanide gives 2-acetoxypropionitrile [8], hydrogen peroxide gives hydroxyacetaldehyde [9], and dienes, such as butadiene or cyclopentadiene, give Diels–Alder products [10].

Transesterification with carboxylic acids produces the corresponding vinyl carboxylate and acetic acid [11] (see also Section 2.3) and with alcohols gives the corresponding acetate and acetaldehyde. Thermal cleavage gives ketene and acetaldehyde [12]. Acid-catalyzed [13] or thermal [14] hydrolysis produces acetaldehyde and acetic acid. Vinyl acetate can be epoxidized with peracetic acid (84% yield) [15]. It undergoes addition of H-active compounds, e.g., dimethyl phosphite gives dimethyl acetoxyethylphosphonate [16].

2.2. Production

There are various possible routes for vinyl acetate production:

1) Addition of acetic acid to acetylene:
   a) in the liquid phase in the presence of homogeneous mercury salt catalysts
   b) in the gas phase in the presence of heterogeneous catalysts containing zinc salts
2) Addition of acetic anhydride to acetaldehyde giving ethylidene diacetate, and subsequent cleavage of the latter to form vinyl acetate and acetic acid
3) Reaction of ethylene with acetic acid and oxygen

a) in the liquid phase in the presence of palladium/copper salts as homogeneous catalysts
b) in the gas phase on heterogeneous catalysts containing palladium
4) Reaction of methyl acetate or dimethyl ether with carbon monoxide and hydrogen in the liquid phase in the presence of homogeneous catalysts, e.g., rhodium salts or noble metals of the platinum group, giving ethylidene diacetate; cleavage of the latter giving acetic acid and vinyl acetate

Acetylene, which is expensive, has mostly been replaced by the cheaper alternative, ethylene; ca. 80% of the available capacity is currently used for process 3b and ca. 20% for process 1b.

Processes 1a, 2, and 3a are no longer used, while process 4, although allegedly developed to an industrial level, has not yet been used industrially. It may become more important because the starting materials can readily be produced from coal or naphtha.

Addition of Acetic Acid to Acetylene (Liquid-Phase Process). The addition of carboxylic acids to acetylene using mercury salts as the catalyst [17]

$$CH_3CO_2H + CH\equiv CH \xrightarrow[\text{Hg salts}]{60-100\,°C} CH_3CO_2CH=CH_2$$

$$\Delta H = -117 \text{ kJ/mol}$$

is now of historical interest only. For more details see [18], [19].

Addition of Acetic Acid to Acetylene (Gas Phase Process).

$$CH\equiv CH + CH_3CO_2H \xrightarrow[\text{zinc acetate on activated charcoal}]{160-210\,°C} CH_3CO_2CH=CH_2$$

$$\Delta H = -117 \text{ kJ/mol}$$

The first process was developed in Munich by Consortium f. Elektrochemische Industrie [20]. It was further developed and used industrially by Wacker Chemie in Burghausen. Until 1965, almost all vinyl acetate was produced by the acetylene gas-phase process. Only two smaller plants used the ethylidene diacetate process [21]. Zinc salts on activated charcoal have proved to be effective catalysts. The development of suitable types of activated charcoal has improved the process [22].

Most of the industrial development work was concerned with carrying out the reaction. In the shaft furnaces used initially as reactors, controlling the heat of reaction was difficult. An occasional runaway of the reaction led to baking of the catalyst. Exothermic autodecomposition of the acetylene could not always be avoided. For better heat removal, other types of reactor employing a cooling medium were used. At Hoechst, Fischer furnaces were initially used, and at Wacker-Chemie tube furnaces.

As far as is known, all producers using the solid bed catalysts have started to use tube reactors because of the defined gas flow, the easier charging and discharging of the catalyst, and the good temperature control. Only at Kurashiki in Japan, Du Pont in the

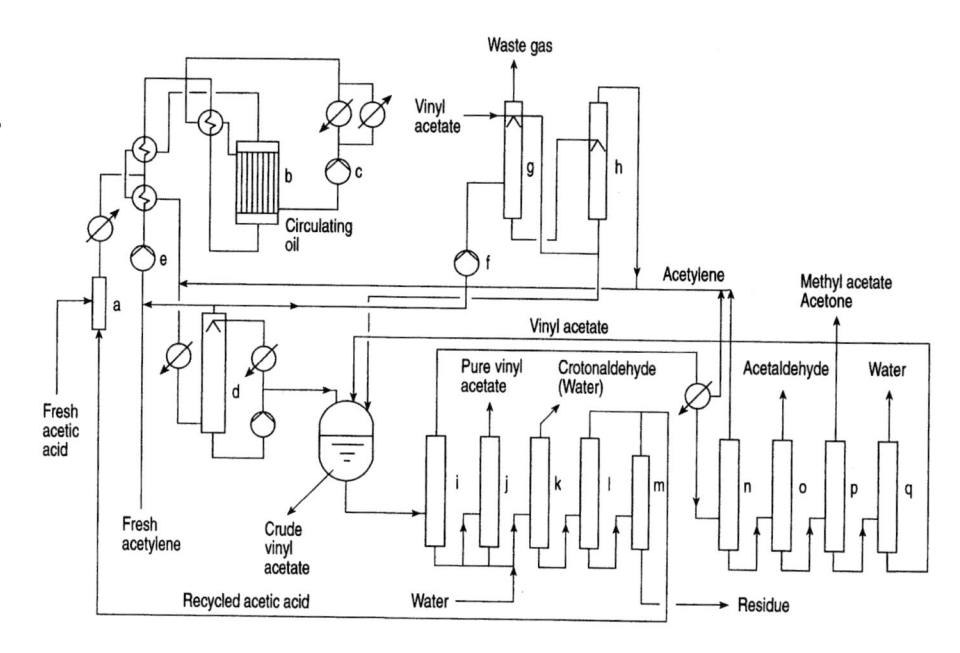

Figure 1. Acetylene gas phase process for vinyl acetate production
a) Acetic acid evaporator; b) Reactor; c) Heat transfer oil, cooling loop; d) Quenching tower; e) Recycle gas blower; f) Liquid ring pump; g) Washing column; h) Regenerating column; i) Lightends column; j) Pure vinyl acetate column; k) Crotonaldehyde column; l) Acetic acid column; m) Residue column; n) Degassing column; o) Acetaldehyde column; p) Acetone column; q) Water removal

United States, and in some plants in the former Soviet Union have fluidized-bed reactors been used. Solid bed and fluidizedbed processes are considered of equal value.

Process Description. The modes of operation of individual producers no longer differ significantly [23]. The process used by Hoechst until 1975 is described as an example (Fig. 1). Process data are given later in this section.

The circulated acetylene is preheated at the exit of the reactor in a countercurrent, and is then mixed with acetic acid vapor. The circulating gas can also be fed directly through the acetic acid evaporator (a). The gas, heated to the reaction temperature, enters the tube reactor (b).

The heat of reaction is removed by heat transfer oil in the cooling loop (c).

The mixture leaving the reactor is cooled in stages. The last cooling stage takes place in the quenching tower (d). In this packed column, the gas mixture is cooled to 0 °C. The condensate itself is used for cooling. It is circulated from the bottom to the top of the quenching tower via a brine – cooled condenser. The liquid product stream is removed; excess acetylene is recycled via a circulating gas blower (e).

The crude vinyl acetate produced at the bottom of the quenching tower is distilled giving acetic acid, which is recycled, and pure vinyl acetate. Some 90% of the circulating gas is acetylene; the remainder is CO_2, CO, and methane formed from thermal decomposition, acetaldehyde, N_2, and other inert substances. To prevent enrichment of the impurities in the circulating gas, a small portion is removed behind the quenching tower (d) and then purified.

The acetylene from the gas stream, which has been brought to ca. 100 kPa overpressure by means of a liquid ring pump (f), is extensively absorbed in a washing column (g) containing brine-cooled vinyl acetate. The inert substances are led as waste-gas to flare. The sump product from (g) is freed from dissolved acetylene in a regeneration column (h) by boiling. The acetylene is recycled. The crude vinyl acetate formed in the bottom of the quenching tower (d) contains ca. 62–63 wt% vinyl acetate and 30–35% acetic acid. It also contains dissolved acetylene, acetaldehyde, crotonaldehyde, acetone, methyl acetate, ethylidene diacetate, and acetic anhydride.

In the lightends column (i) acetaldehyde, acetone, methyl acetate, dissolved acetylene, some vinyl acetate, and water (originating from the starting materials) are first distilled overhead. The sump product is fractionated in distillation columns (j–m). Pure vinyl acetate is removed as the top product of the column (j). In (k) crotonaldehyde distills at the head as an azeotrope after addition of water. The bottom of the acetic acid column (l) also contains, besides acetic acid, the high-boiling ethylidene diacetate and acetic anhydride (unless they have already been hydrolyzed in the crotonaldehyde column), and very small quantities of polymers. Column (m) is operated under vacuum, and to some extent batchwise; the residual acetic acid distills so extensively that the liquid sump product which remains can subsequently be incinerated. The distillate from the lightends column (i) contains mainly acetaldehyde, acetone, methyl acetate, vinyl acetate, and water. It is purified in distillation columns (o–q).

Stabilizers. To avoid polymerization during the distillative work-up of the crude vinyl acetate, polymerization inhibitors are added. The preferred stabilizer is hydroquinone. Copper resinate, phenothiazine, or methylene blue are also used.

Materials and Environmental Aspects. The plant is made from mild steel in the hot area of the reaction section and in places where no liquid acetic acid is present. These areas include the reactor, the gas–gas heat exchanger, and the circulating gas blower. For the distillation section and the equipment in the reaction section which comes into contact with liquid product, stainless steel 316 L is used. The process has almost no polluting waste streams. All gaseous or liquid byproducts (high-boiling, crotonaldehyde, and acetone–methyl acetate fractions) can be incinerated. Waste circulating gas is passed to the excess gas burner.

Process Data. The catalyst consists of zinc acetate on activated charcoal (particle size 3–4 mm) as the carrier material. The zinc content is 10–15 wt%. The catalyst is produced by dipping. As the activated charcoal can contain traces of copper, small quantities of other components are added to the catalyst to prevent the formation of cuprene, which can block the tubes.

The operating time is ca. 5000–7000 h, depending on the type of activated charcoal used. It also depends on the purity of the acetylene and the circulating acetic acid. When using acetylene produced from carbide, the type of carbide is important. Because of the varying contents of phosphorus hydrides, hydrogen sulfide, arsine, and ammonia in the acetylene, it sometimes has to be further purified before being used in vinyl acetate production. This involves several steps.

Acetylene produced from petrochemicals does not contain these components. The decrease in catalyst activity is caused to a small extent by migration of zinc acetate out of the catalyst, but to a greater extent by formation of byproducts or by foreign components, which either act as catalyst poisons or adhere to the catalyst surface and pores [24], [25]. Vinyl acetate itself does not appear to contribute to the deactivation of the catalyst under the reaction conditions.

The space–time yield for the vinyl acetate is normally 60–70 g per liter catalyst per hour.

The reaction temperature is 160–170 °C, depending on the type of activated charcoal used as catalyst, but increases to 205–210 °C as the catalyst activity decreases. At the elevated temperature in the reactor, more byproducts are formed.

The pressure can increase a little during the operating time of a catalyst charge. This is caused by a slight increase in the flow resistance of the catalyst. The maximum pressure is ca. 40 kPa overpressure. Plants are protected against overpressures greater than 40–50 kPa because acetylene can undergo exothermic autodecomposition at high pressures and temperatures:

$$CH \equiv CH \rightarrow H_2 + 2\,C \qquad \Delta H = -229.4 \text{ kJ/mol}$$

The acetylene pipes leading to the plant therefore contain barriers which inhibit the acetylene autodecomposition. The plants are also provided with equipment for automatic flushing with inert gas. The molar ratio acetic acid:acetylene is normally between ca. 1:4 and 1:4.5. The load per m^3 catalyst is ca. 135 m^3 (STP) recycle gas per hour and 76 kg acetic acid per hour. The recycle gas consists of ca. 90% acetylene. Acetylene conversion is ca. 15% and acetic acid conversion ca. 55%. The yields based on acetic acid can reach 99%, or 98% based on acetylene, if the acetaldehyde formed (2–3 kg per 100 kg vinyl acetate) is included in the yield.

Ethylidene Diacetate Process.

$$CH_3CHO + (CH_3CO)_2O \rightarrow CH_3CH(OCOCH_3)_2$$
$$CH_3CH(OCOCH_3)_2 \rightarrow CH_3CO_2CH=CH_2 + CH_3CO_2H$$

The process was developed by Celanese Corporation of America [26] and was operated industrially in the United States from 1953 to 1970, and then replaced by the ethylene gas-phase process. A small plant (Celmex) operated in Mexico until 1991. Acetic anhydride is converted to ethylidene diacetate with acetaldehyde in the liquid phase using catalysts such as iron(III) chloride. The ethylidene diacetate is then cleaved thermally to vinyl acetate and acetic acid, using catalysts such as toluenesulfonic acid. Some of the ethylidene diacetate is converted back to acetic anhydride and acetaldehyde.

Reaction of Acetic Acid with Ethylene and Oxygen (Liquid Phase). The formation of vinyl acetate from ethylene and acetic acid in the presence of palladium chloride and alkali acetate in glacial acetic acid was first described by MOISEEW [27]:

$$C_2H_4 + PdCl_2 + CH_3CO_2H \rightarrow CH_3CO_2CH=CH_2 + 2\,HCl + Pd$$

Addition of benzoquinone to the reaction mixture was said to reoxidize the palladium to palladium chloride. The reaction corresponds to the Wacker–Hoechst process, in which acetaldehyde is obtained from ethylene and water in the presence of palladium chloride:

$$C_2H_4 + PdCl_2 + H_2O \rightarrow CH_3CHO + Pd + 2\,HCl$$

The palladium formed is reoxidized to Pd^{2+} with copper(II) chloride. The copper(I) chloride formed is reoxidized with oxygen ($\rightarrow$ Acetaldehyde). Production of vinyl acetate by a similar route has been widely investigated [28]–[30]. Corresponding production plants were commissioned by ICI in England, Celanese in the United States, and Tokuyama Petrochemical in Japan [31], but later shut down. Now only the ethylene gas-phase process uses ethylene as the starting material (see below).

In the liquid-phase process, a recycle ethylene gas stream is passed through a reaction solution containing acetic acid, water, high-boiling byproducts analogous to those of the acetaldehyde process, $PdCl_2$, and $CuCl_2$. Oxygen is passed into the reaction solution at the same time to reoxidize the palladium and the CuCl. The reaction and regeneration of the catalyst take place in one step:

$$CH_2=CH_2 + CH_3CO_2H + PdCl_2 \rightarrow CH_3CO_2CH=CH_2 + 2\,HCl + Pd$$
$$Pd + 2\,CuCl_2 \rightarrow PdCl_2 + 2\,CuCl$$
$$2\,CuCl + 2\,HCl + 1/2\,O_2 \rightarrow 2\,CuCl_2 + H_2O$$

Water is formed in the reoxidation of CuCl; part of the vinyl acetate formed is hydrolyzed to acetaldehyde and acetic acid. Part of the acetaldehyde is also formed directly from ethylene, as in the acetaldehyde process.

A certain ratio of palladium ions to copper ions and of copper ions to chloride ions is necessary for reoxidation of the palladium. Chlorine, which is lost through formation of chlorinated byproducts, is replaced by hydrochloric acid. To maintain the necessary quantity of chloride ions in solution, alkali metal chloride must be added. A typical reaction solution contains ca. 30–50 mg/L palladium ions and 3–6 g/L copper ions.

Byproducts include CO_2, formic acid, oxalic acid, oxalic acid esters, chlorinated compounds, and butenes. The pressure is 3–4 MPa, and the reaction temperature

110–130 °C. The ratio of acetaldehyde to vinyl acetate can be controlled by adjusting the water concentration and the residence time [32].

Reaction of Ethylene with Acetic Acid and Oxygen (Gas Phase Reaction). The process was developed to an industrial scale only slightly later than the ethylene liquid-phase process, and has been used in industry since 1968. Currently, 80% of world vinyl acetate capacity uses the ethylene gas-phase process. There are two variants: one developed by National Distillers Products (United States) [49], and the other independently by Bayer in cooperation with Knapsack and Hoechst (Germany) [33]–[35]. Most plants employ the Bayer–Hoechst variant, of which there are several versions. The original process has been further developed by various operators.

In the ethylene gas-phase process, ethylene reacts exothermically with acetic acid and oxygen on solid bed catalysts, giving vinyl acetate and water:

$$CH_2=CH_2 + CH_3CO_2H + 1/2\, O_2 \rightarrow CH_3CO_2CH=CH_2 + H_2O \quad \Delta H = -178 \text{ kJ/mol}$$

All catalysts used in industry contain palladium and alkali metal salts on carrier materials, e.g., silicic acid, aluminum oxide, lithium spinel, or activated charcoal. Additional activators can include gold, rhodium, platinum, and cadmium.

The reaction mechanism is assumed to involve either pure metal catalysis [36] or a reaction sequence according to the following equations:

$$CH_2=CH_2 + Pd(CH_3CO_2)_2 \rightarrow Pd + CH_3CO_2CH=CH_2 + CH_3CO_2H$$
$$Pd + 1/2\, O_2 + 2\, CH_3CO_2H \rightarrow Pd(CH_3CO_2)_2 + H_2O$$

The finely divided palladium on the catalyst is thought to be oxidized to divalent palladium. For reoxidation of the palladium, a redox reaction analogous to the liquid-phase process is assumed. Copper, manganese, and iron are cited as redox elements [37].

In the process, which operates above 140 °C and at overpressure 0.5–1.2 MPa, practically no acetaldehyde is formed, even if the acetic acid used as starting material contains water. Byproducts are water, CO_2 and small quantities of ethyl acetate, ethylidene diacetate, and glycol acetates.

Process Description: Reaction Section (Fig. 2). The recycle gas stream, which consists mainly of ethylene, is saturated with acetic acid in the evaporator (a) and is then heated to the reaction temperature. The gas stream is then mixed with oxygen in a special unit.

The allowed oxygen concentration is determined by the flammability limits of the ethylene–oxygen mixture. The flammability limit depends on temperature, pressure, and composition. It is shifted by additional components, such as acetic acid, nitrogen, and argon, which are brought in with the oxygen, or by CO_2. In general, the oxygen concentration at the entry to the reactor is ≤ 8 vol%, based on the acetic-acid-free

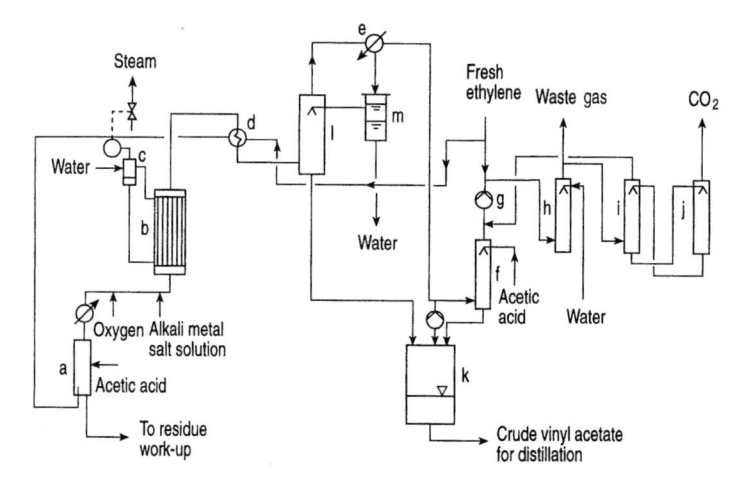

Figure 2. Ethylene gas phase process for vinyl acetate production; reaction section
a) Acetic acid evaporator; b) Reactor; c) Steam drum; d) Countercurrent heat exchanger; e) Water cooler; f) Recycle gas washing column; g) Recycle gas compressor; h) Water wash; i) Potash wash; j) Potash regeneration; k) Crude vinyl acetate collector; l) Predehydration column; m) Phase separator

mixture. It is essential to avoid gas mixtures capable of igniting; great care is taken in mixing in oxygen and measuring the oxygen concentration. If the oxygen stream is switched off, the inlet line must be flushed with nitrogen immediately to avoid back-diffusion of the circulating gas. The mixing chamber is usually installed behind concrete walls. The heat of reaction is removed in the form of steam via (c) from the tube reactor (b) by evaporative cooling in the shell side of the reactor.

The reaction temperature is adjusted by the pressure of the boiling water. The steam formed can be used within the plant itself in the work-up section. The heat of reaction is ca. 250 kJ/mol based on vinyl acetate, because of the simultaneous formation of CO_2. Pressurized water in a circulation loop is used for cooling in some plants.

Ethylene conversion is 8–10%, and that of acetic acid 15–35%. Oxygen conversion can be up to 90%. As small quantities of alkali metal salt on the catalyst migrate under the reaction conditions, traces of alkali metal salt are mixed with the gas at the entry to the reactor [38].

The gas mixture leaving the reactor is first cooled in (d) in countercurrent with the cold recycle gas, which is thus warmed. There is virtually no condensation of the acetic acid, vinyl acetate, or water. The dew point is generally not reached. The gas mixture is then led into the predehydration column (l) and then cooled to about room temperature in (e). The liquid product consists of an acetic-acid-free mixture of vinyl acetate and water. The mixture is separated in a phase separator (m) into an aqueous phase, which is removed, and an organic vinyl acetate phase, which is recycled to the head of the predehydration column.

Between 40 and 50% of the water formed in the reaction is removed in this way without the need to supply extra energy; this quantity of water does not need to be removed in the subsequent distillation of the crude vinyl acetate. Most of the energy

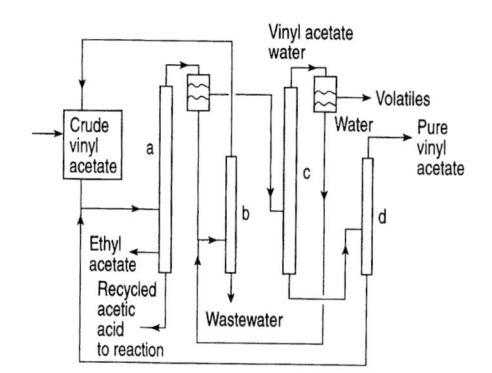

Figure 3. Ethylene gas-phase process for vinyl acetate production; work-up of crude vinyl acetate
a) Azeotrope column; b) Wastewater column; c) Drying/lightends column; d) Pure vinyl acetate column

consumed in the distillation is used for water removal. Crude vinyl acetate, which is low in water, collects in the sump of the predehydration column. Older plants do not have this column [39]. The noncondensed vinyl acetate fraction is washed out of the circulating gas with acetic acid in column (f) [40]. The remaining gas is recycled via compressor (g), after addition of fresh ethylene.

To remove the CO_2 formed in the reaction, a partial stream of recycle gas is first washed with water in column (h) to remove the remaining acetic acid. The CO_2 is then absorbed with potash solution in column (i). The potash solution is regenerated by depressurizing to normal pressure and boiling (j). Depending on the quantity of CO_2 formed in the reactor, the desired CO_2 content of the circulating gas can be adjusted by altering the quantity of circulating gas present in the circulating gas wash, and the degree of absorption in the potash wash. The CO_2 concentration is generally 10–30 vol% [41], [42]. It is also possible to perform water and CO_2 washes in the main gas stream [43].

To remove inert gases (nitrogen, argon) mainly brought in with the oxygen, a small quantity of waste-gas is removed before the CO_2 absorption column (i) and then incinerated. In some plants, part of the ethylene contained in this waste-gas is recovered by additional purification to reduce ethylene loss.

The liquid products formed, i.e., the condensate from the sump of the predehydration column (l) and the sump of the circulating gas wash, are depressurized to almost normal pressure and drained off into a collector for crude vinyl acetate (k). The circulating gas portions dissolved under the pressure of the circulating gas system are degassed and recycled to the circulating gas system after compression.

Process Description: Distillation Section. Distillative work-up of the liquid products to give acetic acid (which is recycled) and pure vinyl acetate is carried out in various ways, depending on the location of the plant and on the relative importance of energy consumption and investment costs. Besides the systems shown in Figures 3 and 4, combinations of both versions are used.

The liquid products contain 20–40 wt% vinyl acetate and ca. 6–10 vol% water. The rest consists of acetic acid and small quantities of byproducts, e.g., ethylidene diacetate

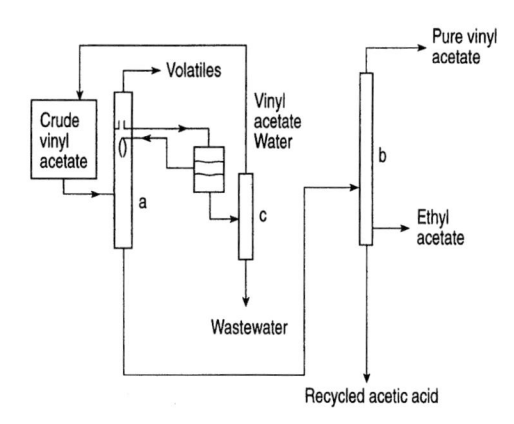

Figure 4. Ethylene gas-phase process for vinyl acetate production; variant for work-up of crude vinyl acetate
a) Dehydration column; b) Pure vinyl acetate column; c) Wastewater column

and ethyl acetate. To work up the liquid products (Fig. 3) a vinyl acetate – water mixture can be distilled at the head in a first distillation (a). This mixture separates into two phases. The dissolved vinyl acetate is distilled from the water phase in wastewater column (b). The remaining water is wastewater. The organic vinyl acetate phase is freed as the top product in a second distillation (c) from dissolved water, other volatile products, and acetaldehyde, formed by vinyl acetate hydrolysis. The sump product is dry vinyl acetate, which is distilled in a third distillation column (d) to give pure vinyl acetate as the top product.

To remove polymers, a small partial stream is removed from the sump of the third distillation column and fed back to the first column. Thus all nonvolatiles and polymers produced in the distillative work-up are contained in the sump of the first column, together with the recycled acetic acid. To remove the polymers and nonvolatiles, a partial stream is removed from the acetic acid evaporator in the reaction section of the plant. From this, the acetic acid for recycling is distilled so that the residue, which is still flowable, can be incinerated.

The small quantities of ethyl acetate formed are removed through a side exit in the first column (a) as a mixture with acetic acid, water, and vinyl acetate [44].

If an additional column is used for the work-up, only the dissolved water is distilled over in the second distillation, in the third the light ends, and in the fourth the pure vinyl acetate [45].

In another version (Fig. 4) the water contained in the crude vinyl acetate is removed as an azeotrope with the vinyl acetate together with volatile products, e.g., acetaldehyde, in a first distillation column (a), which operates at increased pressure. The dry bottom product, which contains vinyl acetate, acetic acid, polymers, and nonvolatiles, is fractionated in a second column (b) into pure vinyl acetate as the top product, and acetic acid and nonvolatiles. The last two are recycled. Ethyl acetate is removed through a side exit in the second distillation column as a mixture with acetic acid and vinyl acetate [46], [47].

To avoid polymer formation during the distillative workup of the crude mixture, polymerization inhibitors, e.g., hydroquinone, benzoquinone, or *tert*-butylcatechol, must be added. Passing in oxygen-containing gases is also said to inhibit polymerization [48].

Process Data. The catalysts used always contain palladium as the metal or a salt, alkali metal salts, and additional activators, e.g., metallic gold or alkali acetoaurate, cadmium acetate, or noble metals of the platinum group. Silicic acid of various structures, aluminum oxide, spinel, or activated charcoal are mainly used as the carrier material [49]–[54], [55].

The space–time yield for vinyl acetate is 200 g L^{-1} h^{-1}, in older plants up to more than 1000 g L^{-1} h^{-1}, depending on the catalyst and the plant layout. The life time of the catalyst is ≤ 4 a.

The reaction pressure is 0.5–1.2 MPa overpressure. The space–time yield for vinyl acetate increases with the reaction pressure and with the oxygen concentration in the reaction gas. However, an increase in pressure shifts the flammability limits of ethylene–oxygen to lower oxygen concentrations, reducing the quantity of oxygen available, and consequently the quantity of vinyl acetate formed, so pressure limits are set; higher pressures also raise equipment costs.

The reaction temperature is generally > 140 °C. It increases to > 180 °C towards the end of the catalyst life. A lower reaction temperature results in the formation of less CO_2, but then the heat produced in the reactor can no longer be used in the plant.

The gas loading of the catalyst is 2–4 m^3 (STP) per liter catalyst per hour. The gas mixture contains 10–20 mol% acetic acid, 10–30% CO_2, and ca. 50% ethylene. The maximum oxygen content is ca. 1.5% below the flammability limit, which varies with the composition of the gas mixture and the reaction conditions. The nitrogen and argon contents are adjusted according to the quantity of waste gas. They are generally ca. 10%, but depend on the purity of the oxygen used.

For old plants, energy consumption is ca. 3 t of heating steam per tonne of vinyl acetate produced. As a result of process improvements, modern plants have a heating steam consumption of 1.2 t per tonne of vinyl acetate.

The yields are up to 99% based on acetic acid, and up to 94% based on ethylene, if the acetaldehyde, formed in small quantities by hydrolysis of vinyl acetate during the distillative work-up, is included in the yield.

The process has not yet posed any environmental problems. Volatile and nonvolatile liquid products are incinerated. The water produced in the reaction can contain traces of acetic acid formed by hydrolysis of vinyl acetate in the wastewater column. It is subjected to biological wastewater treatment. To remove inert gases brought in with the oxygen, some of the waste-gas containing nitrogen and argon is burned after partial recovery of the ethylene it contains. There are small quantities of residual ethylene in the CO_2, formed in the regeneration column of the potash absorber. Removal of ethylene, e.g., by subsequent catalytic incineration, is now necessary in Germany because of more stringent regulations. In addition, special measures are required for waste-gas incineration.

Plants are generally made from stainless steel 316 L, apart from the potash washer, which is made from normal steel or stainless steel 321.

Proposals for Process Improvement. Since the first plants for the ethylene gas-phase process were commissioned, numerous suggestions have been made for improving the process; some of these are now being introduced industrially. Proposals regarding the catalyst include: simplification of production methods [56]; activity improvement (increase in vinyl acetate space–time yield [57]–[60], [55], [61]; improvement in selectivity (less CO_2 formation) [62], [63]; increase in life span [64]–[66]; reduction in flow resistance in the reactor (form of carrier material) [67]; better utilization of the noble metal content (distribution on the carrier) [68]; regeneration of spent catalysts [69]–[74].

Proposals regarding the mode of operation include: increasing the vinyl acetate space–time yield by raising the oxygen concentration in the gas at the entry to the reactor by adding desensitizers [75]; facilitating separation of ethyl acetate from vinyl acetate [76]–[79]; increasing the selectivity of the reaction by adding other components to the reaction gas [80]; carrying out the reaction on a fluidized bed [81]; lowering the energy consumption, e.g., by making use of the heat of condensation of the gas coming out of the reactor for removal of the water formed (see Section 2.2) [39]; simplified work-up of the polymer-containing residue [82].

Reaction of Methyl Acetate with CO and H_2. Recently, several companies have attempted to develop a reaction which, in principle, has been known for a long time. It involves the conversion of methanol, acetic anhydride, dimethyl ether, or preferably methyl acetate to ethylidene diacetate with carbon monoxide and hydrogen [83]–[95], [96]–[98]:

$$2\,CH_3CO_2CH_3 + 2\,CO + H_2 \rightarrow (CH_3CO_2)_2CHCH_3 + CH_3CO_2H$$

The reaction takes place in the liquid phase at ca. 7 MPa and ca. 150 °C. Rhodium salts together with methyl iodide and amines can be used as catalyst. The ethylidene diacetate formed can be cleaved to vinyl acetate and acetic acid, using the process previously employed by Celanese (see Section 2.2). As the acetic acid formed in the process can be reesterified to methyl acetate using methanol, the process essentially involves production of vinyl acetate from methanol and synthesis gas:

$$2\,CH_3CO_2CH_3 + 2\,CO + H_2 \rightarrow (CH_3CO_2)_2CHCH_3 + CH_3CO_2H$$
$$(CH_3CO_2)_2CHCH_3 \rightarrow CH_3CO_2CH{=}CH_2 + CH_3CO_2H$$
$$2\,CH_3OH + 2\,CH_3CO_2H \rightarrow 2\,CH_3CO_2\,CH_3 + 2\,H_2O$$

Dimethyl ether can also be used to prepare methyl acetate:

$$CH_3OCH_3 + CO \rightarrow CH_3CO_2CH_3$$

It is said to be possible to produce dimethyl ether by partial oxidation of natural gas:

$2\,CH_4 + O_2 \rightarrow CH_3OCH_3 + H_2O$

Byproducts are formed in all the steps; recovering them in pure form is expensive [99]–[102].

The process has not yet been used in industry. It could become more important, depending on the further development of raw material prices.

2.3. Quality Specifications, Analysis, Storage, Transport, and Toxicology

Quality. The specifications of individual producers often differ only slightly with regard to the content of impurities. One specification requires: vinyl acetate ≥ 99.9 wt%, acid (as acetic acid) ≤ 0.005 wt%; carbonyl (as acetaldehyde) ≤ 0.02 wt%, water ≤ 0.04 wt%; distillation range 95% at 72–73 °C within 0.5 °C; peroxide free; polymer free; capacity for polymerization (see Section 2.4)

Analysis. The following methods are used for quality testing:

Carbonyl groups: titration of the hydrochloric acid formed after conversion to the oxime using hydroxylamine hydrochloride solution
Water content: titration with Karl-Fischer solution
Acid content: neutral to litmus
Peroxide: test with potassium iodide solution
Polymer content: mixtures with petroleum ether must not produce turbidity or precipitation of solids
Polymerization test: carried out differently by individual producers; the time taken for the onset of polymerization following the addition of a defined quantity of dibenzoyl peroxide and warming to a defined temperature

Storage. As a flammable liquid, vinyl acetate is assigned to VbF, group A, class 1, and ignition group T 2 according to VDE 0165. It can be stored in steel, aluminum, or stainless steel containers under nitrogen. It is not necessary to add stabilizers at lower temperatures. If the vinyl acetate is to be warmed, stabilizers, such as hydroquinone, hydroquinone monomethyl ether, or diphenylamine are added. The quantity of stabilizer used is small, e.g., 3–20 ppm hydroquinone, so that it does not generally need to be removed during the later polymerization.

Transport. Vinyl acetate is transported in practically all quantities: small containers, drums, tankers, tank cars, and ships. Stabilizers are usually added for transportation.

Transport Regulations
EEC guidelines: No. 607-023-00-0
Land transport ADR/RID and GGVS/GGVE
GGVS/GGVE class 3, cipher 3 b
ADR/RID class 3, cipher 3 b
Warning notice no. 339, substance no. 1301
Symbol F flammable liquid
Sea transport IMDG/UN class 2, U.N. no. 1086
IMOG/UN class 3.2 and no. 1301 II
EmS no. 3-07 MFAG 330
Symbol F flammable liquid/3
Air transport ICAO/IATA-DGR
ICAO/IATA-DGR class 3, UN no. 1301 PG II
Dispatch symbols FL: Double arrows FL   flammable liquid/3
Packaging regulations
   PAC: 305 max. net 5 L
   CAO: 307 max. net 60 L
Air mail: not permitted

Safety Data

CAS [108-05-4]	
Flashpoint	−8 °C
Ignition temperature (DIN)	385 °C
Hazard class (VbF)	AI
Explosion limits in air	
lower	2.6 vol%
upper	13.4 vol%
Explosion group (DIN)	IIA
Exothermic decomposition dTA	from 500 °C
Risk assessment depending on the European Seveso guidelines no. 2	
Emergency Response Planning Guidelines	75 ppm (ERPG-2)

Toxicological Data

Acute toxicity	
LD_{50}, oral, rat	2920 mg/kg
LC_{50}, inhalation, rat	14.2 mg/L × 2 h
LD_{50}, dermal, rabbit	2335 mg/kg

Health hazards
F easily inflammable
R 11
S 16-23-29-33
Classification according to the regulations of hazardous chemicals/TRGS 900
Carcinogenic: category K 3
Teratogenic: no (TRGS 905)
Alters genotype: no (TRGS 905)
Skin resorption: no
Sensitization: no
Effect on skin: slightly irritant
Other health hazards, e.g., possible damage to organs; narcosis; damage to liver and kidneys; 4000 ppm causes death in rats in 4-h inhalation test [103].

Protection at Work

MAK/TKR value	10 ppm 35 mg/m^3
Peak limiting category	I
Warning symptoms:	
Sweetish, not unpleasant odor	
Odor threshold	1 mg/m^3
Odor perception threshold	1.8 mg/m^3

Compatibility with water

Water hazard class	2
Fish toxicity LC$_{50}$	20 mg/L
Bacteria toxicity	100–1000 mg/L
Biodegradability	>70% good
Toxicity rating number	fish 4.6; bacteria 5.2; mammals 1

2.4. Use, Economic Importance

Vinyl acetate is used mainly for the production of polymers and copolymers, e.g., for paints (mainly dispersions), adhesives, textile and paper processing, chewing gum, and for the production of poly(vinyl alcohol) and polyvinylbutyral.

Vinyl acetate – ethylene copolymers are processed to give resins, paints, and sheeting. Floor coverings and gramophone records are made from vinyl acetate – vinyl chloride copolymers. Vinyl acetate is also used in small quantities as a comonomer in polyacrylic fiber production.

Uses differ according to the region. In Japan, ca. 70% of vinyl acetate is used in the production of poly(vinyl alcohol), while in the United States and Europe more than half is processed to give poly(vinyl acetate).

The total capacity for vinyl acetate production was $< 10^6$ t/a in 1965. In 1994 it was ca. 3.8×10^6 t/a. The rapid increase has been promoted significantly by the develop-

Table 1. World vinyl acetate capacities

Country	Process	Capacity, 10^3 t/a
United States	Ethylene Bayer – Hoechst and National Distillers	1429
Canada	Ethylene Bayer – Hoechst	54
Mexico	Ethylene Bayer – Hoechst	100
South America	Acetylene, Ethylene National Distillers	100
Australia	Ethylene Bayer – Hoechst	14
Germany	Ethylene Bayer – Hoechst	280
France	Acetylene	150
Spain	Ethylene Bayer – Hoechst, Acetylene	100
United Kingdom	Ethylene Bayer – Hoechst	100
Italy	Acetylene	60
Eastern Europe	Ethylene Bayer – Hoechst, Acetylene	288
Japan	Ethylene Bayer – Hoechst and National Distillers, Acetylene	612
Near East	Ethylene Bayer – Hoechst and National Distillers	90
India	Ethylene Bayer – Hoechst	50
Korea	Ethylene Bayer – Hoechst	150
Taiwan	Ethylene Bayer – Hoechst	100
China	Ethylene Bayer-Hoechst, Acetylene	224

ment of the ethylene gas-phase process. After ethylene had become one of the cheapest raw materials in the chemical industry, it was obvious that vinyl acetate should be produced from ethylene instead of acetylene. The vinyl acetate capacities of individual countries are listed in Table 1.

A plant with a capacity of 150 000 t/y is to be commissioned by Hoechst Celanese in South East Asia in 1997.

The economic optimum of individual process variants depends on location. Investment costs, material consumption (acetic acid, ethylene, acetylene), and energy consumption need to be compared. The weighting depends on the prices of energy and starting materials, and on official regulations specific to a particular country.

3. Vinyl Propionate

3.1. Properties

Physical Properties. Vinyl propionate [105-38-4], $CH_3-CH_2-COO-CH=CH_2$, M_r 100.1, is a clear, colorless, flammable liquid with an ester-like odor; bp 95 °C; d_4^{20} 0.917; n_D^{20} 1.404; solidification point −80 °C; viscosity 0.56 mPa · s at 20 °C; heat of combustion 26 922 kJ/kg; flashpoint according (Abel-Pensky) −2 °C; specific heat (liquid) 2.12 kJ/kg; heat of evaporation 340 kJ/kg at bp; heat of polymerization 882 kJ/kg; vapor pressure 1.34 kPa at 0 °C, 6.45 kPa at 20 °C; explosion limits (vapor in air at 20 °C) 2.4 – 15.5 vol%; ignition temperature 360 °C; solubility in water 6.5 mL/L at 25 °C; solubility of water in vinyl propionate 8.5 mL/L at 25 °C.

Chemical Properties. Vinyl propionate closely resembles vinyl acetate in its chemical reactions and on polymerization. The greatest differences are the more difficult saponification and its activity as an internal plasticizer in copolymers.

3.2. Production

Vinyl propionate is produced on an industrial scale at BASF from acetylene and propionic acid in the gas phase. The catalyst is zinc propionate on activated charcoal.

Process Description (Fig. 5). The recycle gas from blower (a) with an overpressure of up to 40 kPa is driven through the heat exchangers (b), connected in series, where it is heated to ca. 160 °C. It consists of ca. 95 % acetylene together with inert gases and traces of vinyl propionate, propionic acid, and other reaction products. After the second heat exchanger, propionic acid vapor from (j) is mixed into the circulating gas, and the resulting gas mixture is heated to the reaction temperature in (c). Depending on the activity of the zinc propionate catalyst, the temperature in the contact reactors (d) is 160–190 °C. The heat of reaction is removed via a cooling medium (l) by means of evaporative cooling, and is used for steam production in (k). The reaction gases from the reactor outlet (d) are cooled to 80 °C in the heat exchangers (b). The reaction products and unreacted propionic acid are condensed by further cooling in (e), (b), and (f). The recycle gas is washed with propionic acid in (g) to remove vinyl propionate and acetaldehyde. A small quantity is removed via the waste–gas scrubber (h).

The crude mixture in (i) contains vinyl propionate, propionic acid, and, besides dissolved acetylene, small quantities of the byproducts acetaldehyde, crotonaldehyde, diethyl ketone, and propionic anhydride. The latter is hydrolyzed by addition of water. Diethyl ketone and crotonaldehyde are removed by special methods. Water, acetaldehyde, and acetylene are removed at the head of the lightends column.

The sump product is pumped to the purification column where pure vinyl propionate is obtained at the head. The sump product here is at a temperature of 150 °C, and contains propionic acid and higher-boiling products. The propionic acid is recovered 99.5 % pure in an acid recycling column.

The yield is over 98 % based on acetylene, and over 99 % based on propionic acid.

Quality. The following quality specifications are in use:

Purity (by GLC)	99.9 %
Ester content (from saponification number)	99.8 %
Water content	≤ 0.1 %
Acid content (calculated as propionic acid)	≤ 0.03 %
Color value	≤ 10 APHA

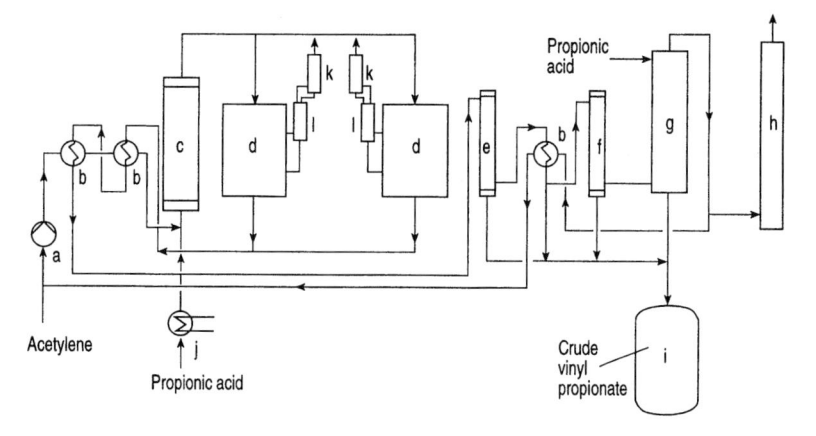

Figure 5. Production of vinyl propionate (BASF)
a) Recycle gas blower; b) Heat exchangers; c) Superheater; d) Reactors; e) Condenser; f) Brine-cooled condenser; g) Recycle gas wash; h) Wastegas wash; i) Crude product receiver; j) Propionic acid evaporator; k) Cooling medium; l) Steam generator

3.3. Use

Vinyl propionate is the starting material for the production of homopolymers and copolymers with common monomers, such as (meth)acrylic compounds, vinyl chloride, and other vinyl esters. The polymers are marketed as Propiofan brands (BASF).

4. Vinyl Esters of Higher Carboxylic Acids

Apart from vinyl acetate and vinyl propionate, the vinyl esters of only a few higher carboxylic acids are important (see Section 4.5.). The properties of polymers, e.g., poly(vinyl acetate), poly(vinyl chloride), polystyrene, and esters of poly(acrylic acid), are modified using these compounds as comonomers. With this type of modification paints hydrolyze less readily and absorb less moisture. The various production methods depend on the physical and chemical properties of the carboxylic acids to be vinylated.

4.1. Reaction with Acetylene in the Gas Phase

The reaction of a thermally stable higher carboxylic acid with acetylene in the gas phase is possible only if the *bp* is sufficiently low for it to be evaporated in an acetylene stream below the reaction temperature. Catalysts containing zinc or cadmium are suitable for the vinylation itself. The production process generally does not run as smoothly as that for vinyl acetate because the zinc salts of higher carboxylic acids, particularly the branched ones, are volatile and migrate from the catalyst more readily. This problem can be circumvented to a certain extent by modification of the catalyst [104].

4.2. Reaction with Ethylene and Oxygen

This process, analogous to that used for the production of vinyl acetate, has not been used industrially, although it is possible in principle [105]. As the molecular mass of the carboxylic acid increases, the quantity required per mole of vinyl ester also increases; the higher equipment expenditure for the ethylene process is no longer compensated for by the cheapness of ethylene compared with acetylene.

4.3. Transvinylation and Other Processes

The transvinylation of the corresponding carboxylic acid with vinyl acetate can be applied to almost all carboxylic acids.

$$CH_3CO_2CH=CH_2 + RCO_2H \rightleftharpoons RCO_2CH=CH_2 + CH_3CO_2H$$

Mercury, palladium, or silver salts are suitable catalysts [106]–[109]. Methods involving mercury salts are appropriate only for very special and expensive vinyl esters because of the associated work-up costs. The disadvantage of the transvinylation method lies in the fact that a large excess of vinyl acetate must be used. Also, working up the mixtures of vinyl acetate, acetic acid, the carboxylic acid, and its vinyl ester by distillation is not always straightforward. Vinyl esters of higher carboxylic acids have therefore been used for transvinylation instead of vinyl acetate.

In special cases, the reaction of a carboxylic anhydride with acetaldehyde or paraldehyde, and subsequent cleavage of the ethylidene dicarboxylate have proved successful [111]. This is analogous to the production of vinyl acetate from ethylidene diacetate. The dehydration of ethylene glycol monocarboxylates has also been proposed [112].

4.4. Reaction with Acetylene in the Liquid Phase

The vinylation of thermally stable carboxylic acids can also be achieved by passing acetylene into a melt formed by heating the corresponding zinc carboxylate or a zinc carboxylate–carboxylic acid mixture to ca. 200 °C. The process is described in more detail below (Wacker and Shell processes). The reaction can also be carried out with acetylene–nitrogen mixtures at pressures of up to 2.5 MPa [112], [113]. Carboxylic anhydrides may also be present [114].

If the vinyl esters of the higher carboxylic acids are thermally unstable, the reaction with acetylene can be carried out in the presence of mercury catalysts, enabling the reaction temperature to be lowered to 60–70 °C.

Wacker and Shell Processes. By these processes, nearly all vinyl esters of higher saturated carboxylic acids up to ca. C_{18} can be produced in the same plant. The Shell process is particularly suitable for vinylating branched higher carboxylic acids [115]–[117]. In the Wacker process, excess acetylene is passed through a melt consisting of the zinc salt of the corresponding carboxylic acid and the acid itself at ca. 200–220 °C. In the Shell process, the reaction mixture contains a high-boiling mineral oil fraction. The vinyl ester formed is carried out with the excess acetylene.

Process Description. A recycle acetylene gas stream is preheated and then passed into the vinylation reactor. This reactor can be a short bubble column, a normal stirred vessel, or a stirred vessel with a gas entrainment stirrer because of the low initial pressure of the acetylene. Trickle bed columns can also be used. The reactor is heated or cooled by a heat transfer oil in the jacket. It contains a melt of the carboxylic acid to be vinylated mixed with its zinc salt (Wacker) or a solution of the mixture in a solvent with $bp > 300$ °C, e.g., a mineral oil fraction with a low aromatic content (Shell). The reactor must be heated or cooled depending on the mode of operation and the nature of the carboxylic acid. At equilibrium, it is also possible to use adiabatic conditions. The reaction temperature is 200–230 °C and depends to some extent on the carboxylic acid used. The vinyl ester formed and the unreacted carboxylic acid are stripped with the acetylene stream. The conditions described for the vinyl acetate process regarding removal of inert gases and safeguarding against acetylene decomposition also apply here (see Section 2.2). The crude mixture, which condenses on cooling, contains up to 90 wt % vinyl ester. The rest consists of unreacted carboxylic acid, carboxylic anhydride, and small quantities of other byproducts. For high-boiling vinyl esters, the distillative work-up of the crude mixture is carried out in vacuo. The vinyl ester is distilled after removal of volatile products, e.g., acetaldehyde and crotonaldehyde. The sump product of the ester distillation, consisting of acid, acid anhydride, and high-boiling products is recycled to the reactor. The carboxylic anhydride formed can first be hydrolyzed thermally to the carboxylic acid. The hydrolysis can also be effected in the reactor itself by addition of small quantities of water, e.g., brought in with the acetylene as steam. Addition of water is not necessary with α-branched carboxylic acids. To avoid enrichment of high-boiling products, a small partial stream of the reactor liquid is removed [118], [119].

A large number of vinyl esters have been produced by the methods described [120]. They are not quite as pure as vinyl acetate or vinyl propionate because the starting acids are less pure than acetic or propionic acid. Plants for the vinylation of higher carboxylic acids are operated by Wacker in Burghausen and by Shell in Moerdijk and Pernis (Holland). Only a few of these vinyl esters are still produced in large quantities. For data on other vinyl esters which have been produced in the past see [121].

4.5. Individual Vinyl Esters

Vinyl laurate [2146-71-6] (Wacker), M_r 226.3, colorless, oily liquid with a weak odor, purity 99.9 ± 0.1%, acid number 1 ± 0.2 mg KOH/g, density at 20 °C 0.87 g/mL, solidification point 6.5 °C, refractive index $n_D^{20} = 1.437$, viscosity 2.617 mPa · s at 20 °C, heat of evaporation 146 kJ/kg at 5.5 kPa, vapor pressure 10 Pa at 50 °C, bp (550 Pa) 123 °C/(101.3 kPa) 254 °C, flashpoint 136 °C, ignition temperature 300 °C, lower explosion limit in air 0.5% (calculated) [114], solubility in water < 1 g/L, solubility of water in the ester < 1 g/L. The ester is not stabilized during transport or storage. Biodegradability, good; water hazard class WGK 2. Transport: not classed as hazardous goods.

Vinyl isononanoate (Wacker), M_r 184.3, colorless liquid with a weak, ester-like odor, acid content calculated as isononanoic acid ≤ 0.3%, carbonyl compounds calculated as acetaldehyde ≤ 0.1%, water content ≤ 0.05%, bp 145 °C at 100 kPa, 44 °C at 50 Pa, solidification point −70 °C, density 0.872 g/mL at 20 °C, flashpoint 68 °C, ignition temperature 390 °C.

Vinyl Esters of Versatic 911, 10, and 1519 Acids (Shell). Versatic acids are produced by the Koch synthesis from olefins, carbon monoxide, and water. 911 refers to mixtures with 9–11 carbons, 1519 to mixtures with 15–19 carbons. The vinyl ester of versatic acid 10 with general formula

$$R^2-\underset{\underset{R^3}{|}}{\overset{\overset{R^1}{|}}{C}}-CO_2CH=CH_2$$

is the most important. R^1, R^2, and R^3 are straight-chain alkyl groups, one of which is always a methyl group.

The vinyl ester of versatic 10 acid, $C_{12}H_{22}O_2$, M_r 198, is a colorless liquid with a slight, ester-like odor, color value ≤ 15 APHA.

Physical Data, Properties

Density at 20 °C	0.88 g/mL
Refractive index, n_D^{25}	1.435
Kinematic viscosity:	
at 20 °C	2.2 mm^2 s^{-1}
at 40 °C	1.3 mm^2 s^{-1}
Vapor pressure:	
at 30 °C	130 Pa
at 110 °C	4.25 kPa
at 210 °C	101 kPa
Solubility in water	< 0.1 wt%
Solubility of water in the ester	0.05 wt%
Acid number	≤ 1
Bromine number	80 g Br/100 g
Heat of evaporation	49 kJ/mol
Heat of polymerization	96.3 kJ/mol
Specific heat	1.97 J/g
Boiling range	133 – 136 °C at 13.3 kPa
Ignition temperature	309 °C
Solidification point	– 20 °C
TA-Luft class III	
VbF AIII	

Regulations governing hazardous substances: not hazardous as understood by these regulations.

Toxicology

LD$_{50}$, oral, rat	> 8800 mg/kg
LD$_{50}$, dermal, rat	> 3500 mg/kg
LD$_{50}$, inhalation, rat, 4 h	> 2.6 mg/L

Ecology

Fish toxicity (rainbow trout) LC$_{50}$, 96 h	14 mg/L
Daphnia toxicity LC$_{50}$, 48 h	110 mg/L
Green algae LC$_{50}$, 96 h	31 mg/L
Not readily biodegradable	

5. References

General References

[1] Beilstein, **II$_1$** p.63, **II$_2$** p.147, **II$_3$** p.268.

[2] C. E. Schildknecht: *Vinyl and Related Polymers*, J. Wiley, New York 1952.

[3] R. P. Arganbright, R. I. Evans: "Vinyl Acetate Still Growing Fast," *Hydrocarbon Process. Pet. Refin.* **43** (1964) no. 11, 149 – 164.

[4] R. B. Stabaugh, W. C. Allen, Jr., R. H. Van Sternberg: "Vinylacetate: How, Where, Who – Future," *Hydrocarbon Process.* **51** (1972) no. 5, 153 – 161.

[5] W. Schwerdtel, *Chem. Ing. Tech.* **40** (1968) 781 – 784.

Specific References

[6] Rhône Poulenc, DE 531 336, 1929.
[7] Chem. Fabrik Griesheim; DE 313 696, 1915.
[8] I. G. Farbenind., DE 712 373, 1938 (P. Kurtz) Du Pont, US 2 397 341, 1940 (E. K. Ellingboe).
[9] N. A. Miles, H. S. Mason, S. Sussmann, *J. Am. Chem. Soc.* **61** (1939) 1844 – 1847.
[10] K. Alder, H. F. Rickert, *Justus Liebigs Ann. Chem.* **543** (1940) 8, 21.
[11] R. L. Adelmann, *J. Org. Chem.* **14** (1949) 1057.
[12] Consortium Elektrochem. Ind. DE 515 307, 1927 (H. Deutsch, W. O. Herrmann).
[13] Consortium Elektrochem. Ind., DE 555 224, 1927 (H. Deutsch, W. O. Herrmann).Hoechst, DE 1 618 426, 1967 (G. Roscher, H. Schmitz).
[14] Bayer, DE-AS 1 283 226, 1967 (W. Krönig, G. Scharfe).
[15] Snamprogetti, DE-OS 2 521 470, 1975 (L. Re, V. Caciagli, G. E. Bianchi).
[16] Hoechst, DE 2 127 821, 1971 (M. Finke, H.-J. Kleiner).
[17] Chem. Fabrik Griesheim, DE 271 381, 1912.
[18] *Ullmann,* 3rd ed., **18,** 77 – 78.
[19] *Ullmann,* 4th ed., **23,** 598.
[20] Consortium Elektrochem. Ind., DE 403 784, 1921 (E. Baum, H. Deutsch, W. O. Herrmann, M. Mugdan).
[21] R. P. Arganbright, R. H. Evans, *Hydrocarbon Process. Pet. Refin.* **43** (1964) no. 11, 159 – 164.
[22] Monsanto, GB 651 072, 1947 (W. H. Gehrke).
[23] *Pet. Refin.* **38** (1959) no. 11, 304.
[24] K. Gerlach, G. Emig, H. Hofmann, *Chem. Ing. Tech.* **49** (1977) 969 – 970.
[25] S. Sacharjan, *Chem. Tech. (Leipzig)* **20** (1968) 617, 620.
[26] Celanese, US 2 859 241, 1956 (A. W. Schnitzer); US 2 425 389, 1945 (H. F. Oxley, E. B. Thomas). *Pet. Refin.* **40** (1961) no. 11, 308.
[27] J. J. Moiseew, M. N. Vargaftik, J. K. Syrkin, *Dokl. Akad. Nauk SSSR* **133** (1960) 377 – 380.
[28] ICI, GB 964 001, 1960 (D. Clark, W. D. Walsh, W. E. Jones).
[29] Hoechst, DE-AS 1 191 366, 1961 (W. Riemenschneider, T. Quadflieg).
[30] Union Carbide, DE 629 885, 1962 (J. E. McKeon, P. S. Starcher).
[31] *Eur. Chem. News* **51** (1973) Feb. 2, 22.
[32] *Ullmann,* 3rd ed., suppl. vol., 137 – 139.
[33] Bayer, BE 627 888, 1962 (H. Holzrichter, W. Krönig, B. Frenz).
[34] Knapsack, DE-AS 1 244 766, 1965 (K. Sennewald, W. Vogt, H. Glaser).
[35] Hoechst, DE-AS 1 296 138, 1967 (H. Fernholz, H.-J. Schmidt, F. Wunder).
[36] S. Nakamura, T. Yasui, *J. Catal.* **23** (1971) 315 – 320.
[37] Hoechst, DE-AS 1 191 366, 1961 (W. Riemenschneider, T. Quadflieg).
[38] Bayer, DE 1 568 657, 1966 (K. Sennewald, G. Roscher, W. Krönig, W. Schwerdtel). DE 1 296 621, 1966 (W. Krönig, W. Schwerdtel, K. Sennewald, G. Roscher).
[39] *Ullmann,* 4th ed. **23,** 602.
[40] Bayer, DE 1 768 078, 1968 (W. Krönig, W. Schwerdtel).
[41] Knapsack, DE 1 277 249, 1966 (K. Sennewald et al.).
[42] Nat. Distillers, DE-AS 1 593 159, 1965 (I. L. Mador, I. L. McClain).
[43] *Eur. Chem. News* **51** (1973) Feb. 2, 21 – 22.
[44] Bayer, DE 1 768 412, 1968 (W. Krönig, W. Schwerdtel).
[45] *Chem. Age Int.* 1973, (Feb.) 9, 15.
[46] Hoechst, DE 1 668 063, 1967 (G. Roscher, H. Schmitz).

[47] Hoechst, DE 1 282 014, 1967 (G. Roscher, K. Günther, O. Probst, J. Scholderer).
[48] Bayer, DE 1 807 738, 1968, (W. Schwerdtel, W. Krönig, G. Scharfe, W. Swodenk).
[49] Nat. Distillers, US 3 190 912, 1962 (R. E. Robinson).
[50] Bayer, DE-AS 1 568 645, 1964 (W. Krönig, W. Schwerdtel).
[51] Hoechst, DE-AS 1 296 138, 1967 (H. Fernholz, H.-J. Schmidt, F. A. Wunder).
[52] Bayer, DE 1 185 604, 1962 (H. Holzrichter et al.).
[53] Knapsack, DE 1 244 766, 1965 (K. Sennewald, W. Vogt, H. Glaser).
[54] Hoechst, DE-OS 2 315 037, 1973 (H. Fernholz et al.).
[55] UCC, US 5 185 308, 1991 (W.-J. Bartley, S. Jobson, G. G. Harkreador, M. Kitson, M. Lemanski).
[56] Nat. Distillers, DE-OS 1 944 933, 1968 (D. Lum).
[57] Knapsack, DE-AS 1 230 784, 165 (K. Sennewald et al.).
[58] Nat. Distillers, DE-OS 1 944 933, 1968 (D. Lum).
[59] Hoechst, DE-OS 2 100 778, 1971 (H. Fernholz, F. Wunder, H.-J. Schmidt).
[60] Hoechst, DE-OS 2 315 037, 1973 (H. Fernholz et al.).
[61] UCC, US 5 179 057, 1991 (W.-J. Bartley et al.).
[62] Hoechst, DE-OS 2 057 087, 1970 (H. Fernholz, F. A. Wunder, H.-J. Schmidt).
[63] Knapsack, DE 1 668 352, 1967 (K. Sennewald et al.).
[64] Nat. Distillers, DE-OS 2 657 442, 1975 (R. C. Bartsch).
[65] Nat. Distillers, DE-AS 1 817 213, 1967 (D. W. Lum).
[66] Knapsack, DE 1 258 863, 1966 (K. Sennewald, W. Vogt, H. Glaser).
[67] Hoechst, DE-OS 2 811 211, 1978 (F. A. Wunder, T. Quadflieg, G. Roscher).
[68] Bayer, DE-OS 2 528 363, 1975 (W. Biedermann, H. Kösler, K. Wedenmeyer).
[69] Nat. Distillers, DE-OS 1 667 206, 1966 (D. W. Lum, I. L. Mador).
[70] Nat. Distillers, DE-AS 1 667 212, 1967 (F. D. Miller).
[71] Nat. Distillers, US 3 634 497, 1969 (C. C. Budke).
[72] Knapsack, DE 1 542 257, 1966 (K. Sennewald, W. Vogt, H. Glaser).
[73] Hoechst, DE-AS 2 513 125, 1975 (H. Fernholz, H. Krekeler, H.-J. Schmidt, F. Wunder).
[74] Nat. Distillers, DE-OS 2 420 374, 1973 (S. Schott, D. W. Lum, I. L. Mador).
[75] Celanese, DE-OS 2 361 098, 1972 (J. Severs).
[76] Nat. Distillers, DE-OS 2 114 751, 1970 (J. Feldmann, F. Lerman, F. D. Miller).
[77] Bayer, DE-AS 1 768 412, 1968 (W. Krönig, W. Schwerdtel).
[78] Du Pont, US 3 591 463, 1968 (H. B. Copelin).
[79] Celanese, DE-AS 1 618 240, 1966 (G. J. Fischer).
[80] Wacker, DE-OS 3 036 421, 1980 (D. Dempf et al.).
[81] Knapsack, DE 1 793 481, 1968 (K. Sennewald et al.).
[82] Celanese, DE-OS 2 359 286, 1973 (G.-J. Fischer, E. N. Wheeler).
[83] Halcon, DE-OS 2 610 035, 1975 (N. Rizkalla, C. N. Winnick).
[84] Halcon, DE-OS 2 261 562, 1971 (R. Fernandez).
[85] Union Oil, US 3 579 566, 1967 (D. M. Fenton).
[86] Ajinomoto, DE-OS 2 016 061, 1969 (H. Wakamatsu, N. Yamagami, J. Furukawa).
[87] Hoechst, DE-OS 2 450 965, 1974 (H. Kuckertz).
[88] Halcon, DE-OS 2 610 035, 1975 (N. Rizkalla, C. N. Winnick).
[89] Halcon, DE-OS 2 856 791, 1977 (C.-G. Wan).
[90] Mitsubishi Gas, EP 0 025 702, 1979 (T. Isshiki, Y. Kijima, Y. Miauchi, T. Jasunaga).
[91] Hoechst, DE-OS 2 941 232, 1979 (H.-K. Duebbeler, H. Erpenbach, K. Gehrmann, G. Kohl).
[92] Mitsubishi Gas, EP 0 028 447, 1979 (K. Jasuhiko, J. Akira, J. Kenji).
[93] Mitsubishi Gas, EP 0 028 515, 1979, T. Issihiki, J. Kijima, T. Kondoh).

[94] Mitsubishi Gas, EP 0 034 062, 1980 (T. Issihiki, J. Kijima, T. Watanabe).
[95] Mitsubishi Gas, EP 0 035 860, 1980 (T. Issihiki, A. Ito, J. Kijima, T. Watanabe).
[96] Halcon, US 4 341 741, 1981 (W. C. Davidson).
[97] Shell, EP 58 442, 1981 (E. Drent).
[98] Daicel, JP 100 135, 1979 (N. Negawa, T. Yokoyama).
[99] Halcon, US 267 960, 1981 (R. V. Porcelli).
[100] Hoechst, DE 134 347, 1981 (H. Erpenbach, K. Gehrmann, W. Lork, P. Prinz).
[101] Halcon, US 241 180, 1981 (W. C. Davidson, R. V. Porcelli).
[102] Daicel, JP 177 671, 1984 (Y. Harano).
[103] *Ullmann*, 4th ed., **23,** 615.
[104] Hoechst, DE 1 233 387, 1964 (G. Roscher, W. Riemenschneider).
[105] Bayer, DE 1 568 657, 1966 (W. Krönig, W. Schwerdtel, K. Sennewald, G. Roscher).
[106] R. L. Adelmann, *J. Org. Chem.* **14** (1949) 1057, 1073.
[107] Consortium Elektrochem. Ind., DE 654 282, 1938 (W. O. Herrmann, E. Baum, W. Haehnel).
[108] Consortium Elektrochem. Ind., DE 1 127 888, 1960, (J. Schmidt, A. Abel).
[109] Sekisui Chem. Ind. JP 19 146/64, 1961 (S. Jwai, Y. Ohzuka).
[110] A. M. Sladkow, G. S. Petrow, *Zh. Obshch. Khim.* **24** (1956) 151.
[111] Wacker, DE-AS 2 142 419, 1971 (H. Eck, H. Spes).
[112] I. G. Farbenind., DE 588 352, 1932 (W. Reppe).
[113] H. M. Teeter, *J. Am. Oil. Chem. Soc.* **40** (1963) 143–155.
[114] H. Hopff, CH 324 667, 1954.
[115] Shell, BE 634 642, 1962 (W. F. Engel, G. E. Rumscheidt).
[116] Shell, DE 1 237 557, 1962 (W. F. Engel, G. E. Rumscheidt).
[117] Shell, DE 1 238 322, 1963.
[118] Wacker, DE 1 542 574, 1965 (H. Kainzmaier, G. Hübner).
[119] Shell, DE-OS 1 768 289, 1967 (A. C. H. Borsboom, P. A. Gautier, D. Medema).
[120] G. Hübner, *Fette Seifen Anstrichm.* **68** (1966) 290–292.
[121] *Ullmann,* 3rd ed., **18,** 84–86.

Vinyl Ethers

ERNST HOFMANN, BASF Aktiengesellschaft, Ludwigshafen, Federal Republic of Germany (Chaps. 1–8)
HANS-JOACHIM KLIMISCH, BASF Aktiengesellschaft, Ludwigshafen, Federal Republic of Germany (Chap. 9)

1.	Introduction	4865	3.3.1.	Production of Vinyl Ethers under Pressure	4870
2.	Properties	4866	3.3.2.	Pressure-free Production of Vinyl Ethers	4872
2.1.	Physical Properties	4866			
2.2.	Chemical Properties	4866	4.	Environmental Protection, Waste Disposal	4872
3.	Production	4866			
3.1.	Reaction of Acetylene with Alcohols	4869	5.	Quality Specifications	4873
			6.	Chemical Analysis	4873
3.2.	Reaction Conditions	4869			
3.2.1.	Introduction	4869	7.	Storage and Transport	4873
3.2.2.	Dependence of Reaction Conditions on the Nature of the Alcohol	4870	8.	Use	4874
			9.	Toxicology	4874
3.3.	Process Description	4870	10.	References	4875

1. Introduction

Vinyl ethers were first described by WISLICENUS in 1878. He obtained ethyl vinyl ether by treating chloroacetal with sodium.

Vinyl ethers (CH_2=CH–O–R; R = alkyl, cycloalkyl, or aryl) may be regarded as derivatives of vinyl alcohol, the unstable enol form of acetaldehyde. Vinyl ethers with substituted vinyl groups are of laboratory interest only, and are not dealt with here. In the 1930s, alkyl vinyl ethers achieved industrial importance as the starting materials for polymers, after they had been made readily accessible through the work of REPPE at I. G. Farbenindustrie, Ludwigshafen [1]–[4].

2. Properties

2.1. Physical Properties

The lower aliphatic vinyl ethers ($C_1 - C_4$) are colorless, combustible, mobile, volatile liquids with an ethereal odor. Methyl vinyl ether is gaseous under normal conditions, but can be liquefied under pressure (pressurized gas). Octadecyl vinyl ether, one of the higher vinyl ethers, is a low-melting solid of low volatility. Vinyl ethers are only slightly soluble in water, but readily miscible with most organic solvents (with the exception of glycol or glycerol). Some physical properties of industrially important vinyl ethers are given in Table 1.

2.2. Chemical Properties

Unlike the chemically very stable saturated ethers (→ Ethers, Aliphatic), vinyl ethers are extremely reactive substances because of the presence of the enol ether group. They can be readily polymerized in the presence of Friedel–Crafts catalysts, such as BF_3, $AlCl_3$, or $SnCl_4$. They are hydrolyzed by dilute aqueous acid at room temperature, to acetaldehyde and an alcohol [5, p. 391]. However, they are stable to alkalis, even hot aqueous caustic solutions.

Ethyl ethers are obtained by catalytic hydrogenation of vinyl ethers. In the presence of acid catalysts they react with alcohols to give acetals. Chlorine and bromine add to vinyl ethers giving 1,2-dihaloethyl ethers, but iodine causes polymerization, even in very small quantities. The reaction of halogens with vinyl ethers in the presence of alcohols gives 2-haloacetals [5, pp. 391–392]. Siggia's method of determination of vinyl ethers is based on this reaction [6]. Vinyl ethers undergo addition reactions with hydrogen halides giving 1-haloethyl ethers [4, pp. 92–93]. The reaction of vinyl ethers with *ortho* esters gives tetraacetals of dialdehydes. With dienes and 1,2-unsaturated aldehydes, vinyl ethers undergo a Diels–Alder reaction [5, pp. 392–395], [7, p. 227]. Vinyl ethers react with numerous other reagents, e.g., acetals, carboxylic acids, hydrogen cyanide, phosgene, carbon tetrachloride, ammonia, and amines [5, pp. 393–396].

3. Production

Methods for preparing vinyl ethers, e.g., reaction of alkali alkoxides with vinyl halides or dihalohydrocarbons [4, pp. 85–86], and transvinylation (transfer of a vinyl group from a vinyl ether or vinyl ester to another alcohol) [5, p. 371], [7, p. 207], [8, p. 508] have not achieved any importance for the production of vinyl ethers beyond laboratory scale.

Table 1. Physical properties of representative vinyl ethers

Vinyl ether	CAS registry number	Formula	M_r	mp, °C	bp, °C	d_4^{20}	n_D^{20}	Heat of vaporization, kJ/kg	Flash point, °C	Ignition temperature, °C	Explosion limits in air, vol% vinyl ether
Methyl	[107-25-5]	$CH_3-O-CH=CH_2$	58.10	−122	6	0.773^a	1.3947^b	417	−60	210	1.9−32.0
Ethyl	[109-92-2]	$C_2H_5-O-CH=CH_2$	72.12	−115	36	0.754	1.3767	379	−45	180	1.35−12.0
Propyl	[764-47-6]	$C_3H_7-O-CH=CH_2$	86.13		65	0.768	1.3908		−25	185	1.8−10.2
Isobutyl	[109-53-5]	$i\text{-}C_4H_9-O-CH=CH_2$	100.18	−112	83	0.769	1.3965	332	−15	195	0.73−7.9
Octadecyl	[930-02-9]	$C_{18}H_{37}-O-CH=CH_2$	296.54	27	$175-190^c$	0.81^d		239	171	265	
Cyclohexyl	[2182-55-0]	$C_6H_{11}-O-CH=CH_2$	126.19	−109	150−152	0.891	1.4550	352	35	190	1.0−7.3
Butanediol-1,4-monovinyl^e	[17832-28-9]	$HO-C_4H_8-O-CH=CH_2$	116.00	−33	189	0.944	1.4440		88	225	1.4−9.9
Butanediol-1,4-divinyl	[3891-33-6]	$CH_2=CH-O-C_4H_8-O-CH=CH_2$	142.20	−8	170	0.898	1.4420		63	205	0.9−13.9
Diethylene glycol divinyl	[764-99-8]	$CH_2=CH-O-C_2H_4-O-C_2H_4-O-CH=CH_2$	158.19	−21	191	0.968	1.4468		83	190	1.1−19.0

^{a} At 0°C.
^{b} At −25°C.
^{c} At 0.6 kPa.
^{d} At 40 °C.
^{e} = Vinyl-4-hydroxybutyl ether.

Production

The production of monovinyl ethers of diols can take place by transfer of a vinyl group from a diol divinyl ether to a free diol:

$CH_2=CH–(CH_2)_4–O–CH=CH_2 + HO–(CH_2)_4–OH \rightleftharpoons$
$2\ CH_2=CH–O–(CH_2)_4–OH$ Butanediol-1,4-monovinyl ether

The equilibrium can be shifted in favor of monovinyl ether formation by excess diol. Compounds of the platinum group metals, in particular those of palladium, are used as catalysts [9].

Vinyloxycarboxylic acid esters (not accessible by Reppe vinylation) can be produced by vinyl group transfer from vinyl ethers to hydroxycarboxylic acid esters, using special palladium compounds as catalysts:

$$HO-CH_2-\underset{CH_3}{\overset{CH_3}{C}}-COOCH_3 + CH_2=CH-O-i\text{-}C_4H_9$$
Methyl hydroxypivalate   Isobutyl vinyl ether

$$\rightleftharpoons CH_2=CH-O-CH_2-\underset{CH_3}{\overset{CH_3}{C}}-COOCH_3 + i\text{-}C_4H_9-OH$$
Methyl vinyloxypivalate   Isobutanol

Here also, the equilibrium can be shifted in favor of vinyloxycarboxylic acid ester formation by excess vinyl ether [10].

An industrial process for vinyl ether production involves the cleavage of acetals at 270–400 °C in the presence of palladium asbestos, thorium oxide, or manganese sulfate on aluminum oxide as catalyst:

$CH_3–CH(OR)_2 \rightarrow CH_2=CH–O–R + R–OH$

Ethyl, butyl, and isobutyl vinyl ethers have been produced industrially by this method (Union Carbide) [5, pp. 369–370], [7, p. 207], [8, p. 508]. Improved processes for vinyl ether production by acetal cleavage use special zeolites as catalysts (mordenite [11], borosilicate – iron silicate zeolites [12]), but these developments have not been exploited industrially.

Vinyl ethers are also obtained by elimination of hydrogen chloride from chloroethyl ethers using alkali [13]. For example, 2-chloroethyl vinyl ether used to be produced industrially from 2,2'-dichlorodiethyl ether by heating with solid NaOH (Union Carbide) [5, p. 371]:

$Cl–CH_2–CH_2–O–CH_2–CH_2–Cl + NaOH \rightarrow CH_2=CH–O–CH_2–CH_2–Cl + NaCl + H_2O$

In a catalytic oxidation process developed by ICI for the production of methyl vinyl ether, ethylene is reacted with methanol in the presence of palladium chloride complexes or palladium metal [14]–[16]. It is not known whether this process has been developed on an industrial scale:

$$CH_2=CH_2 + CH_3-OH \rightarrow CH_2=CH-O-CH_3 + H_2O$$

3.1. Reaction of Acetylene with Alcohols

The most important industrial method for the production of vinyl ethers is the reaction of acetylene with alcohols in the presence of strong alkalis in the liquid phase (Reppe vinylation) [3], [17]:

$$CH\equiv CH + R-OH \rightarrow CH_2=CH-O-R \quad \Delta H = -134 \text{ kJ/mol}$$

The strongly exothermic reaction is carried out in the presence of alkali alkoxide catalysts. The reaction proceeds particularly smoothly in the presence of the corresponding potassium compound [1], [4]. The vinylation temperature is generally 150–180 °C. With methanol, a vigorous reaction occurs at 120 °C.

3.2. Reaction Conditions

3.2.1. Introduction

The choice of operating pressure depends on the boiling point of the alcohol to be vinylated. Lower alcohols (C_1–C_4) are vinylated under pressure, as their boiling points lie below the reaction temperature. Thus, methyl vinyl ether is produced at 1.6 MPa, ethyl vinyl ether at 1.5 MPa, and isobutyl vinyl ether at 0.8 MPa. The vinylation also succeeds at normal pressure if so much KOH is dissolved in the alcohol that the boiling point of the solution rises to the vinylation temperature or above. However, this method gives poor yields because of side reactions.

When working under pressure, the acetylene must be diluted with nitrogen for safety reasons [3] (→ Acetylene). The acetylene concentration must not exceed 55 % for methyl and ethyl vinyl ethers, and 75 % for isobutyl vinyl ether.

Higher alcohols can be vinylated at normal pressure as their boiling points are above the vinylation temperature. In this case, expenditure on equipment is significantly lower than in the pressure process. In the normal pressure process, octadecyl vinyl ether, vinyl ethers of fatty alcohol mixtures, and decalyl vinyl ether, for example, can be produced.

The vinylation of alcohols is also possible in the gas phase. A mixture of alcohol vapor and acetylene is passed over solid-bed catalysts, e.g., soda lime [4, p. 87], KOH on activated charcoal, or magnesium or calcium oxides [18], but such processes have not achieved any practical importance.

3.2.2. Dependence of Reaction Conditions on the Nature of the Alcohol

In the case of primary and secondary aliphatic alcohols, vinylation proceeds very smoothly, even for the highest homologs. Tertiary alcohols react less readily. Polyhydric alcohols, e.g., 1,4-butanediol, can be partially or fully vinylated (to monovinyl or divinyl ethers) depending on the reaction conditions chosen. Where the distance between the hydroxyl groups is small (e.g., with ethylene glycol) the cyclic acetal is formed from the monovinyl ether before the latter can react with the acetylene to give the divinyl ether. Alicyclic alcohols (e.g., cyclohexanol, 2-decalol (2-decahydronaphthol), terpene alcohols, tertiary aminoalcohols, and heterocyclic alcohols, such as hydroxyfuran, can also be vinylated. The vinylation of aminoalcohols with primary or secondary amino groups to give the corresponding vinyl ethers can best be carried out with so-called superbases as catalysts (DMSO with alkali hydroxide/alkoxide, preferably potassium compounds) [19]. Examples are 3-vinyloxy-1-propylamine, *N*-ethyl-2-vinyloxy-1-ethylamine, and 4-vinyloxypiperidine.

Phenols and naphthols react with acetylene less readily than aliphatic alcohols. The corresponding aryl vinyl ethers can be obtained only in the presence of a large quantity of potassium phenoxide and a good solvent for acetylene, e.g., *N*-methylpyrrolidone.

3.3. Process Description

In industry, vinyl ethers are produced in a vertical reaction tower (tube reactor). The alcohol, endowed with KOH, is charged at the top, while the gaseous mixture of acetylene and nitrogen is fed in from below. The acetylene not consumed in the reaction is led back into the reactor (recycling process) and supplemented by pure acetylene (fresh gas).

3.3.1. Production of Vinyl Ethers under Pressure

The pure acetylene is precompressed in the fresh gas compressor (Fig. 1, h) to the required operating pressure and then pumped into the pressure tower (c) from below by means of the circulating gas pump (g). At the same time, the alcohol, to which 2–3 % KOH has been added in the feed vessel (a), is continuously pumped from above toward (c), by pressure pump (b). The vinyl ether formed at 150–160 °C distills together with excess unreacted alcohol from the pressure tower (c), and is condensed

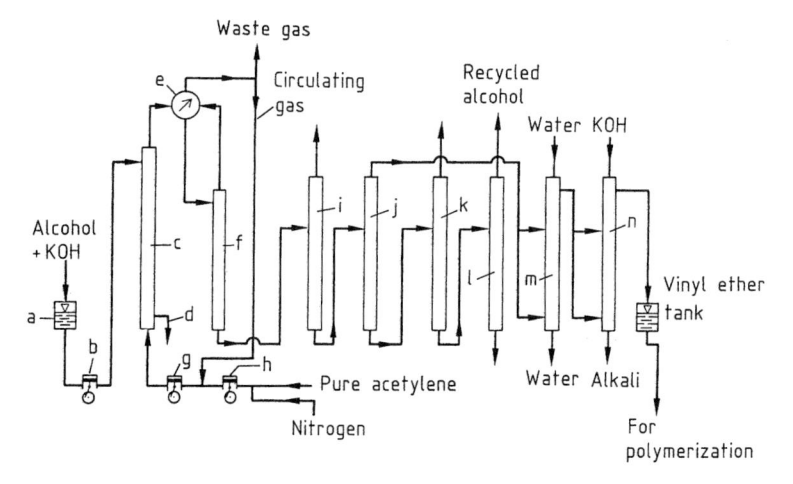

Figure 1. Production of vinyl ethers under pressure
a) Feed vessel; b) Pressure pump; c) Pressure tower; d) Sump exit; e) Condenser; f) Pressurized degasser; g) Circulating gas pump; h) Fresh gas compressor; i) Fores column; j) Purification column; k, l) Alcohol columns; m) Washing column; n) Drying column

out of the circulating gas stream in the condenser (e). The remaining circulating gas is led back to the circulating gas pump after removal of the waste gas by taking out inert components. The heat of reaction is removed by evaporation mainly of the vinyl ether–alcohol mixture and by a cooling coil in the pressure tower (c). A small proportion of the contents of the tower is removed at the lower end of the pressure tower via the drainage exit (d) to remove used catalyst and high-boiling side products from the reaction mixture.

The vinyl ether–alcohol mixture distilling from the pressure tower (c) is freed from dissolved acetylene in the pressure degasser (f) and from low-boiling fractions in the fores column (i). In the purification column (j) the vinyl ether is obtained at the distillation head by fractional distillation, and freed from alcohol residues in the washing column (m) by washing with a countercurrent of water. After drying with solid KOH in the drying tower (n) to remove residual moisture, the vinyl ether is sufficiently pure for further processing (polymerization).

Unreacted alcohol forming the sump in the purification column (j) is purified in two further distillation columns (k, l) and recycled to (a). The whole plant operates completely continuously.

The production of special vinyl ethers under pressure is carried out as a batch operation in stirred autoclaves (vinylation of cyclohexanol, butane-1,4-diol, diethylene glycol, etc.). The crude vinyl ethers produced are purified by vacuum distillation.

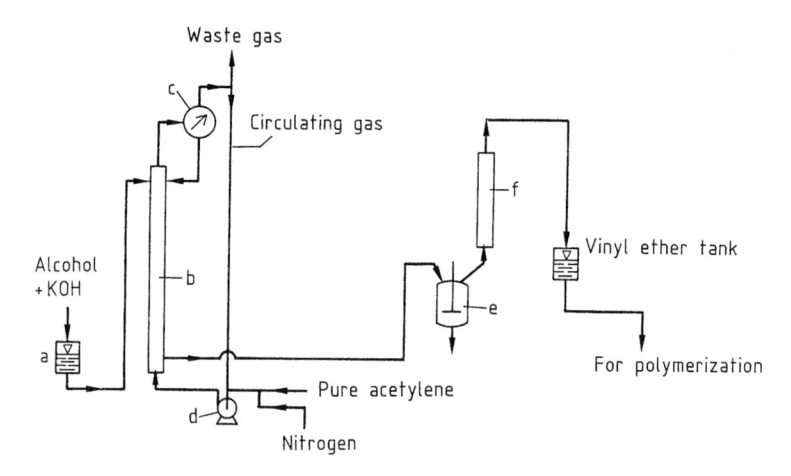

Figure 2. Pressure-free production of vinyl ethers
a) Feed vessel; b) Reaction tower; c) Condenser; d) Circulating gas blower; e) Stirred vessel; f) Column

3.3.2. Pressure-free Production of Vinyl Ethers

Vinylation at normal pressure is carried out as a batch operation so that complete conversion of the alcohol is achieved and the purification of the reaction product is simplified.

The alcohol treated with 2–3% KOH in the feed vessel (Fig. 2, a) is charged in batches to the top of the reaction tower (b). The pure acetylene is fed into the reaction tower from below by means of the circulating gas blower (d). Unreacted acetylene comes out of the top and passes back to the circulating gas blower (d) via the condenser (c). By removal of waste gas, the acetylene concentration in the gas circulation is held at ca. 90% (the remainder is nitrogen). The reaction temperature is 170–180 °C. The heat of reaction is removed from the reaction tower (b) by means of a cooling coil.

After the reaction has finished (acetylene uptake falls to zero) nitrogen is blown through the crude vinyl ether to remove dissolved acetylene, and the vinyl ether is then passed into the stirred vessel (e) where it is purified by vacuum distillation using column (f) in a batch process.

4. Environmental Protection, Waste Disposal

To protect the environment, the residues and waste materials formed in the production of vinyl ethers must be disposed of according to the regulations. Gaseous waste products (process waste gas from the synthesis and displacement gas from the containers) are incinerated in an excess gas burner. Liquid waste and distillation residues are disposed of in an incineration plant for special wastes. Wastewater which is licensed for treatment is subjected to biological purification in a treatment plant. No environmental problems are encountered if disposal measures are carried out properly.

5. Quality Specifications

As precursors for polymer production, vinyl ethers are subject to high quality requirements. The specifications require a purity (vinyl ether content) $>99.8\%$, residual alcohol content $<0.1\%$, other side products $<0.1\%$, and water $<0.01\%$.

6. Chemical Analysis

Vinyl ethers are preferably analyzed by standard GLC methods. They may also be determined quantitatively by addition of iodine to the double bond of the vinyl group in the presence of methanol (Siggia's method) [6]. Absolutely dry vinyl ether (<0.01 water) does not attack freshly pressed sodium wire (no gas formation, no change in the lustrous metal surface within 1 h).

7. Storage and Transport

Vinyl ethers can be stored for only a limited period. They can be kept at 20 °C for several months if water and acids are excluded. They are stored in stainless or normal steel containers under dry nitrogen, which prevents the formation of combustible air–vapor mixtures. The exclusion of atmospheric oxygen inhibits the formation of peroxide in the stored product. Methyl vinyl ether (pressurized gas) is stored in pressurized containers.

High-boiling organic bases have proved to be useful stabilizers for inhibiting polymerization during transport of vinyl ethers, e.g., *N,N*-diethylaniline, or triethanolamine (soluble in the product). The quantity of stabilizer added is 0.05 – 0.1%. The use of solid KOH in pellets as a stabilizer is limited to drums of up to 200 L.

Because vinyl ethers are readily flammable (see Table 1), the relevant safety instructions and explosion protection measures must be adhered to during production, storage, and transport. Dispatch regulations and hazard classes for transport are given in Table 2.

Table 2. Dispatch regulations for vinyl ethers: hazard classes

Vinyl ether	Rail, GGVE/ RID	Trucks, GGVS/ ADR	Ship, IMDG Code
Methyl	2,3 ct	2,3 ct	2, UN 1087
Ethyl	3,2 b	3,2 b	3,1 UN 1302
Propyl	3,3 b	3,3 b	3,1 UN 1993
Isobutyl	3,3 b	3,3 b	3,2 UN 1304
Octadecyl	none	none	none
Cyclohexyl	3,31 c	3,31 c	3,3 UN 1993
Butanediol-1,4-monovinyl *	3,32 c	3,32 c	none
Butanediol-1,4-divinyl	3,32 c	3,32 c	3,3 UN 1993
Diethylene glycol divinyl	3,32 c	3,32 c	

* = Vinyl-4-hydroxybutyl ether.

8. Use

Vinyl ethers are used principally for the production of homo- and copolymers, in the adhesives and varnish industries. Vinyl ethers can also be used as chemical intermediates for the production of glutardialdehyde [7, p. 227], tetramethoxypropane [5, p. 392], γ-picoline [20], and 2-aminopyrimidine [21].

9. Toxicology

Investigations of toxicokinetics and metabolism have shown that, on inhalation, rapid resorption takes place, and that vinyl ethers are relatively stable and metabolized only slightly. They are mainly exhaled unchanged. Epoxidation of the vinyl group has not been described [22]–[25].

The acute toxicity of vinyl ethers is low. In rats the acute oral LD_{50} decreases with increasing chain length:

Methyl vinyl ether	4.900 mg/kg [22], [26]
Ethyl vinyl ether	8.160 mg/kg [27]
Butyl vinyl ether	10.300 mg/kg [28]
Isobutyl vinyl ether	13.100 mg/kg [29]

As LC_{50} indicates, the acute toxicity by inhalation in rats is also very low:

Methyl vinyl ether	150 mg/L (4 h, rat) [26]
Ethyl vinyl ether	21.2 mg/L (4 h, rat) [30]
Isobutyl vinyl ether	21.1 mg/L (4 h, rat) [31]

The symptoms are characterized by the narcotic action of the vinyl ethers at high concentration. Compared with saturated alkyl ethers, the vinyl group reinforces the narcotic properties. Correspondingly, divinyl ether is a stronger narcotic than, for example, ethyl vinyl ether or diethyl ether.

On repeated inhalation (28-day test on rats) only slight systemic toxicity was observed at high concentration (25 000 ppm). With methyl vinyl ether there was an increase in the relative liver weight with no histopathological correlation and a local irritant effect on the mucosa of the nasal cavity in the form of atrophy of the olfactory epithelium. The no adverse effect concentration (NOAEC) was 1500 ppm (3.6 mg/L) for male rats and 3500 ppm (8.3 mg/L) for female rats. No other histopathological change was observed [22]. In experiments with dogs no changes in the liver, no electrocardiographic changes (dogs, monkeys), and no effect on blood pressure were found [32].

On contact with skin and with the mucous membranes of the eye, highly volatile vinyl ethers are not irritant, and those with low volatility slightly irritant [22], [29], [33], [34].

No data from skin sensitization experiments on animals are available. However, short-chain vinyl ethers are not expected to be hazardous because of their high volatility. In humans no cases of sensitization have been published, either in production of the compounds or for their earlier use as an inhaled narcotic.

In vitro investigations on genotoxicity are contradictory. On microorganisms (*Salmonella typhimurium*: Ames test) and with mammalian cells (HPRT test: [35]) vinyl ethers were not mutagenic, with or without metabolic activation. Chromosome aberrations could not be detected on somatic cells with isobutyl vinyl ether [36]. However, with ethyl vinyl ether and 2,2,2-trifluoroethyl vinyl ether possible indications of an increase in the sister chromatid exchange rate (SCE test; [37]) have been described. In animal experiments no genotoxic activity was observed (micronucleus test; [22]).

No investigations of carcinogenicity and reproductive toxicity in animals have been reported.

From the use of ethyl vinyl ether as an inhaled narcotic in the 1950s, a number of effects on humans have been reported. In one out of 220 narcoses spasm of the larynx and flow of saliva occurred. In 12 of 220 cases there was generalized cramp and respiratory and cardiac arrest. Postnarcotic effects are: nausea/vomiting (26.7%), reduction in circulation (4.1%), excitation (1.6%), and hypoxia (0.8%). Liver function tests gave no indication of damage [38]. In another investigation indications of slight leukocytosis and slight myocardial depression were observed, but liver function tests showed no change [38].

10. References

[1] I. G. Farbenind., DE 584 840, 1930; DE 639 843, 1933; DE 640 150, 1934 (W. Reppe).
[2] W. Reppe: *Neue Entwicklungen auf dem Gebiet der Chemie des Acetylens und Kohlenoxids*, Springer-Verlag, Berlin 1949.
[3] W. Reppe: *Chemie und Technik der Acetylen-Druckreaktionen*, 2nd ed., Verlag Chemie, Weinheim 1952.
[4] W. Reppe et al., *Justus Liebigs Ann. Chem.* **601** (1956) 81–110.

[5] N. D. Field et al.: "Vinyl Ethers," in *High Polymers*, vol. **24**, Wiley, New York 1971, pp. 365–411.
[6] S. Siggia, R. L. Edsberg, *Anal. Chem.* **20** (1948) 762.
[7] S. A. Miller: "Vinyl Ethers," in *Acetylene, Its Properties, Manufacture and Uses*, vol. 2, Ernest Benn Ltd., London 1966, pp. 198–231.
[8] D. H. Lorenz: *Encyclopedia of Polymer Science and Technology*, vol. **14**, Interscience, New York 1971, pp. 504–510.
[9] BASF, DE 4 134 807 A1, 1991 (M. Karcher, J. Henkelmann).
[10] BASF, EP 0 538 685 A2, 1992 (M. Karcher, J. Henkelmann).
[11] Degussa, DE 3 535 128 A1, 1985 (D. Arntz et al.).
[12] BASF, EP 0 299 286 A1, 1988 (W. Holderich et al.).
[13] C. E. Rehberg, *J. Am. Chem. Soc.* **71** (1949) 3247.
[14] ICI, NL 6 411 879, 1965.
[15] ICI, NL-Appl. 6 611 559, 1967.
[16] *Ullmann*, 3rd ed., **18**, 96.
[17] *Ullmann*, 4th ed., **7**, 45.
[18] R. L. Zimmerman et al., *Ind. Eng. Chem. Prod. Res. Dev.* **2** (1963) no. 4, 293–296.
[19] BASF, EP 0 514 710 A2, 1992 (M. Karcher, J. Henkelmann).
[20] BASF, DE 944 250, 1952 (W. Reppe et al.).
[21] BASF, DE 1 117 587, 1958 (H. Maisack et al.).
[22] Vinylmethylether (BG no. 63), Toxikologische Bewertung, Berufsgenossenschaft der chemischen Industrie, ed. 02/92.
[23] M. E. Andersen, M. L. Gargas, R. A. Jones, L. J. Jenkins, Jr.: "Determination of the Kinetic Constants for Metabolism of Inhaled Toxicants in vivo Using Gas Uptake Measurements," *Toxicol. Appl. Pharmacol.* **54** (1980) 100–116.
[24] M. E. Andersen: "A Physiologically Based Toxicokinetic Description of the Metabolism of Inhaled Gases and Vapors: Analysis at Steady State," *Toxicol. Appl. Pharmacol.* **60** (1981) 509–526.
[25] R. T. Williams: *Detoxication Mechanism*, 2nd ed., John Wiley & Sons, New York 1959.
[26] W. B. Deichmann: *Toxicology of Drugs and Chemicals*, Academic Press, New York 1969.
[27] G. D. Clayton, F. E. Clayton (eds.): *Patty's Industrial Hygiene and Toxicology*, 3rd ed., vol. **2 A**, John Wiley & Sons, New York 1981, p. 2503.
[28] H. F. Smyth, Ch. P. Carpenter, C. S. Weil, H. C. Pozzani: *AMA Arch. Ind. Hyg. Occup. Med.* **10** (1954) 61.
[29] H. F. Smyth, Ch. P. Carpenter, C. S. Weil, H. C. Pozzani: *Am. Ind. Hyg. J.* **23** (1962) 95–98.
[30] BASF AG, Toxicology department: unpublished results (88/193).
[31] BASF AG, Toxicology department: unpublished results (13I0192/887035).
[32] J. R. Krantz et al.: *J. Pharmacol. Exp. Ther.* **90** (1947) 88.
[33] BASF AG, Toxicology department: unpublished results (79/406).
[34] BASF AG, Toxicology department: unpublished results (88/193).
[35] BASF AG, Toxicology department: unpublished results (50M0234/924205).
[36] BASF AG, Toxicology department: unpublished results (32M0234/924195).
[37] A. E. White, T. Shim, E. Edmond, W. Sheldon, S. Wendell: *Anesthesiology* **50** (1979) 426–430.
[38] Vinylethylether (BG no. 263), Toxikologische Bewertung, Berufsgenossenschaft der chemischen Industrie, unpublished data, 09/95.

Waxes

UWE WOLFMEIER, Hoechst Aktiengesellschaft, Werk Gersthofen, Augsburg, Federal Republic of Germany (Chap. 1)

HANS SCHMIDT, Hoechst Aktiengesellschaft, Werk Gersthofen, Augsburg, Federal Republic of Germany (Chap. 2)

FRANZ-LEO HEINRICHS, Hoechst Aktiengesellschaft, Werk Gersthofen, Augsburg, Federal Republic of Germany (Chap. 3)

GEORG MICHALCZYK, DEA Mineralöl AG, Hamburg, Federal Republic of Germany (Chap. 4, Chap. 7 in part)

WOLFGANG PAYER, Hoechst Aktiengesellschaft, Werk Ruhrchemie, Oberhausen, Federal Republic of Germany (Chap. 5)

WOLFRAM DIETSCHE, BASF Aktiengesellschaft, Ludwigshafen, Federal Republic of Germany (Sections 6.1.1, 6.1.2, and 6.1.4)

KLAUS BOEHLKE, BASF Aktiengesellschaft, Ludwigshafen, Federal Republic of Germany (Sections 6.1.1, 6.1.2, and 6.1.4)

GERD HOHNER, Hoechst Aktiengesellschaft, Werk Gersthofen, Augsburg, Federal Republic of Germany (Sections 6.1.3, 6.1.5, 6.2, and 6.3)

JOSEF WILDGRUBER, Hoechst Aktiengesellschaft, Werk Gersthofen, Augsburg, Federal Republic of Germany (Chap. 7)

1.	Introduction	4878	2.2.6. Jojoba Oil	4898
1.1.	History of Waxes and Their Applications	4879	2.2.7. Other Vegetable Waxes	4899
			2.3. **Animal Waxes**	4900
1.2.	**Definition**	4880	2.3.1. Beeswax	4900
1.3.	**Classification**	4881	2.3.2. Other Insect Waxes	4904
			2.3.3. Wool Wax	4906
1.4.	**Properties and Uses**	4884	3. **Montan Wax**	4907
1.5.	**World Market Volume for Waxes**	4885	3.1. **Formation and Occurrence**	4907
1.6.	**International Wax Organizations**	4886	3.2. **Extraction**	4907
			3.3. **Properties and Composition**	4908
2.	**Recent Natural Waxes**	4886	3.4. **Refining and Derivatization**	4910
2.1.	**Introduction**	4886	3.4.1. Deresinification	4910
2.2.	**Vegetable Waxes**	4887	3.4.2. Bleaching	4910
2.2.1.	Carnaúba Wax	4888	3.4.3. Derivatization	4912
2.2.2.	Candelilla Wax	4892	3.5. **Uses and Economic Aspects**	4913
2.2.3.	Ouricury Wax	4895	4. **Petroleum Waxes**	4914
2.2.4.	Sugarcane Wax	4896	4.1. **Introduction**	4914
2.2.5.	Retamo Wax	4898		

4.2.	Macrocrystalline Waxes (Paraffin Waxes)	4915	4.4.	Legal Aspects............. 4939
4.2.1.	Chemical Composition and General Properties	4915	4.5.	Ecology................. 4940
			4.6.	Economic Aspects 4941
4.2.2.	Division into Product Classes...	4916	4.7.	Candles................. 4941
4.2.3.	Occurrence of Raw Materials and Processing	4920	5.	Fischer–Tropsch Paraffins... 4942
			6.	Polyolefin Waxes 4944
4.2.3.1.	Dewaxing Lubricating Oil Distillates	4920	6.1.	Production and Properties... 4944
			6.1.1.	Polyethylene Waxes by High-Pressure Polymerization 4944
4.2.3.2.	Deoiling Slack Waxes	4922		
4.2.3.3.	Refining Deoiled Slack Waxes ..	4924	6.1.1.1.	Production 4945
4.2.4.	Storage, Transportation, Commercial Forms, and Producers	4927	6.1.1.2.	Properties................ 4948
			6.1.2.	Copolymeric Polyethylene Waxes by High-Pressure Polymerization 4949
4.2.5.	Quality Specifications and Analysis	4928	6.1.3.	Polyolefin Waxes by Ziegler–Natta Polymerization .. 4950
4.2.6.	Uses....................	4930	6.1.3.1.	Production 4950
4.3.	Microcrystalline Waxes (Microwaxes).............	4931	6.1.3.2.	Properties................ 4951
			6.1.4.	Degradation Polyolefin Waxes .. 4953
4.3.1.	Chemical Composition and General Properties	4931	6.1.4.1.	Production 4953
			6.1.4.2.	Properties................ 4955
4.3.2.	Division into Product Classes...	4932	6.1.5.	Polar Polyolefin Waxes 4956
4.3.3.	Occurrence of Raw Materials and Processing................	4934	6.2.	Uses 4957
			6.2.1.	Nonpolar Polyolefin Waxes 4957
4.3.4.	Commercial Products and Producers	4936	6.2.2.	Polar Polyolefin Waxes 4958
			6.3.	Economic Importance 4959
4.3.5.	Quality Specifications and Analysis	4937	7.	Toxicology............... 4959
4.3.6.	Uses....................	4938	8.	References............... 4963

1. Introduction

Waxes are among the oldest worked materials used by humans. Their value as versatile construction materials ("man's first plastic") was discovered very early. Today, waxes are used mostly as additives and active substances. The use of waxes is expected to increase in the future because of their generally favorable toxicological and ecological properties.

The historic prototype of all waxes is beeswax. Since it could be obtained with relatively little effort, it was popular in antiquity, and even now the term wax is occasionally used in everyday speech as a synonym for beeswax. However, the scientific and commercial definition of wax covers a much wider area.

1.1. History of Waxes and Their Applications

Utilization of waxes was probably begun in prehistoric times, but because of their transitory nature, no definite archaeological evidence exists. Thus, the utilization of wax and related substances in mummification and as protective coverings in ancient Egypt (from ca. 3000 B.C.) represents the earliest scientific proof of the use of waxes. Ancient written documents contain many indications that waxes found many different applications. The most well known is the story of Daedalus and Icarus, who used wax as an adhesive to make wings by attaching feathers to each other. In antiquity, wax was used as a raw material for modeling, in the production of casting molds, as a pigment carrier, and for surface protection.

In colonial times, hitherto unknown waxes, such as carnaúba, candelilla, and Chinese insect wax, were introduced in Europe. From the occurrence of wax, COLUMBUS inferred the riches of the Caribbean islands: "Where there is wax, there are also thousands of other things" [10].

For a long time, not much was known about the chemical nature of wax. Only in the 18th century did the discovery occur that beeswax is an animal secretion and not a plant product collected by bees. Wax research was established as a scientific discipline in 1823 [11]. It became part of the new research area of soaps, oils, fats, and waxes. The real breakthrough of wax as an important raw material, in terms of quantity as well, occurred at the beginning of the Industrial Revolution. Ozocerite (fossil wax) was mined and refined to give ceresin (1875), montan wax was obtained from Eocene lignite (1897), and paraffin waxes were obtained from crude petroleum. In 1935 the first fully synthetic waxes were produced by the Fischer–Tropsch process. Polyethylene wax has been synthesized by the high pressure process since 1939, and became available by the low-pressure Ziegler process after 1953. On a laboratory scale polyolefin waxes can also be synthesized by using modern metallocene catalysts.

Degradation processes for the production of waxes, which start from high molar mass plastics (mainly polypropylene), have also achieved a certain degree of importance. Finally, low molar mass substances, which would otherwise have to be disposed of in plastics production, can be processed or refined to give industrially utilizable waxes.

A large and still increasing number of natural and synthetic waxes and related substances exists, as well as applications for these materials. No decrease in overall demand is apparent, so waxes are likely to remain important during the coming decades. The history of waxes has been described in great detail in [12].

1.2. Definition

No generally accepted definition exists for the term wax. All attempts to formulate a precise, comprehensive, and scientifically verifiable definition of wax must take the large number of waxlike products and the chemical complexity of individual types into account. Nevertheless, use of the term wax for different chemical species with common properties is still reasonable.

Typically waxes do not consist of a single chemical compound, but are often very complex mixtures. Being oligomers or polymers in many cases, the components differ in their molar mass, molar mass distribution, or in the degree of side-chain branching. Functional groups (e.g., carboxyl, alcohol, ester, keto, and amide groups) can be detected in waxes, sometimes several different groups.

The academic definition still quoted in chemistry text books—that waxes are esters of long-chain carboxylic acids with long-chain alcohols—is no longer useful. It applies fairly well only to some classical waxes, such as beeswax; others (e.g., petroleum waxes) do not fall in this category. Today, *physical* and *technical* definitions are preferred. Several attempts were made to differentiate between waxes and other classes of substance, particularly fats, resins, and high molar mass polymers, by using several criteria. These primarily *physical* definitions are to some extent arbitrary and are not generalled accepted. Waxes can also be classified according to their *applications*.

Probably the most conclusive definition has been drawn up in Europe by the Deutsche Gesellschaft für Fettwissenschaft (DGF, German Association for Fat Science). It was used in modified form in the customs tariff of the European Union [13], [14]. According to this definition, waxes must have:

1) A drop point (mp) $>40\,°C$
2) Their melt viscosity must not exceed 10 000 mPa · s at 10 °C above the drop point
3) They should be polishable under slight pressure and have a strongly temperature-dependent consistency and solubility
4) At 20 °C they must be kneadable or hard to brittle, coarse to finely crystalline, transparent to opaque, but not glassy, or highly viscous or liquid
5) Above 40 °C they should melt without decomposition
6) Above the *mp* the viscosity should exhibit a strongly negative temperature dependence and the liquid should not tend to stringiness
7) Waxes should normally melt between ca. 50 and 90 °C (in exceptional cases up to 200 °C)
8) Waxes generally burn with a sooting flame after ignition
9) Waxes can form pastes or gels and are poor conductors of heat and electricity (i.e., they are thermal and electrical insulators).

In the case of higher molar mass polyolefin waxes, differentiating between a wax and a thermoplastic polymer is sometimes difficult. This differentiation is particularly important for applications where a food approval is essential (FDA, EC, MITI in Japan, etc.). In case of doubt, relevant national regulations must be consulted.

1.3. Classification

Waxes can be classified according to various criteria such as origin; chemical, physical, and engineering properties; or applications (see Section 1.2). The primary differentiation is usually made according to origin or occurrence and synthesis. A proposal for a classification is shown in Figure 1. Here, waxes are divided into two main groups: natural and synthetic. Natural waxes exhibit their wax character without chemical treatment, whereas synthetic waxes generally acquire their waxy nature in the course of synthesis. There are no clear borderlines.

Natural waxes are formed through biochemical processes and are products of animal or plant metabolism. Natural waxes formed in earlier geological periods are known as *fossil waxes*. All petroleum, lignite (montan wax), and peat waxes belong to this largest category. Fossil waxes occur predominantly as minor components of oil, coal, and peat. Some deposits with a high wax content were formed by sedimentation. *Ozocerite* separated from oil, and *pyropissite*, one of the few organic minerals, is a mixture of montan wax and resins. Whereas petroleum waxes are predominantly nonpolar hydrocarbons, lignite and peat waxes consist of oxygen-containing compounds.

Biological wax synthesis is still taking place in nature. Many plants, particularly carnaúba palm, and animals—most importantly insects, such as bees—produce waxes. These are known as *nonfossil* or *recent natural waxes*. Natural waxes are seldom used industrially in their original form. They are generally converted into *modified natural waxes* by refining, e.g., by distillation, or extraction. Chemical processes, such as hydrogenation, bleaching, and oxidation, can also be used. Physical and chemical treatment can be combined; the objective of all these processes is to obtain a wax that is as pure as possible.

Partially Synthetic Waxes. If natural waxes or waxlike materials are modified by chemical reactions such as esterification, amidation, or neutralization of acidic waxes (with aqueous alkali or alkaline-earth metal hydroxides), partially synthetic waxes are obtained. These can be tailored to particular applications. In *esterified waxes* the relationship to natural waxes is still easily recognizable. For example, natural montan waxes, which consist mainly of esters of long-chain acids with long-chain alcohols, can be converted into acid waxes by cleavage of the esters and oxidation of the alcohols. Subsequent reaction with dihydric alcohols gives esters that have chemical structures comparable to those of the starting materials. *Amide waxes* do not occur naturally to any significant extent. This group consists of reaction products of fatty acids with

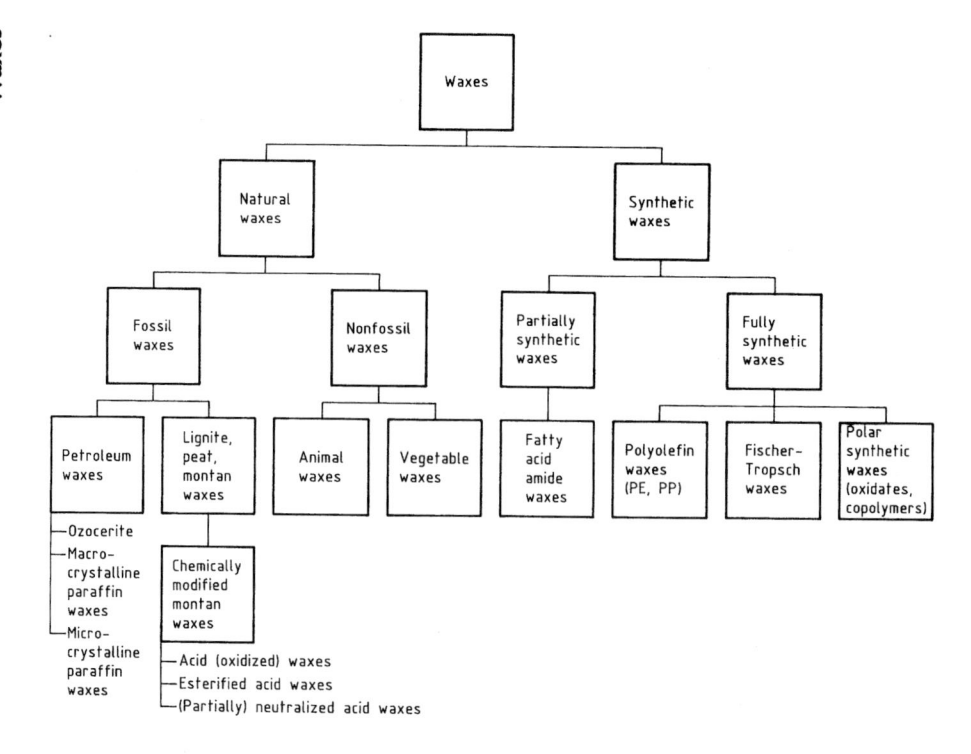

Figure 1. Classification of waxes

ammonia, amines, and diamines. An industrially important amide wax is distearoyl ethylenediamine.

Alcohol waxes (*lanette waxes*) also belong to the partially synthetic waxes. They are mixtures of long-chain alcohols that are optimized for their main area of use, in ointments and creams for pharmaceuticals and cosmetics, by the addition of fatty acid esters and emulsifiers. Alcohol waxes can also be used for industrial emulsions with good long-term stability.

Wool wax (*lanolin*) is obtained in crude form from sheep's wool. It is refined by chemical and adsorptive bleaching, and sometimes by subsequent hydrogenation or separation of the acidic fractions after hydrolysis. Wool wax is a mixture of esters, some being derived from branched or cyclic waxy or fatty acids and alcohols.

Fully synthetic waxes were developed only in the 20th century. Low molar mass compounds are used as starting materials. The products can be waxes in the narrower sense of the definition (see Section 1.2), or substances with only partial wax character. The two main groups of fully synthetic waxes are the Fischer–Tropsch and the polyolefin waxes. These waxes can be classified according to the starting material used for production (C_1 or C_2 chemistry).

C_1 Building Blocks. The Fischer–Tropsch (FT) synthesis is described elsewhere in detail. The oldest FT variant is the ARGE process (Arbeitsgemeinschaft Ruhrchemie–Lurgi). Synthesis gas is obtained by reaction of steam with carbon and is then converted catalytically into a broad spectrum of saturated and unsaturated hydrocarbons. Among these products are the Fischer–Tropsch waxes. They are characterized by low branching of the hydrocarbon chain and a narrow molar mass distribution. They are relatively hard products. The ARGE process has been used commercially by SASOL in the Republic of South Africa.

Besides carbon, natural gas can also be used as a raw material for C_1 syntheses. The Shell Middle Distillate Synthesis (SMDS) process utilizes synthesis gas produced by steam reforming of natural gas [15]. Fischer–Tropsch wax fractions are produced and marketed at the new SMDS plant in Bintulu (Sarawak/Malaysia). This process is a variant of the classical FT process and employs specially developed catalysts.

Polyolefin waxes are, in most cases, products of C_2 or C_3 chemistry. Low molar mass α-olefins, usually ethylene, are polymerized. The other route to polyethylene waxes—thermomechanical degradation of polyethylene plastic—is now of minor importance. The two most important production processes are the high-pressure and the Ziegler processes. Recently, polyolefin waxes have also been synthesized by using modern metallocene catalysts.

Polyolefin waxes can be homopolymers or copolymers of C_2, C_3, C_4 (ethylene, propene, and butene), or higher α-olefins. Polypropylene (PP) waxes are produced by the Ziegler process or by thermomechanical degradation of polypropylene plastic. Polypropylene waxes are generally partially crystalline. In addition, predominantly amorphous and atactic low molar mass polypropylene (APO or APAO) with a waxlike character is now produced and marketed by several companies. Besides this deliberate production, atactic PP (APP) is still produced as a byproduct of PP plastics.

All FT and polyolefin waxes are nonpolar. Polar waxes are obtained by subsequent reactions, such as oxidation by air or grafting (e.g., with maleic anhydride), optionally followed by neutralization to metallic soaps, esterification, or amide formation. Polar polyolefin waxes are also accessible via direct copolymerization or addition reactions (e.g., of olefins with unsaturated carboxylic acids or esters).

A considerable proportion of industrially processed waxes is used not in pure form but as wax compounds. For example, the viscosity and *mp* of paraffin waxes can be increased to the desired degree by mixing with polyolefin waxes.

1.4. Properties and Uses

The number of possible and actual uses of waxes is extraordinarily high. The physical and chemical properties of various types of wax can be combined fairly freely through mixtures and preparations. Therefore, waxes are represented in almost all possible areas of use, although to differing extents. Any selection of applications thus has to be somewhat arbitrary.

For the user, the application-oriented function of the wax is important, not so much its chemical structure. The latter is defined primarily by the characteristic hydrocarbon chain, the basic building block of almost all waxes. It imparts hydrophobing properties to waxes. The carbon chain length of fatlike waxes is $n = 16-18$ (e.g., amide wax); for esterified and paraffin waxes, $n = 20-60$; and for polyethylene waxes, less than hundred to several thousand. Besides the chain length itself (and thus the molar mass), molar mass distribution and degree of branching also affect the properties of waxes. The melting and softening points, melt viscosity, and degree of hardness increase with increasing chain length. High degrees of branching lower the *mp* and the hardness. Polar groups, such as carboxyl, ester, and amide groups, strengthen intermolecular forces and generally raise the *mp* and hardness.

Particular features of many waxes are their good absorption capacities and their ability to bind solvents. Hot wax solutions form stable, homogeneous pastes on cooling. These can be applied to surfaces and polished after evaporation of the solvent. The resulting wax films are shiny or glossy, hard, and resistant to mechanical stress.

When a suitable emulsifier is added, *polar waxes* can form fine stable dispersions in water. Mixed aqueous–organic emulsions can also be produced. These mostly stable dispersions and emulsions are used for surface protection. After application, the wax-containing phase dries and a protective, sometimes even glossy, coating is formed.

In addition to surface protection, hydrophobing, and gloss-giving, other functions of waxes include surface protection, hydrophobing, gloss-giving, matting, mold releasing, lubricating, binding, imparting compatibility or flexibility, regulating viscosity, adjusting consistency, emulsifying, dispersing, and adjusting drop points. Waxes are also used as combustible and illuminating materials.

For special applications only a few — frequently only one — of these functions are important. Some occasionally opposite effects are noteworthy: wax gives polish and gloss to various surfaces but has a matting effect as an additive in paints and coatings. Waxes are used as binders in masterbatches but as release agents in molding plastics. Table 1 gives some idea of the versatility of waxes but does not claim to be complete. For the specific suitability of waxes to individual applications, monographs, company brochures, and prospectuses should be consulted.

Table 1. Branches of industry in which waxes are used

Branch	Examples of applications
Adhesives, hot melts	viscosity regulation, lubricants, surface hardening
Building	modification of bitumen, antigraffiti treatment
Candles	fuel, drop point regulation
Ceramics and metal	binders for sintering
Cosmetics	binders and consistency regulators for ointments, pastes, creams, lipsticks
Electrical and electronics industries	release agents, insulating materials, etching bases
Explosives	stabilization
Foods	citrus fruit and cheese coating, chewing gum base, confectionery
Matches, pyrotechnics	impregnation, fuel
Medicine and pharmaceuticals	molding and release agents in dental laboratories, retardants, surface hardening of pills
Office equipment	dispersing agents and binders for carbon paper and selfduplicating paper; antioffset for toners for photocopiers
Paints and coatings	matting, surface protection
Paper and cardboard	surface hardening
Plastics	lubricants (PVC *), release agents (PA **), pigment carriers (masterbatch)
Polishes	surface protection of leather, floors, cars
Printing inks	improvement of rub resistance, slip
Recycling	compatibilizing
Rubber industry	release agents enhancing rigidity, surface hardening

* PVC = poly(vinyl choride).
** PA = polyamide.

1.5. World Market Volume for Waxes

No reliable global capacity or production data are available for waxes. The main reasons for this are the many types of wax and the difficulties in differentiating among them; the comparatively low transparency of the market; and the dependence of wax production on external factors, such as crude petroleum consumption and annual harvest of natural waxes. Some statistical data exist for certain markets (e.g., the United States [16]). Thus, the following figures can be regarded only as rough indications, with a correspondingly large degree of fluctuation, of the magnitude of annual worldwide wax consumption data:

Paraffin waxes, including microcrystalline waxes	3 000 000 t/a
Polyolefin waxes including oxidates and copolymers	200 000 t/a
Fischer–Tropsch waxes	100 000 t/a
Amide wax	50 000 t/a
Montan wax including crude montan wax	25 000 t/a
Carnaúba wax	15 000 t/a
Beeswax	10 000 t/a
Candelilla and other plant waxes	3 000 t/a

Forecasts for future market development anticipate positive growth factors. Specialties, such as special waxes or compounds for critical applications (e.g., in the electrical, electronics, or ceramics industries), will probably profit more from this development than waxes, that are tend to be continuously replaced in their applications. The decreasing demand for waxes in the packaging sector, for example, is becoming noticeable. The proportion of waxes used in some other classical applications (e.g., carbon paper) is continually declining. However, waxes generally have favorable ecological and toxicological properties, in addition to the variety of applications described, so their market chances may outweigh the risks in the long term.

1.6. International Wax Organizations

In Europe the interests of all wax producers have been represented by the European Wax Federation since 1979 (EWF, Avenue van Nieuwenhuyse 4, Brussels, Belgium). The EWF is a member of the Conseil Européen des Federations de l'Industrie Chimique (CEFIC) and is open to all European wax producers and some outside Europe. It represents the interests of its members, with regard to both technical questions such as specifications and legal matters referring in particular to the European Union. In other regions, no organizations exist that cover the entire wax spectrum.

2. Recent Natural Waxes

2.1. Introduction

Many plant and animal organisms produce waxes with extraordinarily complex compositions. The main components are usually those that correspond to the classical definition of waxes—i.e., esters of long-chain aliphatic alcohols and acids, which belong to homologous series in the ca. $C_{16}-C_{36}$ (predominantly $C_{22}-C_{36}$) range, those with an even number of carbon atoms predominating. Recent natural waxes also contain smaller quantities of esters derived from acids and alcohols with an uneven number of carbon atoms as well as bifunctional components such as diols, hydroxycarboxylic, and dicarboxylic acids. Natural waxes contain very small proportions of branched carboxylic acids, mainly those with ω-methyl branching. Aromatic acids, such as cinnamic, 4-hydroxycinnamic, and ferulic (4-hydroxy-3-methoxycinnamic) acids, have also been found in some waxes. Because of the presence of bifunctional components, natural waxes can contain not only the simple, classical wax esters, but also oligoesters with two or more ester functions and other functional groups in the molecule.

Free acids and alcohols are present in natural waxes in widely varying quantities. They are generally identical to those present as ester components.

Homologous n-alkanes in the ca. $C_{15}-C_{37}$ range, mainly those with an uneven number of carbon atoms, are also components of natural waxes in proportions of

1–50%. Branched-chain and olefinic hydrocarbons are present in insignificant quantities. Glycerides (fats), phytosterols, terpenes, resins, long-chain carbonyl compounds, and flower pigments are also present in very small proportions. Because of the complex mixtures involved and the relatively simple analysis techniques of earlier years, the inconsistency of literature data on the composition of natural waxes is understandable.

Modern analytical instruments are required for complete analysis of a natural wax. High-temperature capillary GC, possibly coupled to a mass spectrometer, is currently the best analytical method.

Beeswax and carnaúba and candelilla waxes have a total annual market volume of ca. 28 000 t worldwide. These are economically the most important natural waxes [17].

Among the large number of other known waxes, ouricury, esparto, bamboo, and Japan wax (actually a special vegetable tallow) are of regional or local importance. Their availability is estimated to be 3000–4000 t/a [17], [18].

In all areas of application for waxes, recent natural waxes are now mostly used together with modified fossil and synthetic waxes. Applications include preparations for cleaning, polishing, and preserving floors, furniture, and car bodywork surfaces; the candle industry; craft work; production of pharmaceuticals, cosmetics, and confectionery (observing appropriate legal approval); metal and ceramic sintering technology; coating techniques; the paint and coating industries; and the production of carbon paper compounds. The latter, which was important in terms of quantity, has been replaced almost completely by new duplicating processes. Despite the market recession this has caused [19], an annual growth of ca. 4% is expected in the U.S. market for natural and synthetic waxes [20].

Special, ion-free or low-ion natural wax raffinates have also been used recently in electronics.

In his standard book *The Chemistry and Technology of Waxes*, A. J. WARTH included a comprehensive chapter on all known recent natural waxes. The second edition of the book was published in 1956 and is currently out of print [21].

2.2. Vegetable Waxes

In contact with the environment, plants form a compressed epidermis on their surfaces by depositing secondary layers of cellulose, except in the case of inflorescence [22]. This mechanical reinforcement is supplemented by deposition of cutin in the outer epidermal layers and an outer coating of cutin (cuticula) to regulate water content. As a result of climatic conditions, many plants in tropical regions also store waxes in cutinized membrane layers in the cuticula as an additional protection against evaporation of water. In many cases, waxes exude outward and form wax coatings, which can be up to several millimeters thick. Xerophytes, plants found in exceptionally arid regions, are however not the greatest wax producers [23]. Specific soil types, temperature, dry periods, and heavy rainfall, as well as temporal changes in these, are responsible for enhanced wax formation. Vegetable waxes can be classified according to their origin as deciduous and coniferous tree waxes, shrub and herb waxes, and grass and flower waxes, and within these groups into leaf, needle, stem, and root waxes; bark and skin waxes; and seed and fruit waxes.

2.2.1. Carnaúba Wax

The leaf wax of the carnaúba palm is by far the most important recent vegetable wax both economically and with regard to applications [21], [22], [24], [25]. The carnaúba palm is a fan palm found mainly in Brazil and given the botanical name *Copernicia cerifera* by v. MARTIUS in 1819 in honor of the astronomer COPERNICUS. It was apostrophized as the life tree of Brazil by v. HUMBOLDT.

History. The existence of carnaúba wax was first reported in 1648 [26], and some data concerning its composition were published in 1811 in *Transactions of the Royal Society, London*.

In 1836 the Brazilian MACEDO arranged for chemical investigations to be carried out on carnaúba wax at the Sorbonne in Paris. He began small-scale extraction and purification of carnaúba wax in 1856. In 1862, 1280 t of carnaúba wax was exported from Brazil.

Occurrence and Isolation. The carnaúba palm occurs mainly in north eastern Brazil in the provinces of Bahia, Rio Grande do Norte, Paraiba, Pernambuco, and particularly Ceara and Piaui, where there is a total stand of ca. 90×10^6 palms in a surface area of 10^6 km² [27], [28]. Smaller areas exist in southern Brazil, northern Argentina, Paraguay, and Bolivia. Different climatic conditions in other tropical and subtropical countries, in which carnaúba palms were planted as an experiment, allow satisfactory growth but markedly reduced formation of wax deposits on the leaves.

The palms grow to a height of 6 – 12 m, occasionally to 20 m, and are said to achieve a lifetime of up to 200 years. When the palm has reached an age of ca. eight years, one leaf gives a wax yield of 5 – 7 g, which is worth harvesting. The annual yield of wax per palm is ca. 150 g.

To isolate the wax, ca. 20 – 30 leaves at different stages of development are removed from each palm twice a year (September and December) [21].

The unopened heart leaves, of which only a few are removed per palm, carry particularly light-colored wax (prime yellow). Wax on the outer palm leaves is yellow, gray-green, or gray-brown, depending on climatic conditions during the vegetation period and on leaf age.

After drying the leaves in the sun or on steam-heated racks, the wax is chipped off manually or nowadays mostly by machine. The primary purification process involves melting over water and gives a crude wax that can be purified further by centrifugation, filtration with additives, or solvent extraction. Additional bleaching with hydrogen peroxide is also carried out.

The central ribs are often removed from the leaves before drying. The dried leaves are then chopped and the chips separated by air classification to give a mixture of ca. 60% wax and 40% chopped plant, which is stirred with water and oxalic acid to form a paste. The latter is heated to its boiling point and pressed through a filter while hot. Centrifugation of the filtrate gives virtually anhydrous wax. The remaining wax is extracted from the dried filter cake with aliphatic hydrocarbons (e.g., heptane).

The crude product is normally supplied in lumps that are obtained by breaking up the wax blocks into which the melt was cast. Carnaúba wax is also sold now in other forms.

Types. Internationally recognized classifications and specifications for genuine, unfalsified types of carnaúba wax were worked out by the American Wax Importers and Refiners Association according to established methods of determination [29].

The classification used in Brazil, shown in Table 2, has been valid worldwide since the beginning of the 1980s.

The highest-value type, *T 1*, is a prime yellow wax purified by filtration. To obtain *T 3*, suitable crude waxes (colored) are filtered and then bleached with hydrogen peroxide. The *T 4* wax is exported predominantly in the form of centrifuged or filtered variants.

Chemically refined or derivatized carnaúba waxes are no longer available commercially because of lack of demand [17]. Chemical refining processes, such as heating with alkali in the presence of paraffins, gave carnaúba wax residues in addition to the corresponding raffinates [30]–[36]. Where these processes are used, they account for <1% of the total volume of carnaúba wax.

Properties. Carnaúba wax is one of the hardest and highest-melting natural waxes. Because of its fine crystalline structure, its fracture edges are smooth and have a macroscopically homogeneous structure. At room temperature the wax has a weakly aromatic odor and a characteristic haylike scent (similar to coumarin) in the molten state [37]. Carnaúba wax is compatible with almost all natural and synthetic waxes and a number of natural and synthetic resins. It is readily soluble in most nonpolar solvents on warming [21] and is miscible with them in all proportions above its *mp*. On cooling, the wax precipitates from solution to form a solid paste. Carnaúba wax is only partially soluble in polar solvents, even on warming, and is generally sparingly soluble at room temperature (0.15–0.6%) [21].

Melting, solidification, and drop points as well as hardness of other waxes increase on addition of small quantities of carnaúba wax. With paraffins, the addition of carnaúba wax suppresses their tendency toward coarse crystallinity and reduces possible tack in the case of soft waxes.

Table 2. Carnaúba wax types and specifications

Characteristic	T 1 (prime yellow)	T 3 (fatty gray/ light)	T 4 (fatty gray/ crude)	T 4 (fatty gray/ centrifuged)	T 4 (fatty gray/ filtered)
Minimum mp, °C	83.0	82.5	82.5	82.5	82.5
Minimum flash point, °C	310	299	299	299	299
Moisture content, %	0	0	2.0 (max.)	1.0 (max.)	0.1 (max.)
Minimum acid number, mg KOH/g	2	4	4	4	4
Maximum acid number, mg KOH/g	6	10	10	10	10
Minimum saponification number, mg KOH/g	78	78	78	78	78
Maximum saponification number, mg KOH/g	88	88	88	88	88
Ester number, mg KOH/g	75–85	75–85	75–85	75–85	75–85
Maximum ash content, %	0.04	0.15	0.20	0.20	0.10
Solidification point, °C	80	79	79	79	79
Drop point, °C	84	83	83	83	83
Minimum particle size, mesh	95	95	95	95	95

Composition. Many analyses of carnaúba wax has been carried out, with conflicting results [21], [23], [24]. According to a thorough investigation, carnaúba wax prime yellow (T 1) has the composition given below [38]:

Aliphatic esters	40.0 wt%
Diesters of 4-hydroxycinnamic acid	21.0 wt%
Esters of ω-hydroxycarboxylic acids	13.0 wt%
Free alcohols	12.0 wt%
Diesters of 4-methoxycinnamic acid	7.0 wt%
Free aliphatic acids	4.0 wt%
Free aromatic acids	1.0 wt%
Hydrocarbons (paraffins)	1.0 wt%
Free ω-hydroxycarboxylic acids	0.5 wt%
Triterpene diols	0.5 wt%
Unsaponifiable components	56.4 wt%
Saponifiable components	39.2 wt%
Aromatics and/or resins	4.4 wt%

The aliphatic esters contain monocarboxylic acids of average chain length C_{26} and monohydric alcohols of average chain length C_{32}. The ω-hydroxyesters contained are mixtures of ca. 90% esters of ω-hydroxyacids (C_{26}) and monohydric alcohols (C_{32}) and 10% esters of monocarboxylic acids (C_{28}) and α,ω-diols (C_{30}).

Esters containing 4-hydroxy- and 4-methoxycinnamic acids are present mainly as oligomers and polymers. The monomer units of these are diesters of the above-named cinnamic acids with mono- and polyhydric alcohols and ω-hydroxycarboxylic acids.

The free alcohols are similar in composition to those in aliphatic esters. The wax also contains a small proportion of secondary alcohols.

Table 3. Chain-length distributions for straight-chain monocarboxylic acids and monohydric alcohols, and paraffins in carnaúba wax

Chain length (C-number)	Acids*, % (approx.)	Alcohols*, % (approx.)	Paraffins*, % (approx.)
16	2.0		
18	3.0		
20	8.5		0.4
21			0.5
22	7.5	0.5	0.7
23			1.3
24	23.2	0.7	1.6
25			2.6
26	10.5	0.5	3.5
27			8.7
28	16.5	2.0	5.4
29			13.1
30	5.5	10.6	13.9
31			29.3
32	2.0	60.8	9.6
33			7.3
34	0.5	16.4	1.5
35			0.3
36		0.6	0.4

* Values based on 100% total acid, total alcohols, and total paraffins in each case (deviations from 100% = sum of all other acids and alcohols).

The even-numbered *n*-monocarboxylic acids, *n*-monohydric alcohols, and *n*-alkanes present in all carnaúba wax types have typical, distinctive chain-length distributions (see Table 3).

The important difference between Type 1 carnaúba wax and the other types is that the latter contain no triterpenediols and the esters of 4-hydroxy- and 4-methoxycinnamic acids have higher degrees of polymerization compared with Type 1. This may be attributed partly to differences in climate but also to the fact that Type 1 is harvested from genetically younger leaves, whereas the others come from older leaves that have been exposed to more sunshine.

Uses. *T 4 Variants.* The wide use of carnaúba wax in the production of *preserving and cleansing agents* for floors (floor polishes, ionogenic and nonionogenic day-bright emulsions, wiping waxes and cleaners, sealing emulsions, spray cleaners, etc.); and polishes for furniture, cars, and shoes (solvent–water or straight solvent products) is based on its ability to form pastes and glossy and repolishable films and its ability to be dispersed. In this classical area a large proportion of T 4 variants of carnaúba wax are used.

Type 1 carnaúba wax and some of Type 3 (after absorptive purification to remove peroxide components arising from hydrogen peroxide bleaching) are approved for use in the *pharmaceutical industry* (e.g., as polishes for pills) and the *cosmetics industry* (as lipsticks and lip salves).

In the *food sector*, T 1 and T 3 waxes are used as release agents for bakery and confectionery products and as additives in chewing gum production.

In *polymer processing*, carnaúba wax is used in release agent preparations and, to a small extent, as a lubricant. The *varnish industry* uses carnaúba wax as an additive in the interior coating of food containers [17].

In the *leather industry*, carnaúba wax is used for cutting waxes, cleaning waxes, and edging inks. It is also approved for coating citrus fruit.

Economic Aspects. About 10 000 – 16 000 t/a carnaúba wax is available worldwirde, of which ca. one-fifth is Type 1. Brazil, the sole exporter, supplies mainly Europe (ca. 2000 t/a), the United States (3000 – 4000 t/a), and Japan [17]. About 3000 – 3500 t/a is for indigenous supply.

Carnaúba wax consumption in the United States decreased markedly from 1955 to 1976 and then remained constant [18]. Brazil is increasingly exporting purified or bleached products in flaked form to achieve higher prices. Harvests, exports, and price structuring are subject to government control.

Waxes with Similar Compositions. *Copernicia australis*, a palm that produces large quantities of leaf wax, is found in Paraguay and around its border with Argentina (Gran Chaco). It is similar to *Copernicia cerifera* in many respects. A few small factories for isolating this palm wax are said to have been erected since 1945 [39].

The Brazilian *cauacu bush*, found mainly in Pará province and by the upper Amazon, produces a wax that is said to be similar to carnaúba wax in chemical composition. However, the wax is isolated only sporadically and to a limited extent [40].

2.2.2. Candelilla Wax

Occurrence and Isolation. The climatic conditions of the semideserts in southern California, Arizona, south western Texas, northern Mexico, and parts of Central and South America favor enhanced wax formation in *Euphorbia* (*E. cerifera*, *E. antisyphilitica*) and *Pedilanthus* (*P. pavonis*, *P. aphyllus*) species. The wax deposited on the stalks and leaf stems of these plants, which grow as bushes or shrubs, is known as candelilla wax [21], [23], [41], [42].

Numerous and dense stands of these plants are found mainly in northern Mexico in the states of Chihuahua, Coahuila, Durango, Nuevo León, San Luis Potosi, Tamaulipas, and Zacatecas over a total area of more than 100 000 km^2. These are the only areas in which candelilla wax is obtained [43].

In manual harvesting the plants are pulled out, leaving most of the root system in the soil. Together with scattered seeds, roots remaining partially in the soil guarantee the growth of new stands, which takes several years.

In collecting and processing locations, plant material is boiled with ca. 0.2 % sulfuric acid in open vessels, which are heated using already dewaxed and dried stalks. Metal grids keep the plants under the surface of the liquid, and addition of sulfuric acid inhibits emulsification of the wax. The latter collects in molten form on the surface of the liquid. After the wax is skimmed off into open vats or barrels, it solidifies to form the crude wax known as cerote. Yields based on plant material are 3–4%.

Wax from outer areas of the solidified wax blocks is removed with impurities that have concentrated there and is subjected to the boiling process again. The remainder of the wax is purified in a central refinery located in Saltillo by melting with sulfuric acid and filtering in filter presses. The crude wax broken down into lumps is the Mexican commercial product (Mexican Standard Grade Candelilla Wax).

Candelilla wax raffinates are obtained from crude wax by adsorptive purification or hydrogen peroxide bleaching. They are produced and sold in the United States and Europe. The properties of these products, apart from color, are identical to those of crude candelilla wax.

Following the transfer of Mexican candelilla wax production from state monopoly to private enterprise, candelilla wax raffinates produced in Mexico will also be put on the market [17].

By deresinification and subsequent oxidative chromic acid bleaching, products that are almost white can be obtained. Because of the high price of genuine wax, these processes have no practical importance.

Properties and Composition. Candelilla wax is a hard, brittle wax that is very similar to carnaúba wax with regard to solubility in polar and nonpolar organic solvents. A large proportion of the resin present in the wax can be extracted at room temperature with hot 70% ethanol, acetone, or low-boiling, chlorinated organic solvents.

The *physical and chemical properties*, the color (brown – brownish yellow – pale yellow), and the degree of purity vary depending on climate, time of harvest, region, and age of harvested plants. The recognized specification for genuine crude candelilla wax is given in the following [43].

mp	68.5 – 72.5 °C
Refractive index	1.4550 – 1.4611
d_4^{15}	0.950 – 0.990
Acid number	12 – 22 mg KOH/g
Saponification number	43 – 65 mg KOH/g
Hydrocarbon content	30.6 – 45.6%
Total acid content	20.6 – 29.0%
Color	amber
Flash point	235.4 – 248.4 °C
Components insoluble in xylene/toluene (1:1)	0.0 – 0.1%

The data agree with those previously laid down by the American Wax Importers and Refiners Association for prime crude candelilla wax [29].

The average *composition* of candelilla wax gives below is derived from the large number of values found in the literature.

Hydrocarbons (ca. 98% paraffins + 2% alkenes)	42.0 wt%
Wax + resin + sitosteroyl esters	39.0 wt%
Lactones	6.0 wt%
Free wax and resin acids	8.0 wt%
Free wax and resin alcohols (terpene alcohols)	5.0 wt%
Saponifiable components	23.0–29.0 wt%
Unsaponifiable components	71.0–77.0 wt%

Candelilla wax differs fundamentally from carnaúba wax in its high hydrocarbon content of 45% (max.) and resin content of 20%. Whereas carnaúba wax contains >80% wax according to the classical chemical definition, candelilla wax consists of <35% of these components.

As in all natural waxes, in candelilla wax the paraffins and wax acids or alcohols with even-numbered carbon chains have typical and distinctive chain-length distributions (Table 4).

Uses. Candelilla wax is used in classical *cleaning preparations and polishes* (e.g., as a component of shoe and other leather polishes) and in furniture, car, and floor polishes. Because of its particular composition, this wax can be used in the applications only in combination with other suitable natural and synthetic waxes. Its use in floor polish is very limited because the high resin content leads to increased dirt pickup.

Candelilla wax is also used in the production of *candles*, *coatings* for paper and cardboard, *hotmelt adhesives*, and in *polymer processing*.

It is also approved for use in the *cosmetics, pharmaceutical,* and *food industries*. The main use of candelilla wax, principally in the United States and Europe, is by the cosmetics industry in the production of lipstick. Here, properties such as outstanding framework formation with castor oil, gloss, and uniform shrinkage on cooling (very good demolding after casting lipsticks) are very important [17]. Candelilla wax is also used as a polish for pills and in the production of chewing gum and confectionery.

Economic Aspects. For nomadic peasants in the areas of Mexico mentioned, proceeds from the candelilla harvest were an important part of their income for a long time. Because of continuing improvements in infrastructure, the uneconomical isolation of candelilla wax has decreased since 1950 and been replaced increasingly by agricultural production. Exports to the main consumer, the United States, fell from 2500 t in 1950 [23] to 1800 t in 1974 [42], 1100 t in 1980, and ca. 310 t in 1992 [44]. About 800–1000 t of candelilla wax is currently available on the world market, of which ca. 200 t is consumed in Europe [17].

In the United States, demand for candelilla wax is increasing because of the shift of use from classical areas to those of foods, cosmetics, and pharmaceuticals [44]. Privatization of candelilla wax isolation in Mexico and efforts to produce raffinates there will probably lead to greater availability of these waxes.

Table 4. Chain-length distribution of even-numbered carboxylic acids and alcohols, and paraffins in candelilla wax

Chain length (C-number)	Acids *, % (approx.)	Alcohols *, % (approx.)	Paraffins *, % (approx.)
22	0.3		
23			
24	0.1		
25			
26	0.3	2.4	0.1
27			0.2
28	3.2	17.4	0.5
29			5.6
30	36.4	58.0	1.3
31			80.0
32	48.6	19.0	1.8
33			9.0
34	8.0		0.2

* Values based on resin-free total acid, total alcohol, and total paraffin (deviations from 100% = sum of all other acids, alcohols, and hydrocarbons).

2.2.3. Ouricury Wax

Occurrence and Isolation. Ouricury wax is deposited on the leaf stems of feather palms (*Syagros coronata*), which grow mainly in the Brazilian state of Bahia. There it is also known as urucury, licuri, aricuri, nicuri, and coqueiro.

The isolation process for ouricury wax is very similar to that used for carnaúba wax, but it is made more difficult by strong adhesion of the wax to the plant, and mechanical scratching or scraping off is necessary. The annual yield per palm is ca. 1 kg of crude wax.

Preliminary purification involves melting crude wax in hot water and filtering the melt through filter cloths or suitable sieves [21], [23], [45]. Additional purification in filter presses is used to obtain *USA Pure Refined* wax. The crude wax is sold in lumps (Brazilian Crude).

Properties and Composition. Ouricury wax resembles carnaúba wax very closely with regard to its hardness, gloss, solubility in polar and nonpolar solvents, and hardening capacity when used in combination with soft waxes. Its color is somewhat darker than that of dark carnaúba wax types. The *mp* is 82.5 °C. The most important properties for genuine commercial products have been established [29]. Thorough investigation of ouricury wax gave the following composition [46]:

Simple aliphatic esters	23.5 wt%
Monoesters of hydroxycarboxylic acids	22.4 wt%
Diesters of hydroxycarboxylic acids	17.2 wt%
Acidic hydroxypolyesters	5.4 wt%
Free acids	8.7 wt%

Free alcohols	3.0 wt%
Resins	14.8 wt%
Water	1.4 wt%
Ash	0.4 wt%

Alkaline saponification of ouricury wax yields varying quantities of total acids (69–74%) and unsaponifiable components (26–31%), as usual in the case of natural waxes. Despite these facts, ouricury wax, like other vegetable waxes, can be characterized definitely by the chainlength distributions of its n-paraffin and acids and alcohol components.

Uses and Importance. Ouricury wax was previously used in floor and shoe polishes and in carbon paper coloring agents, instead of carnaúba wax, mainly because of its low price. As a result of increasing price, the importance of ouricury wax has decreased sharply. According to Brazilian statistics, 39 t was exported in 1980. In 1991 the consumption in the United States was ca. 4.5 t/a [47].

2.2.4. Sugarcane Wax

Occurrence and Isolation. Sugarcane wax is formed as a powdery, pale to dark yellow deposit on sugarcane stalks (particularly in the thickened nodes) in quantities of 0.1–0.25% based on cane weight, depending on country of origin. The main sugarcane-growing countries are India (ca. 96×10^6 t/a), Brazil (ca. 80×10^6 t/a), and Cuba (ca. 40×10^6 t/a).

Sugarcane wax is generally extracted from the dried filter cake (ca. 1% based on sugarcane) [21] formed during sugarcane processing by using alcohols, carbon disulfide, carbon tetrachloride, or aromatic and aliphatic hydrocarbons [48].

The crude wax content of the filter cake (in weight percent) varies between the following limits, depending on sugarcane origin and method of processing:

India	8.0–18.0 wt%
Louisiana	4.4–17.9 wt%
Puerto Rico	12.0–14.0 wt%
Hawaii	9.6–11.0 wt%
Philippines	10.5–11.0 wt%
Cuba	12.4–22.0 wt%
Java	5.0–15.0 wt%
Republic of South Africa	6.9–14.6 wt%
Argentina	11.9–15.5 wt%
Brazil	8.9–17.8 wt%

Crude wax can be freed from most glycerides (fats and oils) by treatment with acetone, methyl ethyl ketone, methanol, ethanol, 1-propanol and isopropanol, 1-butanol, or ethyl acetate. In a subsequent recrystallization step, preferably in isopropanol, resinous and colored components can be removed. Further treatment with bleaching earths or oxidative bleaching of the wax melt with air gives waxes ready for use in a variety of applications. Bleaching processes for prepurified wax using potassium chlorate – sulfuric acid or chromic acid – sulfuric acid would be effective but are not used for cost reasons.

Properties and Composition. *Crude sugarcane wax* is a black-brown to green-brown semisolid with an unpleasant, rancid odor.

Semi- and completely purified products are brown- to yellow-colored hard waxes, whose characteristics vary considerably. Melting points of 68 – 81 °C, acid numbers of 7 – 22 mg KOH/g saponification numbers of 32 – 65 mg KOH/g and proportions of unsaponifiable material of 52 – 62 % have been published in the literature. The following values are typical for a semiraffinate [21]:

Wax esters (aliphatic and sterol esters)	78 – 82 wt %
Free fatty and wax acids	ca. 14 wt %
Free alcohols (wax alcohols and sterols)	6 – 7 wt %
Hydrocarbons	3 – 5 wt %

On comparing crude wax with a raffinate sample [49] in terms of chain-length distributions of the even-numbered acids and wax alcohols and of the paraffins, characteristic differences can be seen, particularly with the acids.

Importance and Uses. From an estimated annual worldwide harvest of ca. 510×10^6 t of sugarcane, ca. 5.1×10^6 t/a of dry filter cake material is theoretically obtained. From this, ca. 510 000 t of crude sugarcane wax could be extracted, given an average wax content of ca. 10 % in the filter cake. If refining losses are assumed to be 50 %, ca. 255 000 t of sugarcane wax raffinate per year remains theoretically available worldwide.

In the past, many attempts have been made to produce crude sugarcane wax on an industrial scale. As early as 1841, isolation of the wax from filter cakes was begun, and in 1909 a corresponding extraction process was patented.

Around 1916, crude sugarcane wax was extracted as a commercial product in South Africa and Java. In 1958 the wax was produced in two Cuban and one Australian plant. However, in 1960, production in Australia was halted and only one plant was still operating in Cuba [48]. In India, an extraction plant is thought to have been operating continuously since 1950 [48]. This plant is attached to a sugar refinery with a daily throughput of 1200 t of sugarcane.

Difficulties resulting from the necessary centralization of production locations, the necessarily rapid and complete drying of large quantities of filter cake material (which has a tendency to ferment), and last, but not least, the increasing costs of extraction have made sugarcane wax raffinate incapable of competing with carnaúba wax on the world market in terms of price. Consequently, sugarcane wax raffinates have only regional importance.

In producer countries, crude sugarcane wax can be used in floor and shoe polishes, carbon paper production, impregnating agents, and production of wax compounds.

2.2.5. Retamo Wax

Occurrence and Isolation. In arid areas of Argentina, *Bulnesia retama* grows wild as trees and shrubs. Branches and twigs are coated with a wax deposit that can be collected mechanically after drying the plant material harvested in summer. The crude wax is purified by melting with dilute sulfuric acid and subsequent filtration through cloths. It is sold exclusively on the home market as retamo wax [21].

Properties and Composition. Retamo wax is a light to medium brown, odorless, hard wax. It is very similar to other vegetable waxes, such as carnaúba wax, with regard to its solubility in polar and nonpolar organic solvents.

Data found in the literature, such as *mp* (76–78 °C), acid number (ca. 49 mg KOH/g), saponification number (ca. 87 mg KOH/g), proportion of hydrocarbons (15–25%), and content of unsaponifiable components (38–42%), characterize the wax less precisely than the typical chain-length distributions of even-numbered wax acids and alcohols and of paraffins [49].

Uses and Importance. With a decreasing annual production of several hundred tonnes (1975, ca. 500 t/a), retamo wax is of limited importance on the Argentinian market. It is used for floor, car, and shoe polishes and in the production of colored polishing inks.

2.2.6. Jojoba Oil

According to the DGF definition, jojoba oil is not a wax. However, its chemical composition allows it to be classified as a "liquid wax" in the sense of the classical definition of waxes.

Occurrence and Isolation. The jojoba bush (*Simmondsia chinensis* = *Simmondsia californica* = *Buxus chinensis*) grows wild over an area of ca. 250 000 km^2, most of which lies in the Sonora desert. The population density varies from a few to 500 plants per hectare. In certain areas, millions of jojoba bushes can be found.

The bushes carry fruit capsules that can contain one to three nuts of differing sizes. The yield from the harvest of wild jojoba bushes can range from a few nuts to 14 kg of nuts per plant. The average yield is ca. 1.8 kg per bush.

The nuts contain 50–60 wt% oil, which can be extracted almost completely from the deshelled, granulated fruit. Around 35–40% of a particularly high-value oil is obtained by cold pressing and a further 15–20% of a lower-value oil by extraction of the pressed cake. The oil is purified by simple filtration [50].

Properties and Composition. Jojoba oil is a virtually colorless to golden yellow, odorless, unsaturated oil, for which the most important characteristics are as follows:

mp	6.8–7.0 °C
Acid number	2 mg KOH/g
Saponification number	92 mg KOH/g
Iodine number	82 g I$_2$/100 g

A highly crystalline, relatively hard wax is obtained from jojoba oil by hydrogenation. Saponification of the oil gives 51 % unsaponifiable material and 52 % total acid.

Jojoba oil has the following composition with regard to substance groups:

Wax esters	ca. 97.0 wt %
Free alcohols	ca. 1.1 wt %
Free acids	ca. 1.0 wt %
Sterols	ca. 0.9 wt %

Unlike other vegetable waxes, the acids and alcohols in jojoba oil are monounsaturated, straight-chain representatives of the chain-length range C_{16}–C_{24}. Among the acids the C_{18} (ca. 12.5 % of the total acids), C_{20} (ca. 67 % of the total), and C_{22} components (ca. 14.5 % of the total) predominate, and among the alcohols, the C_{20} (ca. 48 % of the unsaponifiable components) and C_{22} components (ca. 41 % of the unsaponifiable components). With this unusual composition, jojoba oil is very similar to sperm oil contained in the skull bones of the sperm whale.

Uses and Importance. An extraordinarily high market demand exists for a sperm oil substitute, which can be satisfied to a great extent by jojoba oil. This demand is a result of the worldwide prohibition of whaling, even though the latter is not obeyed by all countries.

The first experimental jojoba plantations were begun in the United States (Arizona, Texas, and California) to effect infrastructural improvements. Many other jojoba oil plantations were started in Mexico, Argentina, Australia, Brazil, Africa, Egypt, Israel, and countries whose climatic conditions allowed desert farming.

Because management of these plantations is difficult and the time required to achieve economical harvests is relatively long (eight to ten years), jojoba oil is available in insufficient quantities at a very high price. Its use is therefore limited to cosmetics and pharmaceuticals. As soon as reliable quantities can be supplied at acceptable prices, jojoba oil may attract extraordinarily great interest as a raw material, particularly in the production of high-performance lubricants and in modified form in plasticizers and stabilizers. *Hydrogenated jojoba oil*, which corresponds to fatty esters such as stearyl stearate in its properties, will remain noncompetitive with these esters in terms of price and will therefore not achieve market importance.

2.2.7. Other Vegetable Waxes

Many other vegetable waxes are known, and the composition of some of them has been investigated [21]. However, they are either of very limited local importance or of purely academic interest. These waxes include

Caranday wax *
Raffia wax *
Columbia wax *
Esparto wax
Alfalfa wax
Bamboo wax
Hemp wax
Douglas fir wax
Cork wax
Sisal wax
Flax wax
Cotton wax
Dammar wax
Cereal wax
Tea wax
Coffee wax
Rice wax
Ocatilla wax
Oleander wax

* Palm waxes

Japan tallow wax, which is available commercially, fulfills some application-oriented criteria as a wax but, because of its chemical nature, is a vegetable tallow and thus belongs to the group of other tallows such as myrica, bayberry, ucuhuba, Borneo, Malabar, and illipe.

2.3. Animal Waxes

Animal waxes are obtained mainly from bees (Apoidae), particularly the honeybee (Apidae), stingless bee (Meliponinae), and bumblebee (Bombinae). Some scale insects (Coccidae) also produce waxes.

Waxes obtained from land mammals (sheep, goats, llamas, and dromedaries) and marine animals, as well as those of microbiological origin, are also known.

2.3.1. Beeswax

Beeswax is the oldest natural wax used by humans and still the most important animal wax.

Occurrence. Beeswax is an end product of the metabolism of a honeybee class (*Apis mellifica, A. carnica*), which belongs to the *Apis* genus. This species of honeybee is found worldwide and domesticated to a high degree. Wax is secreted from the wax glands in the lowest abdominal segments of the worker bee as white, slightly transparent flakes (virgin wax). From there it is taken to jaw, plasticized by chewing, and then used in this form as a building material for incubating combs and honeycombs.

If plenty of food is available and honey production is increasing, the wax glands are stimulated to enhance production for building new honeycombs. If the beekeeper makes preformed honeycombs available to the bees, honey production is increased [22], [45], [51], [52].

A bee population of ca. 30 000 worker bees produces ca. 270 g of wax from 3 kg of sugar or 3.6 kg of honey in one-fifth of a generation — ca. 50 d. A bee population consumes ca. 50 kg of pollen and 62 kg of honey per year.

Beeswax is produced in almost all countries of the world. In Europe, wax produced in a particular country is consumed there and supplemented by imports. Suppliers are African countries, such as Tunisia, Morocco, Kenya, Tanzania, Zambia, and the Central African Republic. North America, Australia, New Zealand, Russia, and the People's Republic of China also export beeswax.

Isolation. To obtain wax, the combs are freed from honey as far as possible by centrifuging. Then the wax is isolated either thermally by using sunlight or with boiling water. Particularly heavy impurities become enriched in the aqueous phase. Insoluble impurities remaining in the wax melt can be removed by filtration through a fine mesh sieve of filter cloths, or by pressure filtration or centrifuging.

Colored, wax-soluble impurities can often be removed by treating the wax melt with activated carbon, aluminum or magnesium silicates, or other bleaching earths. As in pressure filtration and centrifugation, wax-containing residues are formed from which lower-quality waxes can be isolated by solvent extraction. The oldest method of lightening the color, which is said to have been practiced in 1000 B.C., involves bleaching beeswax, that has been processed to thin strips or shavings (large surface), in air and sunshine.

In the aforementioned purification processes, beeswax remains chemically unchanged and its characteristic, honeylike odor is retained. The success of the process depends mainly on the source of the beeswax.

Posttreatment of beeswax is now limited to washing with water or adsorptive purification with bleaching or diatomaceous earths. Bleaching with hydrogen peroxide is used out because peroxy compounds are formed [17].

If the process is carried out skillfully, chemical bleaching processes, such as refining with chromic acid – sulfuric acid, give very light, almost chemically unchanged products, which have lost the aromatic odor of the starting material. However, these processes are not employed. The same holds true for those involving mild bleaching with permanganate and sulfuric acid or the use of chlorine-containing bleaching agents such as chlorine bleach, sodium chlorite, and chloramine. The latter give waxes that do not have stable colors and often contain chlorine.

Beeswax is sometimes derivatized by ethoxylation or esterification. The products obtained do not have any market importance.

Properties. Crude, mechanically purified beeswax contains, besides wax, flower and pollen pigments and propolis resin, which acts as a cement in honeycomb construction. Therefore, depending on the source, the wax can have a yellow, orange, or dark brown color.

Beeswax is moderately hard. It becomes plastic and kneadable on warming in the hands, without sticking. It has a noncrystalline, finely grained fracture pattern. On cutting, matt surfaces are produced. Beeswax has a honeylike odor that intensifies on melting. It is moderately soluble in polar and nonpolar organic solvents in the cold, and completely soluble when heated to its boiling point.

Chemically bleached beeswaxes are cream to ivory in color and practically odorless and tasteless. They have a somewhat harder and more brittle consistency than chemically untreated products.

The following limiting values have been established for the properties of mechanically purified and bleached beeswax [29]:

mp	62–65 °C
Acid number	17–24 mg KOH/g
Ester number	72–79 mg KOH/g
Ester number/acid number	3.3–4.2
Saponification turbidity point	65 °C

Extensive investigations show that properties vary even among waxes from the same source [53].

Composition. The following values are given in the literature for the approximate composition of beeswax with regard to wax classes:

Wax esters	70–80%
Free acids	10–15%
Paraffins	10–20%

Alkenes, isoparaffins, cholesteryl esters, and pollen pigments are present in minor amounts [21], [54].

Since the typical data for beeswax vary between relatively wide limits, misrepresentation of beeswax is difficult to detect by using these data [55]. A large number of publications exist on the analysis of beeswax [56]. Beeswax can be separated into the various substance classes by chromatographic methods [57]–[62]. Beeswax of different origin has been shown to have similar composition.

Comparative GC studies, in which ca. 65% of the waxes investigated are analyzed (residues mostly involatile di- and triesters), also show great similarity among beeswax of widely varying origin [63] (see Table 5).

Table 5. Composition of beeswax of different origins *

Components	Eastern Switzerland	Brazil	China	Canada	Czech Republic
Hydrocarbons, %	14	15	16	14	13
Fatty (wax) acids, %	12	14	9	(8)–12	13
Fatty alcohols, %	1	1	1	1	
Palmitates, %	22	20	24	35	31
				35**	31**
Oleates, %	4	3	5	35	31
Hydroxymonoesters, %	12	12	11	(4)–8	13
Total, %	65	65	66	(62)–70	70

* Values in parantheses were found only in a few cases.
** Palmitates and oleates

Table 6. Percentages of individual substance classed in a hydrolyzed wax sample from the honeybee *Apis mellifica* (Switzerland)

Chain length (C-number)	Hydrocarbons	Fatty acids	Hydroxy fatty acids	Fatty alcohols
16		17.98	11.60	
17				
18		2.44	0.71	
19				
20		0.15		
21				
22		0.33	1.66	0.12
23	0.25			
24	0.10	4.94	0.59	5.59
25	0.58			
26	0.11	1.75	1.38	3.37
27	4.42			
28	0.08	1.78	0.10	4.37
29	2.64			
30	0.07	1.58		9.58
31	3.27			
32		1.05		7.98
33	2.24			
34		0.94		0.88
35	0.12			
36		0.08		
Total	14.23	33.02	16.04	31.89

Parallel GC investigations on a genuine beeswax sample and the same sample after hydrolysis and silylation show that C_{22}–C_{36} wax acids are present predominantly as the free acid, while all C_{12}–C_{20} fatty acids are esterified [63].

Gas chromatography of hydrolyzed and silylated beeswax allows a total analysis in which 95% of its components can be identified [63] (Table 6).

Studies using GC show that a spectrum of wax samples from *Apis mellifica* has the same characteristic composition independent of the breed of bee. Small differences exist only in the percentage of individual components and substance classes, so that in general the following ranges of composition are correct for various classes of substance [63]:

Hydrocarbons	13 – 16%
Fatty acids	30 – 35%
Palmitic acid	15 – 20%
Hydroxy fatty acids	13 – 17%
Fatty alcohols	30 – 35%
Other substances	4 – 8%

The remaining substances are present mainly in concentrations of $<0.05\%$.

By using these facts, beeswax misrepresentation can be detected by a corresponding total analysis of the wax even in the presence of small quantities of blending materials [63].

Uses. Beeswax, mainly the yellow variety, is one of the raw materials used in the production of household *candles*, expensive decorative candles, and candles for religious purposes. For the latter, the Roman Catholic Church prescribes a minimum beeswax content of 10%. Beeswax is also used for artistic and handicraft purposes (wax flowers, wax sculptures, painting, and batik). Most bleached and purified beeswax is used in *cosmetics and pharmaceuticals* (to regulate the consistency of lipsticks, creams, ointments, and suppositories) and in the *food sector* (as a release agent and a polish in confectionery production). Its use in these areas is favored by corresponding licensing regulations.

A significant proportion of beeswax, ca. 40% of the total market volume, is sold back to beekeepers for the production of *synthetic honeycombs*, in which blends with paraffins and ceresin, which are not allowed for pharmaceuticals, are often used.

Economic Aspects. From the worldwide honey production figures for 1976, production of 11 000 – 19 000 t of beeswax was estimated. Whether the quantities used in the candle and beekeeping industries were included in the worldwide consumption figure of ca. 10 000 t for 1981, is not known. At that time, total imports for the Federal Republic of Germany amounted to 1000 t.

At present (1994), ca. 15 000 t of beeswax is available on the world market. Ca. 6000 t is used in industry, and ca. 4000 t is sold back to beekeepers [17].

The annual availability of beeswax depends strongly on climatic variations in exporter countries and bee diseases that break out occasionally. In 1993, for example, a marked drop in exports from the United States occurred because of catastrophic weather conditions in the Mississippi area [17].

2.3.2. Other Insect Waxes

The following waxes are not market products and only in some cases have limited regional importance in the countries of origin.

Ghedda Wax. Ghedda wax is produced by some types of wild bee bound in Asia (*Apis indica, A. florea, A. dorsata*). It was formerly exported as such but more frequently now is blended with beeswax from the East Indies.

Ghedda wax has a yellow-brown color and varying characteristic data (e.g., acid number 4–11 mg KOH/g, saponification number 86–130 mg KOH/g, iodine number 5–11 g I_2/100 g, *mp* 60–66 °C). It has a consistency similar to beeswax with a fatty feel, which can be attributed to its glyceride content. Ghedda wax consists of ca. 78–80% wax esters, 4% glycerides (fats), 5–6% free acids, 8–9% hydrocarbons, and 1–2% partly resinous impurities [21].

Among the acids are 7- and 16-hydroxypalmitic acids and 7,16-dihydroxymargaric acid, which can be used for identification of ghedda wax or other beeswaxes blended with ghedda wax in modern umpire assays.

Shellac Wax. The resinous exudate of the scale insect *Laccifer lacca* (formerly *Tachardia lacca*) of the Coccidae family is an important commercial product, known as shellac. Shellac contains 4–5% wax, which can be obtained by treatment with dilute soda solution or 90–95% ethanol (spirit extraction) as an insoluble component.

Shellac wax is a hard, yellow to brown product. Very light, bleached types are also known. The widely varying data found in the literature (e.g., acid number 12–24 mg KOH/g, saponification number 63–126 mg KOH/g, iodine number 6–9 g I_2/100 g, solidifying point 58–80 °C) may be attributed to the different sources of shellac and different methods of extraction and posttreatment of the wax component. Thus, the composition with regard to substance classes also varies as follows [21]:

Wax esters	70–82%
Free acids	10–24%
Free alcohols	<1%
Hydrocarbons	1– 6%
Shellac	1– 4%

Recent investigations [64] of bleached shellac waxes show that the total acid content of the wax, similar to that of beeswax, consists essentially of a mixture of even-numbered fatty acids (C_{12}–C_{18}, 21–26% of the total acid) and waxy acids (C_{28}–C_{34}, mainly C_{32} and C_{34}). A significant component is trihydroxypalmitic acid (aleuritic acid), even though it occurs only in quantities of 0.5–1.0%. Among the alcohols (C_{28}–C_{32}) the C_{28} component predominates (62–65% of the total alcohol). The main hydrocarbon components are paraffins with 27, 29, and 31 carbon atoms.

Chinese Insect Wax. The East Asian wax scale insect (*Ericerus pela*, subgenus Lecaniinae, family Coccidae) secretes a wax on the twigs and leaves of trees and shrubs on which it is a parasite. The wax can be extracted by melting with hot water. It is hard, brittle, crystalline, and odorless, with a light color, and consists of 95–97% wax esters, 0.5–1.0% free acids, and 2–3% hydrocarbons [21]. The high proportion of wax esters is significant. The East Indian wax scale insect (*Ceroplastes ceriferus*) is said to produce a wax similar to beeswax in composition [21].

2.3.3. Wool Wax

Sheep secrete a substance from their sebaceous glands to protect the epidermis and the wool from adverse weather conditions and from their own acidic perspiration. This substance is incorrectly referred to in the literature as wool fat. Because of its chemical composition it should definitely be classified as a wax.

Wool wax is obtained in emulsified form from raw wool by machine washing with soap liquor. Extraction of wool with suitable organic solvents is not employed because it would lead to complete degreasing of the wool and thus make it nonelastic.

Many processes exist for obtaining crude wool wax from the emulsion, some of them patented [45]. Wax is first precipitated along with the acids of the soap solution by addition of sulfuric acid. The fatty acids remain in the crude product regardless of whether the precipitates are pressed out, centrifuged, or extracted with solvents. They can be removed almost completely by dissolving the crude product (e.g., in hexane) and subsequent extraction with a mixture of alcohol and dilute aqueous sodium hydroxide. Further purification steps (e.g., bleaching with hydrogen peroxide in the presence of phosphoric acid [65]) give neutral wool wax, which is usually treated subsequently with adsorbents such as bleaching earth or activated carbon and sold in a highly pure form as Adeps lanae.

Crude wool wax is a greasy, glutinuous, brown-yellow to brown-black substance with a penetrating goatlike odor (*mp* 34–38 °C). *Neutral wool wax* is yellow to light brown in color (*mp* 38–42 °C), with a milder odor, whereas *Adeps lanae* is a pale yellow, almost odorless substance (*mp* 40–42 °C).

The *composition* of wool wax varies depending on the origin of the wool, breed of sheep, climatic conditions, farming method, and type of pasture.

Wool wax usually consists of a mixture of ca. 48% wax esters, 33% sterol esters, 6% free sterols, 3.5% free acids, 6% lactones, and 1–2% hydrocarbons [21]. The low *mp* and appearance of wool wax may be attributed to the high proportion of branched acids.

Uses. Wool wax is used mainly as the crude product, as neutral wool wax, or as hydrogenated neutral wool wax and in the form of wool wax alcohols (obtainable by hydrolysis) in leather polishes, cosmetics (e.g., creams, baby-care products, and toilet soap), pharmaceuticals (e.g., plasters, ointments, and suppositories), and lubricants. The purified, commercial product is known as wool wax or lanolin.

3. Montan Wax

3.1. Formation and Occurrence

Montan wax is a vegetable fossil wax. It forms part of the extractable, bituminous components of lignite and peat. The composition of bitumen depends on the carbonized plant material and on the geology and conditions of the deposit. The main components of bitumen are waxes, resins, asphaltenes, and dark residues. If the proportion of wax is $>60\%$, the bitumen is generally known as crude montan wax. Prerequisites for formation of this wax were the same or similar to those now necessary for formation of recent natural waxes (e.g., carnaúba wax [66]–[69]). Lignite, which originates from plant material of the Eocene period, has a high and economically utilizable proportion of wax components. Deposits of lignite, which are used for wax extraction, are found mainly in eastern Germany (Röblingen), Ukraine (Alexandrija), Russia (Baschkiren), the United States (California), and China. Other deposits are of only local importance or are not even exploited. The largest crude montan wax producers are the Montanwerke Romonta in Röblingen (Germany) and the American Lignite Co. in the United States.

3.2. Extraction

The currently most economically important bitumen-rich lignite deposits are found in the area around the city of Halle in Germany. An industrial process for extracting montan wax from this type of coal was based on investigations bei RIEBECK. The original patent [70], [71] for this process, published in 1880, describes how montan wax can be obtained by solvent extraction. In 1900 the production of montan wax was begun in Völpke, and later in other locations. Romonta in Amsdorf [71] is currently the most important production site. There, lignite with a bitumen content between 10 and 20% is used for industrial extraction of montan wax. The process used involves the following steps: size reduction, drying, extraction, evaporation of the extract, and formulation [66], [69]. The yield and composition of the extract are determined by coal quality (clay and mineral content); physical parameters such as water content, particle size, and particle-size distribution; and properties of the solvent.

Granulating and Drying. Run-of-mine, pit-wet lignite contains ca. 50% water. Before extraction it is ground in roll mills and dried to a residual moisture level of ca. 15% in steam heated tubular dryers. The particle-size distribution is ca. 7% >8 mm, 75% 1–8 mm, and 18% fines.

Lignite Extraction and Wax Production. Various machines such as band, chamber, carousel, or bucket chain conveyor extractors are used for extraction. Many solvents and solvent mixtures have been tested and used to extract wax from lignite. They include aliphatic, aromatic, and unsaturated hydrocarbons and hydrocarbon mixtures, mono- and polyhydric alcohols, ketones, esters, ethers, and mixtures of hydrocarbons and alcohols. The yield of extract is highest when polar solvents (e.g., ethanol) or solvent mixtures (e.g., ethanol–toluene) are used. However, the extract then contains a high proportion of polar substances, such as resin acids, sterols, triterpenes, or dark residues, and only a small proportion of wax. With hydrocarbons the overall yield is lower, but the proportion of wax in the extract is significantly higher. In industry, only hydrocarbons or hydrocarbon mixtures are therefore used (e.g., toluene at Romonta in Röblingen). Mixtures of aromatic and aliphatic hydrocarbons and naphthenic benzine are used by other producers.

If toluene is used, extraction is carried out at 85 °C. After extraction, the solvent contains ca. 8–10 % dissolved crude montan wax. The mixture is worked up in a three-stage distillation. The bitumen content is increased to 50 % in a continuous evaporator and then to 90 % in a circulation evaporator. The remaining solvent is removed by blowing in steam directly or in a falling-film vaporizer. The wax melt is pelletized or poured into molds. The extracted coal granulates are freed from solvent by steam stripping and used for electricity production in a power station. The toluene–water mixture is worked up, and toluene is recycled to the extraction process.

3.3. Properties and Composition

Crude montan wax is a black-brown, hard, brittle product with a conchoidal fracture pattern. It cannot be scratched by fingernails at room temperature. Crude montan wax has a penetrating terpene-like odor, which becomes very intense on melting. The wax is soluble in many organic solvents, particularly aromatic or chlorinated hydrocarbons, even on moderate heating. Crude montan wax consists of a mixture of wax acids, wax esters, resins, asphaltenes, and dark residues. The ester and acid components can be saponified and emulsified by treatment with aqueous alkali [66], [68], [69]. The waxes form the crystalline component of montan wax. They effect the relatively sharp transition from solid to liquid state and vice versa. The resins and asphaltenes have a more amorphous nature and exhibit glassy behavior. The qualitative and quantitative composition of crude montan wax is determined by the carbonized plants, extent of carbonization, the solvent, and extraction conditions.

A typical crude montan wax contains ca.

Wax acids	35 %
Wax alcohols	20 %
Hydroxycarboxylic acids	10 %
Dicarboxylic acids	3 %
Wax ketones	1 %

Hydrocarbons	2%
Resin acids	15%
Sterols	10%
Dark residues	4%

Characteristic properties are

Acid number	20–40 mg KOH/g
Saponification number	70–120 mg KOH/g
Iodine number	30–45 g I_2/100 g
Drop point	ca. 85 °C
Density (100 °C)	ca. 0.86 g/cm^3
Thermal expansion (20–100 °C)	18%

Wax Components. Crude montan wax contains 70–75% waxlike components, of which the following have been isolated:

Straight-chain carboxylic acids	C_{16}–C_{34}
Straight-chain alcohols	C_{18}–C_{34}
Straight-chain ω-hydroxycarboxylic acids	C_{16}–C_{32}
Straight-chain 2-hydroxycarboxylic acids	C_{16}–C_{32}
Straight-chain ω-dicarboxylic acids	C_{22}–C_{32}
Esters of C_{16}–C_{34} carboxylic acids and C_{18}–C_{34} alcohols	
Straight-chain aliphatic hydrocarbons	C_{25}–C_{33}

These compounds form homologous series, in which chain lengths in the range of C_{28}–C_{30} predominate. The chain-length distribution for straight-chain carboxylic acids and alcohols found by GC analysis is shown in Figure 2.

Resins and Dark Residues. Besides the wax components, crude montan wax also contains ca. 25–30% resins and dark residues. The quantity and composition of these components are determined mainly by the properties of the lignite, the extraction solvent, and the extraction temperature. The resin can be removed from crude montan wax by exhaustive extraction with methanol. In industry, this method is not used to determine resin content because it is expensive and time-consuming. Instead, the proportion of soluble compounds (i.e., resins) is determined by a standardized procedure such as solubility in acetone or in a toluene–ethanol or benzene–ethanol mixture. The resin is characterized by its chemical and physical properties, of which the following are typical:

Drop point	ca. 75 °C
Acid number	ca. 33 mg KOH/g
Saponification number	ca. 75 mg KOH/g
OH number	ca. 70 mg KOH/g
Iodine number	ca. 90 g I_2/100 g
Sulfur content	ca. 2–2.5%

Determination of all individual components in resin extracts has not yet proved possible. However, substance classes and typical compounds in these classes have been

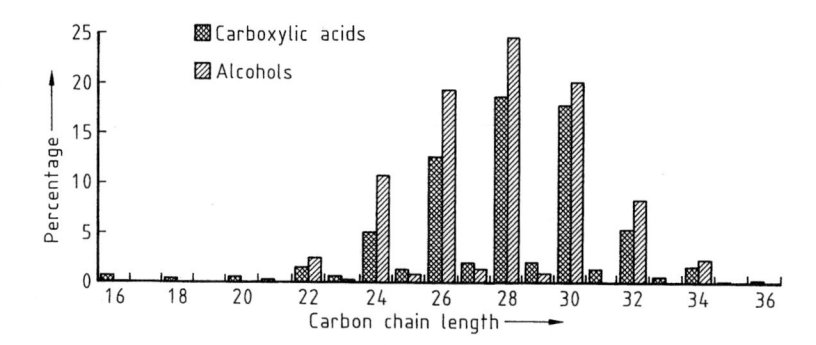

Figure 2. Chain-length distribution of acids and alcohols in montan wax

identified: e.g., triterpenes such as betulinol, sterols, diterpenoid acids and their esters, and diterpenoid alcohols and their esters.

3.4. Refining and Derivatization

Because of its dark color and its resin content, which sometimes lead to problems, direct use of crude montan wax is limited. For most applications the wax must be refined. Refining involves three steps: extractive deresinification, oxidative bleaching, and subsequent derivatization [66]–[69], [72].

3.4.1. Deresinification

The resin components of the crude wax are soluble in many solvents even at low temperature (10–20 °C). Treatment of pulverized crude wax with methylene chloride, ethyl acetate, or methanol reduces the resin content to 5–10%. Other solvents, such as ketones, ethers, hydrocarbons, mixtures of alcohols and aromatics, liquid SO_2 or supercritical CO_2, can be used but have not achieved industrial importance. After removal of the solvent, the extracted resin contains ca. 10–20% wax. Deresinified wax can be used directly or as a mixture with other wax components for various applications. Such products are sold by Romonta, Amsdorf, and the American Lignite Co. Most deresinified crude wax is subjected to oxidative bleaching.

3.4.2. Bleaching

A purification process for the crude wax was sought very early on. Of the many processes investigated, only three have actually been used in production [66], [69], [73]–[76].

Distillation was the first method for obtaining light, hard waxes from crude montan wax. The crude wax was heated to 420 °C in stills and thus partially cracked. The distillate therefore consisted of 50–80% hydrocarbons and 20–50% wax acids. The hydrocarbons were removed in a further workup step, leaving the montan wax acids.

Oxidation with Nitric and Sulfuric Acids. In this process, paraffin had to be added as a diluting agent. The resin was precipitated as the nitroresin and removed. The dark residues were coked and the wax acids extracted from this coke. However, the wax still had a certain amount of dark residues and nitrate-containing products.

Oxidation with Chromic Acid or Chromates in Sulfuric Acid. In processes described thus far, the wax is damaged considerably and the chainlength distributions of its constituents are altered. Oxidative bleaching with aqueous chromic–sulfuric acid mixtures, developed by I.G. Farben, is much more suitable. The starting material is deresinified crude montan wax. The resin components and dark residues are oxidized to carbon dioxide or low molar mass, water-soluble compounds in a multistage oxidation process. The wax esters are simultaneously hydrolyzed; the wax alcohols formed are oxidized to wax acids; and the hydroxycarboxylic acids are oxidized to dicarboxylic acids. Some esters remain unaltered.

$$CH_3-(CH_2)_n-COO-CH_2-(CH_2)_n-CH_3 \xrightarrow{H_2O}$$
$$CH_3-(CH_2)_n-COOH + CH_3-(CH_2)_n-CH_2OH$$
$$CH_3-(CH_2)_n-CH_2OH \xrightarrow{CrO_3} CH_3-(CH_2)_n-COOH$$
$$HOCH_2-(CH_2)_n-COO-CH_2-(CH_2)_n-CH_3 \xrightarrow{H_2O}$$
$$HOCH_2-(CH_2)_n-COOH + CH_3-(CH_2)_n-CH_2OH$$
$$HOCH_2-(CH_2)_n-COOH \xrightarrow{CrO_3} HOOC-(CH_2)_n-COOH$$

where n is an even number between 24 and 32.

A mixture of light wax acids, which are very similar in chain-length distribution to the crude wax, is obtained in a yield of 80–85% based on crude wax. Minor components of the mixture are unchanged esters and long-chain dicarboxylic acids. The spent oxidizing agent is removed and worked up, and the bleached wax is freed from residual chromium salts in a subsequent step.

Depending on the quantity of chromic acid, the acid number of the bleached wax is between 100 and 135 mg KOH/g. Oxidation can be performed batchwise or continuously. In the *Gersthofen process* used by Hoechst, the oxidizing agent is regenerated electrochemically. BASF, Völpke, and other producers use the spent oxidizing agent in an associated plant for the production of chromium-containing tanning agents.

Development is still being carried out on refining processes [77] such as alkaline dehydrogenating oxidation with anhydrous alkali hydroxide. Apparently, these pro-

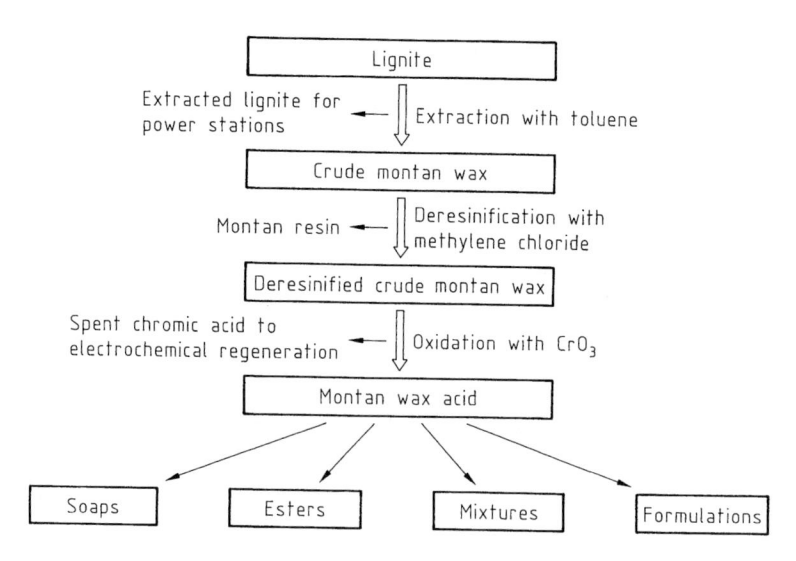

Figure 3. Flow sheet for extraction and refining of montan waxes

cesses have not been used in industry up to now. An overview of the current processes used for extraction and refining of montan wax is given in Figure 3.

3.4.3. Derivatization

Reconstruction of the original wax structure is attempted by treating the wax acids obtained with dihydric alcohols or $Ca(OH)_2$. By varying the proportions of free acid and soap and the degree of esterification, the reaction can be carried out in such a way that the products formed are optimized for a particular application [66], [73], [74], [78].

$$CH_3-(CH_2)_n-COOH + HOCH_2-CH_2OH \rightarrow CH_3-(CH_2)_n-COO-CH_2-CH_2-OR$$

$R = H$ or $CH_3-(CH_2)_n-CO-$

$$CH_3-(CH_2)_n-COOH + Ca(OH)_2 \rightarrow (CH_3-(CH_2)_n-COO)_2^- Ca^{2+}$$

The most important products in this series are the esters of ethylene glycol (e.g., Hoechst Wax E or Hoechst Wax KSL) and mixtures of esters of 1,3-butanediol with calcium montanate (e.g., Hoechst Wax OP). Broad variation in properties can be achieved by using tri- and polyhydric alcohols, aromatic alcohols, aminoalcohols, and ethoxylated or fluorinated alcohols. Reactive waxes can be produced from partial esters of polyols by treatment with dicarboxylic acids, isocyanates, epoxides, or unsaturated carboxylic acids.

$$\text{ROCH}_2-\underset{\underset{\text{CH}_2\text{OOC}-\text{R}^2}{|}}{\overset{\overset{\text{CH}_2\text{CH}_3}{|}}{\text{C}}}-\text{CH}_2\text{OR}^1$$

$$R^1 = -\text{CONH}-(\text{CH}_2)_6-\text{NCO}$$
$$-\text{CO}-(\text{CH}_2)_6-\text{COOH}$$
$$-\text{CH}_2-\text{CHOH}-\text{R}-\overset{O}{\overset{\|}{\text{CH}}}-\text{CH}_2$$
$$-\text{CO}-\text{CH}=\text{CH}_2$$

R^2 = Alkyl group of the montan wax acid
$R = H, R^2CO, R^1$

These waxes exhibit interesting properties and have been produced and used for special applications. If the reaction is not stopped at the reactive wax stage and a suitable stoichiometry is chosen, oligomeric waxes are produced.

3.5. Uses and Economic Aspects

Montan wax acid and its derivatives are versatile materials [73], [74], [78]–[80]. They are used in the form of flakes, powders, pastes with solvents, or aqueous emulsions. The main areas of use are in polishes for floors, cars, and leather; in plastics processing as lubricants, release agents, and nucleating agents; in the paper and building industries; and in wood and metal processing. Montan wax derivatives are also used in small quantities in pharmaceutical formulations; cosmetics; and the production of adhesives, resins, and office equipment. In addition to their technical properties, the low toxicity of montan wax derivatives and the fact that BGA or FDA approval has already been granted for some derivatives are important criteria favoring their use.

Production figure of montan wax for 1994 are given below, together with a breakdown of montan wax consumption in different application areas:

Crude montan wax
Romonta	ca. 19 000 t/a
Alpco	ca. 2 500 t/a
Ukraine	ca. 1 500 t/a
China	ca. 1 000 t/a

Bleached montan waxes
Hoechst Gersthofen	ca. 11 000 t/a
Völpker Montanwachs	ca. 3 500 t/a
China	ca. 500 t/a
Poland	< 500 t/a
Czechia	< 500 t/a

Application areas
Plastics, lubricant	ca. 7 000 t/a
Polishes, coatings	ca. 5 500 t/a
Technical applications	ca. 2 500 t/a

Table 7. Classification of waxes from crude petroleum

	Origin		
	light, medium, heavy lubricating oil distillates	residues from vacuum distillation	crude oil
Group	(macrocrystalline) paraffin waxes	microcrystalline waxes (microwaxes)	settling waxes
Subgroup	paraffin waxes intermediate waxes	residue waxes	pipe waxes tank bottom waxes
Crude products	crude waxes (slack waxes)	petrolatum	raw waxes
Deoiled and refined products	scale waxes deoiled slack waxes filtered (decolorized) waxes fully refined waxes		plastic microwaxes hard microwaxes
Side products from deoiling	soft waxes	soft petrolatum (microwax slacks)	

4. Petroleum Waxes

4.1. Introduction

The quantity of waxes obtained from crude petroleum has increased continuously for two reasons: (1) the demand for lubricating oils with low pour points and (2) the large proportion of paraffinic crudes in total crude oil production that have to be dewaxed for the production of lubricating oils. This development has increased the need to find more industrial applications for petroleum waxes. When the processing of lubricating oil began in the second half of the 19th century, waxes were still inconvenient side products. Since then, worldwide consumption of petroleum waxes has increased to 3×10^6 t/a. This consumption corresponds to more than 95 % of the waxes of all types produced worldwide. Depending on their natural occurrence and their crystallinity, petroleum waxes are divided into

1) Macrocrystalline waxes (paraffin waxes)
2) Microcrystalline waxes (microwaxes)

Table 7 gives a detailed classification of paraffin and microcrystalline waxes based on origin and method of refining [81].

Paraffin waxes are obtained from light and middle lubricating oil cuts of vacuum distillation. Paraffin waxes also include waxes from heavy lubricating oil distillates, which are intermediates between macrocrystalline and microcrystalline waxes with regard to structure and composition (*intermediate waxes*).

Microcrystalline waxes originate from vacuum residues and from the sediments of paraffinic crude oil (settling waxes). Waxes that are liquid at room temperature are mostly contained in diesel oil or gas oil fractions and can be isolated from them. These are not dealt with here.

4.2. Macrocrystalline Waxes (Paraffin Waxes)

4.2.1. Chemical Composition and General Properties

Chemical Composition. *Paraffin waxes* consist predominantly of mixtures of straight-chain alkanes in a typical distribution of the homologous series whose molar masses depend on the boiling range of the lubricating oil distillate from which they are obtained. Long-chain, weakly branched isoalkanes are present in a much lower proportion, along with a very small fraction of monocyclic alkanes.

The *intermediate waxes* have a similar composition, but the molar masses of the *n*- and isoalkanes are higher. Intermediate waxes contain a higher proportion of cycloalkanes and isoalkanes; the latter more strongly branched than those in paraffin waxes.

According to the European Wax Federation (EWF), paraffin waxes have a C-number distribution of *n*-alkanes from 18 to 45 and a total content of iso- and cycloalkanes of 0 – 40 %. Typical data for the intermediate waxes are an *n*-alkane C-number of 22 to 60 and a total content of iso- and cycloparaffins of 30 – 60 %.

General Physical Properties. Paraffin waxes are insoluble in water and sparingly soluble in low molar mass aliphatic alcohols and ethers. They are more soluble in ketones, chlorohydrocarbons, petroleum spirit, solvent naphtha, benzene, toluene, xylene, and higher aromatics, especially at elevated temperature. The solubility decreases markedly with increasing molar mass (higher melting point) of the waxes.

Chemical Properties. Paraffin waxes are extremely unreactive under normal conditions. Oxidation reactions occur only at elevated temperatures (e.g., on storage and processing above 100 °C), particularly in the presence of oxygen and catalytically active metals. These reactions can be recognized from the burnt odor produced and the yellow to brown coloration of the waxes. Nevertheless, under certain thermally and catalytically controlled conditions, these waxes can undergo chemical reactions such as chlorination, oxidation, dehydrogenation, and cracking, of which chlorination and cracking are important in industry.

Table 8. Variation in physical data with degree of refining of paraffin wax (starting material: slack wax from a medium machine oil)

Characteristics	Slack wax	Deoiled wax	Filtered (decolorized) wax	Fully refined wax
Congealing point, °C	59	62	62	62
Needle penetration, 0.1 mm				
at 25 °C	57	18	18	17
at 30 °C	80	26	26	25
at 35 °C	105	36	36	36
Oil content, %	9.9	0.5	0.5	0.4
Viscosity (at 100 °C), mm^2/s	6.8	6.1	6.0	5.8
Color	brown	brown	whitish	white
Fluorescence	very strong		weak	none

4.2.2. Division into Product Classes

Depending on the degree of refining, paraffin waxes are divided into the following product classes:

1) Crude waxes, also known as slack waxes
2) Slack wax raffinates (scale waxes)
3) Deoiled slack waxes
4) Soft waxes
5) Semirefined waxes
6) Filtered (decolorized) waxes
7) Fully refined waxes

Table 8 gives an overview of the change in physical characteristics with degree of refining.

Crude waxes (slack waxes) [64742-61-6] consist of a mixture of alkanes, which can be solid, semisolid, or liquid at room temperature; alkyl-substituted cyclopentanes and cyclohexanes (naphthenes); and alkyl-substituted aromatics. They also contain, as impurities, the typical contents of the lubricating oil cuts from which they originate, such as asphaltenes, resins, olefins, and sulfur and nitrogen compounds. The oil content of slack waxes is usually between 5 and 12%, but can be as high as 25%.

n-Paraffins and isoparaffins predominate in slack waxes in terms of quantity; both are present in continuously distributed homologous series. The chain-length spectrum (C-number distribution) is determined from the width of the boiling range, the distribution maximum (C-number maximum) from the boiling level of the lubricating oil fraction. In isoparaffins, compounds with terminal methyl branches predominate, followed by other methyl-substituted alkanes. The concentration of these compounds decreases as the branching point moves toward the middle of the chain. Other components are compounds with terminal ethyl branches and multiply branched

Table 9. Composition of slack waxes from various lubricating oil distillates of the same origin (80% Arabian, 10% German, and 10% Norwegian crude oil)

Characteristic	Slack waxes from			
	Heavy spindle oil	Light machine oil	Medium machine oil	Heavy machine oil
Congealing point, °C	47	50	58	62
Oil content, %	9.5	9.0	9.8	8.0
n-Paraffin content, %	73	58	52	39
C-number range	16–36	19–43	21–48	22–58
C-number max.	24	29	34	41
C_{max} content, %	17	14	10	8

structures, which can be detected only in slack waxes from heavy vacuum distillates [82], [83].

Table 9 shows the composition of some slack waxes as a function of origin.

Physical Properties. Depending on the origin and production process, slack waxes are light- to dark-brown, soft, unctuous to semisolid materials without a clear crystal structure because of their high oil and soft wax content. The congealing points lie between 35 and 65 °C; the drop points, between 35 and 68 °C; the needle penetration at 25 °C (a reciprocal measure of the hardness) is between 40 and 80 (at greater oil content, even higher); the viscosities at 100 °C are between 3 and 10 mm^2/s; the densities at 70 °C, between 775 and 815 kg/m^3; and the flash points between 190 and 250 °C.

Scale Waxes. Normally being physically decolorized slack waxes, scale waxes [*90669-78-6*] essentiall have the same composition as the raw materials from which they are obtained. Only the dark material contents of the slack waxes (asphaltenes and resins) are removed by the decolorization process, and the content of alkylaromatics, and of sulfur and nitrogen compounds, is reduced. In physical properties, scale waxes are similar to slack waxes. Decolorization renders the products white to pale yellow.

Deoiled Slack Waxes (Crude Hard Waxes, Raw Waxes). In deoiled slack waxes [*8002-74-2*] the predominant proportion of the hydrocarbons that are liquid (oils) under normal conditions, and a certain proportion of the semisolid ones (soft waxes), have been removed. Liquid and semisolid waxes consist mainly of low molar mass n- and isoparaffins and of isoparaffins with centrally located and strongly branched side chains, as well as naphthenes and alkylaromatics. Therefore, in deoiled slack waxes the n- and weakly branched isoparaffins are enriched. By removal of the low molar mass portion a narrower C-number range results along with a more pronounced C-number maximum.

Table 10 shows these relationships with two waxes from heavy spindle oil and medium machine oil used as examples.

Physical Properties. In structure and consistency, deoiled slack waxes are coarse to medium crystalline, brittle to weakly plastic and depending on the degree of deoiling, hard to very hard. These waxes are light to dark brown and darken on heating. The

Table 10. Composition of deoiled slack waxes from two lubricating oil distillates (origin – see Table 9)

Characteristic	Slack wax*	Deoiled	Slack wax**	Deoiled
Congealing point, °C	47	50	58	67
Oil content, %	9.5	0.3	9.8	0.4
n-Paraffin content, %	73	84	52	64
C-number range	16–36	19–36	21–48	23–48
C-number max.	24	24	34	34
C_{max} content, %	17	19	10	11

* From heavy spindle oil.
** From medium machine oil.

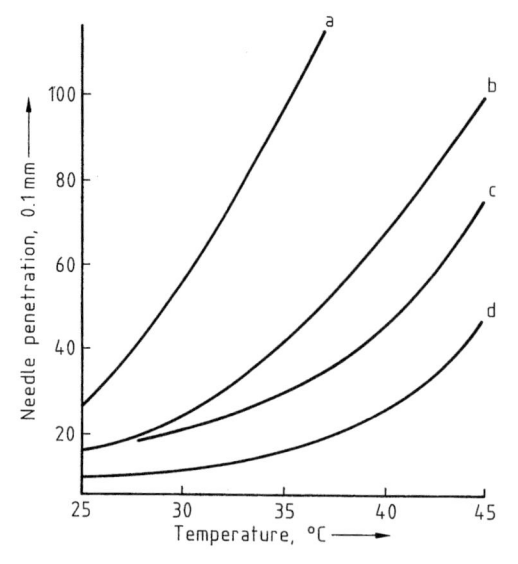

Figure 4. Variation of needle penetration with temperature for deoiled slack waxes of different origins
a) From heavy spindle oil (North Sea); b) From light machine oil (North Sea); c) From heavy machine oil (North Sea); d) Wax from Indonesia

congealing points are between 48 and 72 °C; the drop points between 48 and 75 °C; and the needle penetrations at 25 °C between 10 and 30. At the same degree of deoiling, the course of the penetration—temperature curve (increase in penetration in the penetrogram [84] essentially depends on the origin of the waxes (Fig. 4).

Crude soft waxes [64742-67-2] are sometimes known as foot oils (see Section 4.2.3.2) and are formed in the deoiling of slack waxes. They consist predominantly of low molar mass n-paraffins and strongly branched isoparaffins; higher molar mass, weakly branched isoparaffins; naphthenes; and alkylaromatics. The n-paraffin content of soft waxes is generally < 25 wt%.

Soft waxes are very soft, unctuous to inhomogeneous materials at room temperature because of their high oil content, which can be 30 wt% or greater. The oil is included in voids of the crystal lattice of the solid hydrocarbons. Soft waxes have a light- to dark-brown color and congealing points < 40 °C. Their viscosities at 100 °C are between 3 and 12 mm²/s; their flash points, between 180 and 220 °C.

Table 11. Composition of fully refined waxes (congealing point 62 °C in each case) from different origins

Characteristic	Wax from machine oil distillate		
	Saudi Arabia	Russia	East Asia
Congealing point, °C	62	62	62
Oil content, %	0.3	0.6	0.2
n-Paraffin content, %	52	78	92
C-number range	23–48	23–41	25–38
C-number max.	34	32	30
C_{max} content, %	10	14	21

Semi-refined, filtered (decolorized), and fully refined waxes are similar to deoiled slack waxes in the composition of their main components. Through refining they are freed from more highly condensed alkylaromatics and naphthenes, olefins, and sulfur and nitrogen compounds. Depending on the origin, their content of n-paraffins varies between 92 wt% (East-Asian waxes) and 45 wt% (waxes from heavy machine oil distillates of German origin). The composition of the refined waxes is determined not only by the origin of the crude (Table 11), but also by the boiling range of the starting distillate. As the boiling ranges of the lubricating oils rise, the molar masses of the hydrocarbons contained in waxes obtained from them increase. The concentration ratio of iso- and cycloparaffins to n-paraffins also increases. Thus, for example, a refined wax from Tuismasy (Ural) crudes contains 81.6% n-paraffins with an average molar mass of 360 g/mol (average C-number 25.7) and 9.3% isoparaffins (average molar mass 384 g/mol; average C-number 27.4). The naphthenes (9.1%) consist of 81.5% monocyclic and 15.5% bicyclic alkanes (average C-number 28.9) [85].

Physical Properties. Waxes in this group differ from deoiled slack waxes and from each other in their residual oil content and, thus, in hardness, color, and color stability.

Semirefined waxes have an oil content of 1.5–3%, and the needle penetration at 25 °C can be 20–60. They are somewhat plastic and kneadable, colorless to white, have good color stability under light, and are virtually odorless and tasteless.

Filtered (decolorized) waxes have an oil content of 1.5% (max.) and a needle penetration at 25 °C of 10–26. They have a whitish color and are relatively color stable and generally odorless.

Fully refined waxes (oil content and penetration similar to decolorized waxes) are crystalline, no longer kneadable, pure white, transparent to slightly opaque, color stable, light-resistant, odorless, and tasteless. They fulfill purity criteria for the production of packaging materials for foods and for the formulation of cosmetics and pharmaceuticals. Detailed information on the chemical composition of fully refined waxes and methods for their analytical determination can be found in the CONCAWE report [86].

According to a more recent definition, fully refined wax products are those that fulfill various national purity requirements (see Section 4.2.5). The oil content, which was once limited to 0.5% (max.), is no longer a criterion since the oils contained in the wax also consist of *n*- and isoparaffins of the required purity. Depending on the literature, the oil content of fully refined waxes can be 1.5% (max.) [87] or 2.0% (max.) [88].

4.2.3. Occurrence of Raw Materials and Processing

Paraffin waxes are contained in crude petroleum and are obtained during oil refining. Depending on the source of the petroleum, the content of solid waxes can vary between 2% (Rumanian origin) and 30% (Indonesian origin); generally, however, it is between 3 and 15% in crude petroleum from the main production areas.

Because of their higher molar masses and thus higher boiling ranges, waxes become enriched in the residues from atmospheric distillation (long residues). In processing the residues to lubricating oils, waxes are obtained in all fractions of the vacuum distillation. Because of their poor solubility, they give rise to the high pour points of the basic oils and must therefore be removed. The yields of wax vary depending on the origin and quality requirements of the oils. For a wax content of crude petroleum of ca. 5%, 8–18% solid waxes in the individual lubricating oil fractions can be expected.

Slack waxes are produced in the dewaxing of lubricating oil distillates. The type of process used and process parameters used are according to the desired quality of the lubricating oils.

The process steps necessary for the production of refined waxes are shown in Figure 5.

4.2.3.1. Dewaxing Lubricating Oil Distillates

The principle of dewaxing lubricating oil distillates is based on the different crystallization temperatures of straight-chain and weakly branched paraffins and the oil phase. The wax-containing solvent-neutral oils (i.e., the raffinates from solvent refining) are mixed with suitable solvents such as propane, naphtha, chlorohydrocarbons, ketones, or (most frequently) a toluene–methyl ethyl ketone mixture. The oil–solvent mixture is subsequently warmed to obtain a homogeneous solution and then cooled continuously in scraping chillers to obtain the wax crystals in a loose suspension. In the subsequent filtration, crystallized wax is separated in vacuum rotary filters, and washed with fresh solvent; the low-wax solvent-neutral oil is then freed from the solvent by distillation. The solvent-containing wax is either separated from the solvent by distillation if it is to be obtained as such or fed directly into the deoiling process for production of hard waxes.

Process conditions are controlled in such a wax that high throughput rates (i.e., short filtration times) and as low an oil content of the slack waxes as possible are achieved [89].

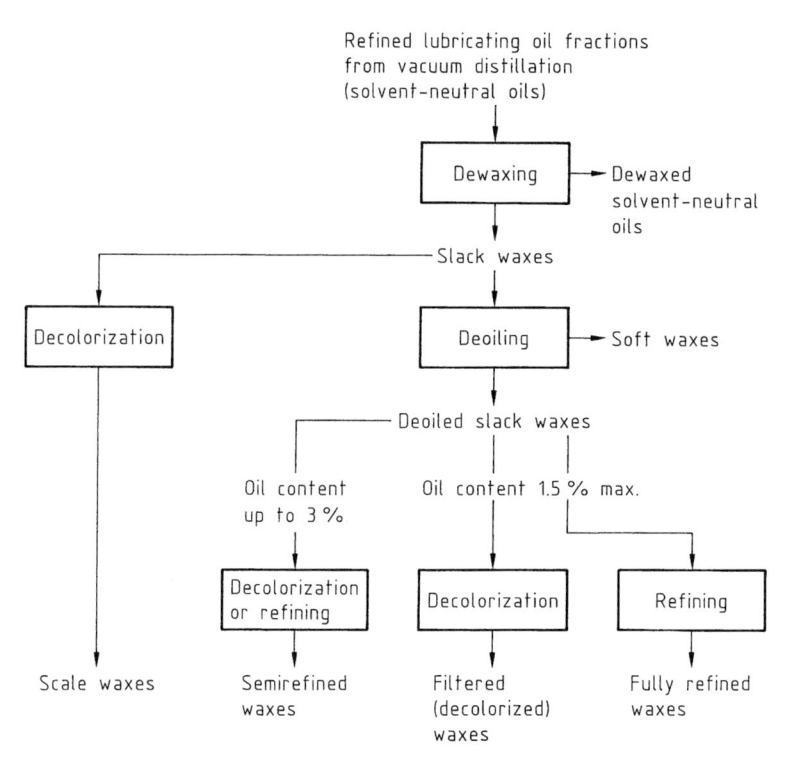

```
                    Refined lubricating oil fractions
                    from vacuum distillation
                    (solvent-neutral oils)
                                │
                                ▼
                         ┌─────────────┐
                         │  Dewaxing   │──▶ Dewaxed
                         └─────────────┘     solvent-neutral
                                │            oils
                    ── Slack waxes ──
                    │           │
                    │           ▼
┌──────────────┐    │    ┌─────────────┐
│Decolorization│    │    │  Deoiling   │──▶ Soft waxes
└──────────────┘    │    └─────────────┘
       │            │           │
       │       ── Deoiled slack waxes ──
       │            │                  │
       │     Oil content      Oil content 1.5 % max.
       │     up to 3 %                  │
       │            ▼            ▼            │
       │    ┌──────────────┐ ┌─────────────┐ ┌─────────┐
       │    │Decolorization│ │Decolorization│ │Refining │
       │    │  or refining │ │             │ │         │
       │    └──────────────┘ └─────────────┘ └─────────┘
       ▼            ▼              ▼              ▼
   Scale waxes  Semirefined     Filtered      Fully refined
                  waxes       (decolorized)      waxes
                                waxes
```

Figure 5. Process schematic for the production of paraffin waxes

The selectivity of the dewaxing process is controlled by the cooling temperature in the scraping chiller and by the oil–solvent ratio. For common lubricating oil fractions the cooling temperature is between –50 and –20 °C at oil yields of 65–85%. The most important process parameter is the targeted pour point of the dewaxed oil. Figure 6 shows a typical dewaxing process.

In dewaxing, all hydrocarbons with crystallization temperatures above the chosen cooling temperature are removed. n-Paraffins are thus almost completely separated, along with a suitable fraction of the long-chain, weakly branched isoparaffins. Even under optimum filtration and washing conditions, impurities consisting of highly branched isoparaffins and cycloparaffins are present in slack waxes, thus giving rise to their oil content [90].

A second process for producing lubricating oils with the required low-temperature properties is *catalytic dewaxing*. n-Paraffins are preferentially cracked on zeolite catalysts to give low molar mass hydrocarbons. Wax thus cannot be obtained from this process. Catalytic dewaxing is said to be very selective in the case of heavy distillates, whereas in light distillates the branched paraffins also react (loss of yield). Since no solvents are used in this process it does not require solvent recovery and is therefore environmentally friendly [91].

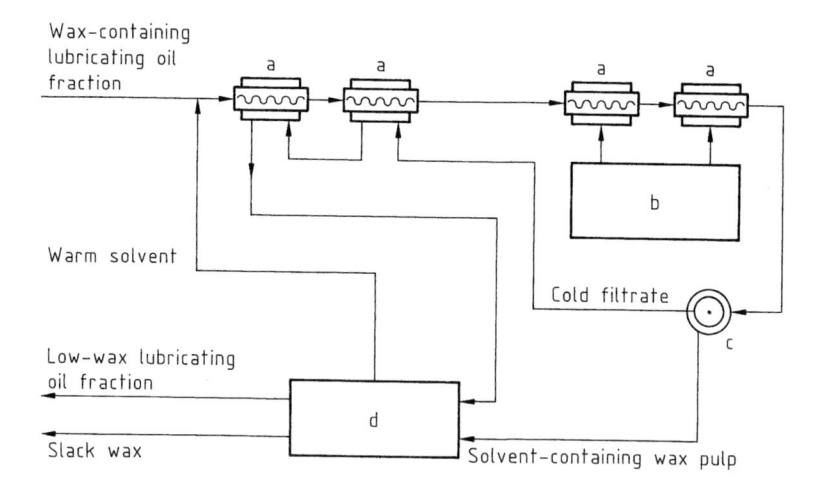

Figure 6. Dewaxing of lubricating oil distillates
a) Scraping chiller; b) Cooler; c) Vacuum rotating filter; d) Solvent recovery

4.2.3.2. Deoiling Slack Waxes

Refining slack waxes begins with deoiling to obtain harder, higher-value products (hard waxes). Two main processes are currently used: solvent and sweat deoiling. The oils produced as byproducts, which contain a higher or lower proportion of soft waxes, are known as foot oils or soft waxes, depending on the consistency and the process used.

Solvent deoiling is the most commonly used process, involving one of three possible techniques: pulping, crystallization, or spray deoiling. Starting materials for all deoiling processes are either molten slack waxes (if deoiling is carried out in a location other than that of slack wax production) or slack wax – solvent mixtures, formed in dewaxing lubricating oil distillates (see Table 12).

Slack waxes from spindle oils to heavy machine oils, and to some extent petroleum as well (see Section 4.3), can be deoiled by the three solvent deoiling methods. The yields of deoiled slack waxes depend on their origin, the process used, and the degree of deoiling. Yields are between 80 % for spindle oil distillates and ca. 60 % for heavy machine oil distillates, based on the slack wax used.

Pulping Process. The inhomogeneous mixture of crystallized wax, oil, and solvent is diluted or repulped by addition of the same solvent or a solvent mixture (normally by using part of the wash filtrates produced later). The oils, soft waxes, and part of the low molar mass hard waxes thus dissolve. The pulp is cooled with stirring in scraping chillers, and the temperature is adjusted to that necessary for the desired degree of

Table 12. Processes for deoiling slack waxes

Company	Process	Solvent	Starting material
Exxon	crystallization process (Dilchill deoiling)	methyl ethyl ketone – methyl isobutyl ketone (or – toluene)	slack wax – solvent mixture *
Edeleanu	pulping process (Di-Me deoiling)	1,2-dichloroethane – methylene chloride	slack wax – solvent mixture *
Edeleanu	spray deoiling	1,2-dichloroethane	slack wax
Texaco	pulp process (one- and two-step)	benzene (or benzene – toluene)	slack wax – solvent mixture *
Texaco	crystallization process (wax fractionation)	methyl ethyl ketone	slack wax and slack wax – solvent mixture *
Union oil	crystallization process	water-saturated methyl isobutyl ketone	slack wax
	sweat deoiling		slack wax

* From the dewaxing step.

deoiling. Hard waxes, which are partly undissolved and have partly recrystallized, are removed on vacuum rotary filters and washed with solvent; the solvent is subsequently distilled off.

The quality of the hard waxes produced is determined by the deoiling temperature and the quantity of solvent used. With a 1,2-dichloroethane – methylene chloride mixture as solvent (Edeleanu Process), for example, typical deoiling temperatures are −10 °C for slack waxes from heavy spindle oil, −5 °C for those from a light machine oil, and +15 °C for those from medium to heavy machine oil.

Crystallization Process. The molten slack wax or solvent-containing wax crystallizate from the dewaxing process is dissolved completely, or almost completely, in solvents by warming, and the mixture is cooled to a given temperature in one or several steps, depending on the desired degree of deoiling. The temperature is chosen such that only the hard wax crystallizes. The crystallizate is filtered and washed, and the solvent is removed by distillation.

In both the pulping and the crystallization processes, technical problems arise in attempting to produce readily filterable hard wax crystals that allow high filtration capacity with good washability. These problems can be solved in various ways: e.g., through the nature and quantity of the solvent; the construction of the crystallization apparatus; successive solvent dosing during cooling; addition of cold solvent to the warm, intensively stirred wax – solvent mixture [92]; and use of filter aids [89]. Other improvements have been brought about by the development of economical cooling and filtration sysems [93].

Spray Deoiling. Molten slack wax is sprayed into a countercurrent of cold air, and precipitated wax particles are washed with solvents in mixers whereby the oil is largely diluted. After settling the washed wax particles are centrifuged and washed, and the adhering solvent is distilled off. The washing temperature is between 5 and 15 °C [94].

Sweat Deoiling. Molten slack wax is charged to chambers equipped with sieve bottoms and heating coils. The wax is solidified by cooling and then warmed very slowly. The oil—and, at higher temperatures, the low-melting soft waxes, as well—sweat from the wax block and run through the perforated bottom (run-off slack wax). At the end of the process the remaining hard wax is melted to remove it from the sweating chamber.

No new plants are being built for this classical deoiling process because of the low selectivity (poor yields of hard wax), time-consuming warmup, need to use a batch process, and inapplicability of the method to strongly oil-binding slack waxes from medium and heavy machine oil distillates. In existing plants an attempt is being made to improve yields of hard waxes by partially recycling the run-off slack [95].

Overview of Processes in Current Use. The deoiling processes most widely used in industry are listed in Table 12.

Of the deoiling plants operating in Germany, half employ solvent deoiling (using mainly the Edeleanu dichloroethane – methylene chloride process) and the others use spray deoiling (Edeleanu) and sweat deoiling.

In the United States, more than 90% of the existing plants use solvent deoiling. The most common solvents are methyl ethyl ketone and methyl ethyl ketone – aromatics mixtures. Propane and methyl isobutyl ketone are also used.

4.2.3.3. Refining Deoiled Slack Waxes

Deoiled slack waxes whose maximum oil content is 0.5%, 0.5 – 1.5%, or 2 – 3%, depending on the use anticipated, still contain impurities and are mostly dark in color. To improve their quality they are purified further by adsorbents (decolorizing) or chemically. The choice of refining process depends on economic factors and the quality of raffinates required. Process combinations have also been introduced into industry. The preferred process in new plants is hydrotreating.

Refining with adsorbents (decolorizing) can be performed batchwise, semicontinuously, or continuously. The decolorizing temperature can be between 70 and 120 °C, depending on starting material.

In *batch decolorizing*, molten wax is stirred in heated vessels with decolorizing clays (chemically activated clays, bentonite, bauxite) in one or more steps until the desired lightening of color is achieved. The used clay is removed in heated filter presses and thermally regenerated.

Percolation Process. In the percolation process, molten wax flows downward through a tower containing 10 – 50 t of decolorizing clay, depending on plant size. After the clay has been exhausted, the wax flow is stopped; wax adhering to the clay is washed off with naphtha; the remaining solvent is driven off with steam; and the clay is burnt off in a separate tower, activated, and reused [97].

Table 13. Wax qualities obtained by decolorization

Starting material	Oil content, %	Decolorized product
Slack wax	> 3	scale wax
Deoiled slack wax	1.5 – 3	semirefined wax
Deoiled slack wax	1.5 (max.)	filtered wax [64742-43-4]

The *semicontinuous decolorizing process* operates with two or three percolator towers, which are in different phases of the adsorption cycle. While the adsorption process occurs in one tower, in the others the used decolorizing clay is washed, stripped, removed for regeneration, and replaced by new activated clay.

In the *continuous process* the regenerated clay flows downward through the adsorption tower in countercurrent to the molten wax. The loaded clay passes in the form of a sludge to a washing tower where adsorbed wax is extracted with naphtha and then to a second tower where naphtha is driven off wich steam. The clay is then burnt off in a rotary kiln, activated, and fed back to the top of the adsorption tower after cooling [98].

The decolorized waxes are then stripped with steam in packed columns at 110 – 170 °C, either under vacuum (0.6 – 1.6 kPa), under reduced pressure (up to 80 kPa), or under pressure (up to 0.8 MPa) to remove odorous substances. The weight ratio of steam to wax varies between 0.5 : 1 and 1 : 1 at detention times up to 15 min. To improve the aging resistance of the waxes, apparatus with aluminum or stainless steel lining is recommended for stripping [99].

Depending on the starting material, the wax qualities listed in Table 13 are obtained by the decolorizing process.

The quantity of decolorizing clay required depends on the origin and desired quality of the waxes and is normally between 2 and 4 %. Refining with adsorbents removes dark substances, condensed alkyl aromatics, compounds containing hetero atoms, and metallic impurities.

Nevertheless, producing fully refined waxes fulfilling all of the purity requirements (see Section 4.2.5) by adsorption is possible only using extremely complex processes (multistage decolorizing) and large quantities of decolorizing clay, particularly in the case of the high molar mass wax fractions. To produce these types of high-purity waxes, chemical refining is the process of choice.

Chemical Refining. Concentrated sulfuric acid, fuming sulfuric acid (oleum), or hydrogen is used as chemical refining agent.

Sulfuric acid refining can only be carried out batchwise because of the slow separation of the wax and acid sludge phases. Molten wax is mixed with sulfuric acid at 80 – 140 °C for some time (between 0.5 and 2 h, depending on the origin of the wax). After separation of the heavier acid sludge, the wax is washed with alkali, treated with decolorizing clay, and stripped with steam to remove the last residues of impurities and odorous substances.

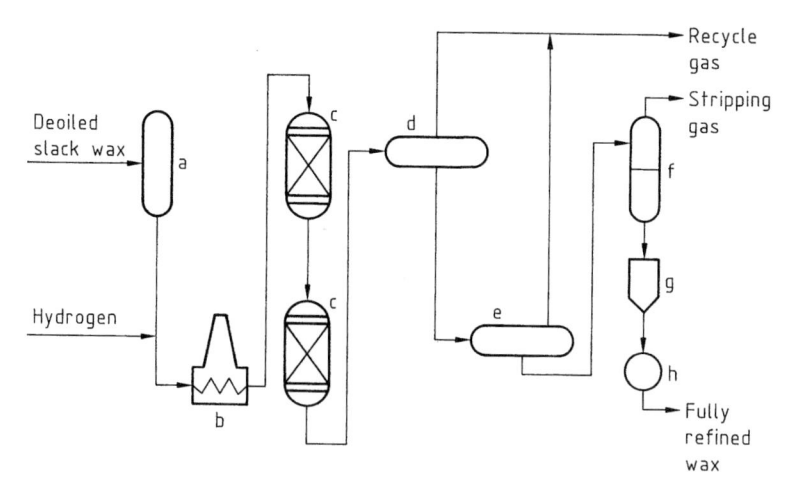

Figure 7. Two-step hydrotreating of waxes
a) Deaerator; b) Preheater; c) Reactors; d) High-pressure gas separator; e) Low-pressure gas separator; f) Stripper; g) Vacuum dryer; h) Filter

All compounds that react chemically with the aggressive acid, such as unsaturated aliphatic and aromatic hydrocarbons, metal compounds, and those containing hetero atoms, are removed by sulfuric acid refining. Labile quaternary carbon atoms are also attacked, with bond cleavage and formation of aliphatic sulfonic acids.

The process can be optimized by variation of the number of reaction steps, degree of acidity, reaction temperature, ratio of wax to acid used, reaction time, and intensity of mixing. The higher the average molar mass of the wax fraction, the slower is the separation between the wax and the acid sludge phase.

The process has several disadvantages such as the necessity for batch operations, poor yields of raffinates, formation of polluting byproducts, and waste gas and corrosion problems. For these reasons, almost all new plants are based on hydrotreating.

Hydrotreating (Fig. 7). Deoiled slack wax is heated to the required temperature together with fresh and recycled hydrogen in the preheater (b) and passed over a sulfur-resistant fixed-bed catalyst (c). After cooling, the reaction mixture and hydrogen are separated in the high-pressure gas separator (d), and hydrogen is recycled into the process. After depressurizing in the low-pressure gas separator (e), the wax passes into the stripper (f), where all the light crack products, odorous substances, and reaction gases are stripped off completely with compressed steam under vacuum. The wax is subsequently dried with nitrogen (g). A combination of metals from groups 6 and 10 on an inert carrier is generally used as the catalyst (e.g., nickel – tungsten or nickel – molybdenum on neutral aluminum oxide).

Reaction conditions can vary within wide limits, depending on starting material, degree of refining required, and composition of the catalyst—e.g., between 200 and 350 °C, 20 and 200 bar, and throughputs of 0.2 – 0.81 L of wax per liter of catalyst and hour. Yields are almost 100 %, based on starting wax [100], [101].

Under the severe reaction conditions of high-pressure hydrogenation, all aromatic compounds are hydrogenated to naphthenes, all dark substances are decomposed, and all sulfur and nitrogen atoms are removed as hydrogen sulfide and ammonia. Metallic impurities become bonded to the catalyst.

The process can be used to refine all classes of wax (paraffin waxes, intermediate waxes, and microwaxes).

4.2.4. Storage, Transportation, Commercial Forms, and Producers

Paraffin waxes are transported in liquid form (molten) in road tank trucks, rail tankers, or liquid containers. Solid waxes are marketed in slabs in small and large cartons and sacks, and as pastilles, flakes, and powder in sacks or big bags. They are also supplied in fiberboard and steel drums and in molded cartons.

Even refined waxes can become yellow or decompose to form odorous substances on prolonged heating, particularly in the presence of air or catalytically active metals. On storage, transportation, and processing, the waxes are therefore preferably heated in vessels, fitted with warm water equipment, and the containers and piping are well insulated. Storage tanks are often blanketed with inert gas. In some cases, antioxidants are added to fully refined waxes in quantities up to 0.01%. Deoiled and refined waxes are differentiated according to melting range (gradation). Some examples of commercial waxes are given below.

Europe (Melting gradation in °C)	United States (AMP* gradation in °F)
48/50	120–122
50/52	122–124
52/54	126–130
54/56	130–132
56/58	132–134
58/60	134–136
60/62	136–140
62/64	143–145

* AMP = American Paraffin Wax

Important producers of paraffin waxes include: Shell (United Kingdom, United States, Germany), Texaco (United States), BP (France, Germany), Mobil (United States, United Kingdom), Lützendorf (Germany), DEA (Germany), Wintershall (Germany), CFP (France), Total (France), Empetrol (Spain), Petrogal (Portugal). In Germany, for example, the companies either process the waxes to refined waxes (DEA, Wintershall) themselves or sell them to wax refineries (H. O. Schümann). Waxes are *sold* mostly by DEA, H. O. Schümann, and Wintershall.

Table 14. Standardized test methods for determination of physical data

Determination	Method	
mp (cooling curve), °C	DIN ISO 3841	ASTM D 87
Congealing point, °C	DIN ISO 2207	ASTM D 938
Needle penetration, 0.1 mm	DIN 51579	ASTM D 1321
Oil content, %	DIN ISO 2908	ASTM D 721
Viscosity, mm^2/s	DIN 51562/T1	ASTM D 445
Density, kg/m^3	DIN 51757	ASTM D 1298
Refractive index	DIN 51423/T2	ASTM D 1747
Color		
ASTM color index	DIN ISO 2049	ASTM D 1500
Lovibond color	IP * 17	
Saybolt color	DIN 51411	ASTM D 156
Odor		ASTM D 1833

* IP = Institute of Petroleum, United Kingdom.

4.2.5. Quality Specifications and Analysis

The quality specifications of paraffin waxes are determined mainly by their uses. They are differentiated in terms of physical characteristics, chemical composition, purity requirements, and application properties. The following standard specifications have been published for waxes that come in contact with foods:

1) Food and Drug Administration: Code of Federal Regulations (United States) [102]
2) FAO Food and Nutrition Paper (EC) [103]
3) Specifications of the European Wax Federation (EC) [87]
4) Kunststoffe im Lebensmittelverkehr (Plastics which come into contact with foods, Recommendations of the German BGA) [104]
5) Hydrocarbons in Food Regulations (United Kingdom) [105]

For waxes in cosmetic and pharmaceutical preparations, standards are found in the following publications: DAB (Germany), Codes Français (CF) (France), B.P. (United Kingdom), ISCI Monograph (Japan), U.S. Formulary (United States), and CTFA International Cosmetics Ingredient Dictionary (United States).

The most common test standards for the determination of physical data are summarized in Table 14. The most important test standards for determining chemical composition can be found in Table 15. Some examples of important test methods for determination of the purity of waxes are listed below:

Fluorescence	49th communication of the BGA
Alkali- or acid-reacting substances	DAB, B.P., U.S.P.
Behavior toward sulfuric acid	DAB, BGA, CF, U.S.P.
Presence of polycyclic aromatics	DAB, BGA, FDA

A looseleaf collection of national and international purity regulations for waxes has been issued by the European Wax Federation and is supplemented continuously [106].

Table 15. Standardized test methods for determining chemical composition

Determination	Method	
Ash content, %	DIN EN 7	ASTM D 482
Sulfur content, ppm	DIN 51400/T7	ASTM D 2622
Heavy metal content (As, Cr, Cd, Pb), ppm	XRF ^{a} analysis	
C-number distribution	DGF ^{b} M-V9	EWF ^{c} method
C-number max.	DGF ^{b} M-V9	IP ^{d} draft
n-Paraffin content, %	DGF ^{b} M-V9	ASTM draft

^{a} XRF = X-ray fluorescence spectromety.
^{b} DGF Einheitsmethoden der Deutschen Gesellschaft für Fettwissenschaften [Standardized methods of the Deutsche Gesellschaft für Fettwissenschaften (German Fat association)].
^{c} EWF = European Wax Federation.
^{d} IP = Institute of Petroleum, United Kingdom.

Some of the methods cited are basic test procedures for gaining accreditation according to the series of standards DIN EN 45 000 and the regulations of the Deutsche Akkreditierungsstelle Mineralöl GmbH (DASMIN). With accreditation, DASMIN recognizes the competence of testing laboratories to test the quality of waxes.

The determination of *application properties* is described mostly in in-house methods and includes, for example, oil binding properties, contraction, polishability, abrasion resistance, solvent retention, blocking point, steam permeability, flexibility, extendability, and color stability. Melting and congealing characteristics and latent heats are determined by using differential thermoanalysis [standardized DGF method M-III 14 (1989); ASTM-D 4419-84].

Special analytical methods have been developed to determine the *chemical composition* of paraffin waxes.

The content of *n*- and isoparaffins is determined most accurately by using computer-supported high-temperature gas chromatography [107], [108]. Here, the DGF method M-V9 (1988) is currently the universal standard. This method allows not only sufficient separation of the *n*- and isoparaffin signals (depending on the column length and injector temperature, waxes up to a C-number of ca. 80 can be determined quantitatively), but also determination of the C-number distribution and position of the C-number maximum.

The determination of *n*- and isoparaffins using urea adduct formation and molecular sieve adsorption has been shown to be a nonquantitative separation method. Only enrichment of the *n*-paraffins results. This, however, clearly diminishes with increasing molar mass and complexity of the waxes investigated [109].

For individual analysis of the different, paraffinic substance groups, coupled GC–MS and GC–NMR methods have been described [86], [110].

The determination of individual polycyclic aromatic hydrocarbons in waxes is currently possible into the μg/kg (ppb) region [86], [111], [112].

Table 16. Uses of paraffin waxes worldwide, 1992 (in 1000 t)

Use	United States	Asia	Western Europe	of which Germany
Candles and decorative modeling	100	450	230	55
Paper, cardboard, wood	450	220	120	25
Hot melt adhesives	60	20	20	10
Rubber and cables	40	20	30	10
Industrial products	50	30	50	
Pharmaceuticals and cosmetics	20	30	10	
Other uses	130	130	40	50 *

*Including industrial products, pharmaceuticals and cosmetics

4.2.6. Uses

Paraffin waxes are used in very different fields of industry. They not only are used as such, but are also important components of wax compounds (i.e., mixtures of paraffin waxes with other waxes, plastics, resins, etc.). The consumption figures listed in Table 16 cover the most important areas of use.

Slack waxes that are not processed further to refined waxes or subjected to dehydrogenation or cracking processes—for example, to give α-olefins —are used mostly in the wood and mineral oil processing industries.

Wood processing requires slack waxes, either molten or in the form of emulsions, mainly for the production of particle board and fiberboard. In the United States the production of artificial logs for barbecues and fires is also important.

In the *mineral oil processing industry*, slack waxes are used to produce petroleum jellies, wire-drawing agents, lubricants, spray emulsions, and anticorrosives.

In addition to wax industry products with a wide area of applications, slack waxes are also used to produce chlorinated paraffins, papers and cardboards for technical uses, cable coverings, solid anticorrosives, fire lighters, and wax torches.

Soft waxes have only a limited market in the production of industrial petroleum jellies and emulsions (wood and building industry), torches, and fire lighters and for conditioning fertilizers. The deoiling of soft waxes gives highly plastic and extendable products (isowaxes).

Refined Waxes [Scale Waxes, Filtered (Decolorized), Semi- and Fully Refined Waxes]. In Germany the producers of candles and wax goods (pictures, sculptures, ornaments) are the main users of refined waxes, which can be processed by all common production methods for candles (molding, drawing, extrusion molding, and powder compressing). In terms of quantity, paraffin wax is the most important raw material for the candle industry.

The second most important consumer of refined waxes is the wax industry, which requires paraffin waxes as an important raw material for compounding wax products for the most widely varying applications.

In the paper, cardboard, and packaging industries, refined waxes are used in the form of emulsions for paper sizing as antifoams in the production of paper, cardboard, and corrugated cardboard; and as impregnating and coating waxes for paper and packaging materials [113].

The viscosity of heat-seal waxes and hot-melt adhesives is adjusted with paraffin waxes. In rubber articles, tires, and solid cable coverings, waxes act as external lubricants/plasticizers, mold-release agents, and rapidly migrating protection agents of rubber against sunlight. They are therefore used predominantly for statically stressed rubber articles (profiles, gaskets, linings, and cable jackets).

Other important applications are in cleansers and polishes; cosmetics (creams, ointments, lipsticks); matches; plastics (external lubricants); office papers; and waterproof textiles (dry impregnation).

In the United States, by far the greatest proportion of refined waxes is used for paper and cardboard finishing, mainly for deep-freeze packaging and wrapping for meat, candy, and flowers; paper cups; yogurt and soft cheese pots; and corrugated cardboard [114].

In the former Soviet Union, refined waxes still play an important role in the production of synthetic fatty acids via catalytic oxidation processes. Fatty acids are mainly used in the soap industry.

4.3. Microcrystalline Waxes (Microwaxes)

4.3.1. Chemical Composition and General Properties

Like paraffin waxes, microcrystalline waxes consist of a mixture of saturated hydrocarbons that are predominantly solid at room temperature, such as n- and isoalkanes, naphthenes, and alkyl- and naphthene-substituted aromatics. Unlike paraffin waxes, isoparaffins and naphthenic compounds predominate here. The microcrystalline structure can be explained by the presence of strongly branched isoparaffins and naphthenes, which inhibit crystallization.

General Physical Properties. Microcrystalline waxes are insoluble in water and most organic solvents at room temperature. They are moderately to readily soluble in solvents such as chlorohydrocarbons, benzene, toluene, xylene, solvent naphtha, and turpentine oil, especially at elevated temperature. Solubility decreases markedly as molar mass increases. Solvents and oils are retained very strongly by microcrystalline waxes and therefore evaporate very slowly. The quality of some consumer products, such as petroleum jellies or floor and shoe polish, is determined by this retention capacity of microwaxes.

Chemical Properties. Microwaxes are more reactive than paraffin waxes because of the higher concentration of complex branched hydrocarbons with tertiary and quaternary carbon atoms. These C–C bonds are not very thermally stable (i.e. the waxes darken and resinify) on prolonged heating. In addition, they form black tarlike substances on contact with aggressive chemicals such as concentrated sulfuric acid or antimony pentachloride.

The reaction of microwaxes with oxygen at elevated temperature and in the presence of catalytically acting heavy-metal soaps is used for the production of oxidized microwaxes.

4.3.2. Division into Product Classes

Depending on the degree of refining, microwaxes are divided into the following classes:

1) Bright stock slack waxes (petrolatum)
2) Plastic microwaxes
3) Hard microwaxes

Table 17 and Figure 8 show some typical characteristics of microwaxes.

Table 17. Typical physical characteristics of microwaxes

Characteristic	Bright stock slack wax	Plastic microwax		Hard microwax
Congealing point, °C	67	72	75	87
Drop point, °C	76	78	80	91
Needle penetration, 0.1 mm				
at 25 °C	65	30	28	8
at 40 °C	140	85	66	18
Oil content, %	3.8	1.8	1.6	0.4
Viscosity (at 100 °C), mm^2/s	18.7	15.7	14.9	16.2
Refractive index (at 100 °C)	1.4505	1.4458	1.4412	1.4410
Color	dark brown	yellow	white	white

Bright stock slack waxes (petrolatum) constitute the most complex mixture in the group of microcrystalline waxes. The molar masses of the hydrocarbons can be between 320 and 1200 g/mol. The n-alkane content is seldom >10%. Asphaltenes, hydrocarbon resins, and sulfur, nitrogen, and oxygen compounds are present as impurities. The oil content is normally between 3 and 15%, but it can be as high as 30%. Bright stock slack waxes are white or yellow to dark brown, greasy to softly plastic, unctuous, smooth

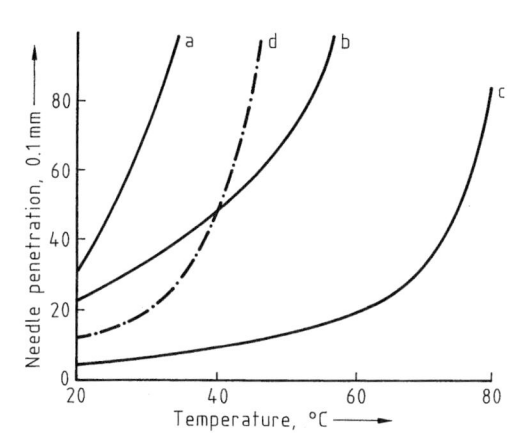

Figure 8. Variation of needle penetration with temperature for various microwaxes compared with a paraffin wax
a) Bright stock slack wax; b) Plastic microwax; c) Hard microwax; d) Paraffin wax 60/62

products with a typical odor, depending on the refining process. They have a very good oil-staining capacity, and some grades have a pronounced stickiness. The congealing points are between 60 and 75 °C; the drop points between 62 and 80 °C; and the hardness (needle penetration at 25 °C), between 45 and 160, depending on the oil content. The viscosities (10–30 mm^2/s at 100 °C), densities (810–860 kg/m^3 at 80 °C), and flash points (290–320 °C) are clearly higher than those of paraffin waxes.

Plastic Microwaxes. In plastic microwaxes the higher molar mass n- and isoparaffins and the weakly branched isoparaffins and naphthenes are enriched compared with the petrolatum from which they are obtained by deoiling. Molar masses are between 450 and 1200 g/mol. The n-paraffin content can be as high as 40 %. Depending on the degree of refining the impurities present in petrolatum are removed completely or to a great extent. The oil content is usually between 1.5 and 5 %.

The plastic microwaxes, although relatively hard, are definitely plastic, readily moldable, and kneadable. The plasticity is due to the presence of very long carbon chains in the n-paraffins and terminally branched isoparaffins. The waxes feel dry to slightly adhesive, and are sold as white to brown products. Congealing points are 65–80 °C, and the hardness (needle penetration at 25 °C) is 10 to 60. Highly refined microwaxes are odorless and tasteless.

Hard microwaxes also essentially consist of high molar mass alkanes (predominantly n-alkanes and isoalkanes with short side chains statistically distributed over the main carbon chain) and naphthenes. Average molar masses are between ca. 500 and 1200 g/mol. Thus, for example, a refined, hard microwax of North American origin contains 89 % nonpolar hydrocarbons (n-, iso-, and cycloparaffins) and has an average C-number of 48 [86]. The n-paraffin content can be as high as 70 %.

The extremely microcrystalline to almost amorphous structure of hard microwaxes is associated with the formation of supercooled melts during the setting process.

The waxes are tough and hard, barely ductile and weakly adhesive. Depending on the degree of refining, hard white, yellow and brown microwaxes are on the market, the white to pale yellow products being of highest purity. They solidify between 80 and 94 °C. The needle penetration at 25 °C is <10. The hardness of hard waxes decreases only slightly with temperature until close to their softening point (Fig. 8).

4.3.3. Occurrence of Raw Materials and Processing

Like paraffin waxes, microwaxes are components of crude petroleum. Because of their high molar masses, and thus high boiling range and poor solubility, they either are enriched in the vacuum residues (short residues) from lubricating oil distillation (residual waxes) or separate during the transportation and storage of crude oils (settling waxes). Settling waxes form pasty to solid deposits on the walls of pump tubes and pipelines and on the bottom of storage tanks (pipe and tank bottom waxes). Vacuum residues and pipe and tank bottom waxes are the raw materials for recovery of microcrystalline waxes.

Figure 9 shows schematically the process steps involved in the production of microcrystalline waxes.

In the United States, the most important producer country for microwaxes, the vacuum residues are subjected to deasphalting, refining, and dewaxing to produce high-value bright stock oils. Bright stock slack waxes (petrolatum) are produced in the process as economically important byproducts. However, only in the rarest cases they are processed further to microwaxes in the lube oil refineries themselves. Petrolatum is either marketed directly or used, for example, for the production of petroleum jellies. For further processing it is sold to companies specializing in the production of microwaxes. The same companies also process settling waxes purchased from oil companies.

Deasphalting is the first step in the processing of tank bottom waxes, pipe waxes, and residues, which have been freed from water, mechanical impurities, and volatile components. When propane is used as the deasphalting agent, considerable proportions of resins, polycyclic aromatics, and nitrogen and sulfur compounds are removed. To improve the selectivity with as low a propane throughput as possible, rotating disk contactors are now used instead of packed or sieve bottom columns. In addition, the extraction process is carried out in two steps [115].

Deasphalting of residues is followed by solvent extraction and dewaxing. The products formed here are the bright stock oils and bright stock slack waxes (petrolatum).

Deoiling. The crude and bright stock slack waxes thus obtained are deoiled by fractional crystallization (see p. 4923) if they are to be processed further to microwaxes. Preferred solvents are methyl ethyl ketone and methyl isobutyl ketone, often as a

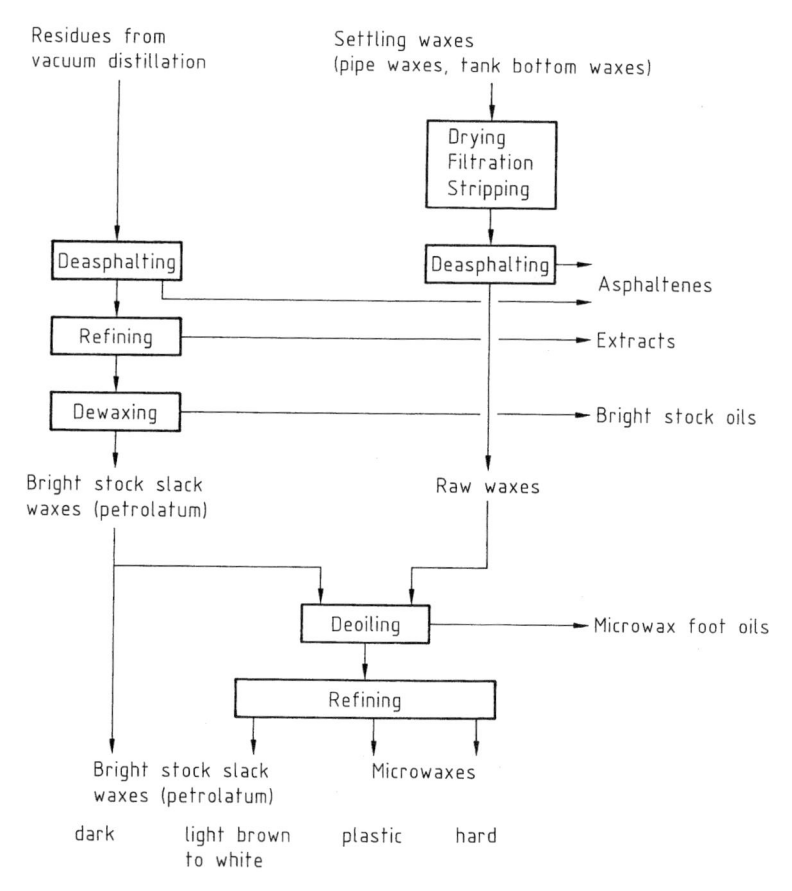

Figure 9. Process schematic for the production of microwaxes

mixture with benzene or toluene. Propane and naphtha, frequently used earlier, have become less important because of their low selectivity.

The crude waxes (deasphalted settling waxes) and bright stock slack waxes are mixed either molten or as a pulp (if the deoiling is to follow immediately after dewaxing the brightstock oils), without isolation of the slack wax with fresh, warm solvents and warmed until only the less soluble (high molar mass) waxes remain undissolved. These waxes are then filtered, washed, and freed from solvent (hard microwax fraction). To obtain plastic microwaxes the filtrate is cooled further. The properties of reprecipitated and refiltered waxes are determined by the filter temperature. The low molar mass, very plastic, and soft microwax foot oils [64742-67-2] remains in the filtrate. If the hard microwaxes are not extracted, a one-step crystallization process is sufficient for obtaining plastic microwaxes.

Deoiled microwaxes are still dark yellow to dark brown and not completely odorless, so they must subsequently be refined to satisfy a broad application range.

Table 18. Producers of microwaxes in the United States and commercial products

Producer	Production site	Commercial products
Bareco Div. of Petrolite Corp.	Kilgore, Tex. Barnsdale, Okla.	Petrolite Crown, Be Square, Mekon, Stawax, Cardis, Victory, Ultraflex
Mobil Oil Co.	Beaumont, Tex. Paulsboro, N.J.	Wax Rex, Mobilwax
Quaker State Oil Ref. Corp.	Emlenton, Pa. Farmers Valley, Pa. Newell, W.V.	Microwax, Superflex
Shell Oil Co.	Deer Park, Tex. Wood River, Ill.	Shellwax
Sun Oil Co.	Tulsa, Okla.	Sunoco Wax, Sunolite
Witco Chem. Corp.	Bradford, Pa.	Multiwax

Refining. Refining is carried out either with adsorbents (decolorizing) or with hydrogen (hydrotreating). Refining with concentrated sulfuric acid, which is possible in principle, has not become important economically because of the high loss of material.

For refining microwaxes with *adsorbents*, the same process principles apply as for decolorizing paraffin wax (see p. 4924). Since microwaxes are more difficult to decolorize than paraffin waxes, countercurrent percolation is used predominantly. Here, the wax melt continually comes in contact with fresh decolorizing clay [116]. Bauxite is also frequently used as a decolorizing agent.

Hydrotreating is carried out as described for paraffin waxes, in one or two steps in reactors packed with sulfur-resistant, heavy-metal-doped mixed catalysts (see p. 4925). Reaction conditions are ca. 350 °C and 150 bar, depending on the properties of the catalyst, at a space velocity of 0.3 L of wax per liter of catalyst and hour. Hydrogenated products are treated further by adsorptive decolorizing or stripping to completely remove odorous materials and volatile components.

In the United States, a whole range of microwaxes is produced according to the processed described. Products differ in their congealing point, drop point, hardness, viscosity, and color. They are sold in the liquid (molten) form and as solids in the form of slabs, pellets, and powder.

4.3.4. Commercial Products and Producers

Table 18 shows the most important producers of microwaxes in the United States and the corresponding commercial products.

In Europe, H. O. Schümann (Germany), Mobil (United Kingdom), CFR/Total, Mobil, and BP (France), Shell and Witco (Holland), Repsol (Spain), and Petrogal (Portugal), for example, produce microwaxes by deoiling bright stock slack waxes.

Table 19. Microwaxes for food packaging

Determination	Method	Specification
Congealing point, °C	DIN ISO 2207 ASTM D 938	65 (min.)
Needle penetration, 0.1 mm	DIN 51 579 ASTM D 1321	<50
Extractables, %	ASTM D 721	2.5 (max.)
Viscosity at 100 °C, mm^2/s	DIN 51 562/T1 ASTM D 445	>13
Density at 100 °C, kg/m^3	DIN 51 757 ASTM D 1298	790–840
Refractive index at 100 °C	DIN 51 423/T2 ASTM D 1747	1.430–1.450
Color	DIN ISO 2049 ASTM D 1500	2.0 (max.)
Odor	ASTM D 1833	weak
Ash content, %	DIN EN 7 ASTM D 482	<0.05
Sulfur content, ppm	DIN 51 400/T7 ASTM D 2622	<300 [a] <4000 [b]
Heavy-metal content (As, Cd, Cr, Pb), ppm	XRF analysis [c]	<1
C-number distribution of n-paraffins	DGF M-V9 (88) IP 372-85	23–85
Non-n-paraffin content, %	DGF M-V9 (88)	25–75
UV-Absorption	$ 172.886 FDA	
at 280–289 mm		0.15 (max.)
at 290–299 mm		0.12 (max.)
at 300–359 mm		0.08 (max.)
at 360–400 mm		0.02 (max.)

[a] Hydrotreated.
[b] Refined with sulfuric acid/decolorizing clay.
[c] XRF = X-ray fluorescence spectrometry.

4.3.5. Quality Specifications and Analysis

The quality required depends on the industrial use. In principle, the specifications given for paraffin waxes apply here.

Basic product specifications can be found in the following sources:

1) Food and Drug Administration; Code of Federal Regulations (United States) [102]
2) European Wax Federation (EC) [87]
3) Kunststoffe im Lebensmittelverkehr; Empfehlungen des Bundesgesundheitsamtes Deutschland (Plastics for Contact with Foods; Recommendations of the Federal Health Authority, Germany) [104]

The EWF specifies microwaxes for use in packaging materials for foods, as shown in Table 19.

In the use of microwaxes to produce food packaging materials, and cheese and chewing gum waxes, purity and physiological nontoxicity are particularly important. For microwaxes the absence of specific polycyclic aromatics is required. These substances can occur as impurities in insufficiently refined waxes, and some of them have been found to be carcinogenic. Polycyclic aromatics can be detected by UV extinction after a special enrichment process (detection limit in microwaxes 10^{-7} g per gram of wax substance). Microwaxes have been shown to be toxicologically and physiologically harmless in extensive animal experiments, when UV adsorption values specified according to test regulations of the FDA [102] and the BGA [104] are not exceeded.

The physical characteristics of microwaxes are determined according to the same standardized methods used for paraffin waxes (see Table 14). In addition, some special analytical methods have been introduced, for example:

Determination of gloss and gloss stability	ASTM D 2895
Determination of blocking point	ASTM D 3234
Determination of sealing stability	TAPPI [Tech. Assoc. Pulp & Paper Ind. (United States)] T 642
Determination of water vapor permeability	ASTM D 988
	DIN 53 413
Determination of specific penetration resistance	DIN 53 482
Determination of dielectric loss factor	DIN 53 483

Oil-binding capacity, adhesiveness, and light stability are additional quality factors.

4.3.6. Uses

Bright Stock Slack Waxes. The greatest proportion of bright stock slack waxes is processed further to microwaxes. Important consumers are the wax-producing industry, which requires crude and refined bright stock slack waxes for compounding adhesive waxes, cheese waxes, chewing gum base, cosmetic preparations, sealing and cable compounds, impregnating insulating materials and pesticide sprays; and the mineral oil processing industry, which produces anticorrosives, cavity and underfloor protection agents for motor vehicles, and petroleum jellies. Other uses include the production of artificial fire logs and oxidized petrolatum, which is suitable for the formulation of rust preventives.

Oxidized petrolatum and microwaxes are mainly produced industrially in the United States. Air at normal or at slight overpressure is blown through the wax, which is heated to 150–180 °C until the desired degree of oxidation is reached. Darkening of the wax is avoided by adding fresh wax to the reaction zone. Oxidized microwaxes are specified and traded according to their melting range and degree of oxidation (expressed in terms of acid and saponification numbers). The acid and ester groups, which are distributed statistically over the hydrocarbon chains, can be reacted with inorganic and organic bases to form soaps that can be dispersed in water (emulsified waxes). These compounds also impart to the oxidized waxes an excellent adhesion to metals.

Plastic Microwaxes. The main consumer of plastic microwaxes is the wax industry. Plastic microwaxes are added to wax compounds that require either pronounced plastic or oil- and solvent-staining properties for their industrial applications.

Typical applications include the following:

1) Paper and cardboard coating for packaging meat and sausage products, cereals, bread, candy, ice cream, cookies, yogurt, and frozen food
2) External lubricant (sliding) for extrusion of plastics
3) Formulation of cheese waxes and chewing gum bases
4) Decorative for candles, production of wax pictures and sculptures

5) Polishes and grinding agents
6) Lost wax casting and dental compounds
7) Foam regulators for detergents
8) Protectives against crack formation in tires and rubber articles (sun-proof waxes) [117]
9) Additives for explosives and propellant materials.

Hard microwaxes have numerous applications based on their high melting points and great hardness; favorable hardness–temperature behavior; and good solvent retention. They are used as release agents for pressing plastics and fiberboard, in the production of ceramic articles and in molding polyurethane foam, in solvent-containing cleansing agents, in grinding and polishing pastes, and flat varnishes and paints.

The Committee on Toxicity (COT) in the United Kingdom has recommended that only hard microwaxes be approved for use in the production of cheese waxes, poultry defeathering compounds, and chewing gum [118].

On an industrial scale, hard microwaxes are *oxidized* to an acid number of 30 and are used in this form to produce emulsions (polishes and release agents) and as additives for paints and coatings, printing inks, and office transfer papers.

4.4. Legal Aspects

Apart from environmental effects, petroleum waxes can be absorbed by the human organism via foods (paraffins either are contained in them naturally, have been added to them, or have migrated there from the packaging materials), from chewing gum (only on swallowing or the simultaneous consumption of fat-solubilizing chocolates), and through the use of lipstick. To protect human health, national and international regulations and recommendations have been enacted to limit the oral consumption of waxes.

The use of waxes in packaging materials is subject of *EEC legislation* [119]. In the appendix lists (lists of additives), waxes are permitted, provided that specific migration limits of 0.3 mg per kilogram food for waxes treated with sulfuric acid or decolorizing clay and 3 mg per kilogram for high-pressure hydrotreated waxes are not exceeded.

Extensive literature on the migration of paraffins into foods has been published [120]–[123].

The permitted paraffins must correspond to published specifications (see Section 4.2.5).

In *Germany* the addition of petroleum waxes to food and animal feed is not allowed. The use of waxes as cheese coatings and in chewing gum is controlled by supplementary regulations.

In the *United Kingdom* the Ministry of Agriculture, Fisheries, and Food (MAFF) proposes a prohibition of petroleum waxes as food additives in England, Scotland, and Northern Ireland [118]. Only specified microwaxes may still be added to foods for a limited period.

In the *United States*, not indication has been given of any intention to ban paraffins in food.

4.5. Ecology

Paraffin and microcrystalline waxes are insoluble in water and can be separated as solids, together with sludge, in oil separators and sewage treatment plants. In the German catalogue of substances that are hazardous to water, they are therefore listed in water hazard class 0 (Wassergefährdungsklasse, WGK 0). Since paraffins have a water solubility of < 1 µg/L, they have no toxic effects on aquatic organisms. Used paraffin-containing packaging materials can be combusted without any problem in waste incineration plants. No polluting substances are formed, and the high calorific value of the paraffins improves the energy balance of the plants.

The *biodegradability* of petroleum waxes satisfies OECD guideline 301 B (1981) [124]. With the exception of microwaxes, biodegradabilities of more than 60% after 28 d were measured. Bacterial degradation mechanisms are described in [125]. According to this, paraffin-containing packaging materials can also be composted [126]–[128]. No substances that are toxic or hazardous to groundwater are formed.

The biodegradation of waxes under anaerobic conditions (covered landfills) proceeds much more slowly.

A large number of investigations have been carried out on the *recycling* of used wax-containing packaging materials. The water-insoluble, hydrophobic petroleum wax tends to form particles in the paper mills and give rise to blockages in the sizing equipment [129].

In Europe, ca. 20×10^6 t of waste paper was recycled as a raw material for paper production in 1991. This figure includes 0.35×10^6 t of wax-containing paper and cardboard with average wax content of 12%. Assuming an even distribution, this gives a wax content of 0.21% wax in wastepaper [130]. The practice has shown that in modern wastepaper processing plants, wax-containing packaging materials do not interfere with the dilution process, but nearly increase the quantity of waste products. These waste products can be used to produce water-resistant cardbord.

4.6. Economic Aspects

In 1992, 9.5×10^5 t of paraffin and microcrystalline waxes were *produced* in the United States, 8×10^5 t in Asia, 6.5×10^5 t in Western Europe, 2.6×10^5 t in Germany alone, 2.0×10^5 t in Eastern Europe. and 4.0×10^5 t in the rest of the world.

In industrialized countries, a stagnation (probably a decrease) in petroleum wax production is anticipated for the coming years, independent of the growth in demand. This is because of the further development of car engine and tribology technology (reduced specific lubricant consumption and longer periods between oil changes), an increase in the proportion of partially or fully synthetic and biodegradable lubricants, the recycling of used oils, and the closing down of excess capacity in distillation plants despite an increase in the number of motor vehicles and the distances traveled [131].

However, in Asia, particularly China and Malaysia, an increase in wax production to 1.5×10^6 t in 2000 is predicted, because these waxes can be inexpensively produced from indigenous wax-rich crude petroleum by direct dewaxing. According to careful estimates, worldwide petroleum wax production will be ca. 3.8×10^6 t in 2000 (40% in Asia, 25% in the United States, 17% in Western Europe).

The worldwide *consumption* of petroleum waxes was ca. 2.8×10^6 t in 1992, of which (in million tonnes) 0.5 was in Western Europe, 0.2 in Eastern Europe, 0.85 in the United States, and 0.9 in Asia.

4.7. Candles

According to German commercial quality class RAL 040 A2 (Issue 02/1993), candles are sources of heat and light and consist of a wick and a solid combustible material surrounding the wick. Therefore tea lights, lights for graveyards, and oil lamps are also included in this category.

The first candles have been traced to shortly after the birth of Christ. With Christianity, and later with the rise of principalities and the upper middle class, the use of candles developed rapidly. Whereas initially, tow wicks, beeswax, tallow, and spermaceti were used for candle production, from the 19th century onward the use of cotton wicks, petroleum wax, and stearin permitted the problem-free handling of candles.

Today, refined (petroleum) wax is the main raw material used in candle production. Candles are also made from stearin, beeswax, composites (mixtures of various waxes), and ceresin (mixtures of paraffin and hard waxes). Candles are dyed with wax-soluble organic aniline dyes (full dyeing), organic pigments, and (less often) natural pigments (dip dyeing). Fashionable candles with a colored dipping-paint coating are also available. Wicks consist of twisted cotton threads and are sold as round or flat wicks of different strengths. The wicks are twisted and prepared so as to guarantee uniform burning, good wick curling position, and good suction capacity for the molten combustible material and to prevent afterglow of the wick once the candle has been extinguished. Candles are sometimes perfumed.

A range of mechanical and hand-worked processes exist for industrial production of candles. In the *drawing process* a candle strand is produced by passing the wick several times through the hot wax bath. The strand then passes onto the cutting machine, where it is cut to the desired length of the candle. With continuously operating candle-drawing machines, production of more than 10 000 candles per hour is possible. In the *molding process*, molten wax is poured into molds, in the middle of which the wicks are held taut. Molding machines are available in different sizes. Modern rotating molding machines have an hourly capacity up to 10 000 candles. The *powder pressing process* uses powdered wax as the starting material, which is either molded together with the wick in an extruder to form a candle strand or pressed into molds in a stamp press. At the same time, the wick is introduced with a tubular needle. Another process exists, known as the *dunking process*. Here, the wick is dipped repeatedly into molten wax.

Blank candles are made pointed by using milling cutters and are coated with dyed dip-coating material if necessary.

The quality control of candles involves, for example, determining their weight and dimensions, burning period, and burning properties (no drips or soot). The production of candles in Germany (East and West) was 80 000 t in 1988, and 17 000 t were imported. Thus, with an export of 22 000 t, a total of 75 000 t was consumed within Germany.

5. Fischer–Tropsch Paraffins

Production. The sole producer of Fischer–Tropsch paraffins is South African Coal, Oil and Gas (SASOL) in the Republic of South Africa. The formerly used ARGE (Ruhrchemie/Lurgi consortium) process, which employed iron catalysts arranged as a fixed bed in multitubular flow reactors, was changed to the SASOL slurry bed process (SSBP), which afforded considerable cost savings. Production capacity has been doubled [132].

In the slurry or bubble column reactor, finely divided iron catalyst (average particle size 50 µm) is suspended in a Fischer–Tropsch wax with low viscosity at the reaction temperature, and synthesis gas (a mixture of carbon monoxide and hydrogen) is bubbled through. At reaction temperatures of 220–240 °C and pressures around 2 MPa, *n*-alkanes and *n*-alkenes are formed preferentially, with up to 40 % crude wax in the synthesis product.

Crude wax is separated from the other products of synthesis by fractional condensation. The low-boiling constituents (gasoline and diesel fuel) are removed by distillation under atmospheric pressure, followed by vacuum distillation to separate the *soft waxes*. Besides *n*-alkanes, distillation bottoms also contain alkenes, hydrocarbons with hydroxyl and carbonyl groups, and colored components. For purification and stabilization,

Table 20. Typical properties of Fischer–Tropsch soft and hard paraffins

Characteristic	Sasol Wax M	Sasol Wax H2 Vestowax SP 1002
Drop point, °C		106–112
Needle penetration, 0.1 mm		1–3
Color	white	white
Molar mass, g/mol	400	700
Density (at 23 °C), g/cm^3	0.94	0.94
Viscosity (120 °C), mPa · s	<20	<20

the bottom products are subjected to a hydrofining step employing a nickel catalyst to yield a white *hard wax* that is practically free of alkenes, aromatics, functional hydrocarbons, and sulfur compounds.

Soft waxes are also subjected to hydrogenation.

Properties. Fischer–Tropsch waxes consist essentially of *n*-paraffins with chain lengths between 20 and 50 carbon atoms. Products with an average molar mass of 400 g/mol are marketed as soft, and those with an average molar mass of 700 g/mol as hard waxes. The paraffins have a fine crystalline structure and, because of the narrow molar mass distribution, a small melting range and very low melt viscosities. Congealing point, density, and hardness increase with increasing mean molar mass. The low molar mass compared with polyolefin waxes is the reason for a certain displaceability of the crystal layers relative to each other and, associated with this, the polishability.

Synthetic paraffins are fully compatible with refined waxes, polyolefin waxes, and most vegetable waxes. They are soluble at elevated temperature in the usual wax solvents (e.g., naphtha, turpentine, and toluene) to give clear solutions. Addition of 10–20% synthetic paraffin to other waxes increases their congealing point and hardness without significantly influencing melt viscosity. In wax pastes, the tendency of Fischer–Tropsch waxes to form microcrystals increases solvent retention. Some typical data for Fischer–Tropsch waxes are listed in Table 20.

Uses. Fischer–Tropsch waxes are used in plastics processing as lubricants for poly(vinyl chloride) and polystyrene, as well as mold-release agents; as melting point improvers, hardeners, and viscosity reducers in hot melts and candles; and, because of their good polishability, for the production of cleaning agents and polishes.

Micronized waxes improve the abrasion resistance of paints and printing inks.

Oxidized waxes containing fatty acids and fatty acid esters are produced from synthetic paraffin by oxidation and partial saponification. Major areas of application are as mold-release agents in plastics processing, in polishes and cleaning agents, and as auxiliaries in the textile and paper industries.

Trade Names. Fischer–Tropsch waxes are marketed, e.g., as Sasol Wax (Sasol Marketing Co., Johannesburg, South Africa) and Vestowax SH/SP (Hüls AG, Marl, Germany).

6. Polyolefin Waxes

6.1. Production and Properties

6.1.1. Polyethylene Waxes by High-Pressure Polymerization

High-pressure polyethylene (PE) waxes are produced, like high-pressure polyethylene plastic, at high pressure and elevated temperature in the presence of radical formers. As waxes, their molar masses are considerably lower than those of plastics. The molar mass range is adjusted during polymerization by the addition of regulators.

High-pressure polyethylene waxes are partially crystalline and therefore consist mainly of branched molecular chains in which shorter side chains, such as ethyl and butyl, predominate. They generally have low densities (low-density polyethylene waxes, LDPE waxes). Less branched polyethylene waxes with higher crystallinity and density (high-density polyethylene waxes, HDPE waxes) can be produced by increasing the pressure.

High-pressure polyethylene waxes with molar masses between 3000 and 20 000 g/mol [weight-average molar mass $\bar{M}_w$, determined by gel permeation chromatography (GPC)] currently dominate the market.

The development of high-pressure PE waxes proceeded in parallel to that of LDPE plastics. In 1939, waxlike ethylene polymers were formed by using a variant of the ICI high-pressure polymerization process [133]. Industrial production of high-pressure ethylene waxes was started in the 1940s [134], [135]. Parallel to this, the technology of thermal depolymerization—the thermal degradation of polyethylene plastics to polyethylene waxes—was developed (see Section 6.1.4).

With the rapid growth of petrochemistry and the expansion of polyethylene plant capacities, the economic importance of PE waxes also increased rapidly. Waxes with very versatile property profiles could be synthesized by varying the density and molar mass of the homopolymers and functionalizing polyethylene by copolymerization with various monomers (e.g., vinyl acetate or acrylic acid) or by melt oxidation.

Because PE waxes are rather inexpensive, have improved applicability and consistency in quality, and are in constant supply they have displaced expensive natural waxes (e.g., carnaúba and montan waxes) in many areas.

The high-pressure process is not applicable for production of polypropylene waxes, because only soft or oily products can be obtained. These waxes are therefore produced by the Ziegler process (see Section 6.1.3) or by depolymerization of high-density polypropylene (see Section 6.1.4).

6.1.1.1. Production

The technology involved in the production of high-pressure PE waxes (see Fig. 10) is analogous to that used for high-pressure polyethylene. The only differences are in the product finishing process. Unlike viscous plastic melts, which must be granulated underwater in granulators, for example, the mobile wax melt can be converted into powder by spraying or into granules by using dicers.

The reaction vessels are stirred autoclaves or tubular reactors, the design of which can vary considerably.

Reaction Mechanism. The high-pressure polymerization of ethylene is a highly exothermic radical chain reaction, which is initiated by reaction of ethylene with radicals formed by decomposition of an initiator [136]. Chain propagation occurs through addition of further ethylene units. Chain growth is terminated, for example, by reaction with a regulator molecule such as propene:

$$R-CH_2\dot{C}H_2 + CH_2=CH-CH_3 \rightarrow R-CH=CH_2 + CH_3-\dot{C}H-CH_3$$

Polymerization is terminated by recombination or disproportionation of two macroradicals:

$$2R-CH_2-\dot{C}H_2 \begin{cases} \rightarrow R-CH_2-CH_2-CH_2-CH_2-R \\ \rightarrow R-CH_2-CH=CH-CH_2-R \end{cases}$$

Long-chain branching occurs by intermolecular chain-transfer reactions:

$$R^1-CH_2\dot{C}H_2 + R^2-CH_2CH_2R^3$$
$$\longrightarrow R^1-CH_2CH_3 + R^2-\dot{C}H-CH_2R^3$$
$$\xrightarrow{nCH_2=CH_2} \begin{array}{c} R^2-CH-CH_2R^3 \\ | \\ (CH_2CH_2)_{n-1}-CH_2\dot{C}H_2 \end{array}$$

A particular feature of polyethylene wax synthesis by the high-pressure process is the increased formation of short ethyl and butyl side chains. These are formed by intramolecular radical transfer:

$$\begin{array}{c} H_2 \\ C \\ R-CH \quad CH_2 \\ | \quad \quad | \\ H \quad \, CH_2 \\ \cdot CH_2 \end{array} \longrightarrow \begin{array}{c} H_2 \\ C \\ R-\dot{C}H \quad CH_2 \\ | \quad \quad | \\ \quad \quad CH_2 \\ CH_3 \end{array}$$

$$\xrightarrow{nCH_2=CH_2} \begin{array}{c} R-CH-CH_2CH_2CH_3 \\ | \\ (CH_2CH_2)_{n-1}CH_2\dot{C}H_2 \end{array}$$

At the high pressures necessary for polymerization, ethylene is in a supercritical state. Polymerization therefore takes place in a one-phase system. After leaving the reactor,

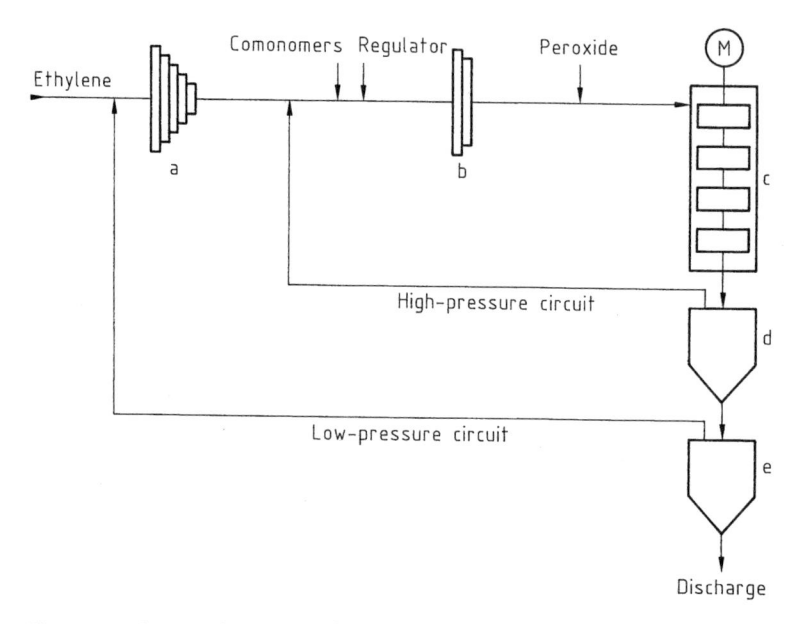

Figure 10. Schematic of production of polyethylene waxes by the high-pressure process in stirred autoclaves
a) Precompressor; b) Postcompressor; c) Autoclave reactor; d) High-pressure separator; e) Low-pressure separator

the reaction mixture is decompressed in high- or low-pressure separators, and unreacted ethylene evaporates. The wax remains as a melt and ethylene is recycled.

Reaction Conditions. The structure, molar mass, and thus properties of polyethylene wax are determined mainly by reaction pressure, reaction temperature, type and quantity of initiator and molar mass regulator, and reactor type and geometry [137]–[141].

In the homo- and copolymerization of ethylene the *reaction pressure* is usually 150–320 MPa. Somewhat lower pressures are used in autoclaves than in tubular reactors. Other process variants involve much lower pressures of 70 MPa (max.) with isopropanol as the molar mass regulator [139], [140].

Higher pressure favors chain propagation and thus leads to very short residence times in the reactor. It also inhibits chain-transfer reactions so that with increasing pressure the degree of branching decreases and the density, crystallinity, hardness, and *mp* increase. In tubular reactors, rising pressure increases the polymer yield, whereas in autoclaves this effect is smaller.

For homopolymerizaion the *reaction temperature* is 200–350 °C and for copolymerization 200–300 °C. High reaction temperature favors chain-transfer reactions. For this reason the polyethylene waxes formed have many short-chain and only a few long-chain branches [142] and thus low densities. Higher-density waxes with lower degrees of branching are formed at lower reaction temperature.

Organic peroxides and molecular oxygen (the latter exclusively in tubular reactors) are used as *initiators*. They decompose into radicals under polymerization conditions. To ensure that the peroxides are metered reproducibly, they are dissolved in organic solvents.

Hydrogen and almost all organic compounds can act as *molar mass regulators* (chain terminators). Hydrogen, lower alkanes (e.g., propane), lower alkenes (e.g., propene or butene), alkyl aromatics, lower aldehydes (e.g., propionaldehyde), and lower alcohols (e.g., isopropanol) or mixtures of these substances are mainly employed. The concentration and reactivity of these regulators determine the average molar mass, the degree of branching, and thus the density of polyethylene wax. The activity of the regulator increases with increasing temperature and decreases with increasing pressure [136].

Reactor type and geometry also significantly influence polymer structure and properties. Because of the backmixing occurring in *autoclaves*, spherical molecules with many long-chain branches are predominantly formed. In *tubular reactors* with plug flow, long, straight molecules with little long-chain branching predominate [143]. High-density, highly crystalline polyethylene waxes can therefore be produced only in tubular reactors.

Just as the geometries of the autoclave and the tubular reactor differ, so do operating procedures. In the case of autoclaves the initiator solution and the molar mass regulator are charged directly to the preheated pressurized vessel by using high-pressure pumps. In tubular reactors the initiator (e.g., air) and the molar mass regulator are added to the reaction mixture before the compression stage or at the entrance to the reactor. With autoclaves, comonomers can be charged to different parts of the reaction vessel (see Fig. 10).

Product Finishing. Shaping of the molten polyethylene can be performed directly after removal of gaseous products in high- and low-pressure separators and subsequent fine degassing and filtration. Since no metal-containing catalysts are used in high-pressure polymerization of ethylene, no catalyst removal is necessary. Low molar mass fragments and residues of the initiator and molar mass regulator evaporate mainly in the separators. Only an extremely small proportion is incorporated into the polymer.

Granules can be produced by using strip dicers. The wax melt flows in thin strips on a continuous, water-cooled metal band and, after solidification, is chopped with a rotating knife. *Pastilles* can be obtained by metered dripping of the melt from a distributor head onto this type of cooling band.

A particular characteristic of wax melts is their low melt viscosity. Thus, spraying the melt to form *powders* of spherical particles is also possible. By using special nozzle arrangements (binary nozzles) with nitrogen as the atomizing gas, *micronized waxes* (very fine powders) with particle sizes in the micrometer range can be produced.

Powders and micronized waxes can also be obtained by grinding granules in jet mills. With soft or very viscoelastic waxes, cooling with dry ice or liquid nitrogen is necessary. Coarse particles must be removed by subsequent classification.

Like all organic dusts (e.g., coal or flour), wax powders, and particularly micronized waxes, are highly susceptible to dust explosions. Appropriate safety precautions must be taken during processing, in particular grounding all installations to avoid spark formation through electrostatic charge buildup.

6.1.1.2. Properties

The transition from polyethylene plastic to polyethylene wax is flexible. If the molar mass is lowered, the thermoplastic materials is gradually changed to a wax. The DGF definition of wax (see Section 1.2) gives an approximate boundary: For waxes, an upper limit to the melt viscosity of ca. 20 000 mm^2/s at 120 °C is defined, which corresponds to an average molar mass (weight-average molar mass $\bar{M}_w$) of ca. 37 000 g/mol.

The properties of polyethylene waxes are determined strongly by $\bar{M}_w$ as a measure of the average chain length and by the degree of branching as a measure of the shape of the molecules. Melt viscosity increases with increasing molar mass. Crystallinity, hardness, *mp*, and solidification point increase as the degree of branching decreases. These data are important for application-oriented properties.

The molar mass can be determined by GPC. This method gives the weight-average ($\bar{M}_w$) and the number-average ($\bar{M}_n$) molar mass. The ratio $\bar{M}_w/\bar{M}_n$ is a measure of molar mass distribution and is known as the polydispersity index. For high-pressure polyethylene waxes currently produced, this value lies between 2.0 and 2.5. The waxes have a much narrower molar mass distribution than high-pressure polyethylene, for which the values are 3 – 8 (tubular reactor) and 12 – 16 (autoclave) [136].

Table 21 gives a series of structure and property data for three high-pressure polyethylene waxes with increasing density from two different producers (some of this represents unpublished data supplied by the authors).

The following conclusions can be drawn from these data:

1) The degree of branching is generally low with ca. ten branches (max.) per molecule. The side chains are mostly ethyl and butyl groups with a very small proportion of long-chain branches [144] – [146].
2) The density, crystallinity, hardness, solidification point, and drop point increase as degree of branching decreases [147]. All density-dependent properties vary correspondingly.
3) High-pressure PE waxes contain a small proportion of double bonds, whose distribution in the macromolecule depends on the production process, as the large differences between the levels of central double bonds show. The formation of double bonds is attributed to depolymerization of chain radicals at high reaction temperature [136], [148], e.g.:

$$\underset{R^2}{\overset{R^1}{\diagdown}}CH-\dot{C}H-R^3 \longrightarrow R^1-CH=CH-R^3 + \dot{R}^2$$

Table 21. Structure and properties of high-pressure polyethylene waxes A – C and D – F from two different producers

Characteristic	A	B	C	D	E	F
Density (23 °C), g/cm^3 (DGF-M-III 2a)	0.92	0.92	0.945	0.92	0.92	0.94
CH$_3$ per 1000 C atoms	26	22	17	32	23	20
Branches per molecule	3.4	1.8	1.0	3.7	1.9	1.8
Crystallinity (from IR data), %	54	59	67	52	62	64
Central double bonds per 1000 C atoms	0.1	0.4	0.4	0	0	0
Terminal double bonds per 1000 C atoms	0.1	0.1	0	1.3	1.2	1.3
Side-chain double bonds per 1000 C atoms	0.6	0.4	0.2	0.9	0.5	0.4
Ball indentation hardness (23 °C), bar (DGF-M-III 9a)	200	370	550	314	413	482
Solidification point, °C (DGF-M-III 4a)	96	99	106	93	101	104
Drop point, °C (DGF-M-III 3)	106	112	116	106	111	113
Melt viscosity (120 °C), mm^2/s (DGF-M-III 8)	1310	1160	1200	780	925	890
Weight-average molar mass $\bar{M}_w$, g/mol	6140	5930	6000	5200	5500	5500
Number average molar mass $\bar{M}_n$, g/mol	2920	2390	2440	2490	2380	2660
$\bar{M}_w/\bar{M}_n$ (polydispersity index)	2.1	2.5	2.5	2.1	2.3	2.1

High-density polyethylene waxes are colorless, white to transparent and form clear melts. Like other waxes, they dissolve in nonpolar solvents (e.g., aliphatic, aromatic, and chlorinated hydrocarbons) on heating and generally crystallize as very fine particles on cooling. Depending on the type and concentration of the wax, they then form mobile dispersions or paste-like gels, which frequently exhibit thixotropic properties.

6.1.2. Copolymeric Polyethylene Waxes by High-Pressure Polymerization

In the high-pressure process, many other monomers can copolymerize with ethylene giving rise to considerable changes in product properties [139], [141]. For copolymerization, autoclaves with their intensive mixing, constant temperature, and consequently stable reaction process are particularly suitable. In industry, vinyl acetate and acrylic acid are used mainly as comonomers, giving waxes with higher polarity and lower crystallinity. Here also, the degree of branching decreases with increasing pressure and decreasing temperature. Density, *mp*, and hardness increase accordingly.

Table 22 gives characteristic data for two important copolymeric high-pressure polyethylene waxes with vinyl acetate and acrylic acid as comonomers.

Ethylene – vinyl acetate copolymer is a relatively polar but nevertheless hydrophobic wax that can be dispersed in organic solvents particularly well. It is used as an additive in metallic automotive paints, as a dispersing agent in pigment concentrates, and as a component of hot melts.

Ethylene – acrylic acid copolymer is suitable for aqueous systems because it can be emulsified readily and is chemically and thermally intensive. Its emulsions are used mainly in modern floor polishes and mold-release agents.

Table 22. Physical and chemical data for copolymeric high-pressure polyethylene waxes

Characteristic	Ethylene–vinyl acetate copolymer wax	Ethylene–acrylic acid copolymer wax
Weight-average molar mass $\bar{M}_w$, g/mol	ca. 6800	ca. 6100
Number-average molar mass $\bar{M}_n$, g/mol	3000	3000
Melt viscosity (120 °C), mm^2/s (DGF-M-III 8)	2000	1300
Solidification point, °C (DGF-M-III 4a)	85	93
Drop point, °C (DGF-M-III 3)	96	102
Ball indentation hardness (120 °C), bar (DGF-M-III 9a)	120	410
Acid number, mg KOH/g		45
Vinyl acetate content, %	10	

6.1.3. Polyolefin Waxes by Ziegler–Natta Polymerization

Like high molar mass polyethylene plastics, PE waxes can be produced by the Ziegler low-pressure process using organometallic catalysts, in addition to the radical process at high pressure and temperature. Like high-density polyethylene, Ziegler waxes have a mainly linear molecular structure. They can contain short side chains but not long ones like high-pressure polyethylene waxes.

For direct synthesis of polymer waxes from propylene or higher α-olefins, only the Ziegler–Natta process is suitable.

Ziegler PE waxes currently on the market have a molar mass between 800 and 8000 g/mol ($\bar{M}_n$). They thus bridge the gap between Fischer–Tropsch and paraffin waxes and HDPE thermoplastics. Besides controlling chain length, the polymerization process allows adjustment of the degree of crystallinity so that both hard–brittle, high-melting products and softer, flexible products with low melting points can be obtained.

The soluble, soft wax fractions inevitably produced in the suspension process for HDPE are more similar to paraffin waxes than to PE waxes because of their low molar masses. For normal hard wax applications, they can be used only after processing by distillation or extraction [149].

6.1.3.1. Production

For the production of Ziegler–Natta polyolefin waxes, Ziegler catalysts that give polymers with low average chain lengths and narrow chain-length distributions, and are sufficiently active under the special conditions of wax synthesis, are particularly suitable. The low degree of polymerization typical of waxes is achieved by carrying out polymerization in the presence of hydrogen as a molar mass regulator at comparatively high temperature (usually between 100 and 200 °C, sometimes even higher) [150]–[154]. Aliphatic hydrocarbons are used as reaction medium. Because of the high reaction temperature the waxes are formed in solution (solution polymerization). If the product viscosity is sufficiently low the molten polymer formed can itself function as the solvent (bulk polymerization) [155], [156]. Unlike modern HDPE processes in which the catalyst is not removed, in the case of wax synthesis the catalyst must usually be decomposed and filtered off [157], [158]. This applies at least to the heterogeneous titanium catalysts still used today (see below) and is necessary because waxes have higher purity requirements than PE plastics as a consequence of the applications for which they are used. After the solvent has been distilled off, the wax melt is shaped (e.g., by spraying or pastille formation).

Catalyst Systems. Classical Ziegler catalysts, consisting of titanium tetrachloride and alkylaluminum compounds, can be used for wax synthesis, but their activity is low [155], [156], [159]–[161]. The degree of branching and thus also the density, drop point, and hardness can be adjusted by copolymerization, by the special type of catalyst preparation [160], and by varying the polymerization temperature [155], [161].

The state of the art for production of *PE homo- and copolymer waxes* involves, as in plastics production, the use of supported catalysts, which contain titanium atoms as the active species and magnesium compounds as the carrier material. Catalysts derived from titanium tetrachloride and magnesium chloride, oxide, hydroxide [151]–[154], or alkoxide [150], for example, are suitable.

Catalysts based on Ti–Mg compounds can also be used in the production of *polypropylene waxes*. Through an appropriate choice of catalyst, the degree of crystallinity (isotacticity) can be varied within wide limits and thus adapted to a particular application. Flexible products with average crystallinity can be obtained [162]. Highly crystalline polypropylene waxes can now be produced economically by using stereoregulating silanes as additional catalyst components [163], [164]. Formerly, these waxes could be produced only by thermal degradation of high molar mass polypropylene (Section 6.1.4).

Indications are that the homogeneous metallocene–alumoxane catalyst systems [165], [166], which became known at the end of the 1970s and nowadays are used for plastics synthesis in industry, will also become important in wax synthesis. Since the 1980s, many patent applications have appeared in this field, relating to both PE [167]–[172] and PP [173]–[175] waxes. Metallocene catalysts have a number of advantages compared with titanium catalysts. They are more active and can thus be employed in lower quantities, so catalyst removal is not necessary. Polymerization can occur at lower temperature, and the process is therefore more economical. Above all, the specific polymerization properties and great structural variability of metallocene systems are likely to give polyolefin waxes with optimized or completely new property profiles.

6.1.3.2. Properties

The most important molecular structure parameters, which determine the macroscopic property profile of PE waxes, are the average molar mass (average chain length)

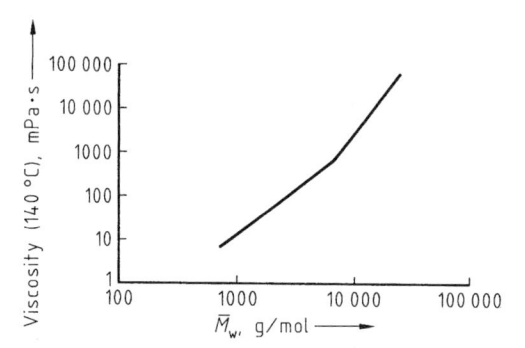

Figure 11. Relationship between average molar mass $\bar{M}_w$ and melt viscosity for PE waxes

Table 23. Structure and properties of typical commercial Ziegler polyolefin waxes

	A Ethylene homopolymer wax	B Ethylene copolymer wax	C Ethylene homopolymer wax	D Ethylene homopolymer wax	E Propylene homopolymer wax
Number-average molar mass $\bar{M}_n$, g/mol (GPC)	1600	2200	850	5000	3000
Weight-average molar mass $\bar{M}_w$, g/mol (GPC)	4800	6300	920	17 000	18 000
Molar mass distribution ($\bar{M}_w/\bar{M}_n$)	3.0	2.9	1.1	3.4	6.0
Density (23 °C), g/cm³ (DGF-M-III 2a)	0.97	0.93	0.96	0.97	0.88
Heat of fusion, J/g (DSC)	240	120	230	200	60
Viscosity (140 °C), mPa · s	400	600	10	25 000	1700 *
Drop point, °C (DGF-M-III 3)	125	118	107	135 **	158 **
Needle penetration, 0.1 mm (DGF-M-III 9b)	<1	3	2	<1	<1

* Temperature of measurement 170 °C.
** Softening point ring/ball (DGF-M-III 13).

and the degree of branching. Melt viscosity rises exponentially with increasing molar mass (Fig. 11). Unbranched, *homopolymeric PE waxes* have a high degree of crystallinity as a result of their undistorted linear molecular structure. They are brittle and hard (needle penetration numbers <1) and have high densities and melting points (up to ca. 127 °C). These products can be obtained only by Ziegler polymerization. Radical polymerization always gives branched waxes.

By using *1-alkenes* (generally propene) *as copolymers*, defects can be introduced into the polymer chain in the form of short-chain branches, which lead to a decrease in crystallinity. The density, hardness, drop point, and heat of fusion are also decreased, while flexibility and plasticity increase.

In Table 23, some important physical properties of typical commercially available Ziegler polyolefin waxes are compared.

Ziegler waxes are colorless and odorless and form clear, transparent melts with high thermostability. Like most other waxes, they are soluble on warming in nonpolar aliphatic and aromatic solvents without leaving a residue, and on cooling often crystallize out as pastes. Commercial products have a melt viscosity range from ca. 10 to 30 000 mPa · s (at 140 °C). Their densities are between 0.92 and 0.97 g/cm^3.

6.1.4. Degradation Polyolefin Waxes

By heating in the absence of air, the molecular chain of high molar mass polyolefins can be cleaved into smaller molecules with waxlike character. The starting materials are usually low- and high-density polyethylene, isotactic polypropylene, and polybutene. The production capacities of degradation polyolefin waxes have diminished since ca. 1980 in favor of the built products (see Sections 6.1.1, 6.1.2, 6.1.3). While the proportion of degradation PE waxes has diminished, that of PP waxes has increased. Despite the relatively unfavorable energy balance of the degradation process, it is still used because of its technological simplicity and, for PE waxes, occupies third place with regard to the production quantity after the high-pressure and low-pressure PE waxes. The degradation process was developed approximately in parallel to high-pressure polymerization [176]–[178].

6.1.4.1. Production

Polyethylene is a thermally stable polymeric hydrocarbon. In the absence of oxyen it can withstand temperatures up to ca. 290 °C [179]. At higher temperature, thermal degradation begins with a decrease in the molar mass; from 400 °C, degradation occurs with high rate and yield [179]. The cleavage products consist almost exclusively of a wax fraction with a molar mass distribution that is barely wider than that of high-pressure PE waxes. Low-boiling and gaseous crack products are formed only in small quantities. Individual process steps are shown in Figure 12.

The starting materials for thermal degradation to polyolefin waxes are LDPE and HDPE [180] and HD polypropylene [181]–[184]. Degradation of copolymeric polyolefins [185] and polybutenes [186] has also been described but has not achieved great economic importance.

In the industrial process [180], polyolefin pellets are melted under nitrogen in a single-screw extruder and extruded into a steel tube heated to ca. 400 °C. The residence time is 15–30 min. The product is then cooled by evaporative cooling in ca. 5 min to <250 °C. It is freed from low molar mass components by degassing in a separator. The product can then be finished as described for high-pressure PE wax (Section 6.1.4.1).

The degree of branching of degradation waxes formed, and thus all structure-dependent properties such as crystallinity, density, hardness, and *mp*, are controlled by the choice of starting material (LDPE or HDPE, stabilizer free). The average molar mass depends on residence time in the hot-tube reactor and the average molar mass of the starting material. The degradation temperature affects mainly the reaction time and thus the throughput rate.

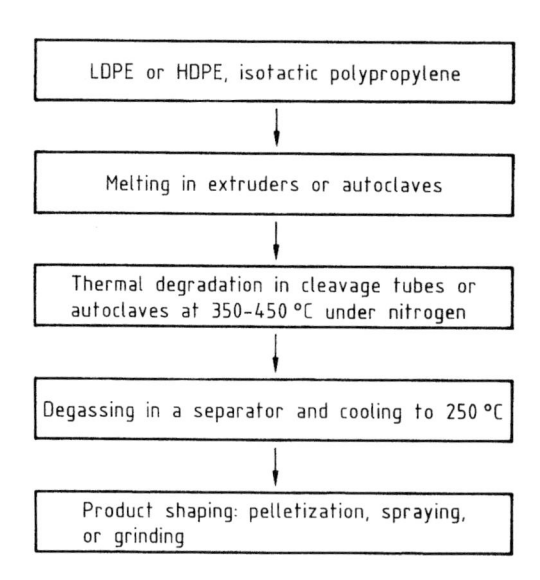

Figure 12. Process schematic for the production of degradation polyolefin waxes

```
┌─────────────────────────────────────────┐
│  LDPE or HDPE, isotactic polypropylene  │
└─────────────────────────────────────────┘
                    │
                    ▼
┌─────────────────────────────────────────┐
│     Melting in extruders or autoclaves  │
└─────────────────────────────────────────┘
                    │
                    ▼
┌─────────────────────────────────────────┐
│  Thermal degradation in cleavage tubes or│
│  autoclaves at 350–450 °C under nitrogen │
└─────────────────────────────────────────┘
                    │
                    ▼
┌─────────────────────────────────────────┐
│ Degassing in a separator and cooling to 250 °C │
└─────────────────────────────────────────┘
                    │
                    ▼
┌─────────────────────────────────────────┐
│   Product shaping: pelletization, spraying, │
│   or grinding                           │
└─────────────────────────────────────────┘
```

Batchwise, thermal degradation in autoclaves has also been described [182], [185]. It is performed in certain cases in the presence of hydrogen and hydrogenation catalysts to saturate double bonds formed on chain cleavage [176].

Like other polyethylene waxes, degradation waxes can be modified by air oxidation and thus rendered emulsifiable (see Section 6.1.5). Preliminary hydrogenation of double bonds can be necessary to avoid cross-linking reactions during oxidation.

Because of their double bonds, degradation waxes are particularly suited to grafting with unsaturated carboxylic acids (e.g., with maleic anhydride [187], [188]).

Little information is available in the literature on the mechanism of thermal degradation of polyolefins to form waxes [179]. Presumably, radicals form from the few oxygen-containing groups in the polyolefin and initiate a radical chain reaction, in which the chain is cleaved through radical transfer and a double bond is formed:

$$\dot{R} + \begin{array}{c} R^1 \\ \diagdown \\ CH-CH_2R^3 \\ \diagup \\ R^2 \end{array} \longrightarrow \begin{array}{c} R^1 \\ \diagdown \\ \dot{C}-CH_2R^3 + RH \\ \diagup \\ R^2 \end{array} \quad \text{(Start)}$$

$$\begin{array}{c} R^1 \\ \diagdown \\ C=CH_2 + \dot{R}^3 \\ \diagup \\ R^2 \end{array} \quad \text{Side-chain double bond}$$

$$\begin{array}{c} R^1 \\ \diagdown \\ CH-\dot{C}H-R_3 \\ \diagup \\ R^2 \end{array} \longrightarrow R^1-CH=CH-R^3 + \dot{R}^2 \quad \text{Central double bond}$$

$$R^1-\dot{C}H-CH_2R^2 \longrightarrow R^1-CH=CH_2 + \dot{R}^2 \quad \text{Terminal double bond}$$

Since tertiary hydrogen atoms are particularly susceptible to radical transfer, the degradation reaction begins at the carbon atoms of the branches, to form fragments

Table 24. Characteristics of three degradation and three high-pressure polyethylene waxes

Characteristic	Degradation PE wax			High-pressure PE wax		
	G	H	I	A	B	C
Density (23 °C), g/cm^3 (DGF-M-III 2a)	0.92	0.93	0.95	0.92	0.93	0.945
CH$_3$ per 1000 C atoms	32	16	8	26	22	17
Branches per molecule	3.5	0.5	0	3.4	1.8	1.0
Crystallinity (from IR data), %	43	63	74	54	59	67
Central double bonds per 1000 C atoms	1.2	1.5	2.0	0.1	0.4	0.4
Terminal double bonds per 1000 C atoms	3.3	2.6	2.6	0.1	0.1	0
Side-chain double bonds per 1000 C atoms	1.7	1.1	0.5	0.6	0.4	0.2
Ball indentation hardness (23 °C), bar (DGF-M-III 9a)	450	500	600	200	370	550
Solidification point, °C (DGF-M-III 4a)	96	100	110	96	99	106
Drop point, °C (DGF-M-III 3)	105	112	121	106	112	116
Melt viscosity (120 °C), mm^2/s (DGF-M-III 8)	930	1230	1300	1310	1160	1200
Weight-average molar mass $\bar{M}_w$, g/mol	5400	6040	6100	6140	5930	6000
Number average molar mass $\bar{M}_n$, g/mol	2400	2160	2440	2920	2390	2440
$\bar{M}_w/\bar{M}_n$ (polydispersity index)	2.3	2.8	2.5	2.1	2.5	2.5

containing side-chain double bonds. Only later are linear chains cleaved, to form central and terminal double bonds.

6.1.4.2. Properties

In Table 24, properties (in some cases unpublished results supplied by the authors) of three degradation and three high-pressure polyethylene waxes, with about the same average molar mass to decreasing branching and increasing density, are compared. The two types of wax can be seen not to differ significantly. For both, the same relationships are valid between molar mass and melt viscosity and between degree of branching and density, crystallinity, hardness, and solidification and drop points. The width of the average molar mass distribution (polydispersity) is also comparable. A significant difference is that degradation polyethylene waxes contain considerably more double bonds than comparable polyethylene waxes formed by high-pressure polymerization. Determination of the type and level of double bonds from the IR spectrum can therefore be used for analytical identification of degradation polyethylene waxes. The technical properties are affected only slightly by double bonds.

Compared with degradation polyethylene waxes, *degradation polypropylene waxes* usually have extremely high melting points and hardness because they are produced from highly crystalline, isotactic high-density polypropylene. Some characteristics of a relatively low molar mass degradation polypropylene wax are listed below:

mp (DSC)	130 – 160 °C
Ball indentation hardness (23 °C) (DGF-M-III 9a)	500 – 1900 bar
Melt viscosity (200 °C) (DGF-M-III 8)	800 – 1900 m^2/s
Weight-average molar mass (GPC, standard: PP)	17 000 – 37 000 g/mol

4955

Degradation polyolefin waxes are colorless, white to transparent, and form clear melts. On warming, they dissolve in nonpolar solvents such as aliphatic, aromatic, and chlorinated hydrocarbons, and on cooling, they crystallize as dispersions or gels (pastes).

6.1.5. Polar Polyolefin Waxes

Polar polyolefin waxes generally contain carboxyl and ester groups, which give them special technical properties, in particular emulsifiability in aqueous media. They can be produced by a variety of methods:

1) Oxidation of nonpolar PE waxes (melt oxidation) or PE plastics (oxidative degradation) with air
2) Radical high-pressure polymerization of ethylene with oxygen-containing comonomers (acrylic acid, acrylates, vinyl acetate, etc.) (see Section 6.1.1.1)
3) Radical grafting of polar unsaturated monomers onto nonpolar PE and PP waxes

To a great extent, PE waxes are converted into wax oxidates by treatment with atmospheric oxygen in the melt at ca. 150–160 °C [189]–[191]. Oxygen functions (carboxyl, ester, carbonyl, hydroxyl groups, etc.) are introduced into the wax molecule by complex mechanisms involving peroxy intermediates and chain-shortening reactions [192]. The usual commercial products with acid numbers between 25 and 30 mg of KOH per gram contain ca. 3–4 wt% oxygen, mainly in the form of carboxyl and ester groups.

Oxidation of the molten wax is limited to waxes whose molar mass is not too high. In the case of longer-chain starting materials, intensive dispersion of air in the melt, which is necessary for the reaction to run smoothly, is inhibited by the melt's high viscosity. In this case, waxes can be oxidized in *inert dispersion agents*. Since a (statistical) chain cleavage is always associated with oxidation (i.e., a lowering of molar mass), PE plastics can also be converted into a polar wax by this route (*oxidative degradation*). For example, molten polyethylene in an aqueous dispersion can be converted into emulsifiable waxes by passing in air at 130–160 °C with intensive stirring. These waxes are harder and have higher drop points than the usual melt oxidates with the same acid number because of their relatively high molar masses [193]. The reaction can be accelerated by addition of peroxides. High-density polyethylene gives hard brittle products. Tough, flexible waxes, which can be used as antislip components of floor polishes, can be obtained from LDPE containing vinyl acetate [194].

Oxidative degradation of polyethylene can also be carried out as a *gas–solid reaction*. In this case, no dispersing agent is required. Air, oxygen-enriched air, or pure oxygen in the presence of radical initiators is reacted with the substrate in powder form a few degrees below the softening point [195]–[198]. Because a low temperature (<130 °C) must be maintained, the oxidation period is comparatively long. For example, it requires at least 20 h to achieve acid numbers of 25–30 mg of KOH per gram. Use

Table 25. Properties of typical commercial oxidized polyolefin waxes

Characteristic	A	B	C	D	E
Raw material	high-pressure PE wax	Ziegler PE wax	HDPE	HDPE	ethylene–vinyl-acetate copolymer
Acid number, mg KOH/g (DGF-M-IV 2)	22	18	16	60	21
Saponification number, mg KOH/g (DGF-M-IV 2)	50	38		103	80
Viscosity (120 °C), mPa · s	400	200	8500 (150 °C)	300	3000
Drop point, °C (DGF-M-III 3)	100	116	140	111	100
Needle penetration, 0.1 mm (DGF-M-III 9b)	5	5	<0.5	1	2
Density, g/cm^3 (DGF-M-III 2a)	0.96	0.98	0.98	1.02	0.96

of air–ozone mixtures as the oxidizing agent has also been described [199], [200]. Another process variant involves oxidation of polyethylene powder with air in *aqueous suspension* under pressure [201]. The properties of some representative commercial wax and plastic oxidates are given in Table 25.

Oxidized PE waxes can be modified further by esterification [202], amidation [203]–[205], saponification, and other derivatization reactions [206] and thus adapted to special technical requirements.

Polyolefin waxes can also be converted into emulsifiable products by *grafting unsaturated polar compounds* (maleic anhydride, etc.) [161], [207]–[210]. For *polypropylene wax* [211]–[213], this is the only method of introducing oxygen functions that is used in practice, since air oxidation would lead to low molar mass, colored, soft degradation products.

6.2. Uses

6.2.1. Nonpolar Polyolefin Waxes

Polyolefin waxes are used in widely differing applications, mostly as auxiliaries in products and production processes. The areas of use of high-pressure, Ziegler, and degradation waxes overlap extensively. More details can be found in leaflets and brochures of various production companies. Important areas of use include the following:

1) *In Additive and Pigment Masterbatches.* PE and PP waxes are carrier materials and binders in concentrates used as plastics additive and for coloration of plastics.
2) *As Additives to Improve Rub Resistance and as Slip Agents* [214]. PE waxes are incorporated into printing inks after micronizing by milling or spraying, or in the form of dispersions.

3) *In paints and coatings*, PE waxes are used for matting effects and to increase resistance to scratching [215], [216].
4) *In plastics processing*, PE waxes are used as lubricants and release agents in molding.
5) *In Polishes*. PE waxes are components of heat-resistant pastes for floor, shoe, and car polishes.
6) *In Hotmelt Coatings*. PE and PP waxes are heat-sealable components of coatings and adhesives for paper and metal sheeting because they improve insulation, gloss properties) and heat resistance.
7) *For Corrosion Protection*. PE waxes are used as hydrophobing components for temporary corrosion protection of motor vehicles, machines, and instruments.
8) *Photocopying*. PE and highly crystalline PP waxes are used in toner preparations for photocopying machines.
9) *In the rubber industry*, PE waxes are used as release agents in processing.

Also, PE and PP waxes are used as components of insulating compounds, as carriers and binders for mechanically and chemically reacting self-duplicating paper and wax crayons, to coat granular fertilizer, and to increase the heat resistance of candles.

6.2.2. Polar Polyolefin Waxes

Polar polyolefin waxes are used mainly in the form of aqueous emulsions (dispersions). Depending on the nature of the emulsifier system used, nonionic, nonionic – ionic, anionic, and cationic formulations can be produced. Emulsions are produced by mixing molten waxes and emulsifiers with hot water. On rapid cooling with intensive stirring, wax droplets solidify and form stable, finely divided dispersions. High-melting waxes must be emulsified under pressure. After the emulsion is applied, water evaporates and leaves a homogeneous wax film, often with a high gloss, on the treated surface. This effect is used in floor polishes where polar waxes are applied in combination with acrylate- and styrene-based polymer dispersions. Waxes obtained by oxidative degradation of plastics are primarily used in this area because they form harder-wearing films due to their greater hardness than lower molar mass melt oxidates. Polyethylene wax emulsions are also used as leather and shoe polishes and in many other areas of industry. In textiles, they are used for wash and wear finishes and to improve sewing properties. They are used as hydrophobing agents in paint dispersions, as release agents in the building industry, and in surface finishing of paper products. Citrus fruit is provided with a glossy surface and protected from drying out through the application of wax films [217]. In the plastics processing industry, polar PE waxes are used as lubricants and release agents.

6.3. Economic Importance

As indispensable additives with a secure raw material supply and broad application potential, polyolefin waxes play an important role, mainly in the industrial, but also in the consumer market sectors.

The quantity of polyolefin waxes currently annually consumed worldwide is ca. 200 000 t, including oxidized and grafted products. High-pressure waxes (PE and polar copolymers) account for ca. 70%. The remainder consists essentially of Ziegler products (PE, PP). Degradation waxes (mainly PP) are less important in terms of quantity.

Ca. 12 000 t of polar waxes are produced worldwide by oxidative degradation of PE plastics.

The main consumers of nonpolar polyolefin waxes are the producers of pigment concentrates and of plastic and rubber articles, as well as the paint and printing ink industry. Polar waxes are used mainly in the form of emulsions in the textile, polish, and paper industries.

Producers and Trade Names. *High-Pressure Waxes.* Allied-Signal, United States (A-C Polyethylene). Eastman Kodak, United States (Epolene Wax). BASF, Germany (Polyethylenwachs BASF, Luwax). Leuna, Germany (LE-Wachs).

Ziegler Waxes. Hoechst, Germany (Hoechst-Wachs). Mitsui Petrochemical Ind., Japan (Hi-wax). Hüls, Germany (Vestowax). Petrolite, United States (Polywax).

Degradation Waxes. Sanyo Chemical Industries, Japan (Viscol, Youmex). Mitsui Petrochemical Ind., Japan (NP-Types). Lion Chemical, Korea (L-C Wax). Eastman Kodak, United States (Epolene). Ceralit, Brazil (Cerit Poly).

7. Toxicology

On the basis of experience accumulated over thousands of years, waxes are generally considered harmless, nontoxic, environmentally friendly products. The results of scientific toxicological and ecotoxicological tests on natural waxes, montan waxes, polyethylene waxes, and other wax products generally support this.

Natural Waxes. From the group of animal and vegetable waxes, *carnaúba wax* was subjected to toxicological tests. In a two-generation feeding experiment, 1% wax (max.) based on the feed, was administered first to pregnant rats and then to their first-generation offspring over a period of 90 d. The results gave no indication of dose-dependent effects in the rats treated compared to the untreated control group [218].

Montan Waxes and Their Derivatives. The toxicological properties of acid, ester, and partially saponified ester waxes (partially neutralized with calcium hydroxide) derived from montan wax are determined mainly by the indigestibility of the products,

Table 26. Acute oral, subchronic, and chronic toxicity of montan wax derivatives

Wax	Chemical composition	Acute oral toxicity, LD_{50}, mg/kg	Subchronic toxicity, 90-d no-effect level *, mg/kg	Subchronic toxicity, 140-d no-effect level *, mg/kg	Chronic toxicity, 2-a no-effect level *, mg/kg
Acid wax (Hoechst wax S)	montanic acid C_{16}–C_{36}	>15.000 (rat)			
Ester wax (Hoechst wax E)	esters of montanic acid with ethanediol	>20 000 (mouse)		50 000 (dog)	50 000 (rat)
Ester wax (Hoechst wax KPS)	esters of montanic acid with ethanediol and 1,3-butanediol	>15 000 (rat)	50 000 (dog)		50 000 (rat)
Ester wax (Hostalub WE 4)	esters of montanic acid with glycerol		50 000 (rat)		
Partially saponified ester wax (Hoechst wax OP)	esters of montanic acid with 1,3-butanediol and calcium salts of montanic acid	>20 000 (mouse)		50 000 (dog)	50 000 (rat)

* Proportion by weight in the feed.

as in the case of natural waxes. Besides determining acute oral toxicity, extensive feeding experiments on rats and dogs over periods of 90 d, 140 d, and 2 a have been carried out with montan wax derivatives [219]. The LD_{50} values are >15 000 and 20 000 mg per kilogram of body weight, respectively. According to the usual classification, these products are considered practically nontoxic and harmless [220].

Long-term feeding experiments led to no macroscopically or microscopically recognizable toxicological damage to the animals, even at very high dosages of 5 % in the feed (Table 26).

Dermatological examination of montan wax derivatives in the percutaneous test and the mucous membrane test on rabbits (Hoechst waxes S, E, and OP) did not lead to any irritation of the skin or mucous membranes.

In the Ames mutagenicity test with and without external metabolic activation, a montanic acid–ethylene glycol ester wax (Hoechst wax E) did not show any mutagenicity (tested dosages 4–5000 µg per plate) [221].

The *ecotoxicological behavior* of montan waxes is attributed to their water insolubility. Mixtures with water are attacked very slowly by microorganisms and cause no perceptible biological oxygen demand (BOD) in water bodies and sewage treatment plants. The waxes are separated from wastewater as ballast materials together with sludge. In the determination of biodegradability according to OECD test method 302 B (static test/Zahn–Wellens test), the finely dispersed montan wax derivatives in wastewater show good elimination of chemical oxygen demand. They are separated with sludge and are nontoxic to fish and bacteria.

In the case of wax emulsions, when the usual test methods for determination of COD and BOD are used, a very slow oxygen consumption can be detected, which is higher

Table 27. Acute oral, subchronic, and chronic toxicity of polyolefin waxes

Wax	Acute oral toxicity, LD_{50}, mg/kg	Subchronic toxicity, 90-d (rat) no-effect level *, mg/kg	Subchronic toxicity, 9-month (rat) no-effect level *, mg/kg
Polyethylene wax (Hoechst wax PE 520)	>15 000		85 000
Polyethylene wax oxidates			
(Hoechst wax PED 121)		16 000	
(Hoechst wax PED 522/PED 136)	>15 000	50 000	
Polypropylene wax (Hoechst wax PP 230)	>5000		

* Proportion by weight in the feed.

than the oxygen demand of the emulsifiers [219]. In wastewater, these emulsions break up as a result of great dilution, altered pH, and the presence of calcium and magnesium ions. The findings described for pure wax–water mixtures also apply to the agglomerates formed.

Polyolefin Waxes. Polyethylene waxes, oxidized polyethylene waxes, and polypropylene waxes are similar to those based on montan wax and natural waxes with regard to toxicology. The LD_{50} values for rats are >5000 and 15 000 mg per kilogram of body weight, respectively. In a long-term toxicity test (a 9-month feeding experiment on rats), polyethylene waxes caused no toxic damage to the animals even at the highest dosage of 8.5 % in the feed. In a 90-d feeding experiment on rats, oxidized polyethylene wax up to acid numbers 70 in dosages of 0.2, 1, 1.6, and 5 % in the feed produced no adverse effects in the animals. Histopathological tests showed no morphologically recognizable damage or changes caused by the waxes. No accumulation was observed in tissues, liver, or lymph glands. Fat and iron metabolism and spermatogenesis and oogenesis were unaffected (Table 27) [219]. Practically no negative effects were found in dermatological tests on rabbits. Polyethylene waxes and oxidized polyethylene wax show good COD elimination under sewage treatment plant conditions and no toxicity to fish or bacteria.

Paraffins and Microwaxes. Petroleum waxes have been used for decades as food additives, as wax coatings for cheese, in chewing gum, for upgrading food packaging, and in cosmetics and pharmaceutical products, without damaging human health. The toxicological properties of paraffins and microwaxes extracted from crude petroleum depend essentially on their composition and purity, particularly their residual aromatics content.

In a study of the toxicity of petroleum waxes, rats were fed with several paraffin products for 2 a. At a dosage of 10 % wax in the feed, no toxicological or carcinogenic effects were observed. Even on repeated application of petroleum waxes to the skin of mice and rabbits, no carcinogenic effects were detected [222].

In recent years, toxicological tests have been carried out with paraffins and microwaxes, mainly 90-d feeding experiments on rats. In these, varying toxicological effects

Table 28. Registration according to EINECS, TSCA, AICS, DSL

Type of wax	CAS registry no.
Natural waxes	
Beeswax	[8012-89-3]
Carnaúba wax	[8015-86-9]
Candelilla wax	[8006-44-8]
Waxes based on montan wax	
Montan wax fatty acids	[68476-03-9]
Ethylene esters of montan wax fatty acids	[73138-45-1]
1-Methyl-1,3-propanediyl esters of montan wax fatty acids	[73138-44-0]
Calcium salts of montan wax fatty acids	[68308-22-5]
Montan wax glycerides	[68476-38-0]
Montan wax (crude montan wax)	[8002-53-7]
Polyolefin waxes	
Polyethylene wax	[9002-88-4] *
Ethylene	[74-85-1]
Polypropylene wax	[9003-07-0] *
Propylene	[115-07-1]
Polyethylene wax oxidates	[68441-17-8] *
Petroleum waxes	
Paraffin and hydrocarbon waxes	[8002-74-2]
Microcrystalline waxes	[63231-60-7]
Amide wax	
N,N'-Ethylenedistearamide	[110-30-5]

* In EINECS, polymers are registered only under the CAS registry no. of the monomer.

were observed. For low molar mass paraffins, accumulation in the tissues and increased organ weights of the liver, kidney, spleen and lymph glands were observed. The most important histopathological findings were granulomata in the liver and histiocytosis in the lymph glands. These effects were not found with high molar mass waxes.

No toxic effects were found at a dose of 0.6 mg per kilogram of body weight and day for low molar mass paraffins (C-number maximum 25) and 1.8 g per kilogram of body weight per day for high molar mass paraffin (microwaxes, C-number maximum 42).

The latest results are published in BIBRA reports [223], [224], CONCAWE reports [225], and reports of the Toxicology Forum [226], [227].

Amide Waxes. The most well known amide wax, bis(stearoyl)- bis(palmitoyl)ethylenediamide, has an LD_{50} of $>20\,000$ mg/kg (mouse). In determination of the chronic toxicity in a 2-a feeding experiment on rats, a non-effect level of 5% in the feed was established [218].

Approval for Food Contact. On the basis of their low toxicity, a range of natural waxes, montan wax derivatives, amide waxes, polyethylene waxes, paraffins, and microcrystalline waxes have been approved for food contact by the health authorities of many countries [e.g., the German BgVV (formerly BGA), the U.S. FDA, and in Australia and

Japan]. The products are also contained in the draft version of the EC additives list (Synoptic Document no. 7 of May, 15, 1994). Some of the waxes are approved in the European Directives 95/2/EC and 95/3/EC for food contact.

Registration According to EINECS/TSCA/AICS/DSL. Waxes are registered under the CAS numbers given in Table 28 in the European Inventory of Existing Chemical Substances (EINECS), the TSCA inventory in the United States, the Australian AICS inventory, and the Canadian DSL inventory.

Waxes are not in the *MAK value list* for 1995.

Labeling According to GefStoffV, ChemG, and EC Guidelines. If they are pure and are not in the form of preparations with components for which labeling is compulsory (e.g., emulsifiers), waxes do not have to be labeled according to GefStoffV (October 26, 1993), the ChemG, Technische Regeln für Gefahrstoffe (technical regulations governing hazardous substances, TRGS) 200, and the EC Guideline 67/548/EC and adaptations and amendments.

Status According to OSHA. Paraffins are regulated by OSHA. Labeling is prescribed on packaging of paraffins (right-to-know label), in which warning of paraffin vapors is given. Paraffin vapors are classified as eye and respiratory irritants; they have the following exposure limits:

OSHA-PEL	2 mg/m^3
ACGIH-TLV	2 mg/m^3

Other waxes are not regulated by OSHA.

8. References

General References

[1] DGF-Einheitsmethoden: *Untersuchung von Fetten, Fettprodukten und verwandten Stoffen*, Abteilung M: "Wachse und Wachsprodukte," Wissenschaftl. Verlagsges., Stuttgart 1975.
[2] H. Bennet: *Industrial Waxes*, vol. 2, *Compounded Waxes*, Chem. Publ. Comp., New York 1963.
[3] A. H. Warth: *The Chemistry and Technology of Waxes*, Reinhold Publ. Co., New York 1956.
[4] J. Teubel, W. Schneider, R. Schmiedel: *Erdölparaffine*, VEB Dtsch. Verlag f. Grundstoffind., Leipzig 1965.
[5] F. Asinger: *Die petrochemische Industrie*, Akademie-Verlag, Berlin 1971, part I, pp. 33–70.
[6] C. Zerbe: *Mineralöle und verwandte Produkte*, 2nd ed., Springer Verlag, Berlin 1969, part 1, pp. 469–488; part 2, pp. 581–593.
[7] R. Sayers: *Wax – An Introduction*, Gentry Books Ltd., London 1983.
[8] M. Freund et al.: *Paraffin Products*, Elsevier Sc. Publ. Co., Amsterdam–Oxford–New York 1982.

[9] A. Davidsohn, B. M. Milwidsky: *Polishes,* Leonard Hill, London 1968.A. Renfrew, P. Morgan; *Polyethylene,* 2nd ed., Iliffe & Sons, London; Interscience Publ., New York 1960.

Specific References

[10] Chr. Columbus, Bordbuch, November, 29, 1492. Reprint: edition Erdman, Thienemann, Stuttgart 1983.
[11] M. E. Chevreul: "Recherches chimiques sur les corps gras d'origine animale," Frankreich 1823.
[12] Hoechst AG: "Vom Wachs," Hoechster Beiträge zur Kenntnis der Wachse, Frankfurt 1959 ff.
[13] DGF-Einheitsmethoden, Abteilung M: Wachse und Wachsprodukte, Stuttgart 1975.
[14] Erläuterungen des Rates für die Zusammenarbeit auf dem Gebiet des Zollwesens 3404/1 (1992) und 3404/2 (1988): International Customs Tariff, Harmonized System HS 3404: "Artificial Waxes (Including Water-Soluble Waxes); Prepared Waxes, not Emulsified or Containing Solvents".
[15] J. Haggin: "Asian Natural Gas Pipeline Proposed for Easing Energy Feedstocks Strains," *Chem. Eng. News*(1994) no. 6, 17–18.
[16] "Wax Data," Newsletter for Marketers and Consumers of Petroleum, Natural and Synthetic Waxes, Rauch Associates Inc., Bridgewater, New Jersey (published 13 times annually).
[17] Kahl & Co., Otto-Hahn-Str. 2, 22946 Trittau, personal communication, 1994.
[18] "Animal and Vegetable Waxes – A Review," *Wax Data 1993,* Rauch Associates Inc., P.O. Box 6802, Bridgewater, N.J. 08807.
[19] *Chem. Mark. Rep.* **244**, (1993) no. 12, 7, 11.
[20] "Study Forecasts Growth for Natural and Synthetic Waxes," *Wax Data 1994,* Rauch Associates Inc., P.O. Box 6802, Bridgewater, N.J. 08807.
[21] A. H. Warth: *The Chemistry and Technology of Waxes,* Reinhold Publ. Corp., New York 1956.
[22] R. Büll et al.: "Vom Wachs," vol. **1**; *Hoechster Beiträge zur Kenntnis der Wachse,* vol. **1** (contributions 1–8); vol. **2** (contributions 1–10), Hoechst AG, Werk Gersthofen, Augsburg 1958–1970.
[23] E. B. Kurtz, Jr., *Fette Seifen Anstrichmittel* **61** (1959) no. 6, 505.
[24] E. J. Fischer, W. Presting: *Laboratoriumshandbuch für die Untersuchung technischer Wachs-, Harz- und Ölgemenge,* VEB Wilhelm Knapp Verlag, Halle/Saale 1958.
[25] L. Ivanovszki: *Wachs-Enzyklopädie,* Verlag für Chem. Industrie H. Ziolkowsky, Augsburg, vol. **1**, 1954; vol. **2**, 1960.
[26] W. Piso, *Historia Naturalis Brasiliae* (1648).
[27] J. B. De Moraes: *Ensaios sobre à Carnanheira,* Nat. Ölinstitut, Rio de Janeiro 1942.
[28] J. B. Nogueira, R. D. Machado: *Glossario da Palmeiras,* Nat. Ölinstitut, Rio de Janeiro 1950.
[29] Am. Wax Importers and Refiners Assoc.: *Blue Book,* New York 1973.
[30] Chem. Fabrik Dessau, DE 581 130, 1931.
[31] Dtsch. Hydrierwerke, DE 658 749, 1932.
[32] Degussa, DE 938 501, 1942.
[33] E. Eisenring, CH 271 936, 1950.
[34] W. A. Sprague, US 1 658 062, 1926.
[35] I.G. Farbenind., US 1 989 626, 1932.
[36] S. C. Johnson & Son, US 2 527 481, US 2 531 785, 1948.
[37] G. Kronbach, *Seifen Öle Fette Wachse* **99** (1973) 97.
[38] L. E. Vandenburg, E. A. Wilder, *J. Am. Oil Chem. Soc.* **44** (1967) 659.
[39] S. Markley, *Econ. Bot.* **9** (1955) 39.
[40] *Seifen Öle Fette Wachse* **86** (1960) 422.

[41] E. A. Wilder: "Chemical Specialties Manufacturers Assn.," *61st. Mid-Year Meeting*, Washington 1975.
[42] G. Illmann, *Fette Seifen Anstrichmittel* **81** (1979) no. 8, 322.
[43] The Candelilla Wax Industry FONCAM; Fideicomisos Relacionados con la Cera de Candelilla; Banco Nacional de Crédito Rural, S.A. Mexico, company brochure.
[44] *Chem. Mark. Rep.* **243** (1993) no. 2, 21.
[45] C. Lüdecke, L. Ivanovszky: *Taschenbuch für die Wachsindustrie*, 4th ed., Wissenschaftl. Verlagsgesellschaft, Stuttgart 1958, pp. 33–36.
[46] L. J. N. Cole, J. B. Brown, *J. Am. Oil Chem. Soc.* **37** (1960) 359.
[47] *Chem. Mark. Rep.* **241** (1992) no. 6, 10, 11.
[48] P. J. Manohar Rao, *Coop. Sugar/India* **13** (1981) no. 2, 1.
[49] H. Schmidt: "Analytik von Naturwachsen am Beispiel des Zuckerrohrwachses," *DGF-Vortragstagung*, Mainz 1967.
[50] A. Schwab: *Jojoba, ein hochwertiges Pflanzenöl aus der Wüste*, Ledermann-Verlag, Bad Wörishofen 1981.
[51] E. Zander: *Handbuch der Bienenkunde in Einzeldarstellung*, Verlag E. Ulmer, Stuttgart 1946.
[52] K. V. Frisch: *Aus dem Leben der Bienen*, Springer Verlag, Heidelberg 1956.
[53] G. Titschak, *Fette Seifen Anstrichmittel* **71** (1969) no. 5, 369–379.
[54] R. J. Holloway, *J. Am. Oil Chem. Soc.* **46** (1969) 189–190.
[55] A. P. Tulloch, *J. Am. Oil Chem. Soc.* **50** (1973) 269.
[56] A. P. Tulloch, *Bee World* **61** (1980) 47.
[57] A. P. Tulloch, *Lipids* **5** (1970) 247.
[58] A. P. Tulloch, *Chem. Phys. Lipids* **6** (1971) 235.
[59] A. P. Tulloch, L. L. Hoffmann, *J. Am. Oil Chem. Soc.* **49** (1971) 696.
[60] K. Stransky, M. Streibl, *Collect. Czech. Chem. Commun.* **36** (1971) no. 6, 2267.
[61] K. Stransky, M. Streibl, V. Kubelka, *Collect. Czech. Chem. Commun.* **36** (1971) no. 5, 2281.
[62] K. Stransky, K. Ubik, M. Streibl, *Collect. Czech. Chem. Commun.* **37** (1972) no. 12, 4099.
[63] H. Brüschweiler, H. Felber, F. Schwager, *Fett Wiss. Technol.* **91** (1989) no. 2, 73–79.
[64] H. Schmidt: "Analytische Untersuchungen an Schellackwachs," *DGF-Vortragstagung*, Freiburg 1981.
[65] C. A. Anderson, *J. Am. Oil Chem. Soc.* **40** (1963) 333–336.
[66] V. Vcelák: *Chemie und Technologie des Montanwachses*, Verlag der tschechoslowakischen Akademie der Wissenschaften, Prague 1959.
[67] A. H. Warth: *The Chemistry and Technology of Waxes*, Reinhold Publ. Corp., New York 1956.
[68] E. J. Fischer, W. Presting: *Laboratoriumsbuch für die Untersuchung technischer Wachs-, Harz- und Ölgemenge*, VEB Wilhelm Knapp Verlag, Halle (Saale) 1958.
[69] A. Lissner, A. Thau: *Die Chemie der Braunkohle*, vols. **1** and **2**, VEB Wilhelm Knapp Verlag, Halle (Saale) 1952.
[70] E. Riebeck, Dissertation, Freiberg 1880.
[71] Von Boyen, DE 101 373, 1897.
[72] VEB Braunkohlekombinat Röblingen: *Festschrift*, "75 Jahre Herstellung von Montanwachs," Röblingen 1972.
[73] R. Büll: *Informationsschrift "Vom Wachs,"* 2 vols, Hoechst AG, Frankfurt-Höchst 1959.
[74] BIOS Final Report 1759 and 1831, Technical Information and Document Unit, London 1947. FIAT Final Report 737, Field Information Agency, London 1946.
[75] VEB Braunkohlekombinat Gustav Sobotka, DD 135 089, 1979 (H. J. Grauen et al.).

[76] Hoechst AG, DE 954 903, 1953 (R. Schirmer); DE 957 508, 1955 (R. Schirmer); DE 1 007 918, 1955 (R. Schirmer); DE 2 855 263, 1979 (A. Riedel); DE 2 915 764, 1979 (H. Stetter); DE 2 915 802, 1979 (H. Stetter); DE 2 915 755, 1979 (H. Stetter).
[77] BASF, DE 3 726 514, 1987 (L. Schuster).
[78] O. Malitschek: *Hoechst Wachse, Ein Leitfaden für Wachsanwender,* Hoechst AG, SBU Wachse und Additive, Werk Gersthofen 1990.
[79] R. Gächter, H. Müller: *Kunststoffadditive,* Hanser Verlag, München 1993.
[80] P. Piesold: "Montanwachs als Kunststoffadditiv," in J. Edenbaum (ed.): *Plastics, Additives and Modifiers Handbook,* Van Nostrand Reinhold, New York 1992, pp. 757–772.
[81] European Wax Federation (EWF): *Glossary of Wax Terminology,* Brussels 1981.
[82] C. Kajdas, *Seifen Öle Fette Wachse* **96** (1970) 251–260.
[83] C. Kajdas, *Seifen Öle Fette Wachse* **112** (1986) 471–474.
[84] G. Matscholl, *Coating* **14** (1981) 117–118.
[85] I. A. Michailov, *Chem. Tech. (Leipzig)* **20** (1968) 351–353.
[86] "Assessment and Comparison of the Composition of Food-Grade White Oils and Waxes Manufactured from Petroleum by Catalytic Hydrogenation Versus Conventional Treatment," *CONCAWE Rep.* **84/60** (1984) 18–26, 33–46.
[87] European Wax Federation (EWF): *Specifications of Petroleum Waxes,* Brussels 1991.
[88] F. V. Synder: Presentation at National Petroleum Refiners Assoc., Houston, Texas 1979.
[89] Röhm GmbH, Spektrum 39, company brochure, Darmstadt 1992.
[90] CONCAWE 1991: Report no. 91/33, 58–62 CONCAWE, Brussels.
[91] R. J. Taylor, *Oil Gas J.* **91** (1993) no. 4, 48–50.
[92] D. A. Gudelis et al., *Hydrocarbon Process.* **52** (1973) no. 9, 141.
[93] M. Soudek, *Hydrocarbon Process.* **53** (1974) no. 12, 64, 66.
[94] *Hydrocarbon Process.* **59** (1980) no. 9, 201.
[95] L. M. L. Tobing, *Hydrocarbon Process.* **56** (1977) no. 9, 143–145.
[96] *Hydrocarbon Process.* **59** (1980) no. 9, 180, 181, 192, 200, 201, 206.
[97] *Hydrocarbon Process.* **59** (1980) no. 9, 218.
[98] Shell Oil, US 4 701 254, 1987 (L. W. Maas, R. A. Geimann, C. T. Adams).
[99] F. Wolf et al., *Erdöl Kohle Erdgas Petrochem.* **24** (1971) 79–82, 153–155.
[100] W. Himmel et al., *Erdöl Kohle Erdgas Petrochem.* **39** (1986) 408–414.
[101] P. Polanek et al., *Erdöl Erdgas Kohle* **109** (1993) 324–326.
[102] U.S. Depart. of Health, Education and Welfare; Food and Drug Admin. (FDA), Part 21, Paragraph 172.886.
[103] FAO Food and Nutrition Paper 52, Rome 1992.
[104] BGA-Regulation XXV.
[105] Mineral Hydrocarbons in Food Regulations, SI 1966, No. 1073 (1966).
[106] European Wax Federation: Summaries of National Regulations Affecting Wax Usage, Brussels 1991.
[107] A. Case, *Adhes. Age* **33** (1990) 28–31.
[108] A. Aduan et al., *J. Pet. Res.* **6** (1987) 63–76.
[109] G. Michalczyk, *Fette Seifen Anstrichmittel* **87** (1985) 481–486.
[110] M. P. Goutam et al., *Seifen Öle Fette Wachse* **118** (1992) 1277–1281.
[111] D. T. Kaschani, DGMK-Projekt 242, Hamburg 1992.
[112] G. Grimmer et al., *Chromatographia* **9** (1976) 30–40.
[113] *Chem. Mark. Rep.* **26** (1980) no. 5, 7.
[114] *Chem. Purch.* **5** (1980) 74–85.

[115] R. Hournac, *Hydrocarbon Process.* **60** (1981) no. 1, 207–211.
[116] A. E. Dustan, *Soap Cosmet. Chem. Spec.* **56** (1980) no. 12, 42B–42E.
[117] K. Marche, Cleveland Rubber Group Meeting, Los Angeles 1984.
[118] Ministery of Agriculture, Fisheries and Food News Release 278/93, London 1993.
[119] EEC-Directive 90/128 (EC 1990) and appendix lists.
[120] D. Monzie, Document of Chemistry and Technology of Petroleum Waxes (CTP) No. 776 (1976), Special report of the Institute of Petroleum, London.
[121] R. Franz et al., *Z. Lebensm. Technol. Verfahrenstech.* **43** (1992) 291–296.
[122] D. van Battum, *TNO-Rep.* **B 77/2694 A** (1979) 26–30.
[123] J. Rustige, Presentation at Toxicology Forum; Special Meeting on Mineral Hydrocarbons, Green College, Oxford 1992.
[124] A. O. Hanstveit, *TNO-Rep.* **R 90/1986** (1990) 18–40.
[125] DGMK Forschungsbericht 461-01, Hamburg 1991.
[126] J. Davie: "Polymers, Laminations and Coatings Conference," *TAPPI Proc.* 1992, 179–183.
[127] A. O. Hanstveit, *TNO-Rep.* **R 90/243 a** (1991) 18–46.
[128] R. Wenzel-Hartung, *Erdöl Erdgas Kohle* **108** (1992) 424–426.
[129] J. G. E. McEwen, *Progr. Pap. Recycl.* 1992, 11–20
[130] J. T. E. Lemmen, *TNO-Rep.* **PK 92.1051** (1992) 33, 38–46.
[131] A. Slaughter, *Pet. Rev.* **98** (1993) 258–260.
[132] *Hydrocarbon Process.* **72** (1993) no. 8, 36.
[133] I.G. Farbenind., DE 745 425, 1939 (H. Hopff, S. Goebel).
[134] I.G. Farbenind., DE 864 150, 1942 (R. Kern).
[135] Allied Chemical, US 2 504 400, 1947 (M. Erchak Jr.).
[136] G. Luft, *Chem. Ing. Tech.* **51** (1979) 960–969.
[137] BASF, DE 1 720 273, 1967 (O. Büchner et al.); DE 1 927 527, 1969 (H. Pfannmüller, F. Urban, W. Hensel, O. Büchner); DE 1 951 879, 1969 (K. Stark, H. Gropper, F. Urban, H. Pfannmüller); DE 2 034 534, 1970 (F. Urban, O. Büchner, W. Hensel, K. Stark).
[138] Eastman Kodak, US 2 772 259, 1953 (H. J. Hagemeyer Jr.); GB 861 277, 1959 (D. C. Hull, R. J. Schrader).
[139] Allied Chemical, US 2 683 141, 1952 (M. Erchak Jr.).
[140] Allied Chemical, BE 780 023, 1971 (J. E. Dench, H. Knutson, M. K. Seven).
[141] Allied Chemical, DE 2 233 360, 1971 (J. E. Dench, H. Knutson, M. K. Seven).
[142] M. J. Roedel, *J. Am. Chem. Soc.* **75** (1953) 6110–6112. W. M. D. Bryant, R. C. Voter, *J. Am. Chem. Soc.* **75** (1953) 6113–6118.F. W. Billmeyer, *J. Am. Chem. Soc.* **75** (1953) 6118–6122.
[143] R. Kuhn, H. Krömer G. Rossmanith, *Angew. Makromol. Chem.* **40/41** (1974) 361–389.
[144] G. Baum, J. Boelter, *Fette Seifen Anstrichmittel* **70** (1968) 15–19.
[145] N. Nakajima, C. F. Stark, R. D. Hoffmann, *Trans. Soc. Rheol.* **15** (1971) 647–662.
[146] E. Brauer, H. Wiegleb, M. Helmstedt, W. Eifert, *Plaste Kautsch.* **24** (1977) 463–468.
[147] H. Hendus, G. Schnell, *Kunststoffe* **51** (1961) 69–74.
[148] G. Ehrlich, G. A. Mortimer, *Adv. Polym. Sci.* **7** (1970) 386–448.
[149] Chusei Oil Co., US 5 223 122, 1992 (Y. Katayama).
[150] Hoechst, DE-OS 1 929 863, 1969 (B. Diedrich, K. Rust, R. Klein).
[151] Hoechst, DE-OS 1 942 734, 1969 (B. Diedrich, R. Klein).
[152] Mitsui Petrochemical, DE 2 241 057, 1972 (T. Tomoshige).
[153] Mitsui Petrochemical, EP-A 0 021 700, 1980 (T. Ueda).
[154] Leuna-Werke, DE-OS 4 235 404, 1992 (P. Adler, J. Krüger, M. Gebauer, R.-P. Pfitzer).
[155] Scholven-Chemie, DE-AS 1 301 524, 1965 (K. Schmitt et al.).

[156] VEBA-Chemie, DE-OS 1 520 927, 1963 (K. Schmitt, F. Gude).
[157] VEBA-Chemie, DE-OS 1 645 409, 1965 (K. Schmitt, G. Baum, F. Gude, W. Nagengast).
[158] Hoechst, DE-OS 2 257 917, 1972 (M. Engelmann, R. Höltermann).
[159] VEBA-Chemie, DE-OS 1 645 423, 1965 (K. Schmitt, F. Gude).
[160] VEBA-Chemie, DE-OS 1 520 925, 1963 (K. Schmitt, F. Gude).
[161] Scholven-Chemie, DE-OS 1 520 914, 1962 (K. Schmitt, G. Keller).
[162] Hoechst, DE-OS 2 329 641, 1973 (M. Engelmann).
[163] BASF, EP-A 480 190, 1991 (R. Schlund et al.).
[164] Hoechst, EP-A 584 586, 1993 (G. Hohner).
[165] F. Kübler, *New Sci.* (1993) Aug., 28.
[166] R. Mühlhaupt, *Nachr. Chem. Tech. Lab.* **41** (1993) 1341.
[167] Mitsui Petrochemical, JA 61-236 804, 1985 (T. Tsutsui, M. Kioka, N. Kashiwa).
[168] Mitsui Petrochemical, JA 62-129 303, 1985 (T. Tsutsui, A. Toyota, N. Kashiwa).
[169] Hoechst, DE-OS 3 743 322, 1987 (H. Lüker).
[170] Exxon Chemical Patents, US 4 914 253, 1988 (M. Chang).
[171] Hoechst, DE-OS 4 217 378, 1992 (H.-F. Herrmann et al.).
[172] Hoechst, EP 602 509, 1993 (R. Hess, H. Voigt, L. Böhm, H.-F. Herrmann).
[173] Hoechst, DE-OS 3 743 320, 1987 (A. Winter, M. Antberg, J. Rohrmann).
[174] Hoechst, DE-OS 3 904 468, 1989 (A. Winter et al.).
[175] Mitsui Toatsu Chemicals, EP-A 563 834, 1993 (Y. Sonobe, M. Mizutani, T. Shiomura, N. Hirayama).
[176] BASF, DE 922 618, 1952 (H. Hopff, H. Eilbracht).
[177] Eastman Kodak, FR 1 289 766, 1960 (H. W. Coover Jr., M. A. McCall, J. E. Guillet).
[178] W. G. Oakes, R. B. Richards, *J. Chem. Soc.* 1949, 2929 – 2935.
[179] A. Schwarz, G. Cramer in *Polyethylene*, 2nd ed., Iliffe & Sons, London; Interscience Publ., New York 1960, pp. 232 – 239.
[180] BASF, DE-AS 1 940 686, 1969 (H.-G. Trieschmann et al.).
[181] Allied Chemical, US 3 345 352, 1964 (J. J. Baron, J. P. Rahus).
[182] Eastman Kodak, US 2 835 659, 1957 (J. E. Guillet).
[183] Eastman Kodak, US 3 519 609, 1967 (R. L. McConnell, D. A. Weemes).
[184] J. E. Guillet, H. W. Coover, *SPE J.* **16** (1960) 311 – 314.
[185] Eastman Kodak, US 3 562 788, 1967 (D. A. Weemes, R. L. McConnell).
[186] H. W. Coover, R. L. Combs, J. E. Guillet, D. A. Weemes, *J. Appl. Polym. Sci.* **11** (1967) 1271 – 1281.
[187] BASF, DE 1 364 747, 1962.
[188] Eastman Kodak, US T 869 011, 1966 (M. O. Bronson).
[189] A. Thalhofer in R. Vieweg, A. Schley, A. Schwarz (eds.): *Kunststoff-Handbuch*, vol. **VI**, Polyolefine, Hanser Verlag, München 1969, p. 161.
[190] VEB Leuna-Werke, DD 283 730, 1988 (I. Klimek et al.).
[191] Mitsui Petrochemical, DE-OS 2 126 725, 1971 (A. Mori et al.).
[192] N. M. Emanuel, E. T. Denisov, Z. K. Maizus: *Liquid-Phase Oxidation of Hydrocarbons*, Plenum Press, New York 1967.
[193] Hoechst, DE-OS 2 035 706, 1970 (H. Korbanka, K.-H. Stetter).
[194] Hoechst, EP 28 384, 1980 (H. Korbanka et al.).
[195] Allied Chemical Corp., DE-OS 1 495 938, 1963 (R. W. Bush, C. L. Kehr).
[196] Allied Chemical Corp., DE-OS 1 520 008, 1964 (C. L. Kehr).
[197] Allied Chemical Corp., DE-OS 1 570 652, 1965 (G. L. Braude, S. T. Phoa).

[198] BASF, DE-OS 3 112 163, 1981 (A. Hettche et al.).
[199] BASF, DE-OS 3 720 952, 1987 (L. Schuster et al.).
[200] BASF, DE-OS 3 720 953, 1987 (L. Schuster et al.).
[201] BASF, DE-OS 3 238 652, 1982 (R. W. Reffert et al.).
[202] Hoechst, DE-OS 2 201 862, 1972 (H. Korbanka, K.-H. Stetter).
[203] Th. Goldschmidt, EP 0 257 506, 1987 (E. Ruf).
[204] Th. Goldschmidt, EP 0 258 724, 1987 (E. Ruf).
[205] Th. Goldschmidt, EP 0 259 652, 1987 (E. Ruf).
[206] Hüls, EP-A 0 451 352, 1990 (A. Kühnle, V. Schrenk, M. Vey, E. Wolf).
[207] Mitsui Petrochemical, DE-OS 285 3862, 1978 (A. Tachi, T. Tomoshige, H. Furuta, N. Matsuzawa).
[208] Petrolite Corp., US 3 590 076, 1967 (W. J. Heintzelmann, M. I. Naiman).
[209] Eastman Kodak Co., US 4 028 436, 1975 (R. T. Bogan, C. M. Shelton).
[210] Mitsui Petrochemical, EP 0 024 034, 1980 (T. Tomoshige, H. Furuta, A. Tachi, N. Kawamoto).
[211] Eastman Kodak Co., US 3 480 580, 1965 (F. B. Joyner, R. L. McConnell).
[212] Eastman Kodak Co., US 3 481 910, 1965 (M. O. Brunson).
[213] Eastman Kodak Co., US 5 290 954, 1992 (T. D. Roberts, T. W. Smith, C. A. Galvin).
[214] K. Rieger, *Fette Seifen Anstrichmittel* **73** (1971) 231.
[215] K. Rieger, G. Grimm, *Adhäsion* **11** (1977) 311.
[216] R. Köhler, *Fette Seifen Anstrichmittel* **87** (1985) 214.
[217] J. Wildgruber, *Fette Seifen Anstrichmittel* **84** (1982) 75.
[218] Results of investigation of Food and Drug Research Ltd., private communication of J. Maehlmann & Cie., Fortaleza, Brazil 1982.
[219] J. Lange, J. Wildgruber, *Fette Seifen Anstrichmittel* **78** (1976) 62–69.
[220] W. S. Spector: *Handbook of Toxicology*, W. B. Saunders, Philadelphia–London 1956.
[221] Hoechst, Pharma Research Toxicology, unpublished results, Frankfurt 1986.
[222] P. Shubik et al.: *Toxicology and Applied Pharmacology*, vol. 4, Supplement, Academic Press, New York–London 1962.
[223] BIBRA 1992: "A 90 Day Feeding Study in the Rat, With Six Different Mineral Oils and Three Different Mineral Waxes," BIBRA Project no. 3, 1010, BIBRA Toxicology, Carshalton, Surrey, U.K.
[224] BIBRA 1993: "A 90 Day Feeding Study in the Rat, With Two Mineral Waxes," (an Intermediate and a Micro-paraffin Blend), BIBRA Project no. 3, 1205, BIBRA Toxicology, Carshalton, Surrey, U.K.
[225] CONCAWE 1993: "Report no. 93/56, White Oil and Waxes, Summary of 90 Day Studies," CONCAWE, Brussels.
[226] Toxicology Forum 1992: Special Meeting on Mineral Hydrocarbons Sept. 21–23, 1992, Green College, Oxford, U.K. The Toxicology Forum, Washington, D.C.
[227] Toxicology Forum 1993: Mineral Hydrocarbons, 1993 Winter Meeting, Feb. 15–17, 1993, Toxicology Forum, Washington, D.C.

Xanthates

KATHRIN-MARIA ROY, Langenfeld, Federal Republic of Germany

The author would like to thank Hoechst AG for undertaking the literature research for this article.

1.	Introduction 4971	6.	Storage and Transportation . . 4978
2.	Physical Properties 4971	7.	Analysis 4978
3.	Chemical Properties 4973	8.	Uses 4979
4.	Production 4976	9.	Economic Aspects 4980
4.1.	General 4976	10.	Toxicology and Occupational
4.2.	Production Processes 4977		Health 4981
5.	Quality Specifications. 4978	11.	References 4981

1. Introduction

According to the IUPAC nomenclature, xanthates or xanthogenates (older name in the continental literature) are described as derivatives of *O*-alkyl- or *O*-aryldithiocarbonates (xanthic acids; **1**). The term xanthates refers both to the salts of the general formula **2** and to the corresponding *O*,*S*-diesters **3**.

$$S=C\begin{matrix}O-R^1\\SH\end{matrix} \quad S=C\begin{matrix}O-R^1\\S^-M^+\end{matrix} \quad S=C\begin{matrix}O-R^1\\S-R^2\end{matrix}$$

$$\quad\quad 1 \quad\quad\quad\quad\quad 2 \quad\quad\quad\quad\quad 3$$

The name xanthogenates (derived from the Greek *xanthos*: yellow) was introduced by ZEISE, who discovered this class of compounds in 1822 [8].

2. Physical Properties

Xanthic Acids. The free xanthic acids (**1**) are oily liquids, that are stable only in the pure state. They are strong acids, the dissociation constants of the C_1-C_5 esters in water are in the range of $1.4-3.4\times 10^{-2}$ [9]. Xanthic acids are readily soluble in nonpolar solvents and sparingly soluble in water. Traces of polar substances cause their decomposition into carbon disulfide and the corresponding alcohol. The alcohol formed accelerates the decomposition autocatalytically.

Table 1. Melting points of some industrially important xanthates, $R-O-CS-S^-M^+$

M	R	CAS registry no.	mp, °C
K	CH_3	[2667-20-1]	182–186
	C_2H_5	[140-89-6]	225–226
	n-C_3H_7	[2720-67-4]	233–239
	iso-C_3H_7	[140-92-1]	278–282
	n-C_4H_9	[871-58-9]	255–256
	iso-C_4H_9	[13001-46-2]	260–270
Na	C_2H_5	[140-90-9]	69–69.5
	iso-C_3H_7	[140-93-2]	124
	n-C_4H_9	[141-33-3]	71.5–75
	$(CH_2)_2CH(CH_3)_2$	[2540-36-5]	85–85.5
	$(CH_2)_9CH_3$		99–99.5
	$(CH_2)_{11}CH_3$		106–107.2

Table 2. Solubilities of alkali-metal xanthates, $R-O-CS-S^-M^+$, in water and alcohols

M	R	Solubility, g/100 g solution			
		H_2O (0 °C)	H_2O (35 °C)	R–OH (0 °C)	R–OH (35 °C)
K	n-C_3H_7	43.0	58.0	1.9	8.9
	iso-C_3H_7	16.6	37.2		2.0
	n-C_4H_9	32.4	47.9		36.5
	iso-C_4H_9	10.7	47.7	1.6	6.2
	$(CH_2)_2CH(CH_3)_2$	28.4	53.3	2.0	6.5
Na	n-C_3H_7	17.6	43.3	10.2	22.5
	iso-C_3H_7	12.1	37.9		19.0
	n-C_4H_9	20.0	76.2		39.2
	iso-C_4H_9	11.2	33.4	1.2	20.5
	$(CH_2)_2CH(CH_3)_2$	24.7	43.5	10.9	15.5

Alkali-Metal Xanthates. In contrast to the free acids, alkali-metal xanthates are relatively stable, yellowish solids of unpleasant odor. The potassium xanthates crystallize without water, whereas the sodium salts form dihydrates upon exposure to atmospheric moisture. The melting points of some industrially important alkali-metal xanthates are given in Table 1. Since the potassium salts melt with decomposition, very different melting points are found in the literature [10].

Alkali-metal xanthates are soluble in water and polar organic solvents such as alcohols, lower ketones, pyridine, and acetonitrile, but sparingly soluble in nonpolar solvents, e.g., ether or petroleum ether. The solubilities of some alkali-metal xanthates in water and alcohols are given in Table 2.

Heavy-Metal Xanthates. The heavy-metal xanthates have lower mps than the corresponding alkali-metal salts and are better soluble in organic solvents, e.g., chloroform, tetrahydrofuran, and carbon disulfide, than in strongly polar ones [11]. The solubilities of heavy-metal O-ethyl xanthates in water increase in the order Au^+, Hg^{2+}, Cu^+, Ag^+, Pb^{2+}, Cd^{2+}, Co^{2+}, Ni^{2+}, Zn^{2+}, Fe^{2+} [12].

Aqueous xanthate solutions are relatively stable between pH 8 and 13 with a maximum of stability at pH $\approx$ 10 [13], but they rapidly decompose at pH < 6 and at pH > 13 (see Chap. 3). Their stability tends to increase with increasing molecular mass and branching of the alkyl group.

The electron absorption spectra of aqueous xanthate solutions show characteristic bands at 301 and 226 nm with extinction coefficients ϵ of $\approx$ 17 500 and 8750, respectively [14]. IR spectra of various xanthates have been published, e.g. [15].

3. Chemical Properties

In the presence of polar substances xanthic acids readily *decompose* into carbon disulfide and the corresponding alcohol, which accelerates the decomposition autocatalytically [16].

$$S=C(SH)(O-R) \longrightarrow CS_2 + R-OH$$

This reaction is also the second step in the decomposition of xanthates in aqueous acidic solution [17]:

$$S=C(S^-)(O-R) \xrightarrow[pH<6]{H^+} S=C(SH)(O-R) \longrightarrow CS_2 + R-OH$$

In neutral and alkaline solutions xanthates *hydrolyze* according to the following equation. The trithiocarbonate formed can decompose further into carbon disulfide and hydrogensulfide

$$6\,S=C(S^-)(O-R) + 3\,H_2O \longrightarrow 6\,R-OH + 3\,CS_2 + CO_3^{2-} + 2\,CS_3^{2-}$$

Xanthates are susceptible to *oxidation*, resulting in the formation of dixanthogens, e.g.:

$$2\,S=C(S^-)(O-R) \xrightarrow{\text{Oxidation}} (R-O)(S=)C-S-S-C(=S)(O-R)$$

Oxidation can be accomplished by various oxidants [18], e.g., oxygen, chlorine, bromine, iodine, hydrogen peroxide, nitrous acid, and chloramine T. In addition, oxidative coupling is achieved by anodic oxidation [19]. Diisopropyl dixanthogen, used as polymerization regulator, is produced by oxidation of potassium isopropyl xanthate with persulfate [20].

Condensation of alkali-metal xanthates with sulfur dichlorides gives higher xanthate sulfanes [21].

$$2S=C\begin{matrix}S^-\\O-R\end{matrix} + S_nCl_2 \longrightarrow \begin{matrix}S\\\|\\R-O-C-S-S_n-S-C-O-R\\\|\\S\end{matrix}$$

$$n = 1-4$$

Alkylation of alkali-metal xanthates with alkyl or allyl halides yields the corresponding xanthate esters (*O,S*-diesters of dithiocarbonic acid).

$$S=C\begin{matrix}S^-M^+\\O-R^1\end{matrix} + R^2-X \xrightarrow{-MX} S=C\begin{matrix}S-R^2\\O-R^1\end{matrix}$$

M: Na, K

This widely employed synthetic procedure [22] is usually carried out in organic solvents such as ethanol, ethers, ketones, or benzol under reflux, but reaction in DMSO is also possible [23]. In aqueous solution, alkylation is often conducted under phase transfer conditions [24] or in the presence of an emulsifier [25]. Especially in industrial processes, alkylation is carried out as a one-pot reaction without isolation of the corresponding alkali-metal xanthate [26]. Xanthate esters have also been obtained by alkylation of polymer-supported ammonium xanthates [27]. Treatment of lithium enolates with carbon disulfide and subsequent alkylation gives α,β-unsaturated esters [28].

Reaction of alkali-metal xanthates with geminal dihalides affords the corresponding bisxanthates [29].

In contrast to alkylation, *arylation* of alkali-metal xanthates is possible only with aryldiazonium salts [30]. In the Leuckart reaction, the arylesters thus obtained are converted either to thiophenols by hydrolysis [31], or thermally to thioethers.

$$S=C\begin{matrix}S^-K^+\\O-C_2H_5\end{matrix} + Ar-N_2^+Cl^- \longrightarrow$$

$$S=C\begin{matrix}S-Ar\\O-C_2H_5\end{matrix} \xrightarrow{H_2O} Ar-SH$$

Mild *pyrolysis* of *S*-methylxanthates results in elimination of carbonyl sulfide and methanethiol to give the corresponding alkenes without rearrangement (Chugaev reaction) [22], [32]. This reaction is widely used to convert alcohols into alkenes via their *S*-methyl xanthates.

$$\underset{H}{\overset{|}{\underset{|}{C}}}-\underset{|}{\overset{|}{C}}-O-\underset{S-CH_3}{\overset{S}{\underset{\|}{C}}} \xrightarrow{\Delta} \overset{\backslash}{\underset{/}{C}}=\overset{/}{\underset{\backslash}{C}} + COS + H_3C-SH$$

Applications in peptide chemistry [33] and in the terpene field have also been described [34].

Treatment of xanthate esters with tributyltin hydride gives *deoxygenated products* by radical reduction (Barton–McCombie deoxygenation) [35].

$$R\underset{O}{\diagdown}\underset{S}{\overset{S}{\underset{\|}{C}}}\diagdown CH_3 \xrightarrow[Cat.]{Bu_3SnH} R-H$$

This method is efficiently applied in the deoxygenation of secondary alcohols [36] and of secondary hydroxyl groups of carbohydrates and nucleosides [37].

Other reagents effecting radical deoxygenation of xanthates include various silanes [37], [38], dialkyl phosphites [39] and hypophosphorus acid or its salts [40]. Analogously, deoxygenations of primary [41] and tertiary alcohols [42] have been achieved via the corresponding xanthates.

Application of this deoxygenation procedure to bisxanthate esters derived from vicinal diols affords the corresponding alkenes [43].

$$\underset{HO\ \ OH}{-\overset{|}{\underset{|}{C}}-\overset{|}{\underset{|}{C}}-} \longrightarrow \underset{\underset{Xan:\ -CS-S-CH_3}{Xan-O\ \ O-Xan}}{-\overset{|}{\underset{|}{C}}-\overset{|}{\underset{|}{C}}-} \xrightarrow{Bu_3SnH} \overset{\backslash}{\underset{/}{C}}=\overset{/}{\underset{\backslash}{C}}$$

Several interesting *radical cyclizations* involving xanthates have been described. Thus, five- to eight-membered cyclic ethers [44], five- and six-membered *N*-heterocycles [45], and thionolactones [46] have been synthesized by this procedure.

Xanthate esters undergo thermally induced *rearrangement* to give *S,S*-dialkyldithiocarbonates [23], [47].

$$R^1\diagdown_O\underset{S}{\overset{S}{\underset{\|}{C}}}\diagup^{S}\diagdown R^2 \xrightarrow{\Delta} O=C\overset{S-R^2}{\underset{S-R^1}{\diagdown}}$$

This isomerization can also be effected catalytically with a Lewis acid or trifluoroacetic acid [48]. The isomerization of *O*-allyl xanthate esters occur with simultaneous allyl rearrangement [49].

Alkali-metal xanthates react with alkyl chloroformates to give alkyl xanthogen formates (mixed anhydrosulfides) (**I**), together with dialkyl xanthic anhydrides (**II**), dialkoxy carbonyl sulfides (**III**), and dialkyl carbonates (**IV**) as coproducts.

$$R^1{-}O{-}C(S){-}S^-Na^+ + Cl{-}C(O){-}O{-}R^2 \xrightarrow{-NaCl} \underset{\mathbf{I}}{R^1{-}O{-}C(S){-}S{-}C(O){-}O{-}R^2}$$

$$+ \underset{\mathbf{II}}{R^1{-}O{-}C(S){-}S{-}C(S){-}O{-}R^1} + \underset{\mathbf{III}}{R^2{-}O{-}C(O){-}C(O){-}O{-}R^2} + \underset{\mathbf{IV}}{R^1{-}O{-}C(O){-}O{-}R^2}$$

These reaction mixtures are used as collectors in the flotation of copper–molybdenum sulfide ores [50]. Efficient methods for obtaining the xanthogen formates (**I**) as the sole products in nearly quantitative yield have been also described [51].

4. Production

4.1. General

Alkali-metal xanthates are principally synthesized by reaction of an alcohol with sodium or potassium hydroxide and carbon disulfide. This reaction consists of two steps: (1) the formation of the alkali-metal alkoxide; and (2) reaction of the alkoxide with carbon disulfide to give the xanthate.

$$R{-}OH + NaOH \longrightarrow R{-}ONa + H_2O$$

$$R{-}ONa + CS_2 \longrightarrow R{-}O{-}C(=S){-}SNa$$

This reaction is applicable to all types of alcohols, including diols, polyols, and carbohydrates, including starch and cellulose. Xanthate salts of primary and secondary alcohols are prepared in aqueous medium, whereas the corresponding derivatives of tertiary alcohols are readily hydrolyzed and are therefore synthesized under anhydrous conditions [52]. Analogous reactions of solid alcohols [52] and of phenol and its derivatives [53] are carried out in aqueous dioxane.

For several purposes (when they are, for example, used as intermediates in the synthesis of xanthate esters) isolation of the alkali-metal xanthates is not necessary. Here, the xanthate is obtained in high yield by converting the alcohol into the corresponding alkoxide with metallic sodium or sodium hydride in an inert solvent and subsequent addition of carbon disulfide. This procedure is also suitable for unsaturated alcohols, e.g., allyl and propargyl alcohol [54].

Another preparation method for alkali-metal xanthates is the reaction of a dialkyl ether with alkali hydroxide and carbon disulfide [55].

$$R{-}O{-}R + 2\,CS_2 + 2\,NaOH \longrightarrow 2\,R{-}O{-}C(=S){-}S^-Na^+ + H_2O$$

Water-insoluble *heavy-metal xanthates* can be prepared by reaction of dissolved alkali-metal xanthates with heavy-metal salts [5].

Reaction of carbon disulfide with epoxides affords *cyclic dithiocarbonates* (1,3-oxathiolane-2-thiones) [56].

$$R-\overset{O}{CH\!-\!CH_2} + CS_2 \longrightarrow \underset{R}{\text{(1,3-oxathiolane-2-thione)}}$$

The corresponding 4,5-unsaturated heterocyclic derivatives are obtained by rhodium-catalyzed decomposition of α-diazocarbonyl compounds in the presence of carbon disulfide [57].

$$R-CO-CH=N_2 + CS_2 \xrightarrow{\text{Cat.}} \underset{R}{\text{(unsaturated heterocycle)}}$$

4.2. Production Processes

In the industrial production of alkali-metal xanthates a primary or secondary C_2–C_6 alcohol is reacted first with anhydrous sodium or potassium hydroxide and subsequently with carbon disulfide. To avoid side reactions, e.g., thermal decomposition, the temperature of the reaction mixture is generally maintained below 40 °C by cooling. Since their introduction into flotation in 1925, detailed reaction conditions of the manufacture of xanthates have been evaluated and described in numerous patents. In principle, the reaction can be carried out with or without a solvent.

Production without solvent is carried out in reactors preferentially made from steel that contain cooling and stirring apparatus. The reactors are also equipped with a heating device so that the reaction and the drying of the product can be carried out in the same vessel. Several improvements of the existing methods are reported in the patent literature. Thus, a reactor for the production of alkali-metal alkyl xanthates in one step with economic raw materials utilization and without environmental pollution is described [58]. Another patent reports increased yields and improved quality of xanthates by carrying out the process in a rotor-plate reactor [59]. Potassium butyl xanthate can be produced more efficiently by recycling of the plant mother liquor [60]. Higher reaction yields of potassium amyl xanthate can be obtained by removing part of the water formed with pervaporation membranes [61].

Production in solvents is carried out either in an excess of the corresponding alcohol [62] or in an inert solvent such as petroleum ether [63]. The precipitated or crystallized xanthate is separated by filtration, washed, and dried. The continuous production of xanthates using aqueous alkali-metal hydroxide, an alcohol, and carbon disulfide has also been reported [64]. Moreover, reaction in aqueous solution is used to produce alkali xanthates in the presence of additives, giving dust-free, noncaking, hydrophobic powders [65].

Safety Aspects. The low flash point and ignition temperature and the toxicity of carbon disulfide require appropriate safety measures for the production plants. Dusting of xanthate powders can be prevented by additives [65] or by shaping processes [66].

5. Quality Specifications

The commercially available xanthates from C_2-C_6 alcohols are of technical grade only and usually 90–95% pure. Another publication reports commercial xanthates to be rarely more than 90% and frequently only 60–90% pure [67]. The content of inorganic byproducts in technical xanthates varies, depending on their exposure to air [68]. Thus, alkali-metal sulfide, sulfate, trithiocarbonate, thiosulfate, or sulfite may be present. In addition, xanthates may contain some residual alcohol or alkali hydroxide. No general specifications for xanthates exist. They can be obtained as powder, granules, flakes, or sticks, and are also available in solution.

6. Storage and Transportation

Xanthates are usually transported and stored in steel drums. Copper, brass, zinc, and similar metals are not suitable. Aqueous xanthate solutions can be transported in vessels made from steel, steel alloys, or plastics, e.g., poly(vinyl chloride) or polyethylene. If the drums are kept cool and dry, solid xanthates are stable for years, but they may ignite spontaneously upon combined action of moisture and warming. Xanthate solutions may form carbon disulfide.

According to GeStoffV xanthates must be labeled as substances that are hazardous to health.

7. Analysis

Xanthates can be analytically determined by different methods [7]. Since many of the published analytical procedures have certain sources of error, the method of assay should be noted together with the specification of the xanthate content.

In general, the xanthate content is determined by a procedure based on decomposition of the xanthate with an excess of acid and subsequent back-titration with a base [3], [69].

Solutions of technical xanthates are usually assayed by *UV spectrometry*. This technique is also used for monitoring the concentration of xanthates in industrial flotation circuits [70]. Furthermore, xanthate solutions can be analyzed *potentiometrically* with silver nitrate [71].

Xanthates of C_1-C_5 alcohols have been separated successfully by *chromatographic methods* such as thin-layer chromatography or HPLC [72].

8. Uses

Cellulose xanthates are the most important xanthates. They are used as starting material in the manufacture of cellophane.

Flotation Reagents. The principal use for alkali-metal xanthates is as flotation collectors for metal-sulfide ores. Since their introduction in flotation in 1925 many patents and articles dealing with this field of application have been published.

Other applications of xanthates are only of minor importance. A selection of recent publications on the use of xanthates, drawn mainly from the extensive patent literature, is given below.

Polymerization. Xanthate derivatives are employed in polymerization processes as regulators [73], photoinitiators [74], and for the synthesis of block and graft copolymers [75]. Cyclic dithiocarbonate derivatives are suitable as cross-linking agents [76].

Additives for Lubricating Oils. Various dialkyl xanthates are claimed to be active as additives for lubricating oils [25], [77]. Moreover, xanthate complexes of molybdenum and vanadium are suitable as lubricant additives [78].

Color Photography. Xanthate derivatives are used in the preparation of electrophotographic material [79], as color developer, and photographic couplers [80].

Pesticides and Agricultural Chemicals. Xanthates and their derivatives have been recommended as fungicides [81], acaricides [82], and insecticides [83]. Microbicidal activity is also described for xanthate derivatives of benzeneacetic acid [84] and for quaternary ammonium salts [85] and *N*-methylpyrroles [86] containing dithiocarbonate residues. Halonitroaryl esters of *O*-butylxanthic acid exhibit nematocidal activity [87].

Other xanthate compounds effective as plant growth regulators are derivatives of dipropylene or triethylene glycol [88] and 3-benzothiazoline [89]. Various xanthate derivatives are claimed to be active as herbicides [90].

Pharmaceuticals. Pharmaceutical effects of xanthates include virustatic [91], antihypertensive [92], anthelmintic [93], analgetic and antipyretic [94], and antibiotic activity [95]. Pharmacological uses are also claimed for xanthate derivatives of aminoalkyl glycerols [96]. Potassium and gallium ethyl xanthate are effective antitumor agents [97] and 1,3-benzoxathiol-2-thiones (cyclic dithiocarbonates) can be used for treatment of liver diseases [98]. A number of phenylacyl xanthate esters and xanthogen formates (mixed anhydrosulfides) have been synthesized as potential radiosensitizers [99]. In addition, xanthates have been described as valuable intermediates for the preparation of drugs [100].

Rubber. Xanthates are used as antioxidants and vulcanization accelerators for rubber [101].

Environmental Chemistry. Traces of heavy metals can be removed from wastewaters by precipitation with xanthates [102]. Similarly, silver can be recovered from spent fixing baths using xanthates [103]. Cadmium can be extracted from soil by complexation with long-chain alkyl xanthates [104]. The recovery of xanthates from production mother liquors has been described [105].

Analytics. DNA suitable for polymerase chain reactions is isolated from plants, yeast, and bacteria using potassium or sodium ethyl xanthate [106]. Xanthates are also useful chelating reagents for the spectrophotometric detection of metals [107].

Other Uses. Substituted noble-metal xanthates are used in catalysts systems for removing impurities, e.g., S and N compounds, from hydrocarbons by hydrotreating [108].

Alkoxycarbodithiochloroformates are suitable condensing reagents in peptide synthesis [109].

9. Economic Aspects [110]

Various xanthates are produced by a large number of companies worldwide. The major mining chemical companies producing xanthates in Northern America are American Cyanamid, C-I-L (a wholly owned subsidiary of ICI), Kerley Enterprises, and Minerec Mining Chemicals. The Charles Tennant Co. plans (1993) to build a 12 000 – 14 000 t/a xanthate facility in Montreal, Canada. UCB in Belgium is the principal supplier of xanthates in Western Europe, but also sells to Africa, Canada, and the Philippines. Other producers in Western Europe are Hoechst (Germany), Shell (United Kingdom), and General Quimica (Spain). Robinson Brothers (United Kingdom) produces diisopropylxanthogen disulfide and zinc isopropylxanthate as rubber processing chemicals. Industrias Quimicas (Mexico), owned by Rhône-Poulenc (France), dominates the large Mexican market for xanthates and also sells to other South and Central American countries. Shell Chemicals produces xanthates and xanthogen for-

Table 3. Toxicity of potassium xanthates, $R-O-CS-S^- \ K^+$, to mice

R	LD$_{50}$, mg/kg	
	Oral	Intravenous
Ethyl	683	199
Isopropyl	583	
Butyl	465	158
Isobutyl	480	158
Isopentyl	480	158

mates in Chile. Karbochem (Republic of South Africa) manufactures a range of alkali-metal xanthates, marketing them in the Republic of South Africa and other African nations.

In 1989 prices for technical grade xanthates were in the range of $ 1.20 – 3.60 per kilogram for potassium amyl xanthate and $ 1.00 – 2.54 per kilogram for sodium isopropyl xanthate.

10. Toxicology and Occupational Health [111]

Data on the toxicity of xanthates and their derivatives in the literature are rather rare.

The alkali-metal xanthates are moderately toxic by ingestion or inhalation, but poisonous by intravenous route, as indicated in Table 3. In the body, xanthates act similarly to carbon disulfide.

Animal experiments on the *chronic toxicity* of potassium amyl xanthate revealed that liver damage occured in dogs exposed to an aqueous aerosol of 23 mg/m^3.

Because of their basicity, prolonged contact of alkali-metal xanthates and their concentrated solutions with the skin and especially with the eyes must be avoided. Skin contact may cause dermatitis.

Xanthates emit toxic fumes of SO$_x$ and metal oxide when heated to decomposition. In the preparation and handling of xanthate solutions the possible formation of carbon disulfide requires protection against fire and explosions.

11. References

General References

[1] *Ullmann,* 4th ed., **24**, 511.
[2] *Houben-Weyl,* **E 4**, 425; **9**, 810.
[3] *Kirk-Othmer,* 3rd ed., **24**, 645.

[4] A. D. Dunn, W.-D. Rudorf: *Carbon Disulphide in Organic Chemistry*, Ellis Horwood, Chichester 1989, pp. 316–339.

[5] G. Gattow, W. Behrendt in A. Senning (ed.): "Carbon Sulfides and Their Inorganic and Complex Chemistry," *Topics in Sulfur Chemistry*, vol. **2**, Georg Thieme, Stuttgart 1977, pp. 119–154.

[6] E. E. Reid: *Organic Chemistry of Bivalent Sulfur*, vol. **4**, Chemical Publishing, New York 1962, p. 131.

[7] S. R. Rao: *Xanthates & Related Compounds*, Marcel Dekker, New York 1971.

Specific References

[8] W. C. Zeise, *Schweiggers J. Chem. Phys.* **35** (1822) 173; **36** (1822) 1.

[9] H. Majima, *Sci. Rep. Res. Inst. Tohoku Univ., Ser. A* **13** (1961) 183.

[10] F. Drawert, K. H. Reuter, F. Born, *Chem. Ber.* **93** (1960) 3056. I. S. Shupe, *J. Assoc. Off. Agric. Chem.* **25** (1942) 495; *Chem. Abstr.* **36** (1942) 4670.

[11] F. Galsbøl, C. E. Schäffer, *Inorg. Synth.* **10** (1967) 42.

[12] I. A. Kakovski, *Proc. Int. Congr. Surf. Act. 2nd* 1957, no. 4, 225. A. T. Pilipenko, *Zh. Anal. Khim.* **4** (1949) 227.

[13] J. Dyer, L. H. Phifer, *Macromolecules* **2** (1969) 111. R. J. Millican, C. K. Sauers, *J. Org. Chem.* **44** (1979) 1964.

[14] M. L. Shankaranarayana, C. C. Patel, *Acta Chem. Scand.* **19** (1965) 1113.

[15] L. H. Little, G. W. Poling, J. Leja, *Can. J. Chem.* **39** (1961) 745. F. G. Pearson, R. B. Stasiak, *Appl. Spectrosc.* **12** (1958) 116. A. T. Pilipenko, N. V. Melnikova, *Russ. J. Inorg. Chem. (Engl. Transl.)* **15** (1970) 608.

[16] J. Berger, *Acta. Chem. Scand.* **6** (1952) 1564. J. R. Millican, C. K. Sauers, *J. Org. Chem.* **44** (1979) 1664.

[17] I. Iwasaki, S. R. B. Cooke, *J. Am. Chem. Soc.* **80** (1958) 285.

[18] Bayer, DE 2533989, 1975 (W. Behr, H. Königshofen). H. Dautzenberg, B. Philipp, M. Frede, *J. Prakt. Chem.* **312** (1970) 1051. Toyo Soda, JP 61210067, 1985 (S. Aman, H. Watanabe, A. Nakanishi, Y. Nonaka). G. Bulmer, F. G. Mann, *J. Chem. Soc.* 1945, 677. E. I. Stout, B. S. Shasha, W. M. Doane, *J. Org. Chem.* **39** (1974) 562.

[19] R. Woods, *Aust. J. Chem.* **25** (1972) 2329. Academy of Sciences Kazakh, SU 1116035, 1981 (L. G. Ili'chenko et al.). Irkut. Poly., SU 1244142, 1984 (S. A. Bogidaev, A. N. Baranov, S. B. Leonov).

[20] W. Winkler, W. Gruneberg, DD 118599, 1966. Orgsintez, Volzhskii, SU 1518338, 1987 (Y. P. Polin et al.).

[21] F. Fehér, W. Becher, R. Kreutz, F. R. Minz, *Z. Naturforsch. B* **18 B** (1963) 507. W. Knoth, G. Gattow, *Z. Anorg. Allg. Chem.* **552** (1987) 181.

[22] H. R. Nace, *Org. React.* **12** (1962) 57.

[23] P. Meurling, K. Sjöberg, B. Sjöberg, *Acta Chem. Scand.* **26** (1972) 279. M. B. Diana, M. Marchetti, G. Melloni, *Tetrahedron: Asymmetry* **6** (1995) 1175.

[24] I. Degani, R. Fochi, *Synthesis* 1978, 365. I. Degani, R. Fochi, V. Regondi, *Synthesis* 1981, 149.

[25] A. S. Nakhmanov, S. I. Krakhmalev, L. I. Nazarova, SU 910613, 1980; SU 910614, 1980.

[26] A. W. M. Lee, W. H. Chan, H. C. Wong, M. S. Wong, *Synth. Commun.* **19** (1989) 547. Phillips Petroleum, US 5239104, 1993 (J. E. Shaw). Inst. Cercetari, Ing. Tehnol., Baia Mare, RO 102036, 1992 (G. Duda et al.); *Chem. Abstr.* **120** (1994) 191136.

[27] A. Le Minor, I. E. Kanjo, D. Villemin, *Polym. Bull. (Berlin)* **21** (1989) 445.

[28] S. M. Ali, S. Tanimoto, *J. Org. Chem.* **54** (1989) 5603. S. M. Ali, S. Tanimoto, *Bull. Inst. Chem. Res. Kyoto Univ.* **68** (1990) 199. S. M. Ali, S. Tanimoto, *J. Chem. Soc. Chem. Commun.* 1989, 684.

[29] A. Schönberg, E. Frese, W. Knöfel, K. Praefke, *Chem. Ber.* **103** (1970) 938. Establ. Ind. Quim. Oxiquim., Chile, US 5 011 965, 1991 (E. Saphores, A. G. Gleisner, J. C. Vega, W. A. Mardones). Indian Oil, IN 171 122, 1992 (Y. K. Gupta et al.); *Chem. Abstr.* **120** (1994) 216722.

[30] D. S. Tarbell, D. K. Fukushima, *Org. Synth. Coll. Vol.* **3** (1955) 809.

[31] J. R. Cox, C. L. Gladys, L. Field, D. E. Pearson, *J. Org. Chem.* **25** (1960) 1083.

[32] J. March: *Advanced Organic Chemistry*, 4th ed., Wiley-Interscience, New York 1992, p. 1015.

[33] J.-R. Dormoy et al., *Angew. Chem. Int. Ed. Engl.* **19** (1980) 742.

[34] D. H. R. Barton, *J. Chem. Soc.* 1949, 2174.

[35] D. H. R. Barton, S. W. McCombie, *J. Chem. Soc. Perkin Trans. 1* 1975, 1574. D. H. R. Barton, D. Crich, A. Loebberding, S. Z. Zard, *Tetrahedron* **42** (1986) 2329.

[36] S. Jacono, J. R. Rasmussen, *Org. Synth.* **64** (1985) 57. K. Nozaki, K. Oshima, K. Utimoto, *Tetrahedron Lett.* **29** (1988) 6125. K. Nozaki, K. Oshima, K. Utimoto, *Bull. Chem. Soc. Jpn.* **63** (1990) 2578.

[37] Univ. Georgia Res. Found., WO 92/15 598, 1991 (Y. Chen, C. K. Chu).

[38] D. H. R. Barton, D. O. Jang, J. C. Jaszberenyi, *Synlett* 1991, 435. J. N. Kirwan, B. P. Roberts, C. R. Willis, *Tetrahedron Lett.* **31** (1990) 5093. D. H. R. Barton, D. O. Jang, J. C. Jaszberenyi, *Tetrahedron* **49** (1993) 7193. S. J. Cole, J. N. Kirwan, B. P. Roberts, C. R. Willis, *J. Chem. Soc. Perkin Trans. 1* 1991, 103.

[39] D. H. R. Barton, D. O. Jang, J. C. Jaszberenyi, *Tetrahedron Lett.* **33** (1992) 2311.

[40] D. H. R. Barton, D. O. Jang, J. C. Jaszberenyi, *Tetrahedron Lett.* **33** (1992) 5709.

[41] D. H. R. Barton, D. O. Jang, J. C. Jaszberenyi, *Tetrahedron Lett.* **32** (1991) 7187. D. H. R. Barton, D. O. Jang, J. C. Jaszberenyi, *Tetrahedron* **49** (1993) 2793.

[42] D. H. R. Barton, S. I. Parekh, C. L. Tse, *Tetrahedron Lett.* **34** (1993) 2733.

[43] A. G. M. Barrett, D. H. R. Barton, R. Bielski, *J. Chem. Soc. Perkin Trans. 1* 1979, 2378. D. H. R. Barton, D. O. Jang, J. C. Jaszberenyi, *Tetrahedron Lett.* **32** (1991) 2569. D. H. R. Barton, D. O. Jang, J. C. Jaszberenyi, *J. Org. Chem.* **58** (1993) 6838.

[44] J. H. Udding, J. P. M. Giesselink, H. Hiemstra, W. N. Speckamp, *J. Org. Chem.* **59** (1994) 6671.

[45] J. H. Udding, J. P. M. Giesselink, H. Hiemstra, W. N. Speckamp, *Bull. Soc. Chim. Belg.* **103** (1994) 392. J. Axon et al., *Tetrahedron Lett.* **35** (1994) 1719.

[46] M. Yamamoto et al., *J. Chem. Soc. Chem. Commun.* 1989, 1265. S. Iwasa, M. Yamamoto, S. Kohmoto, K. Yamada, *J. Org. Chem.* **56** (1991) 2849.

[47] K. G. Rutherford, B. K. Tang, L. K. M. Lam, D. P. C. Fung, *Can. J. Chem.* **50** (1972) 3288. Courtaulds, EP 202 054, 1985 (P. G. Urben).

[48] K. Komaki, T. Kawata, K. Harano, T. Taguchi, *Chem. Pharm. Bull.* **26** (1978) 3807. M. W. Fichtner, N. F. Haley, *J. Org. Chem.* **46** (1981) 3141. J. R. Pougny, *J. Carbohydr. Chem.* **5** (1986) 529.

[49] K. Harano, T. Taguchi, *Chem. Pharm. Bull.* **20** (1972) 2357. S. N. Suryawanshi, A. Rani, D. S. Bhakuni, *Synth. Commun.* **20** (1990) 625. F. Chatzopoulos-Ouar, G. Descotes, *J. Org. Chem.* **50** (1985) 118. D. Villemin, M. Hachemi, *Synth. Commun.* **25** (1995) 2311. T. Taguchi et al., *Tetrahedron Lett.* **6** (1965) 2717.

[50] R. D. G. Crozier, US 4 354 980, US 4 454 051, US 4 605 518, 1981.

[51] Pfizer, EP 274 876, 1986 (C. Sklavounos). M. A. Palominos, J. G. Santos, J. C. Vega, M. A. Martinez, *Phosphorus Sulfur* **25** (1985) 91.

[52] K. A. Jensen, U. Anthoni, A. Holm, *Acta Chem. Scand.* **23** (1969) 1916.

[53] A. K. Kadyrov, L. B. Alavidinov, D. V. Islamova, M. G. Ismatuellaeva, *Dokl. Akad. Nauk. USSR* 1989, 40; *Chem. Abstr.* **112** (1990) 216390.

[54] K. Harano, T. Taguchi, *Chem. Pharm. Bull.* **20** (1972) 2357. K. Tomota, M. Nagano, *Chem. Pharm. Bull.* **20** (1972) 2302.

[55] Standard Oil, US 2 534 085, 1950 (B. M. Vanderbuilt, J. P. Thorn). J. Sejbl, *Rudy* **31** (1983) 163; *Chem. Abstr.* **99** (1983) 175192.

[56] Agency of Industrial Sciences and Technology, JP 63 218 672, 1987 (Y. Taguchi, K. Yanagiya, I. Shibuya, Y. Suhara). N. Kihara, Y. Nakawaki, T. Endo, *J. Org. Chem.* **60** (1995) 473.

[57] T. Ibata, H. Nakano, *Chem. Express* **4** (1989) 93. T. Ibata, H. Nakano, *Bull. Chem. Soc. Jpn.* **63** (1990) 3096.

[58] Inst. Cercetari, Ing. Tehnologica si Proiectari pentru Minereuri Neferoase, RO 89 624, 1984 (G. Duda et al.); *Chem. Abstr.* **106** (1987) 175787; RO 89 625, 1984 (G. Duda et al.); *Chem. Abstr.* **107** (1987) 25115 g.

[59] G. A. Anokhin et al., SU 1 214 662, 1983; *Chem. Abstr.* **105** (1986) 171855 y.

[60] Central Ural Copper-Smelting Plant, SU 1 351 925, 1986 (B. N. Driker et al.); *Chem. Abstr.* **108** (1988) 223445 y. Central Ural Copper-Smelting Plant, SU 1 468 899, 1986 (Y. K. Achaev et al.); *Chem. Abstr.* **111** (1989) 96669 j.

[61] M. M. Cipriano, V. Geraldes, M. Norberta de Pinho, *J. Membr. Sci.* **62** (1991) 103.

[62] Degussa, DE 765 196, 1955 (H. Gunther).

[63] Hoechst, DE 1 934 175, 1971 (K. Baessler, G. Folz).D. I. Mendeleev Chem.-Tech. Inst., SU 727 641, 1980 (V. P. Savel'yanova, R. T. Savel'yanova, N. P. Utrobin, T. V. Rudakova).

[64] O. Leminger et al., CS 97 929, 1961.P. P. Gnatyuk et al., *Khim. Prom-st. (Moscow)* 1983, 402.

[65] Volgograd Polytechn. Inst., SU 1 066 988, 1980 (A. P. Khardin et al.). Volgograd Polytechn. Inst., AU 8 290 043, 1982.

[66] Zhuzhon Dressing Chemical Plants, CN 1 074 902, 1992 (L. Xu, D. Chen, C. Lu); *Chem. Abstr.* **120** (1994) 298077. Qixia County Ore Dressing Agent Factory, CN 1 072 354, 1991 (C. H. Zhang, S. Jian); *Chem. Abstr.* **121** (1994) 113856.

[67] J. Leja: *Surface Chemistry of Froth Flotation*, Plenum Press, New York 1982, p. 229.

[68] S. N. Danilov, N. M. Grad, E. I. Geine, *Zh. Obshch. Khim.* **19** (1949) 826.

[69] W. Hirschkind, *Eng. Min. J.* **119** (1925) 968.

[70] M. H. Jones, J. T. Woodcock, *Proc. Austr. Inst. Min. Metall.* **238** (1973) 1.

[71] C. Du Rietz, *Sven. Kem. Tidskr.* **69** (1957) 310.

[72] A. Messina, D. Corradini, *J. Chromatogr.* **207** (1981) 152.J. G. Eckhardt, K. Stetzenbach, M. F. Burke, J. L. Moyers, *J. Chromatogr. Sci.* **16** (1978) 510.R. T. Honeyman, R. R. Schrieke, *Anal. Chim. Acta* **116** (1980) 345.

[73] Bayer, EP 53 319, 1980 (G. Arend, H. Königshof, P. Müller, R. Musch).Desoto, US 4 537 667, 1984 (T. E. Bisho, G. Pasternack, J. M. Zimmerman).

[74] Tosoh Corp., EP 450 492, 1991 (H. Yamakawa, S. Ozoe). Fuji Photo Film, JP 6 083 079, 1993 (E. Kato, S. Oosawa). A. Ajayaghosh, *Polymer* **36** (1995) 2049.

[75] Moscow Textile Inst., SU 821 453, 1979 (G. A. Gabrielyan, M. S. Kislyuk, V. S. Lebedev, R. P. Shul'ga).M. Niwa, N. Higashi, M. Shimizu, T. Matsumoto, *Makromol. Chem.* **189** (1988) 2187.

[76] Nippon Zeon, JP 7 145 164, 1993 (K. Watanabe, T. Matsura).Nippon Zeon, JP 7 062 190, 1993 (Y. Takao, K. Watanabe, H. Mori).

[77] A. S. Nakhmanovich, SU 996 410, 1981.Inst. Chem. Additives, Acad. Sci. Azerbaidzhan, SU 1 505 931, 1988 (A. B. Kuliev, M. M. Dzhavadov, B. A. Abushova).Inst. Chem. Additives, Acad. Sci, Azerbaidzhan, SU 1 404 504, 1986 (A. M. Levshina, M. A. Agaeva, D. S. Gamidova, I. A. Buniyat-Zade).

[78]   Exxon, US 4 995 996, 1989 (C. L. Coyle). Exxon, US 4 861 958, 1985 (T. R. Halbert, E. I. Stiefel).
[79]   Fuji Photo Film, EP 599 660, 1993 (E. Kato). Fuji Photo Film, WO 92/18 907, 1992 (E. Kato, K. Ishii). Yamada, JP 7 076 568, 1993 (F. Kiryama, I. Nakanishi).
[80]   Konishiroku Photo, JP 57 096 335, 1982.
[81]   Ciba-Geigy, CH 636 610, 1977 (W. Kunz, A. Hubele, W. Eckhardt). P. C. Joshi, Jr., K. B. Karnatak, M. M. Sah, A. K. Pant, *J. Indian Chem. Soc.* **62** (1985) 164. W. Podkoscielny et al., *Przem. Chem.* **69** (1990) 495; *Chem. Abstr.* **114** (1991) 184935.
[82]   Begin-Say, FR 2 543 555, 1983 (J. Mentech).
[83]   Sandoz, EP 169 169, 1985 (C. A. Henrick).Takeda, JP 64 000 063, 1988 (H. Uneme, H. Mitsudera, T. Uekado).
[84]   BASF, EP 582 902, 1993 (H. Winger, H. Sauter, E. Ammermann, G. Loren).
[85]   Degussa, EP 386 508 and EP 386 509, 1989 (P. Werle, M. Trageser, S. Weiss).
[86]   American Cyanamid, EP 545 103, 1992 (D. G. Kuhn, V. Kameswaran).
[87]   P. Dureja, S. P. Gupta, D. Prasad, *Pesticides* **18** (1984) 43.
[88]   CPC International, US 4 806 149, 1987 (M. J. Danzig, A. M. Kinnersley).
[89]   Monsanto, US 4 427 436, 1982 (J. J. D'Amico).
[90]   Monsanto, US 4 556 415, 1985 (J. E. Franz, K. E. Zwikelmaier). Ciba-Geigy, EP 139 612, 1984 (W. Toepfl).
[91]   Merz, US 4 602 037, 1983 (A. Scherm, K. Hummel). Merz, BE 891 245, 1981 (K. Hummel, A. Scherm).
[92]   Boehringer, EP 255 075, 1986 (M. Frigerio et al.). Mitsubishi Kasei, JP 4 021 695, 1990 (F. Takaku et al.).
[93]   Y. D. Kulkarni, B. Kumar, *J. Indian Chem. Soc.* **60** (1983) 882.
[94]   Sandoz, DE 3 320 746, 1982 (R. Achini, L. Revesz).
[95]   Otsuka, WO 84/00 964, 1982 (J. Nogami et al.).
[96]   Ono Pharmaceutical Co., EP 94 586, 1983 (H. Itoh, T. Okada, T. Miyamoto).
[97]   D. Khadzhiolov, N. Frank, N. Khadzhiolov, S. Yanev, *Arch. Geschwulstforsch.* **58** (1988) 313. Ajinimoto, JP 6 197 292, 1984 (T. Cho, T. Tsuchiya, T. Yoshihama, T. Shiio).
[98]   Shinnippon, JP 6 239 856, 1993;JP 6 239 640, 1994 (T. Ueyama et al.).
[99]   D. Skwarski, H. Sobolewski, *Acta Pol. Pharm.* **49** (1992) 71, 77, 81; *Chem. Abstr.* **120** (1994) 191269, 191270; *Chem. Abstr.* **121** (1994) 8835. D. Skwarski, H. Sobolewski, *Acta Pol. Pharm.* **50** (1993) 213; *Chem. Abstr.* **121** (1994) 108121. J. Tulecki, J. Dudzinska-Usarewicz, P. Radola, *Pol. J. Pharmacol. Pharm.* **37** (1985) 709.
[100] Pfizer, EP 131 374, 1984 (E. S. Hamanaka). Sagami Chem. Res. Center, JP 62 132 857, 1985 (M. Matsumoto, N. Watanabe). Leo Pharmaceutical Prod., WO 94/14 766, 1993 (G. Grue-Soerensen, E. R. Ottosen).
[101] P. Rajalingam, G. Radhakrishnan, *Polym. Plast. Technol. Eng.* **30** (1991) 405. Asahi Chem. Ind., JP 62 127 330, 1985 (Y. Kitagawa, Y. Hattori). Tashkent Poly., SU 1 065 412, 1982 (A. Safaev, L. Alavidinov, N. B. Niyazoc).
[102] VEB Werke Buna, DD 148 044, 1979 (K. D. Rauchstein et al.).R. Braun, R. Rafflenbeul, DE 4 009 589, 1990.
[103] M. Nanjo, Y. Itoh, *Tohoku Daigaku Senku Seiren Kenkyusho Iho* **42** (1986) 127; *Chem. Abstr.* **106** (1987) 186253.
[104] N. Wolf, D. M. Roundhill, *Polyhedron* **13** (1994) 2801.
[105] `Kazmekhanobr' Inst., SU 1 449 545, 1986 (A. L. Droyaronov et al.); *Chem. Abstr.* **110** (1989) 195647 m.

[106] C. E. Williams, P. C. Ronald, *Nucleic Acids Res.* **22** (1994) 1917. Pioneer Hi-Bred Int., EP 493 115, 1991 (A. K. Jhingan).

[107] D. Yamamoto, M. Aoyama, *Meiji Daigaku Nogakubu Kenkyu Hokoku* 1985, 69; *Chem. Abstr.* **104** (1986) 198955.

[108] Exxon Res. & Eng. US 5 068 375, 1990 (G. H. Singhal).

[109] Tadzhik State Univ., SU 1 146 302, 1983 (S. K. Khalikov, M. M. Sobirov, S. V. Alieva, A. A. Akhtyamova); *Chem. Abstr.* **103** (1985) 123921 d.S. K. Khalikov, S. V. Alieva, M. M. Sobirov, *Khim. Prir. Soedin.* 1985, 844; *Chem. Abstr.* **105** (1986) 79339.

[110] SRI International (ed.): Chemical Economics Handbook, Directory of Chemical Producers, Europe 1995. E. A. Thiers in SRI International (ed.): Chemical Economics Handbook, Specialty Chemicals, Mining Chemicals-Worldwide, 1989.

[111] R. J. Lewis, Sr. (ed.): *Sax's Dangerous Properties of Industrial Materials*, 8th ed., vol. III, Van Nostrand Reinhold, New York 1992.

Xylenes

Jörg Fabri, Veba AG, Düsseldorf, Federal Republic of Germany
Ulrich Graeser, Veba Öl AG, Gelsenkirchen, Federal Republic of Germany
Thomas A. Simo, Lurgi Öl, Gas, Chemie GmbH, Frankfurt, Federal Republic of Germany

1.	Introduction	4987	4.2.3.	Combination of the Technologies in the Aromatics Complex 5003
2.	Properties	4990	5.	Integration into Refinery and Petrochemical Complexes 5005
3.	Occurrence and Raw Materials	4994		
3.1.	Natural Occurrence	4994	5.1.	Backward Integration into Petroleum Refining 5005
3.2.	Raw Materials for Xylene Production	4994	5.2.	Forward Integration into Chemical Processing 5005
4.	Production, Separation, and Further Processing	4996	6.	Economic Aspects 5006
4.1.	Production of C_8 Aromatics	4996	7.	Quality Specifications and Analysis 5010
4.2.	Separation and Further Processing	4998	8.	Storage, Transport, and Safety 5012
4.2.1.	Separation of the C_8 Aromatics Fraction	4998	9.	Environmental Aspects and Toxicology 5013
4.2.2.	Isomerization of the C_8 Fraction	5002	10.	References 5014

1. Introduction

The benzene homologs of general formula C_8H_{10} are generally known as mixed xylenes. The mixture of isomers with a boiling point range of 135–145 °C mainly consists of the three isomeric dimethylbenzenes and ethylbenzene:

o-Xylene   m-Xylene   p-Xylene   Ethylbenzene

With the exception of xylene production by disproportionation of toluene, the isomeric xylenes and ethylbenzene are always produced as a mixture in all production processes. However, the relative proportions of the C_8 isomers often differ considerably (see Chap. 3).

Because of their high knock resistance (see Chap. 2), xylenes are well suited to the production of motor fuels. In terms of quantity the production of gasoline exceeds that of BTX aromatics (B = benzene, T = toluene, X = xylenes) quite considerably (in Western Europe in 1995 gasoline production was ca. 150×10^6 t and that of BTX aromatics ca. 13.7×10^6 t, of which ca. 2.7×10^6 t/a was mixed xylenes, 0.65×10^6 t/a o-xylene, and 1.4×10^6 t/a p-xylene) [1]. The average aromatics content of motor fuels in Western Europe is ca. 38% [2].

The close association with gasoline production strongly affects the economics of separating xylene mixtures, for example, for use in chemical processes.

In Western Europe the production/use balance in 1995 was as follows [3]:

Production		
from coal	36×10^3 t	(ca. 1.3%)
from pyrolysis gasoline	300×10^3 t	(ca. 11.1%)
from reformate	2165×10^3 t	(ca. 80.0%)
by disproportionation	205×10^3 t	(ca. 7.6%)
Total production	2706×10^3 t	
Use in extraction and chemical processing (incl. imports)	2563×10^3 t	(ca. 88.4%)
solvents	306×10^3 t	(ca. 10.6%)
as gasoline component	30×10^3 t	(ca. 1%)
Total use	2899×10^3 t	
Domestic production	93.3%	
Net impor	6.7%	

The structure of world mixed-xylenes production capacity in 1994 was as follows:

Total capacity	23×10^6 t
Reformer	83%
Disproportionation	10%
Pyrolysis gasoline	6%
Coal	1%

In Western Europe until the year 2000 a significant increase in the proportion of xylene used in gasoline production is predicted, because the growth in production of xylene-containing reformate may exceed its consumption in chemical processes. However, in other regions, such as the United States, the reduction and limitation of the content of aromatics, and thus of xylenes, in gasoline is being discussed (see also Chap. 2and Section 5.2) [2], [4].

The use of xylenes in chemical processes broke down as follows (data for Western Europe 1995) [1]:

p-Xylene	1761×10^3 t (68.7%)
o-Xylene	650×10^3 t (25.4%)
Ethylbenzene and miscellaneous	152×10^3 t (5.9%)
Total	2563×10^3 t

The dominant importance of *p*-xylene can be seen from the relative proportions of the isolated xylene isomers. *p*-Xylene is mainly oxidized to terephthalic acid, which can be esterified to dimethyl terephthalate (precursor for polyesters). *o*-Xylene is oxidized to phthalic anhydride (precursor for plasticizers) and *m*-xylene to isophthalic acid (precursor for polyesters). Ethylbenzene is dehydrogenated to styrene, which is converted to polystyrene and other polymers [5].

History [6]. Xylene was first discovered in crude wood spirit in 1850 by CAHOURS. The name xylene was derived from the Greek word *xylon* (= wood). Xylene was detected in coal tar in 1855 by RITTHAUSEN and CHURCH. In 1891 ethylbenzene was found in hard coal tar by NOELTING and PALMAR, and MOSCHNER carried out the first synthesis of pure ethylbenzene from ethylbenzenesulfonic acid.

Between 1865 and 1869 ERNST and FITTIG found that the "xylene" in coal tar does not consist of a single compound. They synthesized "xylene" from toluene and named it "methyltoluene." They then established that some properties of "methyltoluene" differed from those of coal-tar xylene, particularly with regard to the nitro compounds. FITTIG et al. concluded that in synthetic "methyltoluene" (*p*-xylene) the second methyl group was in a different position from that in coal-tar xylene.

GLINSER and FITTIG described the production of methyltoluene in 1865. FITTIG obtained a hydrocarbon by dry distillation of the calcium salt of mesitoic acid which resembled both known modifications of dimethylbenzene, ("xylene" and "methyltoluene") but was not completely identical to either. The new hydrocarbon was named "isoxylene" (*m*-xylene). This work elucidated the structures of *p*- and *m*-xylene and proved their occurrence in coal tar.

In a further experiment FITTIG identified *o*-xylene as the third isomer, which was discovered in coal tar by JACOBSON in 1877.

m- and *p*-xylene were initially separated on the basis of their different chemical reactivity. *m*-Xylene reacts with concentrated sulfuric acid. *m*-Xylene is then regenerated from the crystalline *m*-xylenesulfonic acid using superheated steam in the presence of sulfuric acid. The fraction unaffected by sulfonation is then reacted with oleum. The sulfonic acid that crystallizes is cleaved to give pure *p*-xylene.

For decades coke-oven tar and benzole were used as raw materials for production of aromatics. Worldwide their production was associated in location and quantity with coal and steel production. The plants for extraction and refining of coke-oven benzole, a mixture of light oil from coke-oven tar and crude benzene, obtained from coke-oven gas by oil scrubbing, were built adjacent to coking plants. These plants were very small compared with those currently in operation. At that time the refining of crude benzole was carried out exclusively by washing with concentrated sulfuric acid, whereby impurities, such as resin-forming compounds, nitrogen and oxygen compounds, and some of the sulfur compounds, were removed as acid tar. However, valuable substances, such as styrene, were also lost in the acid tar. The xylenes were in demand as solvents and were used as raw materials in the chemical and pharmaceutical industries. However, production of the three isomers in pure form was complicated due to their almost identical physical properties. The quantities available were too small for the development of industrial xylene chemistry.

m- and *p*-xylene could be separated from *o*-xylene by distillation and thus *o*-xylene could be obtained in limited purity. *m*- and *p*-xylene, however, could only be separated laboriously on the basis of their different chemical properties.

With the development of the hydrogenative refining of coke-oven benzole over sulfur-resistant catalysts under pressure, sulfuric acid refining became uneconomical and was replaced by a technically superior process (see → Benzene). Parallel to this development in process technology, aromatics production became independent of steelworks or coking plants and centralization took place. Large refineries and distillation plants provided a basis for modest chemical processing of the C_8 aromatics in the mid-1950s. Between 1955 and 1960 ca. 30 000 t of xylenes were obtained from a coke production level of $(40-45) \times 10^6$ t/a.

The fourth isomer, ethylbenzene, must be considered separately from the historical development of the isomeric xylenes. Its synthesis from benzene and ethylene in a Friedel–Crafts reaction solved the problem of raw materials at an early stage.

With the discovery of the catalytic dehydrogenation of naphthenes and the catalytic dehydrocyclization of paraffins and isoparaffins on noble metal catalysts and their bifunctional systems (platforming and rheniforming processes), the raw material basis for the development of xylene chemistry on a large scale was secured in Europe from about 1960. A large number of processes have now been developed: liquid–liquid extraction, azeotropic distillation, extractive distillation, isomerization, crystallization, adsorption, and absorption. These not only provided pure products on an industrial scale, but also permitted the separation of *m*- and *p*-xylene. Thus the basis for a large-scale growth of polyester fiber production was created. The residual *m*-xylene could be isomerized or oxidized to isophthalic acid, which had a limited market. The market for *o*-xylene was provided by phthalic anhydride production.

2. Properties [6]

Physical Properties. Technical-grade xylene is a mixtrue of C_8 aromatics, also known as the A_8 fraction, consisting of the three xylene isomers, ethylbenzene, and depending on the origin, varying amounts of nonaromatics which boil in the same range (136–145 °C). Table 1 lists some physical data for the aromatic components. Other data can be found in [7], [8]. The relative proportions of these components in a xylene fraction depend on its origin and also which process steps the xylene has already passed through (e.g., hydrogenation).

The vapor pressure curves of the three xylene isomers and ethylbenzene are shown in Figure 1.

The isomeric xylenes and ethylbenzene form azeotropic mixtures with water and numerous organic compounds (Table 2) [9]–[13].

Table 1. Physical data for xylene isomers and ethylbenzene

		o-Xylene	m-Xylene	p-Xylene	Ethylbenzene
M_r		106.16	106.16	106.16	106.16
bp at 1 bar,	°C	144.4	139.1	138.4	136.2
Critical temperature,	°C	357.1	343.6	342.8	344.0
Critical pressure	bar	35.20	35.47	34.45	37.27
Critical compressibility		0.260	0.270	0.250	0.260
Critical molar volume	L/mol	0.380	0.390	0.370	0.371
	g/cm^3	0.281	0.282	0.290	0.284
mp,	°C	−25.182	−47.87	+13.26	−95.00
Surface tension at 15.6 °C,	mN/m	30.70	30.12	28.8	29.5
Dynamic viscosity at 20 °C,	mPa · s	0.809	0.617	0.644	0.6783
Thermal conductivity at 0 °C,	kJ m^{-1} h^{-1} K^{-1}	0.544	0.523	0.511	0.494
Enthalpy of evaporation at the bp and 1 bar,	kJ/mol	36.89	36.45	36.00	36.05
Lower calorific value,	kJ/g				
Gas		41.25	41.24	41.24	41.34
Liquid		40.84	40.83	40.84	40.94
Density at 1 bar,	g/cm^3				
at 25 °C		0.8760	0.8599	0.8567	0.8624
at 20 °C		0.8802	0.8642	0.8610	0.8670
Refractive index n_D^{20}		1.50449	1.49712	1.49575	1.49588
Heat of fusion,	kJ/mol	13.78	11.56	17.02	9.16
	kJ/kg	129.8	108.9	160.4	86.2
Explosion limits in air,	vol%				
lower		1.0			
upper		5.3–7.6			
Ignition temperature,	°C	496–502			

The absorption properties of xylene are of technical interest because of the significant differences in the solubilities of various gases as a function of temperature (see Table 3).

Chemical Properties. Oxidation of the xylene isomers gives the corresponding aromatic dicarboxylic acids. Phthalic acid is produced industrially from o-xylene, isophthalic acid from m-xylene, and terephthalic acid from p-xylene (see → Phthalic Acid and Derivatives, → Terephthalic Acid, Dimethyl Terephthalate, and Isophthalic Acid).

The oxidation reactions have been established as industrial processes in both the gas and liquid phases. Attempts have been made to introduce the co-oxidation of p-xylene with paraformaldehyde (Toray Industries) or acetaldehyde (Eastman Kodak) [5].

Ammonoxidation of m- and p-xylenes initially gives the corresponding phthalic acid dinitriles, which are important raw materials for the production of isocyanates via reduction to the corresponding xylylenediamines. The dinitriles can also be hydrolyzed to the corresponding phthalic acids. However, this step has only limited industrial and economic importance.

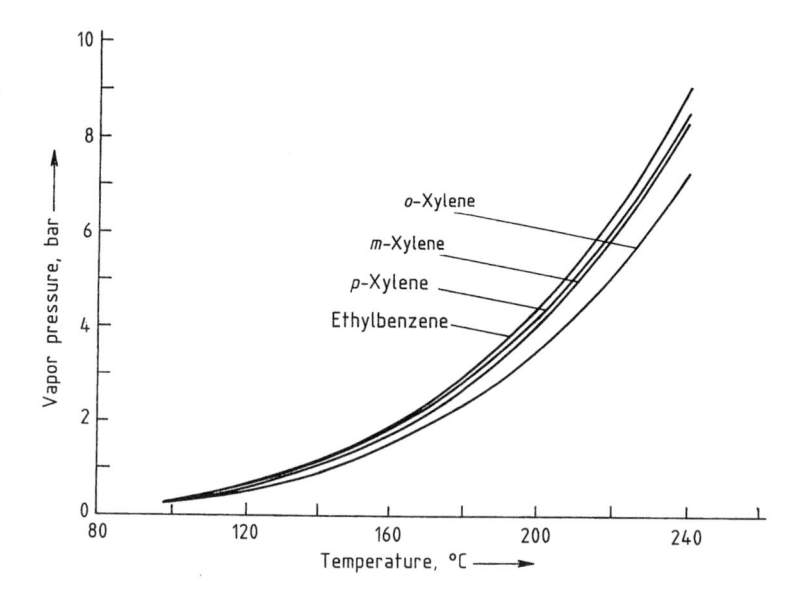

Figure 1. Vapor pressure curves of the xylene isomers and ethylbenzene at 100–240 °C

Table 2. Binary azeotropes

Component A	Aromatic	Azeotrope	
		bp, °C	Proportion of A, %
Water	*m*-xylene	94.5	40
	ethylbenzene	33.5 (80 mbar)	33
Methanol	*p*-xylene	64.0	5
Butanol	ethylbenzene	115.85	65
Isobutanol	ethylbenzene	125.7	49
n-Hexanol	*o*-, *m*-, *p*-xylene	138–143	18–13
Ethylene glycol	*o*-xylene	135.7	7
1,2-Propanediol	*o*-xylene	135.8	10
Glycol	*o*-, *m*-, *p*-xylene and ethylbenzene	139–133	14
4-Heptanone	*m*-xylene	139	10
Formic acid	*m*-xylene	94	70
	o-xylene	95.5	74
Acetic acid	*o*-, *m*-, *p*-xylene and ethylbenzene	114–116	76–66
Propionic acid	*o*-, *m*-, *p*-xylene and ethylbenzene	131–135	42–28
Phenol	*m*-xylene	133	18
Nonane	*o*-xylene	139	10

The nitration of *o*- and *m*-xylenes provides a route to xylidines following hydrogenation of the initially formed dimethylnitrobenzene isomers. Xylidines are used as intermediates in dye and rubber additive production.

The capacity of the xylene isomers to undergo isomerization and disproportionation reactions is also exploited industrially (see Chap. 4).

Table 3. Temperature dependence of the absorption coefficients of a technical-grade xylene (in m^3/m^3 xylene, STP)

T, °C	H_2S	CO_2	CO	H_2	NH_3	Methane	Ethylene	Propene
0	25	1.53	0.30	0.013	8.0	0.6	4.0	24.0
−10	35	1.62	0.33	0.017	12.4	0.68	4.4	33.0
−20	50	1.71	0.36	0.02	18.6	0.78	4.9	46.5
−30	72	1.85	0.39	0.024	27.7	0.92	5.5	66.0
−40	110	2.01	0.44	0.029	42.2	1.08	6.3	95.0
−50	174	2.21						

Table 4. Properties of xylene isomers as motor fuel components compared with finished gasoline

	o-Xylene	m-Xylene	p-Xylene	Ethylbenzene	Super unleaded (DIN 51 607/EN 228)
RON (pure)	107.4	117.5	116.4	113	
MON (pure)	101.5	110	109	98	
RON (typical blending value)	120	145	146	124	95 *
MON (typical blending value)	103	124	127	107	85 *
RVP, bar (pure)	0.028	0.025	0.04	0.045	0.9 (Winter) **
					0.7 (Summer) **

* Minimum values.
** Maximum values.

Sulfonation of *m*-xylene and subsequent decomposition of the sulfonic acid derivatives gives 3,5- and 2,4-xylenols, providing starting materials for insecticides, herbicides, etc.

Of the chemical properties of ethylbenzene as a component of the A_8 fraction, the most important is catalytic dehydrogenation to styrene, although ethylbenzene used for this purpose mainly comes from other sources (alkylation of benzene with ethylene).

Properties as a Motor Fuel Component. The properties of motor fuels are determined by a large number of quality aspects whose minimum requirements are mostly specified. The value which is attributed to xylenes in blending of motor fuels essentially depends on how the xylenes can be combined individually with other available fuel components to produce motor fuel qualities which conform to market requirements. Typically the high knock rating, particularly as a blend component, and the low vapor pressure of xylenes are important factors affecting their use as motor fuel components. Table 4 compares the octane number and vapor pressure values (RVP) of the xylene isomers [14], [15] with the specification of a typical gasoline.

A more detailed account of the overall much more complex interactions with the pool of gasoline components can be found in Section 5.1.

Table 5. Aromatics content of various crude oils

	Libya	Louisiana Gulf	West Texas	Venezuela	Nigerian	Iranian
C_8 aromatics, wt%	0.56	0.50	1.10	1.10	1.47	1.05
$C_6 - C_8$ aromatics, wt%	1.0	1.10	1.79	1.85	2.50	1.80

3. Occurrence and Raw Materials

3.1. Natural Occurrence

Practically the only natural source of xylenes is petroleum. The concentration of xylenes varies considerably depending on location and geological age of the crude oil. Table 5 shows a comparison of the C_8 aromatics and the C_6-C_8 aromatics contents in various types of crude oil [16].

Although in a few cases petroleums (e.g., that of South East Asian origin) can have an even higher aromatics content (up to 35%), the direct isolation of xylenes is not economical.

3.2. Raw Materials for Xylene Production

For the industrial extraction of xylenes, the oil fraction or coal used as the raw material is subjected to thermal or catalytic treatment in which aromatics- and xylene-containing fractions are obtained. These conversion processes determine the quantity of xylenes available and supply them in a form that is enriched sufficiently to make isolation and further processing economical.

Raw Materials based on Petroleum. *Reformates.* Xylenes are mainly isolated from reformates [17]. Naphtha fractions are used as the raw materials for the reforming process, which is now practically only carried out catalytically. Because of the sulfur sensitivity of the Pt–Re catalysts, the naphtha must first be desulfurized. The composition of the reformates and thus the xylene content depends on how the reforming process is carried out (see Section 4.1) and on the quality of the naphtha fraction used. The composition of the naphtha fraction, in particular its naphthene and aromatics content, significantly affects the composition of the reformate [18].

Naphthenic crude oils give naphtha fractions which are particularly suitable for reforming and aromatics production. However, naphtha fractions produced by hydrocracking also generally have high naphthene contents [19] and are therefore suitable starting materials for xylene production by reforming. Besides the composition, the boiling range of the naphtha fraction affects the yield of xylenes (see Table 6). A particularly high yield can be obtained when the naphtha fraction already has a high

Table 6. Dependence of reformate composition on the boiling range of reformer feedstock (Kuwait naphtha)

	Naphtha boiling range, °C		
	60 – 160	107 – 160	90 – 160
Feedstock composition, wt%			
Paraffins	69.6	62.2	64.2
Naphthenes	19.5	21.2	22.2
Aromatics	10.9	16.6	13.6
Aromatics composition of reformate, wt%			
Benzene	9.3	1.6	5.2
Toluene	21.7	19.0	25.1
C_8 aromatics	20.8	34.3	26.2
C_{9+} aromatics	8.8	15.2	11.2

Table 7. Effect of feedstocks on the yields of pyrolysis gasoline and C_8 aromatics from steam cracking (conditions: very high severity with C_2 and C_3 recycle)

Feedstock	Yield of pyrolysis gasoline (C_5 – 200 °C), %	Yield of C_8 aromatics, %	Proportion of C_8 aromatics in pyrolysis gasoline, %
Ethane	1.7		
Propane	6.6		
Butane	7.1	0.4	5.6
Medium-range naphtha	18.7	1.8	9.6
Atmospheric gas oil	18.4	2.2	12.0
Vacuum gas oil	19.3	1.9	9.8

proportion of the corresponding xylene precursors (e.g., C_8 aromatics or C_8 naphthenes) in a suitable boiling range [20].

Pyrolysis Gasoline. The yield of pyrolysis gasoline and its xylene content is decisively affected by the raw materials used, and by the mode of operation of the plant (see Chap. 4). In steam cracking very different types of raw material can be used, which give rise to widely varying yields and compositions of the pyrolysis gasoline (Table 7) [21].

Only on using higher boiling hydrocarbons feedstocks can appreciable xylene yields and concentrations in pyrolysis gasoline be attained. Therefore, in general, isolation of xylenes is only economical when using these types of raw materials.

Raw Materials based on Coal. The isolation of xylenes from coke-oven tar, coke-oven benzole, or from hydrocarbon mixtures produced by hydrogenation of coal is of minor importance [17] (see also Section 4.1). The effect of the composition of the coal is described in [22]. Processes in which xylene-rich aromatics fractions can be produced from LPG or methanol are described in Chapter 4.

Table 8. Typical operating conditions of various reformer types

Parameter	Semiregenerative reformer	Fully regenerative reformer	Continuously regenerative reformer
Pressure, MPa	1.5–2.5	0.7–1.5	1
Operating temperature, °C	510–540	510–540	510–540
LHSV*, h^{-1}	2–3.5	3.5–4	1.5–4

* Liquid hourly space velocity.

4. Production, Separation, and Further Processing

4.1. Production of C$_8$ Aromatics

Catalytic Reforming. Besides the nature of the raw materials, process engineering parameters also significantly determine the composition of the reformate and its content of C$_8$ aromatics [23]. The design of the catalytic reformer [24], its mode of operation, and the catalyst used are also very important. The three following types of catalytic reformer are currently in common use:

1) Semiregenerative reformer, in which the noble metal catalyst is periodically regenerated during a production shutdown (typically every 6 to 12 months)
2) Fully regenerative reformer, in which installation of an additional swing reactor allows the alternating regeneration of one reactor while the other continues to operate
3) Continuously regenerative reformer whose fluidized bed catalyst is continuously regenerated during operation

Operating conditions of the different types of reformer are listed in Table 8 [25].

Since at least six reactions take place in parallel during the reforming process, of which essentially only two lead directly to the formation of aromatics, establishing optimal operation conditions is very important. The reactions which give aromatics (the dehydrogenation of naphthenes and the dehydrocyclization of paraffins) are favored by comparatively high reactor temperatures and low pressures (see Fig. 2). At the same time a simultaneous increase in undesired side reactions (hydrocracking and coke formation) must be reckoned with, so that while increasing the severity leads to an increase in the aromatics content of the reformate, it also decreases reformate yield [26]. Dealkylation reactions are also of specific importance for xylene production. They can lead to decomposition of the xylenes formed, giving toluene, or benzene and methane, particularly at high temperatures.

A further control parameter for xylene production is the space velocity (LHSV). If throughput is lowered to increase the residence time in the reactor, reformates with high aromatics and xylene contents can be produced [25]. The adjustment of the

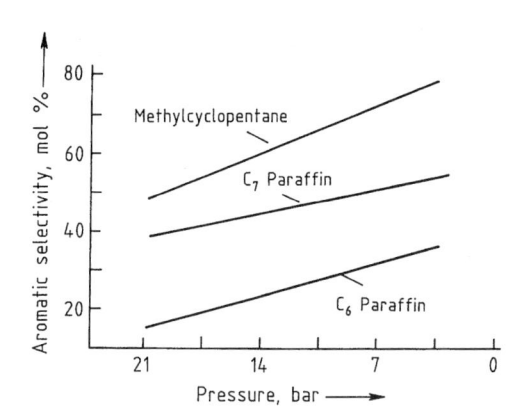

Figure 2. Variation of the selectivity of aromatics production in catalytic reforming with operating pressure

Table 9. Influence of severity in steam cracking on the yield and composition of pyrolysis gasoline (naphtha feedstock)

Severity (% Ethylene)	Xylene + ethylbenzene (all products), wt%	Xylene + ethylbenzene in pyrolysis gasoline, wt%	Pyrolysis benzene yield, wt%
Low (24.4)	0.9	3.5	24.9
Medium (28.5)	1.6	7.2	22.6
High (33.4)	1.7	10.7	16.3

operating parameters to give a maximum aromatics concentration in the reformate is known as aromizing. Within certain limits the quality of the reformate can also be influenced by the choice of catalyst. Besides conventional catalysts, which essentially only contain Pt as the active substances, bimetallic and multimetallic catalyst systems, additionally containing Re, Ge, Ir, or Pb, are used. For aromatics production the selectivity for C_{6+} aromatics is of general importance, while the suppression of dealkylation reactions is particularly important for xylene production. The activity of the catalyst can be increased by regular addition of chlorinated hydrocarbons. Major suppliers of reforming catalysts designed for xylene production include Acreon, Criterion, Exxon, Kataleuna, and UOP [27].

Steam Cracking. In contrast to the reforming process, the xylenes in the pyrolysis gasoline produced by steam cracking are regarded as byproducts of ethylene and propene production. However, their isolation can improve the economics of the process under certain circumstances. Although the operating conditions are not adjusted to maximize the xylene yield in steam cracking, the effect of severity on xylene formation is significant. Increasing the severity—i.e., raising the reaction temperature, lowering the residence time, and reducing the partial pressure by increasing the quantity of steam—significantly decreases the yield of pyrolysis gasoline [28], but increases the concentration of xylene (see Table 9) [29].

Synthesis Processes. In the Cyclar process developed by UOP/BP, propane and butane are converted into aromatics-rich gasoline fractions by cyclization over zeolite catalysts [30], [31]. On using propane the gasoline fraction contains ca. 17.3 wt% C_8 aromatics, and with butane ca. 19.8 wt%. The xylene yield is ca. 15 wt% based on the total starting material. A pilot plant for the Cyclar process has been commissioned at the BP refinery at Grangemouth, Scotland. For economic reasons, however, this process has not been put into industrial-scale operation. The same applies to the Mobil Oil process for synthesizing aromatics from alcohols using ZSM5 catalysts [31].

4.2. Separation and Further Processing

4.2.1. Separation of the C_8 Aromatics Fraction

The reformates used to produce C_8 aromatics can be processed such that C_8 and heavier aromatics are produced in the following approximate proportions [31]:

Benzene	13.2%
p-Xylene	16.5%
o-Xylene	8.3%
C_{9+} aromatics	0.9%

Besides the 38% aromatics, 51–52% naphtha fractions and 10–11% gaseous products (hydrogen, fuel gases, etc.) are formed. This optimized result involves the following operations:

1) Toluene is dealkylated to benzene.
2) m-Xylene is isomerized to o- and p-xylenes.
3) Ethylbenzene is transalkylated to xylenes.

If the toluene is subjected to transalkylation instead of thermal dealkylation, the yield of xylenes can be further increased at the expense of benzene. The structure of a modern aromatics complex, which produces o- and p-xylene in the highest possible yield, is based on a combination of a series of these processes.

The ethylbenzene is mainly converted to xylenes because obtaining ethylbenzene from reformate is energy intensive. The main source of ethylbenzene, which is almost exclusively dehydrogenated to styrene, is therefore now the more economical alkylation of benzene with ethylene (see → Ethylbenzene).

Distillative removal of ethylbenzene, which for styrene production must be toluene-free, from reformates can take place after removal of toluene from the sump of the toluene column. Figure 3 shows a flow sheet of the ethylbenzene distillation. Other products obtained by this separation are the xylene and C_9 aromatics. The major separation problem is caused by a boiling point difference of only 2 °C between ethylbenzene and p-xylene. The distillative separation of this mixture requires high reflux ratios (1:80 to 1:120) with 300–360 effective plates. A yield of ca. 95% ethylbenzene with > 99.5 wt% purity is achieved under these conditions [32].

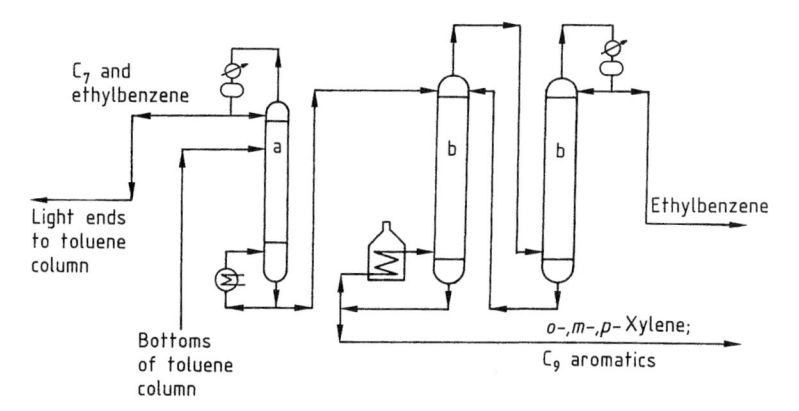

Figure 3. Flow sheet of the ethylbenzene distillation
a) Pre-column; b) Ethylbenzene columns

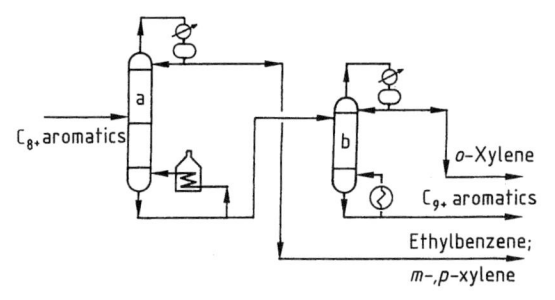

Figure 4. Flow sheet showing separation of
o-xylene
a) Xylene splitter; b) o-Xylene column

The distillative separation of o-xylene from the C_{8+} stream is also difficult. Although the boiling point difference is only ca. 5 °C, the required o-xylene purity (min. 98%) can be achieved with 120–150 effective plates and a reflux ratio of 1:10 to 1:15, with a yield of > 95%. Figure 4 shows a flow sheet for this separation. A further increase in o-xylene purity can be achieved by subsequent extractive distillation. Another method of separation is crystallization, mainly used to obtain p-xylene. This process utilizes the crystallization behavior of p-xylene near the eutectic point of a C_8 aromatics mixture. Depending on the composition of the mixture, this lies between −60 and −68 °C (see Fig. 5).

The Krupp–Koppers process (Fig. 6) is a typical example of the crystallization process. The actual separation process is a fractional crystallization in which the product of the first separation step in the filter (ca. 65% of the p-xylene from the starting mixture) is mixed with highly concentrated p-xylene and is isolated in high purity in a centrifuge. This process and that of Phillips Petroleum [33] operate using indirect refrigeration.

Direct contact refrigeration involves downward flow of the initially cooled liquid in the vertical crystallization vessel. This in turn directly cools the remainder of the contents of the vessel (see Fig. 7).

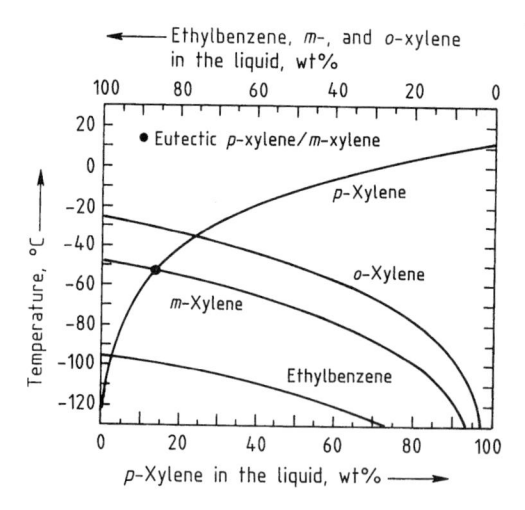

Figure 5. Melting behavior of the C_8 aromatics in their mixtures

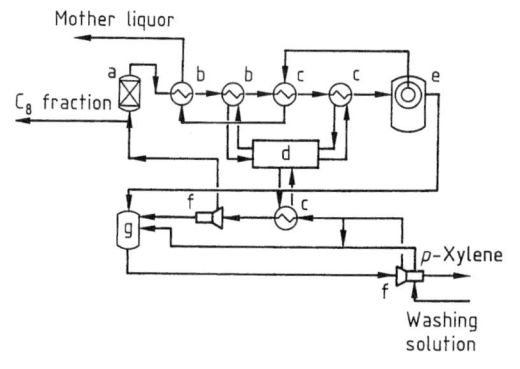

Figure 6. p-Xylene crystallization by indirect refrigeration
a) Drier; b) Precooler; c) Scraped-surface crystallizer; d) Refrigeration plant; e) Filter; f) Centrifuge; g) Mixer

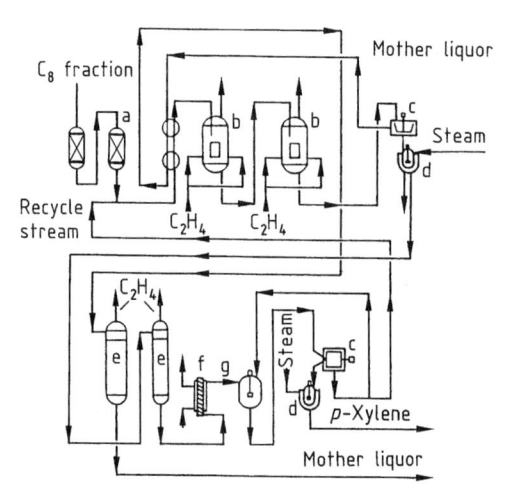

Figure 7. p-Xylene crystallization by direct contact refrigeration
a) Drying; b) Crystallization; c) Centrifuge; d) Melting tank; e) Desorber; f) Scraped-surface crystallizer; g) Mixing tank

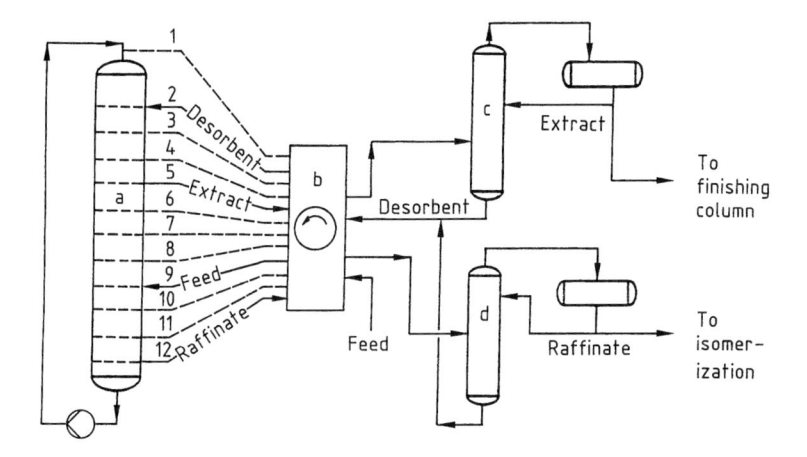

Figure 8. Parex process
a) Adsorbent chamber; b) Rotary valve; c) Extract column; d) Raffinate column

The cooling agent can be evaporating ethylene (Maruzen Oil Co.) [34]. Chevron Research and Development Corp., Sohio/BP, and Arco Technology use other process designs [35].

In the Parex process (UOP) [36] *p*-xylene is obtained in high purity from the isomer mixture by adsorption on a molecular sieve. The attached finishing column removes lighter components (raffinate consisting of ethylbenzene and a xylene mixture low in *p*-xylene) at the head. The *p*-xylene constitutes the bottom product of this column. Adsorption – desorption on the solid bed is carried out in such a way that a moving bed is simulated (Fig. 8). The rotary valve brings segments of the filled adsorption chamber into contact with the feed and with the desorbent in such a way that part of the molecular sieve adsorbs while another part is desorbed. The rotary valve switches the streams in the direction of flow in the bed. This simulates the movement of the adsorbent in a direction opposite to that of the liquid. The adsorbent exhibits high selectivity towards *p*-xylene compared with the other C_8 aromatics, especially ethylbenzene. Ethylbenzene is therefore one of the most suitable desorbents or eluents; this is kept in circulation in the extract column. The heavy part of the raffinate can also be used as the desorbent. The nonadsorbed C_8 aromatics leave the raffinate column at the head and are then subjected to isomerization (Section 4.2.2).

The Parex process is a variant of the Sorbex group of processes. The first process which used this principle of the simulated moving bed was the Molex process developed in 1964 for the separation of *n*-paraffins from isoparaffins and cyclic hydrocarbons. Other modifications of the Sorbex processes are used to separate olefins and paraffins in $C_3 - C_{18}$ hydrocarbon mixtures (Olex process) and for separating carbohydrates (Sarex process).

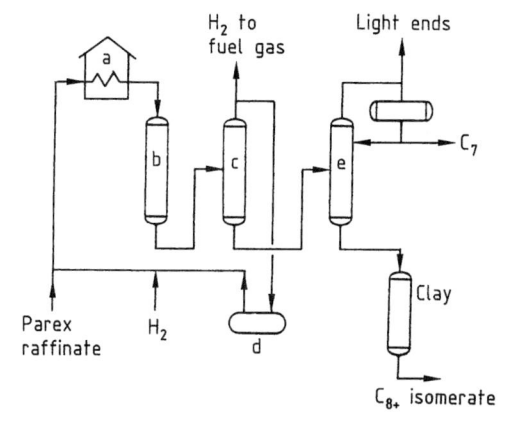

Figure 9. Isomar process
a) Heater; b) Reaction column; c) Hydrogen separation; d) Recycle compressor; e) Deheptanizer column

4.2.2. Isomerization of the C_8 Fraction

The yield of *o*- and *p*-xylenes can be maximized by using catalytic conversion processes. The Isomar process (Fig. 9) [37] enriches the *p*-xylene in the Parex raffinate by re-establishment of the equilibrium. The acidic metal-containing zeolite catalyst used here also isomerizes ethylbenzene selectively to xylene isomers in their equilibrium ratio [38]. Since hydrogenolysis on the metal components must be guaranteed to maintain the activity of this bifunctional catalyst, the Isomar process is operated with a certain partial pressure of hydrogen in the catalytic reactor. Hydrogen is separated from the effluent from the Isomar reactor and then recycled together with fresh hydrogen. The liquid product is separated in a deheptanizer column. A C_7 stream is obtained at the head, which can be recycled to the starting reformate. The bottom product of this column is a *p*-xylene-enriched stream which, after clay treatment to remove unsaturated compounds is fed to the Parex unit to obtain the newly formed *p*-xylene.

In the Tatoray process (Fig. 10) toluene and the C_9 (A_9) bottom fractions from the A_9 column (*o*-xylene splitter column) are mainly converted into xylene isomers by disproportionation of the toluene and transalkylation of the C_9 aromatics (→ Toluene), affording a newly formed $A_6 - A_8$ fraction from A_{9+}. This technology is the product of cooperation between UOP and Toray Industries. Commercial applications have been in operation since 1978. Since then the catalyst has been improved several times with regard to activity, thermal stability, and life span [37].

The Tatoray process consists of a reactor filled with solid catalyst to which are attached a cooling unit, a gas–liquid separation unit, a unit for distillation of the light components and one for clay treating to remove olefinic impurities. As in the Isomar process a partial pressure of hydrogen is maintained in the reactor. By processing different feed qualities, which can be influenced both by the different recycle rates and a varying charge of fresh reformate fractions, the composition of the product can be controlled.

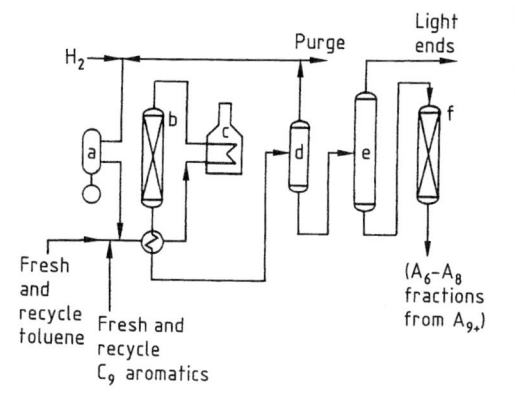

Figure 10. Tatoray process
a) Hydrogen recycle compressor; b) Catalytic reactor; c) Preheater; d) Hydrogen removal; e) Light ends removal; f) Clay tower

Other proven disproportionation processes are the LTD process of Mobil Research of Arco Technology. Alkylation of toluene to increase the xylene yield has been developed by Mobil Chemical (Mobil toluene-to-*p*-xylene process = MTPX) and allows *p*-xylene concentrations of ca. 94% to be achieved [39]. Mobil has announced that it will be using its new MTPX process in two facilities in the United States.

4.2.3. Combination of the Technologies in the Aromatics Complex

The maximization of the yields of *o*- and *p*-xylenes by combining isomerization, transalkylation, disproportionation, and adsorptive extraction of *p*-xylene is exemplified by the UOP technology complex (Fig. 11) [37]. In this complex all processes which are of industrial and commercial importance in this respect are represented. The center of the complex is the Parex process combined with Isomar technology. Tatoray technology is also used for conversion of the toluene and the C_{9+} aromatics fraction (A_9) [40].

Naphtha is used as the feedstock in this complex (cf. Fig. 12). It is obtained by conventional crude oil distillation, sometimes also from an evaporated BTX fraction. The reformer in this case is the UOP–CCR Platformer, which allows particularly high yields of aromatics to be obtained and also provides the hydrogen for subsequent processes, such as the Isomar and Tatoray processes, or for naphtha hydrogenation. The Sulfolane process is a frequently used liquid–liquid extraction, developed by Shell to obtain benzene and toluene from the reformate and from its C_5–C_7 fraction (top product from the deheptanizer; see → Benzene; → Toluene). The bottom products from the reformate deheptanizer and from the toluene and post-Isomar deheptanizer columns are fed to the Parex unit. Consequently the products of the complex are *o*- and *p*-xylene, benzene, aromatics-free raffinate, and the C_{9+} aromatics.

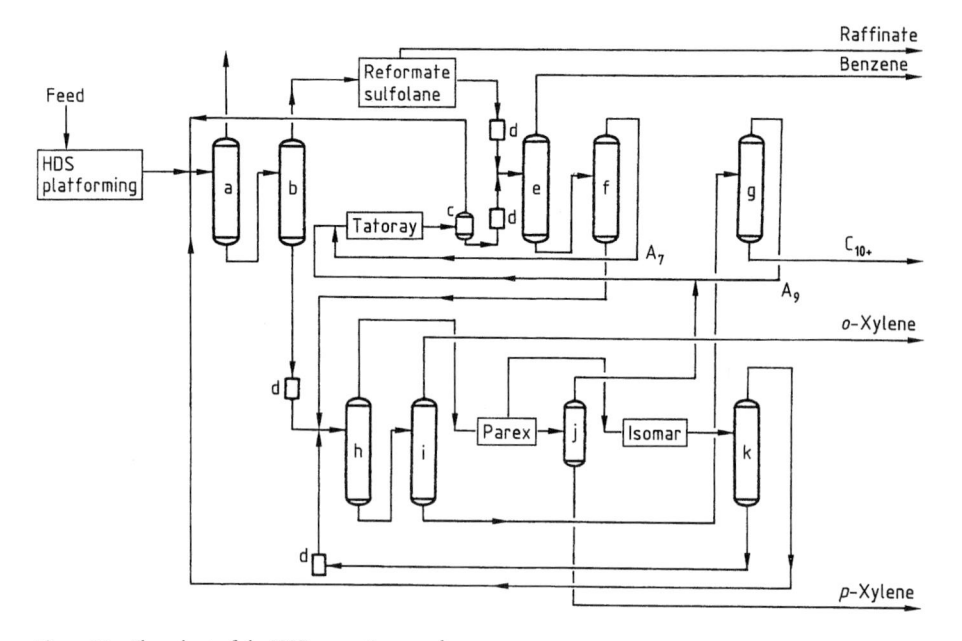

Figure 11. Flow sheet of the UOP aromatics complex
a) Debutanizer; b) Platformate deheptanizer; c) Splitter; d) Clay tower; e) Benzene column; f) Toluene column; g) A$_9$ column; h) Xylene stripper; i) o-Xylene re-run; j) Parex finishing column k) Isomerate deheptanizer

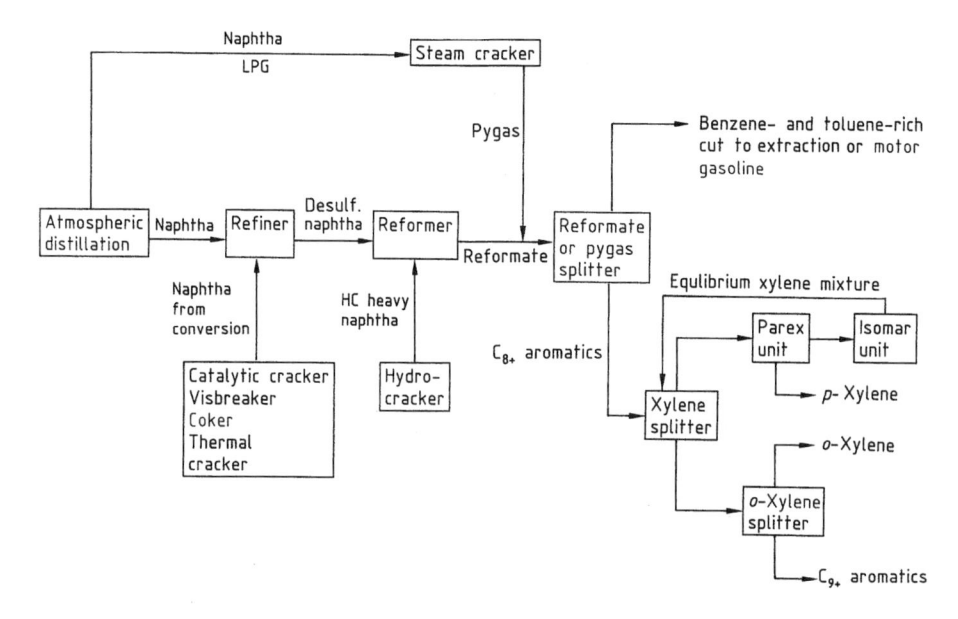

Figure 12. Integration of xylene production into petrochemical refineries
FCCU = Fluid catalytic cracking unit

5. Integration into Refinery and Petrochemical Complexes

5.1. Backward Integration into Petroleum Refining

Usually both the production and the separation and isomerization of the xylenes are integrated into refinery and petrochemical complexes [41], [42]. There are many advantages to this type of incorporation [43]:

1) Availability of a very favorable raw material base
2) Improved economics, especially in the use of byproduct streams unavoidably produced, such as pyrolysis gasoline or raffinates
3) Use of common infrastructure (energy, steam)
4) Possibilities for further processing of xylenes on site
5) Alternative use as fuel components if economical

The integration of xylene production into the structure of a petrochemical refinery is shown schematically in Figure 12.

The separation of xylenes and other aromatics from reformate or pyrolysis gasoline frequently leaves a gap in the boiling curve of the motor fuels produced in the refinery. This phenomenon is typical of petrochemical refineries and is known as the gap fuel characteristic. To comply with motor fuel specifications it is sometimes necessary to add alternative fuel components in the same boiling range [44]. With regard to a possible future limitations on the total aromatics content of motor fuels, as is being discussed in the United States [4], the isolation of xylenes and other aromatics could, however, also help to fulfill the specification requirements [45].

5.2. Forward Integration into Chemical Processing

Apart from their use as solvents, the individual xylene isomers can be regarded as intermediates which are further processed in a series of reactions to give consumer products [5]. While immediate subsequent synthesis steps, such as oxidation of o- and p-xylenes to phthalic anhydride and terephthalic acid, can be integrated into petrochemical complexes to some extent, further processing steps are mainly carried out at purely chemical sites.

The most important further processing steps and downstream products of the various xylene isomers and of ethylbenzene are shown schematically in Figure 13.

```
Ethylbenzene ─(-H₂)→ Styrene ─────────→ Polystyrene
```

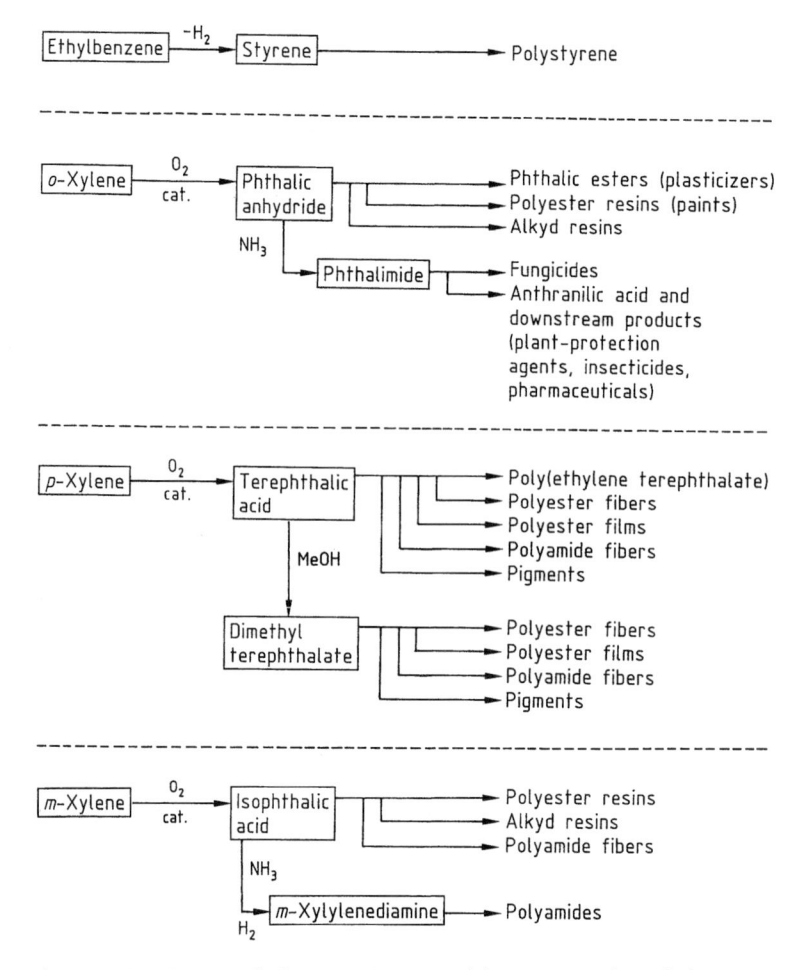

```
o-Xylene ─(O₂/cat.)→ Phthalic anhydride ─┬─→ Phthalic esters (plasticizers)
                         │               ├─→ Polyester resins (paints)
                         │               └─→ Alkyd resins
                         ↓NH₃
                    Phthalimide ─┬─→ Fungicides
                                 └─→ Anthranilic acid and
                                     downstream products
                                     (plant-protection
                                     agents, insecticides,
                                     pharmaceuticals)
```

```
p-Xylene ─(O₂/cat.)→ Terephthalic acid ─┬─→ Poly(ethylene terephthalate)
                         │              ├─→ Polyester fibers
                         │              ├─→ Polyester films
                         │              ├─→ Polyamide fibers
                         ↓MeOH           └─→ Pigments
                    Dimethyl terephthalate ─┬─→ Polyester fibers
                                            ├─→ Polyester films
                                            ├─→ Polyamide fibers
                                            └─→ Pigments
```

```
m-Xylene ─(O₂/cat.)→ Isophthalic acid ─┬─→ Polyester resins
                         │             ├─→ Alkyd resins
                         │             └─→ Polyamide fibers
                         ↓NH₃
                         ↓H₂
                    m-Xylylenediamine ─→ Polyamides
```

Figure 13. Most important further processing steps and downstream products of xylene isomers and ethylbenzene

6. Economic Aspects

Parameters Affecting Economics. The economics of xylene production and isolation are essentially determined by two parameters [3]:

1) The price difference between the proceeds of xylene production and the cost of the naphtha used (see Fig. 14)
2) The value relative to that when used as a motor fuel component

The first aspect is the most important when targeted xylene production by reforming naphtha without the possibility of the alternative use in motor fuel is considered. However, the second aspect determines the economics of the separation of xylene

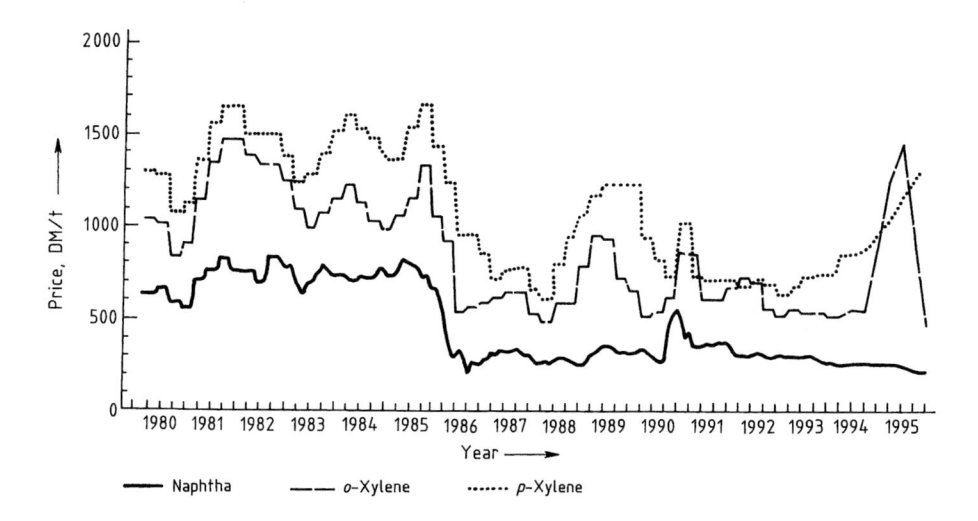

Figure 14. Development of prices of naphtha and xylenes in Western Europe

mixtures, which can originate from reformate or pyrolysis gasoline, if the alternative option of use as a motor fuel component is available. Besides these aspects, other technical parameters, such as the performance of the reformer, the efficiency of the separating and isomerization plants, or the quality of the naphtha, affect the overall economics.

The naphtha/xylene price difference is also controlled by other, overriding factors. While the price of naphtha is closely linked to that of petroleum, the proceeds of xylene production are principally determined by the price and demand for downstream products.

From the historical and expected future development of the quantities of direct downstream products of *p*-xylene produced in Western Europe (see Fig. 15) [3], it may be deduced that conversion to purified terephthalic acid (PTA) will be increasingly favored over dimethyl terephthalate (DMT). On balance this leads to an increased overall demand for *p*-xylene which may essentially be attributed to the increasing demand for poly(ethylene terephthalate) resins for the production of plastic bottles.

In the past the primary downstream product of *o*-xylene, phthalic anhydride, was produced in significant quantities by oxidation of coal-tar naphthalene [3]. Now phthalic anhydride is mainly produced from *o*-xylene so the consumptions of these products are directly associated. They are only increasing slightly because of the relatively low growth rate of plasticizer production.

The main downstream product from ethylbenzene, styrene, is mainly converted to polystyrene and ABS elastomers. Correspondingly, the demand for these products determines the demand for ethylbenzene. Most of the ethylbenzene, however, is not produced within the framework of xylene production, but from benzene and ethylene [46].

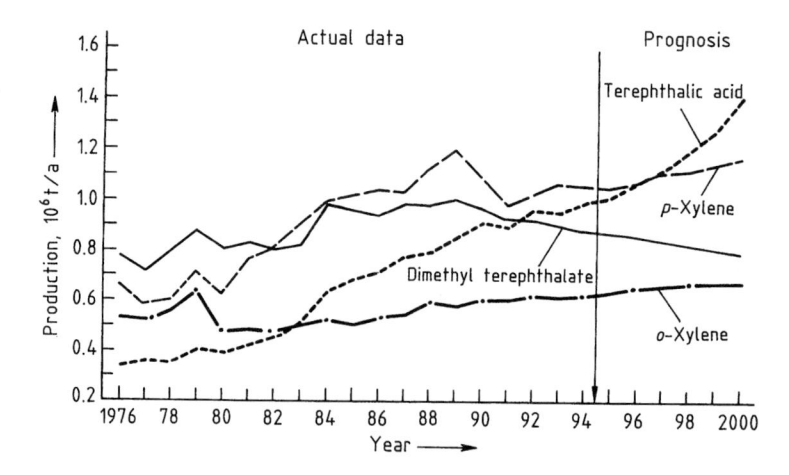

Figure 15. Development of quantities of *o*- and *p*-xylene and primary downstream products produced in Western Europe

The economic importance of *m*-xylene and its primary downstream product, isophthalic acid, is comparatively low. Regional demand for mixed xylenes in 1995 broke down as follows (total demand: 16×10^6 t):

Far East	44 %
North America	28 %
Western Europe	15 %
Eastern Europe	7 %
South America	3 %
Middle East	3 %

Producers. The most important producers and their production capacities for the various xylene isomers in Western Europe are listed in Table 10 [47]. Only in a few cases are additional capacities for conversion to phthalic anhydride or terephthalic acid and dimethyl terephthalate available at the same production site. The largest capacities for mixed xylenes are in North America and the Asia/Pacific region. Production capacities (in 10^3 t) worldwide for mixed xylenes follow [48]:

North America

United States

Amoco Chemical	680
Exxon Chemical	625
Chevron Chemical	573
Koch	550
Phillips Chemical	416
Lyondell	314
Fina	311
Sun	270
BP Chemicals	198
Shell Chemical	198
Ashland Chemical	182
Mobil Chemical	152

Coastal	109
South Western Refining	98
Citgo	68
OxyChem	66
Phibro Energy	45
Marathon Oil	43
Uno-Ven	33

Canada

Kemtec	91
Sunoco	86
Petro-Canada	73
Novacor Chemicals	59 (toluene–xylene mixture)
Shell Canada	32
Imperial Oil	14

South America

Argentina

Pasa	80
YPF	80

Brazil

Petroquimica do Nordeste	250
Petroquimica do Sul	68
Petrobras	25
Petroquimica União	96

Chile

Corpoven	50

Europe

Belgium

Finaneste	35

Germany

DEA Mineraloel	242
Deutsche Shell	200
Ruhr Oel	175
PK Schwedt	138

Italy

EniChem	200
Edison Spa	185

The Netherlands

Exxon Chemical	285
Total	200

Portugal

Petrogal	185

Spain

Petresa	216

United Kingdom

ICI	345

Asia/Pacific

Japan

Idemitsu Petrochemical	550
Tonen Chemical	540
Nikko	510

Mitsubishi Oil	420
Showa Shell Sekiyu	370
Nippon Petrochemicals	340
Koa Oil	250
Fuji Oil	190
General Seikyu	190
Maruzen Petrochemical	170
Kashima Oil	120
Nippon Steel Chemical	115
Mitsubishi Kagaku	100
Mitsui Petrochemical	100
Kyushu Aromatics	90
South Korea	
Yukong	500
Honam Oil	365
Ssangyoung Oil	360
Daelim Industrial	95
Singapore	
Singapore Petrochemical	30
Shell Eastern Chemicals	20
Taiwan	
Chinese Petroleum Corp.	750

7. Quality Specifications and Analysis

Specifications for xylenes are generally laid by a particular company, depending on the intended application. Properties relevant to specification are determined by standard procedures [49].

Some important methods of determination for xylene follow:

1) Distillation range (e.g., DIN 51761, ASTM D 850)
2) Hydrogen sulfide and sulfur dioxide content (e.g., ASTM D 853)
3) Thiol sulfur (e.g., DIN 51765)
4) Density (e.g., DIN 51757, ASTM D 891)
5) Flash point (e.g., DIN 51755, ASTM D 56)
6) Color (e.g., ASTM D 156, ISO 6271)
7) Bromine consumption (e.g., DIN 51774)
8) Purity/composition (e.g., ASTM D 1016; GLC determination)
9) Residue on evaporation (e.g., EN 5).

Typical specifications for *o*- and *p*-xylenes are given in Tables 11 and 12.

The high purity requirement for *p*-xylene are due to the quality requirements for the production of pure terephthalic acid. Provided that further processing takes place via dimethyl terephthalate, sometimes somewhat lower degrees of purity can be tolerated.

Table 10. Western European producers and product capacities for xylene isomers and primary downstream products if produced at the same site (1994, in 10^3 t/a)

Location	Producer	o-Xylene	p-Xylene	m-Xylene	Ethylbenzene from C_8 aromatics	Phthalic anhydride	Terephthalic acid	Dimethyl terephthalate
Geel, Belgium	Amoco			30			250	
Gonfreville, France	Total	90	105					
Schwedt, Germany	PCK	35	25				62	
Gelsenkirchen, Germany	Ruhr Oel	45	105					
Heide, Germany,	RWE/DEA	12			10			
Wesseling, Germany	RWE/DEA	44	110					
Godorf, Germany	Shell	100	75					
Priolo, Italy	Enichem	60	140		70			
Sarroch, Italy	Saras	60	70		50			
Milan, Italy	Sisas			50		45		
Botlek, The Netherlands	EXXON	105	165			50		
Oporto, Portugal	Petrogal	42	115					
Alqeciras, Spain	Cepsa Interquisa	40	34			20	200	70
Wilton, United Kingdom	ICI		330		50		580	

Table 11. Typical product specification for o-xylene

	Method	Value
o-Xylene (min.), wt%	GC	98.0
C_9 and higher aromatics (max.), wt%	GC	0.5
Nonaromatics (max.), wt%	GC	0.5
Density at 15 °C, kg/m^3	DIN 51 757	882–885
APHA color (max.)	ISO 6271	10
Copper corrosion (max.)	DIN 51 759	1
Hydrogen sulfide and sulfur dioxide (max.)	ASTM D 853	free of H_2S and SO_2
Acidity	DIN 51 558 T.1	not detectable
Start of boiling (min.), °C	DIN 51 761	143.0
End of boiling (max.), °C	DIN 51 761	145.0
Residue on evaporation (max.), mg/kg	EN 5	20

For *m*-xylene, which is produced in comparatively small quantities, only purities of 95.4 wt% min. are usually required.

Table 12. Typical specification for *p*-xylene

	Method	Value
p-Xylene (min.), wt%	GC	99.4
C_8 aromatics (max.), wt%	GC	0.55
Other aromatics (max.), wt%	GC	0.05
Nonaromatics	GC	0
Density at 15 °C, kg/m^3	DIN 51 757	865 – 866
Copper corrosion (max.)	DIN 51 709	0
Total sulfur (max.), ppm	DIN 51 768	10 – 15
Acidity	DIN 51 558 T.1	not detectable
Distillation range, °C	DIN 51 761	1
mp, °C		13.1

8. Storage, Transport, and Safety

Instructions for ensuring safe storage, transfer, and transport of flammable liquids can be found in the technical rules for flammable liquids (TRbF), drawn up by the German committee for flammable liquids [50], [51]. For xylene the following TRbF numbers are important:

TRbF 001:	General, form, and application of the rules
TRbF 100:	General safety requirements
TRbF 111:	Filling stations, emptying stations
TRbF 112:	Gasoline stations
TRbF 120/121:	Tanks in a fixed location
TRbF 131/1 and 2:	Piping and tubing
TRbF 141:	Tanks on vehicles
TRbF 142:	Tank containers
TRbF 143:	Movable vessels
TRbF 180:	Operating instructions

The flammability limits for xylenes in air are between 1 and 8%. Because the flash point of xylene mixtures is below 21 °C, they are mostly assigned to hazard class A I. The flash points of pure xylenes (not including ethylbenzene) are, however, > 21 °C so they may be declared as hazard class A II. For storage and transport xylenes must be labeled as readily flammable and harmful [51], [52]. For transport within Germany xylenes are assigned to class 3, cipher 31 c of the GGVS/GGVE regulations (regulations governing transport of hazardous goods by road and rail) [53].

9. Environmental Aspects and Toxicology

Environmental Aspects. In Germany xylenes are assigned to water hazard class 2 (WGK 2) [55]. The solubility of xylenes in water is low (ca. 0.14 g/L). Because of the comparatively low vapor pressure (7 – 8 mbar at 20 °C), the danger of vapor emissions in air is relatively low. However, xylenes can react with other air pollutants to give environmentally damaging products. This applies particularly to the UV-catalyzed photooxidation of xylenes by nitrogen oxides. However, compared with some other hydrocarbons, the reactivity of xylenes is comparatively low, with a reaction rate of ca. 2×10^{-9} min^{-1} [55]. For example, disubstituted internal olefins have a reaction rate of ca. 50×10^{-9} min^{-1}.

Toxicology. Exposure to xylene is possible through inhalation of vapors and resorption through the skin. The MAK value is 100 ppm (ca. 440 mg/m^3) for all three isomers [54].

General Activity. Xylene exhibits acute prenarcotic and narcotic activity. Chronic exposure to xylene leads to disturbance of the CNS (e.g., headaches, sleep disturbance) [56] and damage to the blood picture (dyspepsia). Provided that there is no long-term chronic overexposure, these effects are reversible. Besides developing a certain tolerance, frequent exposure to xylene can also lead to habituation or even addiction (solvent abuse).

Acute Toxicity. The LD$_{50}$, LC$_{50}$, and TCLo values for oral administration and inhalation vary widely, depending on the animal investigated and the isomer composition. The following values give an indication of the toxicity of xylene isomer mixtures [54], [57]: LD$_{50}$ (rat, oral) 4300 mg/kg, LDL$_0$ (rat, i.p.) 2000 mg/kg, TCLo (humans, inhalation) 200 ppm.

A very high exposure to xylene of ca. 10 000 ppm caused by an accident led to lung edema and subsequent death in one person [58]. In other cases severe damage to the CNS, kidneys, and liver were observed.

Irritant Effects. On repeated application xylenes can cause irritation of the respiratory passages and mucous membranes of the eye. Frequent skin contact can lead to blister formation and dermatitis [57], [59]–[62].

Subchronic and Chronic Toxicity. At a concentration range of 100 to over 1000 ppm xylene in inhaled air, damage to the CNS with disturbance of balance or slowing of reactions is observed. Inhalation of xylene vapors can also cause nausea and headaches [63]–[65].

A change in the blood picture is also frequently observed. In ca. 10 % of persons who had been exposed to xylene vapors in concentrations of up to ca. 100 ppm for ca. 5 years, a leucocyte level of < 4500 mm^{-3} was established [66]. A decrease in the immunobiological activity has occasionally been observed [58].

Carcinogenicity, Mutagenicity, and Embryotoxicity. The asessment of the carcinogenicity of xylenes is not consistent [67]–[69]. While in one study no indications of carcinogenicity or cocarcinogenicity were found, another investigation indicated that xylenes act as tumor promoters for skin tumors in rats.

None of the three xylene isomers showed mutagenicity in the Ames test [70]. However, slight mutagenicity was detected in a recessive lethal test on *drosophila* [71].

In animal experiments long-term exposure through inhalation of xylenes causes small changes to fetuses [72].

Pharmacokinetics and Metabolism. All three xylene isomers are resorbed in the same way. Various investigations have shown that ca. 60–70% of the xylene reaching the organism via the lungs is retained [73]–[76]. This percentage remains approximately constant over the whole exposure period. The ratio of the concentration in the air in the alveoli (mg/m^3) and in the blood (mg/kg) changes with the degree of bodily activity. When the body is at rest the ratio is ca. 15:1 and when moving ca. 30–40:1 [77]. Xylene can be resorbed at a rate of ca. 2.5 (0.7–4.3) mg/m^3 per minute through intact skin [78].

Xylenes are deposited rapidly in body fat (up to ca. 5%) and remain there for hours after exposure. The half-life in fat deposits is ca. 0.5–1.0 h [78]. The metabolism of the individual xylene isomers is identical. The main biotransformation pathway initially involves oxidation to methylbenzoic acid, which forms the corresponding methylhippuric acid by conjugation with glycine (A) [74], [79]–[81]. The methylhippuric acid can be excreted rapidly via the kidneys. Another, less favored biotransformation pathway involves the hydroxylation of the xylene on the aromatic ring, forming xylenols (B) [73].

10. References

[1] Parpinelli Technon: "West European Petrochemical Industry," Aromatics Report and Tables, Mailand 1995, pp. 18, 31, 38.

[2] J. S. McArragher: Paper to the Institute of Petroleum Conference 16 Oct. 1991: "Making Cleaner Fuels in Europe – Their Need and Cost," Institute of Petroleum, London 1991.

[3] Chem Systems Ltd.: "Petroleum and Petrochemicals Economics," annual reports 1 and 2, London, Aug. 1995.

[4] *Octane Week,* July 19 (1993) 6.

[5]   K. Weissermel, H. J. Arpe: *Industrielle Organische Chemie*, 3rd ed., Verlag Chemie, Weinheim 1988, pp. 407, 426.
[6]   *Ullmann*, 4th ed., **24**, 526.
[7]   C. L. Yaws, *Chem. Eng. (N.Y.)* **82** (1975) no. 15, 113–122; **82** (1975) no. 20, 73–81.
[8]   *Beilstein*, 3rd ed., suppl., vol. 5, 2nd part (1964) 807–823, 823–845, 845–864, 776–807. W. Berghoff: *Erdölverarbeitung und Petrolchemie*, VEB Deutscher Verlag für Grundstoffindustrie, Leipzig 1968, pp. 80–81. W. L. Nelson: *Petroleum Refinery Engineering*, 4th ed., McGraw-Hill, Toronto 1958. H. J. V. Winkler:
       Der Steinkohlenteer und seine Aufarbeitung, Verlag Glückauf, Essen 1951, pp. 70–71.
[9]   F. D. Rossini et al.: *Selected Values of Physical and Thermodynamic Properties of Hydrocarbons and Related Compounds*, Carnegie Press, Pittsburgh, PA, 1953.
[10]   L. H. Horsley, *Anal. Chem.* **19** (1947) no. 8, 508–600.
[11]   L. H. Horsley, *Anal. Chem.* **21** (1949) no. 7, 831–873.
[12]   L. H. Horsley, *Adv. Chem. Ser.* **35** (1962).
[13]   *Beilstein*, 3rd ed., suppl., vol. **1**, 2nd part (1964) 811 (o-Xylol), 828 (m-Xylol), 849 (p-Xylol), 782 (Ethylbenzol).
[14]   K. Owen, T. Coley: *Automotive Fuels Handbook*, Society of Automotive Engineers, Warrendale, PA, 1990, pp. 564–565.
[15]   F. E. Condon, *J. Am. Chem. Soc.* **70** (1948) 1963.
[16]   H. G. Franck, J. W. Stadelhofer in: *Industrielle Aromatenchemie*, Springer Verlag, Berlin 1987, pp. 60–61.
[17]   Parpinelli Technon: "West European Petrochemical Industry", Aromatics Report and Tables, Mailand 1993, p. 231.
[18]   J. A. Weiszmann: "The UOP Platforming Process," in R. A. Meyers (ed.): *Handbook of Petroleum Refining Processes*, Chap. 3, McGraw-Hill, London 1986, p. 6.
[19]   D. G. Tajbl: "UOP HC Unibon Process for Hydrocracking," in R. A. Meyers (ed.): *Handbook of Petroleum Refining Processes*, Chap. 2, Mc-Graw-Hill, London 1986, p. 41.
[20]   A. Chauvel, G. Lefebre in: *Petrochemical Processes*, vol. **1**, Editions Technip IFP, Paris 1989, p. 176.
[21]   A. Chauvel, G. Lefebre in: *Petrochemical Processes*, vol. **1**, Editions Technip IFP, Paris 1989, p. 130.
[22]   H. G. Franck, J. W. Stadelhofer in: *Industrielle Aromatenchemie*, Springer Verlag, Berlin 1987, pp. 35–57.
[23]   J. A. Weiszmann: "The UOP Platforming Process," in R. A. Meyers (ed.): *Handbook of Petroleum Refining Processes*, Chap. 3, McGraw-Hill, London 1986, pp. 11–13.
[24]   N. L. Gilsdorf, A. P. Furano, R. H. Rachford, R. J. Schmidt, D. L. York: "Platforming Technology for Aromatics Production," in: *Petrochemical Technology: Building Success through Teamwork and Technology*, UOP Technology Conference, Des Plaines, Ill., 1992, pp. 20–21.
[25]   A. Chauvel, G. Lefebre in: *Petrochemical Processes*, vol. **1**, Editions Technip IFP, Paris 1989, pp. 170–172.
[26]   J. A. Weiszmann: "The UOP Platforming Process," in R. A. Meyers (ed:): *Handbook of Petroleum Refining Processes*, Chap. 3, McGraw-Hill, London 1986, pp. 6–9.
[27]   *Oil Gas J.*, 2nd Oct. (1995) 35–37.
[28]   W. Keim, A. Behr, G. Schmitt in: *Grundlagen der Industriellen Chemie*, Verlag Salle-Sauerländer, Frankfurt 1986, p. 129.
[29]   A. Chauvel, G. Lefebre in: *Petrochemical Processes*, vol. **1**, Editions Technip IFP, Paris 1989, p. 156.

[30] C. D. Gosling, G. L. Gray, P. J. Kuchar, L. Sullivan: "Produce BTX from LPG," in: *Building Success through Teamwork and Technology*, UOP Technology Conference, Des Plaines, Ill., 1992, pp. 1–17.

[31] H. G. Franck, J. W. Stadelhofer in: *Industrielle Aromatenchemie*, Springer Verlag, Berlin 1987, pp. 88–91.

[32] E. C. Haun, M. W. Golem, P. P. Piotrowski, S. Sapuntzakis: "The Modern Aromatics Complex and Revamp Options for Existing Plants," UOP Technology Conference, Des Plaines, Ill. 1988.

[33] J. G. Jenkins, *Oil Gas J.* **63** (1965) Jan. 18, 78–86.

[34] N. E. Ockerbloom, *Hydrocarbon Process* **50** (1971) July, 113.

[35] J. Morrison, *Oil Gas Int.* **10** (1970) no. 12, 67–69; *Hydrocarbon Process.* **60** (1981) Nov., 240.

[36] Chevron Res., US 3 219 722, 1965. ARCO Techn., *Hydrocarbon Process.* **60** (1981) Nov., 239.

[37] J. A. Johnson, A. P. Furfaro, S. H. Hobbs, R. G. Kabza: "Advances in para-Xylene Recovery," UOP Technology Conference, Des Plaines, Ill. 1988.

[38] A. K. Aboul-Jheit, *Erdöl Erdgas Kohle* **111** (1995).

[39] *Chemical Week*, Aug. 30, 1995, p. 18.

[40] P. J. Kuchar, O. Y. Lin, V. Zukauskas, D. Brkic: "Isomar and Tatoray Aromatics Production Flexibility," UOP Technology Conference, Des Plaines, Ill. 1988.

[41] L. Grimm, *Erdöl Erdgas* **108** (1992) no. 9, 355.

[42] H. Forth, *Oel* **19** (1981) no. 7, 172–178.

[43] B. Hnat: *Wirtschaftspolitische Studien*, vol. **87**, Verlag Vandenhoek and Ruprecht, Göttingen–Zürich 1992, pp. 69–71.

[44] K. Owen, T. Coley: *Automotive Fuels Handbook*, Society of Automotive Engineers Inc., Warrendale, PA, 1990, pp. 144–145.

[45] UOP: "The Clean Air Act And The Refining Industry," Des Plaines, Ill., 1 Sep. 1991, p. 17. UOP: "Process Solutions for Reformulated Gasoline," Des Plaines, Ill., 1992.

[46] Parpinelli Technon: "West European Petrochemical Industry," Aromatics Report and Tables, Mailand 1993, p. 19.

[47] Parpinelli Technon: "West European Petrochemical Industry," Aromatics Report and Tables, Mailand 1993, pp. 229–311.

[48] *Chemical Week*, Oct. 11, 1995, p. 58.

[49] H. G. Franck, J. W. Stadelhofer in: *Industrielle Aromatenchemie*, Springer Verlag, Berlin 1987, p. 133.

[50] VbF/TRbF, vol. 5 (Regulations), 16. Lfg., June 1989.

[51] Kühn, Birett – Merkblätter Gefährliche Arbeitsstoffe, 49th suppl., 6/90 – X03, X06, X07, X08, X19, 1990.

[52] BArbBl, no. 9/1990, pp. 65 ff.

[53] Unfallmerkblatt für den Straßentransport, MED-Verlagsgesellschaft, Landsberg, edition 5/87.

[54] MERCK Sicherheitsdatenbank MS-Safe, 1 Jul. 1991.

[55] J. Fabri, A. Reglitzky, M. Voisey, *Erdöl, Ergas* **107** (1991) no. 1, 24–29.

[56] K. Savolainen, V. Riihimäki, E. Vaheri, M. Linnoila, *Scand. J. Work Environ. Health* **6** (1980) 94.

[57] M. Wolf, *Arch. Ind. Health* **14** (1956) 387.

[58] R. Morley, D. W. Eccleston, C. P. Douglas, W. E. J. Greville, D. J. Scott, J. Anderson, *Brit. Med. J.* **3** (1970) 442.

[59] E. Browning in: *Toxicity and Metabolism of Industrial Solvents*, Elsevier, Amsterdam 1956.

[60] H. W. Gerarde: *Toxicology and Biochemistry of Aromatic Hydrocarbons*, Elsevier, Amsterdam 1965.

[61] E. Schmidt, *Arch. Gewerbepath. Gewerbehyg.* **15** (1956) 37.

[62] W. Matthaus, *Klin. Monatsbl. Augenheilk.* **144** (1964) 713.
[63] K. Savolainen, V. Riihimäki, A. M. Seppäläinen, M. Linnoila, *Int. Arch. Occup. Environ. Health* **45** (1980) 105.
[64] F. Gamberale, G. Annwall, M. Hultengren, *Scand. J. Work Environ. Health* **4** (1978) 204.
[65] K. Savolainen, M. Linnavuo, *Acta Pharmacol. Toxicol.* **44** (1979) 315.
[66] V. I. Boiko, *Gig. Tr. Prof. Zabol.* 1970, no. 6, 23.
[67] A. Pound, *N. Engl. J. Med.* **67** (1968) 88.
[68] A. Pound, *Pathology* **2** (1970) 269.
[69] I. Berenblum, *Cancer Res.* **1** (1941) 44.
[70] R. P. Bos, R. M. E. Brouns, R. van Doorn, J. L. G. Theuws, P. T. Henderson, *Mutat. Res.* **88** (1981) 273.
[71] M. Donner, *Mutat. Res.* **74** (1980) 171.
[72] I. Krotov, N. Chebotar, *Gig. Tr. Prof. Zabol.* 1972, no. 16, 40.
[73] V. Sedivec, J. Flek, *Int. Arch. Occup. Environ. Health* **37** (1976) 205.
[74] I. Astrand, I. Engström, P. Övrum, *Scand. J. Work Environ. Health* **4** (1978) 185.
[75] V. Riihimäki, P. Pfäffli, K. Savolainen, K. Pekari, *Scand. J. Work Environ. Health* **5** (1979) 217.
[76] W. Senczuk, J. Orlowski, *Br. J. Ind. Med.* **35** (1978) 50.
[77] I. Engström, R. Bjurström, *Scand. J. Work Environ. Health* **4** (1978) 195.
[78] K. Engström, K. Husman, V. Riihimäki, *Int. Arch. Occup. Environ. Health* **39** (1977) 181.
[79] H. Bray, B. Humphris, W. Thorpe, *Biochem. J.* **85** (1949) 241.
[80] H. Bray, B. Humphris, W. Thorpe, *Biochem. J.* **87** (1950) 395.
[81] G. Bienik, T. Wilczok, *Br. J. Ind. Med.* **38** (1981) 304.

Xylidines

MICHAEL MEYER, EMS-Dottikon, Dottikon, Switzerland

1.	Introduction 5019	6.	Use 5023
2.	Physical Properties 5020	6.1.	Dye Production 5023
3.	Chemical Properties. 5020	6.2.	Plant Protection 5024
4.	Industrial Production 5021	6.3.	Production of Pharmaceutical Products, Pharmaceutical Products for Veterinary Medicine, and Vitamins 5024
4.1.	Reduction of Nitroxylenes . . . 5021		
4.2.	Amination of Xylenols 5022		
4.3.	Other Processes. 5022	6.4.	Other Uses 5025
5.	Quality Specifications, Storage, and Transport 5022	7.	Toxicology. 5025
		8.	References. 5026

1. Introduction

Xylidines are homologs of aniline and toluidine and are therefore very similar in chemical behavior.

Xylidines are used mainly in dye production. Their use in the production of vitamins, pharmaceuticals, and plant-protection agents is also important. Xylidine mixtures (crude xylidines), which were previously mixed with aircraft fuels in large quantities as antiknock agents, are still processed as antioxidants in lubricating oils and rubber products, e.g., car tires.

The xylidines **1–6**, $C_8H_{11}N$, molecular mass 121.18, are derived from the three isomeric xylenes:

2. Physical Properties

Xylidines are readily soluble in most organic solvents, but are not very soluble in water. At room temperature, xylidines are colorless liquids or congealed melts, each with its characteristic odor. Xylidines become colored within hours in light and air. Physical properties of xylidines are listed in Table 1.

3. Chemical Properties

Like aniline, xylidines are weak bases. The positions of the methyl groups have little influence on the basicity, which increases in the following order in aqueous media: **2 < 3 < 1 < 2 < 6 < 5** [1]. Xylidines form stable salts with strong mineral or organic acids. The differing solubilities of these salts can be used to separate isomers in some cases. However, this method is unimportant in industry [2].

Electrophilic substitution reactions on the aromatic nucleus, e.g., chlorination, sulfonation, nitration, nitrosation, or coupling with diazonium salts, take place very readily because of the high activation by the amino group and the two methyl groups. The positions *ortho* and *para* to the amino group are preferentially attacked. The amino group often needs to be protected. In strongly acidic media, the amino group is protonated and the ring is deactivated. In this case electrophilic substitution is directed to positions *meta* to the amino group [3].

The most important reactions of the amino group are acylation, alkylation, arylation, and diazotization. Xylidines can also be converted into the corresponding isocyanates [4].

Table 1. Physical properties of xylidines

	Other names	CAS registry no.	mp, °C	bp, °C	d_4^{20}	n_D^{20}	Flash point, °C
2,3-Xylidine (1)	2,3-dimethylaniline, 3-amino-o-xylene	[87-59-2]	3.5	222.7 – 222.9	0.9931	1.5652	97
2,4-Xylidine (2)	2,4-dimethylaniline, 4-amino-m-xylene	[95-68-1]	−12.5 to −14.3	214.3	0.9763	1.5576	100
2,5-Xylidine (3)	2,5-dimethylaniline, 5-amino-p-xylene	[95-78-3]	14.2 – 15.0	217	0.9755	1.5596	97
2,6-Xylidine (4)	2,6-dimethylaniline, 2-amino-m-xylene	[87-62-7]	11.2	214.8 – 215.1	0.9796	1.5612	98
3,4-Xylidine (5)	3,4-dimethylaniline, 4-amino-o-xylene	[95-64-7]	50 – 51	225	1.0755		97
3,5-Xylidine (6)	3,5-dimethylaniline, 5-amino-m-xylene	[108-69-0]	9.8 – 10.0	218	0.9704	1.5581	103

4. Industrial Production

4.1. Reduction of Nitroxylenes

For all xylidines, apart from 3,5-xylidine and, to some extent, 2,6-xylidine, reduction of nitroxylenes is the most important industrial production route. The corresponding nitroxylenes can be obtained by nitration of the pure xylenes. If mixtures of isomers are obtained, as with o- and m-xylenes, the nitro compounds can be easily separated by rectification.

Crude xylidines, previously obtained in large quantities by nitration of xylene mixtures (crude xylene) and subsequent reduction, can be further processed to give the pure xylidines, but at great expense [5].

For the reduction of nitroxylenes, processes used in the production of aniline and toluidines can, in principle, be considered. The classical Béchamp reduction with iron and hydrochloric acid [6] has now been widely replaced by catalytic hydrogenation with molecular hydrogen. Usually continuous or batch processes in the liquid phase, generally without solvents, are used [7], [8]. If solvents are used, these are aromatic hydrocarbons, lower aliphatic alcohols, and acetic acid [9], [10]. Palladium on activated charcoal is the most important catalyst. Other possible catalyst systems are platinum or ruthenium on activated charcoal, aluminum oxide, or kieselguhr and Raney nickel [11].

In 1943 and 1944, xylidines were produced in the United States in large-scale industrial plants. Hydrogenation in the gas phase was mostly used [12], [13]. Nickel, nickel–copper, or sulfidic catalysts based on copper, nickel, chromium, molybdenum, or tungsten were used [14], [15].

4.2. Amination of Xylenols

3,5-Xylidine cannot be prepared at all by the nitration of *m*-xylene, as the corresponding 5-nitro-*m*-xylene is formed in < 1% yield . The production of 2-nitro-*m*-xylene, formed in 16% yield in the nitration of *m*-xylene, depends on the market for 4-nitro-*m*-xylene. 2,6-Xylidine can therefore be produced by the nitration of *m*-xylene only to a limited extent. To produce both these xylidines, processes starting from the generally readily accessible xylenols are more suitable.

In the Halcon process, (also used for the production of aniline) xylenols can be converted to xylidines with excess ammonia at high pressure and temperature on aluminum-oxide-containing catalysts [16], [17]; a disadvantage is that extensive dealkylations and isomerizations can occur, in particular when xylenols with *ortho* methyl groups or spent catalysts are used [18]. In the Ethyl Corporation process these disadvantages are avoided, if sterically hindered xylenols in the presence of catalytic quantities of a cyclohexanone are converted to xylidines with ammonia using a hydrogen transfer catalyst [19]. In another process developed by the same company, 2,6-xylenol can be aminated under anhydrous conditions via the aluminum salt [20].

4.3. Other Processes

3,4-Xylidine can be obtained by catalytic hydrogenation of 2-chloromethyl-4-nitrotoluene [21]. 3,4- and 3,5-Xylidines can be produced by amination of the corresponding substituted bromoxylenes under high pressure with a copper catalyst [22], [23].

Another process, which can be used to obtain 3,5-xylidine, is the reaction of α,β-unsaturated cyclohexenone oximes with dehydrating agents [24]. Ciba-Geigy describes the condensation of acrolein with the enamine from diethyl ketone and morpholine to give 2,6-dimethylcyclohexenone. The latter can be converted to 2,6-xylidine in a second step with ammonia in the presence of hydrogen-transfer catalysts using a method similar to that of Ethyl Corporation [25].

5. Quality Specifications, Storage, and Transport

Technically pure xylidines are rectified products, generally of purity > 99%. The content can be determined simply by titration, GLC [26], or HPLC [27]. Other isomers and side products with hydrogenated aromatic nuclei can be present as impurities. To identify other impurities, e.g., hydroxylamines, azo and azoxy compounds, or phenols, TLC methods are suitable, particularly when these substances are present in trace

Table 2. Hazard classes and registrations of xylidenes

	EINECS* number	TSCA status	RID/ADR number	UN number	RCRA**
2,3-Xylidine	201-755-0	listed	6.1/12 b	UN 1711	
2,4-Xylidine	202-437-4	listed	6.1/12 b	UN 1711	
2,5-Xylidine	202-451-0	listed	6.1/12 b	UN 1711	8 EHQ-0990-1080 S
2,6-Xylidine	201-758-7	listed	6.1/12 b	UN 1711	
3,4-Xylidine	202-437-4	listed	6.1/12 b	UN 1711	
3,5-Xylidine	203-607-0	listed	6.1/12 b	UN 1711	

* European Inventory of Existing Chemical Substances.
** Resource Conservation and Recovery Act.

amounts. The residual moisture is < 0.1 % and is determined titrimetrically by the Karl Fischer method. In the case of crude xylidines, the isomer distribution should usually be specified. This can vary within wide limits, depending on the supplier.

The freshly distilled product is pale yellow. On storage over longer periods, particularly in the presence of light and air, and depending on the other storage conditions, the product increasingly becomes red with no detectable change in the xylidine content. The Hazen color index provides a measure of the coloration. The coloration can be suppressed for a certain period by addition of various stabilizers in quantities of 0.05 – 0.5 %. (For the mode of action of structurally different stabilizers see [28].) The following are effective stabilizers: hydrazine hydrate [29], aliphatic polyamines [30], aromatic biguanides [31], alkylphenols [32], and sulfur-containing compounds [33].

Iron containers are used for storage and transport. Table 2 gives the hazard classes and registrations of the individual xylidines [34].

6. Use

6.1. Dye Production

A large number of azo dyes are derived from 2,4- and 2,5-xylidines, whereby the xylidines can function both as diazo and coupling components. Instead of pure 2,4-xylidines, crude xylidines are frequently used to produce red and orange acid dyes.

Yellow and orange solvent dyes and some red acid and direct dyes are produced from 2,4-xylidine. An example of an azo dye based on 2,5-xylidine is Solvent Red 26 (C.I. 26120).

Of the three theoretically possible 2,4-xylidinesulfonic acids, two are important as diazo components in the production of azo dyes. 4-Amino-1,3-dimethylbenzene-6-sulfonic acid is formed by sulfonation of 2,4-xylidine with oleum [35], and is preferentially used for the production of acid azo dyes. Baking sulfonation of 2,4-xylidine gives 4-amino-1,3-dimethyl-5-sulfonic acid [35], which is further processed to give Acid Orange 9 (C.I. 17 925) and Direct Violet 7 (C.I. 27 855). The reaction of xylidines with diketene [36], [37] is an important first step in the production of pigment dyes. An interesting example of this class of compound is the 2,4-xylidide of acetoacetic acid, used principally as the coupling component for Pigment Yellow 13 (C.I. 21100).

2,5-Xylidine is also used to produce triphenylmethane dyes, such as Solvent Yellow 30 (C.I. 21240), Solvent Red 22 (C.I. 21250), and Acid Red 65 (C.I. 24830).

3,5-Xylidines are currently almost exclusively used to produce the perylene pigment dye Pigment Red 149 (C.I. 71137). Only a few areas of use are known for 2,3-xylidine; e.g., in the form of the amide of 2-hydroxy-3-naphthoic acid, it is a coupling component for azo dyes used in printing inks.

The condensation product of bromamine acid with 2,6-xylidine is an intermediate in the production of acidic blue dyes for wool [38] and reactive dyes.

6.2. Plant Protection

2,3-Xylidine is a component of the herbicide xylachlor (Combat) [39]. The insecticide and acaricide amitraz is obtained from 2,4-xylidine.

5-Acetamido-2,4-dimethylaniline, which can be obtained by nitration and subsequent catalytic hydrogenation of N-acetyl-2,4-xylidine, is an intermediate in the production of the herbicide and growth regulatormethafluoridamid, mefluidide (Embark, Vistar) [40]. Prowl (Pendimethalin, Penoxalin) is also used as a herbicide and growth regulator [41], [42]; it is obtained from 3,4-xylidine by alkylation and dinitration.

2,6-Xylidine is an interesting precursor for a large number of plant-protection agents. Derivatives of N-chloroacetyl-2,6-xylidine are of interest as herbicides, growth regulators, and fungicides. Commercial products based on 2,6-xylidine are the fungicides metalaxyl (Ridomil) [43] and furalaxyl (Fongarid) [44].

6.3. Production of Pharmaceutical Products, Pharmaceutical Products for Veterinary Medicine, and Vitamins

The antiphlogistic, antirheumatic, and analgesic mefenamic acid is produced from 2,3-xylidines [45].

2,4-Xylidine is the starting material for the active substance cymiazole (Tifatol, Besuntol), used in veterinary medicine as an ectoparasiticide for combatting ticks [46], [47]. 2,6-Xylidine is a component of a range of similarly structured local anesthetics, such as lidocaine, bupivacaine, mepivacaine, and etidocaine [48], [49]. 2,6-Xylidine is also used in a range of other medicinal products, e.g., the diuretic xipamide [50], the amidinourea derivative lidamidine hydrochloride [51], which combats diarrhoea, the arylurea derivative xilobam, which has a tranquilizing effect on the central nervous system [52], and the coronary vasodilator lidoflazine [53]. The active substance xylazine hydrochloride (Rompun) is also produced from 2,6-xylidine [54], and is used in veterinary medicine as a narcotic and sedative.

3,4-Xylidine is used in the synthesis of riboflavin (vitamin B_2) [55]. It is also a precursor in the biosynthesis of vitamin B_{12} [56].

6.4. Other Uses

The use of xylidines as antiknock agents for gasoline [57] and as antioxidants for lubricating oils and rubber products [58], [59] has already been mentioned. Xylidines are still added to gasoline as coloring agents; 2,5-xylidine is used as an additive for liquid crystal mixtures [60]. The use of xylidines as corrosion inhibitors in metallic materials has also been described [61]. Xylidines can be used as absorption agents for sulfur dioxide in desulfurizing exhaust gas [62].

The reaction of formaldehyde with xylidines is analogous to that with aniline, and gives diaminotetramethyldiphenylmethanes [63] and, under hydrogenation conditions, diaminotetramethyldicyclohexylmethanes [64]. These products are used as cross-linking agents and hardeners for epoxy resins, or can be further processed to polyurethanes.

7. Toxicology

Xylidines are readily resorbed by the skin and mucous membranes. Their ability to bind hemoglobin is known, in good agreement with the established maximum methemoglobin level.

Detailed test have been carried out with 3,4-xylidine and 2,4-xylidine [65]. Very high xylidine concentrations attack the central nervous system directly. Xylidines also lead to induction of microsomal enzyme systems in the liver. In animal experiments, toxic effects on the lungs and kidneys have also been established.

All the isomers of xylidine are identified as mutagenic, 2,4-, 2,5-, and 2,6-xylidines have also been found to cause tumors. The LD_{50}(oral) for rats is 0.1 – 1 g. The MAK value has been set at 5 mg/kg for any single xylidine. Table 3 gives an overview of the toxicological data described.

There are reports of tests for carcinogenesis on 2,4-, 2,5-, and 2,6-xylidines, which have been found to cause tumors. High doses of 2,6-xylidine, which has been identified as a component of tobacco smoke, leads to adenomas and carcinomas in the nasal cavities in rats. Mesenchymal tumor and rhabdomysarcoma have been observed as

Table 3. Toxicological data for xylidines

	Microsomal mutagenicity on *Salmonella typhimurium*, µmol/plate	LD_{50}, rat oral, mg/kg	References
2,3-Xylidine	5	836	[66]
2,4-Xylidine	1	467	[67]
2,5-Xylidine	10	1297	[66]
2,6-Xylidine	50	707 [b]	[67]
3,4-Xylidine	5	812	[66], [69]
3,5-Xylidine	1 [a]	707	[68], [69]

[a] mg/plate.
[b] Mouse.

malignant carcinomas. Subcutaneous tumors and neoplastic lumps in the liver have also been found. In rats 2,6-xylidine is classed as definitely carcinogenic [70].

8. References

[1] R. N. Beale, *J. Chem. Soc.* 1954, 4494.
[2] G. T. Morgan, W. J. Hickinbottom, *J. Chem. Soc. Ind. London Trans. Commun.* **45** (1926) 221.
[3] J. Bergman, P. Sand, *Tetrahedron* **46** (1990) 6085.
[4] Sumitomo Chemical, JP 58 08, 052 [83 08, 052], 1983.
[5] Anglo-Iranian Oil, GB 599 252, 1944.
[6] BIOS Final Report, **1146** (1947) 65.
[7] Clayton Aniline, EP 2308, 1978.
[8] ICI, GB 982 902, 1962.
[9] Teijin, JP 70 490, 1966.
[10] Mitsui Toatsu Chem., JP 7 0 31 652, 1967.
[11] Texaco, US 3 832 401, 1971.
[12] Shell, US 2 458 214, 1944.
[13] Standard Oil, GB 599 252, 1944.
[14] J. Werner, *Ind. Eng. Chem.* **41** (1949) 1841.
[15] Standard Oil, US 2 455 713, 1945.
[16] General Aniline Works, US 2 013 873, 1934.
[17] Bayer, DE-OS 3 023 487, 1980.
[18] Rütgerswerke, DE-OS 2 516 316, 1975.
[19] Ethyl Corp., DE-OS 2 208 827, 1972.
[20] Ethyl Corp., DE-AS 1 933 636, 1969.
[21] Y. P. Sobolev, V. M. Berezivskii, SU 118 129, 1959.
[22] American Home Products, US 2 347 652, 1941.
[23] V. I. Rogovik, I. I. Gulianyuk, SU 287 958, 1969.
[24] Hoechst, DE-OS 2 654 852, 1978.
[25] Ciba-Geigy, US 4 188 341, 1975.
[26] A. Ono, *Chromatographia* **14** (1981) 39, 692.
[27] F. K. Chow, E. Grushka, *Anal. Chem.* **49** (1977) 1756.
[28] J. F. Kung, Jr., W. C. Howell, Jr., C. E. Starr, Jr., *Ind. Eng. Chem.* **40** (1948) 1534.

[29] Du Pont, DE-AS 1 518 107, 1965; GB 1 047 607, 1965.
[30] Standard Oil, US 2 447 615, 1944.
[31] Standard Oil, US 2 469 745, 1946.
[32] Standard Oil, US 2 510 849, 1946.
[33] Du Pont, US 2 686 809, 1952.
[34] Sigma-Aldrich Library of Regulation and Safety Data, vol. **I**, Sigma-Aldrich Corp., Milwaukee, Wis., 1993.
[35] A. Courtin, H. R. v. Tobel, P. Doswald, *Helv. Chim. Acta* **61** (1978) 3079.
[36] Hoechst, DE-OS 2 749 327, 1977; DE-OS 2 519 036, 1975.
[37] Ciba-Geigy, EP 4611, 1979.
[38] Geigy, DE-AS 1 148 341, 1959.
[39] Amer. Cyanamid, DE-OS 2 632 437, 1976.
[40] Minnesota Mining, DE-OS 2 406 475, 1974.
[41] Amer. Cyanamid, DE-OS 2 429 958, 1973.
[42] Amer. Cyanamid, DE-OS 2 232 263, 1971.
[43] Ciba-Geigy, DE-OS 2 515 091, 1975.
[44] Ciba-Geigy, DE-OS 2 513 732, 1975.
[45] Parke Davis, DE-AS 1 186 073, 1963.
[46] Ciba-Geigy, DE-OS 2 619 724, 1976.
[47] Bayer, DE-OS 2 531 606, 1975.
[48] S. M. Waraszkiewicz, W. O. Foye, H. J. Adams, B. H. Takman, *J. Med. Chem.* **19** (1976) 541.
[49] J. Marcus, A. Bardeanu, Z. Theodorescu, *Pharmazie* **14** (1959) 322.
[50] Beiersdorf, DE-OS 1 270 544, 1965.
[51] William H. Rorer Inc., DE-OS 2 433 837, 1973.
[52] Mc Neil Laboratories, DE-OS 2 313 257, 1971.
[53] Janssen, NL 6 507 312, 1965.
[54] Bayer, DE-AS 1 173 475, 1962.
[55] BASF, DE-OS 3 615 834, 1986.
[56] M. H. Kay, C. Radford, W. L. Alworth, *J. Chem. Soc. Chem. Commun.* **78** (1978) 22.
[57] J. E. Brown, F. X. Markley, H. Shapiro, *Ind. Eng. Chem.* **47** (1955) 2141.
[58] Du Pont, FR 1 553 317, 1967.
[59] Goodyear, US 4 269 763, 1980.
[60] Bayer, DE-OS 3 536 267, 1985.
[61] M. L. J. Bernard, DE-AS 1 084 544, 1958.
[62] Metallgesellschaft, DE-AS 1 025 400, 1956.
[63] ICI, GB 852 651, 1958.
[64] Bayer, DE-OS 2 945 614, 1979.
[65] G. Birnes, H. G. Neumann, *Arch. Toxicol.* **62** (1988) 110.
[66] D. Zimmer, J. Mazurek, G. Petzold, B. K. Bhuyan, *Muta. Res.* **77** (1980) 317.
[67] D. Zimmer, J. Mazurek, G. Petzold, B. K. Bhuyan, *J. Agric. Food Chem.* **34** (1986) 157.
[68] E. Zeiger et al., *Environ. Mol. Mutagen.* **11** (1988) 1.
[69] J. M. Mac Ewen, E. H. Vernot, *Aerosp. Med. Res. Lab. Tech. Rep. AMRL-TRUS* **72** (1972) 72–62.
[70] M. Kornreich, C. A. Montgomery, *Nat. Toxicol. Program Tech. Rep. Ser.* **278** (1990).

Author Index

Abaecherli, Claudio, Lonza AG, Visp, Switzerland, *Ketenes*, **5**
Abe, Nobuyuki, Koei Chemical Co. Ltd., Osaka 541, Japan, *Pyridine and Pyridine Derivatives*, **7**
Aguiló, Adolfo, Celanese Chemical Company Technical Center, Corpus Christi, Texas 78469, United States, *Acetic Acid*, **1**
Arntz, Dietrich, Degussa Corporation, Mobile, Alabama, United States, *Acrolein and Methacrolein*, **1**
Astheimer, Hans-Joachim, Rhodia AG, Freiburg, Federal Republic of Germany, *Cellulose Esters*, **2**

Bach, Hanswilhelm, Ruhrchemie AG, Oberhausen, Federal Republic of Germany, *Butanals*, **2**
Bahrmann, Helmut, Ruhrchemie AG, Oberhausen, Federal Republic of Germany, *Alcohols, Aliphatic*, **1**; *2-Ethylhexanol*, **4**
Balser, Klaus, Wolff Walsrode AG, Walsrode, Federal Republic of Germany, *Cellulose Esters*, **2**
Baltes, Herbert, Hoechst Aktiengesellschaft, Frankfurt, Federal Republic of Germany, *Crotonaldehyde and Crotonic Acid*, **3**
Barda, Henry J., Ethyl Corp., Baton Rouge, Louisiana 70821, United States, *Bromine Compounds*, **2**
Bässler, Jürgen, Uhde GmbH, Dortmund, Federal Republic of Germany, *Acetylene*, **1**
Bauer, Wolfgang, Cassella AG, Frankfurt/Main, Federal Republic of Germany, *Thiourea and Thiourea Derivatives*, **8**
Bauer, Jr., William, Rohm and Haas Co., Philadelphia, Pennsylvania 19477, United States, *Methacrylic Acid and Derivatives*, **6**
Beck, Ferdinand, SKW Trostberg AG, Trostberg, Federal Republic of Germany, *Thiourea and Thiourea Derivatives*, **8**
Beck, Uwe, Bayer AG, Leverkusen, Federal Republic of Germany, *Chlorinated Hydrocarbons*, **3**
Behr, Arno, Henkel KGaA, Düsseldorf, Federal Republic of Germany, *Hydrocarbons*, **5**
Behr, Fred, Minnesota Mining and Manufacturing Company, St. Paul, Minnesota, United States, *Fluorine Compounds, Organic*, **5**
Behringer, Hartmut, Hoechst Aktiengesellschaft, Werk Knapsack, Federal Republic of Germany, *Acetylene*, **1**
Benya, Theodore J., Ethyl Corp., Baton Rouge, Louisiana 70821, United States, *Bromine Compounds*, **2**
Besson, Bernard, Rhône-Poulenc Chimie S.A., Paris, France, *Salicylic Acid*, **7**
Betso, Joanne E., Dow Chemical, Midland, Michigan 48640, United States, *Chlorohydrins*, **3**
Beutel, Klaus K., Dow Chemical Europe, Horgen, Switzerland, *Chlorinated Hydrocarbons*, **3**
Bhargava, Naresh, BASF Canada Inc., Cornwall, Ontario, Canada, *Phthalic Acid and Derivatives*, **7**
Biedenkapp, Dieter, BASF Aktiengesellschaft, Ludwigshafen, Federal Republic of Germany, *Hydrocarbons*, **5**
Bipp, Hansjörg, BASF Aktiengesellschaft, Ludwigshafen, Federal Republic of Germany, *Formamides*, **5**
Blanc, Alain, Société Française Hoechst, C.R.A., Stains, France, *Glyoxal*, **5**
Blau, Werner, Hoechst Aktiengesellschaft, Frankfurt, Federal Republic of Germany, *Crotonaldehyde and Crotonic Acid*, **3**
Boehlke, Klaus, BASF Aktiengesellschaft, Ludwigshafen, Federal Republic of Germany, *Waxes*, **8**
Böhm, Siegfried, Bayer AG, Leverkusen, Federal Republic of Germany, *Chloroformic Esters*, **3**
Bolt, Hermann M., Institut für Arbeitsphysiologie an der Universität Dortmund, Dortmund, Federal Republic of Germany, *Acetaldehyde*, **1**
Booth, Gerald, Booth Consultancy Services, Thorpe House, Uppermill, Oldham OL3 6DP, United Kingdom, *Naphthalene Derivatives*, **6**; *Nitro Compounds, Aromatic*, **6**
Bosche, Horst, BASF Aktiengesellschaft, Ludwigshafen, Federal Republic of Germany, *Butanediols, Butenediol, and Butynediol*, **2**
Boullard, Olivier, Rhône-Poulenc Chimie S.A., Paris, France, *Salicylic Acid*, **7**
Bowers, Jr., Joseph S., First Chemical Corporation, Pascagoula, MS 39568, United States, *Toluidines*, **8**
Brandt, Lothar, Hoechst AG, Niederlassung Kalle, Wiesbaden-Biebrich, Federal Republic of Germany, *Cellulose Ethers*, **2**
Bringer-Meyer, Stefanie, Institut für Biotechnologie der KFA Jülich GmbH, Jülich, Federal Republic of Germany, *Ethanol*, **4**
Brockmann, Rolf, Henkel KGaA, Düsseldorf, Federal Republic of Germany, *Fatty Acids*, **4**
Brühne, Friedrich, Bayer AG, Krefeld-Uerdingen, Federal Republic of Germany, *Benzyl Alcohol*, **2**; *Benzaldehyde*, **2**
Buckl, Klaus, Linde AG, Höllriegelskreuth, Federal Republic of Germany, *Acetylene*, **1**
Butzert, Hartmut, Hüls Aktiengesellschaft, Marl, Federal Republic of Germany, *Methyl Tert-Butyl Ether*, **6**
Buysch, Hans-Josef, Bayer AG, Krefeld, Federal Republic of Germany, *Carbonic Acid Esters*, **2**

Author Index

Buysch, Hans-Josef, Bayer AG, Leverkusen, Federal Republic of Germany, *Phenol Derivatives*, **6**

Caillard, Liliane, Rhône-Poulenc, Courbevoie, France, *Chlorophenols*, **3**
Campbell, M. Larry, Exxon Chemical Co., Florham Park, New Jersey 07932, United States, *Cyclohexane*, **3**
Castor, William M., Dow Chemical, Freeport, Texas 77 541, United States, *Styrene*, **7**
Chahal, Surinder P., Croda Colloids Ltd., Widnes, United Kingdom, *Lactic Acid*, **5**
Chorlton, Alan P., ICI Specialties, Blackley, Manchester, United Kingdom, *Pyrimidine and Pyrimidine Derivatives*, **7**
Christidis, Yani, Société Française Hoechst, C.R.A., Stains, France, *Glyoxylic Acid*, **5**
Collin, Gerd, DECHEMA e.V., Frankfurt/Main, Federal Republic of Germany, *Acridine*, **1**; *Anthracene*, **2**; *Carbazole*, **2**; *Furan and Derivatives*, **5**; *Hydrocarbons*, **5**; *Indole*, **5**; *Naphthalene and Hydronaphthalenes*, **6**; *Quinoline and Isoquinoline*, **7**; *Tar and Pitch*, **8**
Cook, Richard, ICI Chemicals and Polymers, Runcorn, United Kingdom, *Chlorinated Hydrocarbons*, **3**
Cornils, Boy, Hoechst AG, Frankfurt, Federal Republic of Germany, *Dicarboxylic Acids, Aliphatic*, **3**
Cornils, Boy, Ruhrchemie AG, Oberhausen, Federal Republic of Germany, *Butanals*, **2**
Coty, Robert R., Badger Engineers, Cambridge, Massachusetts 02142, United States, *Ethylbenzene*, **4**
Crews, George M., Melamine Chemicals, Inc., Donaldsonville, Louisiana 70346, United States, *Melamine and Guanamines*, **5**

Dagani, Michael J., Ethyl Corp., Baton Rouge, Louisiana 70821, United States, *Bromine Compounds*, **2**
Dämbkes, Georg, Ruhrchemie AG, Oberhausen, Federal Republic of Germany, *Butanols*, **2**
Daniel, James R., Department of Foods and Nutrition, Purdue University, West Lafayette, Indiana 47907, United States, *Starch*, **7**
Davis, Darwin D., E. I. Du Pont de Nemours & Co., Victoria, Texas 77901, United States, *Adipic Acid*, **1**
de Belder, Anthony N., Pharmacia AB, Uppsala, Sweden, *Dextran*, **3**
Decker, Daniel, Hoechst Aktiengesellschaft, Frankfurt am Main, Federal Republic of Germany, *Resorcinol*, **7**
Demmering, Günther, Henkel KGaA, Düsseldorf, Federal Republic of Germany, *Fatty Acids*, **4**
Diehl, Herbert, Bayer AG, Leverkusen, Federal Republic of Germany, *Butanediols, Butenediol, and Butynediol*, **2**
Dietsche, Wolfram, BASF Aktiengesellschaft, Ludwigshafen, Federal Republic of Germany, *Waxes*, **8**
Disteldorf, Walter, BASF Aktiengesellschaft, Ludwigshafen, Federal Republic of Germany, *Formaldehyde*, **5**
Dreher, Eberhard-Ludwig, Dow Chemical GmbH, Stade, Federal Republic of Germany, *Chlorinated Hydrocarbons*, **3**
Droste, Wilhelm, Hüls AG, Marl, Federal Republic of Germany, *Butenes*, **2**
Duvnjak, Zdravko, University of Ottawa, Ottawa, Ontario K1N 9B4, Canada, *Ethanol*, **4**

Ebel, Klaus, BASF Aktiengesellschaft, Ludwigshafen, Federal Republic of Germany, *Imidazole and Derivatives*, **5**
Ebersberg, Günter, Chemische Werke Hüls AG, Marl, Federal Republic of Germany, *Acetylene*, **1**
Ebertz, Wolfgang, Hoechst Aktiengesellschaft, Frankfurt, Federal Republic of Germany, *Oxocarboxylic Acids*, **6**
Eggersdorfer, Manfred, BASF Aktiengesellschaft, Ludwigshafen, Federal Republic of Germany, *Ketones*, **5**; *Terpenes*, **8**
Eicher, Theo, Stuttgart, Federal Republic of Germany, *Cellulose Esters*, **2**
Eisele, Peter, Linde AG, Werksgruppe Verfahrenstechnik und Anlagenbau, Höllriegelskreuth, Federal Republic of Germany, *Propene*, **7**
Enke, Walter, BASF AG, Ludwigshafen, Federal Republic of Germany, *Phthalic Acid and Derivatives*, **7**

Fabri, Jörg, Veba AG, Düsseldorf, Federal Republic of Germany *Toluene*, **8**; *Xylenes*, **8**
Falbe, Jürgen, Henkel KGaA, Düsseldorf, Federal Republic of Germany *Alcohols, Aliphatic*, **1**; *Aldehydes, Aliphatic and Araliphatic*, **1**
Farkas, Adalbert, Consulting Chemist, Delray Beach, Florida 33483, United States, *Ethanol*, **4**
Fedtke, Norbert, Hüls Aktiengesellschaft, Marl, Federal Republic of Germany, *Maleic and Fumaric Acids*, **5**
Feiring, Andrew, E. I. Du Pont de Nemours & Co., Wilmington, Delaware, United States, *Fluorine Compounds, Organic*, **5**
Fiedler, Eckhard, BASF Aktiengesellschaft, Ludwigshafen, Federal Republic of Germany, *Methanol*, **6**
Fiege, Helmut, Bayer AG, Leverkusen, Federal Republic of Germany, *Cresols and Xylenols*, **3**; *Phenol Derivatives*, **6**
Finley, K. Thomas, State University of New York, Brockport, New York 14420, United States, *Benzoquinone*, **2**
Fleig, Helmut, BASF Aktiengesellschaft, Ludwigshafen, Federal Republic of Germany, *Turpentines*, **8**
Fleischmann, Gerald, Wacker-Chemie GmbH, Werk Burghausen, Federal Republic of Germany, *Acetaldehyde*, **1**
Folkins, Hillis O., (Union Oil Company of California), Claremont, California 91711, United States, *Benzene*, **2**
Forgione, Peter S., American Cyanamid Company, Stamford, Connecticut 06905, United States, *Cyanamides*, **3**
Frank, Dieter, Akzo Chemie America, Chicago, Ill., United States, *Amines, Aliphatic*, **1**

Fuchs, Hugo, BASF Aktiengesellschaft, Ludwigshafen, Federal Republic of Germany, *Caprolactam*, **2**
Fuhrmann, Werner, Hüls Aktiengesellschaft, Bottrop, Federal Republic of Germany, *Maleic and Fumaric Acids*, **5**
Fujita, Yasuhiro, Mitsui Petrochemical Industries, Ltd., Yamaguchi, Japan, *Phenol Derivatives*, **6**

Gamer, Armin O., BASF Aktiengesellschaft, Ludwigshafen, Federal Republic of Germany *Formaldehyde*, **5**; *Propionic Acid and Derivatives*, **7**
Garbe, Dorothea, Haarmann & Reimer GmbH, Holzminden, Federal Republic of Germany, *Cinnamic Acid*, **3**; *Phenylacetic Acid*, **6**; *Phenol Derivatives*, **6**
Garbe, Dorothea, Haarmann & Reimer GmbH, Holzminden, Federal Republic of Germany, *Hydrocarbons*, **5**
Gärtner, Roderich, Ruhrchemie AG, Oberhausen, Federal Republic of Germany, *Butanals*, **2**
Gelbke, Heinz-Peter, BASF Aktiengesellschaft, Ludwigshafen, Federal Republic of Germany, *Nitriles*, **6**
Gerlich, Otto, Phenolchemie GmbH, Gladbeck, Federal Republic of Germany, *Phenol*, **6**
Gerulis, John J., Mobay Chemical Corp., Pittsburgh, Pennsylvania 15205, United States, *Amines, Aromatic*, **1**
Goebel, Otto, CORA Engineering, Chur, Switzerland, *Ethanol*, **4**
Golka, Klaus, Institut für Arbeitsphysiologie an der Universität Dortmund, Dortmund, Federal Republic of Germany, *Acetaldehyde*, **1**
Götz, Peter H., Degussa AG, Hanau, Federal Republic of Germany, *Peroxy Compounds, Organic*, **6**
Gousetis, Charalampos, BASF Aktiengesellschaft, Ludwigshafen, Federal Republic of Germany, *Nitrilotriacetic Acid*, **6**
Graeser, Ulrich, Veba Öl AG, Gelsenkirchen, Federal Republic of Germany, *Toluene*, **8**; *Xylenes*, **8**
Gräfje, Heinz, BASF Aktiengesellschaft, Ludwigshafen, Federal Republic of Germany, *Butanediols, Butenediol, and Butynediol*, **2**
Grambow, Clemens, SKW Trostberg AG, Trostberg, Federal Republic of Germany, *Guanidine and Derivatives*, **5**
Griesbaum, Karl, Universität Karlsruhe (TH), Karlsruhe, Federal Republic of Germany, *Cyclopentadiene and Cyclopentene*, **3**; *Hydrocarbons*, **5**
Grolig, Johann, Bayer AG, Leverkusen, Federal Republic of Germany, *Allyl Compounds*, **1**; *Naphthoquinones*, **6**
Grossmann, Georg, BASF Aktiengesellschaft, Ludwigshafen, Federal Republic of Germany, *Methanol*, **6**
Gscheidmeier, Manfred, Hoechst Aktiengesellschaft, Werk Gersthofen, Augsburg, Federal Republic of Germany, *Turpentines*, **8**

Haberstroh, Hans-Joachim, Biochemie Ladenburg GmbH, Ladenburg, Federal Republic of Germany, *Gluconic Acid*, **5**
Haferkorn, Herbert, Hüls Aktiengesellschaft, Bottrop, Federal Republic of Germany, *Maleic and Fumaric Acids*, **5**
Hagedorn, Ferdinand, Bayer AG, Leverkusen, Federal Republic of Germany, *Nitriles*, **6**
Hahn, Heinz-Dieter, Ruhrchemie AG, Oberhausen, Federal Republic of Germany, *Butanols*, **2**; *2-Ethylhexanol*, **4**
Hamamoto, Toshikazu, Ube Industries, Ltd., Ube, Japan, *Phenol Derivatives*, **6**
Hammer, Hans, BASF Aktiengesellschaft, Ludwigshafen/Rhein, Federal Republic of Germany, *Ethanolamines and Propanolamines*, **4**
Harreus, Albrecht Ludwig, BASF Aktiengesellschaft, Ludwigshafen, Federal Republic of Germany, *Pyrrole*, **7**; *2-Pyrrolidone*, **7**
Hart, J. Roger, W. R. Grace & Co., Lexington, Massachusetts 02173, United States, *Ethylenediaminetetraacetic Acid and Related Chelating Agents*, **4**
Hefner, Werner, BASF Aktiengesellschaft, Ludwigshafen, Federal Republic of Germany, *Acetylene*, **1**
Heilen, Gerd, BASF Aktiengesellschaft, Ludwigshafen, Federal Republic of Germany, *Amines, Aliphatic*, **1**
Heinrichs, Franz-Leo, Hoechst Aktiengesellschaft, Werk Gersthofen, Augsburg, Federal Republic of Germany, *Waxes*, **8**
Heitmann, Wilhelm, Hüls AG, Herne, Federal Republic of Germany, *Ethers, Aliphatic*, **4**
Held, Heimo, Wacker-Chemie GmbH, Werk-Burghausen, Burghausen, Federal Republic of Germany, *Acetic Anhydride and Mixed Fatty Acid Anhydrides*, **1**
Hertel, Hasso, Hoechst Aktiengesellschaft, Werk Offenbach, Federal Republic of Germany, *Diazo Compounds and Diazo Reactions*, **3**
Heslinga, Michiel C., Solvay Pharmaceuticals, Weesp, The Netherlands, *Lactose and Derivatives*, **5**
Hilt, Albrecht, Ultraform GmbH, Ludwigshafen, Federal Republic of Germany, *Formaldehyde*, **5**
Hobbs, Charles C., Celanese Chemical Company Technical Center, Corpus Christi, Texas 78469, United States, *Acetic Acid*, **1**
Hofmann, Ernst, BASF Aktiengesellschaft, Ludwigshafen, Federal Republic of Germany, *Vinyl Ethers*, **8**
Hofmann, Thomas, Hoechst Marion Roussel, Hattersheim, Federal Republic of Germany, *Pentanols*, **6**; *Phosphorus Compounds, Organic*, **6**
Hohner, Gerd, Hoechst Aktiengesellschaft, Werk Gersthofen, Augsburg, Federal Republic of Germany, *Waxes*, **8**

Author Index

Höke, Hartmut, Weinheim, Federal Republic of Germany, *Acridine*, **1**; *Anthracene*, **2**; *Carbazole*, **2**; *Furan and Derivatives*, **5**; *Hydrocarbons*, **5**; *Indole*, **5**; *Naphthalene and Hydronaphthalenes*, **6**; *Quinoline and Isoquinoline*, **7**; *Tar and Pitch*, **8**

Hönicke, Dieter, Universität Karlsruhe (TH), Karlsruhe, Federal Republic of Germany, *Cyclopentadiene and Cyclopentene*, **3**

Höpp, Mathias, Degussa AG, Zweigniederlassung Wolfgang, Hanau, Federal Republic of Germany, *Acrolein and Methacrolein*, **1**

Hoppe, Bernd, Degussa AG, Hanau-Wolfgang, Federal Republic of Germany, *Amino Acids*, **1**

Hoppe, Lutz, Wolff Walsrode AG, Walsrode, Federal Republic of Germany, *Cellulose Esters*, **2**

Höver, Hermann, Union Rheinische Braunkohlen Kraftstoff AG, Wesseling, Federal Republic of Germany, *Dimethyl Ether*, **4**

Hudnall, Phillip M., Tennessee Eastman Company, Kingsport, Tennessee 37662, United States, *Hydroquinone*, **5**

Hustede, Helmut, Joh. A. Benckiser GmbH, Ladenburg, Federal Republic of Germany, *Gluconic Acid*, **5**

Ichimura, Hisao, Koei Chemical Co. Ltd., Osaka 541, Japan, *Pyridine and Pyridine Derivatives*, **7**

Iwata, Tadao, Mitsui Petrochemical Industries, Ltd., Yamaguchi, Japan, *Phenol Derivatives*, **6**

J. Voragen, Alphons C., Wageningen, Agricultural University, Department of Food Science, Wageningen, The Netherlands, *Polysaccharides*, **7**

Jäckh, Rudolf, BASF Aktiengesellschaft, Ludwigshafen, Federal Republic of Germany, *Amines, Aliphatic*, **1**; *Imidazole and Derivatives*, **5**; *Phthalic Acid and Derivatives*, **7**

Jacobi, Sylvia, Degussa AG, Zweigniederlassung Wolfgang, Hanau, Federal Republic of Germany, *Acrolein and Methacrolein*, **1**

Jager, Martin, Hoechst Aktiengesellschaft, Frankfurt/Main, Federal Republic of Germany, *Sorbic Acid*, **7**

Jäger, Peter, BASF Aktiengesellschaft, Ludwigshafen/Rh., Federal Republic of Germany, *Carbamates and Carbamoyl Chlorides*, **2**

Jakobson, Gerald, Deutsche Solvay-Werke GmbH, Solinen and Rheinberg, Federal Republic of Germany, *Allyl Compounds*, **1**; *Glycerol*, **5**

James, Denis H., Dow Chemical, Freeport, Texas 77 541, United States, *Styrene*, **7**

Jira, Reinhard, Wacker-Chemie GmbH, Werk Burghausen, Federal Republic of Germany, *Acetaldehyde*, **1**; *Chloroacetaldehydes*, **3**

John, Peter, Wacker Chemie GmbH, Burghausen, Federal Republic of Germany, *Silicon Compounds, Organic*, **7**

Jordan, Wilfried, Phenolchemie GmbH, Gladbeck, Federal Republic of Germany, *Phenol*, **6**

Kahlich, Dietmar, Dow Deutschland Inc., Stade, Federal Republic of Germany, *Propylene Oxide*, **7**

Kassaian, Jean-Maurice, ETS Legré-Mante, Marseilles, France, *Tartaric Acid*, **8**

Kataoka, Toshiaki, Koei Chemical Co. Ltd., Osaka 541, Japan, *Pyridine and Pyridine Derivatives*, **7**

Kathagen, Friedrich W., Deutsche Solvay-Werke GmbH, Solingen and Rheinberg, Federal Republic of Germany, *Glycerol*, **5**

Kersebohm, Dietrich Burkhard, BASF Aktiengesellschaft, Ludwigshafen, Federal Republic of Germany, *Melamine and Guanamines*, **5**; *Methanol*, **6**

Keuser, Ullrich, BASF Aktiengesellschaft, Ludwigshafen, Federal Republic of Germany, *Propionic Acid and Derivatives*, **7**

Kieczka, Heinz, BASF Aktiengesellschaft, Ludwigshafen, Federal Republic of Germany, *Butanediols, Butenediol, and Butynediol*, **2**; *Butyrolactone*, **2**; *Carbamates and Carbamoyl Chlorides*, **2**; *Caprolactam*, **2**; *Ethanolamines and Propanolamines*, **4**; *Formamides*, **5**; *Formic Acid*, **5**

Kiel, Wolfgang, Bayer AG, Leverkusen, Federal Republic of Germany, *Benzylamine*, **2**

Killpack, Richard, Shell International Chemical Company Ltd., Shell Centre, London, United Kingdom, *Propene*, **7**

Klatt, Martin, Deutsche Solvay-Werke GmbH, Solingen and Rheinberg, Federal Republic of Germany, *Glycerol*, **5**

Kleemann, Axel, Degussa AG, Hanau-Wolfgang, Federal Republic of Germany, *Amino Acids*, **1**

Kleine-Boymann, Michael, Phenolchemie GmbH, Gladbeck, Federal Republic of Germany, *Phenol*, **6**

Kleinschmidt, Peter, Bayer AG, Dormagen, Federal Republic of Germany, *Chlorinated Hydrocarbons*, **3**

Klenk, Herbert, Degussa AG, Frankfurt, Federal Republic of Germany, *Cyanuric Acid and Cyanuric Chloride*, **3**; *Peroxy Compounds, Organic*, **6**; *Thiocyanates and Isothiocyanates, Organic*, **8**

Klimisch, Hans-Joachim, BASF Aktiengesellschaft, Ludwigshafen, Federal Republic of Germany, *Vinyl Ethers*, **8**

Klingler, Franz Dietrich, Boehringer, Ingelheim, Federal Republic of Germany, *Oxocarboxylic Acids*, **6**

Koch, Theodore A., E. I. Du Pont de Nemours & Co., Inc., Wilmington, Delaware 19898, United States, *Cyclododecatriene and Cyclooctadiene*, **3**

Koehler, Hermann, BASF Aktiengesellschaft, Ludwigshafen, Federal Republic of Germany, *Imidazole and Derivatives*, **5**

Koenen, Gunther, Peter Greven Fettchemie GmbH, Bad Münstereifel, Federal Republic of Germany, *Metallic Soaps*, **6**

Koenig, Günter, Hoechst Aktiengesellschaft, Augsburg, Federal Republic of Germany, *Chloroacetic Acids*, **3**

Kohler, Walter, BASF Aktiengesellschaft, Ludwigshafen, Federal Republic of Germany, *Propionic Acid and Derivatives*, **7**

Kohlpaintner, Christian, Celanese GmbH, Werk Ruhrchemie, Oberhausen, Federal Republic of Germany, *Aldehydes, Aliphatic and Araliphatic*, **1**

Kopp, Erwin, Wacker-Chemie GmbH, München, Federal Republic of Germany, *Chloroacetaldehydes*, **3**

Körnig, Wolfgang, BASF Aktiengesellschaft, Ludwigshafen, Federal Republic of Germany, *Butanediols, Butenediol, and Butynediol*, **2**; *Ethanolamines and Propanolamines*, **4**

Kosaric, Naim, The University of Western Ontario, London, Ontario N6A 5B9, Canada, *Ethanol*, **4**

Kosswig, Kurt, Hüls Aktiengesellschaft, Marl, Federal Republic of Germany, *Sulfonic Acids, Aliphatic*, **8**

Krähling, Ludger, Deutsche Solvay-Werke GmbH, Rheinberg, Federal Republic of Germany, *Allyl Compounds*, **1**

Kreutzer, Udo, Henkel KGaA, Düsseldorf, Federal Republic of Germany, *Fatty Acids*, **4**

Krey, Jürgen, Deutsche Solvay-Werke GmbH, Rheinberg, Federal Republic of Germany, *Allyl Compounds*, **1**

Kriebitzsch, Norbert, Degussa-Antwerpen N.V., Antwerp, Belgium, *Cyanuric Acid and Cyanuric Chloride*, **3**

Kuster, Ben F. M., Technische Universiteit Eindhoven, Eindhoven, The Netherlands, *Lactose and Derivatives*, **5**

Langer, Ernst, BASF Aktiengesellschaft, Ludwigshafen, Federal Republic of Germany, *Chlorinated Hydrocarbons*, **3**

Langvardt, Patrick W., The Dow Chemical Co., Midland, Michigan 48640, USA, *Acrylonitrile*, **1**

Lappe, Peter, Ruhrchemie AG, Oberhausen, Federal Republic of Germany, *Aldehydes, Aliphatic and Araliphatic*, **1**; *Dicarboxylic Acids, Aliphatic*, **3**; *Pentanols*, **6**

Lawrence, F. R., E. I. du Pont de Nemours & Co., Deepwater, New Jersey 08023, United States, *Aniline*, **1**

Leblanc, Henri, Rhône-Poulenc Chimie S.A., Paris, France, *Salicylic Acid*, **7**

Lehmann, Jochen, Institut für Organische Chemie und Biochemie, Universität Freiburg, Freiburg, Federal Republic of Germany, *Carbohydrates*, **2**

Lendle, Wilhelm, Hoechst Aktiengesellschaft, Frankfurt/Main, Federal Republic of Germany, *Chlorinated Hydrocarbons*, **3**

Leng, Marguerite L., Dow Chemical, Midland, Michigan 48640, United States, *Chlorophenoxyalkanoic Acids*, **3**

Lenz, Hans H., Knoll AG Ludwigshafen, Federal Republic of Germany, *Purine Derivatives*, **7**

Leuchtenberger, Wolfgang, Degussa AG, Hanau-Wolfgang, Federal Republic of Germany, *Amino Acids*, **1**

Levy, Alan B., Allied Signal Inc., Morristown, New Jersey 07962, United States, *Acetone*, **1**

Lindemann, Manfred, Henkel KGaA, Düsseldorf, Federal Republic of Germany, *Fatty Acids*, **4**

Lindner, Jörg, Dow Deutschland Inc., Stade, Federal Republic of Germany, *Propylene Oxide*, **7**

Lindner, Otto, Bayer AG, Leverkusen, Federal Republic of Germany, *Benzenesulfonic Acids and Their Derivatives*, **2**

Lipper, Karl-August, Bayer AG, Krefeld, Federal Republic of Germany, *Chlorinated Hydrocarbons*, **3**

Lipps, Wolfgang, Naturwissenschaftlich-technische Akademie Prof. Dr. Grübler, Isny, Federal Republic of Germany, *Alcohols, Aliphatic*, **1**

Liu, Gordon Y. T., Dow Chemical, Midland, Michigan 48640, United States, *Chlorohydrins*, **3**

Löbbert, Gerd, BASF Aktiengesellschaft, Ludwigshafen, Federal Republic of Germany, *Phthalocyanines*, **7**

Lohbeck, Kurt, Hüls Aktiengesellschaft, Bottrop, Federal Republic of Germany, *Maleic and Fumaric Acids*, **5**

Lohmar, Elmar, Hoechst Aktiengesellschaft, Köln, Federal Republic of Germany, *Chloroacetic Acids*, **3**

Lorz, Peter M., BASF AG, Ludwigshafen, Federal Republic of Germany, *Phthalic Acid and Derivatives*, **7**

Löser, Eckhard, Bayer AG, Wuppertal, Federal Republic of Germany, *Butadiene*, **2**; *Chlorinated Hydrocarbons*, **3**; *Isoprene*, **5**

Lück, Erich, Hoechst Aktiengesellschaft, Frankfurt/Main, Federal Republic of Germany, *Sorbic Acid*, **7**

Lyday, Phyllis A., US Bureau of Mines, Washington, DC 20241, United States, *Iodine Compounds*, **5**

Maki, Takao, Dia Research Martech Inc., Yokohama, Japan, *Benzoic Acid and Derivatives*, **2**

Marr, Beinta U., The Copenhagen Pectin Factory Ltd., Lille Skensved, Denmark, *Polysaccharides*, **7**

Marten, Klaus, Sichel-Werke GmbH, Hannover, Federal Republic of Germany, *Acrylic Acid and Derivatives*, **1**

Mattioda, Georges, Société Française Hoechst, C.R.A., Stains, France, *Glyoxylic Acid*, **5**; *Glyoxal*, **5**

Mayer, Dieter, Pharma Forschung Toxikologie, Hoechst Marion Roussel Aktiengesellschaft, Frankfurt, Federal Republic of Germany, *Acetic Anhydride and Mixed Fatty Acid Anhydrides*, **1**; *Acetylene*, **1**; *Alcohols, Aliphatic*, **1**; *Benzidine and Benzidine Derivatives*, **2**; *Crotonaldehyde and Crotonic Acid*, **3**; *Dialkyl Sulfates and Alkylsulfuric Acids*, **3**; *Ethanol*, **4**; *Ethers, Aliphatic*, **4**; *Ethylene Glycol*, **4**; *Ethylene Oxide*, **4**; *2-Ethylhexanol*, **4**; *Hydrocarbons*, **5**

Mayr, Wilfried, Degussa AG, Hanau, Federal Republic of Germany, *Peroxy Compounds, Organic*, **6**
McKillip, William J., QO Chemicals Inc., West Lafayette, Indiana 47 906, United States, *Furan and Derivatives*, **5**
McKusick, Blaine C., E. I. Du Pont de Nemours & Co., Wilmington, Delaware, United States, *Aziridines*, **2**; *Chloroacetaldehydes*, **3**; *Fluorine Compounds, Organic*, **5**
Meessen, Jozef H., DSM Stamicarbon, Geleen, The Netherlands, *Urea*, **8**
Meinass, Helmut, Linde AG, Höllriegelskreuth, Federal Republic of Germany, *Acetylene*, **1**
Mercker, Hans Jochen, BASF Aktiengesellschaft, Ludwigshafen, Federal Republic of Germany, *Amines, Aliphatic*, **1**; *Butyrolactone*, **2**
Mertschenk, Bernd, SKW Trostberg AG, Trostberg, Federal Republic of Germany, *Thiourea and Thiourea Derivatives*, **8**
Metzger, Adolf, Hoechst Aktiengesellschaft, Frankfurt/Main, Federal Republic of Germany, *Sulfamic Acid*, **7**
Meyer, Michael, EMS-Dottikon, Dottikon, Switzerland, *Xylidines*, **8**
Michalczyk, Georg, DEA Mineralöl AG, Hamburg, Federal Republic of Germany, *Waxes*, **8**
Miki, Hisaya, Mitsui Petrochemical Industries, Ltd., Yamaguchi, Japan, *Phenol Derivatives*, **6**
Miksche, Leopold, Bayer AG, Leverkusen, Federal Republic of Germany, *Allyl Compounds*, **1**
Miller, Raimund, Lonza Inc., Fair Lawn, New Jersey 07410, United States, *Ketenes*, **5**
Miltenberger, Karlheinz, Hoechst Aktiengesellschaft, Gersthofen, Federal Republic of Germany, *Hydroxycarboxylic Acids, Aliphatic*, **5**
Mitchell, Stephen C., University of Birmingham, Birmingham, United Kingdom, *Aminophenols*, **1**
Moran, William C., BASF Corporation, Freeport, Texas, United States, *Caprolactam*, **2**
Morawietz, Marcus, Degussa AG, Hanau-Wolfgang, Federal Republic of Germany, *Alcohols, Polyhydric*, **1**
Morishita, Sinji, Koei Chemical Co. Ltd., Osaka 541, Japan, *Pyridine and Pyridine Derivatives*, **7**
Muller, François, Rhône-Poulenc, Le Pont-de-Claix, France, *Chlorophenols*, **3**
Müller, Hans Joachim, Erdölchemie GmbH, Köln, Federal Republic of Germany, *Butadiene*, **2**
Müller, Herbert, BASF Aktiengesellschaft, Ludwigshafen, Federal Republic of Germany, *Tetrahydrofuran*, **8**
Müller, Richard, Chemische Werke Hüls AG, Marl, Federal Republic of Germany, *Acetylene*, **1**
Musin, Willy, UCB, Brussels, Belgium, *Methylamines*, **6**
Musser, Michael T., E. I. Du Pont de Nemours & Co., Orange, Texas 77631, United States, *Cyclohexanol and Cyclohexanone*, **3**

Narkofsky, Sheldon B., W. G. Grace & Co., Columbia, Maryland 21044, United States, *Nitro Compounds, Aliphatic*, **6**
Narshall, W. J., E. I. du Pont de Nemours & Co., Deepwater, New Jersey 08023, United States, *Aniline*, **1**
Neier, Wilhelm, Deutsche Texaco AG, Moers, Federal Republic of Germany, *2-Butanone*, **2**
Neumeister, Joachim, Hüls Aktiengesellschaft, Marl, Federal Republic of Germany, *Butenes*, **2**; *Methyl Tert-Butyl Ether*, **6**
Nierlich, Franz, Hüls Aktiengesellschaft, Marl, Federal Republic of Germany, *Methyl Tert-Butyl Ether*, **6**
Noweck, Klaus, Condea Chemie GmbH, Brunsbüttel, Federal Republic of Germany, *Fatty Alcohols*, **5**
Nyns, Edmond-Jacques, Unité de Génie Biologique, Université Catholique de Louvain, Louvain-la-Neuve, Belgium, *Methane*, **6**

Obenaus, Fritz, Hüls AG, Marl, Federal Republic of Germany, *Butenes*, **2**
Ohara, Takashi, Nippon Shokubai Kagaku Kogyo Co., Ltd., Osaka, Japan, *Acrolein and Methacrolein*, **1**; *Acrylic Acid and Derivatives*, **1**
Olson, Michael, General Motors Research Laboratories, Warren, Michigan 48090, United States, *Cyclopentadiene and Cyclopentene*, **3**
Opgenorth, Hans-Joachim, BASF Aktiengesellschaft, Ludwigshafen, Federal Republic of Germany, *Nitrilotriacetic Acid*, **6**

Paetz, Christian, Bayer AG, Leverkusen, Federal Republic of Germany, *Hydrocarbons*, **5**
Papa, Anthony J., Union Carbide Chemicals and Plastics Company, Inc., South Charleston, WV 25303, United States, *Propanal*, **7**; *Propanols*, **7**
Pässler, Peter, BASF Aktiengesellschaft, Ludwigshafen, Federal Republic of Germany, *Acetylene*, **1**
Paulus, Wilfried, Bayer AG, Uerdingen, Federal Republic of Germany, *Phenol Derivatives*, **6**
Payer, Wolfgang, Ruhrchemie AG, Oberhausen, Federal Republic of Germany, *Waxes*, **8**
Petersen, Harro, BASF Aktiengesellschaft, Ludwigshafen, Federal Republic of Germany, *Urea*, **8**
Pfleiderer, Gerhard, Hoechst Aktiengesellschaft, Frankfurt/Main, Federal Republic of Germany, *Chlorinated Hydrocarbons*, **3**
Pilnik, Walter, Wageningen, Agricultural University, Department of Food Science, Wageningen, The Netherlands, *Polysaccharides*, **7**

Plachenka, Jürgen, Henkel KGaA, Düsseldorf, Federal Republic of Germany, *Fatty Acids*, **4**
Pluim, Hendrik, Solvay Pharmaceuticals, Weesp, The Netherlands, *Lactose and Derivatives*, **5**
Pollak, Peter, Lonza AG, Basel, Switzerland, *Malonic Acid and Derivatives*, **5**; *Nitriles*, **6**
Pratt, John Wesley, Synthetic Chemicals Limited, Wolverhampton, West Midlands, United Kingdom, *Thiophene*, **8**
Prescher, Günter, Degussa AG, Zweigniederlassung Wolfgang, Hanau, Federal Republic of Germany, *Acrolein and Methacrolein*, **1**; *Acrylic Acid and Derivatives*, **1**

Raczek, Nico, Hoechst Aktiengesellschaft, Frankfurt/Main, Federal Republic of Germany, *Sorbic Acid*, **7**
Rademacher, Hans, Hüls AG, Marl, Federal Republic of Germany, *Cyclododecanol, Cyclododecanone, and Laurolactam*, **3**
Ram, Sanjeev, Badger Engineers, Cambridge, Massachusetts 02142, United States, *Ethylbenzene*, **4**
Rapp, Knut, Südzucker AG, Grünstadt-Obrigheim, Federal Republic of Germany, *Carbohydrates*, **2**
Rassaerts, Heinz, Chemische Werke Hüls AG, Marl, Federal Republic of Germany, *Chlorinated Hydrocarbons*, **3**
Rebsdat, Siegfried, Hoechst Aktiengesellschaft, Gendorf, Federal Republic of Germany, *Ethylene Glycol*, **4**; *Ethylene Oxide*, **4**
Reck, Richard A., Akzo Chemie America, Chicago, Ill, United States, *Amines, Aliphatic*, **1**
Reiß, Wolfgang, BASF Aktiengesellschaft, Ludwigshafen, Federal Republic of Germany, *Butanediols, Butenediol, and Butynediol*, **2**
Reitmeier, Rudolf, Wacker Chemie GmbH, Burghausen, Federal Republic of Germany, *Silicon Compounds, Organic*, **7**
Rengstl, Alfred, Wacker-Chemie GmbH, Burghausen, Federal Republic of Germany, *Acetic Anhydride and Mixed Fatty Acid Anhydrides*, **1**
Rentzea, Costin N., BASF Aktiengesellschaft, Ludwigshafen/Rh., Federal Republic of Germany, *Carbamates and Carbamoyl Chlorides*, **2**
Reuss, Günther, BASF Aktiengesellschaft, Ludwigshafen, Federal Republic of Germany, *Formaldehyde*, **5**
Reutemann, Werner, BASF Aktiengesellschaft, Ludwigshafen, Federal Republic of Germany, *Formic Acid*, **5**
Richey, W. Frank, Dow Chemical, Midland, Michigan 48640, United States, *Chlorohydrins*, **3**
Ridder, Heinz, Condea Chemie GmbH, Brunsbüttel, Federal Republic of Germany, *Fatty Alcohols*, **5**
Riemenschneider, Wilhelm, Hoechst Aktiengesellschaft, Frankfurt, Federal Republic of Germany, *Carboxylic Acids, Aliphatic*, **2**; *Esters, Organic*, **4**; *Oxalic Acid*, **6**
Rieth, Robert, Peroxid-Chemie GmbH, Höllriegelskreuth, Federal Republic of Germany, *Epoxides*, **4**
Rippel, Robert, Hoechst AG, Frankfurt, Federal Republic of Germany, *Mercaptoacetic Acid and Derivatives*, **6**
Ripperger, Willi, BASF Aktiengesellschaft, Ludwigshafen, Federal Republic of Germany, *Melamine and Guanamines*, **5**
Ritz, Josef, BASF Aktiengesellschaft, Ludwigshafen, Federal Republic of Germany, *Caprolactam*, **2**
Ritzer, Edwin, Bayer AG, Leverkusen, Federal Republic of Germany, *Hydroxycarbolic Acids, Aromatic*, **5**
Roark, David N., (currently with Ethyl Corp.), Cordova Chemical Co., North Muskegon, Michigan 49445, United States, *Aziridines*, **2**
Röderer, Gerhard, Wacker-Chemie GmbH, Burghausen, Federal Republic of Germany, *Chloroacetaldehydes*, **3**
Roelfsema, Willem A., ccFriesland Cooperative Company, Leeuwarden, The Netherlands, *Lactose and Derivatives*, **5**
Röhrscheid, Freimund, Hoechst Aktiengesellschaft, Frankfurt/Main, Federal Republic of Germany, *Carboxylic Acids, Aromatic*, **2**
Rolin, Claus, The Copenhagen Pectin Factory Ltd., Lille Skensved, Denmark, *Polysaccharides*, **7**
Romanowski, Frank, Degussa AG, Frankfurt, Federal Republic of Germany, *Thiocyanates and Isothiocyanates, Organic*, **8**
Romeder, Gérard, Lonza AG, Basel, Switzerland, *Nitriles*, **6**
Röper, Harald, Cerestar, Eridania Béghin-Say, Vilvoorde, Belgium, *Starch*, **7**
Rösch, Lutz, Wacker Chemie GmbH, Burghausen, Federal Republic of Germany, *Silicon Compounds, Organic*, **7**
Roscher, Günter, Hoechst Aktiengesellschaft, Frankfurt/Main, Federal Republic of Germany, *Vinyl Esters*, **8**
Rossberg, Manfred, Hoechst Aktiengesellschaft, Frankfurt/Main, Federal Republic of Germany, *Chlorinated Hydrocarbons*, **3**
Rowbottom, Kenneth T., Laporte Industries Ltd., Widnes, Cheshire WA8 OJU, United Kingdom, *Epoxides*, **4**
Roy, Kathrin-Maria, Langenfeld, Federal Republic of Germany, *Sulfones and Sulfoxides*, **7**; *Thiols and Organic Sulfides*, **8**; *Xanthates*, **8**
Rupprich, Norbert, Bundesanstalt für Arbeitsschutz, Dortmund, Federal Republic of Germany, *Butanols*, **2**; *Chloroacetic Acids*, **3**

Sahm, Hermann, Institut für Biotechnologie der KFA Jülich GmbH, Jülich, Federal Republic of Germany, *Ethanol*, **4**
Said, Adel, Lonza AG, Basel, Switzerland, *Ketenes*, **5**
Sakata, Gozyo, Central Research Institute, Nissan Chemical Ind., Ltd., Chiba, Japan, *Chloroamines*, **3**

Samel, Ulf-Rainer, BASF Aktiengesellschaft, Ludwigshafen, Federal Republic of Germany, *Propionic Acid and Derivatives*, **7**
Sato, Takahisa, Nippon Shokubai Kagaku Kogyo Co., Ltd., Osaka, Japan, *Acrolein and Methacrolein*, **1**; *Acrylic Acid and Derivatives*, **1**
Sauer, Jörg, Degussa AG, Zweigniederlassung Wolfgang, Hanau, Federal Republic of Germany, *Acrolein and Methacrolein*, **1**
Schinzig, Elisabeth, Biochemie Ladenburg GmbH, Ladenburg, Federal Republic of Germany, *Gluconic Acid*, **5**
Schmidt, Hans, Hoechst Aktiengesellschaft, Werk Gersthofen, Augsburg, Federal Republic of Germany, *Waxes*, **8**
Schmiedel, Klaus W., Hoechst Aktiengesellschaft, Frankfurt am Main, Federal Republic of Germany, *Resorcinol*, **7**
Schneider, Kurt, BASF Aktiengesellschaft, Ludwigshafen, Federal Republic of Germany, *Butanediols, Butenediol, and Butynediol*, **2**
Scholz, Bernhard, Hüls Aktiengesellschaft, Marl, Federal Republic of Germany, *Methyl Tert-Butyl Ether*, **6**
Schubart, Rüdiger, Bayer AG, Leverkusen, Federal Republic of Germany, *Dithiocarbamic Acid and Derivatives*, **4**; *Sulfinic Acids and Derivatives*, **7**
Schulte, Markus, Celanese GmbH, Werk Ruhrchemie, Oberhausen, Federal Republic of Germany, *Aldehydes, Aliphatic and Araliphatic*, **1**
Schwarz, Michael W., E. Merck OHG, Darmstadt, Federal Republic of Germany, *Saponins*, **7**
Schwenecke, Hans, Hoechst Aktiengesellschaft, Frankfurt/Main, Federal Republic of Germany, *Benzidine and Benzidine Derivatives*, **2**
Schwertfeger, Werner, Hoechst Aktiengesellschaft, Frankfurt, Federal Republic of Germany, *Fluorine Compounds, Organic*, **5**
Schwind, Helmut, Degussa AG, Zweigniederlassung Wolfgang, Hanau, Federal Republic of Germany, *Acrolein and Methacrolein*, **1**; *Acrylic Acid and Derivatives*, **1**
Seeholzer, Josef, SKW Trostberg, Trostberg, Federal Republic of Germany, *Melamine and Guanamines*, **5**
Sheehan, Richard J., Amoco Research Center, Amoco Chemical Company, Naperville, Illinois, United States, *Terephthalic Acid, Dimethyl Terephthalate, and Isophthalic Acid*, **8**
Shimizu, Noboru, Nippon Shokubai Kagaku Kogyo Co., Ltd., Osaka, Japan, *Acrolein and Methacrolein*, **1**; *Acrylic Acid and Derivatives*, **1**
Shimizu, Shinkichi, Koei Chemical Co. Ltd., Osaka 541, Japan, *Pyridine and Pyridine Derivatives*, **7**
Shoji, Takayuki, Koei Chemical Co. Ltd., Osaka 541, Japan, *Pyridine and Pyridine Derivatives*, **7**
Siegel, Hardo, BASF Aktiengesellschaft, Ludwigshafen, Federal Republic of Germany, *Ketones*, **5**
Siegemund, Günter, Hoechst Aktiengesellschaft, Frankfurt, Federal Republic of Germany, *Fluorine Compounds, Organic*, **5**
Siegmeier, Rainer, Degussa AG, Hanau, Federal Republic of Germany, *Peroxy Compounds, Organic*, **6**
Sienel, Guenter, Peroxid-Chemie GmbH, Höllriegelskreuth, Federal Republic of Germany, *Epoxides*, **4**
Sifniades, Stylianos, Allied Signal Inc., Morristown, New Jersey 07962, United States, *Acetone*, **1**
Simo, Thomas A., Lurgi Öl-Gas-Chemie GmbH, Frankfurt/Main, Federal Republic of Germany, *Toluene*, **8**, *Xylenes*, **8**
Singh, Jasbir, Badger Engineers, Cambridge, Massachusetts 02142, United States, *Ethylbenzene*, **4**
Smart, Bruce, E. I. Du Pont de Nemours & Co., Wilmington, Delaware, United States, *Fluorine Compounds, Organic*, **5**
Smiley, Robert A., Wilmington, Delaware 19880-0336, United States, *Hexamethylenediamine*, **5**; *Phenylene- and Toluenediamines*, **6**
Steffan, Guido, Bayer AG, Leverkusen, Federal Republic of Germany, *Butanediols, Butenediol, and Butynediol*, **2**
Steinberner, Udo, Henkel KGaA, Düsseldorf, Federal Republic of Germany, *Fatty Acids*, **4**; *Glycerol*, **5**
Strack, Heinz, Rheinfelden, Federal Republic of Germany, *Chlorinated Hydrocarbons*, **3**
Strehlke, Guenter, Deutsche Texaco AG, Moers, Federal Republic of Germany, *2-Butanone*, **2**; *Ethers, Aliphatic*, **4**
Sullivan, Carl J., ARCO Chemical Company, Newtown Square, Pennsylvania 19073, United States, *Propanediols*, **7**
Sundermann, Rudolf, Bayer AG, Leverkusen, Federal Republic of Germany, *Hydroxycarboxylic Acids, Aromatic*, **5**
Suzuki, Yoshihisa, Central Research Laboratories, Ajinomoto Co., Inc., Kawasaki, Japan, *Choline*, **3**
Svara, Jürgen, Hoechst Aktiengesellschaft, Werk Knapsack, Hürth-Knapsack, Federal Republic of Germany, *Phosphorus Compounds, Organic*, **6**
Szczepanek, Alfred, Faistenlohestr. 5, München, Federal Republic of Germany, *Metallic Soaps*, **6**

Takeda, Kazuo, Mitsubishi Chemical Safety Institute Ltd., Yokohama, Japan, *Benzoic Acid and Derivatives*, **2**
Tanifuji, Minoru, Ube Industries, Tokyo, Japan, *Oxalic Acid*, **6**
Tanner, Herbert, Degussa AG, Hanau-Wolfgang, Federal Republic of Germany, *Amino Acids*, **1**
Tanno, Hiroyuki, Central Research Laboratories, Ajinomoto Co. Inc., Kawasaki, Japan, *Lecithin*, **5**
Thomas, Alfred, Unimills International, Hamburg, Federal Republic of Germany, *Fats and Fatty Oils*, **4**

Tögel, Adolf, Hoechst Aktiengesellschaft, Frankfurt/Main, Federal Republic of Germany, *Chlorinated Hydrocarbons*, **3**
Torkelson, Theodore R., Dow Chemical, Midland, Michigan 48674, United States, *Chlorinated Hydrocarbons*, **3**
Towae, Friedrich K., BASF Aktiengesellschaft, Ludwigshafen, Federal Republic of Germany, *Phthalic Acid and Derivatives*, **7**

Ullrich, Jochen, Phenolchemie GmbH, Gladbeck, Federal Republic of Germany, *Phenol*, **6**
Ulrich, Henri, Dow Chemical Company, North Haven, Connecticut 06473, United States, *Isocyanates, Organic*, **5**
Umemura, Sumio, Ube Industries, Ltd., Ube, Japan, *Phenol Derivatives*, **6**
Ura, Yasukazu, Central Research Institute, Nissan Chemical Ind., Ltd., Chiba, Japan, *Chloroamines*, **3**

van Barneveld, Heinrich, Phenolchemie GmbH, Gladbeck, Federal Republic of Germany, *Phenol*, **6**
van Gysel, August B., UCB, Brussels, Belgium, *Methylamines*, **6**
Verhage, Marinus, Solvay Pharmaceuticals, Weesp, The Netherlands, *Lactose and Derivatives*, **5**
Verhoff, Frank H., Miles Laboratories, Elkhart, Indiana 46515, United States, *Citric Acid*, **3**
Vogel, Axel, Bayer AG, Leverkusen, Federal Republic of Germany, *Anthraquinone*, **2**
Vogel, Herward, Minnesota Mining and Manufacturing Company, St. Paul, Minnesota, United States, *Fluorine Compounds, Organic*, **5**
Voges, Heinz-Werner, Hüls Aktiengesellschaft, Marl, Federal Republic of Germany, *Hydrocarbons*, **5**; *Phenol Derivatives*, **6**
Vogt, Peter F., Mobay Chemical Corp., Pittsburgh, Pennsylvania 15205, United States, *Amines, Aromatic*, **1**

Wagner, Rudolf, Bayer AG, Leverkusen, Federal Republic of Germany, *Naphthoquinones*, **6**
Walzl, Roland, Linde AG, Höllriegelskreuth, Federal Republic of Germany, *Ethylene*, **4**
Wandel, Martin, Bayer AG, Leverkusen, Federal Republic of Germany, *Cellulose Esters*, **2**
Waring, Rosemary H., University of Birmingham, Birmingham, United Kingdom, *Aminophenols*, **1**
Watanabe, Nanao, Koei Chemical Co. Ltd., Osaka 541, Japan, *Pyridine and Pyridine Derivatives*, **7**
Weber, Edwin, Institut für Organische Chemie und Biochemie, Universität Bonn, Federal Republic of Germany, *Crown Ethers*, **3**
Weber, Jürgen, Ruhrchemie AG, Oberhausen, Federal Republic of Germany, *Aldehydes, Aliphatic and Araliphatic*, **1**
Weber, Theodor, BASF Aktiengesellschaft, Ludwigshafen/Rhein, Federal Republic of Germany, *Ethanolamines and Propanolamines*, **4**
Weferling, Norbert, Hoechst Aktiengesellschaft, Werk Knapsack, Hürth-Knapsack, Federal Republic of Germany, *Phosphorus Compounds, Organic*, **6**
Weiberg, Otto, Degussa AG, Zweigniederlassung Wolfgang, Hanau, Federal Republic of Germany, *Acrolein and Methacrolein*, **1**; *Acrylic Acid and Derivatives*, **1**
Weisenberger, Karl, Hoechst AG, Frankfurt, Federal Republic of Germany, *Dialkyl Sulfates and Alkylsulfuric Acids*, **3**
Weiss, Günther, BASF Aktiengesellschaft, Ludwigshafen, Federal Republic of Germany, *Methanol*, **6**
Weiss, Stefan, SKW Trostberg AG, Trostberg, Federal Republic of Germany, *Guanidine and Derivatives*, **5**
Weitz, Hans-Martin, BASF Aktiengesellschaft, Ludwigshafen, Federal Republic of Germany, *Butanediols, Butenediol, and Butynediol*, **2**; *Isoprene*, **5**
Welch, Vincent A., Badger Engineers, Cambridge, Massachusetts 02142, United States, *Ethylbenzene*, **4**
Werle, Peter, Degussa AG, Hanau-Wolfgang, Federal Republic of Germany, *Alcohols, Polyhydric*, **1**
Wernicke, Hans-Jürgen, Linde AG, Höllriegelskreuth, Federal Republic of Germany, *Acetylene*, **1**
Whistler, Roy L., Whistler Center for Carbohydrate Research, Purdue University, West Lafayette, Indiana 47907, United States, *Starch*, **7**
Wiechern, Uwe, Dow Deutschland Inc., Stade, Federal Republic of Germany, *Propylene Oxide*, **7**
Wildgruber, Josef, Hoechst Aktiengesellschaft, Werk Gersthofen, Augsburg, Federal Republic of Germany, *Waxes*, **8**
Witte, Claus, BASF Aktiengesellschaft, Ludwigshafen, Federal Republic of Germany, *Methanol*, **6**
Wolfmeier, Uwe, Hoechst Aktiengesellschaft, Werk Gersthofen, Augsburg, Federal Republic of Germany, *Waxes*, **8**
Wright, Elaine, General Motors Research Laboratories, Warren, Michigan 48090, United States, *Benzaldehyde*, **2**; *Benzyl Alcohol*, **2**

Youngman, Richard, SKW Trostberg AG, Trostberg, Federal Republic of Germany, *Guanidine and Derivatives*, **5**

Zey, Edward G., Celanese Chemical Company Technical Center, Corpus Christi, Texas 78469, United States, *Acetic Acid*, **1**

CAS Registry Number Index

[50-0-0]   **1**: 352; **4**: 2309, 2311, 2315; **5**: 2641, 2812, 3037; **7**: 4263
[50-1-1]   **5**: 2825
[50-14-6]   **4**: 2382
[50-21-5]   **5**: 3120
[50-29-3]   **3**: 1531
[50-32-8]   **5**: 2927
[50-44-2]   **7**: 4166
[50-65-7]   **6**: 3547
[50-70-4]   **2**: 1074
[50-71-5]   **7**: 4231
[50-78-2]   **4**: 2042; **7**: 4278
[50-79-3]   **2**: 832, 838
[50-84-0]   **2**: 832, 838
[50-85-1]   **5**: 2980
[50-99-7]   **2**: 1062; **6**: 3607; **7**: 4397
[51-17-2]   **6**: 3810
[51-18-3]   **2**: 669
[51-19-4]   **1**: 603
[51-21-8]   **5**: 2620
[51-28-5]   **6**: 3563
[51-35-4]   **1**: 544
[51-56-9]   **5**: 2983
[51-58-1]   **6**: 3542
[51-66-1]   **1**: 506, 510
[51-78-5]   **1**: 603
[51-79-6]   **2**: 1047
[51-81-0]   **1**: 603
[51-83-2]   **3**: 1635
[51-84-3]   **3**: 1633, 1635
[52-51-7]   **6**: 3495
[52-67-5]   **1**: 583; **8**: 4650
[52-86-8]   **5**: 2616
[52-89-1]   **1**: 579
[52-90-4]   **1**: 543, 579
[53-19-0]   **3**: 1529
[54-11-5]   **7**: 4219
[54-12-6]   **1**: 545
[54-21-7]   **7**: 4277
[54-85-3]   **7**: 4218
[55-21-0]   **2**: 828
[55-22-1]   **7**: 4202, 4218
[55-55-0]   **1**: 618; **5**: 2944
[55-56-1]   **5**: 2840
[55-63-0]   **5**: 2804
[55-91-4]   **5**: 2622
[56-3-1]   **5**: 2826, 2839
[56-18-8]   **1**: 488
[56-23-5]   **3**: 1252, 1376, 1472; **5**: 2622, 2702, 2978
[56-38-2]   **6**: 3562, 3852
[56-40-6]   **1**: 543, 580; **3**: 1542
[56-41-7]   **1**: 543, 578; **6**: 3600
[56-45-1]   **1**: 545, 581

[56-73-5]   **2**: 1070
[56-81-5]   **5**: 2788
[56-82-6]   **1**: 206
[56-84-8]   **1**: 543, 578
[56-85-9]   **1**: 543, 580
[56-86-0]   **1**: 543, 579; **6**: 3608
[56-87-1]   **1**: 544, 580
[56-89-3]   **1**: 543, 579; **8**: 4661
[56-95-1]   **5**: 2841
[57-0-1]   **5**: 2835
[57-6-7]   **8**: 4636
[57-10-3]   **3**: 1909; **4**: 2485
[57-11-4]   **3**: 1909; **4**: 2485
[57-13-6]   **5**: 2818, 2826, 3198; **8**: 4784
[57-48-7]   **2**: 1062; **6**: 3607
[57-50-1]   **2**: 1067; **4**: 2091
[57-53-4]   **1**: 330; **3**: 1580
[57-55-6]   **6**: 3601; **7**: 4047
[57-57-8]   **4**: 2033; **5**: 3067
[57-67-0]   **3**: 1750; **5**: 2830
[57-68-1]   **5**: 2830, 3067
[57-88-5]   **4**: 2382
[58-8-2]   **5**: 2693, 2697; **7**: 4166
[58-14-0]   **5**: 2830
[58-27-5]   **2**: 844; **6**: 3355, 3450
[58-55-9]   **5**: 2693, 2697; **7**: 4166
[58-73-1]   **7**: 4171
[58-74-2]   **6**: 3762
[58-90-2]   **3**: 1606
[59-2-9]   **4**: 2382
[59-5-2]   **5**: 2830
[59-30-3]   **2**: 834; **5**: 2830
[59-31-4]   **7**: 4254
[59-43-8]   **7**: 4231
[59-46-1]   **4**: 2162
[59-50-7]   **6**: 3785
[59-51-8]   **1**: 213, 544, 581
[59-52-9]   **8**: 4650
[59-56-3]   **2**: 1070
[59-67-6]   **7**: 4202, 4218, 4254
[59-87-0]   **6**: 3571
[59-92-7]   **1**: 577; **6**: 3762
[60-0-4]   **4**: 2295, 2297; **5**: 2667
[60-9-3]   **3**: 1889; **6**: 3812
[60-18-4]   **1**: 545, 581
[60-23-1]   **2**: 666
[60-24-2]   **8**: 4654
[60-29-7]   **4**: 2053, 2187, 2189
[60-33-3]   **4**: 2460, 2487
[60-35-5]   **1**: 56
[61-54-1]   **5**: 3003
[61-57-4]   **6**: 3573
[61-82-5]   **5**: 2833
[61-90-5]   **1**: 544, 580

[62-23-7]	**2**: 835, 838; **6**: 3519	[69-81-8]	**6**: 3762
[62-44-2]	**1**: 620; **6**: 3563	[69-89-6]	**7**: 4166
[62-49-7]	**3**: 1633	[69-93-2]	**7**: 4166
[62-51-1]	**3**: 1635	[70-26-8]	**1**: 544
[62-52-3]	**5**: 2945	[70-30-4]	**6**: 3790
[62-53-3]	**1**: 527, 529, 627; **5**: 2946; **6**: 3812	[70-34-8]	**5**: 2616; **6**: 3536
[62-54-4]	**1**: 89	[70-47-3]	**1**: 543, 578
[62-56-6]	**3**: 1748; **8**: 4708	[70-53-1]	**1**: 544, 580
[62-74-8]	**5**: 2622	[70-54-2]	**1**: 580
[62-76-0]	**6**: 3594	[70-55-3]	**2**: 763
[63-25-2]	**2**: 1047; **3**: 1580; **6**: 3374	[71-0-1]	**1**: 544, 580
[63-42-3]	**2**: 1067; **5**: 3135	[71-23-8]	**1**: 281, 421; **2**: 1008; **4**: 2116; **7**: 4038, 4061
[63-56-9]	**7**: 4232		
[63-68-3]	**1**: 544, 581	[71-36-3]	**1**: 281; **2**: 1008; **7**: 4388
[63-74-1]	**2**: 772	[71-36-8]	**4**: 2116
[63-91-2]	**1**: 544, 581	[71-41-0]	**1**: 281
[64-2-8]	**4**: 2297	[71-43-2]	**1**: 93; **2**: 693, 695; **3**: 1813; **5**: 2899; **6**: 3694, 3704; **7**: 4263, 4434
[64-10-8]	**8**: 4819		
[64-17-5]	**1**: 281, 53; **4**: 2048; **7**: 4105, 4391	[71-44-3]	**1**: 491
[64-18-6]	**4**: 2309; **5**: 2711; **7**: 4102, 4108	[71-55-6]	**3**: 1309, 1473
[64-19-7]	**1**: 30; **3**: 1649; **4**: 1932, 2057, 2059; **5**: 2718, 3065; **6**: 3704; **7**: 4102, 4105; **8**: 4576	[72-17-3]	**5**: 3130
		[72-18-4]	**1**: 545, 581
		[72-19-5]	**1**: 545, 581
[64-55-1]	**1**: 330	[72-54-8]	**3**: 1529
[64-67-5]	**3**: 1854	[72-56-0]	**3**: 1529
[64-69-7]	**3**: 1542	[73-22-3]	**1**: 545, 581
[65-45-2]	**7**: 4278	[73-24-5]	**7**: 4166
[65-49-6]	**1**: 618; **6**: 3562	[73-32-5]	**1**: 544, 580
[65-71-4]	**7**: 4231	[73-40-5]	**5**: 2823; **7**: 4166
[65-85-0]	**2**: 819, 837; **5**: 2975; **6**: 3698; **7**: 4184	[73-48-3]	**5**: 2612
[65-86-1]	**7**: 4230	[74-11-3]	**2**: 832, 838
[66-22-8]	**7**: 4230	[74-31-7]	**5**: 2944
[66-25-1]	**1**: 352	[74-79-3]	**1**: 543, 578
[67-20-9]	**6**: 3571	[74-82-8]	**1**: 92; **5**: 2855; **6**: 3269
[67-43-6]	**4**: 2297	[74-83-9]	**2**: 869, 878, 888
[67-45-8]	**3**: 1580; **6**: 3571	[74-84-0]	**4**: 2224; **5**: 2855
[67-47-0]	**2**: 1080; **6**: 3607	[74-85-1]	**4**: 2052, 2221, 2311, 2314, 2329; **5**: 2699, 2871; **7**: 4105, 4187; **8**: 4962
[67-48-1]	**3**: 1634; **4**: 2161		
[67-56-1]	**1**: 281, 34; **2**: 958; **4**: 2052, 2058, 2315; **5**: 2647, 3037; **6**: 3288; **7**: 4067, 4105; **8**: 4573	[74-86-2]	**1**: 118, 92; **5**: 2699; **7**: 4368
		[74-87-3]	**2**: 1215; **3**: 1252, 1470
		[74-88-4]	**1**: 36; **4**: 2059; **5**: 2719, 3008
[67-63-0]	**1**: 281, 92, 97; **7**: 4061	[74-89-5]	**2**: 1007; **6**: 3325
[67-64-1]	**1**: 89; **5**: 2900; **6**: 3694; **7**: 4183, 4266	[74-90-8]	**1**: 253; **5**: 2697
[67-66-3]	**1**: 91; **3**: 1252, 1471	[74-95-3]	**2**: 869, 880
[67-68-5]	**2**: 722	[74-96-4]	**2**: 869, 888
[67-72-1]	**3**: 1328, 1474	[74-97-5]	**2**: 869, 880, 888
[67-97-0]	**4**: 2382	[74-98-6]	**4**: 2224; **5**: 2699, 2855
[68-11-1]	**3**: 1542, 1546; **6**: 3221	[74-99-7]	**1**: 185
[68-12-2]	**5**: 2691, 2698, 2737	[75-0-3]	**2**: 1223; **3**: 1290, 1472; **4**: 2223
[68-26-8]	**4**: 2382	[75-1-4]	**1**: 124; **3**: 1330, 1474
[68-35-9]	**5**: 2830	[75-2-5]	**5**: 2584, 2624
[68-94-0]	**7**: 4166	[75-3-6]	**5**: 3008
[69-23-8]	**5**: 2612	[75-4-7]	**1**: 454, 6
[69-53-4]	**3**: 1580	[75-5-8]	**1**: 253; **6**: 3460; **7**: 4187
[69-72-7]	**5**: 2974; **7**: 4271	[75-7-0]	**1**: 124, 2, 352, 43; **4**: 2056; **5**: 2812; **8**: 4576
[69-74-9]	**7**: 4232		
[69-79-4]	**2**: 1062	[75-9-2]	**3**: 1252, 1471; **5**: 2679

[75-10-5]	**5**: 2623	[75-98-9]	**2**: 955; **4**: 2016
[75-11-6]	**5**: 3007	[75-99-0]	**2**: 1116; **7**: 4120
[75-12-7]	**5**: 2691, 2718	[76-1-7]	**3**: 1325, 1474
[75-13-8]	**3**: 1770	[76-2-8]	**3**: 1554
[75-16-1]	**2**: 879	[76-3-9]	**3**: 1552
[75-18-3]	**4**: 1932; **8**: 4658	[76-5-1]	**5**: 2602
[75-19-4]	**5**: 2855	[76-6-2]	**6**: 3494
[75-21-8]	**2**: 1224, 666; **4**: 2223, 2305, 2311, 2329; **7**: 4407	[76-9-5]	**1**: 337
		[76-11-9]	**5**: 2576, 2623
[75-25-2]	**2**: 869, 888	[76-12-0]	**5**: 2576, 2623
[75-26-3]	**2**: 869	[76-13-1]	**5**: 2576, 2589, 2623
[75-28-5]	**2**: 959; **4**: 2224; **5**: 2855	[76-14-2]	**5**: 2576, 2623
[75-29-6]	**3**: 1373, 1477	[76-15-3]	**5**: 2576
[75-31-0]	**1**: 454, 92; **7**: 4065	[76-16-4]	**5**: 2568
[75-34-3]	**3**: 1295, 1472	[76-38-0]	**5**: 2589, 2596, 2625
[75-35-4]	**3**: 1348, 1475	[76-39-1]	**6**: 3497
[75-36-5]	**1**: 55	[77-25-8]	**5**: 3184
[75-37-6]	**5**: 2587, 2623	[77-47-4]	**5**: 2923
[75-38-7]	**5**: 2584, 2624	[77-49-6]	**6**: 3496
[75-43-4]	**5**: 2576, 2624	[77-54-3]	**4**: 2041
[75-43-40]	**5**: 2618	[77-58-7]	**7**: 4348
[75-44-5]	**7**: 4118	[77-61-2]	**6**: 3754
[75-45-6]	**5**: 2576, 2623	[77-73-6]	**3**: 1826
[75-47-8]	**5**: 3007	[77-75-8]	**1**: 309; **2**: 975
[75-50-3]	**5**: 2700, 3073; **6**: 3325	[77-78-1]	**3**: 1853, 1855; **4**: 1932
[75-52-5]	**5**: 2697; **6**: 3488	[77-85-0]	**1**: 339; **7**: 4039
[75-54-7]	**7**: 4311	[77-86-1]	**6**: 3494
[75-55-8]	**2**: 663	[77-91-8]	**3**: 1635
[75-56-9]	**1**: 331; **2**: 1228; **3**: 1375, 1592; **7**: 4129, 4407	[77-92-9]	**3**: 1645; **6**: 3608
		[77-99-6]	**1**: 338, 344; **5**: 2667, 2728
[75-61-6]	**2**: 888; **5**: 2581	[78-10-4]	**7**: 4322
[75-62-7]	**2**: 869	[78-19-3]	**1**: 203
[75-63-8]	**2**: 869, 881, 888; **5**: 2581, 2624	[78-24-0]	**1**: 343
[75-64-9]	**1**: 454	[78-26-2]	**1**: 330
[75-65-0]	**1**: 281; **5**: 3038	[78-40-0]	**5**: 3065
[75-66-1]	**6**: 3853	[78-42-2]	**6**: 3850
[75-68-3]	**5**: 2576, 2587	[78-44-4]	**1**: 330
[75-69-4]	**5**: 2576, 2622	[78-50-2]	**1**: 47
[75-71-8]	**5**: 2576, 2623	[78-59-1]	**1**: 105; **3**: 1819; **5**: 3105
[75-72-9]	**5**: 2576, 2624	[78-62-6]	**7**: 4322
[75-75-2]	**8**: 4504	[78-63-7]	**6**: 3641
[75-76-3]	**7**: 4311, 4322	[78-75-1]	**2**: 869, 879
[75-77-4]	**7**: 4311, 4322, 4346	[78-76-2]	**2**: 869
[75-78-5]	**7**: 4311, 4322, 4346	[78-78-4]	**5**: 2855
[75-79-6]	**7**: 4311, 4322	[78-79-5]	**5**: 3033
[75-83-2]	**5**: 2855	[78-81-9]	**1**: 454
[75-84-3]	**1**: 281	[78-82-0]	**6**: 3461
[75-85-4]	**1**: 281	[78-83-1]	**1**: 281; **4**: 2116; **5**: 2738
[75-86-5]	**1**: 109	[78-84-2]	**1**: 336, 352; **2**: 925
[75-87-6]	**3**: 1531; **5**: 2598	[78-85-3]	**1**: 200, 368; **7**: 4039
[75-88-7]	**5**: 2576	[78-87-5]	**1**: 414; **3**: 1374, 1477, 1592
[75-89-8]	**5**: 2590, 2624	[78-89-7]	**3**: 1586; **7**: 4135
[75-90-1]	**5**: 2598	[78-90-0]	**1**: 484
[75-91-2]	**2**: 959; **4**: 1997; **6**: 3634, 3636	[78-92-2]	**1**: 281, 41; **2**: 974
[75-93-4]	**3**: 1854	[78-93-3]	**1**: 41; **2**: 955, 971; **3**: 1718; **5**: 3085; **8**: 4584
[75-94-5]	**7**: 4318		
[75-97-8]	**5**: 3090	[78-94-4]	**5**: 3101

CAS Registry Number Index

[78-96-6]	**4**: 2164	[81-4-9]	**6**: 3369
[78-97-7]	**6**: 3463	[81-5-0]	**6**: 3415
[78-98-8]	**6**: 3601	[81-6-1]	**6**: 3410
[79-0-5]	**3**: 1317, 1473	[81-7-2]	**1**: 492
[79-1-6]	**1**: 129; **3**: 1356, 1475, 1543	[81-10-7]	**6**: 3549
[79-2-7]	**3**: 1529	[81-11-8]	**1**: 505, 514; **6**: 3554
[79-3-8]	**7**: 4118	[81-16-3]	**6**: 3414
[79-4-9]	**3**: 1541, 1548; **5**: 3067	[81-20-9]	**6**: 3522
[79-6-1]	**1**: 225, 243; **2**: 1234; **5**: 2819	[81-33-4]	**2**: 1132
[79-7-2]	**3**: 1542, 1549	[81-64-1]	**3**: 1613; **5**: 2944
[79-8-3]	**2**: 877	[81-83-4]	**2**: 1132
[79-9-4]	**2**: 1008; **7**: 4038, 4064, 4102	[82-5-3]	**2**: 652
[79-10-7]	**1**: 126, 223	[82-12-2]	**5**: 2982
[79-11-8]	**2**: 1230; **3**: 1540	[82-34-8]	**2**: 652; **6**: 3451
[79-14-1]	**3**: 1542; **4**: 2309, 2315; **5**: 2818	[82-35-9]	**2**: 652
[79-17-4]	**3**: 1749; **5**: 2833	[82-45-1]	**1**: 524; **2**: 652
[79-20-9]	**1**: 53; **4**: 2014, 2060; **5**: 3066	[82-47-3]	**6**: 3439
[79-21-0]	**1**: 43, 66; **6**: 3643	[82-48-4]	**2**: 652
[79-22-1]	**3**: 1576	[82-49-5]	**2**: 652
[79-24-3]	**6**: 3488	[82-68-8]	**6**: 3534
[79-27-6]	**2**: 888; **8**: 4576	[82-75-7]	**6**: 3413
[79-28-7]	**2**: 869	[82-86-0]	**5**: 2915
[79-29-8]	**5**: 2855	[83-13-6]	**5**: 3184
[79-33-4]	**6**: 3600	[83-30-7]	**3**: 1866
[79-34-5]	**1**: 129; **3**: 1321, 1474	[83-32-9]	**5**: 2913
[79-35-6]	**5**: 2589	[83-34-1]	**5**: 3002
[79-36-7]	**3**: 1551	[83-38-5]	**2**: 684
[79-38-9]	**5**: 2624	[83-40-9]	**5**: 2979
[79-39-0]	**1**: 104, 110; **6**: 3250	[83-41-0]	**6**: 3522
[79-41-4]	**1**: 110; **6**: 3249	[83-42-1]	**2**: 785; **6**: 3535
[79-43-6]	**3**: 1549	[83-46-5]	**4**: 2382
[79-46-9]	**6**: 3488	[83-55-6]	**6**: 3426
[79-51-6]	**5**: 2597	[83-56-7]	**6**: 3379
[79-52-7]	**5**: 2597	[83-64-7]	**6**: 3433
[79-53-8]	**5**: 2597	[83-65-8]	**6**: 3399
[79-72-1]	**6**: 3554	[83-66-9]	**6**: 3744
[79-94-7]	**2**: 879, 882	[83-67-0]	**5**: 2697; **7**: 4166
[80-4-6]	**1**: 333	[83-72-7]	**6**: 3451
[80-5-7]	**1**: 105, 334; **2**: 975; **5**: 2947	[83-83-0]	**2**: 811
[80-8-0]	**6**: 3530	[84-11-7]	**5**: 2922
[80-10-4]	**7**: 4315	[84-15-1]	**5**: 2910
[80-11-5]	**3**: 1876	[84-48-0]	**2**: 652
[80-15-9]	**1**: 95; **5**: 2900; **6**: 3634, 3637, 3694	[84-49-1]	**2**: 652
[80-17-1]	**2**: 759	[84-58-2]	**2**: 841
[80-18-2]	**2**: 758	[84-61-7]	**3**: 1816
[80-37-5]	**6**: 3550	[84-65-1]	**2**: 649
[80-40-0]	**2**: 763	[84-66-2]	**4**: 2015, 2040; **7**: 3886
[80-43-3]	**1**: 94, 96; **6**: 3640	[84-67-3]	**2**: 806
[80-46-6]	**6**: 3747	[84-68-4]	**2**: 807
[80-48-8]	**2**: 763	[84-69-5]	**7**: 3886
[80-54-6]	**1**: 382	[84-74-2]	**4**: 2040, 2121; **7**: 3886
[80-56-8]	**5**: 2584	[84-86-6]	**1**: 512; **6**: 3411
[80-62-6]	**1**: 103, 110, 429; **4**: 2014, 2039; **6**: 3250; **7**: 4039	[84-87-7]	**6**: 3390, 3450
		[84-89-9]	**1**: 512; **6**: 3412
[80-68-2]	**1**: 545	[84-92-4]	**6**: 3423
[80-70-6]	**5**: 2836	[85-0-7]	**7**: 4192
[80-82-0]	**6**: 3549	[85-1-8]	**5**: 2920

[85-18-7]	**7**: 4171	[88-20-0]	**2**: 762
[85-41-6]	**7**: 3868, 3880	[88-21-1]	**1**: 637; **2**: 769; **6**: 3549
[85-44-9]	**7**: 3866; **8**: 4584	[88-22-2]	**2**: 778
[85-47-2]	**6**: 3367	[88-23-3]	**2**: 783
[85-48-3]	**7**: 4254	[88-26-6]	**6**: 3743
[85-49-4]	**6**: 3372	[88-27-7]	**6**: 3741
[85-52-9]	**2**: 657, 831	[88-39-1]	**2**: 767
[85-68-7]	**7**: 3886	[88-42-6]	**2**: 768
[85-74-5]	**6**: 3417	[88-44-8]	**1**: 512; **2**: 778; **8**: 4752
[85-75-6]	**6**: 3417	[88-45-9]	**6**: 3555
[86-28-2]	**2**: 1059; **3**: 1867	[88-50-6]	**2**: 784; **6**: 3533
[86-48-6]	**6**: 3383	[88-51-7]	**2**: 786; **6**: 3535; **8**: 4754
[86-50-0]	**3**: 1894	[88-53-9]	**2**: 785; **8**: 4753
[86-53-3]	**6**: 3473	[88-60-8]	**6**: 3737
[86-57-7]	**6**: 3523	[88-61-9]	**2**: 765
[86-60-2]	**6**: 3415	[88-63-1]	**2**: 774; **6**: 3552
[86-61-3]	**6**: 3436	[88-69-7]	**6**: 3732
[86-65-7]	**6**: 3420	[88-72-2]	**2**: 835; **6**: 3515
[86-66-8]	**6**: 3371	[88-73-3]	**6**: 3526, 3811
[86-74-8]	**2**: 1057	[88-74-4]	**1**: 637; **6**: 3540, 3811
[86-77-1]	**3**: 1899	[88-75-5]	**6**: 3560
[86-86-2]	**6**: 3355	[88-85-7]	**6**: 3566
[86-87-3]	**6**: 3355	[88-88-0]	**6**: 3531
[86-88-4]	**1**: 529; **6**: 3402	[88-90-4]	**6**: 3555
[87-2-5]	**6**: 3434	[88-91-5]	**6**: 3543, 3558
[87-3-6]	**6**: 3435	[88-99-3]	**7**: 3866
[87-10-5]	**6**: 3793	[89-0-9]	**7**: 4202, 4254
[87-13-8]	**5**: 3184	[89-2-1]	**2**: 774; **6**: 3551
[87-17-2]	**7**: 4278	[89-21-4]	**2**: 809
[87-20-7]	**7**: 4277	[89-32-7]	**5**: 2896
[87-44-5]	**8**: 4611	[89-39-4]	**6**: 3564
[87-52-5]	**5**: 3002	[89-56-5]	**5**: 2980
[87-56-9]	**2**: 938	[89-57-6]	**1**: 620
[87-59-2]	**1**: 636	[89-58-7]	**6**: 3522
[87-60-5]	**1**: 519; **8**: 4750	[89-59-8]	**6**: 3535
[87-62-7]	**1**: 636, 639; **6**: 3522	[89-60-1]	**6**: 3535
[87-63-8]	**6**: 3535; **8**: 4750	[89-61-2]	**6**: 3533
[87-65-0]	**3**: 1606	[89-62-3]	**6**: 3545
[87-66-1]	**5**: 2981; **6**: 3763	[89-63-4]	**6**: 3546
[87-67-2]	**3**: 1634	[89-64-5]	**6**: 3561
[87-68-3]	**3**: 1395, 1478	[89-68-9]	**6**: 3787
[87-69-4]	**6**: 3601	[89-69-0]	**6**: 3534
[87-73-0]	**5**: 2775	[89-72-5]	**6**: 3735
[87-78-5]	**2**: 1074	[89-83-8]	**6**: 3732
[87-85-4]	**5**: 2894	[89-84-9]	**5**: 2980
[87-86-5]	**3**: 1606	[89-86-1]	**5**: 2980
[87-88-7]	**2**: 842	[89-87-2]	**6**: 3522
[87-90-1]	**3**: 1566, 1568	[89-98-5]	**2**: 684
[88-4-0]	**6**: 3786	[90-0-6]	**6**: 3728, 3730
[88-5-1]	**6**: 3525	[90-2-8]	**2**: 686
[88-6-2]	**3**: 1606	[90-4-0]	**1**: 505, 612, 636; **6**: 3561
[88-12-0]	**1**: 125; **7**: 4246	[90-5-1]	**5**: 2819; **6**: 3761, 3780
[88-14-2]	**5**: 2748	[90-11-9]	**2**: 870
[88-16-4]	**5**: 2609	[90-12-0]	**6**: 3355
[88-17-5]	**5**: 2611	[90-13-1]	**3**: 1441
[88-18-6]	**6**: 3737	[90-15-3]	**6**: 3357, 3373
[88-19-7]	**2**: 762	[90-20-0]	**6**: 3439

[90-40-4]	**6**: 3442	[93-69-6]	**5**: 2840
[90-44-8]	**2**: 651	[93-72-1]	**3**: 1620
[90-51-7]	**6**: 3436	[93-76-5]	**3**: 1620
[90-66-4]	**6**: 3744	[93-89-0]	**4**: 2015
[90-80-2]	**5**: 2775	[93-97-0]	**2**: 826
[91-2-1]	**7**: 4218	[93-99-2]	**2**: 827, 837
[91-15-6]	**7**: 3868, 3883	[94-9-7]	**2**: 834; **4**: 2042
[91-16-7]	**6**: 3761	[94-13-3]	**5**: 2978
[91-17-8]	**6**: 3358	[94-18-8]	**5**: 2978
[91-19-0]	**6**: 3811	[94-26-8]	**5**: 2978
[91-20-3]	**6**: 3351	[94-36-0]	**2**: 828
[91-22-5]	**1**: 509; **7**: 4253	[94-59-7]	**6**: 3762
[91-23-6]	**2**: 809; **6**: 3561	[94-60-0]	**1**: 335
[91-25-8]	**2**: 767	[94-68-8]	**8**: 4749
[91-29-2]	**6**: 3556	[94-70-2]	**1**: 612, 636; **6**: 3561
[91-57-6]	**6**: 3355	[94-71-3]	**6**: 3781
[91-58-7]	**3**: 1441	[94-74-6]	**3**: 1620
[91-59-8]	**1**: 527, 529; **6**: 3403	[94-75-7]	**3**: 1620
[91-63-4]	**7**: 4255	[94-81-5]	**3**: 1620
[91-66-7]	**1**: 637	[94-82-6]	**3**: 1620
[91-68-9]	**1**: 617	[94-85-9]	**6**: 3565
[91-76-9]	**2**: 829; **5**: 3214	[95-13-6]	**5**: 2915
[91-81-6]	**7**: 4207	[95-14-7]	**6**: 3811
[91-88-3]	**8**: 4749	[95-20-5]	**5**: 3004
[91-94-1]	**1**: 525, 527, 529; **2**: 807; **6**: 3528	[95-21-6]	**1**: 616
[91-95-2]	**2**: 805	[95-46-5]	**2**: 870
[92-6-8]	**5**: 2910	[95-47-6]	**8**: 4582
[92-27-3]	**6**: 3400	[95-48-7]	**5**: 2980
[92-28-4]	**6**: 3419	[95-49-8]	**2**: 832
[92-40-0]	**6**: 3393	[95-50-1]	**6**: 3812
[92-41-1]	**6**: 3371	[95-51-2]	**1**: 519, 636, 638; **6**: 3528
[92-44-4]	**6**: 3381	[95-52-3]	**5**: 2613
[92-52-4]	**5**: 2906	[95-53-4]	**1**: 529, 636; **8**: 4746
[92-55-7]	**6**: 3571	[95-54-5]	**5**: 2818; **6**: 3540, 3810
[92-59-1]	**1**: 637	[95-55-6]	**1**: 602; **6**: 3561
[92-67-1]	**1**: 527, 529	[95-57-8]	**3**: 1606
[92-70-6]	**6**: 3383	[95-63-6]	**5**: 2894
[92-81-9]	**1**: 197	[95-64-7]	**1**: 636; **6**: 3522
[92-84-2]	**1**: 509	[95-68-1]	**1**: 636, 639; **6**: 3522
[92-86-4]	**2**: 870	[95-69-2]	**8**: 4750
[92-87-5]	**1**: 527, 529; **2**: 793, 803	[95-70-5]	**6**: 3809
[92-88-6]	**6**: 3743	[95-73-8]	**2**: 832
[92-94-4]	**5**: 2910	[95-74-9]	**6**: 3535; **8**: 4751
[93-0-5]	**6**: 3415	[95-75-7]	**3**: 1607
[93-1-6]	**6**: 3392	[95-76-1]	**1**: 519, 636, 638; **6**: 3532
[93-4-9]	**6**: 3378	[95-77-2]	**3**: 1606
[93-14-1]	**6**: 3762	[95-78-3]	**1**: 636; **6**: 3522
[93-18-5]	**6**: 3378	[95-79-4]	**1**: 519; **8**: 4750
[93-26-5]	**1**: 614	[95-80-7]	**6**: 3520, 3809
[93-50-5]	**6**: 3562	[95-81-8]	**6**: 3535; **8**: 4750
[93-53-8]	**1**: 382	[95-82-9]	**1**: 519, 636; **6**: 3533
[93-55-0]	**5**: 3111	[95-84-1]	**1**: 614
[93-58-3]	**2**: 826, 837; **4**: 2015	[95-85-2]	**1**: 614; **6**: 3561
[93-59-4]	**6**: 3643	[95-86-3]	**1**: 613
[93-60-7]	**7**: 4218	[95-92-1]	**6**: 3595, 3608
[93-61-8]	**5**: 2716	[95-93-2]	**5**: 2894
[93-65-2]	**3**: 1620	[95-95-4]	**3**: 1606

[96-5-9]	**1:** 426, 435	[97-96-1]	**1:** 352
[96-9-3]	**4:** 2003	[97-97-2]	**3:** 1524
[96-12-8]	**2:** 874, 888	[97-99-4]	**5:** 2747, 2762
[96-13-9]	**1:** 421	[98-0-0]	**5:** 2747, 2756
[96-14-0]	**5:** 2855	[98-1-1]	**2:** 1079; **5:** 2747, 2750
[96-17-3]	**1:** 352; **5:** 3038	[98-6-6]	**5:** 2902
[96-18-4]	**1:** 415; **3:** 1376	[98-7-7]	**2:** 828; **3:** 1458
[96-22-0]	**5:** 3095	[98-8-8]	**5:** 2609
[96-23-1]	**3:** 1586	[98-9-9]	**2:** 757
[96-24-2]	**3:** 1586	[98-10-2]	**2:** 758
[96-31-1]	**8:** 4819	[98-11-3]	**2:** 755
[96-33-3]	**1:** 226; **4:** 2014, 2039; **7:** 4266	[98-13-5]	**7:** 4318
[96-34-4]	**3:** 1541, 1548; **4:** 2016	[98-16-8]	**5:** 2612; **6:** 3538
[96-35-5]	**5:** 2966	[98-18-0]	**2:** 771
[96-37-7]	**3:** 1797	[98-19-1]	**5:** 2902
[96-45-7]	**8:** 4719	[98-25-9]	**6:** 3555
[96-47-9]	**5:** 2747, 2761	[98-27-1]	**6:** 3737
[96-48-0]	**2:** 1005; **4:** 2033; **7:** 4242	[98-29-3]	**6:** 3762
[96-49-1]	**2:** 1086; **4:** 2313, 2334, 2350	[98-31-7]	**2:** 769
[96-53-7]	**4:** 2153	[98-32-8]	**2:** 782; **6:** 3557
[96-54-8]	**7:** 4238	[98-33-9]	**2:** 779; **8:** 4751
[96-67-3]	**6:** 3558	[98-34-0]	**2:** 777
[96-69-5]	**6:** 3744	[98-36-2]	**2:** 784; **6:** 3557
[96-73-1]	**6:** 3555	[98-37-3]	**2:** 781; **6:** 3557
[96-75-3]	**6:** 3555	[98-42-0]	**2:** 782
[96-76-4]	**6:** 3737	[98-43-1]	**2:** 779; **6:** 3553
[96-77-5]	**2:** 776	[98-44-2]	**2:** 773; **6:** 3557
[96-82-2]	**5:** 3146	[98-46-4]	**5:** 2610; **6:** 3537
[96-91-3]	**1:** 613; **3:** 1890	[98-47-5]	**6:** 3550
[96-96-8]	**6:** 3563	[98-48-6]	**2:** 759; **7:** 4263
[96-99-1]	**2:** 835	[98-51-1]	**5:** 2902
[97-0-7]	**6:** 3531	[98-52-2]	**6:** 3743
[97-2-9]	**1:** 523, 637; **6:** 3542	[98-53-3]	**6:** 3743
[97-5-2]	**2:** 780	[98-54-4]	**6:** 3737
[97-6-3]	**2:** 811; **6:** 3552	[98-56-6]	**5:** 2609
[97-8-5]	**6:** 3557	[98-59-9]	**2:** 763
[97-9-6]	**6:** 3557	[98-60-2]	**2:** 768
[97-18-7]	**6:** 3791	[98-64-6]	**2:** 768
[97-23-4]	**6:** 3789	[98-66-8]	**2:** 768
[97-24-5]	**6:** 3791	[98-67-9]	**2:** 775
[97-39-2]	**5:** 2838	[98-69-1]	**2:** 765
[97-44-9]	**1:** 616	[98-70-4]	**2:** 766
[97-50-7]	**6:** 3534	[98-73-7]	**2:** 831, 838
[97-52-9]	**1:** 614; **6:** 3561	[98-74-8]	**6:** 3551
[97-53-0]	**6:** 3762	[98-79-3]	**1:** 581
[97-56-3]	**3:** 1889	[98-82-8]	**1:** 93; **5:** 2899; **6:** 3694, 3704
[97-59-6]	**5:** 2818	[98-83-9]	**6:** 3697; **7:** 4436
[97-62-1]	**4:** 2014, 2016	[98-86-2]	**5:** 3110; **6:** 3704
[97-62-3]	**6:** 3250	[98-87-3]	**2:** 828; **3:** 1456, 1640, 1642
[97-64-3]	**5:** 3130	[98-88-4]	**2:** 827, 837
[97-65-4]	**3:** 1648	[98-89-5]	**2:** 836, 838
[97-67-6]	**5:** 2970	[98-92-0]	**7:** 4202, 4218
[97-85-8]	**4:** 2014	[98-94-2]	**1:** 461
[97-86-9]	**6:** 3250	[98-95-3]	**6:** 3509
[97-88-1]	**6:** 3250	[98-96-4]	**5:** 2814
[97-93-8]	**5:** 2700	[98-98-6]	**7:** 4202, 4218
[97-95-0]	**1:** 281, 295	[99-4-7]	**2:** 831, 837

[99-5-8]	**2**: 834, 838	[100-46-9]	**2**: 829, 863
[99-6-9]	**2**: 836; **5**: 2975	[100-47-0]	**2**: 829, 837
[99-7-0]	**1**: 617	[100-48-1]	**7**: 4202
[99-8-1]	**6**: 3518	[100-50-5]	**1**: 202
[99-9-2]	**1**: 514, 637; **6**: 3541	[100-51-6]	**2**: 849
[99-10-5]	**5**: 2981	[100-52-7]	**2**: 837; **3**: 1641; **5**: 2982
[99-20-7]	**2**: 1067	[100-54-9]	**7**: 4202, 4218
[99-24-1]	**5**: 2982	[100-55-0]	**7**: 4211
[99-29-6]	**6**: 3547	[100-60-7]	**1**: 461
[99-30-9]	**6**: 3541	[100-61-8]	**1**: 529, 637
[99-35-4]	**6**: 3514	[100-64-1]	**3**: 1817
[99-50-3]	**5**: 2980	[100-66-3]	**6**: 3779
[99-51-4]	**6**: 3522	[100-69-6]	**7**: 4190, 4218
[99-52-5]	**6**: 3545	[100-70-9]	**7**: 4195, 4202
[99-54-7]	**6**: 3532	[100-73-2]	**1**: 200
[99-55-8]	**6**: 3520, 3545	[100-74-3]	**1**: 468
[99-56-9]	**6**: 3543	[100-83-4]	**2**: 687; **5**: 2977
[99-57-0]	**1**: 612; **6**: 3564	[100-97-0]	**1**: 470, 493; **3**: 1863; **5**: 2667
[99-59-2]	**1**: 615; **6**: 3561	[101-14-4]	**1**: 527, 529
[99-61-6]	**2**: 685, 835	[101-18-8]	**1**: 617
[99-62-7]	**5**: 2901; **7**: 4263	[101-37-1]	**1**: 426
[99-63-8]	**6**: 3815	[101-39-3]	**1**: 382
[99-64-9]	**2**: 834	[101-53-1]	**6**: 3754
[99-65-0]	**6**: 3513, 3812	[101-54-2]	**1**: 527
[99-71-8]	**6**: 3735	[101-56-4]	**3**: 1881, 1901
[99-75-2]	**8**: 4580	[101-68-8]	**1**: 634; **5**: 2667
[99-76-3]	**5**: 2978	[101-69-9]	**3**: 1881, 1884, 1901
[99-81-0]	**6**: 3568	[101-72-4]	**6**: 3530
[99-82-1]	**8**: 4601	[101-77-9]	**1**: 512, 527, 529; **5**: 3016
[99-85-4]	**8**: 4602	[101-81-5]	**5**: 2904
[99-87-6]	**5**: 2902; **8**: 4603	[101-83-7]	**1**: 462, 512
[99-89-8]	**6**: 3732	[101-84-8]	**6**: 3779
[99-94-5]	**2**: 831, 838; **8**: 4577	[101-86-0]	**1**: 382
[99-96-7]	**5**: 2975, 2978	[101-89-3]	**3**: 1901
[99-97-8]	**8**: 4749	[101-96-2]	**7**: 4247
[99-99-0]	**2**: 835; **6**: 3519	[102-1-2]	**1**: 637; **5**: 3075
[100-0-5]	**1**: 522; **6**: 3529	[102-6-7]	**5**: 2838
[100-1-6]	**1**: 505, 522, 529, 637; **6**: 3541, 3812	[102-7-8]	**8**: 4820
[100-2-7]	**6**: 3562	[102-9-0]	**2**: 1086
[100-14-1]	**6**: 3536	[102-20-5]	**4**: 2041
[100-16-3]	**6**: 3542	[102-25-0]	**5**: 2894
[100-17-4]	**6**: 3563	[102-27-2]	**8**: 4749
[100-18-5]	**1**: 100; **5**: 2901, 2945	[102-28-3]	**6**: 3541
[100-19-6]	**6**: 3568	[102-56-7]	**6**: 3564
[100-20-9]	**6**: 3815	[102-69-2]	**1**: 454
[100-21-0]	**4**: 2310; **7**: 3868; **8**: 4573	[102-70-5]	**1**: 432, 436
[100-28-7]	**6**: 3542	[102-71-6]	**4**: 2150
[100-29-8]	**6**: 3563	[102-76-1]	**4**: 2014
[100-32-3]	**6**: 3551	[102-82-9]	**1**: 454
[100-35-6]	**1**: 455	[102-84-1]	**1**: 428
[100-37-8]	**4**: 2159	[102-87-4]	**1**: 481; **5**: 2731
[100-40-3]	**2**: 904; **7**: 4429	[102-92-1]	**3**: 1640
[100-41-4]	**2**: 695; **4**: 2207, 2223; **7**: 4417	[102-94-3]	**3**: 1640
[100-42-5]	**2**: 904; **3**: 1640; **4**: 2207, 2223; **7**: 4145, 4417; **8**: 4588	[102-96-5]	**6**: 3495
		[103-9-3]	**1**: 54; **4**: 2014
[100-43-6]	**7**: 4190, 4218	[103-11-7]	**1**: 226; **4**: 2014, 2039, 2366
[100-44-7]	**2**: 1235, 828; **3**: 1448; **5**: 2647	[103-23-1]	**1**: 271; **4**: 2014, 2040, 2366

[103-41-3]	**3**: 1640; **4**: 2015	[106-23-0]	**1**: 368
[103-50-4]	**2**: 858	[106-26-3]	**1**: 368
[103-63-9]	**2**: 876	[106-27-4]	**4**: 2041
[103-64-0]	**3**: 1640	[106-35-4]	**5**: 3097
[103-69-5]	**1**: 637	[106-36-5]	**7**: 4040, 4071, 4119
[103-70-8]	**1**: 637; **5**: 2693, 2705	[106-37-6]	**2**: 870
[103-72-0]	**8**: 4636	[106-38-7]	**2**: 870
[103-73-1]	**6**: 3779	[106-39-8]	**2**: 870
[103-74-2]	**7**: 4211	[106-41-2]	**2**: 870
[103-79-7]	**5**: 3112	[106-42-3]	**8**: 4573
[103-82-2]	**1**: 57; **6**: 3805	[106-43-4]	**2**: 832
[103-84-4]	**1**: 505, 637, 639	[106-44-5]	**5**: 2978, 2980
[103-89-9]	**1**: 511	[106-45-6]	**8**: 4662
[103-90-2]	**1**: 619	[106-46-7]	**6**: 3812
[103-95-7]	**1**: 382	[106-47-8]	**1**: 519, 636, 638; **6**: 3530
[104-3-0]	**6**: 3536	[106-48-9]	**3**: 1606
[104-4-1]	**6**: 3541	[106-49-0]	**1**: 512, 522, 636; **8**: 4746
[104-9-6]	**1**: 382	[106-50-3]	**1**: 529; **6**: 3541, 3810
[104-15-4]	**2**: 763	[106-51-4]	**2**: 841; **5**: 2942, 2981; **6**: 3811
[104-38-1]	**5**: 2943	[106-54-7]	**8**: 4662
[104-53-0]	**1**: 382	[106-63-8]	**1**: 226
[104-55-2]	**1**: 382; **3**: 1642	[106-68-3]	**2**: 974
[104-57-4]	**5**: 2738	[106-69-4]	**1**: 200
[104-62-1]	**5**: 2738	[106-75-2]	**3**: 1576
[104-75-6]	**1**: 455	[106-88-7]	**2**: 1215, 947
[104-76-7]	**1**: 282, 298; **2**: 955; **4**: 2361; **5**: 2551	[106-89-8]	**2**: 664; **3**: 1594; **4**: 1998; **5**: 2797
[104-78-9]	**1**: 485	[106-92-3]	**1**: 431, 435
[104-87-0]	**2**: 688; **8**: 4585	[106-93-4]	**2**: 869, 873, 888; **4**: 2305
[104-88-1]	**2**: 684	[106-94-5]	**2**: 869
[104-90-5]	**7**: 4178, 4218	[106-95-6]	**2**: 869, 879
[104-92-7]	**2**: 870	[106-97-8]	**1**: 39; **5**: 2699, 2727, 2855
[104-93-8]	**6**: 3780	[106-97-81]	**4**: 2224
[104-94-9]	**1**: 619, 636; **6**: 3563	[106-98-9]	**2**: 954, 959, 987; **5**: 2871
[105-5-5]	**5**: 2894	[106-99-0]	**2**: 1008, 899; **3**: 1783; **5**: 2699; **7**: 4368
[105-8-8]	**1**: 334, 344; **4**: 2019	[107-1-7]	**2**: 959; **5**: 3038
[105-30-6]	**1**: 281, 293; **5**: 2551; **7**: 4039	[107-2-8]	**1**: 199, 330, 368; **3**: 1594; **5**: 2799
[105-31-7]	**1**: 309	[107-5-1]	**1**: 410, 433; **3**: 1374; **5**: 2797; **7**: 4368
[105-34-0]	**5**: 3187, 3193	[107-6-2]	**2**: 664; **3**: 1298, 1473
[105-37-3]	**4**: 2014; **7**: 4119	[107-7-3]	**3**: 1586
[105-38-4]	**8**: 4855	[107-10-8]	**1**: 454
[105-39-5]	**3**: 1541, 1548	[107-11-9]	**1**: 432, 436, 455
[105-45-3]	**4**: 2015	[107-12-0]	**6**: 3461
[105-46-4]	**1**: 46, 54	[107-13-1]	**1**: 126, 249; **2**: 1234; **5**: 2848; **7**: 4266
[105-53-3]	**5**: 3179, 3193	[107-14-2]	**6**: 3462
[105-54-4]	**4**: 2014	[107-15-3]	**1**: 482, 493; **2**: 664; **4**: 2151
[105-56-6]	**5**: 3187, 3193	[107-16-4]	**5**: 2646, 2681; **6**: 3463
[105-58-8]	**2**: 1086	[107-18-6]	**1**: 205, 309, 331, 419, 434; **3**: 1594; **5**: 2799
[105-59-9]	**4**: 2159		
[105-60-2]	**2**: 1014	[107-19-7]	**1**: 126, 309; **2**: 938
[105-64-6]	**1**: 428; **3**: 1580	[107-20-0]	**3**: 1522
[105-76-0]	**4**: 2039	[107-21-1]	**4**: 2062, 2305, 2319, 2329, 2342, 2349; **5**: 2812; **6**: 3601; **8**: 4586
[105-83-9]	**1**: 488		
[105-85-1]	**4**: 2041; **5**: 2738	[107-22-2]	**1**: 390; **2**: 1221; **4**: 2309; **5**: 2809; **6**: 3811
[105-87-3]	**4**: 2041		
[105-97-5]	**1**: 271	[107-25-5]	**1**: 10
[105-99-7]	**1**: 271	[107-30-2]	**4**: 2202; **5**: 3038
[106-19-4]	**1**: 271	[107-31-3]	**4**: 2014; **5**: 2718, 2736

[107-32-4]	**5:** 2717; **6:** 3643	[108-73-6]	**5:** 2754; **6:** 3766
[107-35-7]	**8:** 4505	[108-75-8]	**7:** 4184, 4218
[107-36-8]	**8:** 4505	[108-77-0]	**3:** 1773
[107-39-1]	**6:** 3749	[108-78-1]	**3:** 1735, 1747, 1768; **5:** 2828, 3197; **8:** 4818
[107-40-4]	**6:** 3749		
[107-41-5]	**1:** 105	[108-80-5]	**3:** 1768; **5:** 3201; **8:** 4818
[107-43-7]	**1:** 579	[108-82-7]	**1:** 282, 301
[107-46-0]	**7:** 4324	[108-83-8]	**5:** 3097
[107-47-1]	**8:** 4658	[108-86-1]	**2:** 870
[107-68-6]	**3:** 1888; **8:** 4505	[108-87-2]	**3:** 1797
[107-75-5]	**1:** 375	[108-88-3]	**2:** 695; **6:** 3698; **8:** 4584, 4728
[107-83-5]	**5:** 2855	[108-89-4]	**7:** 4178, 4218
[107-85-7]	**1:** 454	[108-91-8]	**1:** 459, 492, 512
[107-87-9]	**5:** 2760	[108-93-0]	**1:** 265; **3:** 1807
[107-88-0]	**1:** 344; **2:** 946	[108-94-1]	**1:** 265; **3:** 1807
[107-89-1]	**1:** 375, 5	[108-95-2]	**1:** 90, 93; **3:** 1609, 1809; **5:** 2900, 2945, 2975; **6:** 3689, 3694
[107-92-6]	**2:** 1007, 955		
[107-93-7]	**3:** 1716	[108-98-5]	**8:** 4662
[107-94-8]	**2:** 1116; **7:** 4120	[108-99-6]	**1:** 204; **7:** 4178, 4218
[107-97-1]	**3:** 1888	[109-0-2]	**7:** 4209
[107-99-3]	**1:** 455	[109-2-4]	**1:** 468; **4:** 2159
[108-1-0]	**4:** 2159	[109-4-6]	**7:** 4199
[108-3-2]	**6:** 3488; **7:** 4105	[109-5-7]	**7:** 4197
[108-5-4]	**1:** 125; **4:** 2014, 2039; **7:** 4405; **8:** 4840, 4853	[109-6-8]	**7:** 4178, 4218
		[109-9-1]	**7:** 4195, 4199, 4218
[108-8-7]	**5:** 2855	[109-21-7]	**4:** 2014
[108-10-1]	**1:** 105, 293; **5:** 3087; **7:** 4265	[109-55-7]	**1:** 485
[108-11-2]	**1:** 105, 281, 293; **5:** 3088	[109-60-4]	**1:** 55; **4:** 2014; **7:** 4064
[108-18-9]	**1:** 454; **3:** 1609; **7:** 4065	[109-61-5]	**3:** 1576
[108-19-0]	**8:** 4818	[109-63-7]	**4:** 2024
[108-20-3]	**4:** 2187, 2193; **7:** 4065, 4265	[109-64-8]	**2:** 869
[108-21-4]	**1:** 55; **4:** 2014; **7:** 4064	[109-65-9]	**2:** 869
[108-22-5]	**5:** 3067	[109-66-0]	**5:** 2855
[108-23-6]	**3:** 1576	[109-67-1]	**5:** 2871
[108-24-7]	**1:** 55, 63; **3:** 1609; **4:** 2060; **7:** 4404	[109-70-6]	**2:** 875
[108-27-0]	**6:** 3607	[109-73-9]	**1:** 454; **3:** 1868
[108-29-2]	**6:** 3606	[109-74-0]	**6:** 3461
[108-30-5]	**1:** 268; **2:** 1008	[109-75-1]	**1:** 425
[108-31-6]	**2:** 1007; **5:** 3161; **8:** 4588	[109-76-2]	**1:** 484
[108-32-7]	**2:** 1086	[109-77-3]	**5:** 2824, 3188, 3193
[108-36-1]	**2:** 870	[109-78-4]	**6:** 3464
[108-38-3]	**2:** 695; **8:** 4573	[109-86-4]	**4:** 2320
[108-39-4]	**5:** 2980	[109-87-5]	**5:** 3038
[108-41-8]	**2:** 832; **3:** 1898	[109-89-7]	**1:** 454, 6
[108-42-9]	**1:** 519, 636, 638; **6:** 3529	[109-93-3]	**4:** 2199
[108-43-0]	**3:** 1606	[109-94-4]	**4:** 2014; **5:** 2738
[108-44-1]	**1:** 636; **8:** 4746	[109-96-6]	**7:** 4236
[108-45-2]	**6:** 3810; **7:** 4266	[109-97-7]	**5:** 2749; **7:** 4235
[108-46-3]	**1:** 523; **3:** 1869; **5:** 2980; **7:** 4261	[109-99-9]	**2:** 1008, 903, 944; **3:** 1381; **4:** 2059; **8:** 4621
[108-47-4]	**7:** 4218		
[108-48-5]	**7:** 4178, 4218	[110-0-9]	**5:** 2746
[108-55-4]	**1:** 268	[110-5-4]	**6:** 3640
[108-59-8]	**5:** 3179	[110-12-3]	**5:** 3092
[108-60-1]	**3:** 1592	[110-15-6]	**1:** 268; **3:** 1905
[108-67-8]	**5:** 2894	[110-16-7]	**5:** 3159
[108-69-0]	**1:** 636	[110-17-8]	**5:** 3170
[108-71-4]	**6:** 3809	[110-19-0]	**1:** 54; **4:** 2014

[110-22-5]	**1:** 66, 68	[111-65-9]	**5:** 2855
[110-30-5]	**8:** 4962	[111-66-0]	**5:** 2871
[110-41-8]	**1:** 353	[111-69-3]	**1:** 270; **3:** 1381; **5:** 2847
[110-43-0]	**5:** 3091	[111-70-6]	**1:** 281, 295; **5:** 2535
[110-44-1]	**5:** 3066; **7:** 4365	[111-71-7]	**1:** 352
[110-45-2]	**5:** 2738	[111-76-2]	**1:** 431; **4:** 2320
[110-46-3]	**2:** 974	[111-77-3]	**4:** 2320
[110-49-6]	**4:** 2014, 2321	[111-78-4]	**2:** 904
[110-53-2]	**2:** 869	[111-83-1]	**2:** 869
[110-54-3]	**5:** 2855	[111-84-2]	**5:** 2855
[110-57-6]	**3:** 1381	[111-86-4]	**1:** 455, 481
[110-58-7]	**1:** 454	[111-87-5]	**1:** 282, 295, 298; **5:** 2535
[110-60-1]	**1:** 486, 491	[111-90-0]	**4:** 2014, 2320
[110-61-2]	**6:** 3461	[111-92-2]	**1:** 454
[110-62-3]	**1:** 352	[111-96-6]	**4:** 2320
[110-63-4]	**1:** 331; **2:** 1007, 941; **3:** 1381; **5:** 2667	[112-5-0]	**3:** 1908; **4:** 2485
[110-64-5]	**2:** 940	[112-7-2]	**4:** 2014, 2321
[110-65-6]	**1:** 126; **2:** 1008, 937; **5:** 2646	[112-15-2]	**4:** 2321
[110-68-9]	**1:** 455; **3:** 1868	[112-18-5]	**1:** 481
[110-71-4]	**4:** 2320	[112-24-3]	**1:** 487, 493
[110-74-7]	**4:** 2015	[112-27-6]	**4:** 2319, 2342, 2350
[110-75-8]	**4:** 2199	[112-29-8]	**2:** 869
[110-80-5]	**4:** 2320	[112-30-1]	**1:** 282, 301; **5:** 2535
[110-81-6]	**8:** 4658	[112-31-2]	**1:** 353
[110-82-7]	**1:** 265; **3:** 1795, 1797, 1810	[112-34-5]	**4:** 2014, 2320
[110-83-8]	**3:** 1797, 1813	[112-35-6]	**4:** 2320
[110-85-0]	**1:** 468, 492; **4:** 2151	[112-36-7]	**4:** 2320
[110-86-1]	**1:** 204; **5:** 3073; **7:** 4178, 4218	[112-37-8]	**4:** 2485
[110-87-2]	**5:** 2747, 2763	[112-38-9]	**4:** 2487
[110-88-3]	**5:** 2677	[112-40-3]	**5:** 2855
[110-89-4]	**1:** 465; **7:** 4197, 4219	[112-42-5]	**1:** 282; **5:** 2535
[110-91-8]	**1:** 467, 493; **4:** 2151	[112-43-6]	**1:** 309; **5:** 2549
[110-94-1]	**1:** 268; **3:** 1905	[112-44-7]	**1:** 353
[110-96-3]	**1:** 454	[112-45-8]	**1:** 368
[110-97-4]	**4:** 2164	[112-47-0]	**1:** 333; **5:** 2553
[110-98-5]	**7:** 4048	[112-48-1]	**4:** 2320
[110-99-6]	**3:** 1542	[112-50-5]	**4:** 2320
[111-2-4]	**8:** 4613	[112-53-8]	**1:** 282; **5:** 2535
[111-13-7]	**5:** 3093	[112-54-9]	**1:** 353
[111-14-8]	**4:** 2485	[112-56-1]	**8:** 4632, 4634
[111-15-9]	**4:** 2014, 2039, 2321	[112-57-2]	**1:** 488
[111-16-0]	**3:** 1905	[112-58-3]	**4:** 2187
[111-20-6]	**3:** 1905	[112-60-7]	**4:** 2319
[111-25-1]	**2:** 869	[112-61-8]	**4:** 2014
[111-26-2]	**1:** 454	[112-69-6]	**1:** 481
[111-27-3]	**1:** 281, 293; **5:** 2535	[112-70-9]	**1:** 282; **5:** 2535
[111-29-5]	**1:** 326, 344	[112-71-0]	**2:** 869
[111-30-8]	**1:** 201, 390	[112-72-1]	**1:** 282; **5:** 2535
[111-40-0]	**1:** 487, 493	[112-73-2]	**4:** 2320
[111-42-2]	**4:** 2150; **5:** 2811	[112-75-4]	**1:** 481
[111-43-3]	**4:** 2187, 2195; **7:** 4065	[112-79-8]	**4:** 2487
[111-44-4]	**4:** 2187, 2199	[112-80-1]	**3:** 1908; **4:** 2487
[111-46-6]	**2:** 718, 720; **4:** 2306, 2319, 2342, 2349	[112-82-3]	**2:** 869
[111-47-7]	**8:** 4658	[112-85-6]	**4:** 2485
[111-49-9]	**1:** 466, 493; **5:** 2847	[112-86-7]	**3:** 1918; **4:** 2487
[111-55-7]	**1:** 68; **4:** 2014, 2314, 2321	[112-88-9]	**5:** 2871
[111-61-5]	**4:** 2014	[112-89-0]	**2:** 869

[112-90-3]	**1**: 481	[118-75-2]	**1**: 510; **2**: 841
[112-92-5]	**1**: 282; **5**: 2535	[118-79-6]	**2**: 870, 879
[112-95-8]	**5**: 2855	[118-82-1]	**6**: 3741, 3743
[112-99-2]	**1**: 481	[118-83-2]	**5**: 2616; **6**: 3537
[113-0-8]	**5**: 2823	[118-88-7]	**2**: 779; **6**: 3554; **8**: 4752
[114-26-1]	**6**: 3762	[118-90-1]	**2**: 831, 837
[115-7-1]	**1**: 252, 93, 97; **2**: 957; **5**: 2699, 2797, 2848, 2871, 2899; **6**: 3694; **7**: 4081, 4265; **8**: 4962	[118-91-2]	**2**: 832, 838
		[118-92-3]	**2**: 833, 838; **7**: 3880
		[119-0-6]	**6**: 3557
[115-10-6]	**4**: 1931, 2187	[119-6-2]	**4**: 2040
[115-11-7]	**2**: 954, 959, 987; **3**: 1858; **5**: 2699, 2871, 3037	[119-21-1]	**6**: 3534
		[119-23-3]	**6**: 3564
[115-19-5]	**1**: 309, 92	[119-26-6]	**6**: 3531
[115-25-3]	**5**: 2623	[119-27-7]	**1**: 523; **6**: 3564
[115-69-5]	**6**: 3496	[119-28-8]	**6**: 3412
[115-77-5]	**1**: 205, 340, 344; **5**: 2647, 2667, 2728; **6**: 3840	[119-32-4]	**6**: 3520
		[119-33-5]	**6**: 3565
[115-78-6]	**6**: 3830	[119-36-8]	**4**: 2015; **7**: 4277
[115-84-4]	**1**: 330	[119-40-4]	**6**: 3435
[116-2-9]	**3**: 1915	[119-42-6]	**6**: 3754
[116-14-3]	**5**: 2584, 2624	[119-47-1]	**6**: 3744
[116-15-4]	**5**: 2584, 2587, 2624	[119-61-9]	**5**: 3111
[116-16-5]	**1**: 92; **5**: 2597	[119-64-2]	**6**: 3358
[116-44-9]	**5**: 2814	[119-65-3]	**7**: 4256
[116-54-1]	**3**: 1552; **4**: 2016	[119-70-0]	**6**: 3556
[116-63-2]	**6**: 3429	[119-79-9]	**6**: 3412
[117-14-6]	**2**: 652	[119-90-4]	**2**: 808; **6**: 3561
[117-22-6]	**6**: 3391	[119-93-7]	**1**: 529; **2**: 805
[117-42-0]	**6**: 3421	[120-12-7]	**2**: 643
[117-43-1]	**6**: 3394	[120-18-3]	**6**: 3368
[117-46-4]	**6**: 3441	[120-32-1]	**6**: 3785
[117-55-5]	**6**: 3417	[120-36-5]	**3**: 1620
[117-56-6]	**6**: 3395	[120-47-8]	**5**: 2978
[117-59-9]	**6**: 3390	[120-51-4]	**2**: 827, 837, 859
[117-61-3]	**2**: 810; **6**: 3550	[120-54-7]	**7**: 4198
[117-62-4]	**6**: 3418	[120-55-8]	**2**: 827, 837
[117-80-6]	**6**: 3450	[120-57-0]	**6**: 3762
[117-81-7]	**4**: 2015, 2026, 2040, 2366; **7**: 3886	[120-61-6]	**1**: 335; **4**: 2015, 2039; **6**: 3812; **8**: 4573
[117-82-8]	**7**: 3886	[120-71-8]	**1**: 511, 614, 637; **6**: 3535
[117-83-9]	**7**: 3886	[120-72-9]	**5**: 3001
[117-84-0]	**7**: 3886	[120-73-0]	**7**: 4165
[117-92-0]	**7**: 4255	[120-80-9]	**3**: 1607; **5**: 2980; **6**: 3758; **7**: 4263
[117-98-6]	**4**: 2041	[120-83-2]	**3**: 1606
[118-3-6]	**6**: 3422	[120-92-3]	**1**: 264; **5**: 3099; **7**: 4184
[118-10-5]	**5**: 2983	[120-93-4]	**8**: 4820
[118-28-5]	**6**: 3432	[120-95-6]	**6**: 3747
[118-30-9]	**6**: 3426	[120-98-9]	**2**: 782
[118-32-1]	**6**: 3397	[121-0-6]	**5**: 2944
[118-33-2]	**6**: 3419	[121-2-8]	**6**: 3554
[118-41-2]	**5**: 2982	[121-3-9]	**6**: 3553
[118-46-7]	**6**: 3426	[121-4-0]	**2**: 764
[118-48-9]	**7**: 3880	[121-14-2]	**6**: 3519
[118-52-5]	**3**: 1566, 1569	[121-17-5]	**5**: 2610; **6**: 3537
[118-55-8]	**4**: 2015, 2042; **7**: 4277	[121-18-6]	**6**: 3556
[118-58-1]	**2**: 859; **7**: 4277	[121-25-5]	**7**: 4194
[118-61-6]	**4**: 2015	[121-30-2]	**2**: 785
[118-74-1]	**5**: 2702	[121-32-4]	**6**: 3761

[121-33-5]	**5:** 2819, 2980; **6:** 3761	[123-54-6]	**5:** 2599, 3067, 3106
[121-34-6]	**5:** 2980	[123-62-6]	**7:** 4117
[121-44-8]	**1:** 454, 6; **5:** 2730, 3073	[123-63-7]	**1:** 20
[121-47-1]	**1:** 637; **2:** 771	[123-68-2]	**4:** 2041
[121-50-6]	**5:** 2611; **6:** 3538	[123-72-8]	**1:** 352; **2:** 925; **4:** 2361
[121-51-7]	**6:** 3550	[123-73-9]	**1:** 368; **3:** 1711, 1717; **5:** 3066; **7:** 4368
[121-52-8]	**6:** 3550	[123-75-1]	**1:** 464, 492; **7:** 4237
[121-53-9]	**2:** 836; **5:** 2976	[123-76-2]	**5:** 2757; **6:** 3606
[121-57-3]	**1:** 512, 637; **2:** 771	[123-86-4]	**1:** 54; **4:** 2014
[121-60-8]	**2:** 772	[123-91-1]	**4:** 1937, 2310
[121-62-0]	**2:** 772	[123-92-2]	**4:** 2041
[121-66-4]	**6:** 3572	[123-93-3]	**3:** 1542
[121-69-7]	**1:** 529, 637	[123-95-5]	**4:** 2014, 2040
[121-72-2]	**8:** 4749	[123-96-6]	**1:** 282, 298
[121-73-3]	**6:** 3529	[123-99-9]	**3:** 1905
[121-75-5]	**6:** 3853	[124-2-7]	**1:** 432, 436, 455
[121-79-9]	**5:** 2982	[124-4-9]	**1:** 263; **3:** 1905; **5:** 2845
[121-82-4]	**1:** 70	[124-7-2]	**3:** 1909; **4:** 2485
[121-86-8]	**6:** 3535	[124-9-4]	**2:** 903; **5:** 2845
[121-87-9]	**6:** 3546	[124-11-8]	**5:** 2871
[121-88-0]	**1:** 612	[124-13-0]	**1:** 352
[121-89-1]	**6:** 3568	[124-17-4]	**4:** 2321
[121-91-5]	**8:** 4573	[124-18-5]	**5:** 2855
[121-92-6]	**2:** 810, 835, 838; **6:** 3518	[124-19-6]	**1:** 353
[121-880-0]	**6:** 3561	[124-20-9]	**1:** 491
[122-4-3]	**2:** 835	[124-22-1]	**1:** 481
[122-14-5]	**6:** 3565	[124-28-7]	**1:** 481
[122-20-3]	**4:** 2164	[124-29-8]	**1:** 282
[122-39-4]	**1:** 529, 637, 639	[124-30-1]	**1:** 481
[122-40-7]	**1:** 382	[124-40-3]	**5:** 2737; **6:** 3325
[122-42-9]	**3:** 1580	[124-41-4]	**5:** 2720
[122-51-0]	**4:** 2033	[124-58-9]	**1:** 344
[122-59-8]	**6:** 3780	[124-64-1]	**6:** 3828
[122-62-3]	**4:** 2042	[124-68-5]	**6:** 3498
[122-63-4]	**4:** 2041; **7:** 4119	[124-73-2]	**5:** 2581
[122-68-9]	**3:** 1640	[126-11-4]	**6:** 3494
[122-69-0]	**3:** 1640	[126-30-7]	**1:** 326, 344; **5:** 2667, 2728
[122-78-1]	**1:** 382	[126-33-0]	**2:** 719, 904
[122-80-5]	**6:** 3541	[126-58-9]	**1:** 342
[122-87-2]	**1:** 621	[126-98-7]	**1:** 433; **6:** 3253, 3462
[122-99-6]	**6:** 3780	[126-99-8]	**2:** 902; **3:** 1383, 1478
[123-3-5]	**7:** 4194	[127-0-4]	**3:** 1586
[123-5-7]	**1:** 353	[127-8-2]	**1:** 52
[123-7-9]	**6:** 3728, 3731	[127-9-3]	**1:** 52; **2:** 1132
[123-8-0]	**2:** 687	[127-17-3]	**6:** 3600
[123-15-9]	**1:** 352; **7:** 4040	[127-18-4]	**1:** 129; **3:** 1361, 1376, 1476
[123-17-1]	**1:** 282	[127-20-8]	**2:** 1116
[123-25-1]	**6:** 3608	[127-21-9]	**5:** 2597
[123-30-8]	**1:** 515, 603; **6:** 3562	[127-40-2]	**4:** 2382
[123-31-9]	**1:** 100, 127; **3:** 1607; **5:** 2941, 2981; **6:** 3812; **7:** 4263	[127-52-6]	**3:** 1571
		[127-65-1]	**3:** 1566, 1571
[123-35-3]	**5:** 3050; **8:** 4598	[127-95-7]	**6:** 3593
[123-38-6]	**1:** 352; **7:** 4037, 4107	[128-9-6]	**3:** 1566, 1570
[123-39-7]	**5:** 2705	[128-37-0]	**4:** 2190; **6:** 3737
[123-42-2]	**1:** 105, 111	[128-39-2]	**6:** 3737
[123-44-4]	**1:** 282	[128-42-7]	**6:** 3553
[123-51-3]	**1:** 281; **4:** 2116	[128-70-1]	**5:** 2923

[128-97-2]	**5**: 2923	[139-66-2]	**3**: 1609
[129-0-0]	**5**: 2922	[139-89-9]	**4**: 2297
[129-39-5]	**2**: 652	[140-1-2]	**4**: 2297
[129-91-9]	**6**: 3416	[140-7-8]	**4**: 2299
[129-93-1]	**6**: 3423	[140-10-3]	**3**: 1640
[130-0-7]	**6**: 3413	[140-11-4]	**2**: 858; **4**: 2014
[130-15-4]	**6**: 3448	[140-29-4]	**6**: 3469
[130-23-4]	**6**: 3441	[140-31-8]	**1**: 492
[131-11-3]	**4**: 2015, 2040; **7**: 3886	[140-40-9]	**6**: 3569, 3572
[131-17-9]	**1**: 426, 435	[140-56-7]	**3**: 1887, 1900
[131-27-1]	**6**: 3419	[140-66-9]	**6**: 3748
[131-91-9]	**6**: 3377	[140-67-0]	**8**: 4772
[131-99-7]	**7**: 4167	[140-76-1]	**7**: 4191
[132-57-0]	**6**: 3393	[140-80-7]	**1**: 486
[132-64-9]	**5**: 2765	[140-88-5]	**1**: 127, 226; **4**: 2014, 2039
[132-66-1]	**6**: 3402	[140-90-9]	**3**: 1580
[132-86-5]	**6**: 3380	[140-93-2]	**7**: 4064
[133-8-4]	**5**: 3184	[140-95-4]	**8**: 4823
[133-37-9]	**8**: 4562	[141-12-8]	**4**: 2041
[133-53-9]	**6**: 3788	[141-22-0]	**3**: 1909
[133-59-5]	**2**: 762	[141-27-5]	**1**: 368
[133-73-3]	**2**: 777	[141-28-6]	**1**: 271; **4**: 2014
[133-74-4]	**2**: 784	[141-31-1]	**1**: 200
[133-78-8]	**2**: 779; **8**: 4751	[141-32-2]	**1**: 226; **4**: 2014, 2039
[134-20-3]	**4**: 2015	[141-43-5]	**2**: 666; **4**: 2150, 2350
[134-32-7]	**1**: 512, 527; **6**: 3401	[141-46-8]	**1**: 375; **4**: 2309
[134-34-9]	**6**: 3441	[141-53-7]	**5**: 2700, 2718, 2728, 2738
[134-47-4]	**6**: 3436	[141-78-6]	**1**: 53; **2**: 1006; **4**: 2014, 2057, 2059
[134-54-3]	**6**: 3411	[141-79-7]	**1**: 105; **5**: 3104; **6**: 3697
[134-62-3]	**2**: 831	[141-82-2]	**3**: 1905; **5**: 3178, 3193; **7**: 4368
[134-81-6]	**5**: 3113	[141-86-6]	**7**: 4206
[135-1-3]	**5**: 2894	[141-93-5]	**5**: 2894
[135-19-3]	**6**: 3354, 3377	[141-97-9]	**4**: 2015, 2028, 2039; **5**: 3073
[135-51-3]	**3**: 1891	[142-3-0]	**1**: 51
[135-88-6]	**1**: 529	[142-4-1]	**1**: 636
[136-35-6]	**3**: 1900; **6**: 3812	[142-8-5]	**7**: 4209, 4218
[136-40-3]	**7**: 4208	[142-17-6]	**6**: 3237
[136-47-0]	**4**: 2162	[142-22-3]	**1**: 426; **2**: 1086
[136-51-6]	**6**: 3237	[142-29-0]	**3**: 1832
[136-52-7]	**6**: 3245	[142-30-3]	**1**: 333
[136-53-8]	**6**: 3243	[142-47-2]	**1**: 579
[136-60-7]	**2**: 827, 837	[142-62-1]	**4**: 2485; **5**: 2775
[137-5-3]	**1**: 242	[142-77-8]	**4**: 2040
[137-7-5]	**6**: 3549; **8**: 4662	[142-82-5]	**5**: 2855
[137-9-7]	**1**: 613	[142-83-6]	**7**: 4368
[137-32-6]	**1**: 281; **4**: 2116	[142-84-7]	**1**: 454
[137-40-6]	**7**: 4116	[142-90-5]	**6**: 3250
[137-50-8]	**2**: 774	[142-96-1]	**2**: 955; **4**: 2187, 2196
[137-51-9]	**2**: 773	[143-7-7]	**4**: 2485
[137-99-5]	**6**: 3748	[143-8-8]	**1**: 282; **5**: 2535
[138-15-8]	**1**: 579	[143-15-7]	**2**: 869, 874
[138-42-1]	**6**: 3551	[143-23-7]	**5**: 2847
[138-86-3]	**5**: 2584	[143-27-1]	**1**: 481
[139-12-8]	**1**: 51	[143-28-2]	**1**: 309; **5**: 2549
[139-13-9]	**3**: 1542; **4**: 2295; **5**: 2667; **6**: 3479	[144-16-1]	**3**: 1650
[139-33-3]	**4**: 2297	[144-19-4]	**1**: 336, 344
[139-45-7]	**4**: 2014	[144-49-0]	**5**: 2625

CAS #	Refs	CAS #	Refs
[144-62-7]	**3**: 1649, 1905; **4**: 2309, 2315; **5**: 2738, 2811, 2818; **6**: 3577, 3601	[298-12-4]	**4**: 2309; **5**: 2811, 2817; **6**: 3599
[144-83-2]	**7**: 4207	[299-27-4]	**5**: 2782
[146-54-3]	**5**: 2612	[299-28-5]	**5**: 2782
[147-47-7]	**1**: 505, 508	[299-29-6]	**5**: 2782
[147-82-0]	**2**: 870	[300-92-5]	**6**: 3239
[147-85-3]	**1**: 544, 581	[301-11-1]	**8**: 4634
[147-93-3]	**8**: 4662	[302-1-2]	**7**: 4410
[148-24-3]	**7**: 4255	[302-17-0]	**3**: 1532
[148-25-4]	**6**: 3400	[302-72-7]	**1**: 543
[148-75-4]	**6**: 3396	[302-84-1]	**1**: 545, 581
[149-30-4]	**1**: 505	[303-4-8]	**5**: 2624
[149-44-0]	**7**: 4465	[303-7-1]	**5**: 2980
[149-46-2]	**2**: 776	[303-38-8]	**5**: 2980
[149-74-6]	**7**: 4318	[303-45-7]	**4**: 2384
[149-91-7]	**5**: 2981	[304-55-2]	**8**: 4650
[150-13-0]	**1**: 515; **2**: 834, 838	[306-83-2]	**5**: 2576
[150-25-4]	**5**: 2811	[306-94-5]	**5**: 2575, 2626
[150-30-1]	**1**: 544	[307-24-4]	**5**: 2602
[150-39-0]	**4**: 2297	[307-55-1]	**5**: 2602
[150-69-6]	**6**: 3563; **8**: 4821	[311-89-7]	**5**: 2608
[150-75-4]	**1**: 618; **5**: 2944	[317-34-0]	**7**: 4170
[150-76-5]	**5**: 2943	[319-88-0]	**5**: 2619
[150-78-7]	**5**: 2943	[320-60-5]	**5**: 2609
[150-84-5]	**4**: 2041	[321-14-2]	**7**: 4278
[150-86-7]	**8**: 4613	[324-74-3]	**5**: 2621
[150-97-0]	**5**: 3033	[327-54-8]	**5**: 2617
[151-18-8]	**6**: 3464	[327-92-4]	**5**: 2617
[151-32-6]	**1**: 271	[327-98-0]	**6**: 3847
[151-56-4]	**2**: 663; **4**: 2151; **7**: 4409	[328-39-2]	**1**: 544
[151-67-7]	**2**: 888; **5**: 2581	[328-42-7]	**6**: 3608
[154-42-7]	**7**: 4166	[328-50-7]	**6**: 3608
[156-10-5]	**1**: 524	[328-74-5]	**5**: 2611
[156-43-4]	**1**: 620, 636; **6**: 3563	[328-84-7]	**5**: 2610
[156-59-2]	**3**: 1353	[328-99-4]	**5**: 2609
[156-60-5]	**3**: 1353	[333-20-0]	**5**: 2699
[156-62-7]	**3**: 1735; **5**: 2827	[334-48-5]	**4**: 2485
[198-55-0]	**6**: 3352	[334-88-3]	**3**: 1876; **5**: 2982
[206-44-0]	**5**: 2919	[335-58-0]	**5**: 2582
[208-96-8]	**5**: 2915	[335-67-1]	**5**: 2568, 2602
[244-36-0]	**5**: 2917	[335-76-2]	**5**: 2602
[260-94-6]	**1**: 197	[336-8-3]	**5**: 2604
[271-89-6]	**5**: 2764	[338-83-0]	**5**: 2608, 2626
[280-57-9]	**1**: 470	[338-84-1]	**5**: 2608
[287-23-0]	**5**: 2855	[344-4-7]	**5**: 2618
[287-92-3]	**5**: 2855	[344-7-0]	**5**: 2619
[288-32-4]	**5**: 2693, 2987	[345-18-6]	**5**: 2616
[289-95-2]	**5**: 2693; **7**: 4227	[345-92-6]	**5**: 2617
[292-64-8]	**3**: 1790	[348-54-9]	**5**: 2621
[293-30-1]	**5**: 2680	[349-78-0]	**6**: 3550
[294-62-2]	**3**: 1784, 1790	[350-30-1]	**5**: 2616
[294-93-9]	**3**: 1725; **4**: 2335	[350-46-9]	**5**: 2616; **6**: 3536
[296-39-9]	**3**: 1725	[352-32-9]	**5**: 2613
[297-97-2]	**5**: 2814	[352-70-5]	**5**: 2613
[298-0-0]	**6**: 3562, 3850	[352-93-2]	**8**: 4658
[298-6-6]	**6**: 3853	[353-42-4]	**4**: 1932
[298-7-7]	**6**: 3837, 3850	[353-54-8]	**5**: 2581, 2618
		[353-55-9]	**5**: 2581

CAS Registry Number	Reference
[353-59-3]	**2**: 869, 881; **5**: 2581
[354-6-3]	**5**: 2581
[354-21-2]	**5**: 2576
[354-58-5]	**5**: 2576
[354-64-3]	**5**: 2582
[354-65-4]	**5**: 2582
[354-88-1]	**5**: 2605; **8**: 4504
[355-43-1]	**5**: 2582
[355-46-4]	**5**: 2605; **8**: 4504
[357-83-5]	**5**: 2590
[359-13-7]	**5**: 2624
[359-29-5]	**5**: 2584, 2589
[359-70-6]	**5**: 2608
[360-52-1]	**5**: 2597
[363-72-4]	**5**: 2617
[363-80-4]	**5**: 2617
[365-33-3]	**3**: 1901
[366-18-7]	**7**: 4193, 4218
[367-23-7]	**5**: 2617
[367-25-9]	**5**: 2617, 2621
[368-48-9]	**5**: 2620
[368-94-5]	**5**: 2610
[371-40-4]	**5**: 2617
[371-41-5]	**5**: 2613
[371-62-0]	**5**: 2624
[372-9-8]	**3**: 1542; **5**: 3185, 3193
[372-18-9]	**5**: 2613
[372-31-6]	**5**: 2597, 3067
[372-38-3]	**5**: 2617
[372-39-4]	**5**: 2617
[372-47-4]	**5**: 2619
[372-48-5]	**5**: 2619; **7**: 4199
[372-75-8]	**1**: 543, 579
[373-25-1]	**5**: 2622
[373-44-4]	**1**: 486
[373-49-9]	**4**: 2487
[374-7-2]	**5**: 2576
[375-2-0]	**5**: 2568
[375-17-7]	**5**: 2568
[375-22-4]	**5**: 2602
[375-73-5]	**8**: 4504
[375-85-9]	**5**: 2602
[375-95-1]	**5**: 2602
[376-73-8]	**5**: 2604
[377-38-8]	**5**: 2604
[381-73-7]	**5**: 2625
[382-10-5]	**5**: 2589
[382-21-8]	**5**: 2583, 2622
[382-85-4]	**5**: 2611; **6**: 3538
[384-22-5]	**5**: 2610
[387-89-1]	**5**: 2625
[389-8-2]	**7**: 4207
[392-56-3]	**5**: 2618
[393-9-9]	**5**: 2616
[393-75-9]	**5**: 2610, 2612; **6**: 3537
[396-1-0]	**5**: 2830
[398-23-2]	**5**: 2621
[402-31-3]	**5**: 2609
[402-54-0]	**5**: 2610
[406-90-6]	**5**: 2595
[407-25-0]	**6**: 3599
[420-4-2]	**3**: 1735, 1745; **5**: 2824, 2835; **7**: 4409
[420-26-8]	**2**: 1114
[420-46-2]	**5**: 2587
[420-52-0]	**6**: 3820
[421-50-1]	**5**: 2597
[421-53-4]	**5**: 2598
[422-64-0]	**5**: 2602
[423-39-2]	**5**: 2582
[423-41-6]	**8**: 4504
[423-46-1]	**5**: 2568
[423-55-2]	**5**: 2581
[423-62-1]	**5**: 2582
[425-88-7]	**5**: 2595
[428-59-1]	**5**: 2586, 2592
[430-92-2]	**2**: 1114
[431-3-8]	**2**: 945, 973; **5**: 3108
[431-46-9]	**5**: 2598
[432-3-1]	**5**: 2608
[432-8-6]	**5**: 2608
[433-19-2]	**5**: 2609
[433-27-2]	**5**: 2598
[434-64-0]	**5**: 2618
[439-14-5]	**3**: 1899
[443-48-1]	**5**: 2814, 2988, 2992; **6**: 3570
[443-79-8]	**1**: 544
[445-3-4]	**5**: 2611; **6**: 3538
[446-35-5]	**5**: 2616
[452-58-4]	**7**: 4218
[455-90-3]	**3**: 1901
[456-22-4]	**5**: 2617
[456-39-3]	**3**: 1901
[456-59-7]	**5**: 2983
[457-68-1]	**5**: 2613
[458-24-2]	**5**: 2612
[459-57-4]	**5**: 2617
[460-0-4]	**2**: 870; **5**: 2613
[460-19-5]	**6**: 3578
[460-96-8]	**6**: 3820
[461-42-7]	**3**: 1860; **8**: 4506
[461-58-5]	**3**: 1735, 1755; **5**: 2826, 3198
[462-6-6]	**5**: 2613
[462-8-8]	**7**: 4206, 4208, 4218
[462-94-2]	**1**: 486
[463-40-1]	**4**: 2487
[463-49-0]	**1**: 185
[463-51-4]	**1**: 92; **5**: 3062; **7**: 4118, 4368
[463-57-0]	**5**: 2642
[463-82-1]	**5**: 2855
[464-6-2]	**5**: 2855
[470-82-6]	**8**: 4604
[471-46-5]	**6**: 3595
[472-93-5]	**4**: 2382
[473-29-0]	**3**: 1571

[473-34-7]	**3**: 1566	[504-60-9]	**5**: 3033
[473-54-1]	**8**: 4607	[504-61-0]	**1**: 309; **3**: 1716
[473-55-2]	**8**: 4605	[504-63-2]	**1**: 202, 330, 344; **7**: 4055
[475-20-7]	**8**: 4612	[505-48-6]	**3**: 1905
[475-38-7]	**6**: 3451	[505-52-2]	**3**: 1905
[476-66-4]	**5**: 2982	[505-60-2]	**5**: 2622
[479-18-5]	**7**: 4170	[505-73-7]	**6**: 3222
[479-27-6]	**6**: 3405	[506-12-7]	**4**: 2485
[480-96-6]	**1**: 509	[506-21-8]	**4**: 2487
[481-39-0]	**6**: 3451	[506-30-9]	**4**: 2485
[483-63-6]	**3**: 1719	[506-33-2]	**4**: 2487
[486-25-9]	**5**: 2918	[506-42-3]	**1**: 309; **5**: 2549
[487-89-8]	**5**: 3002	[506-43-4]	**5**: 2549
[488-23-3]	**5**: 2894	[506-44-5]	**5**: 2549
[488-69-7]	**2**: 1070	[506-46-7]	**4**: 2486
[488-70-0]	**6**: 3728	[506-48-9]	**4**: 2486
[489-1-0]	**6**: 3744	[506-50-3]	**4**: 2486
[489-78-1]	**6**: 3433	[506-51-4]	**5**: 2535
[489-98-5]	**1**: 637; **6**: 3544	[506-52-5]	**1**: 282; **5**: 2535
[490-20-0]	**3**: 1640	[506-87-6]	**5**: 2827
[490-79-9]	**5**: 2981	[506-93-4]	**3**: 1757; **5**: 2825
[491-33-8]	**7**: 4256	[507-40-4]	**3**: 1609
[491-35-0]	**7**: 4255	[512-69-6]	**2**: 1067
[492-61-5]	**2**: 1064	[513-13-3]	**3**: 1854
[494-44-0]	**6**: 3415	[513-31-5]	**2**: 869
[495-15-1]	**5**: 3003	[513-42-8]	**1**: 424
[496-3-7]	**1**: 375	[513-74-6]	**4**: 1949
[496-11-7]	**5**: 2916; **6**: 3757	[513-85-9]	**2**: 945
[496-46-8]	**8**: 4830	[513-86-0]	**2**: 945
[496-72-0]	**6**: 3809	[515-46-8]	**2**: 758
[496-78-6]	**6**: 3728, 3730	[515-69-5]	**8**: 4611
[497-3-0]	**1**: 368	[515-84-4]	**3**: 1555
[497-4-1]	**3**: 1586	[516-6-3]	**1**: 545
[497-39-2]	**6**: 3737	[519-37-9]	**7**: 4170
[498-15-7]	**8**: 4604	[521-74-4]	**7**: 4256
[498-23-7]	**3**: 1649; **5**: 3169	[523-87-5]	**7**: 4171
[498-24-8]	**3**: 1649; **5**: 3169	[524-42-5]	**6**: 3451
[498-94-2]	**7**: 4197	[525-6-4]	**5**: 3077
[498-95-3]	**7**: 4197	[525-37-1]	**6**: 3370
[499-75-2]	**6**: 3732	[526-7-8]	**4**: 2384
[499-80-9]	**7**: 4202	[526-55-6]	**5**: 3003
[499-83-2]	**7**: 4202	[526-73-8]	**5**: 2894
[500-8-3]	**6**: 3572	[526-85-2]	**6**: 3728
[500-12-9]	**4**: 2385	[526-95-4]	**5**: 2773
[500-22-1]	**7**: 4213, 4218	[527-7-1]	**5**: 2782
[500-92-5]	**5**: 2840	[527-20-8]	**6**: 3534
[501-52-0]	**3**: 1640	[527-35-5]	**6**: 3728, 3730
[502-42-1]	**3**: 1915; **5**: 3100	[527-53-7]	**5**: 2894
[502-49-8]	**5**: 3101	[527-54-8]	**6**: 3728
[502-65-8]	**4**: 2382	[527-60-6]	**6**: 3728
[502-97-6]	**3**: 1542	[527-62-8]	**1**: 614; **6**: 3561
[503-40-2]	**8**: 4503	[527-73-1]	**6**: 3570
[503-64-0]	**3**: 1716	[527-84-4]	**5**: 2902
[504-2-9]	**7**: 4266	[529-20-4]	**2**: 688; **7**: 3868
[504-20-1]	**5**: 3104	[529-23-7]	**2**: 687
[504-24-5]	**6**: 3569; **7**: 4206, 4208, 4218	[529-86-2]	**2**: 651
[504-29-0]	**7**: 4206, 4218	[530-57-4]	**5**: 2982

[530-78-9]	**5**: 2611
[531-85-1]	**2**: 804
[531-86-2]	**2**: 804
[532-28-5]	**5**: 2982; **6**: 3470
[532-32-1]	**2**: 825, 837
[533-31-3]	**4**: 2384
[533-37-9]	**7**: 4184
[533-73-3]	**6**: 3765
[534-22-5]	**5**: 2747, 2760
[534-52-1]	**6**: 3566
[534-73-6]	**2**: 1062
[535-15-9]	**3**: 1552
[535-77-3]	**5**: 2902
[535-80-8]	**2**: 832, 838
[536-25-4]	**1**: 616
[536-71-0]	**3**: 1889
[536-90-3]	**1**: 636; **6**: 3562
[538-93-2]	**5**: 2902
[539-86-6]	**7**: 4449
[540-10-3]	**4**: 2022
[540-11-4]	**5**: 2553
[540-36-3]	**5**: 2613
[540-49-8]	**2**: 869
[540-59-0]	**1**: 129
[540-61-4]	**6**: 3463
[540-73-8]	**3**: 1868
[540-84-1]	**5**: 2855
[540-88-5]	**1**: 54
[541-2-6]	**7**: 4324
[541-5-9]	**7**: 4324
[541-41-3]	**3**: 1576
[541-50-4]	**6**: 3608
[541-85-5]	**5**: 3098
[541-88-8]	**3**: 1541
[542-2-9]	**5**: 3214
[542-5-2]	**3**: 1649; **6**: 3607
[542-10-9]	**1**: 10
[542-16-5]	**1**: 636
[542-55-2]	**5**: 2738
[542-59-6]	**4**: 2314, 2321
[542-75-6]	**1**: 415; **3**: 1376
[542-76-7]	**6**: 3464
[542-78-9]	**1**: 390; **7**: 4367
[542-88-1]	**4**: 2187, 2199; **5**: 2647
[542-90-5]	**8**: 4632
[542-92-7]	**3**: 1825
[543-28-2]	**2**: 1047
[543-49-7]	**1**: 281
[544-0-3]	**1**: 454
[544-1-4]	**4**: 2187, 2197
[544-17-2]	**5**: 2718, 2728, 2738
[544-40-1]	**8**: 4658
[544-63-8]	**4**: 2485
[544-64-9]	**4**: 2487
[544-73-0]	**4**: 2487
[544-76-3]	**5**: 2855
[544-86-5]	**5**: 2535
[545-6-2]	**6**: 3463
[546-63-4]	**3**: 1635
[546-68-9]	**1**: 312; **7**: 4064
[547-63-7]	**4**: 2014
[547-64-8]	**5**: 3130
[548-39-0]	**5**: 2923
[550-33-4]	**7**: 4167
[550-97-0]	**6**: 3375
[551-62-2]	**5**: 2617
[551-92-8]	**5**: 2988
[552-16-9]	**2**: 810; **6**: 3515
[552-30-7]	**5**: 2895
[552-86-3]	**5**: 2751
[552-89-6]	**2**: 685; **6**: 3515
[553-26-4]	**7**: 4193, 4218
[553-90-2]	**4**: 2016; **6**: 3595
[553-97-9]	**2**: 841
[554-0-7]	**1**: 636; **6**: 3533
[554-12-1]	**4**: 2014; **7**: 4119
[554-70-1]	**6**: 3822
[554-84-7]	**6**: 3562
[554-95-0]	**5**: 2893
[555-3-3]	**6**: 3562
[555-16-8]	**2**: 685; **6**: 3519
[555-21-5]	**6**: 3536
[555-30-6]	**6**: 3762
[555-31-7]	**7**: 4064
[555-35-1]	**6**: 3239
[555-36-2]	**6**: 3245
[555-48-6]	**6**: 3540
[555-84-7]	**5**: 2975
[556-3-6]	**1**: 545
[556-18-3]	**1**: 514; **2**: 687
[556-52-5]	**1**: 421; **4**: 1999, 2003; **5**: 2799
[556-61-6]	**8**: 4636
[556-64-9]	**8**: 4632
[556-67-2]	**7**: 4324
[556-88-7]	**5**: 2824
[556-99-0]	**8**: 4818
[557-4-0]	**6**: 3235
[557-5-1]	**6**: 3243
[557-7-3]	**6**: 3243
[557-17-5]	**4**: 2187
[557-31-3]	**1**: 436
[557-40-4]	**1**: 421, 436
[557-59-5]	**4**: 2486
[557-61-9]	**5**: 2535
[557-91-5]	**2**: 869
[557-98-2]	**1**: 414
[558-13-4]	**2**: 869, 888
[558-38-3]	**3**: 1587
[558-42-9]	**3**: 1587
[562-49-2]	**5**: 2855
[563-45-1]	**3**: 1858
[563-46-2]	**1**: 92
[563-47-3]	**3**: 1390
[563-80-4]	**2**: 974; **5**: 3085

[563-84-8]	**3:** 1587	[589-53-7]	**5:** 2855
[564-2-3]	**5:** 2855	[589-55-9]	**1:** 281
[565-20-8]	**2:** 759	[589-81-1]	**5:** 2855
[565-59-3]	**5:** 2855	[589-82-2]	**1:** 281
[565-61-7]	**5:** 3089	[589-87-7]	**2:** 870
[565-75-3]	**5:** 2855	[589-92-4]	**3:** 1818
[565-80-0]	**5:** 3096	[589-96-8]	**1:** 425
[567-13-5]	**6:** 3433	[590-1-2]	**7:** 4119
[567-18-0]	**6:** 3389	[590-18-1]	**2:** 954, 987; **5:** 2871
[567-47-5]	**6:** 3392	[590-19-2]	**2:** 899
[568-80-9]	**5:** 2982	[590-21-6]	**3:** 1374
[569-42-6]	**6:** 3380	[590-35-2]	**5:** 2855
[571-60-8]	**6:** 3380	[590-46-5]	**1:** 579
[574-0-5]	**6:** 3380	[590-66-9]	**3:** 1798
[575-5-3]	**6:** 3397	[590-86-3]	**1:** 352
[575-38-2]	**6:** 3380, 3452	[590-88-5]	**1:** 486
[575-44-0]	**6:** 3380	[591-9-3]	**1:** 67
[576-24-9]	**3:** 1606, 1896	[591-12-8]	**6:** 3606
[576-55-6]	**6:** 3788	[591-22-0]	**7:** 4178
[577-55-9]	**5:** 2901	[591-24-2]	**3:** 1818
[577-59-3]	**6:** 3567	[591-27-5]	**1:** 603; **6:** 3562; **7:** 4268
[578-66-5]	**6:** 3570	[591-35-5]	**3:** 1606
[578-68-7]	**6:** 3570	[591-50-4]	**5:** 3008
[578-85-8]	**6:** 3394	[591-76-4]	**5:** 2855
[578-95-0]	**1:** 198	[591-87-7]	**1:** 426, 435
[579-44-2]	**5:** 3113	[592-27-8]	**5:** 2855
[581-42-0]	**6:** 3357	[592-34-7]	**3:** 1576
[581-43-1]	**6:** 3381, 3451	[592-35-8]	**2:** 1047
[581-46-4]	**7:** 4193	[592-41-6]	**5:** 2871
[581-47-5]	**7:** 4193	[592-76-7]	**5:** 2871
[581-50-0]	**7:** 4193	[592-84-7]	**4:** 2015
[581-75-9]	**6:** 3370	[592-88-1]	**8:** 4658
[581-89-5]	**6:** 3524	[593-39-5]	**4:** 2487
[582-17-2]	**6:** 3381	[593-40-8]	**4:** 2487
[582-25-2]	**2:** 826	[593-45-3]	**5:** 2855
[583-39-1]	**6:** 3814	[593-50-0]	**5:** 2535
[583-52-8]	**6:** 3593	[593-54-4]	**6:** 3821
[583-53-9]	**2:** 870	[593-60-2]	**2:** 869, 874, 888
[583-60-8]	**3:** 1818	[593-68-0]	**6:** 3821
[583-61-9]	**7:** 4184	[593-70-4]	**5:** 2618
[583-63-1]	**2:** 841	[593-84-0]	**5:** 2825
[583-78-8]	**3:** 1606	[593-85-1]	**5:** 2825
[583-91-5]	**1:** 213	[593-87-3]	**5:** 2825
[584-2-1]	**1:** 281	[593-92-0]	**2:** 869
[584-3-2]	**1:** 344; **2:** 947	[594-9-2]	**6:** 3822
[584-45-2]	**5:** 3178	[594-14-9]	**5:** 2825
[585-34-2]	**6:** 3737	[594-27-4]	**1:** 425
[585-38-6]	**2:** 775	[594-45-6]	**8:** 4504
[585-47-7]	**2:** 761	[594-61-6]	**3:** 1869
[585-71-7]	**2:** 876	[594-82-1]	**5:** 2855
[586-62-9]	**8:** 4602	[595-46-0]	**5:** 3077, 3193
[586-95-8]	**7:** 4211	[597-31-9]	**1:** 327, 375
[586-98-1]	**7:** 4196, 4211, 4218	[597-32-0]	**3:** 1587
[587-4-2]	**2:** 684	[598-5-0]	**3:** 1855
[587-23-5]	**5:** 2983	[598-12-9]	**5:** 2835
[587-90-6]	**6:** 3542	[598-16-3]	**2:** 869
[589-34-4]	**5:** 2855	[598-21-0]	**2:** 878; **5:** 3061

[598-26-5]	**5**: 3077	[612-44-2]	**2**: 810
[598-32-3]	**1**: 309	[612-82-8]	**2**: 806
[598-50-5]	**8**: 4819	[612-83-9]	**2**: 807
[598-53-8]	**4**: 2187	[613-17-2]	**6**: 3384
[598-55-0]	**2**: 1047	[613-20-7]	**6**: 3451
[598-56-1]	**1**: 455	[613-35-4]	**2**: 804
[598-75-4]	**1**: 281	[613-46-7]	**6**: 3473
[598-78-7]	**7**: 4119	[613-73-0]	**6**: 3470
[598-99-2]	**3**: 1555	[613-76-3]	**6**: 3405
[599-52-0]	**3**: 1608	[613-90-1]	**6**: 3470
[599-64-4]	**6**: 3697, 3754	[614-33-5]	**2**: 827
[600-14-6]	**5**: 3109	[614-45-9]	**2**: 828
[600-30-6]	**3**: 1716	[614-80-2]	**1**: 612
[600-36-2]	**1**: 281	[615-5-4]	**6**: 3564
[602-38-0]	**6**: 3524	[615-71-4]	**6**: 3543
[602-94-8]	**5**: 2618	[616-2-4]	**3**: 1648; **5**: 3169
[603-0-9]	**7**: 4170	[616-23-9]	**3**: 1586
[603-34-9]	**1**: 529	[616-38-6]	**2**: 1086
[603-35-0]	**6**: 3822	[616-45-5]	**1**: 432; **7**: 4241
[603-71-4]	**6**: 3525	[616-47-7]	**5**: 2988
[603-83-8]	**6**: 3545	[616-55-7]	**6**: 3737
[604-88-6]	**5**: 2894	[616-84-2]	**6**: 3557
[605-1-6]	**5**: 2894	[616-85-3]	**6**: 3557
[605-65-2]	**6**: 3410	[616-91-1]	**1**: 577; **8**: 4656
[605-71-0]	**6**: 3451, 3524	[617-35-6]	**4**: 2016
[605-91-4]	**7**: 4255	[617-45-8]	**1**: 543, 578
[605-94-7]	**2**: 844	[617-78-7]	**5**: 2855
[606-20-2]	**6**: 3520	[617-94-7]	**6**: 3697
[606-21-3]	**6**: 3543	[617-97-0]	**2**: 762
[606-22-4]	**6**: 3543	[618-3-1]	**2**: 779; **6**: 3552; **8**: 4751
[606-55-3]	**3**: 1533	[618-45-1]	**6**: 3732
[607-34-1]	**6**: 3570	[618-46-2]	**2**: 832
[607-35-2]	**6**: 3570	[619-10-3]	**3**: 1897
[607-80-7]	**4**: 2384	[619-27-2]	**6**: 3541
[608-27-5]	**1**: 636	[619-33-0]	**3**: 1529
[608-31-1]	**1**: 511, 519, 636	[619-60-3]	**1**: 619
[609-8-5]	**5**: 3184	[619-66-9]	**8**: 4577
[609-19-8]	**3**: 1606	[620-14-4]	**5**: 2894
[609-36-9]	**1**: 544	[620-17-7]	**6**: 3728, 3731
[609-46-1]	**2**: 774	[620-23-5]	**2**: 688
[609-54-1]	**2**: 765	[620-64-2]	**5**: 2982
[609-60-9]	**2**: 765	[621-62-5]	**3**: 1522, 1524
[609-72-3]	**8**: 4749	[621-77-2]	**1**: 454
[609-89-2]	**6**: 3561	[621-82-9]	**3**: 1639
[609-93-8]	**6**: 3566	[622-46-8]	**2**: 1047
[610-30-0]	**6**: 3520	[622-57-1]	**8**: 4749
[610-54-8]	**6**: 3564	[622-96-8]	**5**: 2894
[610-67-3]	**2**: 809; **6**: 3561	[622-97-9]	**5**: 2898
[610-92-4]	**3**: 1865	[623-3-0]	**6**: 3471
[611-6-3]	**6**: 3533	[623-8-5]	**8**: 4749
[611-14-3]	**5**: 2894	[623-36-9]	**1**: 368; **7**: 4039
[611-21-2]	**8**: 4749	[623-42-7]	**4**: 2014
[611-24-5]	**1**: 611	[623-43-8]	**3**: 1716
[611-34-7]	**6**: 3570	[623-50-7]	**5**: 2966
[611-71-2]	**5**: 2983	[623-70-1]	**3**: 1716
[611-72-3]	**5**: 2982	[623-73-4]	**3**: 1876, 1900
[612-23-7]	**6**: 3517	[623-84-7]	**1**: 425

[623-93-8]	**1**: 282	[630-10-4]	**3**: 1748
[624-29-3]	**3**: 1798	[630-19-3]	**1**: 352
[624-38-4]	**8**: 4584	[630-20-6]	**3**: 1320
[624-64-6]	**2**: 954, 987; **5**: 2871	[631-61-8]	**1**: 52
[624-92-0]	**8**: 4658	[631-64-1]	**2**: 878
[625-22-9]	**3**: 1855	[632-22-4]	**8**: 4821
[625-30-9]	**1**: 454	[632-79-1]	**2**: 885
[625-35-4]	**3**: 1716	[633-47-6]	**3**: 1719
[625-36-5]	**7**: 4120	[634-55-9]	**1**: 67
[625-37-6]	**3**: 1716	[634-93-5]	**1**: 510
[625-38-7]	**3**: 1715	[635-81-4]	**5**: 2894
[625-72-9]	**5**: 2968	[636-30-6]	**1**: 636; **6**: 3534
[626-26-6]	**8**: 4658	[636-61-3]	**5**: 2970
[626-39-1]	**2**: 870	[637-12-7]	**6**: 3239
[626-43-7]	**1**: 636	[637-32-1]	**5**: 2840
[626-48-2]	**5**: 3075	[637-89-8]	**8**: 4662
[626-55-1]	**7**: 4199	[638-4-0]	**3**: 1798
[626-56-2]	**7**: 4185, 4197	[638-21-1]	**6**: 3821
[626-58-4]	**7**: 4197	[638-23-3]	**1**: 577
[626-60-8]	**7**: 4199	[638-37-9]	**1**: 390
[626-61-9]	**7**: 4195	[638-42-6]	**2**: 1047
[626-64-2]	**7**: 4209	[638-53-9]	**4**: 2485
[626-67-5]	**7**: 4197	[638-68-6]	**5**: 2855
[626-86-8]	**1**: 271	[638-79-9]	**5**: 2582
[626-89-1]	**1**: 281	[642-32-0]	**5**: 2894
[627-12-3]	**2**: 1047	[643-28-7]	**1**: 529
[627-20-3]	**5**: 2871	[643-38-9]	**1**: 197
[627-30-5]	**3**: 1586	[644-97-3]	**6**: 3825
[627-91-8]	**1**: 271	[645-41-0]	**1**: 454
[627-93-0]	**1**: 271; **4**: 2014	[645-62-5]	**1**: 368
[627-97-9]	**7**: 4434	[645-92-1]	**3**: 1757
[628-13-7]	**7**: 4218	[646-4-8]	**5**: 2871
[628-20-6]	**6**: 3464	[646-6-0]	**5**: 3038
[628-28-4]	**4**: 2187	[646-25-3]	**1**: 486
[628-34-2]	**4**: 2187	[646-30-0]	**4**: 2485
[628-37-5]	**6**: 3640	[650-51-1]	**1**: 92; **3**: 1555
[628-55-7]	**4**: 2187	[652-67-5]	**2**: 1074
[628-63-7]	**1**: 55; **4**: 2014	[653-37-2]	**5**: 2618
[628-81-9]	**4**: 2187; **5**: 2730	[655-86-7]	**6**: 3811
[628-94-4]	**1**: 272	[657-27-2]	**1**: 544, 580
[629-11-8]	**1**: 331, 344; **5**: 2553, 2848	[659-70-1]	**4**: 2041
[629-14-1]	**4**: 2320	[661-19-8]	**5**: 2535
[629-20-9]	**1**: 128	[669-90-9]	**5**: 2775
[629-30-1]	**5**: 2553	[670-2-0]	**2**: 783
[629-38-9]	**4**: 2321	[670-54-2]	**6**: 3462
[629-41-4]	**5**: 2553	[670-96-2]	**5**: 2988
[629-45-8]	**8**: 4658	[671-89-6]	**2**: 784
[629-50-5]	**5**: 2855	[673-6-3]	**1**: 581
[629-59-4]	**5**: 2855	[673-84-7]	**8**: 4600
[629-62-9]	**5**: 2855	[674-82-8]	**4**: 2020; **5**: 3068
[629-73-2]	**5**: 2871	[675-14-9]	**5**: 2621
[629-76-5]	**1**: 282; **5**: 2535	[676-22-1]	**3**: 1790
[629-78-7]	**5**: 2855	[676-83-5]	**6**: 3825
[629-92-5]	**5**: 2855	[676-97-1]	**6**: 3842, 3847
[629-96-9]	**1**: 282; **5**: 2535	[676-98-2]	**6**: 3842, 3847
[629-98-1]	**5**: 2549	[677-21-4]	**5**: 2584
[630-8-0]	**1**: 34; **2**: 956; **4**: 2315; **7**: 4106	[677-69-0]	**5**: 2582

CAS Registry Number Index

[678-45-5]	**5**: 2604	[787-77-9]	**2**: 809
[680-31-9]	**3**: 1611	[793-24-8]	**1**: 508
[684-16-2]	**5**: 2597, 2625	[802-93-7]	**5**: 2591
[684-93-5]	**3**: 1876	[811-96-1]	**5**: 2598
[687-48-9]	**2**: 1047	[811-98-3]	**1**: 53
[689-12-3]	**1**: 227	[813-60-5]	**1**: 330
[689-97-4]	**5**: 2699	[814-68-6]	**1**: 225
[690-2-8]	**6**: 3640	[814-78-8]	**5**: 3103
[693-23-2]	**3**: 1783, 1791, 1905	[814-80-2]	**5**: 3129
[693-57-2]	**3**: 1791	[815-51-0]	**3**: 1718
[693-65-2]	**4**: 2187, 2197	[815-68-9]	**1**: 67
[693-98-1]	**5**: 2988	[817-83-4]	**6**: 3239
[694-17-7]	**5**: 2593	[818-61-1]	**1**: 227
[694-52-0]	**5**: 2619	[820-71-3]	**1**: 426
[694-59-7]	**7**: 4196, 4218	[821-6-7]	**2**: 869
[694-83-7]	**5**: 2847	[821-38-5]	**3**: 1905
[696-23-1]	**5**: 2988; **6**: 3570	[821-60-3]	**6**: 3811
[696-29-7]	**3**: 1798	[821-95-4]	**5**: 2871
[696-44-6]	**8**: 4749	[822-6-0]	**2**: 666; **5**: 2847
[696-54-8]	**7**: 4213	[822-36-6]	**5**: 2988
[696-59-3]	**7**: 4238	[822-68-4]	**6**: 3821
[697-82-5]	**6**: 3728, 3730	[823-40-5]	**6**: 3809
[697-83-6]	**5**: 2620	[824-72-6]	**6**: 3847
[698-63-5]	**6**: 3571	[826-62-0]	**7**: 4341
[698-71-5]	**6**: 3728	[830-13-7]	**3**: 1783; **5**: 3101
[700-12-9]	**5**: 2894	[831-59-4]	**7**: 4265
[700-16-3]	**5**: 2620	[836-30-6]	**1**: 522; **6**: 3530
[701-97-3]	**3**: 1640	[838-88-0]	**1**: 527
[706-31-0]	**3**: 1790	[849-99-0]	**3**: 1817
[717-44-2]	**2**: 764	[865-33-8]	**5**: 2720
[721-26-6]	**7**: 4266	[867-44-7]	**5**: 2835
[727-81-1]	**2**: 844	[868-14-4]	**8**: 4567
[732-26-3]	**6**: 3737	[868-54-2]	**6**: 3465
[738-70-5]	**5**: 2830; **6**: 3762	[868-77-9]	**6**: 3250
[756-80-9]	**6**: 3853	[869-24-9]	**2**: 1234
[757-44-8]	**7**: 4343	[869-29-4]	**1**: 204, 425
[760-23-6]	**2**: 943; **3**: 1381	[870-24-6]	**2**: 666
[760-93-0]	**6**: 3253	[871-76-1]	**2**: 666
[761-65-9]	**5**: 2705	[872-5-9]	**5**: 2871
[762-50-5]	**5**: 2588	[872-50-4]	**1**: 121, 47; **2**: 1007, 721; **7**: 4243
[762-75-4]	**5**: 2719	[872-85-5]	**7**: 4213
[764-1-2]	**1**: 309	[873-32-5]	**6**: 3471
[764-41-0]	**2**: 943	[873-69-8]	**7**: 4213, 4218
[765-4-8]	**5**: 2553	[873-94-9]	**3**: 1819
[765-34-4]	**1**: 206	[874-42-0]	**2**: 684
[766-7-4]	**3**: 1810; **6**: 3634	[877-44-1]	**5**: 2894
[766-17-6]	**7**: 4197	[877-83-8]	**3**: 1776
[766-94-9]	**1**: 125	[879-97-0]	**6**: 3737
[767-0-0]	**6**: 3472	[883-40-9]	**3**: 1876
[767-54-4]	**3**: 1819	[885-62-1]	**6**: 3552
[771-3-9]	**5**: 3075	[919-30-2]	**7**: 4335
[771-56-2]	**5**: 2618	[920-46-7]	**6**: 3253
[771-60-8]	**5**: 2618	[920-66-1]	**5**: 2590, 2625
[771-61-9]	**5**: 2618	[923-9-1]	**1**: 578
[771-97-1]	**6**: 3405	[923-26-2]	**6**: 3250
[776-19-2]	**3**: 1570	[925-16-6]	**1**: 426
[777-37-7]	**5**: 2610; **6**: 3537	[925-54-2]	**1**: 352

[925-60-0]	**1**: 227	[1072-63-5]	**5**: 2988
[926-3-4]	**3**: 1867	[1076-22-8]	**7**: 4166
[926-39-6]	**2**: 667	[1076-97-7]	**8**: 4588
[926-41-0]	**4**: 2016	[1077-28-7]	**8**: 4661
[928-51-8]	**3**: 1587	[1079-66-9]	**6**: 3825
[931-19-1]	**7**: 4196	[1083-48-3]	**3**: 1863
[931-36-2]	**5**: 2988	[1087-21-4]	**1**: 426
[931-87-3]	**3**: 1790	[1113-38-8]	**6**: 3594
[931-89-5]	**3**: 1790	[1115-20-4]	**1**: 329
[933-48-2]	**3**: 1818	[1115-63-5]	**1**: 578
[933-75-5]	**3**: 1606	[1115-84-0]	**1**: 582
[933-78-8]	**3**: 1606	[1116-40-1]	**1**: 454
[934-32-7]	**5**: 2838	[1116-76-3]	**1**: 481; **5**: 2731
[935-95-5]	**3**: 1606	[1117-86-8]	**1**: 337, 344
[936-89-0]	**6**: 3728	[1118-61-2]	**6**: 3465
[937-14-4]	**6**: 3643, 3648	[1119-16-0]	**1**: 352
[938-25-0]	**6**: 3405	[1119-34-2]	**1**: 543, 578
[939-23-1]	**7**: 4182, 4218	[1120-21-4]	**5**: 2855
[939-95-7]	**2**: 768	[1120-36-1]	**5**: 2871
[939-97-9]	**2**: 831	[1120-46-3]	**6**: 3241
[943-39-5]	**6**: 3643	[1120-48-5]	**1**: 481
[944-22-9]	**6**: 3847	[1120-49-6]	**1**: 481
[946-89-4]	**3**: 1784	[1120-71-4]	**8**: 4505
[947-4-6]	**3**: 1783, 1786	[1121-31-9]	**7**: 4196
[947-42-2]	**7**: 4324	[1121-60-4]	**7**: 4213
[955-1-1]	**6**: 3743	[1121-78-4]	**7**: 4209
[976-56-7]	**6**: 3846	[1122-17-4]	**6**: 3451
[987-78-0]	**3**: 1635; **7**: 4232	[1122-58-3]	**7**: 4206, 4208, 4218
[992-94-9]	**7**: 4330, 4351	[1122-62-9]	**7**: 4218
[993-0-0]	**7**: 4351	[1123-73-5]	**6**: 3728
[993-13-5]	**6**: 3844	[1123-94-0]	**6**: 3728
[993-43-1]	**6**: 3843, 3847	[1124-33-0]	**6**: 3569; **7**: 4219
[998-30-1]	**7**: 4323	[1124-64-7]	**7**: 4194
[998-40-3]	**6**: 3822	[1129-42-6]	**1**: 5; **8**: 4832
[999-21-3]	**1**: 426, 435	[1129-52-8]	**3**: 1528
[999-55-3]	**1**: 227, 426	[1129-89-1]	**3**: 1790
[999-61-1]	**1**: 227	[1131-60-8]	**6**: 3754
[999-81-5]	**3**: 1636	[1138-52-9]	**6**: 3737
[999-97-3]	**7**: 4346	[1141-38-4]	**2**: 1126; **6**: 3355
[1002-35-3]	**2**: 903	[1163-19-5]	**2**: 885
[1002-84-2]	**4**: 2485	[1174-72-7]	**7**: 4323
[1003-67-4]	**7**: 4196	[1184-84-5]	**8**: 4505
[1003-73-2]	**7**: 4196	[1187-42-4]	**6**: 3465
[1004-99-5]	**3**: 1587	[1187-93-5]	**5**: 2594
[1006-59-3]	**6**: 3728, 3730	[1190-63-2]	**4**: 2014
[1007-49-4]	**7**: 4196	[1193-92-6]	**7**: 4213
[1008-89-5]	**7**: 4182	[1194-65-6]	**6**: 3471
[1025-15-6]	**1**: 429	[1195-59-1]	**7**: 4211
[1066-35-9]	**7**: 4311	[1196-57-2]	**5**: 2818
[1066-40-6]	**7**: 4324	[1197-34-8]	**6**: 3728
[1066-42-8]	**7**: 4324	[1198-27-2]	**6**: 3451
[1070-78-6]	**5**: 2588	[1198-61-4]	**5**: 2619
[1071-83-6]	**6**: 3845	[1198-69-2]	**5**: 2775
[1072-21-5]	**1**: 390	[1204-28-0]	**7**: 4341
[1072-35-1]	**6**: 3241	[1207-69-8]	**1**: 198
[1072-52-2]	**2**: 666	[1228-53-1]	**6**: 3550
[1072-62-4]	**5**: 2988	[1300-73-8]	**1**: 529

5061

CAS Registry Number Index

[1301-96-8]	**2**: 663	[1571-33-1]	**6**: 3844
[1310-58-3]	**2**: 1132, 663	[1572-52-7]	**6**: 3462
[1310-73-2]	**2**: 664	[1573-91-7]	**6**: 3384
[1315-9-9]	**5**: 2730	[1576-87-0]	**1**: 368
[1315-11-3]	**5**: 2730	[1582-9-8]	**5**: 2612; **6**: 3538
[1319-77-3]	**3**: 1656	[1592-23-0]	**6**: 3237
[1321-64-8]	**3**: 1442	[1596-9-4]	**6**: 3754
[1321-65-9]	**3**: 1442	[1599-68-4]	**5**: 2871
[1321-74-0]	**7**: 4435	[1606-80-8]	**8**: 4506
[1331-22-2]	**3**: 1818	[1615-75-4]	**5**: 2588
[1331-46-0]	**6**: 3786	[1620-75-3]	**7**: 4202
[1333-74-0]	**2**: 956	[1623-5-8]	**5**: 2594
[1333-82-0]	**1**: 91	[1623-19-4]	**1**: 426
[1335-87-1]	**3**: 1442	[1634-4-4]	**4**: 2187; **5**: 2719; **6**: 3337; **7**: 4145
[1335-88-2]	**3**: 1442	[1635-61-6]	**6**: 3533
[1336-93-2]	**6**: 3245	[1637-25-8]	**1**: 112
[1338-2-9]	**6**: 3241	[1641-41-4]	**5**: 2917
[1344-28-1]	**5**: 2749	[1643-19-2]	**1**: 432
[1398-61-4]	**2**: 1065	[1649-8-7]	**5**: 2576, 2587
[1401-55-4]	**5**: 2982	[1653-19-6]	**3**: 1383, 1389
[1420-7-1]	**6**: 3566	[1656-44-6]	**6**: 3552
[1421-49-4]	**6**: 3743	[1663-39-4]	**1**: 227
[1429-50-1]	**6**: 3845	[1677-46-9]	**2**: 834
[1445-23-4]	**6**: 3737	[1678-82-6]	**3**: 1798
[1450-14-2]	**7**: 4344, 4346	[1678-91-7]	**3**: 1798
[1452-77-3]	**7**: 4202	[1678-92-8]	**3**: 1798
[1453-82-3]	**7**: 4202	[1678-93-9]	**3**: 1798
[1454-84-8]	**1**: 282; **5**: 2535	[1678-98-4]	**3**: 1798
[1454-85-9]	**1**: 282; **5**: 2535	[1679-64-7]	**8**: 4580
[1459-93-4]	**4**: 2015	[1686-14-2]	**4**: 2002
[1470-91-3]	**2**: 1114	[1687-61-2]	**6**: 3728
[1470-94-6]	**5**: 2917; **6**: 3757	[1687-64-5]	**6**: 3728
[1471-18-7]	**1**: 431	[1694-92-4]	**6**: 3549
[1476-11-5]	**3**: 1381	[1702-42-7]	**6**: 3828
[1477-55-0]	**1**: 529	[1708-29-8]	**2**: 944
[1484-13-5]	**1**: 125; **2**: 1059	[1708-32-3]	**2**: 904
[1484-84-0]	**7**: 4197	[1712-79-4]	**6**: 3652
[1486-75-5]	**3**: 1790	[1717-0-6]	**5**: 2576
[1493-13-6]	**5**: 2605; **8**: 4504	[1724-39-6]	**3**: 1783
[1493-27-2]	**5**: 2616; **6**: 3536	[1732-72-5]	**6**: 3821
[1501-82-2]	**3**: 1784	[1732-74-7]	**6**: 3821
[1502-3-0]	**1**: 464	[1746-1-6]	**3**: 1612
[1502-6-3]	**5**: 3101	[1746-3-8]	**6**: 3844
[1502-47-2]	**5**: 3199	[1746-77-6]	**2**: 1047
[1513-65-1]	**5**: 2620	[1754-62-7]	**3**: 1640
[1514-85-8]	**5**: 2604	[1758-73-2]	**7**: 4468
[1518-16-7]	**5**: 3192	[1760-24-3]	**7**: 4336
[1518-83-8]	**6**: 3754	[1763-23-1]	**5**: 2605
[1518-84-9]	**6**: 3754	[1769-88-6]	**6**: 3383
[1520-22-5]	**8**: 4636	[1779-48-2]	**6**: 3827, 3837
[1552-12-1]	**3**: 1790	[1786-87-4]	**6**: 3647
[1561-49-5]	**3**: 1580	[1794-90-7]	**6**: 3494
[1561-92-8]	**8**: 4506	[1795-16-0]	**3**: 1798
[1562-0-1]	**8**: 4505	[1817-73-8]	**6**: 3543
[1562-85-2]	**5**: 2982	[1822-51-1]	**7**: 4199
[1563-66-2]	**6**: 3762	[1824-81-3]	**7**: 4206
[1570-45-2]	**7**: 4218	[1825-62-3]	**7**: 4322

[1829-0-1]	**3:** 1888	[2146-71-6]	**8:** 4860
[1829-32-9]	**7:** 4278	[2163-42-0]	**1:** 331
[1835-49-0]	**5:** 2617	[2164-17-2]	**5:** 2612
[1836-75-5]	**6:** 3530	[2170-3-8]	**3:** 1648
[1841-19-6]	**5:** 2616	[2176-62-7]	**7:** 4199, 4218
[1845-51-8]	**5:** 3049	[2179-57-9]	**8:** 4658
[1852-4-6]	**3:** 1905	[2188-9-2]	**8:** 4820
[1861-40-1]	**5:** 2612	[2198-23-4]	**5:** 2871
[1866-31-5]	**1:** 431	[2203-24-9]	**3:** 1749; **5:** 2834
[1873-25-2]	**3:** 1587	[2203-34-1]	**3:** 1587
[1879-9-0]	**6:** 3737	[2203-35-2]	**3:** 1587
[1883-75-6]	**5:** 2757	[2207-1-4]	**3:** 1798
[1885-14-9]	**3:** 1576	[2207-3-6]	**3:** 1798
[1892-33-7]	**2:** 776	[2207-4-7]	**3:** 1798
[1897-52-5]	**5:** 2616; **6:** 3471	[2208-89-1]	**3:** 1756
[1899-93-0]	**2:** 762	[2216-38-8]	**5:** 2871
[1899-94-1]	**2:** 762	[2216-51-5]	**5:** 3050
[1910-42-5]	**7:** 4192, 4218	[2219-30-9]	**1:** 583
[1929-82-4]	**7:** 4218	[2219-72-9]	**6:** 3737
[1934-75-4]	**5:** 2839	[2219-73-0]	**6:** 3728
[1941-19-1]	**6:** 3828	[2219-82-1]	**6:** 3737
[1941-79-3]	**6:** 3643	[2223-93-0]	**6:** 3243
[1948-33-0]	**5:** 2944	[2223-95-2]	**6:** 3245
[1951-12-8]	**2:** 1116	[2225-18-5]	**2:** 778
[2004-39-9]	**5:** 2535	[2234-13-1]	**3:** 1442
[2004-70-8]	**7:** 4368	[2243-61-0]	**6:** 3405
[2007-20-7]	**6:** 3439	[2243-62-1]	**6:** 3404
[2015-14-7]	**2:** 764	[2243-63-2]	**6:** 3405
[2016-42-4]	**1:** 481	[2243-64-3]	**6:** 3405
[2016-57-1]	**1:** 455, 481	[2243-67-6]	**6:** 3405
[2016-63-9]	**7:** 4170	[2243-78-9]	**2:** 803
[2027-17-0]	**6:** 3357	[2258-42-6]	**1:** 81
[2028-63-9]	**1:** 309	[2274-11-5]	**1:** 227
[2031-62-1]	**7:** 4323	[2274-74-0]	**3:** 1891
[2031-67-6]	**7:** 4322	[2283-8-1]	**6:** 3382
[2033-24-1]	**5:** 3179	[2311-91-3]	**6:** 3643, 3647
[2035-75-8]	**1:** 271	[2314-97-8]	**5:** 2582, 2610
[2039-87-4]	**7:** 4437	[2367-82-0]	**5:** 2617
[2042-14-0]	**3:** 1896	[2370-88-9]	**7:** 4324, 4350
[2045-70-7]	**6:** 3572	[2374-5-2]	**6:** 3787
[2049-95-8]	**5:** 2902	[2378-2-1]	**5:** 2590
[2050-60-4]	**6:** 3595	[2378-8-7]	**5:** 2597
[2050-92-2]	**1:** 454	[2386-47-2]	**8:** 4504
[2051-65-2]	**2:** 781	[2386-87-0]	**4:** 2004
[2051-76-5]	**1:** 225	[2388-12-7]	**6:** 3643, 3648
[2058-94-8]	**5:** 2602	[2389-73-3]	**2:** 764
[2068-80-6]	**1:** 578	[2398-81-4]	**7:** 4196
[2074-50-2]	**7:** 4192	[2399-48-6]	**5:** 2762
[2074-67-1]	**6:** 3837	[2401-24-3]	**1:** 614
[2078-54-8]	**6:** 3732	[2402-77-9]	**7:** 4199, 4218
[2104-64-5]	**6:** 3847	[2402-78-0]	**7:** 4199, 4218
[2109-98-0]	**1:** 352	[2402-79-1]	**7:** 4199
[2114-20-7]	**2:** 1047	[2402-95-1]	**7:** 4196
[2123-89-9]	**6:** 3659	[2409-55-4]	**6:** 3737
[2130-56-5]	**2:** 810	[2416-94-6]	**6:** 3727
[2134-29-4]	**1:** 202, 330, 375	[2420-98-6]	**6:** 3243
[2136-89-2]	**2:** 832	[2425-6-1]	**5:** 2837

CAS Registry Number Index

[2425-77-6]	**5**: 2551	[2725-55-5]	**2**: 768
[2426-54-2]	**4**: 2159	[2741-16-4]	**6**: 3732
[2431-50-7]	**3**: 1382	[2751-90-8]	**6**: 3828
[2432-74-8]	**5**: 2849	[2764-72-9]	**7**: 4218
[2435-53-2]	**3**: 1608	[2765-29-9]	**3**: 1790
[2437-16-3]	**6**: 3382	[2767-77-3]	**2**: 764
[2437-17-4]	**6**: 3382	[2767-80-8]	**6**: 3823
[2437-18-5]	**6**: 3384	[2768-2-7]	**7**: 4348
[2437-56-1]	**5**: 2871	[2783-17-7]	**1**: 487
[2438-20-2]	**7**: 4119	[2807-54-7]	**1**: 426
[2438-80-4]	**2**: 1066	[2809-21-4]	**6**: 3844
[2439-10-3]	**5**: 2837	[2816-11-7]	**4**: 2380
[2439-35-2]	**4**: 2159	[2819-86-5]	**6**: 3728
[2440-60-0]	**5**: 2824	[2834-90-4]	**6**: 3425
[2452-1-9]	**6**: 3243	[2834-92-6]	**6**: 3426
[2455-14-3]	**6**: 3743	[2837-89-0]	**5**: 2576, 2623
[2457-1-4]	**6**: 3237	[2840-26-8]	**1**: 615
[2457-47-8]	**7**: 4199	[2855-13-2]	**5**: 3016, 3106
[2459-10-1]	**4**: 2015	[2855-19-8]	**4**: 2001
[2469-55-8]	**7**: 4337	[2857-97-8]	**4**: 2335
[2469-99-0]	**6**: 3465	[2867-47-2]	**6**: 3250
[2479-47-2]	**1**: 527	[2872-8-4]	**6**: 3744
[2495-39-8]	**8**: 4506	[2878-14-0]	**1**: 433
[2499-58-3]	**1**: 227	[2893-33-6]	**7**: 4202
[2499-95-8]	**1**: 227	[2893-78-9]	**3**: 1568
[2501-94-2]	**6**: 3824	[2904-27-0]	**3**: 1770
[2517-43-3]	**3**: 1714	[2904-28-1]	**3**: 1770
[2524-3-0]	**6**: 3852	[2905-52-4]	**3**: 1580
[2524-4-1]	**6**: 3852	[2934-5-6]	**6**: 3732
[2530-26-9]	**6**: 3569	[2934-7-8]	**6**: 3732
[2530-83-8]	**7**: 4324, 4339	[2935-44-6]	**1**: 344
[2530-85-0]	**7**: 4340	[2937-50-0]	**3**: 1576
[2530-87-2]	**7**: 4351	[2942-59-8]	**7**: 4196, 4199
[2548-85-8]	**4**: 2487	[2971-90-6]	**5**: 3075
[2554-6-5]	**7**: 4324	[2973-10-6]	**3**: 1855
[2565-58-4]	**2**: 1235	[2987-16-8]	**1**: 352
[2576-47-8]	**2**: 663	[2998-4-1]	**1**: 426
[2581-34-2]	**6**: 3565	[2998-8-5]	**1**: 227
[2582-30-1]	**5**: 2831, 2833	[2998-23-4]	**1**: 227
[2605-44-9]	**6**: 3243	[3007-31-6]	**1**: 481
[2613-76-5]	**5**: 2901	[3010-82-0]	**6**: 3812
[2615-25-0]	**6**: 3810	[3017-95-6]	**2**: 876
[2618-96-4]	**2**: 759	[3029-30-9]	**6**: 3473
[2622-14-2]	**6**: 3822	[3031-66-1]	**1**: 332
[2623-37-2]	**6**: 3382	[3031-73-0]	**6**: 3634
[2627-95-4]	**7**: 4324	[3031-74-1]	**6**: 3634
[2628-17-3]	**6**: 3756	[3031-75-2]	**6**: 3634
[2633-64-9]	**2**: 766	[3033-77-0]	**2**: 1229
[2637-34-5]	**7**: 4218	[3034-38-6]	**5**: 2988; **6**: 3570
[2687-25-4]	**6**: 3809	[3034-42-2]	**5**: 2988
[2687-91-4]	**7**: 4247	[3039-83-6]	**8**: 4505
[2687-94-7]	**7**: 4247	[3041-16-5]	**4**: 1938
[2687-96-9]	**7**: 4247	[3058-35-3]	**6**: 3643
[2696-92-6]	**1**: 91	[3068-88-0]	**5**: 3076
[2699-12-9]	**6**: 3668	[3069-29-2]	**7**: 4336
[2706-90-3]	**5**: 2602	[3071-32-7]	**4**: 2209
[2706-91-4]	**8**: 4504	[3073-66-3]	**3**: 1798

[3087-37-4]	**7:** 4064	[3592-12-9]	**2:** 1091
[3115-68-2]	**6:** 3828	[3603-99-4]	**5:** 3101
[3127-44-4]	**5:** 3004	[3605-1-4]	**7:** 4232
[3133-1-5]	**5:** 2535	[3622-84-2]	**2:** 758
[3135-18-0]	**6:** 3846	[3632-91-5]	**5:** 2782
[3137-0-6]	**1:** 426	[3638-97-9]	**5:** 2617
[3147-55-5]	**2:** 870	[3648-20-2]	**7:** 3886
[3155-43-9]	**5:** 2553	[3658-48-8]	**6:** 3850
[3156-7-8]	**5:** 3077	[3658-77-3]	**1:** 333
[3159-62-4]	**6:** 3237	[3658-80-8]	**8:** 4658
[3159-98-6]	**1:** 100	[3695-77-0]	**8:** 4652
[3161-14-6]	**6:** 3791	[3697-42-5]	**5:** 2841
[3173-72-6]	**6:** 3405	[3701-1-7]	**2:** 761
[3178-22-1]	**3:** 1798	[3702-98-5]	**5:** 3067
[3184-13-2]	**1:** 544, 581	[3710-23-4]	**6:** 3357
[3194-55-6]	**2:** 886	[3720-97-6]	**5:** 2813
[3209-22-1]	**6:** 3533	[3724-65-0]	**3:** 1715
[3228-1-1]	**6:** 3732	[3731-51-9]	**7:** 4218
[3228-2-2]	**6:** 3732	[3736-8-1]	**7:** 4171
[3228-3-3]	**6:** 3732	[3741-15-9]	**6:** 3570
[3228-4-4]	**6:** 3732	[3746-67-6]	**2:** 782
[3230-94-2]	**1:** 581	[3771-14-0]	**6:** 3389
[3238-38-8]	**6:** 3728, 3730	[3792-50-5]	**1:** 578
[3268-49-3]	**1:** 203	[3792-59-4]	**6:** 3847
[3277-26-7]	**7:** 4324	[3796-23-4]	**5:** 2620
[3278-35-1]	**3:** 1580	[3796-24-5]	**5:** 2620
[3279-20-7]	**6:** 3747	[3810-35-3]	**6:** 3573
[3279-27-4]	**6:** 3747	[3847-19-6]	**7:** 4195
[3279-76-3]	**7:** 4209	[3848-12-2]	**3:** 1524
[3290-70-8]	**1:** 421	[3848-24-6]	**5:** 3109
[3306-6-7]	**5:** 2983	[3855-26-3]	**6:** 3728
[3306-62-5]	**2:** 770	[3872-25-1]	**5:** 2605
[3312-60-5]	**1:** 485	[3896-11-5]	**6:** 3744
[3316-2-7]	**6:** 3398	[3913-2-8]	**5:** 2551
[3353-5-7]	**6:** 3245	[3917-15-5]	**1:** 436
[3365-90-0]	**2:** 811	[3926-62-3]	**2:** 1230; **3:** 1541, 1547
[3366-95-8]	**5:** 2992	[3933-88-8]	**6:** 3789
[3377-24-0]	**1:** 463	[3934-16-5]	**8:** 4506
[3380-34-5]	**6:** 3788	[3937-56-2]	**5:** 2553
[3385-41-9]	**1:** 271	[3937-59-5]	**3:** 1819, 1905
[3388-4-3]	**7:** 4339	[3938-95-2]	**4:** 2016
[3413-97-6]	**3:** 1716	[3984-22-3]	**1:** 203
[3433-37-2]	**7:** 4197	[4000-16-2]	**5:** 2831, 2834
[3445-11-2]	**7:** 4247	[4009-98-7]	**6:** 3828
[3452-7-1]	**5:** 2871	[4015-58-1]	**3:** 1541
[3452-97-9]	**1:** 282	[4023-53-4]	**6:** 3822
[3482-86-8]	**5:** 2835	[4039-32-1]	**7:** 4329
[3493-11-6]	**1:** 582	[4043-87-2]	**7:** 4197
[3496-32-0]	**5:** 3077	[4050-45-7]	**5:** 2871
[3497-0-5]	**6:** 3847	[4075-81-4]	**7:** 4117
[3506-9-0]	**6:** 3375	[4076-2-2]	**8:** 4650
[3576-88-3]	**5:** 3200	[4088-60-2]	**1:** 309
[3577-63-7]	**2:** 780	[4098-71-9]	**5:** 3016, 3106
[3586-39-8]	**3:** 1819	[4130-42-1]	**6:** 3742
[3588-11-2]	**3:** 1389	[4137-11-5]	**1:** 205
[3588-12-3]	**3:** 1389	[4170-30-3]	**2:** 957; **3:** 1711; **4:** 2056
[3588-13-4]	**3:** 1389	[4181-95-7]	**5:** 2855

[4185-69-7]	**6**: 3531	[5036-48-6]	**5**: 2991
[4206-94-4]	**6**: 3827, 3837	[5064-31-3]	**6**: 3480
[4212-43-5]	**6**: 3643, 3646; **7**: 4039	[5076-19-7]	**4**: 2001
[4214-79-3]	**7**: 4209	[5089-70-3]	**7**: 4336
[4216-2-8]	**2**: 1134	[5104-49-4]	**5**: 2621
[4254-29-9]	**5**: 2917	[5106-98-9]	**7**: 4278
[4286-23-1]	**6**: 3756	[5118-80-9]	**5**: 2666, 3217
[4297-95-4]	**6**: 3837	[5152-99-8]	**2**: 1006
[4316-42-1]	**5**: 2730	[5183-78-8]	**2**: 758
[4320-30-3]	**1**: 578	[5191-97-9]	**1**: 581
[4336-48-5]	**5**: 2830	[5221-53-4]	**5**: 2836
[4337-71-7]	**4**: 2487	[5246-57-1]	**6**: 3550
[4350-9-8]	**1**: 582	[5259-71-2]	**3**: 1790
[4364-78-7]	**5**: 2834	[5284-66-2]	**8**: 4504
[4377-73-5]	**6**: 3811	[5294-5-3]	**2**: 773
[4384-92-3]	**6**: 3426	[5303-25-3]	**4**: 2014
[4394-0-7]	**5**: 2611; **7**: 4196, 4219	[5326-23-8]	**7**: 4199
[4394-11-0]	**7**: 4193	[5329-15-7]	**1**: 615
[4394-85-8]	**1**: 468; **2**: 724; **5**: 2705, 2723, 2731	[5332-73-0]	**1**: 455
[4417-80-5]	**1**: 368	[5333-42-6]	**5**: 2551
[4424-6-0]	**2**: 1134	[5333-44-8]	**5**: 2551
[4427-56-9]	**6**: 3732	[5333-48-2]	**5**: 2551
[4454-5-1]	**1**: 201, 392	[5339-67-3]	**2**: 758
[4468-2-4]	**5**: 2783	[5343-92-0]	**1**: 337
[4482-52-4]	**3**: 1640	[5344-27-4]	**7**: 4211
[4485-12-5]	**6**: 3235	[5363-54-2]	**2**: 767
[4499-40-5]	**7**: 4170	[5366-84-7]	**3**: 1892; **6**: 3431
[4515-26-8]	**6**: 3558	[5371-52-8]	**1**: 421
[4525-33-1]	**3**: 1580	[5378-84-7]	**2**: 765
[4533-95-3]	**6**: 3556	[5382-16-1]	**7**: 4197
[4533-96-4]	**6**: 3556	[5392-40-5]	**1**: 368; **2**: 974
[4543-33-3]	**2**: 1059	[5394-63-8]	**5**: 3072, 3075
[4553-62-2]	**7**: 4185	[5398-69-6]	**3**: 1897
[4588-69-6]	**5**: 3217	[5402-73-3]	**2**: 768
[4593-90-2]	**3**: 1717	[5405-84-5]	**8**: 4606
[4618-18-2]	**5**: 3142	[5408-52-6]	**1**: 580
[4628-47-1]	**1**: 112	[5418-24-6]	**2**: 1009
[4696-56-4]	**6**: 3237	[5421-92-1]	**7**: 4193
[4696-57-5]	**6**: 3237	[5423-22-3]	**5**: 2825
[4710-40-1]	**6**: 3811	[5423-23-4]	**5**: 2825
[4722-59-2]	**4**: 1938	[5424-20-5]	**2974**:
[4724-48-5]	**6**: 3844	[5435-64-3]	**1**: 353
[4726-14-1]	**6**: 3559	[5452-37-9]	**1**: 464
[4736-48-5]	**3**: 1790	[5454-79-5]	**3**: 1818
[4795-29-3]	**5**: 2750, 2762	[5455-59-4]	**6**: 3549
[4812-20-8]	**6**: 3781	[5459-93-8]	**1**: 462
[4839-46-7]	**3**: 1915	[5460-9-3]	**6**: 3439
[4855-68-9]	**6**: 3754	[5461-32-5]	**6**: 3515
[4857-47-0]	**3**: 1884, 1891; **6**: 3431	[5466-91-1]	**2**: 759
[4883-67-4]	**1**: 67	[5486-84-0]	**3**: 1884
[4892-31-3]	**6**: 3737	[5510-99-6]	**6**: 3736
[4901-51-3]	**3**: 1606	[5530-30-3]	**6**: 3742
[4904-61-4]	**2**: 904; **3**: 1783	[5634-26-4]	**5**: 2549
[4939-92-8]	**6**: 3452	[5639-34-9]	**6**: 3431
[4941-53-1]	**5**: 2871	[5675-51-4]	**5**: 2553
[4988-51-6]	**6**: 3451	[5681-36-7]	**4**: 2380
[4995-91-9]	**6**: 3245	[5692-86-4]	**1**: 68

[5724-16-3]	**6:** 3370	[6286-30-2]	**3:** 1640
[5743-12-4]	**7:** 4166	[6290-3-5]	**2:** 946
[5756-24-1]	**8:** 4658	[6291-84-5]	**1:** 485
[5766-67-6]	**4:** 2299	[6292-58-6]	**2:** 765
[5776-28-3]	**6:** 3384	[6294-31-1]	**8:** 4658
[5794-3-6]	**8:** 4608	[6303-21-5]	**6:** 3836
[5794-13-8]	**1:** 578	[6303-58-8]	**2:** 1007
[5796-85-0]	**6:** 3643	[6317-18-6]	**8:** 4632
[5809-59-6]	**1:** 204	[6331-68-6]	**6:** 3552
[5813-77-4]	**3:** 1717	[6331-96-0]	**2:** 784
[5824-51-1]	**6:** 3643	[6334-97-0]	**6:** 3435
[5855-70-9]	**2:** 809	[6357-85-3]	**3:** 1895; **6:** 3392
[5857-42-1]	**2:** 776	[6357-89-7]	**6:** 3424
[5858-18-4]	**8:** 4662	[6357-93-3]	**6:** 3399
[5867-98-1]	**3:** 1855	[6357-94-4]	**6:** 3399
[5877-42-9]	**1:** 309	[6358-7-2]	**1:** 615
[5903-13-9]	**6:** 3402	[6358-9-4]	**1:** 615
[5908-27-0]	**2:** 804	[6358-64-1]	**6:** 3564
[5921-65-3]	**5:** 3214	[6359-40-6]	**7:** 4257
[5928-66-5]	**5:** 3113	[6361-37-1]	**6:** 3395
[5928-67-6]	**5:** 3113	[6361-38-2]	**6:** 3397
[5930-28-9]	**1:** 621	[6361-41-7]	**6:** 3437
[5934-29-2]	**1:** 544, 580	[6361-49-5]	**6:** 3441
[5967-84-0]	**7:** 4166	[6362-5-6]	**6:** 3416
[5989-27-5]	**5:** 3034	[6362-6-7]	**6:** 3424
[5996-22-5]	**1:** 579	[6362-11-4]	**6:** 3424
[6004-24-6]	**7:** 4219	[6362-18-1]	**6:** 3423
[6004-44-0]	**5:** 3077	[6373-15-5]	**6:** 3543
[6023-44-5]	**3:** 1884	[6375-2-6]	**2:** 782
[6025-45-2]	**7:** 4265	[6378-25-2]	**2:** 769
[6027-71-0]	**4:** 2022	[6381-59-5]	**8:** 4567
[6032-29-7]	**1:** 281; **5:** 2760	[6386-38-5]	**6:** 3743
[6061-6-9]	**3:** 1389	[6387-14-0]	**2:** 786; **8:** 4753
[6068-96-8]	**6:** 3634	[6387-27-5]	**2:** 785; **6:** 3559; **8:** 4753
[6069-98-3]	**3:** 1798	[6399-72-0]	**6:** 3434
[6072-57-7]	**3:** 1717	[6409-58-1]	**2:** 776
[6094-26-4]	**6:** 3369	[6416-39-3]	**2:** 658
[6099-90-7]	**3:** 1869	[6419-19-8]	**6:** 3845
[6117-91-5]	**3:** 1712	[6422-86-2]	**4:** 2032; **8:** 4587
[6119-92-2]	**3:** 1719	[6471-78-9]	**1:** 511
[6135-31-5]	**2:** 1047	[6484-52-2]	**5:** 2826
[6145-42-2]	**5:** 2836	[6485-39-8]	**5:** 2782
[6153-39-5]	**7:** 4261	[6492-78-0]	**5:** 2611
[6153-56-6]	**6:** 3577, 3579	[6493-5-6]	**7:** 4171
[6161-67-7]	**6:** 3728	[6506-37-2]	**5:** 2992; **6:** 3570
[6163-66-2]	**4:** 2187	[6535-20-2]	**6:** 3245
[6168-72-5]	**4:** 2164	[6535-70-2]	**6:** 3442
[6168-76-9]	**3:** 1719	[6564-95-0]	**1:** 425
[6250-86-8]	**5:** 3004	[6602-54-6]	**7:** 4196
[6251-3-2]	**6:** 3452	[6606-65-1]	**1:** 242
[6251-7-6]	**6:** 3416	[6617-91-5]	**1:** 310
[6259-19-4]	**2:** 811	[6624-76-6]	**5:** 2535
[6259-53-6]	**6:** 3438	[6624-79-9]	**5:** 2535
[6259-66-1]	**6:** 3398	[6626-92-2]	**6:** 3792
[6264-77-3]	**2:** 809	[6634-82-8]	**6:** 3554
[6268-17-3]	**6:** 3553	[6642-30-4]	**2:** 1047
[6285-57-0]	**6:** 3573	[6654-64-4]	**6:** 3371

CAS Registry Number Index

[6654-67-7]	**6**: 3372	[7446-70-0]	**3**: 1609
[6669-13-2]	**6**: 3737	[7447-41-8]	**5**: 2699
[6728-26-3]	**1**: 368	[7467-12-1]	**2**: 765
[6765-39-5]	**5**: 2871	[7474-97-7]	**6**: 3382
[6816-17-7]	**5**: 2871	[7486-38-6]	**1**: 271
[6863-58-7]	**4**: 2187	[7488-99-5]	**4**: 2382
[6865-33-4]	**6**: 3237	[7491-74-9]	**7**: 4247
[6865-35-6]	**6**: 3237	[7492-55-9]	**7**: 4367
[6876-23-9]	**3**: 1798	[7529-27-3]	**1**: 431, 436
[6920-22-5]	**1**: 337, 344	[7539-12-0]	**7**: 4341
[6940-53-0]	**6**: 3564	[7541-16-4]	**2**: 1047
[6945-23-9]	**5**: 2840	[7560-83-0]	**1**: 463
[6954-48-9]	**6**: 3451	[7572-29-4]	**3**: 1477
[6959-47-3]	**7**: 4199, 4218	[7601-89-0]	**5**: 2699
[6959-48-4]	**7**: 4199, 4218	[7617-31-4]	**6**: 3241
[6986-48-7]	**4**: 2187	[7620-77-1]	**6**: 3235
[7008-81-3]	**3**: 1890	[7623-9-8]	**7**: 4120
[7038-17-7]	**3**: 1524	[7641-77-2]	**5**: 2896
[7038-25-7]	**3**: 1528	[7642-4-8]	**5**: 2871
[7047-84-9]	**6**: 3239	[7642-9-3]	**5**: 2871
[7048-4-6]	**1**: 543, 579	[7642-10-6]	**5**: 2871
[7081-78-9]	**4**: 2187	[7642-15-1]	**5**: 2871
[7085-85-0]	**1**: 242	[7647-1-0]	**1**: 55
[7094-26-0]	**3**: 1798	[7664-39-3]	**4**: 2315
[7094-27-1]	**3**: 1798	[7664-41-7]	**5**: 2849
[7125-73-7]	**5**: 2611	[7664-93-9]	**2**: 667
[7134-9-0]	**2**: 776	[7669-54-7]	**6**: 3549
[7139-89-1]	**2**: 774	[7675-83-4]	**1**: 578
[7149-69-1]	**6**: 3535	[7681-11-0]	**5**: 2699
[7165-13-1]	**1**: 83	[7681-52-9]	**3**: 1609
[7166-19-0]	**6**: 3495	[7681-76-7]	**5**: 2992
[7173-62-8]	**1**: 481	[7681-84-7]	**5**: 2762
[7175-63-5]	**7**: 4365	[7683-64-9]	**4**: 2382
[7206-29-3]	**5**: 2871	[7688-21-3]	**5**: 2871
[7214-18-8]	**6**: 3742	[7705-8-0]	**3**: 1609; **5**: 2699
[7217-59-6]	**8**: 4662	[7719-9-7]	**1**: 55; **7**: 4118
[7227-91-0]	**3**: 1900	[7719-12-2]	**1**: 55
[7235-40-7]	**4**: 2382	[7722-57-8]	**3**: 1855
[7268-65-7]	**5**: 2553	[7722-84-1]	**1**: 91; **7**: 4266
[7324-2-9]	**1**: 242	[7727-21-7]	**5**: 2981
[7327-60-8]	**6**: 3482	[7731-28-4]	**3**: 1818
[7330-31-6]	**8**: 4658	[7731-29-5]	**3**: 1818
[7351-61-3]	**7**: 4340	[7735-42-4]	**5**: 2553
[7379-35-3]	**7**: 4199	[7735-43-5]	**5**: 2553
[7411-49-6]	**2**: 805	[7737-2-2]	**3**: 1522
[7415-31-8]	**3**: 1383	[7757-81-5]	**7**: 4367
[7416-48-0]	**1**: 43	[7757-82-6]	**7**: 4265
[7423-53-2]	**3**: 1770	[7758-29-4]	**7**: 4406
[7429-90-5]	**1**: 48	[7771-44-0]	**4**: 2487
[7433-56-9]	**5**: 2871	[7775-14-6]	**5**: 2738
[7433-78-5]	**5**: 2871	[7782-50-5]	**3**: 1609
[7443-52-9]	**3**: 1818	[7783-18-8]	**2**: 668
[7443-55-2]	**3**: 1818	[7790-94-5]	**3**: 1858
[7443-70-1]	**3**: 1818	[7791-25-5]	**3**: 1609
[7446-8-4]	**5**: 2812	[7803-49-8]	**2**: 975
[7446-9-5]	**2**: 718	[7803-51-2]	**6**: 3820
[7446-11-9]	**7**: 4265	[7803-62-5]	**7**: 4329

[8000-41-7]	**8:** 4777	[9004-62-0]	**2:** 1206
[8001-20-5]	**4:** 2465	[9004-64-2]	**2:** 1206
[8001-21-6]	**4:** 2456	[9004-65-3]	**2:** 1206
[8001-22-7]	**4:** 2461	[9004-67-5]	**2:** 1206
[8001-23-8]	**4:** 2459	[9004-70-0]	**2:** 1138
[8001-26-1]	**4:** 2458	[9005-25-8]	**7:** 4379
[8001-29-4]	**4:** 2452	[9005-27-0]	**7:** 4407
[8001-30-7]	**4:** 2454	[9005-35-0]	**4:** 2089
[8001-31-8]	**4:** 2445	[9005-82-7]	**7:** 4380
[8001-79-4]	**3:** 1909; **4:** 2466	[9005-90-7]	**8:** 4770
[8002-3-7]	**4:** 2462	[9013-34-7]	**2:** 1206
[8002-13-9]	**4:** 2464	[9015-11-6]	**2:** 1206
[8002-24-2]	**4:** 2472	[9015-12-7]	**2:** 1178
[8002-43-5]	**4:** 2380	[9015-14-9]	**2:** 1139, 1164
[8002-53-7]	**8:** 4962	[9016-87-9]	**1:** 634
[8002-74-2]	**8:** 4917, 4962	[9025-71-2]	**5:** 2982
[8002-75-3]	**4:** 2442	[9032-37-5]	**2:** 1139, 1166
[8004-92-0]	**7:** 4255	[9032-39-7]	**2:** 1206
[8006-44-8]	**8:** 4962	[9032-42-2]	**2:** 1206
[8006-64-2]	**8:** 4761, 4770	[9032-43-3]	**2:** 1139, 1163
[8007-40-7]	**4:** 2465	[9037-22-3]	**7:** 4380
[8008-45-5]	**4:** 2449	[9041-56-9]	**2:** 1206
[8008-74-0]	**4:** 2456	[9045-28-7]	**7:** 4404
[8012-89-3]	**8:** 4962	[9049-76-7]	**7:** 4407
[8013-7-8]	**4:** 2004	[9068-8-0]	**4:** 2080
[8015-51-8]	**3:** 1719	[9069-33-4]	**2:** 1206
[8015-86-9]	**8:** 4962	[9082-51-3]	**4:** 2080
[8016-35-1]	**4:** 2465	[9088-4-4]	**2:** 1206
[8023-79-8]	**4:** 2447	[9088-5-5]	**2:** 1206
[8037-20-5]	**4:** 2470	[10025-67-9]	**6:** 3852
[9000-7-1]	**4:** 2089	[10025-78-2]	**7:** 4311, 4329
[9000-11-7]	**2:** 1206; **7:** 4071	[10026-4-7]	**7:** 4311, 4322
[9000-73-1]	**3:** 1641	[10026-13-8]	**7:** 4118
[9001-4-1]	**4:** 2080	[10033-99-5]	**3:** 1381, 1383
[9001-37-0]	**5:** 2778	[10034-85-2]	**1:** 35
[9002-84-0]	**5:** 2586, 2702	[10042-59-8]	**5:** 2551
[9002-85-1]	**5:** 2698	[10043-35-3]	**3:** 1812
[9002-86-2]	**5:** 2698	[10048-32-5]	**7:** 4365
[9002-88-4]	**5:** 2702; **8:** 4962	[10049-7-7]	**4:** 2059
[9002-89-5]	**5:** 2698, 2702	[10075-38-4]	**3:** 1383
[9002-91-9]	**1:** 22	[10094-58-3]	**2:** 1067
[9002-98-6]	**2:** 664	[10191-60-3]	**5:** 2839
[9003-7-0]	**5:** 2702; **6:** 3749; **8:** 4962	[10210-68-1]	**1:** 127, 35
[9003-20-7]	**5:** 2698	[10215-25-5]	**3:** 1880
[9003-53-6]	**5:** 2698	[10308-82-4]	**5:** 2833
[9004-32-4]	**2:** 1206; **3:** 1546; **5:** 2966	[10308-84-6]	**5:** 2830
[9004-34-6]	**7:** 4381	[10332-40-8]	**4:** 2016
[9004-35-7]	**2:** 1167	[10402-16-1]	**6:** 3241
[9004-36-8]	**2:** 1167, 1178	[10412-98-3]	**7:** 4184
[9004-38-0]	**2:** 1167, 1178	[10413-1-1]	**7:** 4184
[9004-39-1]	**2:** 1167, 1178; **7:** 4114	[10416-59-8]	**7:** 4346
[9004-41-5]	**2:** 1206	[10468-30-1]	**6:** 3243
[9004-48-2]	**2:** 1178	[10471-40-6]	**3:** 1770
[9004-54-0]	**3:** 1843	[10479-42-2]	**3:** 1716
[9004-57-3]	**2:** 1206	[10486-19-8]	**1:** 353
[9004-58-4]	**2:** 1206	[10508-9-5]	**6:** 3641
[9004-59-1]	**2:** 1206	[10543-57-4]	**1:** 66

CAS Registry Number	Reference
[10543-60-9]	**5:** 2814
[10574-17-1]	**6:** 3736
[10599-90-3]	**3:** 1566
[10605-21-7]	**5:** 2838
[11072-16-5]	**5:** 2592
[11081-15-5]	**6:** 3748
[12001-85-3]	**6:** 3243
[12054-48-7]	**5:** 2739
[12379-38-3]	**3:** 1571
[12597-68-1]	**1:** 48
[12605-84-4]	**1:** 48
[12605-85-5]	**1:** 48
[12689-13-3]	**4:** 2024
[13001-15-5]	**6:** 3245
[13005-35-1]	**5:** 2782
[13007-90-4]	**1:** 128
[13064-50-1]	**3:** 1524
[13071-79-9]	**6:** 3853
[13080-96-1]	**1:** 83
[13115-21-4]	**3:** 1750
[13122-71-9]	**6:** 3643
[13195-64-7]	**5:** 3184, 3193
[13248-54-9]	**3:** 1576
[13257-95-9]	**6:** 3550
[13269-52-8]	**5:** 2871
[13284-42-9]	**7:** 4187
[13292-87-0]	**8:** 4661
[13296-94-1]	**6:** 3542
[13306-42-8]	**6:** 3418
[13347-42-7]	**6:** 3785
[13360-61-7]	**5:** 2871
[13360-63-9]	**1:** 455
[13360-80-0]	**6:** 3838
[13362-52-2]	**5:** 2553
[13373-32-5]	**5:** 2988
[13382-47-3]	**4:** 1938
[13389-42-9]	**5:** 2871
[13396-80-0]	**6:** 3820
[13457-18-6]	**6:** 3852
[13460-50-9]	**3:** 1812
[13463-39-3]	**1:** 126; **7:** 4106
[13463-41-7]	**7:** 4196
[13483-16-4]	**3:** 1768
[13501-73-0]	**1:** 282
[13511-38-1]	**2:** 1116
[13531-52-7]	**1:** 488
[13547-6-3]	**3:** 1384
[13569-65-8]	**5:** 2719
[13573-16-5]	**3:** 1634
[13586-84-0]	**6:** 3245
[13595-73-8]	**8:** 4504
[13636-32-3]	**5:** 2837
[13682-92-3]	**1:** 580
[13684-56-5]	**3:** 1580
[13684-63-4]	**3:** 1580
[13710-57-1]	**3:** 1712
[13718-94-0]	**2:** 1062
[13738-63-1]	**5:** 2617; **6:** 3530
[13838-16-9]	**5:** 2595, 2625
[13852-81-8]	**6:** 3556
[13871-68-6]	**1:** 603
[13887-2-0]	**6:** 3820
[13952-84-6]	**1:** 454
[14098-44-3]	**3:** 1725
[14103-61-8]	**7:** 3886
[14174-9-5]	**3:** 1725
[14187-32-7]	**3:** 1725
[14205-39-1]	**5:** 3076
[14236-50-1]	**6:** 3239
[14239-23-7]	**3:** 1901
[14239-24-8]	**3:** 1901
[14250-88-5]	**1:** 352
[14255-87-9]	**5:** 2839
[14263-89-9]	**3:** 1901
[14263-92-4]	**3:** 1901
[14263-94-6]	**3:** 1901
[14289-96-4]	**1:** 421
[14414-68-7]	**2:** 804
[14486-58-9]	**2:** 652
[14499-87-7]	**3:** 1382
[14605-55-1]	**6:** 3759
[14666-78-5]	**3:** 1580
[14686-14-7]	**5:** 2871
[14703-69-6]	**1:** 617
[14722-40-8]	**5:** 2553
[14850-22-7]	**5:** 2871
[14850-23-8]	**5:** 2871
[14885-29-1]	**5:** 2988
[14906-64-0]	**7:** 4196
[14918-69-5]	**6:** 3451
[14919-1-8]	**5:** 2871
[14937-45-2]	**6:** 3828
[14984-21-5]	**5:** 3112
[15022-8-9]	**2:** 1086
[15049-85-1]	**6:** 3844
[15124-9-1]	**2:** 668
[15128-82-2]	**7:** 4209
[15171-48-9]	**6:** 3838
[15221-81-5]	**5:** 2612
[15238-0-3]	**1:** 35
[15299-99-7]	**6:** 3374
[15433-79-1]	**5:** 2757
[15435-29-7]	**6:** 3790
[15457-5-3]	**5:** 2612
[15484-44-3]	**2:** 1235
[15506-53-3]	**5:** 3069
[15594-90-8]	**5:** 2535
[15595-61-6]	**3:** 1576
[15595-62-7]	**3:** 1576
[15625-89-5]	**2:** 665
[15658-52-3]	**1:** 603
[15694-70-9]	**5:** 2739
[15721-2-5]	**2:** 808
[15798-64-8]	**3:** 1711

[15827-56-2]	**6:** 3810	[17894-99-4]	**6:** 3421
[15827-60-8]	**6:** 3845	[18162-48-6]	**7:** 4346
[15862-82-5]	**7:** 4244	[18197-26-7]	**5:** 2694
[15877-57-3]	**1:** 352	[18232-3-6]	**2:** 843
[15894-70-9]	**5:** 2839	[18240-93-2]	**5:** 2836
[15909-83-3]	**8:** 4505	[18295-59-5]	**1:** 353
[15945-7-0]	**2:** 769	[18297-63-7]	**7:** 4346
[15950-66-0]	**3:** 1606	[18368-64-4]	**7:** 4195
[15956-58-8]	**6:** 3245	[18368-76-8]	**7:** 4199
[16029-98-4]	**7:** 4346	[18435-53-5]	**5:** 2871
[16069-36-6]	**3:** 1725	[18472-51-0]	**5:** 2841
[16079-88-2]	**3:** 1569	[18598-63-5]	**1:** 582
[16086-14-9]	**3:** 1529	[18621-75-5]	**1:** 425, 68; **2:** 943
[16094-31-8]	**6:** 3743	[18748-9-9]	**5:** 2739
[16110-9-1]	**7:** 4199, 4218	[18850-78-7]	**1:** 271
[16152-65-1]	**6:** 3754	[19009-56-4]	**1:** 353
[16208-32-5]	**3:** 1861	[19040-62-1]	**2:** 765
[16215-49-9]	**3:** 1580	[19044-88-3]	**6:** 3559
[16241-25-1]	**6:** 3762	[19071-44-4]	**6:** 3600
[16288-77-0]	**2:** 770	[19147-16-1]	**1:** 271
[16337-84-1]	**5:** 2739	[19147-46-7]	**7:** 4367
[16503-25-6]	**3:** 1381, 1383	[19172-47-5]	**8:** 4652
[16528-92-0]	**5:** 2681	[19179-44-3]	**6:** 3241
[16588-2-6]	**6:** 3472, 3547	[19224-26-1]	**2:** 827
[16630-91-4]	**1:** 352	[19238-50-7]	**1:** 579
[16646-44-9]	**5:** 2813	[19348-37-9]	**5:** 2871
[16672-87-0]	**6:** 3845	[19387-91-8]	**5:** 2992
[16712-64-4]	**6:** 3357, 3384	[19393-67-0]	**2:** 973
[16724-63-3]	**1:** 481	[19434-65-2]	**1:** 202
[16773-42-5]	**5:** 2992	[19473-49-5]	**1:** 579
[16794-13-1]	**8:** 4504	[19524-6-2]	**7:** 4199
[16856-18-1]	**1:** 578	[19549-79-2]	**1:** 282
[16948-63-3]	**2:** 784	[19550-10-8]	**5:** 3098
[16996-40-0]	**6:** 3241	[19562-30-2]	**5:** 2836
[17058-53-6]	**2:** 843	[19659-81-5]	**6:** 3424
[17102-64-6]	**7:** 4366	[19689-18-0]	**5:** 2871
[17146-86-0]	**1:** 579	[19700-42-6]	**6:** 3382
[17199-29-0]	**5:** 2983	[19768-54-8]	**5:** 2722
[17243-13-9]	**2:** 780	[19780-25-7]	**1:** 368
[17256-62-1]	**5:** 2750	[19803-43-1]	**3:** 1854
[17342-77-7]	**3:** 1717	[19812-64-7]	**5:** 2553
[17352-16-8]	**3:** 1531	[19824-34-1]	**5:** 3077
[17420-30-3]	**6:** 3472, 3547	[20048-27-5]	**6:** 3811
[17422-32-1]	**5:** 3004	[20063-77-8]	**5:** 2871
[17455-13-9]	**3:** 1725; **4:** 2335	[20191-62-2]	**6:** 3391
[17455-23-1]	**3:** 1725	[20298-69-5]	**6:** 3742
[17496-8-1]	**7:** 4116	[20325-40-0]	**2:** 808
[17557-76-5]	**2:** 810	[20334-52-5]	**5:** 3077
[17617-23-1]	**5:** 2616	[20452-67-9]	**5:** 3077
[17639-93-9]	**7:** 4119	[20559-55-1]	**5:** 2839
[17658-63-8]	**5:** 2551	[20680-48-2]	**6:** 3433
[17671-50-0]	**1:** 579	[20845-34-5]	**7:** 4197
[17692-38-5]	**5:** 2621	[20888-43-1]	**2:** 663
[17696-37-6]	**6:** 3737	[20942-99-8]	**2:** 1062
[17788-47-5]	**6:** 3450	[21013-47-8]	**6:** 3438
[17804-35-2]	**5:** 2838; **6:** 3814	[21282-97-3]	**5:** 3076
[17865-32-6]	**7:** 4351	[21544-2-5]	**5:** 2849

5071

[21564-17-0]	**8:** 4634	[25103-9-7]	**6:** 3223
[21645-51-2]	**5:** 2739	[25151-33-1]	**1:** 227
[21963-38-2]	**7:** 4368	[25154-52-3]	**6:** 3748
[22117-79-9]	**6:** 3561	[25167-67-3]	**1:** 46; **7:** 4105
[22118-9-8]	**2:** 878	[25231-46-3]	**2:** 761
[22221-10-9]	**6:** 3241	[25265-75-2]	**1:** 344
[22282-69-5]	**5:** 2620	[25296-66-6]	**6:** 3837
[22302-43-8]	**6:** 3237, 3243	[25322-68-3]	**4:** 2317
[22322-28-7]	**1:** 271	[25323-68-6]	**3:** 1436
[22346-43-6]	**6:** 3435	[25377-83-7]	**6:** 3749
[22480-69-9]	**2:** 773	[25378-22-7]	**5:** 2871
[22494-42-4]	**5:** 2621; **7:** 4279	[25415-76-3]	**3:** 1719
[22513-81-1]	**5:** 2553	[25429-29-2]	**3:** 1436
[22513-82-2]	**5:** 2553	[25512-42-9]	**3:** 1436
[22535-90-6]	**5:** 3217	[25586-43-0]	**3:** 1442
[22839-47-0]	**5:** 2735	[25609-89-6]	**3:** 1719
[22918-74-7]	**5:** 2739	[25620-59-1]	**6:** 3451
[23077-93-2]	**6:** 3452	[25639-42-3]	**3:** 1818
[23103-98-2]	**5:** 2836	[25656-90-0]	**1:** 428
[23128-41-8]	**6:** 3651	[25714-71-0]	**1:** 331, 375, 421; **2:** 1007
[23235-61-2]	**1:** 340	[25790-55-0]	**3:** 1383
[23250-73-9]	**6:** 3245	[25875-50-7]	**5:** 2834
[23335-74-2]	**6:** 3245	[25875-51-8]	**5:** 2834
[23468-31-7]	**7:** 4199	[25896-97-3]	**1:** 390
[23505-41-1]	**5:** 2836	[25956-17-6]	**1:** 511
[23530-40-7]	**6:** 3549	[26040-98-2]	**5:** 2535
[23564-5-8]	**8:** 4643	[26106-95-6]	**3:** 1586
[23585-26-4]	**2:** 652	[26272-90-2]	**3:** 1576
[23593-75-1]	**5:** 2991	[26281-43-6]	**2:** 781
[23815-28-3]	**2:** 769	[26322-14-5]	**3:** 1580; **6:** 3653
[23894-7-7]	**6:** 3400	[26352-16-9]	**7:** 4326
[23947-60-6]	**5:** 2836	[26399-36-0]	**5:** 2612
[23978-10-1]	**3:** 1725	[26523-73-9]	**3:** 1580
[23978-55-4]	**3:** 1725	[26601-64-9]	**3:** 1436
[23996-53-4]	**5:** 2991	[26605-69-6]	**5:** 2917
[24075-24-9]	**6:** 3672	[26675-46-7]	**5:** 2595
[24157-81-1]	**6:** 3357	[26739-53-7]	**5:** 2825
[24261-30-1]	**1:** 51	[26760-64-5]	**5:** 2871
[24264-8-2]	**5:** 3077	[26761-40-0]	**4:** 2040; **7:** 3886
[24309-97-5]	**5:** 3075	[26780-96-1]	**1:** 505
[24356-66-9]	**7:** 4171	[26811-78-9]	**7:** 4368
[24468-13-1]	**3:** 1576	[26864-1-7]	**6:** 3451
[24634-61-5]	**7:** 4366	[26886-5-5]	**6:** 3732
[24653-79-0]	**2:** 778	[26914-33-0]	**3:** 1436
[24771-25-3]	**5:** 2839	[27138-31-4]	**2:** 827
[24800-44-0]	**7:** 4048	[27157-94-4]	**6:** 3854
[24824-28-0]	**6:** 3405	[27178-83-2]	**2:** 835
[24891-71-2]	**6:** 3600	[27193-28-8]	**6:** 3748
[24935-08-8]	**7:** 4247	[27205-24-9]	**3:** 1714
[24937-79-9]	**5:** 2587	[27215-95-8]	**6:** 3749
[24938-60-1]	**1:** 507; **6:** 3811	[27247-96-7]	**4:** 2366
[24938-64-5]	**1:** 507; **6:** 3815	[27314-13-2]	**5:** 2612
[24966-39-0]	**8:** 4662	[27323-18-8]	**3:** 1436
[24981-14-4]	**5:** 2588	[27336-20-5]	**6:** 3747
[25007-86-7]	**1:** 52	[27348-32-9]	**1:** 580
[25013-15-4]	**5:** 2898; **7:** 4434	[27458-92-0]	**1:** 302
[25014-41-9]	**5:** 2697	[27458-93-1]	**1:** 303

[27459-10-5]	**6:** 3748	[33100-27-5]	**3:** 1725; **4:** 2335
[27554-26-3]	**7:** 3886	[33528-41-5]	**1:** 431
[27577-42-0]	**3:** 1528	[33817-63-9]	**3:** 1384
[27636-85-7]	**5:** 2582	[33817-64-0]	**3:** 1384
[27816-23-5]	**1:** 242	[33965-80-9]	**3:** 1524
[27887-43-0]	**2:** 758	[34149-71-8]	**5:** 2589
[28163-0-0]	**6:** 3471	[34202-28-3]	**5:** 2597
[28166-6-5]	**5:** 2617	[34202-29-4]	**5:** 2597
[28188-48-9]	**2:** 766	[34202-30-7]	**5:** 2739
[28290-79-1]	**4:** 2487	[34220-70-7]	**1:** 580
[28300-74-5]	**8:** 4567	[34594-19-9]	**6:** 3666
[28317-46-6]	**6:** 3643	[34662-32-3]	**6:** 3472
[28317-47-7]	**6:** 3643	[34708-8-2]	**8:** 4632
[28388-44-5]	**3:** 1531	[34789-9-8]	**3:** 1522
[28484-70-0]	**5:** 2535	[34885-3-5]	**1:** 335
[28507-96-2]	**3:** 1382	[34937-0-3]	**7:** 4337
[28523-86-6]	**5:** 2596	[34941-90-7]	**5:** 2620
[28553-12-0]	**7:** 3886	[35203-93-1]	**2:** 766
[28699-88-9]	**3:** 1442	[35266-49-0]	**8:** 4636
[28758-94-3]	**2:** 651	[35337-99-6]	**2:** 781
[28843-76-7]	**3:** 1725	[35346-33-9]	**6:** 3825, 3827
[28884-42-6]	**6:** 3660	[35367-38-5]	**5:** 2617
[28933-81-5]	**3:** 1384	[35400-55-6]	**6:** 3433
[28944-41-4]	**6:** 3754	[35452-30-3]	**8:** 4504
[29191-52-4]	**1:** 529	[35554-44-0]	**5:** 2991
[29204-2-2]	**4:** 2487	[35674-68-1]	**6:** 3243
[29232-93-7]	**5:** 2836	[35850-94-3]	**1:** 205
[29480-42-0]	**3:** 1384	[36110-12-0]	**2:** 784
[29636-87-1]	**5:** 2988	[36311-34-9]	**1:** 303
[29696-35-3]	**6:** 3594	[36322-90-4]	**7:** 4207
[30030-25-2]	**7:** 4438	[36400-98-3]	**1:** 303
[30058-79-8]	**1:** 203	[36597-97-4]	**1:** 69
[30169-34-7]	**3:** 1899	[36638-69-4]	**5:** 3077
[30266-58-1]	**6:** 3452	[36653-82-4]	**5:** 2535
[30525-89-4]	**5:** 2674	[36854-53-2]	**5:** 3077
[30618-84-9]	**6:** 3223	[36888-99-0]	**7:** 4233
[31036-27-8]	**1:** 205	[36947-68-9]	**5:** 2988
[31249-95-3]	**3:** 1725	[37051-42-6]	**6:** 3652
[31252-42-3]	**7:** 4197	[37148-27-9]	**6:** 3568
[31330-63-9]	**5:** 2833	[37264-91-8]	**2:** 1165
[31430-15-6]	**5:** 2839	[37330-39-5]	**6:** 3757
[31431-39-7]	**5:** 2839	[37640-71-4]	**5:** 2917
[31442-13-4]	**1:** 227	[37971-36-1]	**6:** 3845
[31570-4-4]	**6:** 3743	[38083-17-9]	**5:** 2991
[31631-13-7]	**6:** 3751	[38194-50-2]	**5:** 2621
[31656-49-2]	**6:** 3473	[38255-39-9]	**1:** 36
[31906-4-4]	**1:** 214	[38360-74-6]	**5:** 2834
[32241-8-0]	**3:** 1442	[38425-52-4]	**5:** 3069
[32289-58-0]	**5:** 2841	[38433-80-6]	**1:** 331
[32333-53-2]	**2:** 778	[38641-94-0]	**6:** 3845
[32360-5-7]	**6:** 3250	[38830-94-3]	**6:** 3452
[32368-14-2]	**2:** 1091	[38842-5-6]	**5:** 2894
[32432-55-6]	**6:** 3525	[38867-16-2]	**7:** 4368
[32518-77-7]	**5:** 3200	[39148-24-8]	**6:** 3841
[32582-32-4]	**5:** 2551	[39169-92-1]	**2:** 764
[32749-94-3]	**1:** 352	[39259-32-0]	**5:** 2839
[33089-36-0]	**3:** 1725	[39277-57-1]	**2:** 1206

5073

CAS Number	Ref
[39300-45-3]	**6:** 3566
[39638-32-9]	**4:** 2187, 2200
[39665-12-8]	**1:** 580
[39978-42-2]	**6:** 3571
[40018-26-6]	**3:** 1524
[40117-41-7]	**2:** 766
[40137-0-6]	**5:** 3050
[40248-84-8]	**7:** 4265
[40372-72-3]	**7:** 4338
[40943-37-1]	**5:** 2839
[41051-88-1]	**2:** 695
[41205-21-4]	**5:** 2617; **6:** 3536
[41479-14-5]	**1:** 49
[41483-43-6]	**5:** 2836
[41505-40-2]	**6:** 3754
[41601-60-9]	**3:** 1389
[41687-8-5]	**2:** 803
[41687-30-3]	**6:** 3550
[42116-76-7]	**5:** 2992
[42125-46-2]	**3:** 1576
[42151-64-4]	**3:** 1587
[42205-8-3]	**5:** 2894
[42347-74-0]	**1:** 353
[42413-3-6]	**2:** 777
[42783-4-0]	**6:** 3572
[42937-13-3]	**2:** 1133; **5:** 2914
[43027-41-4]	**6:** 3450
[43168-33-8]	**6:** 3235
[43210-67-9]	**5:** 2839
[46001-16-5]	**6:** 3452
[46060-27-9]	**2:** 780
[46062-27-5]	**2:** 764
[49719-55-3]	**5:** 2840
[49735-69-5]	**3:** 1901
[49735-71-9]	**3:** 1901
[50594-66-6]	**5:** 2612
[50632-94-5]	**3:** 1855
[50651-39-3]	**1:** 510
[51181-40-9]	**1:** 335
[51331-9-0]	**2:** 1206
[51392-67-7]	**6:** 3640
[51407-46-6]	**1:** 382
[51460-26-5]	**6:** 3762
[51481-61-9]	**5:** 2839, 2992
[51528-20-2]	**5:** 2825
[51575-25-8]	**5:** 2739
[51583-62-1]	**6:** 3451
[51592-6-4]	**3:** 1570
[51632-16-7]	**2:** 877
[51806-69-0]	**6:** 3754
[51896-27-6]	**2:** 777
[51940-44-4]	**5:** 2836
[52101-1-6]	**1:** 578
[52205-85-3]	**2:** 766
[52328-5-9]	**5:** 2835
[52334-81-3]	**7:** 4199
[52549-17-4]	**7:** 4196
[52756-22-6]	**6:** 3537
[52927-22-7]	**6:** 3473
[52993-54-1]	**3:** 1798
[53369-7-6]	**6:** 3837
[53558-25-1]	**6:** 3542
[53716-50-0]	**5:** 2839
[53718-26-6]	**6:** 3757
[53874-67-2]	**2:** 877
[54063-54-6]	**7:** 4171
[54090-41-4]	**6:** 3552
[54322-31-5]	**2:** 765
[54439-52-0]	**5:** 2551
[54481-12-8]	**6:** 3557
[54532-93-33]	**6:** 3451
[54734-85-9]	**6:** 3556
[54739-18-3]	**5:** 2616
[54910-89-3]	**5:** 2611
[54965-21-8]	**5:** 2839
[55034-25-8]	**2:** 782
[55268-74-1]	**7:** 4257
[55524-84-0]	**6:** 3422
[55758-32-2]	**5:** 2620
[56004-61-6]	**2:** 695
[56136-14-2]	**6:** 3449
[56172-46-4]	**3:** 1719
[56189-9-4]	**6:** 3241
[56278-50-3]	**6:** 3470
[56461-98-4]	**6:** 3535
[56537-19-0]	**6:** 3634
[56639-51-1]	**6:** 3239
[57282-49-2]	**1:** 580
[57491-45-9]	**5:** 2617
[57491-99-3]	**6:** 3543
[57520-17-9]	**5:** 2837
[57530-25-3]	**6:** 3572
[57597-62-3]	**6:** 3465
[57609-64-0]	**7:** 4057
[57828-3-2]	**4:** 2336
[58058-72-3]	**6:** 3553
[58166-83-9]	**7:** 4171
[58200-82-1]	**6:** 3450
[58306-30-2]	**3:** 1580; **5:** 2837
[58479-61-1]	**7:** 4315
[58596-7-9]	**6:** 3432
[58622-66-5]	**6:** 3792
[58629-1-9]	**3:** 1570
[58670-89-6]	**5:** 2551
[58677-63-3]	**5:** 2617
[58718-78-8]	**5:** 2871
[58723-2-7]	**2:** 761
[58844-38-5]	**5:** 3077
[58905-32-1]	**5:** 3091
[59277-89-3]	**7:** 4171
[59804-37-4]	**7:** 4207
[59933-66-3]	**5:** 2605
[59945-37-8]	**3:** 1725
[60178-85-0]	**1:** 271

[61213-25-0]	**7:** 4247	[67747-9-5]	**5:** 2991
[60259-81-6]	**1:** 581	[68018-42-8]	**5:** 2705
[60385-32-2]	**2:** 785	[68025-25-2]	**3:** 1901
[60628-96-8]	**5:** 2991	[68153-21-9]	**4:** 2443
[61186-96-7]	**6:** 3759	[68189-28-6]	**6:** 3559
[61759-63-5]	**5:** 2738	[68217-36-7]	**6:** 3451
[61788-44-1]	**6:** 3754	[68227-69-0]	**2:** 770
[61788-45-2]	**1:** 481	[68308-22-5]	**8:** 4962
[61788-46-3]	**1:** 481	[68399-99-5]	**5:** 2611
[61788-63-4]	**1:** 481	[68441-17-8]	**8:** 4962
[61788-71-4]	**6:** 3245	[68476-3-9]	**8:** 4962
[61788-91-8]	**1:** 481	[68476-38-0]	**8:** 4962
[61788-95-2]	**1:** 481	[68515-41-3]	**7:** 3886
[61789-36-4]	**6:** 3237	[68515-42-4]	**7:** 3886
[61789-51-3]	**6:** 3245	[68515-43-5]	**7:** 3886
[61789-67-1]	**6:** 3237	[68515-47-9]	**7:** 3886
[61789-76-2]	**1:** 481	[68515-48-0]	**7:** 3886
[61789-79-5]	**1:** 481	[68515-50-4]	**7:** 3886
[61789-97-7]	**4:** 2469	[68515-51-5]	**7:** 3886
[61789-99-9]	**4:** 2468	[68553-81-1]	**4:** 2455
[61790-14-5]	**6:** 3241	[69121-20-6]	**6:** 3243
[61790-18-9]	**1:** 481	[69335-91-7]	**5:** 2620
[61790-33-8]	**1:** 481	[69409-94-5]	**5:** 2612
[61791-55-7]	**1:** 481	[70170-61-5]	**5:** 2841
[61791-63-7]	**1:** 481	[70179-79-2]	**5:** 2739
[61791-67-1]	**1:** 481	[70233-62-4]	**6:** 3845
[61889-43-8]	**3:** 1885	[70299-48-8]	**6:** 3640
[61919-18-4]	**3:** 1901	[71159-90-5]	**8:** 4653
[62284-99-5]	**7:** 4367	[71353-61-2]	**5:** 2535
[62309-51-7]	**7:** 4105, 4107	[71555-10-7]	**1:** 580
[62477-23-0]	**2:** 803	[71662-46-9]	**7:** 3886
[62519-7-7]	**3:** 1389	[72317-18-1]	**6:** 3749
[62571-86-2]	**8:** 4656	[72388-18-2]	**5:** 2551
[63034-44-6]	**5:** 2598	[72556-58-2]	**1:** 613
[63231-60-7]	**8:** 4962	[73138-44-0]	**8:** 4962
[63270-14-4]	**4:** 2041	[73138-45-1]	**8:** 4962
[63284-71-9]	**5:** 2617; **7:** 4233	[73590-58-6]	**7:** 4196
[63449-39-8]	**3:** 1396	[73761-81-6]	**5:** 2551
[63661-2-9]	**5:** 2611	[74409-42-0]	**7:** 4196
[64044-51-5]	**5:** 3135	[74525-31-8]	**6:** 3438
[64054-77-9]	**6:** 3756	[74738-17-3]	**7:** 4238
[64062-91-5]	**2:** 765	[74829-49-5]	**8:** 4587
[64071-86-9]	**3:** 1901	[74832-35-2]	**6:** 3442
[64165-14-6]	**5:** 2739	[75648-1-0]	**3:** 1719
[64519-82-0]	**2:** 1062	[76530-15-9]	**6:** 3420
[64701-3-7]	**2:** 844	[76843-24-8]	**5:** 2813
[64742-43-4]	**8:** 4925	[77182-82-2]	**6:** 3837
[64742-61-6]	**8:** 4916	[78320-64-6]	**3:** 1580
[64742-67-2]	**8:** 4918, 4935	[78782-47-5]	**5:** 2837
[64969-36-41]	**2:** 806	[78957-7-0]	**3:** 1716
[65053-91-0]	**6:** 3787	[79380-22-6]	**8:** 4753
[65122-21-6]	**3:** 1384	[79554-59-9]	**1:** 488
[65405-39-2]	**8:** 4820	[79864-2-1]	**5:** 2551
[66280-55-5]	**6:** 3643, 3647	[80198-19-2]	**6:** 3555
[66577-59-1]	**5:** 2553	[80450-4-0]	**1:** 214
[66671-80-5]	**1:** 603	[80632-67-3]	**2:** 844
[67373-56-2]	**7:** 4321	[80866-98-4]	**2:** 844

CAS Registry Number Index

[81344-22-1]	**6**: 3400	[87458-73-9]	**6**: 3647
[81405-85-8]	**7**: 4254	[89981-68-0]	**2**: 764
[83021-64-1]	**6**: 3452	[90459-52-2]	**6**: 3241
[83055-99-6]	**7**: 4233	[90669-78-6]	**8**: 4917
[83411-71-6]	**6**: 3837	[91403-50-8]	**5**: 2841
[84324-99-2]	**4**: 2022	[92238-33-0]	**5**: 2553
[84777-6-0]	**7**: 3886	[93042-68-3]	**7**: 4341
[84852-15-3]	**6**: 3748	[95008-70-1]	**5**: 2553
[85222-98-6]	**3**: 1901	[95491-58-0]	**5**: 2553
[85371-64-8]	**7**: 4208	[97141-5-4]	**8**: 4507
[85507-79-5]	**7**: 3886	[104294-11-3]	**6**: 3791
[85535-84-8]	**3**: 1396	[104466-83-3]	**5**: 3191
[85535-85-9]	**3**: 1396	[107667-2-7]	**6**: 3838
[85605-29-4]	**2**: 782	[116088-82-5]	**7**: 4341
[87025-51-2]	**6**: 3837	[117397-61-2]	**6**: 3742
[87244-72-]	**7**: 4352		

Index

A acid **6**: 3400
ABS **1**: 256
Absorption
   of acetylene **1**: 139
Acenaphthene **5**: 2913
Acenaphthenequinone **5**: 2915
Acenaphthylene **5**: 2915
ACES urea process **8**: 4806
Acetaldehyde **1**: 1, 352
   see also aldehydes, aliphatic
   by vinylation with acetylene **1**: 124
   Chemical Properties **1**: 4
   economic aspects **1**: 19
   for production of pyridines **7**: 4181
   Physical Properties **1**: 2
   toxicology **1**: 23
Acetaldehyde cyanohydrin **6**: 3463
Acetaldehyde diethyl acetal **1**: 395
Acetaldehyde dimethyl acetal **1**: 395
Acetaldehyde monoperacetate **1**: 43
Acetaldehyde oxidation
   for acetic acid production **1**: 43
   for acetic anhydride production **1**: 75
   for production of glyoxal **5**: 2812
   industrial operation **1**: 46
Acetaldehyde polymers **1**: 20
Acetaldehyde production **1**: 7
   from acetylene **1**: 9
   from C_1 Sources **1**: 16
   from ethanol **1**: 8
Acetaldol **1**: 375, 378
Acetalization **1**: 203
Acetals **1**: 393
   of carbohydrates **2**: 1073
   production **1**: 396
   uses **1**: 398
Acetamide **1**: 56
7-Acetamido-4-hydroxynaphthalene-2-sulfonic
   acid **6**: 3435, 3437
3-Acetamido-4-hydroxyphenylarsonic acid **1**: 616
4-Acetamido-3-methoxyaniline **1**: 615
4-Acetamidophenetole **1**: 620
2-Acetamidophenol **1**: 612, 619
Acetaminophen **1**: 619
Acetanilide **1**: 637, 639; **2**: 759
Acetarsol **1**: 616
Acetarsone **1**: 616
Acetates **1**: 51
Acetic acid
   chemical properties **1**: 33
   concentration **1**: 47
   economic aspects **1**: 57
   fluorinated **5**: 2600
   for production of vinyl acetate **8**: 4841
   from methanol, economic aspects **6**: 3311
   physical properties **1**: 30
   specifications **1**: 49
   toxicology **1**: 58
   uses **1**: 50
Acetic acid cellulose acetate process **2**: 1171
Acetic acid chlorination **3**: 1544
Acetic acid esters **1**: 52
Acetic acid 2-methyl-2-propenyl ester **1**: 426
Acetic acid production **1**: 34
Acetic acid 2-propenyl ester **1**: 426
Acetic acid pyrolysis **5**: 3065
Acetic acid recovery
   in cellulose acetate production **2**: 1175
Acetic acid solutions **1**: 31
Acetic anhydride
   chemical properties **1**: 65
   physical properties **1**: 64
   specifications **1**: 79
   uses **1**: 80
Acetic anhydride production **1**: 71
   from ketene **5**: 3066
Acetoacetalkylamides
   from diketene **5**: 3075
Acetoacetamides
   production **6**: 3605
   toxicology **6**: 3606
   uses and economic aspects **6**: 3605
Acetoacetanilide **1**: 637
Acetoacetarylamides
   from diketene **5**: 3075
Acetoacetates
   from diketene **5**: 3073
   production **3**: 2028
Acetoacetic acid **6**: 3602
Acetoacetic amides **6**: 3603
Acetoacetic esters **6**: 3603
   production **6**: 3604
   toxicology **6**: 3606
   uses and economic aspects **6**: 3605
Acetoacetoxyethyl methacrylate
   from diketene **5**: 3076
Acetoformic anhydride **1**: 81
Acetoguanamine **5**: 3214
   uses **5**: 3216
Acetone **1**: 89
   chemical properties **1**: 91
   economic aspects **1**: 106
   for production of pyridines **7**: 4183
   physical properties **1**: 90
   specifications **1**: 102
   toxicology **1**: 108
   uses **1**: 103
Acetone condensation products **1**: 105

Acetone cyanohydrin **1**: 91, 109
  for production of methacrylic acid and methyl
    methacrylate **6**: 3254
Acetonedicarboxylic acid **6**: 3607
Acetone-insoluble matter
  in lecithin **5**: 3154
Acetone production **1**: 92
Acetonitrile **6**: 3460
  physical properties **6**: 3457
p-Acetophenetidine **1**: 620
Acetophenone **5**: 3110
Acetoxylation
  of benzene, for phenol production **6**: 3704
Acetoxy silanes **1**: 68
Acetylacetonates
  metal **5**: 3108
Acetylacetone **5**: 3106
Acetylacetone production
  from ketene **5**: 3067
Acetylacetonitrile **6**: 3465
N-Acetyl γ acid **6**: 3437
4-(Acetylamino)benzenesulfonamide **2**: 773
4-(Acetylamino)benzenesulfonic acid **2**: 772
4-(Acetylamino)benzenesulfonyl chloride **2**: 772
N-Acetyl-2-amino-2-deoxy-D-glucopyranose **2**: 1065
5-Acetylamino-2-hydroxybenzenesulfonic acid **2**: 782
4-Acetylamino-5-hydroxynaphthalene-2,7-disulfonic
  acid **6**: 3441
3-Acetylamino-4-methoxybenzenesulfonamide **2**: 782
3-Acetylamino-4-methoxybenzenesulfonyl
  chloride **2**: 782
3-Acetylamino-4-methoxy-N-methylbenzene-sulfona-
  mide **2**: 782
4-(Acetylaminomethyl)benzenesulfonamide **2**: 764
4-(Acetylaminomethyl)benzenesulfonyl
  chloride **2**: 764
Acetylaniline, see N-Phenylacetamide
N-Acetylanthranilic acids **1**: 66
Acetylation
  of arylamines **1**: 506
  of mineral acids with acetic anhydride **1**: 67
  with acetic anhydride **1**: 66, 67
  with ketene **5**: 3063
Acetyl benzoyl peroxide
  physical properties **6**: 3650
N-Acetylcaprolactam
  deacetylation **2**: 1035
Acetyl chloride **1**: 55
Acetyl cyclohexylsulfonyl peroxide
  physical properties **6**: 3650
  toxicology **6**: 3676
Acetylene **1**: 117
  chemical properties **1**: 123
  economic aspects **1**: 183
  for acrylic acid production **1**: 230
  for acrylonitrile production **1**: 254
  for β-chloroprene production **3**: 1387

for production of 1,1-dichloroethane **3**: 1297
for production of pyridines **7**: 4187
for production of vinyl acetate **8**: 4841
for production of vinyl ethers **8**: 4869
for production of 1,1,2,2-tetrachloro-
  ethane **3**: 1324
for tetrachloroethylene production **3**: 1364
for THF production **8**: 4623
for trichloroethylene production **3**: 1357
for vinyl propionate production **8**: 4856
gas phase vinyl acetate process **8**: 4842
in acetaldehyde production **1**: 9
physical properties **1**: 119
recovery from cracked gas in ethylene production **4**: 2277
safety precautions **1**: 171
storage **1**: 177
toxicology **1**: 189
uses **1**: 183
Acetylene–air flames **1**: 182
Acetylene hydrochlorination
  for vinyl chloride production **3**: 1334
Acetylene hydrogenation
  of cracked gas in ethylene production **4**: 2276
Acetylene–methanol plant
  flow sheer **1**: 185
Acetylene oil quench process **1**: 137
Acetylene–oxygen mixtures **1**: 173
Acetylene production **1**: 130
Acetylene water quench process **1**: 135
N-Acetyl-D-glucosamine **2**: 1065
O-Acetylglycosyl halides **2**: 1076
N-Acetyl H acid **6**: 3441
Acetylides **1**: 129
N-Acetyl J acid **6**: 3435
N-Acetylmethionine-S-oxide **1**: 66
β-Acetylpropionic acid **6**: 3606
Acetylsalicylic acid **7**: 4278
  production **3**: 2027
2-Acetylthiophene **8**: 4700
  specifications **8**: 4696
  toxicology **8**: 4704
C acid **6**: 3419, 3436
Acid-gas removal
  from cracked gas in ethylene production **4**: 2269
Acidity
  of alkylphenols **6**: 3715
Acid-modified starch **7**: 4397
Acidolysis **3**: 2018
  of fats and fatty oils **4**: 2429
Acid potassium tartrate **8**: 4567
Acid soaps **6**: 3228
Acid value
  of fats and fatty oils **4**: 2438
Acridine **1**: 197
Acrolein **1**: 368
  chemical properties **1**: 200

for glycerol production **5**: 2799
for production of pyridines **7**: 4181
handling, storage, and transportation **1**: 212
physical properties **1**: 200
production **1**: 207
specifications **1**: 211
toxicology **1**: 215
uses **1**: 213
Acrolein acetals **1**: 203
Acrolein cyanohydrin **1**: 204
Acrolein diacetate **1**: 204
Acrolein hydrogenation
for allyl alcohol production **1**: 424
Acrolein hydrolysis **7**: 4056
Acrolein polymers **1**: 215
Acrylaldehyde, *see acrolein*
Acrylamide **1**: 225, 243
Acrylic acid **1**: 223
by carbonylation of acetylene **1**: 126
physical and chemical properties **1**: 224
specifications **1**: 236
toxicology **1**: 240
uses **1**: 238
Acrylic acid derivatives **1**: 223, 239
Acrylic acid production **1**: 229
Acrylic anhydride **1**: 225
Acrylic esters **1**: 223
production **1**: 234
toxicology **1**: 240
uses **1**: 238
Acrylic fibers **1**: 256
Acrylonitrile **1**: 249
by vinylation with acetylene **1**: 126
chemical properties **1**: 252
economic aspects **1**: 257
for production of pyridines **7**: 4184
from propene **7**: 4097
physical properties **1**: 249
reaction with amines **1**: 447
specifications **1**: 255
toxicology **1**: 258
uses **1**: 256
Acrylonitrile – butadiene – styrene **1**: 256
Acrylonitrile dimer **6**: 3462
Acrylonitrile hydration **1**: 243
Acrylonitrile hydrolysis
for acrylic acid production **1**: 230
Acrylonitrile production **1**: 252
Acrylosilanes **7**: 4340
Acryloyl chloride **1**: 225
Activated charcoal
for bleaching of fats and fatty oils **4**: 2418
Activated earths
for bleaching of fats and fatty oils **4**: 2418
Active primary amyl alcohol **6**: 3612
Acylals
production from acetic anhydride **1**: 69

Acylation
of alcohols with acylhalides or carboxylic anhydrides **3**: 2027
of amino phenols **1**: 604, 604
of ammonia or amines **8**: 4822
of hydrogen peroxide **6**: 3652
of olefins **3**: 2031
of phosphonic acid **6**: 3844
with ketenes **3**: 2028
Acyl halides **2**: 1114
for production of esters **3**: 2027
Acyloxysilanes **7**: 4327
Addition reactions
of acrylic acid and acrylates **1**: 226
of aldehydes **1**: 354
Adenine, *see purine derivatives*
Adhesives
cellulose ethers **2**: 1241
cyanoacrylates **1**: 242
melamine resins **5**: 3211
resorcinol **7**: 4268
Adiabatic dehydrogenation
of ethylbenzene **7**: 4424
Adipate **1**: 429
Adipic Acid **1**: 263
economic aspects **1**: 273
physical properties **3**: 1905
specifications **1**: 269
toxicology **1**: 274
uses **1**: 272
Adipic acid esters **1**: 271
Adipic acid production **1**: 265
Adipic acid salts **1**: 271
Adipic amide **1**: 272
Adipic anhydrides **1**: 271
Adipinaldehyde **1**: 390
Adiponitrile **1**: 270
Adiponitrile hydrogenation
for production of hexamethylenediamine **5**: 2847
Adsorption
for bleaching of fats and fatty oils **4**: 2418
for separation of *m-/p*-cresol mixtures **3**: 1682
Adsorptive dehydration
of ethanol **4**: 2122
Advanced cracking reactor acetylene process **1**: 165
Aerosol propellants
chlorofluoroalkanes **5**: 2579
dimethyl ether **4**: 1935
Agar **7**: 4009
applications **7**: 4012
properties **7**: 4011
structure **7**: 4010
Agar-agar, *see agar*
Agar gels **7**: 4011
Agarose **2**: 1068
Agar production **7**: 4010

Agrochemicals
 chlorophenols **3:** 1613
 from 1-naphthol **6:** 3374
 heterocyclic thiols and sulfides **8:** 4675
 methyl isothiocyanate **8:** 4642
 pyridines **7:** 4213
 pyrimidines **7:** 4233
 thiols **8:** 4656
Aipum
 as raw material for ethanol production **4:** 2098
Alcohol dehydration
 for production of ethers **4:** 2188
Alcohol dimerization
 with Guerbet process **1:** 292
Alcoholic fermentation
 yeasts **4:** 2063
Alcohol mixtures **1:** 294
Alcohols
 alkali fusion **2:** 1103
 for amine production **1:** 448
 saturated aliphatic, C_6 **1:** 293
 saturated aliphatic, C_7 **1:** 295
 saturated aliphatic, C_8 **1:** 298
 saturated aliphatic, C_9 **1:** 300
 saturated aliphatic, C_{10} **1:** 301
 saturated aliphatic, chemical properties **1:** 283
 saturated aliphatic, economic aspects **1:** 303
 saturated aliphatic, physical properties **1:** 280
 saturated aliphatic, production **1:** 284
 saturated aliphatic, specifications **1:** 307
 unsaturated aliphatic **1:** 308
Alcohols, aliphatic **1:** 279
 toxicology **1:** 314
Alcohols, polyhydric **1:** 321
 diols **1:** 326
 esters **1:** 238
 polyols **1:** 342
 synthesis routes **1:** 325
 tetrols **1:** 340
 triols **1:** 338
Alcoholysis **3:** 2017
 of acyl halides and carboxylic anhydrides **3:** 2027
 of glycerides **4:** 2430
 of ketenes **3:** 2028
 of nitriles **3:** 2030
Alcon oilseed extraction process **4:** 2409
Aldehyde hydrogenation
 for alcohol production **1:** 286
Aldehyde oxidation
 for production of saturated monocorboxylic acids **2:** 1101
Aldehydes
 condensation to esters **3:** 2029
 for amine production **1:** 450
 for production of pyridines **7:** 4183
 saturated aliphatic C_5 **1:** 361
 saturated aliphatic C_6 and C_7 **1:** 362
 saturated aliphatic C_8 **1:** 363
 saturated aliphatic C_9 **1:** 364
 saturated aliphatic C_{10} and C_{11} **1:** 365
 saturated aliphatic C_{12} and C_{13} **1:** 366
 saturated aliphatic, chemical properties **1:** 354
 saturated aliphatic, physical properties **1:** 351
 saturated aliphatic, production **1:** 357
 unsaturated aliphatic **1:** 366
 unsaturated aliphatic, C_5–C_{11} **1:** 372
 unsaturated aliphatic, production **1:** 369
Aldehydes, aliphatic **1:** 349
 economic aspects **1:** 401
Aldehydes, araliphatic **1:** 349, 381
 economic aspects **1:** 401
 production **1:** 383
Aldoketenes **5:** 3076
Aldol addition **1:** 5
Aldol chemicals **1:** 105
Aldol condensation
 for crotonaldehyde production **3:** 1713
 for hydroxyaldehyde production **1:** 376
 for production of aliphatic alcohols **1:** 287
 for production of araliphatic aldehydes **1:** 383
 of acetaldehyde **1:** 370
 of propanal **7:** 4039
Aldol dehydration
 for production of unsaturated aldehydes **1:** 370
Aldolization
 of butyraldehyde **4:** 2363
Aldoses **2:** 1063
Alen acid **6:** 3424
Alfa–Laval Centrimeal process **4:** 2410
Alfol alcohol process **1:** 288
Alfol fatty alcohol process **5:** 2542
Alginate production **7:** 3996
Alginates
 applications **7:** 4002
 gel properties **7:** 3999
 occurrence and structure **7:** 3993
 properties **7:** 3998
Alginic acid **2:** 1068
Alkali fusion
 of benzenesulfonic acids **2:** 743
 of naphthalenesulfonic acids **6:** 3386
 of toluenesulfonates **3:** 1664
 of xylenesulfonates for xylenol production **3:** 1696
Alkali metals
 in organosilane synthesis **7:** 4315
Alkalization
 of cellulose **2:** 1212
Alkane oxidation
 for production of carboxylic acids **2:** 1103
Alkanes **5:** 2924
 chemical properties **5:** 2857
 fluorinated **5:** 2572
 oxidation to fatty acids **4:** 2521
 physical properties **5:** 2854

production **5**: 2860
uses **5**: 2863
Alkanesulfonic acids **8**: 4503
Alkar ethylbenzene process **4**: 2213
Alkenals **1**: 372
 for production of aliphatic alcohols **1**: 287
Alkenes **5**: 2870
 see also olefins
 higher, economic aspects **5**: 2888
 higher, production **5**: 2873
 higher, uses **5**: 2884
 properties **5**: 2870
 toxicology **5**: 2925
α-Alkenesulfonates
 from higher olefins **5**: 2885
Alkenylphenols **6**: 3756
Alkoxides **1**: 311
 preparation **1**: 312
 uses **1**: 314
α-Alkoxyalkylureas **8**: 4825
3-Alkoxypropylamines **4**: 2170
Alkoxysilanes **7**: 4321
Alkyd resins
 with fatty acids **4**: 2527
Alkyl acrylates
 production **1**: 234
Alkylamines **1**: 443, 458
 chemical properties **1**: 444
 cyclic amines **1**: 464
 cycloalkylamines **1**: 459
 diamines and polyamines **1**: 482
 fatty amines, *see fatty amines*
 lower **1**: 453
 lower, production and uses **1**: 457
 production from alcohols **1**: 448
 toxicology **1**: 490
Alkylate gasoline **2**: 997
Alkylation
 of alkali-metal xanthates **8**: 4974
 of amines **1**: 447
 of amino phenol **1**: 615
 of arylamines **1**: 508
 of benzene, liquid-phase **4**: 2209
 of benzene, vapor-phase **4**: 2212
 of hydrocarbons **5**: 2863
 of hydrogen peroxide **6**: 3635
 of metal carboxylates **3**: 2026
 of phenols **6**: 3721
 of propylene **7**: 4096
 of sulfinate anions **7**: 4457
 of sulfinic acid derivatives **7**: 4480
 of urea with tertiary alcohols **8**: 4820
 with butanols **2**: 955
 with dialkyl sulfates **3**: 1864
N-Alkylation
 with dialkyl sulfates **3**: 1867

C-Alkylation
 with dialkyl sulfates **3**: 1868
Alkylation rules
 for phenols **6**: 3724
Alkylbenzenes **5**: 2893
 from higher olefins **5**: 2885
 linear secondary **5**: 2903
 toxicology **5**: 2925
α-Alkylcinnamaldehydes **1**: 389
Alkylguanidines **5**: 2835
Alkyl halides
 for amine production **1**: 451
Alkyl hydroperoxides **6**: 3632
 for hydrocarbon oxidation **1**: 289
Alkyl ketones
 methyl alkyl **5**: 3085
Alkylnaphthalenes **6**: 3355
Alkylnaphthalenesulfonic acids **6**: 3372
Alkylphenols
 higher **6**: 3747
 occurrence **6**: 3718
 production **6**: 3721
 properties **6**: 3714
Alkylphenols oxidation
 for production of hydroxybenzoic acids **5**: 2976
Alkylpyridines
 properties **7**: 4176
Alkylsulfuric acids **3**: 1853
 production **3**: 1858
 properties **3**: 1854
Alkylthioformimidic chlorides **4**: 1965
N-Alkyltoluidines **8**: 4749
Allied/UOP cumene hydroperoxide cleavage process **1**: 96
Allied/UOP cumene oxidation process **1**: 95
Allied chemical oxalic acid process **6**: 3583
Alloocimen **8**: 4600
D-Allose **2**: 1065
Alloxan **7**: 4231
Allyl acetate **1**: 425, 426, 427
Allyl acetate hydrolysis
 for allyl alcohol production **1**: 423
Allyl acrylate **1**: 425, 426
Allyl alcohol **1**: 309, 419
 from acrolein **1**: 205
 production **1**: 421
 toxicology **1**: 434
Allylamine **1**: 432, 455
Allylamines
 toxicology **1**: 436
Allyl bromide, *see 3-bromo-1-propene*
Allyl carbonate **1**: 426
Allyl chloride **1**: 409
 production **1**: 412
 toxicology **1**: 433
 uses and economic aspects **1**: 419

Allyl chloride hydrolysis
   for allyl alcohol production **1**: 422
Allyl chloroformate **3**: 1576
Allyl compounds **1**: 409, 431
   toxicology **1**: 433
Allyl cyanide
   physical properties **6**: 3457
Allyl esters **1**: 425
   production **1**: 426
   toxicology **1**: 435
   uses **1**: 428
Allyl ethers **1**: 431
   production **1**: 432
   toxicology **1**: 435
Allyl glycidyl ether **1**: 431
Allyl isothiocyanate
   physical properties **8**: 4636
   toxicology **8**: 4643
Allyl isothiocyanate production **8**: 4640
Allyl methacrylate **1**: 426, 430
(Allyloxymethyl)oxirane **1**: 431
Allylsulfonic acid **8**: 4506
Allylsulfonic acid production **8**: 4507
ALMA process **5**: 3166
Almond oil
   fatty acid content **4**: 2376
Alphabutol olefin process **5**: 2882
Alphamin **6**: 3404
D-Altrose **2**: 1065
Aluminum acetate **1**: 51
Aluminum chloride
   for alkylation of benzene **4**: 2209
Aluminum chloride ethyl benzene process **4**: 2210
Aluminum dilaurate **6**: 3239
Aluminum dimyristate **6**: 3239
Aluminum di-12-oxystearate **6**: 3239
Aluminum dipalmitate **6**: 3239
Aluminum distearate **6**: 3239
Aluminum formate **5**: 2739
Aluminum hydroxide
   as catalyst in esterification of carboxlic
      acids **3**: 2024
Aluminum monopalmitate **6**: 3239
Aluminum monostearate **6**: 3239
γ-Aluminum oxide
   as phenol methylation catalyst **3**: 1673
   as phenol alkylation catalyst **6**: 3724
Aluminum phenolate
   as phenol alkylation catalyst **6**: 3722
Aluminum soaps **6**: 3238
Aluminum tristearate **6**: 3239
Alusuisse – Ftalital LAR phthalic anhydride process **7**: 3870
American Cyanamid adiabatic nitration process **6**: 3512
Amides
   for production of nitriles **6**: 3459

Amide waxes
   toxicology **8**: 4962
Amido F acid **6**: 3415
Amidosilanes **7**: 4328
Amidosulfuric acid, *see sulfamic acid*
Amination
   of phenol **1**: 632
   of xylenols **8**: 5022
Amine hydrochlorides
   phosgenation **5**: 3019
Amines
   aliphatic, *see Alkylamines*
   aromatic, *see Arylamines*
   fatty, *see fatty amines*
   fluorinated tertiary **5**: 2607
   for production of nitriles **6**: 3459
   in diazotization processes **3**: 1878
   phosgenation **5**: 3017
   salt formation **1**: 445
Aminoacetonitrile **6**: 3463
1-Amino-3(3-aminopropyl)-aminopropane **1**: 488
2-Aminobenzaldehyde **2**: 687
4-Aminobenzaldehyde **2**: 687
Aminobenzene, *see aniline*
4-Amino-1,3-benzenedisulfonamide **2**: 773
2-Amino-1,4-benzenedisulfonic acid **2**: 773
4-Amino-1,3-benzenedisulfonic acid **2**: 773
5-Amino-1,3-benzenedisulfonic acid **2**: 773
5-Amino-1,3-benzenedisulfonyl chloride **2**: 773
2-Aminobenzenesulfonamide **2**: 770
3-Aminobenzenesulfonamide **2**: 771
4-Aminobenzenesulfonamide **2**: 772
2-Aminobenzenesulfonic acid **2**: 769
3-Aminobenzenesulfonic acid **2**: 771
4-Aminobenzenesulfonic acid **2**: 771
2-Aminobenzimidazole **5**: 2838
2-Aminobenzoic Acid **2**: 833
3-Aminobenzoic acid **2**: 834
4-Aminobenzoic acid **2**: 834
5-Amino-*N,N*-bis(2-chloroethyl)–1,3-benzene-
      disulfonamide **2**: 773
3-Amino-2-butenenitrile **6**: 3465
1-Amino-1-*tert*-butylperoxycyclohexane
   physical properties **6**: 3670
Amino C acid **6**: 3424
Amino-chloro-alkylbenzenesulfonic acids **2**: 785
1-Amino-4-chlorobenzene, *see p-chloroaniline*
4-Amino-6-chloro-1,3-benzenedisulfonamide **2**: 785
4-Amino-6-chloro-1,3-benzenedisulfonic acid **2**: 784
4-Amino-6-chloro-1,3-benzenedisulfonyl chlor-
      ide **2**: 784
2-Amino-5-chlorobenzenesulfonic acid **2**: 784
3-Amino-4-chlorobenzenesulfonic acid **2**: 784
4-Amino-6-chloro-*N,N*-dimethyl-1,3-benzene-
      disulfonamide **2**: 785
3-Amino-5-chloro-2-hydroxybenzenesulfonic
      acid **2**: 783

3-Amino-5-chloro-4-hydroxybenzenesulfonic acid   **2**: 783
2-Amino-4-chloro-3-methylbenzenesulfonic acid   **8**: 4753
2-Amino-4-chloro-5-methylbenzenesulfonic acid   **2**: 786; **8**: 4754
2-Amino-5-chloro-4-methylbenzenesulfonic acid   **2**: 785; **8**: 4753
3-Amino-5-chloro-4-methylbenzenesulfonic acid   **2**: 785; **8**: 4753
4-Amino-3-chloro-5-methylbenzenesulfonic acid   **8**: 4753
4-Amino-5-chloro-2-methylbenzenesulfonic acid   **2**: 786
4-Amino-5-chloro-3-methylbenzenesulfonic acid   **2**: 786
2-Amino-6-chloro-4-nitrophenol   **1**: 615
2-Amino-4-chlorophenol   **1**: 614
2-Amino-3-chlorotoluene-5-sulfonic acid   **8**: 4753
2-Amino-6-chlorotoluene-3-sulfonic acid   **8**: 4753
2-Amino-6-chlorotoluene-4-sulfonic acid   **8**: 4753
3-Amino-6-chlorotoluene-4-sulfonic acid   **8**: 4753
4-Amino-6-chlorotoluene-3-sulfonic acid   **8**: 4754
3-Aminocrotonate esters from diketene   **5**: 3076
β-Aminocrotononitrile   **6**: 3465
Aminocyclohexane, *see cyclohexylamine*
1-Amino-3-cyclohexylaminopropane   **1**: 485
2,2-Bis(4-aminocyclohexyl)propane   **1**: 463
1-Amino-3,4-dichlorobenzene, *see 3,4-dichloroaniline*
2-Amino-4,5-dichlorobenzenesulfonamide   **2**: 784
2-Amino-4,5-dichlorobenzenesulfonic acid   **2**: 784
4-Amino-2,5-dichlorobenzenesulfonic acid   **2**: 784
2-Amino-4,5-dichlorobenzenesulfonyl chloride   **2**: 784
6-Amino-2,4-dichloro-3-methylbenzenesulfonic acid   **2**: 786
2-Amino-4,6-dichlorophenol   **1**: 614
4-Amino-2,6-dichlorophenol   **1**: 621
1-Amino-3-diethylaminopropane   **1**: 485
1-Amino-3-dimethylaminopropane   **1**: 485
2-Amino-3,5-dimethylbenzenesulfonic acid   **2**: 778
2-Amino-4,6-dinitrophenol   **1**: 613
12-Aminododecanoic acid   **3**: 1791
Amino epsilon acid   **6**: 3416
2-Aminoethanol, *see monoethanolamine*
3-Amino-6-ethoxybenzenesulfonic acid   **2**: 782
1-Amino-2-ethoxynaphthalene   **6**: 3426
3-(2-Aminoethyl)-aminopropylamine   **1**: 488
Aminoethylated starches   **7**: 4409
N-(2-aminoethyl)–1,3-diaminopropane   **1**: 488
3-(2-Aminoethyl)indole   **5**: 3003
Aminoethylisopropanolamine   **4**: 2171
1-(2-Aminoethyl)piperazine toxicity   **1**: 492
Amino G acid   **6**: 3420
Aminoguanidine   **3**: 1749; **5**: 2833

1-Amino-1-hydroperoxycyclohexane physical properties   **6**: 3670
1-Amino-1-hydroperoxycyclohexane production   **6**: 3672
2-Amino-1-hydroxybenzene   **1**: 602
3-Amino-1-hydroxybenzene   **1**: 603
4-Amino-1-hydroxybenzene   **1**: 603
5-Amino-4-hydroxy-1,3-benzenedisulfonic acid   **2**: 782
3-Amino-4-hydroxybenzenesulfonamide   **2**: 782
3-Amino-4-hydroxybenzenesulfonic acid   **2**: 781
4-Amino-2-hydroxybenzoic acid   **1**: 618
5-Amino-2-hydroxybenzoic acid   **1**: 620
3-Amino-4-hydroxybenzoic acid methyl ester   **1**: 616
3-Amino-5-hydroxy-2,7-disulfonic acid   **6**: 3442
1-Amino-2-hydroxynaphthalene-3,6-disulfonic acid   **6**: 3439
1-Amino-8-hydroxynaphthalene-2,4-disulfonic acid   **6**: 3439
1-Amino-8-hydroxynaphthalene-3,6-disulfonic acid   **6**: 3439
1-Amino-8-hydroxynaphthalene-4,6-disulfonic acid   **6**: 3441
2-Amino-5-hydroxynaphthalene-1,7-disulfonic acid   **6**: 3442
2-Amino-5-hydroxynaphthalene-4,8-disulfonic acid   **6**: 3442
2-Amino-8-hydroxynaphthalene-3,6-disulfonic acid   **6**: 3442
3-Amino-8-hydroxynaphthalene-1,5-disulfonic acid   **6**: 3442
4-Amino-3-hydroxynaphthalene-2,7-disulfonic acid   **6**: 3439
4-Amino-5-hydroxynaphthalene-1,3-disulfonic acid   **6**: 3439
4-Amino-5-hydroxynaphthalene-1,7-disulfonic acid   **6**: 3441
4-Amino-5-hydroxynaphthalene-2,7-disulfonic acid   **6**: 3439
1-Amino-2-hydroxynaphthalene-4-sulfonic acid   **6**: 3429
1-Amino-2-hydroxynaphthalene-6-sulfonic acid   **6**: 3431
1-Amino-5-hydroxynaphthalene-6-sulfonic acid   **6**: 3432
1-Amino-8-hydroxynaphthalene-4-sulfonic acid   **6**: 3433
1-Amino-8-hydroxynaphthalene-6-sulfonic acid   **6**: 3433
2-Amino-1-hydroxynaphthalene-4-sulfonic acid   **6**: 3433
2-Amino-3-hydroxynaphthalene-6-sulfonic acid   **6**: 3434
2-Amino-5-hydroxynaphthalene-7-sulfonic acid   **6**: 3434
2-Amino-6-hydroxynaphthalene-8-sulfonic acid   **6**: 3436

2-Amino-8-hydroxynaphthalene-6-sulfonic
  acid **6:** 3436
2-Amino-8-hydroxynaphthalene-7-sulfonic
  acid **6:** 3438
3-Amino-4-hydroxynaphthalene-1-sulfonic
  acid **6:** 3433
4-Amino-3-hydroxynaphthalene-1-sulfonic
  acid **6:** 3429
4-Amino-5-hydroxynaphthalene-1-sulfonic
  acid **6:** 3433
5-Amino-1-hydroxynaphthalene-2-sulfonic
  acid **6:** 3432
5-Amino-4-hydroxynaphthalene-2-sulfonic
  acid **6:** 3433
5-Amino-6-hydroxynaphthalene-2-sulfonic
  acid **6:** 3431
6-Amino-4-hydroxynaphthalene-2-sulfonic
  acid **6:** 3436
6-Amino-7-hydroxynaphthalene-2-sulfonic
  acid **6:** 3434
7-Amino-1-hydroxynaphthalene-2-sulfonic
  acid **6:** 3438
7-Amino-3-hydroxynaphthalene-1-sulfonic
  acid **6:** 3436
7-Amino-4-hydroxynaphthalene-2-sulfonic
  acid **6:** 3434
8-Amino-4-hydroxynaphthalene-2-sulfonic
  acid **6:** 3433
Aminohydroxynaphthalenesulfonic acids
  production and properties **6:** 3427
2-amino-6-hydroxypurine, *see purine derivatives*
Amino J acid **6:** 3419
2-amino-6-mercaptopurine, *see purine derivatives*
Aminomethane, *see monomethylamine*
4-Amino-2-methoxy-acetanilide **1:** 615
4-Amino-6-methoxy-1,3-benzenedisulfonic acid **2:** 783
4-Amino-6-methoxy-1,3-benzenedisulfonyl
  chloride **2:** 783
3-Amino-4-methoxybenzenesulfonic acid **2:** 782
3-amino-4-methoxytoluene **1:** 614
3-Amino-3-methylacrylonitrile **6:** 3465
1-Amino-3-methylaminopropane **1:** 485
1-Amino-3-*N*-methyl-*N*-(3-aminopropyl)-aminopropane **1:** 488
2-Amino-4-methylanisole **1:** 614
2-Amino-*N*-methylbenzenesulfonamide **2:** 770
2-Amino-5-methylbenzenesulfonic acid **2:** 778;
  8: 4752
3-Amino-4-methylbenzenesulfonic acid **2:** 779;
  8: 4751
4-Amino-2-methylbenzenesulfonic acid **2:** 779;
  8: 4751
4-Amino-3-methylbenzenesulfonic acid **2:** 779;
  8: 4751
5-Amino-2-methylbenzenesulfonic acid **2:** 779;
  8: 4752
2-Amino-4-methylphenol **1:** 614

2-Amino-6-methylpyridine **7:** 4207
  physical properties **7:** 4206
1-Aminonaphthalene **6:** 3401
2-Aminonaphthalene **6:** 3403
1-Aminonaphthalene-3,7-disulfonic acid **6:** 3416
1-Aminonaphthalene-3,8-disulfonic acid **6:** 3416
1-Aminonaphthalene-4,6-disulfonic acid **6:** 3417
1-Aminonaphthalene-4,7-disulfonic acid **6:** 3417
1-Aminonaphthalene-4,8-disulfonic acid **6:** 3417
1-Aminonaphthalene-5,7-disulfonic acid **6:** 3418
2-Aminonaphthalene-1,5-disulfonic acid **6:** 3418
2-Aminonaphthalene-3,6-disulfonic acid **6:** 3419
2-Aminonaphthalene-4,8-disulfonic acid **6:** 3419
2-Aminonaphthalene-5,7-disulfonic acid **6:** 3419
2-Aminonaphthalene-6,8-disulfonic acid **6:** 3420
3-Aminonaphthalene-1,5-disulfonic acid **6:** 3419
3-Aminonaphthalene-2,7-disulfonic acid **6:** 3419
4-Aminonaphthalene-1,5-disulfonic acid **6:** 3417
4-Aminonaphthalene-1,6-disulfonic acid **6:** 3417
4-Aminonaphthalene-1,7-disulfonic acid **6:** 3417
4-Aminonaphthalene-2,6-disulfonic acid **6:** 3416
4-Aminonaphthalene-2,7-disulfonic acid **6:** 3416
5-Aminonaphthalene-1,3-disulfonic acid **6:** 3418
6-Aminonaphthalene-1,3-disulfonic acid **6:** 3419
7-Aminonaphthalene-1,3-disulfonic acid **6:** 3420
8-Aminonaphthalene-1,6-disulfonic acid **6:** 3416
Aminonaphthalenes **6:** 3401
1-Aminonaphthalene-sulfonic acid **6:** 3411
1-Aminonaphthalene-2-sulfonic acid **6:** 3410
1-Aminonaphthalene-3-sulfonic acid **6:** 3411
1-Aminonaphthalene-3,6-sulfonic acid **6:** 3416
1-Aminonaphthalene-5-sulfonic acid **6:** 3412
1-Aminonaphthalene-6-sulfonic acid **6:** 3412
1-Aminonaphthalene-7-sulfonic acid **6:** 3412
1-Aminonaphthalene-8-sulfonic acid **6:** 3413
2-Aminonaphthalene-sulfonic acid **6:** 3415
2-Aminonaphthalene-1-sulfonic acid **6:** 3414
2-Aminonaphthalene-5-sulfonic acid **6:** 3415
2-Aminonaphthalene-7-sulfonic acid **6:** 3415
2-Aminonaphthalene-8-sulfonic acid **6:** 3415
4-Aminonaphthalene-1-sulfonic acid **6:** 3411
4-Aminonaphthalene-2-sulfonic acid **6:** 3411
5-Aminonaphthalene-1-sulfonic acid **6:** 3412
5-Aminonaphthalene-2-sulfonic acid **6:** 3412
6-Aminonaphthalene-1-sulfonic acid **6:** 3415
6-Aminonaphthalene-2-sulfonic acid **6:** 3415
7-Aminonaphthalene-1-sulfonic acid **6:** 3415
7-Aminonaphthalene-2-sulfonic acid **6:** 3415
8-Aminonaphthalene-1-sulfonic acid **6:** 3413
8-Aminonaphthalene-2-sulfonic acid **6:** 3412
Aminonaphthalenesulfonic acids
  production and properties **6:** 3406
  toxicity **6:** 3425
1-Aminonaphthalene-2,4,8-trisulfonic acid **6:** 3420
1-Aminonaphthalene-3,6,8-trisulfonic acid **6:** 3421
1-Aminonaphthalene-4,6,8-trisulfonic acid **6:** 3421
2-Aminonaphthalene-1,5,7-trisulfonic acid **6:** 3422

2-Aminonaphthalene-3,6,8-trisulfonic acid **6:** 3422
4-Aminonaphthalene-1,3,5-trisulfonic acid **6:** 3420
6-Aminonaphthalene-1,3,5-trisulfonic acid **6:** 3422
7-Aminonaphthalene-1,3,6-trisulfonic acid **6:** 3422
8-Aminonaphthalene-1,3,5-trisulfonic acid **6:** 3421
8-Aminonaphthalene-1,3,6-trisulfonic acid **6:** 3421
1-Amino-2-naphthalenol **6:** 3426
4-Amino-1-naphthalenol **6:** 3425
5-Amino-1-naphthalenol **6:** 3426
7-Amino-1-naphthalenol **6:** 3426
8-Amino-2-naphthalenol **6:** 3426
1-Amino-1-naphthol **6:** 3426
1-Amino-7-naphthol **6:** 3426
4-Amino-1-naphthol **6:** 3425
5-Amino-1-naphthol **6:** 3426
7-Amino-1-naphthol **6:** 3426
8-Amino-2-naphthol **6:** 3426
2-Amino-5-nitrobenzonitrile **6:** 3472
2-Amino-4-nitrophenol **1:** 612
2-Amino-5-nitrophenol **1:** 612
2-Amino-5-nitrothiazole **6:** 3572
Aminoorganosilanes **7:** 4335
α-Aminoperoxides **6:** 3669
  production **6:** 3671
  uses **6:** 3672
Aminophen, *see aniline*
2-Aminophenol **1:** 602
3-Aminophenol **1:** 603
4-Aminophenol **1:** 603
Aminophenol production **1:** 606
Aminophenols **1:** 601
  chemical properties **1:** 603
  physical properties **1:** 601
  toxicology **1:** 610
  uses **1:** 610
N-(4-Aminophenyl)benzenesulfonamide **2:** 759
2-Amino-*N*-(phenylsulfonyl)benzene-
  sulfonamide **2:** 770
3-Amino-*N*-(phenylsulfonyl)benzene-
  ulfonamide **2:** 771
**Aminopolycarboxylic acid chelating
  agents** **4:** 2295
Aminopolycarboxylic acids
  economic aspects **4:** 2302
  properties **4:** 2296
  uses **4:** 2300
3-Aminopropanenitrile **6:** 3464
1-Amino-2-propanol **4:** 2164, 2170
  see also Isopropanolamines
2-Amino-1-propanol **4:** 2164, 2170
  see also Isopropanolamines
3-Amino-1-propanol **4:** 2170, 2173
  toxicology **4:** 2179
2-Amino-1-propene-1,1,3-tricarbonitrile **6:** 3465
β-Aminopropionitrile **6:** 3464
3-Aminopropionitrile **6:** 3464
N,N'-Bis(3-aminopropyl)–1,2-diaminoethane **1:** 488

N,N'-Bis(3-aminopropyl)ethylenediamine **1:** 488
Aminopropyltriethoxysilane **7:** 4335
6-aminopurine, *see purine derivatives*
2-Aminopyridine **7:** 4206
3-Aminopyridine **7:** 4208
  physical properties **7:** 4206
4-Aminopyridine **7:** 4208
  physical properties **7:** 4206
Aminopyridines **7:** 4206
Amino R acid **6:** 3419
4-Aminosalicylic acid **1:** 618
5-Aminosalicylic acid **1:** 620
Amino-Schaeffer acid **6:** 3431
Aminosilanes **7:** 4328
2-Amino-4-sulfobenzoic acid **2:** 779
2-Amino-5-sulfobenzoic acid **2:** 780
3-Aminotetrahydro-1,3-thiazine-2,4-diones **4:** 1971
2-Aminothiazol-4-yl acetates
  from diketene **5:** 3076
Aminotoluene, *see toluidine*
Aminotriazines
  uses **3:** 1777
2-Amino-1,1,3-tricyanopropene **6:** 3465
2-Amino-*m*-xylene
  physical properties **8:** 5021
3-Amino-*o*-xylene
  physical properties **8:** 5021
4-Amino-*o*-xylene
  physical properties **8:** 5021
5-Amino-*m*-xylene
  physical properties **8:** 5021
5-Amino-*p*-xylene
  physical properties **8:** 5021
Aminoxysilanes **7:** 4327
Ammonia
  for production of pyridines **7:** 4181
  in urea production **8:** 4787
Ammonia-carbon dioxide ratio
  in urea production **8:** 4788
Ammonia – Casale methanol process **6:** 3304
Ammonia – urea production **8:** 4809
Ammonium acetate **1:** 52
Ammonium dithiocarbamate **4:** 1949
Ammonium formate **5:** 2739
Ammonium iron(III) oxalate **6:** 3594
Ammonium oxalate **6:** 3594
Ammonium propionate **7:** 4116
Ammonium sulfamate **7:** 4446
Ammonolysis
  for production of nitriles **6:** 3458
  of esters **3:** 2018
  oxidative, of *o*-xylene **7:** 3884
Ammoxidation **6:** 3458, 3469
  of alkylpyridines **7:** 4203
  of *o*-xylene **7:** 3884
Amoco carboxylic acid procedure **2:** 1129
Amoco terephalic ester purification process **8:** 4578

Amoco xylene oxidation process **8**: 4576
AMP, *see acetaldehyde monoperacetate*
Amyl acetate
   toxicology **3**: 2043
Amyl alcohol **1**: 281
n-Amyl alcohol **6**: 3612
sec-Amyl alcohol **1**: 281; **6**: 3612
tert-Amyl alcohol **1**: 281; **6**: 3612
Amyl alcohols, *see pentanols*
α-Amylcinnamaldehyde **1**: 382
α-Amylcinnamaldehydes **1**: 389
Amyl ethyl ketone **5**: 3098
Amylopectin **7**: 4381
   structure **7**: 4380
Amylose **7**: 4381
   structure **7**: 4380
Amylose complexes **7**: 4388
Anarcardol **6**: 3757
Andrussow hydrogen cyanide process **6**: 3458
Anesthetic ether **4**: 2191
Anesthetics
   fluorinated ethers **5**: 2595
Aniline **1**: 627
   economic aspects **1**: 639
   physical and chemical properties **1**: 628
   specifications **1**: 633
   toxicology **1**: 640
   uses **1**: 505, 634
Aniline hydrochloride **1**: 636
Aniline oxidation
   for production of hydroquinone **5**: 2946
Aniline production **1**: 629
Aniline sulfate **1**: 636
Animal fats **4**: 2467
   manufacture and processing **4**: 2410
Animal waxes **8**: 4900
m-Anisidine **1**: 636
o-Anisidine **1**: 612, 636
   uses **1**: 505
p-Anisidine **1**: 636
Anisidine value
   of fats and fatty oils **4**: 2438
Anobial **6**: 3792
Anomeric configuration
   of monosaccharides **2**: 1064
Anthelmintics
   guanidine derivatives **5**: 2838
Anthracene **1**: 643
Anthracene oil **8**: 4536
Anthracene oxidation **2**: 653
Anthranilic acid **2**: 833
Anthraquinone
   economic aspects **2**: 660
   properties **2**: 650
   reduction **2**: 651
   toxicology **2**: 661
   uses **2**: 659

Anthraquinone production **2**: 652
Anticorrosion protective coatings
   pitches **8**: 4533
Antifreezes
   ethylene glycol **4**: 2321
   methanol **6**: 3315
   propylene glycol **7**: 4048
Antiknock agents
   MTBE **6**: 3344
Antimicrobial activity
   of sorbic acid **7**: 4371
Antioxidants
   for cellulose ester molding compounds **2**: 1194
   from tert-butyl phenols **6**: 3742
   phosphites **6**: 3840
Antiseptics
   chlorhexidine **5**: 2840
Antoine equation **4**: 2490
Apiose **2**: 1066
Arabinan
   as resources for furfural production **5**: 2751
Arabinogalactan
   in gum arabic **7**: 4013
   in gum tragacanth **7**: 4015
L-Arabinose
   fermentation in yeast **4**: 2069
Arachidic acid
   physical properties **4**: 2485
Arachidonic acid
   physical properties **4**: 2487
Arachidyl alcohol
   physical properties **5**: 2535
Aralkylphenols **6**: 3754
Aramid fibers **1**: 507
Arc coal acetylene process **1**: 154
Arc furnace
   in acetylene production **1**: 147
Arco MTBE process **6**: 3340
Armour fatty acid separation process **4**: 2515
Armstrong acid **6**: 3369
Aromatic nitro compounds
   electrochemical reduction **1**: 515
Aromatic oils **8**: 4537
C_8 Aromatics **8**: 4990
   isomerization **8**: 5002
   separation **8**: 4998
   separation from BTX feedstocks **2**: 718
   xylene production **8**: 4996
Aromatics nitration **6**: 3505
Arosolvan aromatic separation process **2**: 721
Arosolvan toluene extraction process **8**: 4732
Arrowroot starch production **7**: 4395
Arylamines **1**: 503
   major uses **1**: 504
   physical and chemical properties **1**: 506
   production **1**: 513
   toxicology **1**: 525

Arylguanidines **5**: 2837
Aryloxysilanes **7**: 4321
Arylphosphines
    for production of tertiary phosphines **6**: 3824
Asahi Glass chloromethane process **3**: 1270
Asphalt emulsification **1**: 479
Asterosaponins **7**: 4299
2AT **8**: 4700
Atmospheric decomposition
    of chloromethanes **3**: 1283
Atochem vinyl chloride process **3**: 1344
Ausind melamine process **5**: 3208
Autoxidation **6**: 3639
    of aldehydes **6**: 3648
    of ethers **6**: 3666
    of unsaturated fatty acids **4**: 2399
AVCO plasma furnace
    for the pyrolysis of coal **1**: 155
Avocado oil **4**: 2445
    fatty acid content **4**: 2376
Azacycloheptane **1**: 466
Azacyclotridecan-2-one, *see laurolactam*
8-Azaisatoic anhydride **3**: 1567
Azelaic acid **3**: 1916
    physical properties **3**: 1905
Azelaic acid production **3**: 1908
Azeotropes
    ammonia-carbon dioxide **8**: 4793
    of peroxycarboxylic acids and water **6**: 3646
    pentanol-water **6**: 3613
    styrene-water **7**: 4419
    ternary systems, liquid gas equilibrium **8**: 4793
    with acrylonitrile **1**: 251
    with aliphatic alcohols **1**: 284
    with aliphatic aldehydes **1**: 354
    with allyl alcohol **1**: 420
    with allyl chloride **1**: 411
    with benzaldehyde **2**: 674
    with benzene **2**: 697
    with benzyl alcohol **2**: 850
    with benzylamine **2**: 864
    with benzyl chloride **3**: 1450
    with bis(2-chloroethyl) ether **4**: 2200
    with butadiene and other C_4-hydrocarbons **2**: 902
    with butanols **2**: 954
    with butyl acetates **1**: 54
    with butyraldehydes **2**: 927
    with 1-chlorobutane **3**: 1379
    with chlorohydrius **3**:
    with cyclohexane **3**: 1799
    with diamyl ethers **4**: 2198
    with di-*n*-butyl ether **4**: 2197
    with 1,2-dichloroethane **3**: 1299
    with 1,2-dichloroethylene **3**: 1353
    with dichloromethane **3**: 1256
    with diethyl ether **4**: 2190
    with diisopropyl ether **4**: 2195

    with dioxane **4**: 1938
    with di-*n*-propyl ether **4**: 2196
    with hexachloroethane **3**: 1329
    with isoprene **5**: 3036
    with methanol **6**: 3290
    with methyl isobutylketone **5**: 3087
    with MTBE **6**: 3338
    with nitrobenzene **6**: 3510
    with pentachloroethane **3**: 1326
    with phenol **6**: 3692
    with piperidine **1**: 465
    with 1-propanol and 2-propanol **7**: 4062
    with propionic acid **7**: 4102
    with propylene glycol **7**: 4050
    with pyridines **7**: 4178
    with 1,1,2,2-tetrachloroethane **3**: 1322
    with tetrachloroethylene **3**: 1363
    with tetrachloromethane **3**: 1258
    with toluene **8**: 4729
    with 1,1,1-trichloroethane **3**: 1311, 1318
    with trichloroethylene **3**: 1358
    with trichloromethane **3**: 1257
    with xylenes and ethylbenzene **8**: 4992
Azeotropic distillation
    of esters **3**: 2025
    of ethanol **4**: 2115
Azeotropes
    formic acid–water **5**: 2714
Aziridine, *see ethylenimine*
Aziridines **2**: 663
    economic aspects, toxicology **2**: 669
    physical and chemical properties **2**: 664
    polyfunctional **2**: 665
    production **2**: 666
    uses **2**: 668
Azo coupling **3**: 1893
Azole, *see pyrrole*
Azurin acid **6**: 3390
Azurol **6**: 3379

Babassu oil **4**: 2448
    fatty acid content **4**: 2376
B acid **6**: 3421
Backward integration
    of toluene production into refineries **8**: 4734
    of xylene production into refineries **8**: 5005
Bacteria
    in ethanol production **4**: 2075
Badger fatty acid distillation process **4**: 2508
Badische acid **6**: 3415
Baeyer–Villiger oxidation
    for production of hydroquinone **5**: 2947
    of methylbenzaldehyde **3**: 1678
Baking process **1**: 512
Balanced vinyl chloride processes **3**: 1341
Ball drop method **2**: 1158
Balz–Schiemann reaction **5**: 2614

Barium 2-ethylhexanoate **6**: 3237
Barium laurate **6**: 3237
Barium naphthenate **6**: 3237
Barium soaps **6**: 3238
Barium stearate **6**: 3237
Barth reaction **3**: 1898
Basaroff reactions **8**: 4787
BASF acetic acid process **1**: 35
BASF acetylene process **1**: 133
BASF acidic oximation caprolactam process **2**: 1022
BASF aniline process **1**: 631
BASF butadiene extraction procedure **2**: 913
BASF caprolactam process **2**: 1019
BASF formaldehyde process **5**: 2650
BASF formamide process **5**: 2695
BASF formic acid process **5**: 2724
BASF melamine process **5**: 3204
BASF phthalate process **7**: 3890
BASF phthalic anhydride process **7**: 3870
BASF propionic acid process **7**: 4106
BASF vinyl propionate process **8**: 4856
Bashkirov oxidation **1**: 288
　of paraffins **5**: 2546
Basic soaps **6**: 3228
Batch deodorizers
　for fats and fatty oils **4**: 2420
Batter wheat starch process **7**: 4392
Baum's acid **6**: 3389
Bayer acetic acid process **1**: 47
Bayer cyclopentene process **3**: 1835
Bayer's acid **6**: 3393
BBP
　physical properties **7**: 3886
Béchamp aniline process **1**: 632
BÉCHAMP arylamine process **1**: 513
Beckmann rearrangement
　of cyclohexanone oxime **2**: 1025
　to caprolactam **2**: 1018
Beef tallow **4**: 2469
　fatty acid content **4**: 2376
Beeswax **8**: 4900
　isolation **8**: 4901
　properties and composition **8**: 4902
　uses **8**: 4904
Behenic acid
　physical properties **4**: 2485
Behenyl alcohol
　physical properties **5**: 2535
Bensulfuron-methyl **7**: 4233
Benzal chloride
　hydrolysis **2**: 677
　physical properties **3**: 1456
　production **3**: 1457
　specifications and uses **3**: 1457
Benzaldehyde **2**: 673
　economic aspects and toxicology **2**: 689
　physical and chemical properties **2**: 674

　specifications **2**: 682
　uses **2**: 683
Benzaldehyde cyanohydrin **6**: 3470
Benzaldehyde hydrogenation
　for benzyl alcohol production **2**: 853
Benzaldehyde production **2**: 677
4-Benzamido-5-hydroxynaphthalene-1,7-disulfonic acid **6**: 3441
4-Benzamido-5-hydroxynaphthalene-2,7-disulfonic acid **6**: 3441
Benzenamine, *see aniline*
Benzene **2**: 693
　alkylation with aluminium chloride **4**: 2209
　chlorinated, *see chlorobenzenes*
　economic aspects **2**: 728
　for anthraquinone production **2**: 656
　for biphenyl production **5**: 2908
　highly fluorinated **5**: 2617
　physical and chemical properties **2**: 694
　separation and purification **2**: 714
　specifications **2**: 726
　toxicology **2**: 733
　uses **2**: 732
Benzenecarboxylic acid, *see benzoic acid*
Benzene chlorination
　continuous **3**: 1413
　discontinuous **3**: 1415
　liquid phase **3**: 1412
1,2-Benzenedicarbonitrile **7**: 3883
1,2-benzenedicarboxylic acid, *see phthalic acid*
1,3-Benzenedicarboxylic acid, *see isophthalic acid*
1,4-Benzenedicarboxylic acid, *see terephthalic acid*
1,2-Benzene dicarboxylic acid di-2-propenyl ester **1**: 426
1,3-Benzene dicarboxylic acid di-2-propenyl ester **1**: 426
1,3-Benzenediol, *see resorcinol*
1,3-Benzenedisulfonamide **2**: 761
1,3-Benzenedisulfonic acid **2**: 759
1,3-Benzenedisulfonohydrazide **2**: 761
1,3-Benzenedisulfonyl chloride **2**: 761
Benzenehexacarboxylic acid, *see mellitic acid*
Benzene oxidation
　for maleic anhydride production **5**: 3162
Benzene oxychlorination **3**: 1416
Benzenepentacarboxyxlic acid **2**: 1121
Benzenepolycarboxylic acid **2**: 1131
Benzene production **2**: 700
Benzene products **2**: 699
Benzenesulfonamides **2**: 753
Benzene sulfonation
　for phenol production **6**: 3701
　for resorcinol production **7**: 4265
Benzenesulfonic acid **2**: 741, 755
　environmental aspects **2**: 750
　physical and chemical properties **2**: 742
　production **2**: 744

uses **2**: 749
Benzenesulfonic acid esters **2**: 754
Benzenesulfonohydrazides **2**: 754
Benzenesulfonyl chlorides **2**: 751
1,2,3,4-Benzenetetracarboxylic acid, *see mellophanic acid*
1,2,3,5-Benzenetetracarboxylic acid, *see prenitic acid*
1,2,4,5-Benzenetetracarboxylic acid, *see pyromellitic acid*
1,2,3-Benzenetricarboxylic acid, *see hemimellitic acid*
1,2,4-Benzenetricarboxylic acid, *see trimellitic acid*
1,2,4-Benzenetriol **6**: 3765
1,2,5-Benzenetriol **6**: 3766
Benzidine
   economic aspects **2**: 812
   environmental protection **2**: 802
   properties **2**: 803
   rearrangement **2**: 797
   safety precautions **2**: 799
   toxicology **2**: 812
   uses **2**: 804
Benzidine cresol adducts **3**: 1683
Benzidine-2,2'-dicarboxylic acid **2**: 810
Benzidine-3,3'-dicarboxylic acid **2**: 810
Benzidine-2,2'-disulfonic acid **2**: 810
Benzidine production **2**: 794
Benzidine sulfate **2**: 804
Benzidinesulfone **2**: 811
Benzidinesulfone-disulfonic acid **2**: 812
Benzoates **2**: 825
Benzo[1,2-c:3,4-c']-difuran-1,3,4,6-tetrone, *see mellophanic dianhydride*
Benzo[1,2-c:4,5-c']difuran-1,3,5,7-tetrone, *see pyromellitic dianhydride*
1,3-Benzodioxole **6**: 3781
Benzofuran **5**: 2764
   toxicology **5**: 2768
Benzoguanamine **5**: 3214
   uses **5**: 3216
Benzoic acid **2**: 819, 1121
   chemical properties **2**: 821
   economic aspects **2**: 825
   physical properties **2**: 820
   specifications **2**: 823
   toxicology **2**: 836
Benzoic acid derivates **2**: 825
   toxicology **2**: 836
Benzoic acid esters **2**: 826
Benzoic acid production **2**: 821
Benzoin **5**: 3113
Benzonitriles **2**: 829; **6**: 3471
Benzophenone **5**: 3111
[2]Benzopyrano[6,5,4-def][2]benzopyran-1,3,6,8-tetrone, *see 1,8:4,5-naphthalenetetracarboxylic dianhydride*
1-Benzo[*b*]pyrrole **5**: 3001
Benzoquinone **2**: 841
   properties **2**: 842
   toxicology **2**: 846

uses **2**: 845
1,2-Benzoquinone **2**: 842
1,4-Benzoquinone **2**: 842
1,4-Benzoquinone production **2**: 845
Benzothiazole-2-thiol **8**: 4668
Benzothiazolyl-2-acetonitrile **6**: 3470
Benzotrichloride
   environmental protection and specifications **3**: 1461
   physical properties **3**: 1458
   toxicology **3**: 1485
Benzotrichloride production **3**: 1460
Benzotrifluorides
   production **5**: 2610
   properties **5**: 2609
   uses **5**: 2611
Benzo[1,2-c:3,4-c':5,6-c'']-trifuran-1,3,4,6,7,9-hexone, *see mellitic trianhydride*
2-Benzoylbenzoic acid **2**: 831
Benzoyl chloride **2**: 827
   toxicology **3**: 1485
Benzoyl cyanide **6**: 3470
N-Benzoyl H acid **6**: 3441
N-Benzoyl K acid **6**: 3441
Benzyl acetate **2**: 858
   physical properties **3**: 2014
Benzyl alcohol **2**: 849
   chemical properties **2**: 851
   economic aspects and toxicology **2**: 860
   physical properties **2**: 850
   specifications **2**: 856
   uses **2**: 857
Benzyl alcohol production **2**: 852
Benzylamine **2**: 863
   specifications **2**: 864
   toxicology **2**: 866
Benzylamine cresol adducts **3**: 1683
Benzylamine production **2**: 864
Benzylbenzene **5**: 2904
Benzyl benzoate **2**: 827, 859
Benzyl cellulose **2**: 1235
Benzyl chloride
   chemical properties **3**: 1449
   physical properties **3**: 1448
   specifications and environmental protection **3**: 1454
   toxicology **3**: 1484
   uses **3**: 1455
Benzyl chloride hydrolysis **2**: 852
Benzyl chloride production **3**: 1450
4-Benzyl-2-chloro-6-methylphenol **6**: 3787
2-Benzyl-4-chlorophenol **6**: 3785
Benzyl cinnamate
   physical properties **3**: 2015
Benzyl cyanide **6**: 3469
Benzylidene chloride, *see benzal chloride*
Benzyl methyl ketone **5**: 3112

2-Benzylphenol
  physical properties **6**: 3754
4-Benzylphenol
  physical properties **6**: 3754
4-Benzylpiperidine
  physical properties **7**: 4197
Benzyl propionate
  properties **7**: 4119
Benzyl salicylate **2**: 859; **7**: 4277
Bergmann–Junk test **2**: 1157
BHT **6**: 3741
Bicyclo[4.4.0]decane, *see decalin*
Bifurandiones
  by carbonylation of acetylone **1**: 127
Biguanide **5**: 2839
Biocides
  chlorophenols **3**: 1613
  organosilicon compounds **7**: 4347
  saponins **7**: 4285
  sorbic acid **7**: 4371
  sulfones **7**: 4485
  sulfoxides **7**: 4493
  xylidines **8**: 5024
Biodegradation
  of fatty acids **4**: 2379
  of organic esters **3**: 2035
  of petroleum waxes **8**: 4940
  of phenol derivatives **6**: 3784
  of sodium carboxylates **2**: 1104
  of terpenes **8**: 4597
Biogas
  economic aspects **6**: 3284
  handling **6**: 3282
  in developing countries **6**: 3283
  properties **6**: 3271
Biomass fermentation
  for acetone production **1**: 100
Biomethanation
  environmental aspects **6**: 3281
  feedstocks **6**: 3277
  process parameters **6**: 3278
  reactors **6**: 3274
  technological **6**: 3273
Biosynthesis
  of alginates **7**: 3996
  of amylose, amylopectin and involved enzymes **7**: 4381
  of dextran **3**: 1846
  of fatty acids **4**: 2375
  of purines **7**: 4167
  of terpenes **8**: 4596
Biphenyl
  properties **5**: 2906
  toxicology **5**: 2926
  uses **5**: 2910
Biphenyl–diphenyl ether mixtures **5**: 2909
4,4′-Biphenyl-diyldiamine, *see benzidine*

Biphenyl production **5**: 2907
Biphenyls **3**: 1435
3,3′,4,4′-Biphenyl-tetrayltetramine **2**: 805
Bipyridines **7**: 4192
Bisabolene **8**: 4610
Bisabolol **8**: 4611
Bis(2-aminoethyl)amine **1**: 487
Bis(2-amino-ethyl)ethylenediamine **1**: 487
Bis(aminothiocarbonyl)disulfanes **4**: 1953
Bis(aminothiocarbonyl)sulfanes **4**: 1955
Bis(aminothiocarbonyl)trisulfanes **4**: 1955
Bis(tert-amylperoxy)-2,5-dimethylhexane
  production **6**: 3641
2,2-Bis(*tert*-butylperoxy)-butane
  physical properties **6**: 3664
1,1-Bis(*tert*-butylperoxy)-cyclohexane
  physical properties **6**: 3664
2,2-Bis(*tert*-butylperoxy)-propane
  physical properties **6**: 3664
N-{2-[Bis(carboxymethyl)amino]ethyl}-N-(2-hydroxyethyl)glycine, *see hydroxyethylethylenediamine triacetic acid*
N,N-Bis{2-[bis(carboxymethyl)amino]ethyl}glycine, *see diethylenetriamine pentaacetic acid*
N,N-bis(carboxymethyl)glycine, *see nitrilotriacetic acid*
Bischler–Napieralski reaction **7**: 4257
Bis(1-chloroethyl) ether
  physical properties **4**: 2187
Bis(2-chloroethyl) ether **4**: 2199
  physical properties **4**: 2187
  toxicology **4**: 2202
Bis(2-chloroisopropyl) ether **4**: 2200
  physical properties **4**: 2187
  toxicology **4**: 2203
Bis(chloromethyl) ether **4**: 2199
  physical properties **4**: 2187
  toxicology **4**: 2202
1,6-Bis(4-chlorophenylbiguanido)hexane **5**: 2840
1,2-Bis-(cyanomethyl)benzene **6**: 3470
Bisdesmosides **7**: 4284
Bis(2-ethylhexyl) adipate
  physical properties **3**: 2014
  specifications **3**: 2037
Bis(2-ethylhexyl) phthalate
  physical properties **3**: 2015
  specifications **3**: 2038
Bis(2-ethylhexyl) terephthalate
  specifications **3**: 2038
Bis(2-ethylhexyl) trimellitate
  specifications **3**: 2038
3(4),8(9)-Bis(formyl)tricyclo[5.2.1.0$^{2.6}$]decane, *see TCD dialdehyde*
α,α′-Bishydroperoxydicyclohexyl peroxide
  physical properties **6**: 3664
Bishydroxyarylalkanes, *see bisphenols*
Bis(2-hydroxy-3-cyclohexyl-5-methylphenyl)methane
  physical properties **6**: 3769

2,2-Bis(4-hydroxycyclohexyl)-propane   **1**: 323, 333
Bis(4-hydroxy-3,5-di-*tert*-butylphenyl)methane
　physical properties   **6**: 3769
Bis(2-hydroxyethyl) terephthalate
　production   **3**: 2032
1,4-Bis(hydroxymethyl)cyclohexane   **1**: 323, 334
　toxicology   **1**: 343
2,2-Bis(4-hydroxy-3-methylphenyl)propane
　physical properties   **6**: 3769
2,2-Bis(hydroxymethyl)-1,3-propanediol   **1**: 340
2,2-Bis(4-hydroxyphenyl)cyclohexane, *see bisphenol Z*
　physical properties   **6**: 3769
α,α'-Bis(4-hydroxyphenyl)-*p*-diisopropylbenzene
　physical properties   **6**: 3769
Bis-(4-hydroxyphenyl)methane
　physical properties   **6**: 3769
2,2-Bis(4-hydroxyphenyl)propane, *see bisphenol A*
　physical properties   **6**: 3769
4,4-Bis(4-hydroxyphenyl)valeric acid
　physical properties   **6**: 3769
Bis(3-methylbutyl)amine   **1**: 454
Bis(1-methyl)propylamine   **1**: 454
Bis(nitrosocyclohexane) caprolactam process   **2**: 1033
Bisphenol A
　physical properties   **6**: 3769
　production from acetone   **1**: 105
Bisphenol A production   **6**: 3772
Bisphenols
　production   **6**: 3771
　properties   **6**: 3768
　toxicology   **6**: 3774
　uses and economic aspects   **6**: 3773
Bisphenol Z
　physical properties   **6**: 3769
2,6-Bis-(thiocarbamoylthiomethyl)pyridine   **4**: 1971
1,3-Bis(trifluoromethyl)benzene
　physical properties   **5**: 2609
1,4-Bis(trifluoromethyl)benzene
　physical properties   **5**: 2609
Bithionol   **6**: 3791
Biuret   **8**: 4818
　as side product in urea production   **8**: 4799
Bleaching
　of fats and fatty oils   **4**: 2417
　of montan wax   **8**: 4910
Bleaching agents
　oxalic acid   **6**: 3591
Boat form
　of cyclohexane   **3**: 1796
Boeniger acid   **6**: 3429
Boiling point elevation
　of citric acid solutions   **3**: 1648
BON acid   **6**: 3383
Boric acid
　for hydrocarbon oxidation   **1**: 288
Boric acid modified oxidation
　of cyclohexane   **3**: 1812

Boron trifluoride
　as catalyst for esterification of carboxylic
　　acids   **3**: 2023
　reaction with acetic anhydride   **1**: 70
Borsig linear quencher   **4**: 2258
Borsig transfer-line exchangers   **4**: 2257
Bottom-circulation distillation
　of coal tar   **8**: 4521
Bouveault–Blanc reduction   **3**: 2019
BP Chemicals propionic acid process   **7**: 4108
Brabender viscoamylograph   **7**: 4386
Brassidic acid
　physical properties   **4**: 2487
Brassidyl alcohol
　physical properties   **5**: 2549
Brassylic acid   **3**: 1918
　physical properties   **3**: 1905
Bright stock slack waxes   **8**: 4932
　uses   **8**: 4938
Bromination of olefins   **2**: 871
Bromine compounds   **2**: 867
　physical and chemical properties   **2**: 868
　production   **2**: 871
　toxicology   **2**: 887
Bromoacetic acid   **2**: 877
　toxicology   **2**: 889
Bromoalkanes
　from higher olefins   **5**: 2886
4-Bromoanisole   **2**: 870
Bromobenzene   **2**: 870
1-Bromobutane   **2**: 869
2-Bromobutane   **2**: 869
1-Bromo-4-chlorobenzene   **2**: 870
Bromochlorodifluoromethane
　properties   **5**: 2581
1-Bromo-3-chloro-5,5-dimethylhydantoin   **3**: 1569
Bromochloromethane   **2**: 869, 880
　toxicology   **2**: 890
Bromochlorophene   **6**: 3790
1-Bromo-3-chloropropane   **2**: 875
　toxicology   **2**: 888
1-Bromo-2-chloro-1,1,2-trifluoroethane
　properties   **5**: 2581
2-Bromo-2-chloro-1,1,1-trifluoroethane
　properties   **5**: 2581
1-Bromodecane   **2**: 869
4-Bromo-2,6-dimethylphenol   **6**: 3787
1-Bromododecane   **2**: 869, 874
　toxicology   **2**: 888
Bromoethane   **2**: 869
4-(2-Bromoethyl)benzenesulfonamide   **2**: 765
4-(2-Bromoethyl)benzenesulfonic acid   **2**: 765
4-(2-Bromoethyl)benzenesulfonyl chloride   **2**: 765
Bromoethylene   **2**: 869
Bromofluoroalkanes   **5**: 2580
　toxicology   **5**:
1-Bromo-4-fluorobenzene   **2**: 870

5091

1-Bromohexadecane **2**: 869
1-Bromohexane **2**: 869
1-Bromo-4-iodobenzene **2**: 870
Bromomethane **2**: 869, 878
   toxicology **2**: 889
1-(Bromomethyl)-3-phenoxybenzene **2**: 877
   toxicology **2**: 889
1-Bromonaphthalene **2**: 870
1-Bromooctadecane **2**: 869
1-Bromooctane **2**: 869
1-Bromopentane **2**: 869
4-Bromophenol **2**: 870
1-Bromo-2-phenylethane **2**: 876
   toxicology **2**: 889
2-Bromopropane **2**: 869, 869
3-Bromo-1-propene **2**: 869, 879
   toxicology **2**: 889
2-Bromopyridine
   physical properties **7**: 4199
3-Bromopyridine
   physical properties **7**: 4199
4-Bromopyridine hydrochloride
   physical properties **7**: 4199
1-Bromotetradecane **2**: 869
2-Bromothiophene **8**: 4700
   specifications **8**: 4696
   toxicology **8**: 4704
3-Bromothiophene **8**: 4700
2-Bromotoluene **2**: 870
4-Bromotoluene **2**: 870
Bromotrifluoromethane
   properties **5**: 2581
4-Bromo-2,6-xylenol **6**: 3787
Bronner acid **6**: 3415
Bruckner cresol separation process **3**: 1685
2BT **8**: 4700
3BT **8**: 4700
BTX aromatics
   production **2**: 700
   separation and refining **2**: 702
Bucherer reaction **1**: 524
Bureau of Mines correlation index of naphtha **4**: 2225
Burner
   for acetylene production **1**: 135
Butadiene **2**: 899
   acetoxylation **8**: 4624
   chlorination **3**: 1385
   for styrene production **7**: 4429
   physical and chemical properties **2**: 900
   production **2**: 906
   specifications **2**: 914
   toxicology **2**: 917
   uses and economic importance **2**: 916
1,2-Butadiene **2**: 899
1,3-Butadiene, *see butadiene*
Butanals, *see butyraldehydes*

Butane, **2**: 911
   one-step dihydrogenation **2**: 908
   oxidative dehydrogenation **2**: 911
   physical properties **5**: 2855
   toxicology **5**: 2924
   uses **5**: 2867
Butanedial, *see succinedialdehyde*
1,4–Butanedicarboxylic acid, *see adipic acid*
Butanedinitrile **6**: 3461
Butanedioic acid, *see succinic acid*
Butanedioic acid di-2-propenyl ester **1**: 426
1,2-Butanediol **2**: 947
1,3-Butanediol **2**: 946
1,4-Butanediol **2**: 941
   uses **2**: 944
2,3-Butanediol **2**: 945
Butane-1-4-diol dehydrogenation
   for production of butyrolactone **2**: 1007
Butanediol-1,4-divinyl ether
   physical properties **8**: 4867
Butanediol-1,4-monovinyl ether
   physical properties **8**: 4867
1,4-Butanediol production **2**: 942
   from butadiene **2**: 903
Butanediols **2**: 937
   toxicology **2**: 947
2,3-Butanedione **5**: 3108
1,4-Butanedisulfinic acid
   synthesis **7**: 4452
Butane liquid-phase oxidation
   for acetic acid production **1**: 38
n-Butane – maleic anhydride THF process **8**: 4625
Butanenitrile **6**: 3461
Butane oxidation
   for acetic acid production **1**: 38
Butane pyrolysis
   cracking yield **4**: 2235
Butane-1-sulfonic acid
   physical properties **8**: 4504
tert-Butanethiol production **8**: 4652
Butanethiols
   toxicology **8**: 4678
n-Butanoic acid, *see n-butyric acid*
1-Butanol **1**: 281, 284; **2**: 956
   toxicology **2**: 964
   uses **2**: 961
2-Butanol **1**: 281; **2**: 956
   toxicology **2**: 965
   uses **2**: 963
tert-Butanol **2**: 956
   from butenes **2**: 997
   toxicology **2**: 966
   uses **2**: 963
Butanols **2**: 951
   chemical properties **2**: 954
   dehydrogenation **2**: 928
   economic aspects **2**: 963

physical properties **2:** 952
production **2:** 956
specifications **2:** 960
toxicology **2:** 964
uses **2:** 961
2-Butenal, *see crotonaldehyde*
trans-2-Butenal, *see crotonaldehyde*
Butene
    oligomerization **5:** 2883
1-Butene **2:** 987
    physical properties **5:** 2871
2-Butene
    for isoprene synthesis **5:** 3038
cis-2-Butene **2:** 987
    physical properties **5:** 2871
trans-2-Butene **2:** 987
    physical properties **5:** 2871
Butene chlorohydrins **3:** 1588
    production **3:** 1594
Trans-Butenedioic acid, *see fumaric acid*
cis-Butenedioic acid, *see maleic acid*
cis-Butenedioic acid di-2-propenyl ester **1:** 426
trans-Butenedioic acid di-2-propenyl ester **1:** 426
Butenediol **2:** 937, 940
    production and uses **2:** 941
    toxicology **2:** 947
2-Butene-1,4-diol **2:** 940
n-**Butenes**
    dehydrogenation **2:** 909
    hydroformylation **6:** 3616
    oxidative dehydrogenation **2:** 909
    physical and chemical properties **2:** 986
    specifications **2:** 995
    toxicology **2:** 1001
    upgrading **2:** 991
    uses and economic data **2:** 997
2-Butenoic acid, *see crotonic acid*
(*E*)-2-Butenoic acid ethyl ester
    physical properties **3:** 1716
(*E*)-2-Butenoic acid methyl ester
    physical properties **3:** 1716
(*E,E*)-2-Butenoic anhydride
    physical properties **3:** 1716
2-Buten-1-ol **1:** 310
3-Buten-2-ol **1:** 309
3-Buten-2-one **5:** 3101
(*E*)-2-Butenoyl chloride
    physical properties **3:** 1716
1-Butoxybutane, *see di-n-butyl ether*
2-(2-Butoxyethoxy)ethyl acetate
    physical properties **3:** 2014
    specifications **3:** 2037
2-(2-Butoxyethoxy)ethyl thiocyanate
    physical properties **8:** 4632
    toxicology **8:** 4635
2-Butoxyethyl acetate
    physical properties **3:** 2014

specifications **3:** 2037
Butterfat
    fatty acid content **4:** 2376
Butyl acetate **1:** 54
    physical properties **3:** 2014
    specifications **3:** 2037
    toxicology **3:** 2043
n-Butyl acrylate **1:** 226
    physical properties **3:** 2014
    specifications **3:** 2037
sec-Butyl alcohol **1:** 284
    for isoprene synthesis **5:** 3038
tert-Butyl alcohol **1:** 284
tert-Butyl alcohol oxidation
    for methacrolein production **1:** 210
Butylamine **1:** 454
Butylamines
    uses **1:** 458
Butylated hydroxytoluene **6:** 3741
tert-Butylbenzene
    physical properties **5:** 2902
N-(Butyl)benzenesulfonamide **2:** 758
n-Butyl benzoate **2:** 827
4-*tert*-Butylbenzoic acid **2:** 831
Butyl benzyl phthalate
    physical properties **7:** 3886
Butyl butyrate
    physical properties **3:** 2014
n-Butyl carbamate **2:** 1047
n-Butyl chloride **3:** 1378
tert-Butyl Chloride **3:** 1379
    from isobutene chlorination **3:** 1393
n-Butyl chloroformate **3:** 1576
n-Butylcyclohexane
    physical properties **3:** 1798
sec-Butylcyclohexane
    physical properties **3:** 1798
tert-Butylcyclohexane
    physical properties **3:** 1798
4-*tert*-Butylcyclohexyl chloroformate **3:** 1576
6-*tert*-Butyl-2-cyclohexyl-4-methylphenol
    physical properties **6:** 3754
n-Butyldiethanolamine **4:** 2160
    see also ethanolamines, N-alkylated
    physical properties **4:** 2160
Butyldiisopropanolamine
    toxicology **4:** 2179
4-*tert*-Butyl-2,6-dimethylbenzenesulfonic acid **2:** 766
tert-Butyl-2,2-dimethylperoxypropionate
    toxicology **6:** 3676
2-*tert*-Butyl-4,5-dimethylphenol **6:** 3737
4-*tert*-Butyl-2,5-dimethylphenol **6:** 3737
4-*tert*-Butyl-2,6-dimethylphenol **6:** 3737
6-*tert*-Butyl-2,4-dimethylphenol **6:** 3737
2,3-Butylene glycol **2:** 945
Butylenes, *see butenes*

n-Butylethanolamine **4:** 2160
  see also ethanolamines, N-alkylated
  physical properties **4:** 2160
Butyl ethyl ketone **5:** 3097
tert-Butyl-2-ethylperoxyhexanoate
  toxicology **6:** 3676
2-Butyl-2-ethyl-1,3-propanediol **1:** 323, 330
n-Butyl glycolate **5:** 2966
  physical properties **5:** 2958
  specifications **3:** 2037
tert-Butyl hydroperoxide production **6:** 3636
tert-Butyl hydroperoxide
  for epoxidation **4:** 1997
  in propylene oxide production **7:** 4146
  physical properties **6:** 3634
  toxicology **6:** 3676
  uses **6:** 3637
sec-Butyl-bis(3-hydroxypropyl)phosphine **6:** 3834
1-n-Butylimidazole
  toxicology **5:** 2994
n-Butyl isocyanate **5:** 3011
O,O-tert-Butyl-O-isopropyl monoperoxycarbonate
  physical properties **6:** 3655
n-Butyl lactate **5:** 3130
Butyl methacrylate
  physical properties **6:** 3252
Butyl methacrylates **6:** 3265
tert-Butyl methyl ketone **5:** 3090
tert-Butyl methyl peroxide
  physical properties **6:** 3640
2-tert-Butyl-4-methylphenol **6:** 3737
2-tert-Butyl-5-methylphenol **6:** 3737
4-tert-Butyl-2-methylphenol **6:** 3737
4-tert-Butyl-3-methylphenol **6:** 3737
5-tert-Butyl-3-methylphenol **6:** 3737
6-tert-Butyl-2-methylphenol **6:** 3737
2-sec-Butyl-2-methyl-1,3-propanediol **1:** 323, 330
tert-Butyl monoperoxysuccinate production **6:** 3660
2-Butyl-1-octanol
  physical properties **5:** 2551
tert-Butyl peracetate
  toxicology **6:** 3676
tert-Butyl perbenzoate
  toxicology **6:** 3676
tert-Butyl peroxyoctadecanoate production **6:** 3659
tert-Butylphenol
  toxicology and handling **6:** 3745
2-tert-Butylphenol **6:** 3737
3-tert-Butylphenol **6:** 3737
4-tert-Butylphenol **6:** 3737
sec-Butylphenol **6:** 3735
tert-Butylphenol **6:** 3736
  uses **6:** 3742
2-tert-Butylphenol production **6:** 3736, 3739
3-tert-Butylphenol production **6:** 3739
4-tert-Butylphenol production **6:** 3737
3-(4-tert-Butylphenyl)-2-methylpropanal, see lilial

n-Butyl phosphine
  physical properties **6:** 3821
Butyl propionate
  properties **7:** 4119
Butyl rubber
  from butenes **2:** 999
Butyl stearate
  physical properties **3:** 2014
4-tert-Butyl toluene
  physical properties **5:** 2902
tert-Butyl urea **8:** 4820
Butyl vinyl ether
  toxicology **8:** 4874
5-tert-Butyl-m-xylene
  physical properties **5:** 2902
Butynediol **2:** 937
  by ethynylation **1:** 126
  toxicology **2:** 947
  uses **2:** 940
2-Butyne-1,4-diol **2:** 937
  by ethynylation **1:** 126
Butynediol production **2:** 938
2-Butyn-1-ol **1:** 309
3-Butyn-2-ol **1:** 309
Butyraldehyde **1:** 352
  chemical reactions and uses **2:** 930
  for production of 2-ethylhexanol **4:** 2363
  hydrogenation **2:** 927
  physical properties **2:** 925
  production **2:** 926
  specifications **2:** 929
  toxicology **2:** 934
Butyraldehyde diethyl acetal **1:** 395
Butyraldehyde dimethyl acetal **1:** 395
Butyraldol **1:** 375, 380
n-Butyric acid **2:** 1107
  specification **2:** 1106
Butyrolactone **2:** 1005; **3:** 2033
  chemical properties **2:** 1006
  from diketene **5:** 3076
  physical properties **5:** 2958
  specifications **2:** 1008
  toxicology **5:** 2971
  uses and toxicology **2:** 1009
Butyrolactone production **2:** 1007
Butyronitrile **6:** 3461
  physical properties **6:** 3457

CAA liquid–liquid butadiene extraction **2:** 912
Cadaverine **1:** 486
Cadmium 2-ethylhexanoate **6:** 3243
Cadmium 12-hydroxystearate **6:** 3243
Cadmium laurate **6:** 3243
Cadmium oleate **6:** 3243
Cadmium soaps **6:** 3243
Cadmium stearate **6:** 3243
Caffeine monohydrate, see purine derivatives

Calcium arachidate **6**: 3237
Calcium carbide
  for acetylene production **1**: 158
Calcium carbide nitrogenation **3**: 1738
Calcium cyanamide
  economic aspects **3**: 1761
  for thiourea production **8**: 4710
  processing **3**: 1743
  properties **3**: 1736
  specifications **3**: 1744
  toxicology **3**: 1762
  uses **3**: 1745
Calcium cyanamide production **3**: 1737
Calcium 2-ethylhexanoate **6**: 3237
Calcium formate
  for production of formic acid **5**: 2728
Calcium hydroxymethanesulfinate
  uses **7**: 4470
Calcium 12-hydroxystearate **6**: 3237
Calcium lactate **5**: 3129
Calcium laurate **6**: 3237
Calcium naphthenate **6**: 3237
Calcium oleate **6**: 3237
Calcium pectate gels **7**: 3987
Calcium propionate **7**: 4117
Calcium ricinoleate **6**: 3237
Calcium soaps **6**: 3236
Calcium sorbate **7**: 4367
Calcium stearate **6**: 3237
Calcium tartrate
  in tartaric acid production **8**: 4563
Camphene **8**: 4608
Candelilla wax **8**: 4892
  properties and composition **8**: 4893
  uses **8**: 4894
Candles **8**: 4941
  from fatty acids **4**: 2527
Capraldehyde **1**: 352, 362
n-capric acid **2**: 1097
  physical properties **4**: 2485
Capric alcohol
  physical properties **5**: 2535
Caprinaldehyde **1**: 353, 365
Caprinoguanamine **5**: 3214
  uses **5**: 3217
n-Capriotic acid **2**: 1097
Caproic acid
  physical properties **4**: 2485
Caproic alcohol
  physical properties **5**: 2535
ε-**Caprolactam** **2**: 1013
  chemical properties **2**: 1015
  economic aspects and toxicology **2**: 1039
  physical properties **2**: 1013
  specifications **2**: 1037
ε-Caprolactam production **2**: 1015
Caprylaldehyde **1**: 352, 353, 363

n-Caprylic acid **2**: 1097
Caprylic acid **2**: 1109
  physical properties **4**: 2485
Caprylic alcohol
  physical properties **5**: 2535
Carbamates **2**: 1045
  toxicology **2**: 1053
Carbamate salts
  phosgenation **5**: 3020
Carbamic acid esters **2**: 1046
Carbamoyl chlorides **2**: 1045, 1048
  alcoholysis **2**: 1046
  reactions **2**: 1050
Carbanilide **8**: 4820
Carbazole **2**: 1057
Carbide-to-water generator **1**: 159
Carbinol, *see methanol*
Carbobitumen **8**: 4533
Carbohydrate allyl ethers **1**: 432
Carbohydrate nomenclature **2**: 1069
Carbohydrate reactions **2**: 1070
Carbohydrates **2**: 1061; **7**: 3972
  see also polysaccharides
  classification and functions **2**: 1062
  fermentable **4**: 2091
  fermentation **4**: 2063
  oxidation, oxalic acid production **6**: 3582
Carbolic acid, *see phenol*
Carbolic oil **8**: 4539
Carbonates, organic, *see carbonic acid esters*
Carbonation
  of calcium cyanamide **3**: 1751
Carbon-black oils **8**: 4537
Carbon bond energies **7**: 4309
Carbon dioxide
  hydrogenation to formic acid **5**: 2729
  in urea production **8**: 4787
  reaction with aliphatic amines **1**: 446
Carbon dioxide extraction
  of ethanol **4**: 2121
Carbon disulfide
  for production of heterocyclic thiols **8**: 4671
  reaction with aliphatic amines **1**: 446
Carbon disulfide chlorination **3**: 1273
Carbonic acid esters **2**: 1085
  production **2**: 1087
  specifications and uses **2**: 1089
  toxicology **2**: 1088
Carbon monoxide
  formation from formic acid **5**: 2717
  for production of oxalic acid **6**: 3586
  reaction with ammonia **5**: 2694
  reaction with water, for formic acid production **5**: 2729
Carbon monoxide malonate process **5**: 3181
Carbonochloridic acid esters, *see chloroformates*
Carbon tetrachloride, *see tetrachloromethane*

5095

Carbonylation
  for cresol production **3**: 1680
  of acetylene **1**: 126
  of ethylene **7**: 4106
  of methanol **4**: 2059; **5**: 2720
  of methanol, for acetic acid production **1**: 34
  of methyl acetate **1**: 76
  of olefins **3**: 2029
Carbonyl J acid **6**: 3436
Carbonyl value
  of fats and fatty oils **4**: 2438
Carboxamides
  synthesis from amines **1**: 445
Carboxyethyl cellulose **2**: 1234
Carboxylates
  alkylation **3**: 2026
Carboxylation
  with carbon dioxide, of aromates **2**: 1125
Carboxylic acid anhydrides **2**: 1115
Carboxylic acid chlorides
  for production of parcarboxylic acids **6**: 3648
Carboxylic acid esters, *see esters*
Carboxylic acid hydrogenation
  for alcohol production **1**: 286
Carboxylic acids
  aliphatic **2**: 1095
  aromatic **2**: 1119
  chemical properties **2**: 1099
  economic aspects and toxicology **2**: 1112, 1134
  esterification **3**: 2022
  ethoxylation **3**: 2032
  fluorinated **5**: 2600
  halogenated **2**: 1116
  in nature **2**: 1100
  peroxidation **6**: 3644
  physical and chemical properties **2**: 1120
  physical properties **2**: 1096
  production **2**: 1100, 1125
  specifications **2**: 1105
  trade names **2**: 1112
  uses **2**: 1105
Carboxylic anhydrides
  for production of esters **3**: 2027
  for production of peroxycarboxylic acids **6**: 3647
Carboxymethyl cellulose **2**: 1230
  properties and uses **2**: 1231
Carboxymethyl hydroxyethyl cellulose **2**: 1233
Carboxymethyl methyl cellulose **2**: 1215
Carboxysilanes **7**: 4341
Cardanol **6**: 3757
3-Carene **8**: 4604
Carlson calcium cyanamide process **3**: 1742
Carnaúba wax **8**: 4888
  composition **8**: 4890
  types and specifications **8**: 4890
  uses **8**: 4891
Carob gum **7**: 4023

Carousel extractor
  for oilseeds **4**: 2408
Carrageenan **7**: 4003
  applications **7**: 4009
  properties **7**: 4006
  structure **7**: 4003
Carrageenan gel
  properties **7**: 4006
Carrageenan production **7**: 4004
Caryophyllene **8**: 4611
Cassada
  as raw material for ethanol production **4**: 2098
Cassava **4**: 2098
Cassella acid **6**: 3393, 3419
Castor oil **4**: 2466
Catalysis mechanism
  ethylene oxidation **4**: 2337
Catalyst deactivation
  in methanol production **6**: 3298
Catalysts
  for acetylene hydrochlorination **3**: 1334
  for aldehyde oxidation **2**: 1101
  for amine production form alcohols **1**: 449
  for Amoco xylene oxidation **8**: 4576
  for anthracene oxidation **2**: 654
  for benzal chloride hydrolysis **2**: 679
  for benzaldehyde reduction **2**: 854
  for benzene chlorination **3**: 1412
  for benzene hydrogenation to cyclohexane **3**: 1798
  for calcium carbide nitrogenation **3**: 1739
  for chloromethyl silane synthesis **7**: 4311
  for cyanogen chloride trimerization **3**: 1776
  for 1,2-dichloroethane cracking **3**: 1340
  for direct oxidation of ethylene to ethylene glycol **4**: 2314
  for drect oxidation of ethylene to ethylene oxide **4**: 2336
  for epoxidation of olefins with hydrogen peroxide **4**: 1996
  for esterification of carboxylic acids **3**: 2023
  for esterification of cellulose **2**: 1168
  for ethylbenzene dehydrogenation **7**: 4423
  for ethylene oligomerization **5**: 2876, 2878
  for ethylene oxichlorination **3**: 1303
  for fat hardening **4**: 2426
  for formaldehyde production **5**: 2648
  for formox formaldehyde process **5**: 2654
  for gas-phase alkylation of benzene **4**: 2213
  for glucose oxidation to D-gluconic acid **5**: 2777
  for hardening of fatty acids **4**: 2518
  for homologation of methanol **4**: 2060
  for hydration of ethylene to ethanol **4**: 2054
  for hydrochlorination of ethylene **3**: 1292
  for hydrogenation of aromatic nitro compunds **1**: 518
  for hydrogenation of esters **3**: 2019
  for hydrosilylation reactions **7**: 4317

for IFP methanol process **1**: 285
for metathesis of olefins **5**: 2892
for methanol production from synthesis gas **6**: 3296
for nitrophenol reduction **1**: 606
for oxidation of benzene to maleic anhydride **5**: 3162
for oxidation of *o*-xylene or naphthalene to phthalic anhydride **7**: 3871
for pentanol oxidation **6**: 3614
for phenol hydrogenation to cyclohexanone **3**: 1809
for phenol methylation **3**: 1673
for phthalic anhydride esterification **7**: 3888
for production of butynediol **2**: 939
for propene oxidation **1**: 231
for propene oxidation to oxalic acid **6**: 3585
for propene oxide isomerization **1**: 422
for pyridine production from acrolein and ammonia **7**: 4181
for sulfide oxidation to sulfoxides **7**: 4487
for synthesis gas conversion to ethanol **4**: 2061
for toluene chlorination **3**: 1427
for toluene oxidation to benzaldehyde **2**: 680
for toluene oxidation to phenol **6**: 3699
for 1,1,2-trichloroethane dehydrochlorination **3**: 1351
for vapor-phase hydrochlorination of methanol **3**: 1266
for vapor-phase oxidation of aromatic componds **2**: 1127
for vinyl acetate production **8**: 4844
for vinyl acetate production from ethylene **8**: 4850
Friedel–Crafts, for alkylphenol production **6**: 3721
in formamide direct synthesis **5**: 2694
in Sohio acrylonitrile process **1**: 253
phthalocyanines **7**: 3932
silanes and siloxanes **7**: 4351
Ziegler–Natta **8**: 4951
Catalytic reduction
in hydrazobenzene production **2**: 797
Catalytic reforming
benzene production **2**: 704
xylene production **8**: 4996
Catechol
economic aspects and toxicology **6**: 3762
properties **6**: 3758
uses **6**: 3761
Catechol production **6**: 3760
Cation diameters **3**: 1727
Cationic starch **7**: 4408
Catofin propene process **7**: 4091
Caustic soda
in propylene oxide production **7**: 4140
C_{19} dicarboxylic acids **3**: 1918

C_{36} dicarboxylic acids
unsaturated **3**: 1922
CEC **2**: 1234
Cell immobilization **4**: 2089
Celluloid **2**: 1071, 1161
Cellulose
as raw material for ethanol production **4**: 2102
as raw material for inorganic esterification **2**: 1145
as raw material for organic esterification **2**: 1169
direct fermentation **4**: 2074
Cellulose acetate **2**: 1168
production **3**: 2027
properties **2**: 1175
Cellulose acetate butyrate **2**: 1178
characteristics **2**: 1180
polymer-modified **2**: 1193
uses **2**: 1197
Cellulose acetate fibers **2**: 1181
economic aspects **2**: 1183
Cellulose acetate production **2**: 1170
Cellulose acetate propionate **2**: 1178
characteristics **2**: 1179
polymer-modified **2**: 1193
uses **2**: 1196
Cellulose borate **2**: 1165
Cellulose esters **2**: 1071, 1137
inorganic **2**: 1138
organic **2**: 1167
organic, uses **2**: 1180
Cellulose ester thermoplastics **2**: 1183
chemical properties **2**: 1193
polymer-modified **2**: 1192
uses **2**: 1196
Cellulose ether esters **2**: 1233
Cellulose ethers **2**: 1072, 1203
analysis **2**: 1236
classification **2**: 1205
economic aspects and trade names **2**: 1243
general properties **2**: 1208
production **2**: 1211
synthesis **2**: 1207
toxicology **2**: 1210
uses **2**: 1238
Cellulose halogenides **2**: 1165
Cellulose mixed esters **2**: 1178
Cellulose nitrate **2**: 1141
specifications **2**: 1158
toxicology **2**: 1162
uses **2**: 1159
Cellulose nitrate films
mechanical properties **2**: 1142
Cellulose nitrate production **2**: 1149
Cellulose nitrate types **2**: 1156
Cellulose nitrite **2**: 1166
Cellulose phosphate **2**: 1164
Cellulose phosphite **2**: 1164
Cellulose sulfates **2**: 1163

5097

Cellulose titanate **2:** 1165
Cellulose xanthate **2:** 1166
Cellulose xanthates **8:** 4979
Cement additives
   ethanolamines **4:** 2158
Cephalin **4:** 2380
Ceramics
   use of cellulose ethers **2:** 1239
Cereal oils **4:** 2455
Cerotic acid
   physical properties **4:** 2486
Ceryl alcohol **1:** 282
   physical properties **5:** 2535
Cetyl alcohol **1:** 282
   physical properties **5:** 2535
Cetyl chloroformate **3:** 1576
C_4 fractions
   from fluid catalytic cracking **2:** 991
   in steam cracking **2:** 911
CH-Acidic compounds
   for sulfinyl chlorides production **7:** 4460
Chair form
   of cyclohexane **3:** 1796
Chaulmoogra oil **4:** 2467
CHDM **1:** 334
Cheese whey
   as raw material for ethanol production **4:** 2111
Chelating agents
   aminopolycarboxylic acids **4:** 2295
   nitrilotriacetic acid **6:** 3479
Chemical equilibrium
   in urea production **8:** 4787
Chemie Linz melamine process **5:** 3204
Chemische Werke Hüls acetic acid process **1:** 46
Cherry-T electrode pitch process **8:** 4528
Cherry-T tar distillation process **8:** 4524
Chevron paraffin cracking process **5:** 2874
Chicago acid **6:** 3439
Chichibabin amination **7:** 4207
Chicken fat **4:** 2471
   fatty acid content **4:** 2376
Chinese insect wax **8:** 4905
Chiral lactones
   from ketene **5:** 3068
Chisso acetaldehyde process **1:** 10
Chloral **3:** 1531
     toxicology **3:** 1535
   uses and economic aspects **3:** 1534
Chloral hydrate **3:** 1532
Chloral production **3:** 1532
Chloramine-B **3:** 1571
Chloramine-T **3:** 1566, 1571
p-Chloranil **2:** 841, 842
Chlorhexidine **5:** 2840
Chlorinated benzenes, *see chlorobenzenes*
Chlorinated biphenyls
   properties and disposal **3:** 1436

   toxicology **3:** 1481
   uses and trade names **3:** 1438
Chlorinated hydrocarbons **3:** 1249
   aliphatic, environmental degradation **3:** 1478
   aliphatic, toxicology **3:** 1467
   aromatic, toxicology **3:** 1480
   nucleus-chlorinated aromatic **3:** 1406
   nucleus-chlorinated aromatic, economic
      facts **3:** 1446
   nucleus-chlorinated aromatic, environmental
      protection **3:** 1445
   production **3:** 1391
   side-chain chlorinated aromatic **3:** 1447
   side-chain chlorinated aromatic, economic aspects **3:** 1465
Chlorinated naphthalene mixtures
   physical properties **3:** 1442
Chlorinated naphthalenes **3:** 1439
   production **3:** 1441
   properties **3:** 1440
   toxicology **3:** 1483
   uses **3:** 1444
Chlorinated paraffins **3:** 1396
   chemical properties and structure **3:** 1398
   physical properties **3:** 1397
   production **3:** 1400
   safety and uses **3:** 1404
   trade names **3:** 1402
Chlorinated toluenes, *see chlorotoluenes*
Chlorinated xylenes
   side-chain **3:** 1462
   side-chain, toxicology **3:** 1485
Chlorination
   multistep **3:** 1276
   of acetaldehyde and paraldehyde **3:** 1525
   of acetic acid **3:** 1544
   of aromatic side-chains, mechanism and reaction
      conditions **3:** 1450
   of benzene **3:** 1412
   of benzene, in phenol production **6:** 3701
   of butadiene **3:** 1385
   of carbon disulfide **3:** 1273
   of ethane **3:** 1292
   of ethanol **3:** 1530
   of ethylene, high-temperature **3:** 1365
   of ethylene, liquid-phase **3:** 1300
   of methane **3:** 1261, 1268
   of phenoxyacetic acid **3:** 1622
   of propene **1:** 412
   of saturated hydrocarbons **5:** 2866
   of toluene **3:** 1427
   of trichloroethylene **3:** 1364
   of vinyl compounds in aqueous media and
      alcohols **3:** 1526
   photochemical **3:** 1312
Chlorination – dehydrochlorination
   of paraffins **5:** 2876

Chlorination–hydrogenation process
  for chloroacetic acid production **3:** 1546
Chlorination mechanism
  isobutylene chlorination **3:** 1391
Chlorine
  for starch oxidation **7:** 4399
Chlorine-fluorine exchange **5:** 2570
  for fluorobenzene production **5:** 2615
Chlorinolysis **3:** 1264
  for production of tetrachloromethane **3:** 1274
Chlorinolysis tetrachloroethylene process **3:** 1365
  high pressure **3:** 1368
Chloroacetaldehyde
  analysis and handling **3:** 1527
  anhydrous, production **3:** 1526
  properties **3:** 1522
  reactions **3:** 1523
  toxicology **3:** 1534
Chloroacetaldehydes **3:** 1521
  polymeric **3:** 1528
  production **3:** 1523
Chloroacetamide **3:** 1549
Chloroacetates
  toxicology **3:** 1560
Chloroacetic acid **3:** 1539
  economic aspects **3:** 1559
  environmental protection **3:** 1556
  for production of malonates **5:** 3180
  properties **3:** 1540
  purification **3:** 1544
  specifications and uses **3:** 1545
  toxicology **3:** 1560
Chloroacetic acid production **3:** 1543
Chloroacetoacetates
  from diketene **5:** 3073
Chloroacetonitrile **6:** 3462
Chloroacetyl chloride **3:** 1548
Chloroacetyl chloride production
  from ketene **5:** 3067
Chloroacetylenes **1:** 130
Chloroalkylbenzenesulfonic acids **2:** 777
Chloroalkyl ethers **4:** 2198
N-Chloroamidines **3:** 1567
N-Chloroamines **3:** 1565
  toxicology **3:** 1571
m-Chloroaniline **1:** 636, 638
o-Chloroaniline **1:** 636, 638
p-Chloroaniline **1:** 636, 638
2-Chlorobenzal chloride
  physical properties **3:** 1466
3-Chlorobenzal chloride
  physical properties **3:** 1466
4-Chlorobenzal chloride
  physical properties **3:** 1466
2-Chlorobenzaldehyde **2:** 684
3-Chlorobenzaldehyde **2:** 684
4-Chlorobenzaldehyde **2:** 684

Chlorobenzene hydrolysis **3:** 1610
Chlorobenzene nitration process **6:** 3526
Chlorobenzenes
  chemical properties **3:** 1411
  for phenol production **6:** 3701
  physical properties **3:** 1407
  production **3:** 1412
  specifications **3:** 1420
  toxicology **3:** 1480
  uses **3:** 1422
4-Chlorobenzenesulfonamide **2:** 768
N-Chlorobenzenesulfonamide **3:** 1571
4-Chlorobenzenesulfonic acid **2:** 768
4-Chlorobenzenesulfonyl chloride **2:** 768
2-Chlorobenzoic acid **2:** 832
3-Chlorobenzoic acid **2:** 832
4-Chlorobenzoic acid **2:** 832
2-Chlorobenzonitrile **6:** 3471
4-Chlorobenzonitrile **6:** 3471
2-Chlorobenzotrichloride
  physical properties **3:** 1466
4-Chlorobenzotrichloride
  physical properties **3:** 1466
2-Chlorobenzotrifluoride
  physical properties **5:** 2609
4-Chlorobenzotrifluoride
  physical properties **5:** 2609
3-Chlorobenzoyl 2-chlorobutyryl peroxide production **6:** 3651
2-Chlorobenzyl chloride
  physical properties **3:** 1466
3-Chlorobenzyl chloride
  physical properties **3:** 1466
4-Chlorobenzyl chloride
  physical properties **3:** 1466
4-Chlorobenzyl-N,N-diethylcarbamoyl sulfoxide
  synthesis **7:** 4489
1-Chloro-1,3-butadiene, see α-chloroprene
2-Chloro-1,3-butadiene, see β-chloroprene
4-Chloro-1,2-butadiene, see isochloroprene
1-Chlorobutane **3:** 1378
  physical properties **3:** 1378
2-Chlorobutane
  physical properties **3:** 1378
4-Chlorobutanenitrile **6:** 3464
Chlorobutanes **3:** 1377
1-Chloro-2-butanol **3:** 1588
2-Chloro-1-butanol **3:** 1588
3-Chloro-1-butanol **3:** 1588
3-Chloro-2-butanol **3:** 1588
4-Chloro-1-butanol **3:** 1588
4-Chloro-2-butanol **3:** 1588
Chlorobutenes **3:** 1380
4-Chlorobutyronitrile **6:** 3464
Chlorocarbonates, see chloroformates
Chlorocarbonic esters, see chloroformates
Chlorocarboxylic acids **2:** 1116

1-Chloro-2-chloromethyl-2-propanol   **3:** 1589
1-Chloro-2-chloromethyl-1-propene
   from isobutene chlorination   **3:** 1393
3-Chloro-2-chloromethyl-1-propene
   from isobutene chlorination   **3:** 1393
4-Chloro-N-[(4-chlorophenyl)sulfonyl]benzenesulfonamide   **2:** 768
p-Chloro-m-cresol   **6:** 3785
2-Chloro-1-cyanobenzene   **6:** 3471
4-Chloro-1-cyanobenzene   **6:** 3471
1-Chloro-2-cyano-4-nitrobenzene   **6:** 3472
1-Chloro-4-cyano-3-nitrobenzene   **6:** 3472
2-Chloro-1-cyano-4-nitrobenzene   **6:** 3471
4-Chloro-2-cyclopentylphenol   **6:** 3785
5-Chloro-2-(2,4-dichlorophenoxy)phenol   **6:** 3788
1-Chloro-1,1-difluoroethane
   properties   **5:** 2576
Chlorodifluoromethane
   properties   **5:** 2576
   pyrolysis   **5:** 2585
3-Chloro-4,4-dimethyl-2-oxazolidinone   **3:** 1570
4-Chloro-3,5-dimethylphenol   **6:** 3786
1-Chloro-2,4-dinitrobenzene   **6:** 3531
2-Chloro-3,5-dinitrobenzenesulfonic acid   **6:** 3558
4-Chloro-3,5-dinitrobenzenesulfonic acid   **6:** 3558
Chlorodiphenyl phosphine
   physical properties   **6:** 3826
1-Chloro-2,3-epoxypropane, *see epichlorohydrin*
2-Chloroethanal, *see chloroacetaldehyde*
Chloroethanes   **3:** 1286
   analysis and quality control   **3:** 1369
   environmental aspects   **3:** 1371
   production scheme   **3:** 1288
   specifications   **3:** 1370
   toxicology   **3:** 1472
2-Chloroethanesulfinic acid
   synthesis   **7:** 4460
Chloroethanoic acid, *see chloroacetic acid*
2-Chloro-1-ethanol   **3:** 1588
2-Chloroethyl benzenesulfonate   **2:** 758
Chloroethylenes   **3:** 1330
   analysis and quality control   **3:** 1369
   environmental aspects   **3:** 1371
   toxicology   **3:** 1472
1-Chloroethyl ethyl ether
   physical properties   **4:** 2187
2-Chloroethyl ethyl ether
   physical properties   **4:** 2187
β-Chloroethylsilanes   **7:** 4334
Chlorofluoroacetones   **5:** 2596
Chlorofluoroalkanes   **5:** 2575
   toxicology   **5:** 2623
Chlorofluoroolefins   **5:** 2589
Chloroform, *see trichloromethane*
Chloroformates   **3:** 1575
   aminolysis   **2:** 1046
   production   **3:** 1577

safety   **3:** 1578
specifications   **3:** 1579
toxicology   **3:** 1581
uses and economic aspects   **3:** 1580
Chloroformic esters, *see chloroformates*
N-Chloroglycolurils   **3:** 1570
N-Chloroguanidines   **3:** 1567
Chlorohydrination
   of allyl chloride   **4:** 2000
Chlorohydrination process
   in propylene oxide production   **7:** 4138
Chlorohydrin propylene oxide process   **7:** 4135
Chlorohydrins   **3:** 1585
   nomenclature   **3:** 1586
   production   **3:** 1590
   toxicology   **3:** 1597
   uses   **3:** 1596
2-Chloro-1-hydroperoxycyclohexanol
   physical properties   **6:** 3664
3-Chloro-4-hydroxybenzenesulfonic acid   **2:** 780
5-Chloro-2-hydroxybenzenesulfonic acid   **2:** 781
3-Chloro-2-hydroxybenzoic acid   **7:** 4278
4-Chloro-2-hydroxybenzoic acid   **7:** 4278
5-Chloro-2-hydroxybenzoic acid   **7:** 4278
5-Chloro-2-hydroxybiphenyl   **6:** 3786
Chloroisocyanurates   **3:** 1568
N-Chloroisocyanuric acids   **3:** 1568
4-Chloro-6-isopropyl-3-methylphenol   **6:** 3787
Chloromethanenitrile   **6:** 3462
Chloromethanes   **3:** 1252
   chemical properties   **3:** 1257
   decomposition in the atmosphere   **3:** 1283
   environmental aspects   **3:** 1282
   physical properties   **3:** 1254
   production   **3:** 1260
   specifications   **3:** 1278
   storage and handling   **3:** 1280
   toxicology   **3:** 1470
   uses and economic aspects   **3:** 1284
Chloromethanesulfinyl chloride
   synthesis   **7:** 4459
3-Chloro-2-methylaniline   **8:** 4750
3-Chloro-4-methylaniline   **8:** 4751
4-Chloro-2-methylaniline   **8:** 4750
5-Chloro-2-methylaniline   **8:** 4750
6-Chloro-2-methylaniline   **8:** 4750
6-Chloro-3-methylaniline   **8:** 4750
Chloromethylbenzene, *see benzyl chloride*
3-Chloro-4-methylbenzenesulfonamide   **2:** 777
3-Chloro-4-methylbenzenesulfonic acid   **2:** 777
4-(Chloromethyl)benzenesulfonic acid   **2:** 764
5-Chloro-4-methylbenzenesulfonic acid   **2:** 777
3-Chloro-4-methylbenzenesulfonyl chloride   **2:** 777
4-(Chloromethyl)benzenesulfonyl chloride   **2:** 764
Chloromethyl cyanide   **6:** 3462
Chloromethyldisilanes   **7:** 4344
Chloromethylethenylbenzene   **7:** 4438

2-Chloromethyl-1,4-naphthalenedione **6**: 3450
2-Chloromethyl-1,4-naphthoquinone **6**: 3450
1-Chloromethyl-4-nitrobenzene **6**: 3536
3-Chloro-4-methyl-5-nitrobenzenesulfonic
  acid **6**: 3559
3-Chloro-2-methyl-5-nitrobenzenesulfonic
  acid **6**: 3559
Chloromethyl-oxirane, see epichlorohydrin
4-Chloro-3-methylphenol **6**: 3785
Chloromethylphthalocyanines **7**: 3948
1-Chloro-2-methylpropane
  physical properties **3**: 1378
2-Chloro-2-methylpropane **3**: 1379
  physical properties **3**: 1378
2-Chloro-2-methyl-1-propanol **3**: 1589
1-Chloro-2-methyl-2-propanol **3**: 1589
3-Chloro-2-methyl-1-propene, see methallyl chloride
2-Chloro-3-methylpyridine
  physical properties **7**: 4199
2-Chloro-6-methylpyridine
  physical properties **7**: 4199
2-Chloromethylpyridine hydrochloride
  physical properties **7**: 4199
3-Chloromethylpyridine hydrochloride
  physical properties **7**: 4199
4-Chloromethylpyridine hydrochloride
  physical properties **7**: 4199
Chloromethylsilanes **7**: 4333
  physical properties **7**: 4322
  production process **7**: 4313
  synthesis **7**: 4310
α-Chloromethylstyrene **7**: 4438
1-Chloronaphthalene
  physical properties **3**: 1441
2-Chloronaphthalene
  physical properties **3**: 1441
2-Chloro-4-nitroaniline **6**: 3546
4-Chloro-2-nitroaniline **6**: 3546
2-Chloronitrobenzene **6**: 3526
3-Chloronitrobenzene **6**: 3529
4-Chloronitrobenzene **6**: 3529
2-Chloro-5-nitrobenzenesulfonic acid **6**: 3555
4-Chloro-2-nitrobenzenesulfonic acid **6**: 3556
4-Chloronitrobenzene-3-sulfonic acid **6**: 3555, 3556
5-Chloro-2-nitrobenzenesulfonic acid **6**: 3557
2-Chloro-4-nitrobenzonitrile **6**: 3471
2-Chloro-5-nitrobenzonitrile **6**: 3472
4-Chloro-2-nitrobenzonitrile **6**: 3472
4-Chloro-2-nitrophenol **6**: 3561
2-Chloro-4-nitrotoluene **6**: 3535
2-Chloro-6-nitrotoluene **6**: 3535
4-Chloro-2-nitrotoluene **6**: 3535
6-Chloro-2-nitrotoluene **6**: 3535
6-Chloro-2-nitrotoluene-4-sulfonic acid **6**: 3559
6-Chloro-4-nitrotoluene-2-sulfonic acid **6**: 3559
Chloroorgano silanes **7**: 4333
Chloropentafluoroacetone **5**: 2597

Chloropentafluoroethane
  properties **5**: 2576
1-Chloro-1,1,3,3,3-pentafluoro-2-propanone **5**: 2597
3-Chloroperbenzoic acid production **6**: 3648
3-Chloroperbenzoic acid
  physical properties **6**: 3643
Chlorophene **6**: 3785
2-Chlorophenol **3**: 1606
  physical properties **3**: 1606
3-Chlorophenol **3**: 1606
4-Chlorophenol **3**: 1606
  physical properties **3**: 1606
2-Chlorophenol hydrolysis
  for catechol production **6**: 3760
Chlorophenols **3**: 1605
  production **3**: 1608
  properties **3**: 1606
  specifications and economic aspects **3**: 1612
  toxicology **3**: 1614
  uses **3**: 1613
Chlorophenoxyalkanoic acids **3**: 1619
  biochemical and environmental aspects **3**: 1628
  toxicology **3**: 1625
Chlorophenoxyalkanoic acids
  production **3**: 1620
  specifications **3**: 1623
  uses **3**: 1624
p-Chlorophenyl benzenesulfonate **2**: 758
3-Chlorophenyl isocyanate **5**: 3011
Chlorophenylsilanes **7**: 4334
Chlorophosphines **6**: 3825
  toxicology **6**: 3856
Chloroprene **5**: 2890
  economic aspects **3**: 1388
  properties **3**: 1383, 1384
  toxicology **3**: 1478
β-Chloroprene production
  from acetylene **3**: 1387
  from butadiene **3**: 1385
2-Chloropropane **3**: 1373
  toxicology **3**: 1477
2-Chloro-1,3-propanediol **3**: 1588
3-Chloro-1,2-propanediol **3**: 1588
  toxicology **3**: 1598
3-Chloropropanenitrile **6**: 3464
Chloropropanes **3**: 1373
1-Chloro-2-propanol **3**: 1588
  toxicology **3**: 1599
2-Chloro-1-propanol **3**: 1588
  toxicology **3**: 1599
3-Chloro-1-propanol **3**: 1588
3-Chloropropene, see allyl chloride
2-Chloropropionic acid **7**: 4119
  toxicology **7**: 4123
  uses **7**: 4121
3-Chloropropionic acid **7**: 4120
  uses **7**: 4123

3-Chloropropionitrile 6: 3464
β-Chloropropionitrile 6: 3464
2-Chloropropionyl chloride 7: 4120
  uses 7: 4122
3-Chloropropionyl chloride 7: 4120
Chloropropylsilanes 7: 4333
2-Chloropyridine
  physical properties 7: 4199
3-Chloropyridine
  physical properties 7: 4199
2-Chloro-3-pyridinecarboxylic acid
  physical properties 7: 4199
6-Chloro-3-pyridinecarboxylic acid
  physical properties 7: 4199
4-Chloropyridine hydrochloride
  physical properties 7: 4199
Chloropyridines 7: 4198
5-Chloro-2-pyridinol
  physical properties 7: 4209
1-Chloro-2,5-pyrrolidinedione 3: 1570
3-Chlorosalicylic acid 7: 4278
4-Chlorosalicylic acid 7: 4278
5-Chlorosalicylic acid 7: 4278
Chlorosilanes 7: 4321
Chlorostyrene 7: 4437
N-Chlorosuccinimide 3: 1566, 1570
Chlorosulfonation
  of aromatics 2: 751
5-(Chlorosulfonyl)-2-hydroxybenzoic acid 2: 780
Chlorosulfuric acid
  reaction with alcohols 3: 1858
1-Chloro-1,2,2,2-tetrafluoroethane
  properties 5: 2576
p-Chlorothiophenol
  toxicology 8: 4678
4-Chlorothymol 6: 3787
2-Chlorotoluene
  physical properties 3: 1424
  specifications 3: 1433
  uses 3: 1434
3-Chlorotoluene
  physical properties 3: 1424
4-Chlorotoluene
  physical properties 3: 1424
  specifications 3: 1433
  uses 3: 1434
α-Chlorotoluene, see benzyl chloride
Chlorotoluene hydrolysis
  alkaline, for cresol production 3:
Chlorotoluenes
  chemical properties 3: 1426
  physical properties 3: 1423
  production 3: 1427
  toxicology 3: 1481
  uses 3: 1434
3-Chloro-o-toluidine 8: 4750
3-Chloro-p-toluidine 8: 4751

4-Chloro-o-toluidine 8: 4750
5-Chloro-o-toluidine 8: 4750
6-Chloro-m-toluidine 8: 4750
6-Chloro-o-toluidine 8: 4750
Chlorotoluidinesulfonic acids 8: 4752
(4-Chloro-o-tolyloxy)acetic acid 3: 1620
4-(4-Chloro-o-tolyloxy)butyric acid 3: 1620
2-(4-Chloro-o-tolyloxy)propionic acid 3: 1620
1-Chloro-2,2,2-trifluoroethane
  properties 5: 2576
Chlorotrifluoroethylene 5: 2589
  physical properties 5: 2584
Chlorotrifluoromethane
  properties 5: 2576
4-Chloro-3-(trifluoromethyl)benzenesulfonyl chloride 2: 778
2-Chloro-5-trifluoro-methylpyridine
  physical properties 7: 4199
N-Chloroureas 3: 1567
Chloroxylene hydrolysis
  alkaline for xylenol production 3: 1696
p-Chloro-m-xylenol 6: 3786
Choline 3: 1633
Choline chloride 3: 1634
Choline dihydrogen citrate 3: 1635
Choline gluconate 3: 1635
Choline hydrogen tartrate 3: 1634
Chrome blue dyes 2: 783
Chromic acid
  for oxidation of anthracene 2: 653
Chromotropic acid 6: 3400
Cigarette smoke
  formaldehyde content 5: 2662
1,8-Cineole 8: 4604
Cinnamaldehyde 1: 382, 389
Cinnamene, see styrene
Cinnamic acid 3: 1639
  properties and occurrence 3: 1639
  uses 3: 1642
cis-Cinnamic acid 3: 1640
trans-Cinnamic acid 2: 1121; 3: 1639
Cinnamic acid production 3: 1641
Citicoline 7: 4232
Citraconic acid 5: 3169
Citral 1: 368, 373
Citric acid 3: 1645
  chemical properties 3: 1648
  physical properties 3: 1647; 5: 2958
  specifications and uses 3: 1653
Citric acid cycle 3: 1646
Citric acid production 3: 1650
Citronellal 1: 373
Claisen condensation 3: 1641, 2020
  of sulfinic acid esters 7: 4463
Claisen rearrangement
  of allyl vinyl ethers 1: 371

5102

Cleaning compounds
  candelilla wax **8:** 4894
  carnaúba wax **8:** 4891
  chelating agents **4:** 2301
1,6-Cleve's acid **6:** 3412
1,7-Cleve's acid **6:** 3412
Clinical products
  dextran **3:** 1849
Cloud point **4:** 2434
CLT Acid **2:** 785
Clupanodonic acid
  physical properties **4:** 2487
CMC, see carboxymethyl cellulose
CMHEC **2:** 1233
CMMC, see carboxymethyl methyl cellulose
CN, see cellulose nitrate
CNEC **2:** 1234
Coagulation
  of hydroxyethyl cellulose **2:** 1226
Coal
  as feedstock for xylene production **8:** 4995
  as raw material in benzene production **2:** 698
  as source of saturated hydrocarbons **5:** 2861
Coalite process **8:**
Coal pyrolsis
  for acetylene production **1:** 154
Coal tar
  as feedstock for naphthalene production **6:** 3352
  hydrocarbons **5:** 2913
  thiophene content **8:** 4694
Coal-tar oils
  specifications **8:** 4538
Coal-tar pitch
  low-temperature **8:** 4548
  processing **8:** 4525
  properties and composition **8:** 4514
Coal tars
  alkylphenols content **6:** 3719
  isolation of cresols **3:** 1661
  low-temperature **8:** 4547
  properties and composition **8:** 4513
Cobald soaps **6:** 3246
Cobalt 2-ethylhexanoate **6:** 3245
Cobalt naphthenate **6:** 3245
Cobalt stearate **6:** 3245
Cocoa butter **4:** 2449
  fatty acid content **4:** 2376
Coco amine **1:** 472
Coconut oil **4:** 2445
  fatty acid content **4:** 2376
Codimerization
  of ethylene and propene **5:** 3039
Coke-oven coal tar
  constituents **8:** 4517
  processing **8:** 4519
  properties and composition **8:** 4515

Coke-oven tar
  properties and composition **8:** 4513
Coking
  production of saturated hydrocarbons **5:** 2861
  in cracking furnaces **4:** 2254
  propylene production **7:** 4089
Cold acid olefin process **5:** 2884
Color number **2:** 1038
Color properties
  of phthalocyanines **7:** 3924
Columbia acid **6:** 3442
Combes quinoline synthesis **1:** 509
Complexes
  of crown ethers **3:** 1726
  of dioxane **4:** 1939
Complex soaps **6:** 3228
Compression
  of cracked gas in ethylene production **4:** 2268
Condensation
  of aldehydes **3:** 2029
  of formaldehyde **5:** 2647
  of ketones or aldehydes with phenols **6:** 3771
  of propanal **7:** 4039
Condensers
  in phthalic anhydride production **7:** 3874
Conjugation
  of polyunsaturated fatty acids **4:** 2520
Conoco alcohol process **1:** 288
Continuous straight distillation
  of fatty acids **4:** 2507
Conversion
  of hydrocarbons **5:** 2862
Copolymers
  of linear α-olefins **5:** 2887
Copper
  as catalyst in acetaldehyde oxidation **1:** 45
  as catalyst in chloromethylsilane synthesis **7:** 4310
Copper 2-ethylhexanoate **6:** 3241
Copper formate **5:** 2739
Copper(I) chloride
  in diazo reactions **3:** 1897
Copper laurate **6:** 3241
Copper naphthenate **6:** 3241
Copper oleate **6:** 3241
Copper oxide–zinc oxide catalysts
  in methanol production **6:** 3296
Copper phthalocyanides **7:** 3952
Copper phthalocyanine
  toxicology **7:** 3957
Copper phthalocyanine disulfonic acid production **7:** 3947
Copper phthalocyanine pigments **7:** 3939
Copper phthalocyanine production **7:** 3936
Copper phthalocyanine tetrasulfonyl tetrachloride production **7:** 3947
Copper soaps **6:** 3241
Copper stearate **6:** 3241

5103

Coproduct propylene oxide processes **7**: 4144
Coproducts
  from ethylene production **4**: 2265
Coproportionation
  of silanes **7**: 4319
Corands **3**: 1723
Corn
  as raw material for ethanol production **4**: 2096
Corn germ oil
  fatty acid content **4**: 2376
Corn oil **4**: 2454
Cornstalks
  as raw materials for ethanol production **4**: 2107
Corn starch production **7**: 4389
Coronands **3**: 1723
Corrosion
  in urea production **8**: 4797
Corrosion inhibition
  with fatty amines **1**: 478
Corrosion inhibitors
  ethanolamines **4**: 2157
  isopropanolamines **4**: 2169
Cosmetics
  use of cellulose ethers **2**: 1241
Cotton linters
  for production of cellulose ethers **2**: 1212
Cottons
  cellulose nitrate lacquer **2**: 1161
Cottonseed
  conditioning **4**: 2403
Cottonseed oil **4**: 2452
  fatty acid content **4**: 2376
Coumarine **5**: 2764
Crack gas composition
  in BASF acetylene process **1**: 138
Crack gasoline
  for isoprene production **5**: 3041
Crack gas processing
  in ethylene production **4**: 2266
Crack gas vinyl chloride processes **3**: 1335
Cracking **4**: 2226
  see also hydrocarbon pyrolysis
Cracking conditions
  in ethylene production **4**: 2227
Cracking furnaces **4**: 2231, 2239
  coking and decoking **4**: 2254
  in 1,2-dichloroethane cracking **3**: 1338
  thermal efficiency **4**: 2252
  tube metallurgy **4**: 2250
Cracking mechanism
  of ethane to ethylene **4**: 2228
Cracking severity
  and product composition of BTX fractions **2**: 707
Cream of tartar
  uses **8**: 4568
Creep modulus
  of cellulose ester thermoplastics **2**: 1191

Creep rupture strength
  of cellulose ester thermoplastics **2**: 1190
Cresex cresol separation process **3**: 1682
p-Cresidine **1**: 637
Cresol **3**: 1655, 1700
  chemical properties **3**: 1658
  economic aspects **3**: 1690
  environmental protection **3**: 1699
  formation and isolation **3**: 1660
  handling **3**: 1687
  production **3**: 1664
  separation of m- and p-Cresol **3**: 1681
  specifications **3**: 1685
m-Cresol
  physical properties **3**: 1657
  uses **3**: 1689
o-Cresol
  physical properties **3**: 1657
  production by vapor-phase methylation of phenol **3**:
  uses **3**: 1688
p-Cresol
  physical properties **3**: 1657
  uses **3**: 1689
Cresol mixtures
  uses **3**: 1689
Cresol-phenol adducts **3**: 1684
m-Cresotic acid **5**: 2980
o-Cresotic acid **5**: 2979
p-Cresotic Acid **5**: 2980
Cresylic acids **3**: 1656
Cresyl phosphates
  toxicology **6**: 3858
Critical data
  of saturated hydrocarbons **5**: 2857
Critical relative humidity
  of urea **8**: 4785
CR 39 monomer **1**: 428
Crocein acid **6**: 3393
Crop protection materials
  from carbamoyl chlorides **2**: 1052
Cross-linked starch **7**: 4402
Cross-linking agents
  glyoxal **5**: 2813
  silanes **7**: 4349
Crotonaldehyde **1**: 368; **2**: 927; **3**: 1711
  production **3**: 1713
  toxicology **3**: 1719
  uses **3**: 1714
Crotonaldehyde hydrogenation **2**: 957
Crotonaldehyde oxidation **3**: 1717
Crotonamide
  physical properties **3**: 1716
Crotonic acid **3**: 1711, 1715
  production **3**: 1717
  toxicology **3**: 1720
  uses **3**: 1719

Crotonic anhydride
   physical properties   **3**: 1716
Crotonyl chloride
   physical properties   **3**: 1716
Crotonylenediurea   **8**: 4832
Crotyl alcohol   **1**: 309, 310
Crown ether   **3**: 1723
   formation from ethylene oxide   **4**: 2334
   properties   **3**: 1726
   toxicology   **3**: 1731
   uses   **3**: 1730
Crown ether complexes   **3**: 1726
Crown ether production   **3**: 1728
Cruciferous oils   **4**: 2463
Crude hard waxes   **8**: 4917
Crude oil
   composition   **2**: 700
Crude soft waxes   **8**: 4918
Crude waxes   **8**: 4916
Cryptands   **3**: 1723
Crystallization
   for *p*-xylene production   **8**: 4999
   of chloroacetic acid   **3**: 1544
   of m-/p-cresol mixtures   **3**: 1681
   of slack waxes   **8**: 4923
Cumene   **5**: 2899
   from propene   **7**: 4097
   specifications, for phenolacetone process   **1**: 94
Cumene hydroperoxide   **1**: 93
   physical properties   **6**: 3634
   toxicology   **6**: 3676
   uses   **6**: 3637
Cumene hydroperoxide cleavage   **1**: 95; **6**: 3696
Cumene hydroperoxide production   **6**: 3637
Cumene oxidation   **6**: 3694
   for acetone production   **1**: 93
Cumene oxidation process   **6**: 3695
Cumene phenol–acetone process   **1**: 94
Cumene–phenol process   **6**: 3694
p-Cumylphenol
   physical properties   **6**: 3754
Cuprene   **1**: 128
Cuprous ammonium acetate, *see CAA*
Cyanamide
   for production of guanidine salts   **5**: 2827
   production and specifications   **3**: 1751
   properties   **3**: 1745
   toxicology   **3**: 1762
   uses   **3**: 1753
Cyanamide starch   **7**: 4409
Cyanides, *see nitriles*
Cyanoacetaldehyde dimethyl acetal   **6**: 3465
Cyanoacetates   **5**: 3187
Cyanoacetic acid   **5**: 3185
   toxicology   **5**: 3193
Cyanoacrylates   **1**: 241
Cyanobenzenes   **6**: 3471

Cyano compounds, *see nitriles*
1-Cyano-2,6-dichlorobenzene   **6**: 3471
1-Cyano-2,6-difluorobenzene   **6**: 3471
Cyanoethyl cellulose   **2**: 1234
Cyanoethylsilanes   **7**: 4335
Cyanoguanidine, *see dicyandiamide*
Cyanoguanidines   **5**: 2839
Cyanohydrins
   formaldehyde   **5**: 2681
1-Cyano-4-hydroxybenzene   **6**: 3472
α-Cyano-α-hydroxymethylbenzene   **6**: 3470
6-Cyano-2-hydroxynaphthalene   **6**: 3473
Cyanol   **6**: 3426
Cyanomethane   **6**: 3460
Cyanomethylamine   **6**: 3463
Cyanomethylation
   of ammonia   **6**: 3481
α-Cyanomethylbenzene   **6**: 3469
2-Cyanomethylbenzothiazole   **6**: 3470
1-Cyanonaphthalene   **6**: 3473
2-Cyanonaphthalene   **6**: 3473
Cyanonaphthalenes   **6**: 3472
6-Cyano-β-naphthol   **6**: 3473
2-Cyano-4-nitroaniline   **6**: 3472, 3547
4-Cyanophenol   **6**: 3472
Cyanuric acid   **3**: 1768; **8**: 4818
   properties   **3**: 1768
   specifications and uses   **3**: 1772
   toxicology   **3**: 1773
Cyanuric acid production   **3**: 1770
Cyanuric chloride
   properties   **3**: 1773
   specifications   **3**: 1776
   toxicology   **3**: 1778
   uses   **3**: 1777
Cyanuric chloride production   **3**: 1775
Cyclamenaldehyde   **1**: 382, 388
Cyclar xylene process   **8**: 4997
Cyclic compounds
   oxidative cleavage   **3**: 1910
Cyclization
   of acetylene   **1**: 127
   of arylamines   **1**: 508
   of diazo compounds   **3**: 1893
   of heterocyclic thiols   **8**: 4671
   of hydrocarbons   **5**: 2862
Cyclization reactions
   of aminophenols   **1**: 605
Cyclizations
   of urea and its derivatives with aldehydes   **8**: 4828
Cycloaddition
   of isocyanates   **5**: 3014
   with ketene   **5**: 3063
Cycloaddition reactions
   of cyclopentadiene   **3**: 1828
   of cyclopentene   **3**: 1833

Cycloalkanes
　uses **5**: 2869
Cycloalkenes
　for production of dialdehydes **1**: 391
Cycloalkylphenols **6**: 3752
Cyclobutane
　physical properties **5**: 2855
Cyclodecanone **5**: 3101
Cyclododecane **3**: 1790
　uses **5**: 2869
Cyclododecane oxidation **3**: 1785
Cyclododecanol **3**: 1783
　properties **3**: 1784
　toxicology **3**: 1787
Cyclododecanol production **3**: 1785
Cyclododecanone **3**: 1783; **5**: 3101
　properties **3**: 1784
　toxicology **3**: 1787
Cyclododecanone production **3**: 1785
Cyclododecatriene **3**: 1789; **5**: 2891
　properties **3**: 1790
　uses **3**: 1791
Cyclododecatriene production **3**: 1791
Cyclododecene **3**: 1790
Cycloheptanone **5**: 3100
Cyclohexane **3**: 1795
　properties **3**: 1795
　separation from naphtha **3**: 1802
　specifications, uses, economic aspects **3**: 1803
　toxicology **3**: 1804
Cyclohexanecarboxylic acid **2**: 836
　in caprolactam production **2**: 1028
1,4-Cyclohexanedimethanol **1**: 334
Cyclohexane oxidation
　boric acid modified **3**: 1812
　liquid-phase **3**: 1810
Cyclohexane production **3**: 1797
Cyclohexanethiol
　toxicology **8**: 4678
Cyclohexanethiol production **8**: 4653
Cyclohexane-1,3,5-trione **6**: 3766
Cyclohexanol **3**: 1807
　economic aspects and toxicology **3**: 1820
　properties **3**: 1808
　specifications **3**: 1815
　uses and trade names **3**: 1816
Cyclohexanol–cyclohexanone mixtures
　for phenol production **6**: 3703
Cyclohexanol dehydrogenation **3**: 1814
Cyclohexanol esters **3**: 1816
Cyclohexanol oxidation
　for adipic acid production **1**: 265
Cyclohexanol production **3**: 1808
Cyclohexanone **3**: 1807
　economic aspects and toxicology **3**: 1820
　properties **3**: 1808
　specifications **3**: 1815
　uses and trade names **3**: 1816
Cyclohexanone-cyclohexanol oxidation **1**: 266
Cyclohexanone oxime **3**: 1817
　catalytic rearrangement **2**: 1024
　for ε-caprolactam production **2**: 1017
Cyclohexanone peroxide
　toxicology **6**: 3676
Cyclohexanone production **3**: 1808
Cyclohexene **3**: 1803
　physical properties **3**: 1797
Cyclohexene hydration **3**: 1813
Cyclohexylamine **1**: 459
　toxicology **1**: 492
1-Cyclohexylamino-1-hydroperoxycyclohexane
　physical properties **6**: 3670
Cyclohexyl chloroformate **3**: 1576
Cyclohexyldiethanolamine **4**: 2160
　see also ethanolamines, N-alkylated
　physical properties **4**: 2160
Cyclohexyldiisopropanolamine **4**: 2171
　toxicology **4**: 2179
Cyclohexylethanolamine **4**: 2160
　see also ethanolamines, N-alkylated
　physical properties **4**: 2160
Cyclohexyl hydroperoxide
　physical properties **6**: 3634
Cyclohexylisopropanolamine **4**: 2171
　toxicology **4**: 2179
2-Cyclohexyl-4-methylphenol
　physical properties **6**: 3754
2-Cyclohexyl-6-methylphenol
　physical properties **6**: 3754
2-Cyclohexylphenol
　physical properties **6**: 3754
4-Cyclohexylphenol
　physical properties **6**: 3754
Cyclohexylphenols **6**: 3752
Cyclohexyl vinyl ether
　physical properties **8**: 4867
Cyclooctadiene **3**: 1789; **5**: 2891
　properties **3**: 1790
　uses **3**: 1791
Cyclooctadiene production **3**: 1791
Cyclooctane **3**: 1790
　uses **5**: 2869
Cyclooctanone **5**: 3101
Cyclooctene **3**: 1790
Cyclooctyldiisopropanolamine **4**: 2171
Cyclooctylisopropanolamine **4**: 2171
Cyclopentadiene **3**: 1825
　properties **3**: 1825
　toxicology **3**: 1837
　uses **3**: 1831
Cyclopentadiene production **3**: 1829
Cyclopentane
　physical properties **5**: 2855
Cyclopentanone **5**: 3099

Cyclopentene **3**: 1825
   properties **3**: 1832
   toxicology **3**: 1838
   uses **3**: 1837
Cyclopentene production **3**: 1835
2-Cyclopentylphenol
   physical properties **6**: 3754
4-Cyclopentylphenol
   physical properties **6**: 3754
Cyclopentylphenols **6**: 3752
Cyclopropane
   physical properties **5**: 2855
Cyclotetradecanone **5**: 3101
Cydohexyl phosphine
   physical properties **6**: 3821
p-Cymene **8**: 4603
Cymene – cresol process **3**: 1668
Cymene hydroperoxide cleavage **3**: 1668
Cymenes **5**: 2902
Cytarabin **7**: 4232
Cytosine **7**: 4231
2,4-D **3**: 1620

DABCO **1**: 470
D acid **6**: 3395, 3415
Dactin **3**: 1569
Dahl's acid **6**: 3395, 3415
Dahl's acid II **6**: 3417
Dahl's acid III **6**: 3417
DAIP, *see diallyl isophthalate*
Dammaranes **7**: 4297
DAMN **6**: 3465
Damping maxima
   of cellulose ester thermoplastics **2**: 1189
Danish Distilleries fermentation process **4**: 2093
Darzens glycidic ester synthesis
   for production of araliphatic aldehydes **1**: 384
Darzens reaction **1**: 356; **4**: 1998
2,4-DB **3**: 1620
DBBPO, *see decabromobiphenyl oxide*
DBEP
   physical properties **7**: 3886
DBP
   physical properties **7**: 3886
DBS **6**: 3792
DCC-Na **3**: 1568
DC-EC process **3**: 1300
DCMX **6**: 3788
DCNB **6**: 3532
DDQ **2**: 841
DEA, *see diethanolamine*
Deacidification
   of fats and fatty oils **4**: 2414
DEAEC **2**: 1234
Dealkylation
   of alkylpyridines **7**: 4185
   of tertiary amines **1**: 448

   of toluene **8**: 4733
Deasphalting
   of tank bottom waxes, pipe waxes, and residues **8**: 4934
Debrisan **3**: 1849
Debutanization
   of cracked gas in ethylene production **4**: 2276
Decabromobiphenyl oxide **2**: 885
   toxicology **2**: 891
Decabromodiphenyl ether, *see decabromobiphenyl oxide*
Decahydronaphthalene, *see decalin*
Decalin **6**: 3358
   toxicology **6**: 3360
   uses **5**: 2869
Decanal, *see caprinaldehyde*
n-Decane
   physical properties **5**: 2855
Decanedioic acid, *see sebacic acid*
Decanedioic acid di-2-propenyl ester **1**: 426
1,10-Decanediol **1**: 323, 333
   physical properties **5**: 2553
Decanoic acid, *see capric acid*
1-Decanol, *see capric alcohol*
   toxicology **5**: 2559
cis-5-Decene
   physical properties **5**: 2871
1-Decene
   physical properties **5**: 2871
2-Decene
   physical properties **5**: 2871
3-Decene
   physical properties **5**: 2871
4-Decene
   physical properties **5**: 2871
trans-5-Decene
   physical properties **5**: 2871
Decoking
   in cracking furnaces **4**: 2255
Decolorizing
   of deoiled slack waxes **8**: 4924
Decomposition
   of diazonium salts **3**: 1879
   of formaldehyde **5**: 2646
Decylamine **1**: 455, 473
Decylbutanediperoxycarboxylic acid
   production **6**: 3647
n-Decylcyclohexane
   physical properties **3**: 1798
1-Decylimidazole
   toxicology **5**: 2994
Decyloxirane **4**: 2001
2-Decyl-1-tetradecanol
   physical properties **5**: 2551
Dediazotization **3**: 1895
Deethanization
   of cracked gas in ethylene production **4**: 2275

Deethanizer
  in ethylene production **4:** 2272
Deflagration **1:** 171
Degradation
  of alginates **7:** 3999
  of chlorophenoxy acids and chlorophenols **3:** 1628
  of monocarboxylic acids **3:** 1908
  of monosaccharides **2:** 1081
Degradation polyolefin waxes
  production **8:** 4953
  properties **8:** 4955
Degradation polypropylene waxes **8:** 4955
Degree of amidation
  of pectin **7:** 3991
Degree of esterification
  of pectin **7:** 3990
Degree of polymerization
  in cellulose **2:** 1144
Degree of substitution
  of cellulose derivates **2:** 1141
Degumming
  of fats and fatty oils **4:** 2413
DEHP
  physical properties **7:** 3886
Dehulling
  of oil seeds **4:** 2401
Dehydration
  of aldols, for production of unsaturated aldehydes **1:** 370
  of butanols **2:** 954
  of fatty acids **4:** 2520
  of pentanols **6:** 3613
  with acetic anhydride **1:** 70
Dehydroacetic acid **5:** 3075
Dehydrochlorination
  in propylene oxide production **7:** 4138
  of 3,4-dichloro-1-butene **3:** 1386
  of 1,2-dichloropropane, for allyl chloride production **1:** 416
  of ethylene chlorohydrin **3:** 1590
  of propylene chlorohydrin **7:** 4137
  of 1,1,2-trichloroethane **3:** 1350
Dehydrogenation
  of butane and butenes **2:** 907
  of butane-1,4-diol **2:** 1007
  of n-butenes **2:** 989
  of 1,2-cyclohexanediol **6:** 3761
  of cyclohexanol **3:** 1814
  of cyclohexanol – cyclohexanone mixtures, for phenol production **6:** 3703
  of ethylbenzene **7:** 4422
  of Δ^2-imidazolines **5:** 2989
  of isopentane or isopentenes, for isoprene synthesis **5:** 3040
  of paraffins **5:** 2875
  of primary alcohols **1:** 358
  of propane **7:** 4089

of 2-propanol **1:** 97; **7:** 4063
Dehydrogenation catalysts
  for amine production **1:** 449
Delayed coker **8:** 4524
Delayed coking pitch coke process **8:** 4531
Delta acid **6:** 3395
Deltamin **6:** 3405
Demethanization
  of cracked gas in ethylene production **4:** 2274
  of isophorone **3:** 1695
Demethanizer
  in ethylene production **4:** 2271
Density
  of cellulose ester thermoplastics **2:** 1185
  of fatty acids **4:** 2484
  of fatty acids and glycerides **4:** 2389
  of urea **8:** 4785
Deodorization
  of fats and fatty oils **4:** 2419
Deoiled slack waxes **8:** 4917
  refining **8:** 4924
Deoiling
  of bright stock slack waxes **8:** 4934
  of slack waxes **8:** 4922
6-Deoxy-L-galactopyranose **2:** 1066
6-Deoxy-D-glucose-6-sulfonic acid **2:** 1066
DEP
  physical properties **7:** 3886
Depolymerization
  of pectin **7:** 3988
Depropanization
  of cracked gas in ethylene production **4:** 2275
Depropanizer
  in ethylene production **4:** 2273
Depsidone acid **5:** 2974
Deresinification
  of montan wax **8:** 4910
Desulfonation
  of benzenesulfonic acids **2:** 742
  of hydroxynaphthalenedi- or trisulfonic acids **6:** 3388
  of hydroxynaphthalene sulfonic acid **6:** 3409
Detergent alcohols
  saturated aliphatic, linear C_{12}–C_{18} **1:** 302
Detergents
  alkylbenzenesulfonates **5:** 2885
  fatty acids **4:** 2527
  from higher olefins **5:** 2884
  low phosphate **6:** 3483
  with cellulose ethers **2:** 1240
Detol benzene process **2:** 708
Detonation
  of acetylene **1:** 172
Dextran **3:** 1843
  economic aspects **3:** 1850
  specifications **3:** 1848
  structure and properties **3:** 1844

toxicology **3**: 1851
uses **3**: 1849
Dextran production **3**: 1846
Dextransucrase **3**: 1846
L(+)-Dextrotartaric acid **8**: 4559
2,5-DHBA **5**: 2981
DHP
   physical properties **7**: 3886
Diacetates **1**: 68
Diacetone alcohol **1**: 111
Diacetonitrile **6**: 3465
Diacetyl **5**: 3108
Diacetyl peroxide
   physical properties **6**: 3650
   toxicology **6**: 3676
Diacids, *see dicarboxylic acids*
Diacyl peroxides
   production **6**: 3651
   properties **6**: 3649
   uses **6**: 3653
Dialdehydes **1**: 390
Dialdehyde starch **7**: 4410
2-Dialkylamino-1,3-dithiolanylium salts **4**: 1965
N,N'-Dialkylaminomethylphenols
   for cresol production **3**: 1680
Dialkyl peroxides **6**: 3638
Dialkyl sulfates **3**: 1853
   production **3**: 1857
   specifications **3**: 1860
   uses **3**: 1864
Diallyl adipate **1**: 426
Diallylamine **1**: 432, 455, 458
Diallyl carbonate **2**: 1086
Diallyl fumarate **1**: 426, 429
Diallyl isophthalate **1**: 426, 429
Diallyl maleate **1**: 426, 429
Diallyl phthalate **1**: 426, 429
Diallyl sebacate **1**: 426, 429
Diallyl succinate **1**: 426, 429
Dialysis fermentor **4**: 2087
Diamines **1**: 482
2,5-Diamino-1,4-benzenedisulfonic acid **2**: 774
4,6-Diamino-1,3-benzenedisulfonic acid **2**: 774
2,4-Diaminobenzenesulfonic acid **2**: 774
4,4′-Diaminobiphenyl-2,2′-dicarboxylic acid, **2**: 810
4,4′-Diaminobiphenyl-3,3′-dicarboxylic acid **2**: 810
4,4′-Diaminobiphenyl-2,2′-disulfonic acid **2**: 810
1,3-Diaminobutane **1**: 486
1,4-Diaminobutane **1**: 486
2,3-Diamino-2-butenedinitrile **6**: 3465
1,10-Diaminodecane **1**: 486
3,7-Diaminodibenzothiophene-5,5-dioxide **2**: 811
3,7-Diaminodibenzothiophene-2,8-disulfonic acid 5,5-dioxide **2**: 812
4,4′-Diamino-5,5′-dimethylbiphenyl-2,2′-disulfonic acid **2**: 811
1,12-Diaminododecane **1**: 487

1,2-Diaminoethane, *see ethylenediamine*
Diaminoguanidine **5**: 2834
1,6-Diaminohexane, *see hexamethylenediamine*
Diaminomaleonitrile **6**: 3465
1,5-Diaminonaphthalene **6**: 3404
1,8-Diaminonaphthalene **6**: 3405
3,8-Diaminonaphthalene-1,5-disulfonic acid **6**: 3424
4,8-Diaminonaphthalene-2,6-disulfonic acid **6**: 3424
1,2-Diaminonaphthalene-5-sulfonic acid **6**: 3423
1,3-Diaminonaphthalene-8-sulfonic acid **6**: 3423
1,4-Diaminonaphthalene-2-sulfonic acid **6**: 3424
1,8-Diaminonaphthalene-4-sulfonic acid **6**: 3423
4,5-Diaminonaphthalene-1-sulfonic acid **6**: 3423, 3424
5,6-Diaminonaphthalene-1-sulfonic acid **6**: 3423
6,8-Diaminonaphthalene-1-sulfonic acid **6**: 3423
1,8-Diaminooctane **1**: 486
1,5-Diaminopentane **1**: 486
2,4-Diaminophenol **1**: 613
1,2-Diaminopropane **1**: 484
1,3-Diaminopropane **1**: 484
2,6-Diaminopyridine **7**: 4207
   physical properties **7**: 4206
4,4′-Diamino-2,2′-stilbenedisulfonic acid
   uses **1**: 505
Diamond Black P 2 B **2**: 783
Diamond Black PV **2**: 781
Di-*n*-amyl ether **4**: 2197
   physical properties **4**: 2187
Di-*tert*–amyl peroxide production **6**: 3641
6,8-Dianilinonaphthalene-1-sulfonic acid **6**: 3423
o-Dianisidine **2**: 808
   toxicology **2**: 815
Diarylamines
   uses **1**: 505
1,4-Diazabicyclo[2.2.2]octane **1**: 470
Diazoamino compounds, *see triazenes*
Diazo compounds **3**: 1875
   aliphatic **3**: 1875
   aromatic **3**: 1876
   environmental protection **3**: 1899
   safety and toxicology **3**: 1900
Diazomethane
   toxicology **3**: 1900
Diazonium arylsulfonates **3**: 1882
Diazonium chlorides **3**: 1881
Diazonium salts **1**: 605
   crystalline **3**: 1880
   decomposition half-life **3**: 1879
   production **3**: 1878
   properties **3**: 1877
   toxicology **3**: 1901
   uses **3**: 1883
Diazonium tetrachlorozincates **3**: 1881
Diazonium tetrafluoroborates **3**: 1882
Diazophenols **3**: 1890

Diazo reaction 3: 1875
  in benzenesulfonic acid production 2: 748
  modified diazo group 3: 1893
  replacement of the diazo group 3: 1895
  retained diazo group 3: 1892
Diazosulfonates 3: 1886
Diazotates 3: 1884
Diazotization
  direct 3: 1878
  for production of fluorobenzenes 5: 2614
  indirect 3: 1880
  of arylamines 1: 509
Dibenzenesulfonylamines 2: 753
Dibenzofuran 5: 2765
  toxicology 5: 2768
Dibenzopyrrole, see carbazole
Dibenzoyl peroxide
  physical properties 6: 3650
  toxicology 6: 3676
Dibenzyl ether 2: 858
DIBP
  physical properties 7: 3886
1,2-Dibromobenzene 2: 870
1,3-Dibromobenzene 2: 870
1,4-Dibromobenzene 2: 870
4,4'-Dibromo-1,1'-biphenyl 2: 870
trans-1,4-Dibromo-2-butene 2: 869
Dibromochlorofluoromethane
  properties 5: 2581
Dibromodifluoromethane
  properties 5: 2581
1,1-Dibromoethane 2: 869
1,2-Dibromoethane 2: 869, 873
  toxicology 2: 887
1,1-Dibromoethene 2: 869
cis-1,2-Dibromoethene 2: 869
1-Dibromoethyl-3,4-dibromocyclohexane 2: 883
Dibromomethane 2: 869, 880
  toxicology 2: 890
Dibromoneopentyl glycol 2: 883
1,2-Dibromopropane 2: 869
1,3-Dibromopropane 2: 869
2,3-Dibromo-1-propene 2: 869
5,4'-Dibromosalicylanilide 6: 3792
3,5-Dibromosalicylic acid 2: 870
1,2-Dibromotetrafluoroethane
  properties 5: 2581
Di(butoxyethyl) phthalate
  physical properties 7: 3886
Dibutylamine 1: 454, 458
4,4'-Di-tert-butylbenzoyl peroxide
  production 6: 3652
Di-n-butyl carbamoyl chloride 2: 1049
Di-sec-butyl carbamoyl chloride 2: 1049
Dibutylethanolamine 4: 2160
   see also ethanolamines, N-alkylated
  physical properties 4: 2160

  toxicology 4: 2178
Di-n-butyl ether 4: 2196
Di-n-butyl ether
  physical properties 4: 2187
  toxicology 4: 2201
Di-sec-butyl ether
  physical properties 4: 2187
Di-tert-butyl ether
  physical properties 4: 2187
Dibutylisopropanolamine 4: 2171
Di-t-butylketene
  physical properties 5: 3077, 3077
2,6-Di-tert-butyl-4-methylphenol 6: 3737
4,6-Di-tert-butyl-2-methylphenol 6: 3737
4,6-Di-tert-butyl-3-methylphenol 6: 3737
2,6-Di-tert-butyl-4-methylphenol production 6: 3741
Di-tert-butyl peroxide
  physical properties 6: 3640
  toxicology 6: 3676
2,4-Di-tert-butylphenol 6: 3737
2,6-Di-tert-butylphenol 6: 3737
3,5-Di-tert-butylphenol 6: 3737
2,6-Di-tert-butylphenol production 6: 3739
3,5-Di-tert-butylphenol production 6: 3739
Di-n-butyl phosphine
  physical properties 6: 3822
Dibutyl phthalate
  physical properties 7: 3886
  specifications 3: 2038
Di-n-butyl sulfate
  physical properties 3: 1855
Dicarboxylic acids, aliphatic 3: 1903
  saturated, production 3: 1908
  saturated, properties 3: 1904
  specifications 3: 1923
  unsaturated, production 3: 1921
  unsaturated, properties 3: 1919
Dicetyl peroxydicarbonate production 6: 3653
Dichlone 6: 3450
Dichloramine-B 3: 1571
Dichloramine-T 3: 1566
Dichloroacetaldehyde 3: 1529
  toxicology 3: 1535
Dichloroacetaldehydes
  polymeric 3: 1531
Dichloroacetate
  toxicology 3: 1560
Dichloroacetic acid 3: 1549
  toxicology 3: 1560
Dichloroacetic acid esters 3: 1552
Dichloroacetic acid production 3: 1550
Dichloroacetyl chloride 3: 1551
Dichloroacetylene
  toxicology 3: 1477
2,3-Dichloroaniline 1: 636
2,4-Dichloroaniline 1: 636
2,5-Dichloroaniline 1: 636

2,6-Dichloroaniline **1:** 636
3,4-Dichloroaniline **1:** 636, 638
3,5-Dichloroaniline **1:** 636
2,5-Dichlorobenzal chloride
  physical properties **3:** 1466
2,6-Dichlorobenzal chloride
  physical properties **3:** 1466
2,4-Dichlorobenzaldehyde **2:** 684
2,6-Dichlorobenzaldehyde **2:** 684
Dichlorobenzene
  chemical properties **3:** 1411
  physical properties **3:** 1407
  production **3:** 1417
1,2-Dichlorobenzene
  specifications **3:** 1421
  uses **3:** 1422
1,3-Dichlorobenzene
  specifications **3:** 1421
  uses **3:** 1423
1,4-Dichlorobenzene
  specifications **3:** 1421
  uses **3:** 1423
Dichlorobenzene CuPc process **7:** 3938
1,3-Dichlorobenzene production **3:** 1418
3,4-Dichlorobenzenesulfonamide **2:** 769; **3:** 1571
2,5-Dichlorobenzenesulfonic acid **2:** 768
3,4-Dichlorobenzenesulfonic acid **2:** 768
2,5-Dichlorobenzenesulfonyl chloride **2:** 768
3,4-Dichlorobenzenesulfonyl chloride **2:** 769
2,2′-Dichlorobenzidine **2:** 807
3,3′-Dichlorobenzidine **2:** 807
  toxicology **2:** 813
2,4-Dichlorobenzoic acid **2:** 832
2,5-Dichlorobenzoic acid **2:** 832
2,6-Dichlorobenzonitrile **6:** 3471
2,4-Dichlorobenzotrichloride
  physical properties **3:** 1466
2,4-Dichlorobenzotrifluoride
  physical properties **5:** 2609
2,6-Dichlorobenzyl chloride
  physical properties **3:** 1466
3,5-Dichlorobenzyl chloride
  physical properties **3:** 1466
2,2′-dichloro-4,4′-biphenyl-diyldiamine **2:** 807
3,3′-dichloro-4,4′-biphenyl-diyldiamine **2:** 807
1,1-Dichloro-1,3-butadiene **3:** 1389
1,3-Dichloro-1,3-butadiene **3:** 1389
1,4-Dichloro-1,3-butadiene **3:** 1389
2,3-Dichloro-1,3-butadiene **3:** 1389
1,4-Dichlorobutane **3:** 1380
  physical properties **3:** 1378
Dichlorobutene
  isomerization **3:** 1386
1,4-Dichloro-2-butene **3:** 1381
3,4-Dichloro-1-butene **3:** 1382
3,4-Dichloro-1-butene dehydrochlorination **3:** 1386
N,N-Dichlorocarbamates **3:** 1567

Dichlorodianisidine **2:** 809
2,3-Dichloro-5,6-dicyano-1,4-benzoquinone **2:** 843
1,2-Dichloro-1,1-difluoroethane
  properties **5:** 2576
Dichlorodifluoromethane
  production **5:** 2575
  properties **5:** 2576
2,2′-Dichloro-5,5′-dimethoxy-4,4′-biphenyl-diyldiamine **2:** 809
Dichlorodimethyl ether **4:** 2199
1,3-Dichloro-5,5-dimethylhydantoin **3:** 1566, 1569
2,4-Dichloro-3,5-dimethylphenol **6:** 3788
2,2-Dichloroethanal, *see dichloroacetaldehyde*
1,1-Dichloroethane
  for production of 1,1,1-trichloroethane **3:** 1312
  properties **3:** 1295
  toxicology **3:** 1472
  uses and economic aspects **3:** 1297
1,2-Dichloroethane
  for production of 1,1,2,2-tetrachloroethane **3:** 1324
  for production of 1,1,2-trichloroethane **3:** 1318
  for tetrachloroethylene production **3:** 1364
  for trichloroethylene production **3:** 1359
  properties **3:** 1298
  specifications **3:** 1370
  toxicology **3:** 1473
  uses and economic aspects **3:** 1309
1,2-Dichloroethane conversion
  for production of 1,1-dichloroethane **3:** 1296
1,2-Dichloroethane cracking **3:** 1336
  by products **3:** 1339
  catalytic and photochemically induced reaction **3:** 1340
  noncatalystic gas-phase reaction **3:** 1337
1,1-Dichloroethane production **3:** 1296
1,2-Dichloroethane production **3:** 1299
2,2-Dichloroethanoic acid, *see dichloroacetic acid*
1,1-Dichloroethylene
  toxicology **3:** 1475, 1475
  uses and economic aspects **3:** 1352
1,1-Dichloroethylene
  properties **3:** 1348
1,2-Dichloroethylene
  properties **3:** 1353
cis-1,2-Dichloroethylene
  properties **3:** 1354
trans-1,2-Dichloroethylene
  properties **3:** 1353
1,1-Dichloroethylene production **3:** 1350
1,2-Dichloroethylene production **3:** 1355
Dichloroethyl ether **4:** 2199
1,1-Dichloro-1-fluoroethane
  properties **5:** 2576
Dichlorofluoromethane
  properties **5:** 2576
3,5-Dichloro-2-hydroxybenzenesulfonamide **2:** 781

3,5-Dichloro-2-hydroxybenzenesulfonic acid **2**: 781
1,2-Dichloroisobutane
    from isobutene chlorination **3**: 1393
Dichloroisopropyl ether **4**: 2200
Dichloromethane **3**: 1253, 1278
    see also chloromethanes
    chemical properties **3**: 1259
    physical properties **3**: 1255
    toxicology **3**: 1471
Dichloromethane production **3**: 1268
Dichloromethylbenzene, *see benzal chloride*
3,5-Dichloro-4-methylbenzenesulfonic acid **2**: 778
3,5-Dichloro-4-methylbenzenesulfonyl chloride **2**: 778
Dichloromethyl phosphine
    physical properties **6**: 3826
1,2-Dichloro-2-methyl-3-propanol **3**: 1589
2,3-Dichloro-1,4-naphthalenedione **6**: 3450
2,3-Dichloro-1,4-naphthoquinone **6**: 3450
1,2-Dichloro-3-nitrobenzene **6**: 3533
1,2-Dichloro-4-nitrobenzene **6**: 3532
1,3-Dichloro-4-nitrobenzene **6**: 3533
1,4-Dichloro-2-nitrobenzene **6**: 3533
2,3-Dichloronitrobenzene **6**: 3533
2,4-Dichloronitrobenzene **6**: 3533
2,5-Dichloronitrobenzene **6**: 3533
3,4-Dichloronitrobenzene **6**: 3532
Dichlorophene **6**: 3789
2,3-Dichlorophenol **3**: 1606
2,4-Dichlorophenol **3**: 1606
2,5-Dichlorophenol **3**: 1606
2,6-Dichlorophenol **3**: 1606
3,4-Dichlorophenol **3**: 1606
3,5-Dichlorophenol **3**: 1606
(2,4-Dichlorophenoxy)acetic acid **3**: 1620
4-(2,4-Dichlorophenoxy)butyric acid **3**: 1620
2-(2,4-Dichlorophenoxy)propionic acid **3**: 1620
2,4-Dichlorophenyl benzenesulfonate **2**: 758
3,4-Dichlorophenyl isocyanate **5**: 3011
Dichlorophenyl phosphine
    physical properties **6**: 3826
1,2-Dichloropropane **3**: 1374
    for allyl chloride production **1**: 416
    toxicology **3**: 1477
1,3-Dichloro-2-propanol
    toxicology **3**: 1599
1,3-Dichloro-2-propanol **3**: 1588
2,3-Dichloro-1-propanol
    toxicology **3**: 1599
2,3-Dichloro-1-propanol **3**: 1588
2,2-Dichloropropionic acid **2**: 1116; **7**: 4120
    uses **7**: 4123
2,3-Dichloropyridine
    physical properties **7**: 4199
2,4-Dichloropyridine
    physical properties **7**: 4199

2,5-Dichloropyridine
    physical properties **7**: 4199
3,5-Dichloropyridine
    physical properties **7**: 4199
5,2'-Dichlorosalicylanilide **6**: 3792
sym-Dichlorotetrafluoroacetone **5**: 2597
1,1-Dichlorotetrafluoroethane
    properties **5**: 2576
1,2-Dichlorotetrafluoroethane
    properties **5**: 2576
1,3-dichloro-1,1,3,3-tetrafluoro-2-propanone **5**: 2597
2,5-Dichlorothiophenol production **8**: 4663
2,3-Dichlorotoluene
    physical properties **3**: 1425
    production **3**: 1430
2,4-Dichlorotoluene
    physical properties **3**: 1425
    production **3**: 1430
    specifications **3**: 1433
    uses **3**: 1434
2,5-Dichlorotoluene
    physical properties **3**: 1425
    production **3**: 1430
2,6-Dichlorotoluene
    physical properties **3**: 1425
    production **3**: 1430
    uses **3**: 1434
3,4-Dichlorotoluene
    physical properties **3**: 1425
    production **3**: 1431
    specifications **3**: 1433
    uses **3**: 1434
3,5-Dichlorotoluene
    physical properties **3**: 1425
    production **3**: 1431
α,α-Dichlorotoluene, *see benzal chloride*
1,3-Dichloro-1,3,5-triazine-2,4,6(1*H*,3*H*,5*H*)trione sodium salt **3**: 1568
Dichloro-*m*-xylenol **6**: 3788
Dichlorprop **3**: 1620
1,1-Dichlor-2,2,2-trifluoroethane
    properties **5**: 2576
Dicoco amine **1**: 474
Dicumyl peroxide **1**: 94
    physical properties **6**: 3640
    production **6**: 3641
    toxicology **6**: 3676
Dicy, *see dicyandiamide*
Dicyandiamide **3**: 1754
    for production of guanidine salts **5**: 2826
    properties **3**: 1755
    specifications **3**: 1759
    toxicology **3**: 1762
    uses **3**: 1760
Dicyandiamide production **3**:
Dicyandiamides **5**: 2839
1,2-Dicyanobenzene **7**: 3883

1,4–Dicyanobutane, see adiponitril
2,4-Dicyano-1-butene **6:** 3462
Dicyanoethane **6:** 3461
Dicyanomethane, see malononitrile
1,4-Dicyanonaphthalene **6:** 3473
2,6-Dicyanonaphthalene **6:** 3473
Dicyclohexyl adipate **3:** 1817
Dicyclohexylamine **1:** 462
N,N-Dicyclohexyl-2-benzo-1,3-thiazolesulfinic acid amide
   synthesis **7:** 4465
Dicyclohexylidene diperoxide
   physical properties **6:** 3664
Dicyclohexyl peroxide
   toxicology **6:** 3676
Dicyclohexyl phosphine
   physical properties **6:** 3822
Dicyclohexyl phthalate **3:** 1816
Dicyclopentadiene
   properties **3:** 1825
   toxicology **3:** 1837
   uses **3:** 1832
Dicyclopentylidene diperoxide
   physical properties **6:** 3664
2,6-Dicyclopentyl-4-methylphenol
   physical properties **6:** 3754
Didecylamine **1:** 474
Didodecylamine **1:** 473, 474
DIDP
   physical properties **7:** 3886
Dieckmann reaction **3:** 1907
Diels–Alder dimerization
   of 1,3-butadiene **7:** 4429
Diels–Alder reaction
   with acrolein **1:**
   with butadiene **2:** 904
   with methacrylic acid **6:** 3251
Diels–Alder ring closure
   of isoprene and vinyl acetate **3:** 1680
Dienes **5:** 2889
   for production of dialdehydes **1:** 391
   production by oligomerization **5:** 2891
   removal from pyrolysis gasolines **2:** 715
Dienophiles
   crotonic acid **3:** 1716
Diepoxides **4:** 2004
Diesel fuel
   methanol **6:** 3314
Diesel fuels **8:** 4537
Diethanolamine **4:** 2152
   see also ethanolamines
   physical properties **4:** 2152
   toxicology **4:** 2177
3,3'-diethoxy-4,4'-biphenyl-diyldiamine, see o-dipheneti-
   dine
1,1-Diethoxybutane **1:** 395
1,1-Diethoxyethane **1:** 395

1,1-Diethoxyhexane **1:** 395
O,O-Diethoxylthiophosphoryl chloride **6:** 3852
1,4-Diethoxy-2-nitrobenzene **6:** 3564
2,5-Diethoxynitrobenzene **6:** 3564
1,1-Diethoxypentane **1:** 395
1,1-Diethoxypropane **1:** 395
Diethyiisopropanolamine **4:** 2171
Diethyl adipate
   physical properties **3:** 2014
Diethylamine **1:** 454, 458
1-Diethylamino-4-aminopentane **1:** 486
2-(N,N-Diethylamino)-ethyl cellulose **2:** 1234
3-(N,N-Diethylamino)phenol **1:** 617
N,N-Diethylaniline **1:** 637, 638
1,2-Diethylbenzene
   physical properties **5:** 2894
1,3-Diethylbenzene
   physical properties **5:** 2894
1,4-Diethylbenzene
   physical properties **5:** 2894
Diethylbenzenes **5:** 2897
   toxicology **5:** 2926
Diethyl carbamoyl chloride **2:** 1049
   toxicology **2:** 1053
Diethyl carbonate **2:** 1086
N,N-Diethyl-2-chloroethylamine **1:** 455
Diethylene dioxide, see dioxane
Diethylene ether, see dioxane
Diethylene glycol **2:** 721
   physical properties **4:** 2319
   specifications **4:** 2318
   toxicology **4:** 2324
Diethylene glycol bis(allyl carbonate) **1:** 426, 428;
   2: 1086
Diethylene glycol bis-chloroformate **3:** 1576
Diethylene glycol butyl ether acetate
   physical properties **4:** 2321
Diethylene glycol dehydration
   in dioxane production **4:** 1940
Diethylene glycol dibenzoate **2:** 827
Diethylene glycol divinyl ether
   physical properties **8:** 4867
Diethylene glycol ethyl ether acetate
   physical properties **4:** 2321
Diethylene glycol methyl ether acetate
   physical properties **4:** 2321
Diethylene oxide, see dioxane
Diethylenetriamine **1:** 487
   toxicity **1:** 493
Diethylenetriaminepentaacetic acid
   physical properties **4:** 2297
Diethylenetriurea **8:** 4820
Diethylethanolamine **4:** 2160
   see also ethanolamines, N-alkylated
   physical properties **4:** 2160
   toxicology **4:** 2178

Diethyl ether
  physical properties **4**: 2187
  properties **4**: 2189
  specifications **4**: 2191
  toxicology **4**: 2201
  uses and economic aspects **4**: 2192
Diethyl ether production **4**: 2190
Di-2-ethylhexyl phthalate
  physical properties **7**: 3886
Diethylketene
  physical properties **5**: 3077
Diethyl ketone **5**: 3095
Diethyl malonate **5**: 3179
  toxicology **5**: 3193
Diethyl malonate production **5**: 3182
Diethyl peroxide
  physical properties **6**: 3640
2,6-Diethylphenol **6**: 3730
Diethyl phosphine
  physical properties **6**: 3822
Diethyl phosphite
  toxicology **6**: 3859
Diethyl phthalate
  physical properties **3**: 2015; **7**: 3886
  specifications **3**: 2037
Diethyl sulfate
  physical properties **3**: 1855
  toxicology **3**: 1871
1,3-Difluorobenzene **5**: 2613
1,4-Difluorobenzene **5**: 2613
2,6-Difluorobenzonitrile **6**: 3471
Difluorobromochloromethane **2**: 869, 881
Difluorobromomethane
  toxicology **2**: 891
4,4′-Difluorodiphenyl methane **5**: 2613
1,1-Difluoroethane
  properties **5**: 2574
1,1-Difluoroethylene **5**: 2587
  physical properties **5**: 2584
Difluoromethane
  properties **5**: 2574
5-(2,4-Difluorophenyl)salicylic acid **7**: 4279
Diformyl, *see glyoxal*
m-Digallic acid **5**: 2974
Digesters
  for biomethanation **6**: 3274
Diglycerides
  in fats and fatty oils **4**: 2374
Diglycerol **5**: 2805
Di(heptyl-nonyl) phthalate
  physical properties **7**: 3886
Di(heptyl-nonyl-undecyl) phthalate
  physical properties **7**: 3886
Diheptyl phthalate
  physical properties **7**: 3886
Di-*n*-heptyl sulfate
  physical properties **3**: 1855

Dihexadecylamine **1**: 474
Di-*n*-hexyl carbamoyl chloride **2**: 1049
Di-*n*-hexyl ether
  physical properties **4**: 2187
Di(hexyl-octyl-decyl) phthalate
  physical properties **7**: 3886
Di-*n*-hexyl sulfate
  physical properties **3**: 1855
1,3-Dihydro-1,3-dioxo-4-isobenzofurancarboxylic acid, *see hemimellitic anhydride*
1,3-Dihydro-1,3-dioxo-5-isobenzofurancarboxylic acid, *see trimellitic anhydride*
Dihydrogenated-tallow amine **1**: 474
2,3-Dihydro-1*H*-indene-5-sulfonamide **2**: 766
2,3-Dihydro-1*H*-indene-5-sulfonic acid **2**: 766
2,3-Dihydro-1*H*-indene-5-sulfonyl chloride **2**: 766
3,4-Dihydro-2*H*-pyran **5**: 2763
2,3-Dihydroindole **5**: 3003
α,α′-Dihydroperoxydicyclohexyl peroxide
  production **6**: 3668
2,3-Dihydropyran **5**: 2763
  physical properties **5**: 2747
  toxicology **5**: 2768
1,2-Dihydroxybenzene, *see catechol*
1,3-Dihydroxybenzene, *see resorcinol*
1,4-Dihydroxybenzene, *see hydroquinone*
p-Dihydroxybenzene, *see hydroquinone*
4,5-Dihydroxy-1,3-benzenedisulfonic acid **2**: 776
2,4-Dihydroxybenzenesulfonic acid **2**: 776
3,4-Dihydroxybenzenesulfonic acid **2**: 776
2,3-Dihydroxybenzoic acid **5**: 2980
2,4-Dihydroxybenzoic acid **5**: 2980
2,5-Dihydroxybenzoic acid **5**: 2981
3,4-Dihydroxybenzoic acid **5**: 2980
3,5-Dihydroxybenzoic acid **5**: 2981
2,2′-Dihydroxybiphenyl
  physical properties **6**: 3775
4,4′-Dihydroxybiphenyl
  physical properties **6**: 3775
2,2′-Dihydroxybiphenyl production **6**: 3776
4,4′-Dihydroxybiphenyl production **6**: 3776
2,3-Dihydroxybutanedioic acid, *see tartaric acid*
Dihydroxycarboxylic acids
  production **5**: 2962
1,1′-Dihydroxy-2,2′-dichlorodiethyl ether **3**: 1522
α,α′-Dihydroxydicyclohexyl peroxide
  physical properties **6**: 3664
N,*N*′-Dihydroxymethyl-2-oxo-4,5-dihydroxyimidazolidine **8**: 4825
N,*N*′-Dihydroxymethyl-2-oxo-4-hydroxy(methoxy)-5,5-dimethylhexahydropyrimidine **8**: 4832
N,*N*′-Dihydroxymethylurea **8**: 4823
1,5-Dihydroxynaphthalene **6**: 3379
1,8-Dihydroxynaphthalene-3,6-disulfonic acid **6**: 3400
3,6-Dihydroxynaphthalene-2,7-disulfonic acid **6**: 3400
4,5-Dihydroxynaphthalene-2,7-disulfonic acid **6**: 3400
1,8-Dihydroxynaphthalene-4-sulfonic acid **6**: 3399

2,3-Dihydroxynaphthalene-6-sulfonic acid **6:** 3400
2,5-Dihydroxynaphthalene-7-sulfonic acid **6:** 3399
2,8-Dihydroxynaphthalene-6-sulfonic acid **6:** 3399
3,5-Dihydroxynaphthalene-2,7-sulfonic acid **6:** 3400
4,5-Dihydroxynaphthalene-1-sulfonic acid **6:** 3399
4,6-Dihydroxynaphthalene-2-sulfonic acid **6:** 3399
4,7-Dihydroxynaphthalene-2-sulfonic acid **6:** 3399
6,7-Dihydroxynaphthalene-2-sulfonic acid **6:** 3400
1,2-Dihydroxypentane **1:** 337
2,6-Dihydroxypurine, see purine derivatives
2,3-Dihydroxysuccinic acid, see tartaric acid
Diimidotricarbonic diamide **8:** 4818
Diiminoisoindolenine CuPc process **7:** 3938
1,2-Diiodotetrafluoroethane
   properties **5:** 2582
Diisoamyl ether **4:** 2197
   physical properties **4:** 2187
Diisobutylamine **1:** 454, 458
Diisobutyl carbamoyl chloride **2:** 1049
Diisobutyl carbinol **1:** 282
Diisobutyl ether
   physical properties **4:** 2187
Diisobutyl ketone **5:** 3097
Diisobutyl phthalate
   physical properties **7:** 3886
Diisononyl phthalate
   physical properties **7:** 3886
Diisooctyl phthalate
   physical properties **7:** 3886
Diisopentyl ether **4:** 2197
Diisopentyl phthalate
   physical properties **7:** 3886
Diisopropanolamine **4:** 2164
   see also Isopropanolamines
   toxicology **4:** 2179
Diisopropylamine **1:** 454
Diisopropylbenzene **5:** 2901
p-Diisopropylbenzene hydroperoxidation
   for production of hydroquinone **5:** 2945
   for resorcinol production **7:** 4266
p-Diisopropylbenzene oxidation
   for acetone production **1:** 100
Diisopropylbenzene peroxide
   toxicology **6:** 3676
Diisopropyl carbamoyl chloride **2:** 1049
Diisopropyl ether **4:** 2193
   physical properties **4:** 2187
   toxicology **4:** 2201
Diisopropylidene diperoxide
   physical properties **6:** 3664
Diisopropyl ketone **5:** 3096
Diisopropyl malonate **5:** 3184
   toxicology **5:** 3193
3,5-Diisopropyl-4-methylphenols **6:** 3735
2,6-Diisopropylnaphthalene **6:** 3357
Diisopropylnaphthalenes **6:** 3357
   toxicology **6:** 3360

Diisopropyl peroxydicarbonate
   physical properties **6:** 3650
   toxicology **6:** 3676
2,6-Diisopropylphenol **6:** 3732, 3733
Diisotridecyl phthalate
   physical properties **7:** 3886
Diisoundecyl phthalate
   physical properties **7:** 3886
Di-J acid **6:** 3435
Diketene
   for production of acetoacetic esters and
      amides **6:** 3604
   handling **5:** 3072
   properties **5:** 3068
   toxicology **5:** 3079
   uses **5:** 3073
Diketene–acetone adduct **5:** 3075
Diketene polymers **5:** 3071
Diketene production
   from ketene **5:** 3066
Diketene pyrolysis **5:** 3065, 3069
Diketones **5:** 3106
   fluorinated **5:** 2599
Dilatation **4:** 2434
   of fats **4:** 2434
Dilauroyl peroxide
   toxicology **6:** 3676
Dimer acids **3:** 1922
Dimerization
   electrochemical **3:** 1912
   of cyclopentadiene **3:** 1826
   of formic acid **5:** 2713
   of isoprene **5:** 3047
   of proylene **7:** 4095
Dimersol olefin process **5:** 2883
Dimersol X butene oligomerization process **2:** 989
3,3′-Dimethoxy-4,4′-biphenyl-diyldiamine, see o-dianisidine
1,1-Dimethoxybutane **1:** 395
1,1-Dimethoxyethane **1:** 395
Di(methoxyethyl) phthalate
   physical properties **7:** 3886
O,O-Dimethoxythiophosphoryl chloride **6:** 3852
1,4-Dimethoxy-2-nitrobenzene **6:** 3564
2,5-Dimethoxynitrobenzene **6:** 3564
3,3-Dimethoxy-propanenitrile **6:** 3465
Dimethylacetaldehyde, see isobutyraldehyde
N,N-Dimethylacetamide **1:** 56
2,2-Dimethylacrylic acid **2:** 1097
Dimethylacrylic acid production **2:** 1104
Dimethyl adipate
   physical properties **3:** 2014
Dimethylamine **6:** 3326, 3325
   see also methylamines
   specifications and handling **6:** 3331
   toxicology **6:** 3334
   uses **6:** 3332

5115

3-(Dimethylaminomethyl)indole   **5**: 3002
3-(*N*,*N*-Dimethylamino)phenol   **1**: 617
4-(*N*,*N*-Dimethylamino)phenol   **1**: 619
3-Dimethylamino-1-propanol   **4**: 2173
   toxicology   **4**: 2179
Dimethylaniline, *see also xylidines*
2,3-Dimethylaniline
   physical properties   **8**: 5021
2,4-Dimethylaniline
   physical properties   **8**: 5021
2,5-Dimethylaniline
   physical properties   **8**: 5021
2,6-Dimethylaniline
   physical properties   **8**: 5021
3,4-Dimethylaniline
   physical properties   **8**: 5021
3,5-Dimethylaniline
   physical properties   **8**: 5021
N,*N*-Dimethylaniline   **1**: 637, 638
Dimethylbenzenes, *see xylenes*
2,4-Dimethylbenzenesulfonamide   **2**: 765
2,5-Dimethylbenzenesulfonamide   **2**: 765
2,4-Dimethylbenzenesulfonic acid   **2**: 765
2,5-Dimethylbenzenesulfonic acid   **2**: 765
Dimethylbenzenesulfonic acids   **2**: 761
2,4-Dimethylbenzenesulfonyl chloride   **2**: 765
2,5-Dimethylbenzenesulfonyl chloride   **2**: 765
2,2′-Dimethyl-4,4′-bi-phenyl-diyldiamine   **2**: 806
3,3′-Dimethyl-4,4′-biphenyl-diyldiamine, *see o-tolidine*
2,5-Dimethyl-2,5-bis(benzoylperoxy)hexane
   physical properties   **6**: 3655
2,3-Dimethylbutadiene   **5**: 2890
2,3-Dimethylbutanal   **1**: 352
3,3-Dimethylbutanal   **1**: 352
2,2-Dimethylbutane
   physical properties   **5**: 2855
2,3-Dimethylbutane
   physical properties   **5**: 2855
2,3-Dimethyl-2,3-butanediol, *see Pinacol*
3,3-Dimethyl-2-butanone   **5**: 3090
Dimethyl carbamoyl chloride   **2**: 1049
   toxicology   **2**: 1053
Dimethyl carbonate   **2**: 1086
Dimethylcumene hydroperoxide cleavage
   for xylenol production   **3**: 1696
Dimethylcyclohexane
   physical properties   **3**: 1798
N,*N*-Dimethylcyclohexylamine   **1**: 461
Dimethyldiphenylurea
   symmetric   **8**: 4821
N,*N*-Dimethyldodecylamine   **1**: 474
Dimethylethanolamine   **4**: 2160
   see also ethanolamines, N-alkylated
   physical properties   **4**: 2160
Dimethyl ether   **4**: 1931
   physical properties   **4**: 2187
   properties   **4**: 1931

   specifications   **4**: 1934
   uses and toxicology   **4**: 1935
Dimethyl ether production   **4**:
1,1-Dimethylethylamine   **1**: 454
N,*N*-Dimethylethylamine   **1**: 455
N,*N*-Dimethylformamide
   economic aspects   **5**: 2704
   handling   **5**: 2702
   properties   **5**: 2698
   specifications   **5**: 2701
   toxicology   **5**: 2706
   uses   **5**: 2703
N,*N*-Dimethylformamide production   **5**: 2700
2,2-Dimethylglutaric acid
   physical properties   **3**: 1905
Dimethylglutaric acids   **3**: 1915
2,6-Dimethyl-4-heptanol   **1**: 282, 301
3,5-Dimethyl-4-heptanol   **1**: 282
2,6-Dimethyl-4-heptanone   **5**: 3097
(*E*, *Z*)-4-(1,5-Dimethyl-1,4-hexadienyl)-1-methyl-1-cyclohexane   **8**: 4610
(*E*, *Z*)-4-(1,5-Dimethyl-4-hexenylidene)-1-methyl-1-cyclohexene   **8**: 4610
2,5-Dimethyl-3-hexyne-2,5-diol   **1**: 323, 333
3,7-Dimethyl-7-hydroxyoctanal, *see hydroxycitronellal*
2,2-Dimethyl-3-hydroxypropanal, *see hydroxypivaldehyde*
1,2-Dimethylimidazole
   toxicology   **5**: 2994
Dimethyl isophthalate
   physical properties   **3**: 2015
Dimethylisopropanolamine   **4**: 2171
   toxicology   **4**: 2179
Dimethylketene
   physical properties   **5**: 3077
Dimethyl ketone, *see Acetone*
Dimethyl malonate   **5**: 3179
   toxicology   **5**: 3193
N,*N*-Dimethylmethanamine, *see trimethylamine*
2,4-Dimethyl-6-(1-methylcyclohexyl)phenol
   physical properties   **6**: 3754
2,2-Dimethyl-3-methylenenorbornane   **8**: 4608
2,6-Dimethylnaphthalene   **6**: 3357
1,2-Dimethyl-3-nitrobenzene   **6**: 3522
1,2-Dimethyl-4-nitrobenzene   **6**: 3522
1,3-Dimethyl-2-nitrobenzene   **6**: 3522
1,3-Dimethyl-4-nitrobenzene   **6**: 3522
1,4-Dimethyl-2-nitrobenzene   **6**: 3522
1,2-Dimethyl-5-nitroimidazole
   physical properties   **5**: 2988
N,*N*-Dimethyloctadecylamine   **1**: 474
(2*Z*)–3,7-Dimethyl-2,6-octadienal, *see neral*
2,6-Dimethyl-2,4,6-octatriene   **8**: 4600
(*Z*,*E*)-2,6-dimethyl-2,5,7-octatriene   **8**: 4600
1,4-Dimethylolcyclohexane   **1**: 334
N,*N′*-Dimethyl-2-oxo-4,5-dihydroxyimidazolidine   **8**: 4825
2,2-Dimethylpentanal   **1**: 352

2,3-Dimethylpentanal **1**: 352, 363
2,2-Dimethylpentane
　physical properties **5**: 2855
2,3-Dimethylpentane
　physical properties **5**: 2855
2,4-Dimethylpentane
　physical properties **5**: 2855
3,3-Dimethylpentane
　physical properties **5**: 2855
2,2 Dimethylpentanedioic acid, *see 2,2-dimethylglutaric acid*
2,4-Dimethyl-3-pentanol **1**: 281
2,4-Dimethyl-3-pentanone **5**: 3096
Dimethyl peroxide
　physical properties **6**: 3640
Dimethylphenol, *see xylenols*
N,N-Dimethylphenylamine, *see N,N-Dimethylaniline*
Dimethyl phosphine
　physical properties **6**: 3822
Dimethyl phthalate
　physical properties **3**: 2015; **7**: 3886
　specifications **3**: 2037
cis-2,6-Dimethylpiperidine
　physical properties **7**: 4197
2,2-Dimethylpropanal, *see pivaldehyde*
2,2-Dimethylpropane
　physical properties **5**: 2855
2,2-Dimethyl-1,3-propanediol, *see neopentyl glycol*
2,2-Dimethylpropanoic acid, *see pivalic acid*
2,2-Dimethyl-1-propanol **1**: 281; **6**: 3612
2,2-Dimethylpropenoic acid, *see 2,2-dimethylacrylic acid*
N,N-Dimethyl-4-pyridinamine **7**: 4208
　physical properties **7**: 4206
2,6-Dimethylpyridine
　physical properties **7**: 4178
3,5-Dimethylpyridine
　physical properties **7**: 4178
Dimethyl sulfate
　economic aspects **3**: 1870
　handling **3**: 1862
　physical properties **3**: 1855
　production **3**: 1857
　specifications **3**: 1860
　toxicology **3**: 1870
Dimethyl sulfide
　toxicology **8**: 4678
　uses **8**: 4661
Dimethyl sulfide production **8**: 4659
Dimethyl sulfoxide **7**: 4494
　toxicology **7**: 4496
Dimethyl terephthalate **8**: 4573
　economic aspects **8**: 4588
　physical properties **3**: 2015; **8**: 4574
　specifications **3**: 2038; **8**: 4586
　toxicology **8**: 4589
　uses **8**: 4587
Dimethyl terephthalate hydrolysis **8**: 4583

Dimethyl terephthalate production **2**: 1128; **8**: 4576
N,N-Dimethyl-*m*-toluidine **8**: 4749
N,N-Dimethyl-*o*-toluidine **8**: 4749
N,N-Dimethyl-*p*-toluidine **8**: 4749
3,3-dimethyl-1,2,4-trioxolane
　physical properties **6**: 3664
N,N'-Dimethylurea
　symmetric **8**: 4819
3,7-dimethylxanthine, *see purine derivatives*
Dinitriles
　for production of pyridines **7**: 4185
　hydrolysis **3**: 1911
4,6-Dinitro-2-aminophenol **1**: 613
2,4-Dinitroaniline **1**: 637; **6**: 3542
2,6-Dinitroaniline **6**: 3543
2,4-Dinitroanisole **6**: 3564
1,3-Dinitrobenzene **6**: 3513
m-Dinitrobenzene **6**: 3513
Dinitrobenzenes **6**: 3514
2,4-Dinitrobenzenesulfonic acid **6**: 3551
2,4-Dinitrochlorobenzene **6**: 3531
4,6-Dinitro-*o*-cresol **6**: 3566
2,6-Dinitro-*p*-cresol **6**: 3566
2,4-Dinitro-1-methylbenzene **6**: 3519
3,5-Dinitro-4-methylbenzenesulfonic acid **6**: 3555
1,5-Dinitronaphthalene **6**: 3524
1,8-Dinitronaphthalene **6**: 3524
2,4-Dinitrophenol **6**: 3563
2,4-Dinitrotoluene **6**: 3519
2,6-Dinitrotoluene-4-sulfonic acid **6**: 3555
2,4-Dinonylphenol **6**: 3748
Di(nonyl-decyl-undecyl) phthalate
　physical properties **7**: 3886
DINP
　physical properties **7**: 3886
Dioctadecylamine **1**: 473, 474
Dioctylamine **1**: 473, 474
Di(octyl-nonyl-decyl) phthalate
　physical properties **7**: 3886
Di-*n*-octyl carbamoyl chloride **2**: 1049
Di-*n*-octyl phthalate
　physical properties **7**: 3886
1,4-Diols **2**: 937
　toxicology **1**: 344
DIOP
　physical properties **7**: 3886
1,4-Dioxacyclohexane, *see dioxane*
Dioxane **4**: 1937
　economic aspects and toxicology **4**: 1943
　properties **4**: 1937
　specifications **4**: 1941
　uses **4**: 1942
1,3-Dioxane **1**: 395
1,4-Dioxane
　formation from ethylene glycol **4**: 2310
Dioxane production **4**: 1940
1,3-Dioxoisoindoline, *see phthalimide*

1,3-Dioxolane **1**: 395
　formation from ethylene glycol **4**: 2309
1,3-Dioxolan-2-one
　as intermediate in ethylene glycol production **4**: 2313
Dioxybenzene, *see resorcinol*
Dioxyethylene ether, *see dioxane*
Dioxy G acid **6**: 3399
Dioxy J acid **6**: 3399
Dioxy R acid **6**: 3400
Dipentaerythritol **1**: 324, 342
Dipentylamine **1**: 454
Di-*n*-pentyl ether **4**: 2197
2,4-Di-*tert*-pentylphenol
　physical properties **6**: 3747
2,6-Di-*tert*-pentylphenol
　physical properties **6**: 3747
Di-*n*-pentyl sulfate
　physical properties **3**: 1855
Diperoxyadipic acid
　physical properties **6**: 3643
Diperoxyadipic acid di-*tert*-butyl ester
　physical properties **6**: 3655
Diperoxycarbonic acid di-*tert*-butyl ester
　physical properties **6**: 3655
Diperoxydecanedioic acid
　physical properties **6**: 3643
Diperoxydodecanedioic acid
　physical properties **6**: 3643
　toxicology **6**: 3676
Diperoxydodecanedioic acid production **6**: 3647
Diperoxyglutaric acid
　physical properties **6**: 3643
Diperoxyisophthalic acid production **6**: 3647
Diperoxynonanedioic acid
　physical properties **6**: 3643
Diperoxyoctanedioic acid
　physical properties **6**: 3643
o-Diphenetidine **2**: 809
4,4'-Diphenoxybenzophenone **5**: 3112
4,4'-Diphenoxydiphenyl ketone **5**: 3112
Diphenyl, *see biphenyl*
N-Diphenylamine **1**: 637, 639
Diphenylamines **1**: 605
Diphenyl carbamoyl chloride **2**: 1049
　toxicology **2**: 1054
Diphenyl carbonate **2**: 1086
Diphenylene oxide **5**: 2765
Diphenyl epsilon acid **6**: 3423
Diphenyl ether **6**: 3779
Diphenyl ethers **1**: 605
N,N'-Diphenylguanidine **5**: 2838
Diphenylketene
　physical properties **5**: 3077
Diphenyl ketone **5**: 3111
Diphenylmethane **5**: 2904

Diphenyl phosphine
　physical properties **6**: 3822
Diphenyl tetramine **2**: 805
N,N'-Diphenylurea **8**: 4820
　asymmetric **8**: 4821
Diphthaloyl peroxide production **6**: 3652
DIPP
　physical properties **7**: 3886
Di-*n*-propylacetic acid **2**: 1097
　specification **2**: 1106
Dipropylamine **1**: 454
Di-*n*-propyl carbamoyl chloride **2**: 1049
Dipropylene glycol
　physical properties **7**: 4048
　toxicology **7**: 4058
　uses **7**: 4054
Dipropylene glycol dibenzoate **2**: 827
Dipropylenetriamine **1**: 488
Dipropylenetriurea **8**: 4820
Di-*n*-propyl ether **4**: 2195
　physical properties **4**: 2187
Di-*n*-propylketene
　physical properties **5**: 3077
Di-*n*-propyl sulfate
　physical properties **3**: 1855
Di-*sec*-propyl sulfate
　physical properties **3**: 1855
Disanyl **6**: 3792
Disilanes **7**: 4344
Disilazanes **7**: 4329
Disinfectants
　formaldehyde **5**: 2668
Disposal
　of chlorinated biphenyls **3**: 1437
Disproportionation
　of alkylated aromatics **2**: 710
　of silanes **7**: 4319
　of toluene **8**: 4733
Distapex toluene extraction process **8**: 4732
Distillation
　of acetic anhydride **1**: 74
　of aqueous formaldehyde solutions **5**: 2658
　of crude coal tar **8**: 4520
　of crude phenol **8**: 4544
　of crude styrene **7**: 4426
　of fatty acids **4**: 2506
　of montan wax **8**: 4911
　of phthalic anhydride **7**: 3876
　of vinyl acetate **8**: 4848
Distillative neutralization
　of fats and fatty oils **4**: 2416
Distribution coefficient
　of citric acid between water and ether **3**: 1649
Disulfides
　for sufinyl chloride production **7**: 4458
　for sulfinic acid ester production **7**: 4462
　toxicology **8**: 4677

Disulfides, aliphatic
   production **8**: 4660
   properties **8**: 4657
   uses **8**: 4661
Disulfides, aromatic
   production and uses **8**: 4666
   properties **8**: 4662
Disulfides, heterocyclic
   production **8**: 4673
   properties **8**: 4667
   uses **8**: 4674
Diterpenes
   acyclic **8**: 4613
   basic structures **8**: 4593
Ditetradecylamine **1**: 474
Dithiocarbamates
   photolysis **4**: 1967
   production **4**: 1949
   toxicology **4**: 1975
   uses **4**: 1967
Dithiocarbamic acid **4**: 1947
   formation of heterocyclic compounds **4**: 1961
   properties **4**: 1948
   toxiocology **4**: 1975
   transformation products **4**: 1953
   uses **4**: 1967
Dithiocarbamic acid esters **4**: 1957
Dithiocarbonic acid
   salts, physical properties **4**: 1950
Dithiolanylium salts **4**: 1965
Dithiols
   aromatic, production **8**: 4663
Dithiophosphates **6**: 3853
Dithiourethanes **4**: 1957
N,N'-Di-o-tolylguanidine **5**: 2838
Di(3,5,5-trimethylhexyl) phthalate
   physical properties **7**: 3886
Ditrimethylolpropane **1**: 324, 340
Diundecyl phthalate
   physical properties **7**: 3886
DIUP
   physical properties **7**: 3886
Divinylbenzene **7**: 4435
DME, see dimethyl ether
DMEP
   physical properties **7**: 3886
DMF, see N,N-Dimethylformamide
DMP
   physical properties **7**: 3886
DMSO, see dimethyl sulfoxide
DMSO aromatic separation process **2**: 722
2,4 DNA **6**: 3542
2,6 DNA **6**: 3543
DNCB **6**: 3531
DNOC **6**: 3566
DNOP
   physical properties **7**: 3886

DNT **6**: 3519
1,22-Docosanediol
   physical properties **5**: 2553
Docosanoic acid, see behenic acid
1-Docosanol, see behenyl alcohol
4,8,12,15,19-Docosapentaenoic acid, see clupanodonic acid
cis-13-Docosenoic acid, see erucic acid
trans-13-Docosenoic acid, see brassidic acid
(Z)-13-Docosen-1-ol, see erucyl alcohol
Docosylamine **1**: 473
Dodecanal, see lauric aldehyde
n-Dodecane
   physical properties **5**: 2855
Dodecanedioic acid **3**: 1791, 1917
   physical properties **3**: 1905
1,12-Dodecanedioic acid production **3**: 1910
1,12-Dodecanediol
   physical properties **5**: 2553
1-Dodecanesulfinic acid
   synthesis **7**: 4453
1-Dodecanethiol production **8**: 4654
Dodecanethiols
   toxicology **8**: 4678
n-Dodecanoic acid, see n-lauric acid
1-Dodecanol, see lauryl alcohol
1-Dodecene
   physical properties **5**: 2871
6-Dodecene
   physical properties **5**: 2871
1-Dodecene oxide **4**: 2001
1-Dodecyl Alcohol
   toxicology **5**: 2559
Dodecylamine **1**: 473
p-Dodecylbenzenesulfinic acid
   synthesis **7**: 4453
2-Dodecyl-1-hexadecanol
   physical properties **5**: 2551
1-Dodecylimidazole
   toxicology **5**: 2994
tert-Dodecyl mercaptan production **8**: 4653
Dodecylphenol **6**: 3748
   production **6**: 3750
Dodecyl stearate
   physical properties **3**: 2014
Doebner sorbic acid synthesis **7**: 4368
DOP
   physical properties **7**: 3886
DOTG **5**: 2838
1-Dotriacontanol, see lacceryl alcohol
Double-column propene–propane separation process **7**: 4086
Double decomposition
   metallic soap production **6**: 3230
Dow acetylene process **1**: 166
Dow aziridine process **2**: 667

Dow–Bayer chlorobenzene–phenol hydrolysis process **3**: 1667
DPG **5**: 2838
Dressel acid **6**: 3415
Dripolene
   for BTX production **2**: 706
Drop point **4**: 2434
Dry content
   of cellulose nitrate **2**: 1157
Dry Generators
   for acetylene production from calcium carbide **1**: 160
DSM caprolactam process **2**: 1020
DTDP
   physical properties **7**: 3886
DTPA, *see diethylenetriaminepentaacetic acid*
Dulcin **8**: 4821
DUP
   physical properties **7**: 3886
Du Pont cellulose nitrate process **2**: 1150
Du Pont *p*-phenylenediamine process **6**: 3812
Durene **5**: 2895
   physical properties **5**: 2894
Dyes
   chlorophenols **3**: 1613
   coupling components **6**: 3396
   see also naphthalene derivates
   from xylidines **8**: 5023
   phthalocyanines **7**: 3955
   pyrene-based **5**: 2923
   pyrimidines **7**: 4233
   use of arylamines **1**: 504
   use of benzotrifluorides **5**: 2611
Dynamic fatigue test
   of cellulose ester thermoplastics **2**: 1192
Dynamit Nobel dimethyl terephtalate process **8**: 4580
Dynamit Nobel–Hercules dimethyl terephthalate process **2**: 1128

EC, *see ethyl cellulose*
EDC, *see 1,2-dichloroethane*
Edeleanu aromatic separation process **2**: 718
Edetic acid, *see ethylenediaminetetraacetic acid*
EDTA, *see ethylenediaminetetraacetic acid*
EHD adiponitrile process **6**: 3460
EHEC, *see ethyl hydroxyethyl cellulose*
1,20-Eicosanediol
   physical properties **5**: 2553
Eicosanoic acid, *see arachidic acid*
1-Eicosanol, *see arachidyl alcohol*
5,8,11,14-Eicosatetraenoic acid, *see arachidonic acid*
cis-9-Eicosenoic acid **4**: 2487
Eicosylamine **1**: 473
2-Eicosyl-1-tetracosanol
   physical properties **5**: 2551
Elaidic acid
   physical properties **4**: 2487

Elaidyl alcohol **1**: 309
   physical properties **5**: 2549
Electrochemical dimerization
   of dicarboxylic acids **3**: 1912
Electrochemical fluorination **5**: 2569
Electrochemical oxidation
   of glucose **5**: 2776
Electrochemical reduction
   of nitro compounds **1**: 515
Electrode pitch production **8**: 4526
Electrode potentials
   for metals and their EDTA complexes **4**: 2299
Electrolytic reduction
   hydrazobenzene production **2**: 796
Electrophilic aromatic substitution
   of alkylphenols **6**: 3717
Electrophilic nuclear substitution
   of cresols **3**: 1659
Electrophilic substitution
   of thiophene **8**: 4693
   sulfonation **2**: 744
Electrothermic acetylene processes **1**: 146
α-Eleostearic acid
   physical properties **4**: 2487
β-Eleostearic acid
   physical properties **4**: 2487
ELF Atochem/Nippon Shokubai phthalic anhydride process **7**: 3871
Elongation
   of cellulose ester thermoplastics **2**: 1185
EMC, *see ethyl methyl cellulose*
Emersol fatty acid separation process **4**: 2514
Enamines **1**: 445
Enanthaldehyde **1**: 352, 362
Enanthic acid
   physical properties **4**: 2485
Enanthic alcohol
   physical properties **5**: 2535
n-Enantic acid **2**: 1097
Energy balance
   of a urea stripping plant **8**: 4796
Energy sources
   methanol **6**: 3312
Enheptin **6**: 3572
Enthalpies
   of saturated hydrocarbons **5**: 2856
Enthalpy of formation
   of saturated hydrocarbons **5**: 2856
Entrainers **3**: 2025
Entraining agents
   for maleic anhydride isolation **5**: 3164
Entramin **6**: 3572
Enzyme immobilization
   silanes **7**: 4346
Epal fatty alcohol process **5**: 2543
Epichlorohydrin **4**: 1999
   for cross-linking of starch **7**: 4402

for epoxidation **4:** 1998
in glycerol production **5:** 2797
Epihydrine alcohol **4:** 2003
Epoxidation
of propene **7:** 4143
with halohydrins **4:** 1997
with hydrogen peroxide **4:** 1995
with hydroperoxides **4:** 1996
with oxygen **4:** 1998
Epoxides **4:** 1987
economic aspects **4:** 2005
formation from chlorohydrins **3:** 1590
production **4:** 1992
reactions **4:** 1988
reaction with aliphatic amines **1:** 446
rearrangements **4:** 1991
toxicology **4:** 2006
Epoxidized polyoils **4:** 2005
Epoxidized soybean oil **4:** 2004
1,2-Epoxy-3-chloropropane, *see epichlorohydrin*
3,4-Epoxycyclohexanecarboxylic acid-3′4′-epoxycyclo-
  hexylmethyl ester **4:** 2004
3,4-Epoxycyclohexylmethyl-3′,4′-epoxycyclohexane carboxylate **4:** 2004
1,2-Epoxydodecane **4:** 2001
1,8-Epoxy-*p*-menthane **8:** 4604
2,3-Epoxy-2-methylbutane **4:** 2001
Epoxyoleanane saponins **7:** 4297
2,3-Epoxypinane **4:** 2002
1,2-Epoxypropane, *see propylene oxide*
2,3-Epoxypropanol **4:** 2003
Epoxysilanes **7:** 4339
3,4-Epoxytetrahydrobenzyl-3′,4′-epoxytetrahydrobenzo-
  ate **4:** 2004
Epsilon acid **6:** 3394
Erucic acid
physical properties **4:** 2487
Erucyl alcohol
physical properties **5:** 2549
ESBO **4:** 2004
Ester–ester interchange **3:** 2018
Esterification
of acrylic acid **1:** 229, 234
of alkylphenols **6:** 3716
of allyl alcohol **1:** 427
of butanols **2:** 955
of carboxylic acids **3:** 2022
of cellulose, inorganic esters **2:** 1140
of cellulose, organic esters **2:** 1168
of pentanols **6:** 3614
of propanols **7:** 4064
of terephthalic acid **8:** 4582
with dialkyl sulfates **3:** 1866
Esterification–hydrolysis ethanol process **4:** 2056
Esterified waxes **8:** 4881
Esters **3:** 2011
ammonolysis **3:** 2018

chemical properties **3:** 2015
hydrolysis **3:** 2016
natural sources **3:** 2021
physical properties **3:** 2013
production **3:** 2022
pyrolysis **3:** 2021
reduction **3:** 2019
specifications **3:** 2034
toxicology **3:** 2042
uses and economic aspects **3:** 2036
Esters hydrogenation
for alcohol production **1:** 286
Estolide formation **5:** 2959
Ethanal, *see acetaldehyde*
Ethane
direct conversion to vinyl chloride **3:** 1345
for production of 1,2-dichloroethane **3:** 1308
for production of 1,1,1-trichloroethane **3:** 1315
physical properties **5:** 2855
toxicology **5:** 2924
uses **5:** 2867
Ethane chlorination **3:** 1292
Ethanedial, *see glyoxal*
Ethanedioic acid, *see oxalic acid*
1,2-Ethanediol, *see ethylene glycol*
N,N′-1,2-Ethanediylbis[N-(carboxymethyl)glycine, *see
  ethylenediaminetetraaceticacid*
Ethanenitrile **6:** 3460
Ethane pyrolysis
cracking yield **4:** 2235
mechanism **4:** 2228
Ethanesulfinic acid
synthesis **7:** 4454
Ethanesulfonic acid
physical properties **8:** 4504
Ethanethiol
toxicology **8:** 4677
Ethanethiol production **8:** 4653
Ethanol **1:** 281, 284; **4:** 2047, 2123
analysis **4:** 2130
as ethylene feedstock **4:** 2283
as motor fuel **4:** 2118
chemical properties **4:** 2050
economic aspects **4:** 2133
for monochloroethane production **3:** 1294
physical properties **4:** 2049
process economics for synthetis and fermentation **4:** 2124
specifications and grades **4:** 2114
tolerance by bacteria **4:** 2077
tolerance by yeast **4:** 2072
toxicology **4:** 2137
uses **4:** 2131
Ethanolamines **4:** 2149
N-alkylated, production **4:** 2161
N-alkylated, properties **4:** 2158

N-alkylated, specifications, uses and economic
  aspects **4:** 2162
economic aspects **4:** 2158
production **4:** 2154
properties **4:** 2150
specifications **4:** 2156
toxicology **4:** 2177
uses **4:** 2156
Ethanol chlorination **3:** 1530
Ethanol dehydrogeneration
  for acetaldehyde production **1:** 8
Ethanol production
  by fermentation, bacteria **4:** 2075
  by fermentation, costs **4:** 2124
  by fermentation, major producers **4:** 2135
  by fermentation, processes **4:** 2082
  by fermentation, raw materials **4:** 2091
  by fermentation, yeast **4:** 2062
  from wood **4:** 2104
Ethanol purification **4:** 2112
  by azeotropic distillation **4:** 2115
  by extraction and adsorption methods **4:** 2121
  by membrane technologies **4:** 2122
Ethanol synthesis **4:** 2052
  costs **4:** 2124
  diethyl ether as byproduct **4:** 2191
Ethanol-water mixtures **4:** 2051
Ethene, *see ethylene*
Ethenesulfonic acid **8:** 4505
Ethenetetracarbonitrile **6:** 3462
Ether, *see dieethyl ether*
Etherification
  of alkylphenols **6:** 3716
  of butenes **2:** 986
  of cellulose, principles **2:** 1207
  of cellulose, process **2:** 1213
  of propanols **7:** 4065
  with dimethyl sulfate **3:** 1865
Ethers, aliphatic **4:** 2185
  properties **4:** 2185
  synthesis **4:** 2188
4-Ethoxyacetanilide **1:** 620
2-Ethoxyaniline **1:** 612
4-Ethoxyaniline **1:** 620
Ethoxyethane, *see dieethyl ether*
2-(2-Ethoxyethoxy)ethylacetate
  physical properties **3:** 2014
  specifications **3:** 2037
2-Ethoxyethyl acetate
  physical properties **3:** 2014
  specifications **3:** 2037
  toxicology **3:** 2043
Ethoxylation **4:** 2332
  of carboxylic acids **3:** 2032
  of ethylene glycol **4:** 2310
Ethoxymethylsilanes
  physical properties **7:** 4322

2-Ethoxynitrobenzene **6:** 3561
4-Ethoxynitrobenzene **6:** 3563
2-Ethoxyphenol **6:** 3781
4-Ethoxyphenylurea **8:** 4821
3-Ethoxypropylamine **4:** 2173
  toxicology **4:** 2179
Ethoxysilanes
  synthesis **7:** 4323
Ethyl acetate **1:** 53
  physical properties **3:** 2014
  specifications **3:** 2037
  toxicology **3:** 2043
Ethyl acetoacetate
  specifications **3:** 2037
Ethyl acetoacetate production **3:** 2028
Ethyl acrylate **1:** 226
  by carbonylation of acetylone **1:** 127
  physical properties **3:** 2014
  specifications **3:** 2037
Ethylal **1:** 395
Ethyl alcohol, *see ethanol*
Ethylamine **1:** 454
  uses **1:** 457
N-Ethylaniline **1:** 637
Ethylbenzene **2:** 695; **4:** 2207
  dehydrogenation to styrene **4:** 2209; **7:** 4422
  physical properties **8:** 4991
  properties **4:** 2208
  separation from C_8 mixtures **4:** 2214
  specifications **4:** 2215
  toxicology **4:** 2218
  uses and economic aspects **4:** 2217
Ethylbenzene distillation **8:** 4998
Ethylbenzene hydroperoxide
  in propylene oxide production **7:** 4147
Ethylbenzene production **4:** 2209
N-(Ethyl)benzenesulfonamide **2:** 758
Ethyl benzenesulfonate **2:** 758
4-Ethylbenzenesulfonic acid **2:** 765
Ethyl benzoate
  physical properties **3:** 2015
Ethylbenzol, *see ethylbenzene*
N-Ethyl-N-benzylaniline **1:** 637
2-Ethylbutanal **1:** 352, 362
2-Ethylbutanoic acid, *see 2-ethylbutyric acid*
2-Ethyl-1-butanol **1:** 281, 294, 295
2-Ethyl-2-butenal **1:** 368, 373
N-Ethylbutylamine **1:** 455
Ethyl *n*-butyl ether
  physical properties **4:** 2187
Ethyl butyrate
  physical properties **3:** 2014
2-Ethylbutyric acid
  specification **2:** 1106
α-ethylcaproic acid **2:** 1097, 1109
  specification **2:** 1106

Ethyl carbamate **2:** 1047
   toxicology **2:** 1053
Ethyl carbamoyl chloride **2:** 1049
N-Ethylcarbazole **2:** 1059
Ethyl cellulose **2:** 1222
Ethyl chloride, *see monochloroethane*
Ethyl chloroformate **3:** 1576
   uses **3:** 1580
Ethyl Corp. alcohol process **1:** 288
Ethyl crotonate
   physical properties **3:** 1716
Ethyl cyanide **6:** 3461
Ethyl cyanoacetate **5:** 3185
   toxicology **5:** 3193
Ethyl α-cyanoacetoacetate **1:** 67
Ethylcyclohexane
   physical properties **3:** 1798
N-Ethylcyclohexylamine **1:** 462
Ethyl dichloroacetate **3:** 1552
Ethyl N,N-dimethyl carbamate **2:** 1047
Ethylene **4:** 2221
   catalytic hydration to ethanol **4:** 2053
   cyclooligomerization **3:** 1729
   direct conversion to vinyl chloride **3:** 1343
   for methacrylic acid production **6:** 3260
   for production of 1,1,2,2-tetrachlor-
      oethane **3:** 1324
   for production of vinyl acetate **8:** 4845
   for tetrachloroethylene production **3:** 1364
   for trichloroethylene production **3:** 1359
   handling **4:** 2288
   indirect hydration to ethanol **4:** 2056
   physical properties **5:** 2871
   properties **4:** 2222
   specifications **4:** 2286
   toxicology **4:** 2290
   uses and economic aspects **4:** 2289
Ethylene–acrylic acid copolymer **8:** 4949
Ethylenebis(dithiocarbamic acid)
   salts **4:** 1950
Ethylene carbonate **2:** 1086
   as intermediate in ethylene glycol produc-
      tion **4:** 2313
Ethylene carbonylation
   propionic acid production **7:** 4106
Ethylene chlorination
   liquid-phase **3:** 1324
Ethylene chlorohydrin **3:** 1588
   toxicology **3:** 1597
Ethylene chlorohydrin production **3:** 1590
Ethylene copolymer wax
   structure and properties **8:** 4952
Ethylene cyanide **6:** 3461
Ethylene cyanohydrin **6:** 3464
   for acrylonitrile production **1:** 254
Ethylene cyanohydrin acrylic acid process **1:** 230

Ethylenediamine **1:** 482
   toxicology **1:** 493
Ethylenediaminetetraacetic acid **4:** 2295
   economic aspects **4:** 2302
   metal complexes, production **4:** 2299
   properties **4:** 2296
   toxicology **4:** 2303
   uses **4:** 2300
γ-(Ethylenediamino)propylalkoxysilanes **7:** 4336
Ethylene dibromide, *see 1,2-dibromoethane*
Ethylene dichloride, *see 1,2-dichloroethane*
Ethylene dicyanide **6:** 3461
1,2-Ethylenediurea **8:** 4820
Ethylene fractionation **4:** 2278
Ethylene furnaces
   thermal efficiency **4:** 2252
Ethylene gas phase vinyl acetate process **8:** 4847
Ethylene glycol **4:** 2305
   chemical properties **4:** 2308
   economic aspects **4:** 2322
   physical properties **4:** 2306, 2319
   specifications **4:** 2316
   toxicology **4:** 2323
   uses **4:** 2321
Ethylene glycol acetate
   toxicology **3:** 2043
Ethylene glycol butyl ether acetate
   physical properties **4:** 2321
Ethylene glycol diacetate **1:** 68
   physical properties **3:** 2014; **4:** 2321
   specifications **3:** 2037
Ethylene glycol diallyl ether **1:** 431
Ethylene glycol esters **4:** 2319
Ethylene glycol ethyl ether acetate
   physical properties **4:** 2321
Ethylene glycol methyl ether acetate
   physical properties **4:** 2321
Ethylene glycol monoacetate
   physical properties **4:** 2321
Ethylene glycol monoethers **4:** 2317
Ethylene glycol oxidation
   for production of glyoxal **5:** 2812
   oxalic acid production **6:** 3583
Ethylene glycol production **4:** 2311
Ethylene glycol–water mixtures **4:** 2307
Ethylene homopolymer wax
   structure and properties **8:** 4952
Ethylene hydrochlorination **3:** 1291
Ethylene hydroformylation **7:** 4066
Ethylene liquid phase chlorination **3:** 1300
Ethylene oligomerization
   with organoaluminum compounds **5:** 2876
   with transition-metal catalysis **5:** 2878
Ethylene oxichlorination
   gas-phase **3:** 1302
Ethylene oxidation
   for acetaldehyde production **1:** 9

mechanism of catalysis **4:** 2337
to ethylene glycol **4:** 2314
Ethylene oxide 4: 2329
economic aspects and toxicology **4:** 2350
handling **4:** 2345
properties **4:** 2330
reaction with ammonia **4:** 2154
reaction with carboxylic acids **3:** 2032
specifications **4:** 2344
uses **4:** 2349
Ethylene oxide decomposition **4:** 2332
Ethylene oxide cyclooligomerization **3:** 1729
Ethylene oxide hydrolysis
for production of ethylene glycol **4:** 2311
Ethylene oxide isomerisation **1:** 16
Ethylene oxide production **4:** 2335
air-based process **4:** 2341
environmental protection **4:** 2343
oxygen-based process **4:** 2340
Ethylene oxide solutions **4:** 2331
Ethylene oxychlorination
liquid-phase **3:** 1308
Ethylene ozonide
physical properties **6:** 3664
Ethylene polymerization **8:** 4945
Ethylene production
energy consumption **4:** 2281
environmental protection **4:** 2285
propylene as byproduct **7:** 4084
pyrolysis of hydrocarbons **4:** 2226
raw materials **4:** 2224
Ethyleneurea **8:** 4820
Ethylene–vinyl acetate copolymer **8:** 4949
Ethylenimine **2:** 663
toxicology **2:** 670
Ethyl formate **5:** 2738
physical properties **3:** 2014
Ethyl glycolate **5:** 2966
2-Ethylhexanal **1:** 353, 363
2-Ethylhexanoic acid, *see* α-*ethylcaproic acid*
2-Ethylhexanol **1:** 284; **4:** 2361
economic aspects and toxicology **4:** 2367
physical properties **5:** 2551
properties **4:** 2362
specifications **4:** 2365
uses **4:** 2366
2-Ethylhexanol production **4:** 2362
2-Ethyl-2-hexenal **1:** 368
2-Ethylhexyl acetate **1:** 54
physical properties **3:** 2014
specifications **3:** 2037
toxicology **3:** 2043
2-Ethylhexyl acrylate **1:** 226
physical properties **3:** 2014
specifications **3:** 2037
2-Ethylhexylamine **1:** 455, 458
2-Ethylhexyl chloroformate **3:** 1576

2-Ethylhexyl diphenyl phosphate
toxicology **6:** 3858
3-(2-Ethylhexyloxy)propylamine **4:** 2173
toxicology **4:** 2179
Ethyl hydroperoxide
physical properties **6:** 3634
Ethylhydroxyethylaniline **4:** 2160
see also ethanolamines, N-alkylated
physical properties **4:** 2160
Ethyl hydroxyethyl cellulose **2:** 1223
N-Ethyl-*N*-(2-hydroxyethyl)-*m*-toluidine **8:** 4749
2-Ethyl-3-hydroxyhexanal, *see butyraldol*
2-Ethyl-2-hydroxymethyl-1,3-propanediol **1:** 338
Ethylidene diacetate vinyl acetate process **8:** 4844
Ethylidenenorbornene **5:** 2892
1-Ethylimidazole
toxicology **5:** 2994
2-Ethylimidazole
physical properties **5:** 2988
toxicology **5:** 2994
Ethyl iodide **5:** 3008
Ethyl isobutyrate
physical properties **3:** 2014
Ethylketene
physical properties **5:** 3077
Ethyl lactate **5:** 3130
1-Ethyl-4-methylbenzene, *see 4-ethyltoluene*
Ethyl 4-methylbenzenesulfonate **2:** 763
Ethyl methyl cellulose **2:** 1215
2-Ethyl-4(5)-methylimidazole
physical properties **5:** 2988
toxicology **5:** 2995
2-Ethyl-3-methylpentanal **1:** 353
Ethyl methyl peroxide
physical properties **6:** 3640
5-Ethyl-2-methylpyridine
physical properties **7:** 4178
5-Ethyl-2-Methylpyridine production **7:** 4185
N-Ethylmorpholine **1:** 468
4-Ethyl-1-octyn-3-ol **1:** 309
Ethyl olefin process **5:** 2877
3-Ethylpentane
physical properties **5:** 2855
2-Ethylphenol **6:** 3730
3-Ethylphenol **6:** 3731
4-Ethylphenol **6:** 3731
Ethylphenol separation
from coal tar cracking fraction **6:** 3720
Ethyl phenyl ketene
physical properties **5:** 3077
Ethyl phenyl ketone **5:** 3111
Ethyl phosphine
physical properties **6:** 3821
Ethyl propionate
physical properties **3:** 2014
production **3:** 2029
properties **7:** 4119

Ethyl salicylate
  physical properties **3**: 2015
Ethyl silicates **7**: 4326
Ethyl stearate
  physical properties **3**: 2014
Ethyl thiocyanate
  physical properties **8**: 4632
Ethyltoluene **5**: 2898
2-Ethyltoluene
  physical properties **5**: 2894
3-Ethyltoluene
  physical properties **5**: 2894
4-Ethyltoluene
  physical properties **5**: 2894
N-Ethyl-m-toluidine **8**: 4749
N-Ethyl-o-toluidine **8**: 4749
N-Ethyl-p-toluidine **8**: 4749
Ethyl trichloroacetate **3**: 1555
Ethyl vinyl ether
  physical properties **8**: 4867
  toxicology **8**: 4874
Ethynylation **1**: 126
Eucalyptol **8**: 4604
Eutectic point
  of binary mixtures of fatty acids **4**: 2484
Expellers
  for oil seeds **4**: 2404
Explosion
  of acetylene **1**: 172
  of ethylene oxide **4**: 2345
  of methane and chlorine **3**: 1262
  of saturated hydrocarbons **5**: 2858
Explosives
  cellulose nitrate **2**: 1159
  guanidine nitrate **5**: 2831
Extractive distillation
  for butadiene isolation **2**: 913
  of isoprene from C_5 cracking fractions **5**: 3044
Extractive pitches **8**: 4532
Extractors
  for oilseeds **4**: 2406
Extreme pressure additives
  chlorinated paraffins **3**: 1406
Exxon olefin process **5**: 2882

Fabric softening
  with fatty amines **1**: 478
F acid **6**: 3393
Farnesenes **8**: 4609
Fats and fatty oils 4: 2385, 2400
  as natural source for esters **3**: 2021
  chemical properties **4**: 2395
  composition **4**: 2374
  economic aspects **4**: 2473
  environmental aspects **4**: 2431
  fractionation **4**: 2423
  handling **4**: 2440

hydrogenation **4**: 2425
interesterification **4**: 2428
physical properties **4**: 2385
refining **4**: 2412
sampling and analysis **4**: 2432
Fat splitting **4**: 2496
  for glycerol production **5**: 2793
Fatty acid anhydrides
  mixed **1**: 63, 81
Fatty acid esters
  conversion into fatty nitriles **1**: 477
Fatty acid methyl esters
  hydrogenation **5**: 2539
Fatty acid production
  fat splitting **4**: 2496
  hydrogenation **4**: 2517
  isomerization **4**: 2519
  separation by crystallization **4**: 2512
  separation by distillation **4**: 2506
Fatty acids 4: 2481, 2502
  analysis **4**: 2523
  chemical properties **4**: 2495
  contents in fats and fatty oils **4**: 2499
  environmental protection **4**: 2526
  for alcohol production **1**: 289
  handling **4**: 2524
  in fats and fatty oils **4**: 2374
  melting points **4**: 2385
  physical properties **4**: 2482
  synthetic **4**: 2521
  toxicology **4**: 2529
  uses **4**: 2527
Fatty alcohols 5: 2533
  bifunctional **5**: 2552
  economic aspects **5**: 2556
  handling **5**: 2555
  saturated, production from natural
    sources **5**: 2536
  saturated, properties **5**: 2534
  saturated, uses **5**: 2547
  specifications **5**: 2552
  Synthesis from petrochemical feedstocks **5**: 2542
  toxicology **5**: 2558
  unsaturated **5**: 2548
Fatty amines **1**: 472
  economic aspects and trade names **1**: 480
  production **1**: 475
  toxicology **1**: 494
  uses **1**: 478
Fatty nitriles
  for fatty amine production **1**: 475
Fatty oils, *see Fats and fatty oils*
FCC offgas
  recovery of ethylene **4**: 2283
Fenoprop **3**: 1620
Fenticlor **6**: 3791

Fermentation
 bacteria **4**: 2075
 batch processes **4**: 2081
 citric acid production **3**: 1651
 continuous processes **4**: 2082
 direct, of polysaccharides **4**: 2074
 of biomass, for acetone production **1**: 100
 of glucose to D-gluconic acid **5**: 2777
 of glucose to lactic acid **5**: 3124
 of hexoses by yeast **4**: 2063
 of pentoses by bacteria **4**: 2080
 of saccharified waste starch **4**: 2078
 of saturated hydrocarbons **5**: 2866
 raw materials **4**: 2091
 tartaric acid production **8**: 4565
 with flocculating cells **4**: 2084
 with immobilized cells **4**: 2089
 yeast **4**: 2062
 yeast, effect of ethanol **4**: 2072
 yeast, effect of pH and temperature **4**: 2073
 yeast, effect of substrate concentrations **4**: 2071
Fermentation reactors **4**: 2085
Fertilization
 urea **8**: 4816
Fiber cellulose acetate process **2**: 1172
Fibers
 cellulose acetate **2**: 1181
Film evaporation
 of fatty acids **4**: 2506
Filter low
 cellulose acetate **2**: 1181
Filtrex oilseed extraction process **4**: 2409
Fire extinguishing agents
 bromofluoromethanes **5**: 2581
Fire point **4**: 2434
Fire retardants
 bromine compounds **2**: 882
 chlorinated paraffins **3**: 1405
 guanidine salts **5**: 2831
Fischer projection **2**: 1063
Fischer–Tropsch paraffins **8**: 4883
 production **8**: 4942
 properties and uses **8**: 4943
Fish oils **4**: 2472
 fatty acid content **4**: 2376
 manufacture and processing **4**: 2411
Fixed-bed reactors
 for oxychlorination of ethylene **3**: 1304
Flameproofing agents
 phosphine oxides **6**: 3834
 trialkylphosphates **6**: 3849
Flame trap
 for acetylene lines **1**: 176
Flammability
 of aliphatic carboxylic acids **2**: 1107
Flash distillation
 of coal tar **8**: 4521

Flash evaporation
 of fatty acids **4**: 2506
Flash point **4**: 2434
 of ethanol–water mixtures **4**: 2051
 of formaldehyde solutions **5**: 2666
 of saturated hydrocarbons **5**: 2858
Flavors
 benzaldehyde **2**: 683
 use of esters **3**: 2041
Flexural stress
 of cellulose ester thermoplastics **2**: 1187
Flotation reagents
 fatty amines **1**: 478
 xanthates **8**: 4979
Flow point **4**: 2434
Fluid-catalytic cracking process
 propylene production **7**: 4089
Fluidized-bed oxidation
 of naphthalene **7**: 3876
Fluidized-bed reactors
 for oxychlorination of ethylene **3**: 1305
Fluoral **5**: 2598
Fluoranthene **5**: 2919
Fluorene **5**: 2917
9-Fluorenone **5**: 2918
Fluorescent dyes
 from fluoranthene **5**: 2920
Fluorinated synthons **5**: 2571
Fluorinating agents **5**: 2569
Fluorination
 addition of hydrogen fluoride **5**: 2571
 substitution of halogen **5**: 2570
 substitution of hydrogen **5**: 2569
 with nonmetal fluorides **5**: 2572
Fluorine compounds, organic **5**: 2565
 nomenclature **5**: 2567
 production processes **5**: 2568
 toxicology **5**: 2622
Fluoroacetic acid
 toxicology **5**: 2625
Fluoroacetic acids **5**: 2600
Fluoroacetones **5**: 2596
Fluoroalcohols **5**: 2590
 toxicology **5**: 2624
Fluoroaldehydes **5**: 2596
Fluoroalkanes **5**: 2573
 toxicology **5**: 2623
Fluoroaromatic compounds **5**: 2613
Fluorobenzene **5**: 2613
 production **5**: 2614
Fluorocarboxylic acids **2**: 1116
Fluoroepoxides **5**: 2592
Fluoroethane
 properties **5**: 2574
Fluoroethers **5**: 2591
Fluoroethylene
 physical properties **5**: 2584

Fluoroketones **5**: 2596
   toxicology **5**: 2625
Fluoromethane
   properties **5**: 2574
Fluoronitrobenzenes **6**: 3536
Fluoronitrotoluenes **6**: 3536
Fluoroolefins **5**: 2583
   toxicology **5**: 2624
Fluoroorganosilanes **7**: 4334
2-Fluoropyridine
   physical properties **7**: 4199
Fluoropyridines **5**: 2619
Fluoropyrimidines **5**: 2620
2-Fluorotoluene **5**: 2613
3-Fluorotoluene **5**: 2613
4-Fluorotoluene **5**: 2613
Fluorotriazines **5**: 2621
Fluorovinyl ethers **5**: 2594
Fluxing oils **8**: 4539
Folic acid **2**: 834
Food additives
   cellulose ethers **2**: 1242
   lactic acid **5**: 3128
Food applications
   of lecithin **5**: 3156
Food preservatives
   propionic acid **7**: 4111
Foot oils **8**: 4918
Formaldehyde **1**: 352; **5**: 2639
   see also aldehydes, aliphatic
   chemical properties **5**: 2646
   economic aspects **5**: 2668
   environment protection **5**: 2660
   for production of pyridines **7**: 4182
   for THF production **8**: 4623
   from methanol, economic aspects **6**: 3311
   handling **5**: 2666
   liquid monomeric **5**: 2659
   physical properties **5**: 2641
   polymers **5**: 2671
   specifications and analysis **5**: 2664
   toxicology **5**: 2670
   trimerization **5**: 2678
   uses **5**: 2667
Formaldehyde cyanohydrin **5**: 2681; **6**: 3463
Formaldehyde diethyl acetal **1**: 395
Formaldehyde dimethyl acetal **1**: 395
Formaldehyde – methanol solutions
   aqueous **5**: 2644
Formaldehyde production **5**: 2647
   process economics **5**: 2656
Formaldehyde solutions
   distillation **5**: 2658
   handling **5**: 2666
   physical properties **5**: 2642
Formamide
   properties **5**: 2692

specifications **5**: 2696
   toxicology **5**: 2705
   uses and economic aspects **5**: 2697
Formamide hydrolysis
   for formic acid production **5**: 2727
Formamide production **5**: 2694
Formamides **5**: 2691
   for production of nitriles **6**: 3459
Formamidinesulfinic acid **7**: 4468
   uses **7**: 4470
Formanilide **1**: 637; **5**: 2705
Formates **5**: 2738
   for production of formic acid **5**: 2728
Formazan formation **3**: 1893
Formex aromatic separation process **2**: 724
Formex toluene extraction process **8**: 4732
Formic acid **5**: 2711
   chemical properties **5**: 2716
   concentration **5**: 2730
   economic aspects **5**: 2735
   environmental protection and specifica-
      tions **5**: 2732
   handling **5**: 2733
   physical properties **5**: 2712
   toxicology **5**: 2740
   uses **5**: 2734
Formic acid esters **5**: 2735
Formic acid production **5**: 2718
   construction materials **5**: 2731
Formox formaldehyde process **5**: 2654
Formylation
   of aromatic amines **6**: 3469
4-Formyl-1,3-benzenedisulfonic acid **2**: 767
2-Formylbenzenesulfonic acid **2**: 767
4-Formylbenzenesulfonic acid **2**: 767
1-Formyl-1-methylethanesulfinyl chloride
   synthesis **7**: 4461
N-Formylmorpholine **1**: 468; **5**: 2705
   for aromatic extraction **2**: 724
2-Formyl-5-nitrofuran **6**: 3571
Forward integration
   of toluene production into chemical pro-
      cesses **8**: 4735
   of xylene production into chemical proces-
      sing **8**: 5005
Fossil waxes **8**: 4881
Foster – Wheeler fat splitting process **4**: 2503
Foundry pitches **8**: 4534
Fractional crystallization of fatty acids **4**: 2513
Fractionation
   for benzene separation **2**: 716
   of C_2 components in ethylene production **4**: 2278
   of cracked gas in ethylene production **4**: 2267
   of fats and fatty oils **4**: 2423
   of fatty acids **4**: 2508
   of hydrocarbons in ethylene production **4**: 2270
Fragrance aldehydes **1**: 359

5127

Fragrances
   use of esters **3:** 2041
Frank–Caro batch oven calcium cyanamide process **3:** 1740
Freezing point
   of acetic and chloroacetic acid mixtures **3:** 1542
   of 1- and 2-propanol–water mixtures **7:** 4063
   of binary systems of fatty acids **4:** 2488
   of citric acid solutions **3:** 1648
   of ethanol–water mixtures **4:** 2051
   of formic acid–water binary mixtures **5:** 2712
   of glycerol–water solutions **5:** 2790
   of mono- and diethylene glycol –water mixtures **4:** 2307
Freund's acid(1,3,6) **6:** 3416
Freund's acid(1,3,7) **6:** 3416
Friedel–Crafts catalysts **6:** 3721
Friedel–Crafts reaction
   for production of araliphatic aldehydes **1:** 383
   for production of chlorophosphines **6:** 3826
Front-end deethanizer **4:** 2272
Front-end demethanizer **4:** 2272
Front-end depropanizer **4:** 2273
Front-end hydrogenation
   of acetylene in ethylene production **4:** 2277
Front-end process
   in ethylene production from hydrocarbon pyrolysis **4:** 2266
Fructose
   fermentation in yeasts **4:** 2068
Fruit crops
   for fermentation **4:** 2095
Fruit ethers **3:** 2022
Fruit pulp fats **4:** 2441
L-Fucose **2:** 1066
Fuel
   methanol **6:** 3312
   propene **7:** 4095
Fuel components
   MTBE as antiknock agent **6:** 3344
   toluene **8:** 4731
   xylenes **8:** 4993
Fuel oils **8:** 4537
Fujiyama calcium cyanamide process **3:** 1742
Fumaric acid **5:** 3170
   toxicology **5:** 3173
Fumaric acids **5:** 3159
Fungicides
   copper soaps **6:** 3242
2-Furaldehyde, *see furfural*
Furan **5:** 2745
   properties **5:** 2745
   specifications and handling **5:** 2749
   toxicology **5:** 2767
2-Furancarboxaldehyde, *see furfural*
2,5-Furandione, *see maleic anhydride*
2-Furanmethanol, *see furfuryl alcohol*

Furan nitro derivatives **6:** 3571
Furanol, *see furfuryl alcohol*
Furanoses **2:** 1063
Furcellaran **7:** 4003
Furfural **2:** 1079
   physical properties **5:** 2747
   properties **5:** 2750
   raw materials **5:** 2752
   specifications **5:** 2754
   toxicology **5:** 2767
   uses **5:** 2755
Furfural alcohol
   physical properties **5:** 2747
Furfuraldehyde, *see furfural*
Furfural production **5:** 2752
Furfuran, *see furan*
Furfuryl alcohol
   properties **5:** 2756
   specifications and handling **5:** 2758
   toxicology **5:** 2767
   uses and economic aspects **5:** 2759
Furnace design
   for hydrocarbon pyrolysis **4:** 2239
Furostanol saponins **7:** 4291
Fusel oils
   composition **4:** 2116

G acid **6:** 3397
Gadoleic acid
   physical properties **4:** 2487
Galactomannans **7:** 4023
   derivates and applications **7:** 4027
   production and properties **7:** 4025
   structure **7:** 4024
4-*O*-β-D-Galactopyranosyl-D-fructose, *see lactulose*
4-*O*-β-D-Galactopyranosyl-α-D-glucopyranose, *see lactose*
α-D-Galactopyranosyluronic acid
   in pectin **7:** 3976
D-Galactose **2:** 1065
Galacturonans, *see pectin*
Galam butter **4:** 2451
Gallic acid **5:** 2981
   decarboxylation **6:** 3764
Gasification of coal
   for production of saturated hydrocarbons **5:** 2861
Gasification tars
   properties and composition **8:** 4513
Gas oil pyrolysis
   cracking yield **4:** 2238
Gas oils
   as raw material for ethylene production **4:** 2225
Gasoline
   1,2-dibromoethane content **2:** 874
Geddyl alcohol
   physical properties **5:** 2535

Gelation
   of cellulose nitrate **2:** 1153
   of methyl cellulose **2:** 1218
   of pectin **7:** 3990
Gellan gum **7:** 4023
Gelose, *see agar*
Gels
   agar **7:** 4011
   alginate **7:** 3999
   carrageenan **7:** 4006
   pectin **7:** 3984
   starch **7:** 4384
   xanthans **7:** 4021
Gentisic acid **5:** 2981
Geranial **1:** 368
Gersthofen montan wax oxidation process **8:** 4911
GfT/Koppers continuous tar distillation process **8:** 4522
Ghedda wax **8:** 4905
Girdler fatty oil deodorizer **4:** 2421
D-Gluconic acid **5:** 2773
   economic aspects and specifications **5:** 2782
   properties **5:** 2774
   uses **5:** 2780
Gluconic acid lactones **5:** 2775
D-Gluconic acid production **5:** 2776
Gluconobacter suboxydans process **5:** 2779
D-Glucose **2:** 1065
   fermentation by yeast **4:** 2065
Glucose oxidase **5:** 2778
Glucose oxidation
   for production of D-gluconic acid **5:** 2776
Glutaraldehyde
   from acrolein **1:** 213
Glutardialdehyde **1:** 390, 392
Glutaric acid **3:** 1914
   as byproduct in adipic acid production **1:** 268
   physical properties **3:** 1905
Glutaronitrile
   physical properties **6:** 3457
Glycan **2:** 1068
(*R,S*)-Glyceric acid
   physical properties **5:** 2958
Glyceride hydrolysis **4:** 2498
Glycerides **4:** 2394
   see also fats and fatty oils; fatty acids
   conversion into fatty nitriles **1:** 477
   in fats and fatty oils **4:** 2374
   melting points **4:** 2385
Glyceride splitting
   for fatty acid production **4:** 2496
Glycerin, *see glycerol*
Glycerin anhydride **4:** 2003
Glycerol **5:** 2787
   chemical properties **5:** 2792
   economic aspects and toxicology **5:** 2806
   environmental protection **5:** 2801

   handling **5:** 2803
   physical properties **5:** 2788
   purification and refining **5:** 2794
   specifications **5:** 2802
   uses **5:** 2804
Glycerol chlorohydrins **3:** 1588
   production **3:** 1593
Glycerol dichlorohydrins **3:** 1588
   production **3:** 1593
Glycerol esters **5:** 2804
Glycerol ethers **5:** 2805
Glycerol fatty esters **5:** 2805
Glycerol production
   from fats and oils **5:** 2793
   from propene **5:** 2797
Glycerol trinitrate **5:** 2804
Glycerol–water solutions **5:** 2789
Glyceryl triacetate
   physical properties **3:** 2014
   specifications **3:** 2037
   toxicology **3:** 2043
Glyceryl tripropionate **3:** 2037
   physical properties **3:** 2014
   toxicology **3:** 2043
Glycide **4:** 2003
Glycidic ester
   for production of araliphatic aldehydes **1:** 383
Glycidol **4:** 2003
   for epoxidation **4:** 1999
Glycidyl alcohol **4:** 2003
Glycinonitrile **6:** 3463
Glycoalkaloids **7:** 4292
Glycogen **2:** 1068
Glycol, *see ethylene glycol*
Glycol aldehyde **1:** 375
Glycol benzoates **2:** 827
Glycol ethers
   from 1-propanol **7:** 4071
Glycolic acid **5:** 2964
   physical properties **5:** 2958
   toxicology **5:** 2970
Glycolic acid esters **5:** 2966
Glycolic nitrile **5:** 2681
Glycolonitrile **5:** 2681; **6:** 3463
Glycolysis
   in yeast **4:** 2065
N-Glycosides **2:** 1076
   synthesis **2:** 1076
Glyoxal **1:** 390; **5:** 2809
   properties **5:** 2809
   toxicology **5:** 2814
   uses **5:** 2813
Glyoxalic acid, *see glyoxylic acid*
Glyoxal oligomers **5:** 2810
Glyoxal oxidation **5:** 2821
Glyoxal production **5:** 2812

Glyoxylic acid 5: 2817
   properties 5: 2817
   toxicology 5: 2821
Glyoxylic acid production 5: 2821
Goldenberg–Géromont method 8: 4566
Goodyear acetone process 1: 100
Goodyear–Scientific Design isoprene synthesis 5: 3040
Goose fat 4: 2471
   fatty acid content 4: 2376
Gorli oil 4: 2467
Gramine 5: 3002
Granulation
   of urea 8: 4812
Grape-seed oil 4: 2460
   fatty acid content 4: 2376
Graphite hydrogen sulfate
   as catalyst in esterification of carboxylic acids 3: 2024
Grignard organosilane synthesis 7: 4314
Grignard reaction
   for production of tertiary phosphine oxides 6: 3832
   reaction with sulfoxides 7: 4492
   with bromine 2: 870
Grinding process
   of CuPc pigments 7: 3942
Groundnut oil 4: 2462
Growth factor
   in oligomerization of ethylene 5: 2880
Guaiacol 6: 3780
Guanamines 5: 3197, 3214
   production 5: 3215
   toxicology 5: 3217
Guanidine 3: 1748; 5: 2823
   organic derivatives 5: 2835
   organic derivatives, toxicology 5: 2841
   properties 5: 2824
   toxicology 5: 2841
   uses 5: 2830
Guanidine bases 5: 2836
Guanidine salts
   biotechnological applications 5: 2831
   economic aspects 5: 2832
   environmental protection and specification 5: 2829
   physical properties 5: 2825
   production 5: 2826
   toxicology 5: 2841
Guanine, see purine derivatives
Guar gum 7: 4023
   applications 7: 4027
   production and properties 7: 4025
Guerbet alcohol 5: 2550
Guerbet alcohol process 1: 292
Gulf oil olefin process 5: 2877
Gulf oxychlorination phenol process 6: 3702

Gulf toluene oxychlorination 3: 1676
D-Gulose 2: 1065
Gum arabic 2: 1068; 7: 4012
   structure and properties 7: 4013
Gum ghatti 7: 4017
Gum karaya 7: 4016
Gum tragacanth 7: 4015
Gum turpentine 8: 4761
   composition 8: 4768
Gum turpentine production 8: 4762
Guncotton 2: 1071

H acid 6: 3439
Halane 3: 1569
Halcon acetic anhydride process 1: 77
Halcon aniline process 1: 632
Halcon arylamine process 1: 524
Halcon propylene oxide process 7: 4147
Haldor Topsøe methanol process 6: 3304
Halex fluorobenzene process 5: 2615
Halide exchange
   nucleophilic aromatic 1: 521
Halides
   activated aromatic 1: 521
Halogenation
   of arylamines 1: 510
Halogen–fluorine exchange 5: 2570
Halogenohydrins
   for epoxidation 4: 1997
Halogenonitroanilines 6: 3546
Halogenophenols 6: 3782
Halogenophosphines 6: 3825
   toxicology 6: 3856
Halogenopyridines 7: 4198
Halogenosilanes 7: 4321
   synthesis 7: 4310
Halogenothiophenols
   production 8: 4663
Hantzsch pyrrole syntheses 7: 4237
Hardening
   of fats and fatty oils 4: 2398
   of fatty acids 4: 2517
Hard microwaxes 8: 4933
   uses 8: 4939
Hard pitch 8: 4530
Haworth projection 2: 1063
Hazen color number 2: 1038
HBCD, see hexabromocyclododecane
HBMC, see hydroxybutyl methyl cellulose
HCN tetramer 6: 3465
HDA benzene process 2: 709
Heat capacities, of saturated hydrocarbons 5: 2857
Heat distortion temperature
   of cellulose ester thermoplastics 2: 1189
Heat of fusion
   of fatty acids 4: 2489
Heat-pumped C_2 fractionation 4: 2278

Heat pump propene–propane separation process **7:** 4087
Heat transfer agents **5:** 2910
    dibenzofuran **5:** 2766
    terphenyl **5:** 2912
Heavy oil **8:** 4549
Heavy-solvent extraction process **1:** 234
HEC, *see hydroxyethyl cellulose*
HEEDTA, *see hydroxyethylethylenediamine tricetic acid*
HEHPC **2:** 1229
HEMC, *see hydroxylethyl methyl cellulose*
Hemimellitene **5:** 2893
    physical properties **5:** 2894
Hemimellitic acid **2:** 1121
Hemimellitic anhydride **2:** 1130
Hempseed oil **4:** 2459
1,21-Heneicosanediol
    physical properties **5:** 2553
1-Heneicosanol
    physical properties **5:** 2535
1-Heneicosene
    physical properties **5:** 2871
Henkel cortsoxylic acid procedure **2:** 1126
Henkel fatty acid distillation process **4:** 2508
Henkel fatty acid separation process **4:** 2514
1-Hentriacontanol, *see melissyl alcohol*
Heparin **2:** 1068
HEPMC **2:** 1235
1-Heptacosanol
    physical properties **5:** 2535
n-Heptadecane
    physical properties **5:** 2855
1,17-Heptadecanediol
    physical properties **5:** 2553
Heptadecanoic acid, *see margaric acid*
1-Heptadecanol, *see margaryl alcohol*
1-Heptadecene
    physical properties **5:** 2871
2-Heptadecyl-1-heneicosanol
    physical properties **5:** 2551
3-(Heptafluorobutyryl)camphor **5:** 2602
1,1,1,2,2,3,3-Heptafluoro-7,7-dimethyl-4,6-octanedione **5:** 2602
3-(2,2,3,3,4,4,4-heptafluoro-1-oxo-butyl)bicyclo[2,2,1]-heptan-2-one **5:** 2602
Heptanal, *see enanthaldehyde*
n-Heptane
    physical properties **5:** 2855
    toxicology **5:** 2924
Heptanedioic acid, *see oxalic acid*
1,7-Heptanediol
    physical properties **5:** 2553
Heptanoic acid, *see enanthic acid*
    production **4:** 2522
1-Heptanol, *see enanthic alcohol*
2-Heptanone **5:** 3091
3-Heptanone **5:** 3097

1-Heptene
    physical properties **5:** 2871
cis-3-Heptene
    physical properties **5:** 2871
cis,trans-2-Heptene
    physical properties **5:** 2871
trans-3-Heptene
    physical properties **5:** 2871
2-Heptyl-1-undecanol
    physical properties **5:** 2551
Herbicides
    based on propionic acid **7:** 4122
    chlorophenoxyalkanoic acids **3:** 1620
    from guanidine **5:** 2830
    from imidazoles **5:** 2991
    of quaternary pyridinium salts **7:** 4194
Hercules cellulose nitrate process **2:** 1151
Hercules cumene hydroperoxide cleavage process **1:** 96
Hercules cumene oxidation process **1:** 95
Heterocycles
    from isothiocyanates **8:** 4637
Heterolactic fermentation **5:** 3124
Hexabromocyclododecane **2:** 886
    toxicology **2:** 892
Hexachlorobenzene
    chemical properties **3:** 1412
    physical properties **3:** 1410
Hexachlorobenzene production **3:** 1420
Hexachlorobutadiene **3:** 1395
    toxicology **3:** 1478
Hexachlorocyclopentadiene
    uses **3:** 1831
Hexachloroethane
    properties **3:** 1328
    toxicology **3:** 1474
Hexachloroparaldehyde **3:** 1531
Hexachlorophene **6:** 3790
Hexachlorophenol **3:** 1608
Hexacosanoic acid, *see cerotic acid*
1-Hexacosanol, *see ceryl alcohol*
Hexadecachloro copper phthalocyanine production **7:** 3945
n-Hexadecane
    physical properties **5:** 2855
1,16-Hexadecanediol
    physical properties **5:** 2553
Hexadecanoic acid, *see palmitic acid*
1-Hexadecanol, *see cetyl alcohol*
    toxicology **5:** 2559
1-Hexadecene
    physical properties **5:** 2871
cis-9-Hexadecenoic acid, *see palmitoleic acid*
Hexadecylamine **1:** 473
2-Hexadecyl-1-eicosanol
    physical properties **5:** 2551

Hexadecyl stearate
 physical properties **3:** 2014
trans,trans-2,4-Hexadienoic acid, *see sorbic acid*
Hexaethylbenzene
 physical properties **5:** 2894
Hexafluoroacetone **5:** 2596
Hexafluorobenzene **5:** 2618
Hexafluoroethane
 properties **5:** 2574
Hexafluoroisobutene **5:** 2589
1,1,1,5,5,5-Hexafluoro-2,4-pentanedione **5:** 2602
1,1,1,3,3,3-Hexafluoro-2-propanol **5:** 2590
1,1,1,3,3,3-Hexafluoro-2-propanone **5:** 2596
Hexafluoropropene **5:** 2586
 physical properties **5:** 2584
Hexafluoropropylene oxide **5:** 2592
Hexahydroazepine **1:** 466
Hexahydrobenzoic acid **2:** 836
Hexahydro-1*H*-azepin-2-one, *see ε-Caprolactam*
Hexahydro-*p*-xylylene glycol **1:** 334
1-Hexaicosanol **1:** 282
Hexamethylbenzene, *see mellitene*
Hexamethyldisilane **7:** 4344
Hexamethylene carbamoyl chloride **2:** 1049
Hexamethylenediamine **5:** 2845
 handling and economic aspects **5:** 2849
 properties **5:** 2846
 specifications **5:** 2848
 toxicology **5:** 2850
Hexamethylenediamine production **5:** 2847
Hexamethylenediurea **8:** 4820
Hexamethylene glycol **1:** 331
Hexamethyleneimine **1:** 466
 toxicology **1:** 493
Hexamethylenetetramine **1:** 470
 toxicology **1:** 493
2,6,10,15,19,23-Hexamethyltetracosa-2,*t*-6,*t*-10,*t*-14,*t*-18,*r*-22-hexaene **8:** 4613
Hexamine, *see hexamethylenetetramine*
Hexanal, *see capraldehyde*
Hexanal diethyl acetal **1:** 395
Hexane
 for solvent extraction of oil seeds **4:** 2405
 physical properties **5:** 2855
 toxicology **5:** 2924
 uses **5:** 2868
Hexanedial, *see adipinaldehyde*
1,6-Hexanediamine, *see hexamethylenediamine*
Hexanedinitrile, *see adiponitril*
Hexanedioic acid, *see adipic acid*
Hexanedioic acid di-2-propenyl ester **1:** 426
1,2-Hexanediol **1:** 337
1,6-Hexanediol **1:** 323, 331
 physical properties **5:** 2553
2,3-Hexanedione **5:** 3109
Hexane-insoluble matter
 in lecithin **5:** 3154

Hexane-1-sulfonic acid
 physical properties **8:** 4504
Hexanoic acid, *see caproic acid*
Hexanol, *see caproic alcohol*
 toxicology **5:** 2558
2-Hexenal **1:** 368
trans-2-Hexenal **1:** 372
1-Hexene
 physical properties **5:** 2871
cis-2-Hexene
 physical properties **5:** 2871
cis-3-Hexene
 physical properties **5:** 2871
trans-2-Hexene
 physical properties **5:** 2871
trans-3-Hexene
 physical properties **5:** 2871
Hexogen (1,3,5-trinitrohexahydro-1,3,5-triazine) **1:** 70
Hexoses **2:** 1063
 fermentation pathways **4:** 2063
Hexuloses **2:** 1063
Hexylamine **1:** 454
n-Hexyl carbamate **2:** 1047
α-Hexylcinnamaldehyde **1:** 382
α-Hexylcinnamaldehydes **1:** 389
2-Hexyl-1-decanol
 physical properties **5:** 2551
 toxicology **5:** 2559
3-Hexyne-2,5-diol **1:** 323, 332
Hexynediols **1:** 332
1-Hexyn-3-ol **1:** 309
HFIB **5:** 2589
HFPO **5:** 2592
High-amylose starches **7:** 4388
High-pressure C_2 fractionation **4:** 2278
High-temperature carbonization
 for benzene production **2:** 701
High-temperature coking
 for benzene production **2:** 701
High-temperature pyrolysis acetylene process **1:** 164
Hilum **7:** 4382
7*H*-Imidazo[4,5-*d*]pyrimidine, *see purine*
7*H*-Imidazo*d*]pyrimidine, *see purine derivatives*
Hinsberg test **1:** 445
Hitachi two-stage quench-cooler system **4:** 2262
HM-pectin gelation **7:** 3985
HMTA, *see hexamethylenetetramine*
HNP
 physical properties **7:** 3886
HNUP
 physical properties **7:** 3886
Hock phenol-acetone process **1:** 93; **6:** 3694
Hock reaction
 of toluene derivatives **3:** 1679
Hoechst acetaldehyde process **1:** 10
Hoechst chloromethane process **3:** 1268

Hoechst high-pressure chlorinolysis tetrachloromethane process **3**: 1275
Hoechst high-temperature pyrolysis acetylene process **1**: 164
Hoechst multistep chlorination tetrachloromethane process **3**: 1276
Hoechst oxidation acetaldehyde process **1**: 10
Hoechst plasma arc acetylene process **1**: 153
Höls plasma arc acetylene process **1**: 153
Hofmann degradation **1**: 477; **2**: 1125
Holothurins **7**: 4300
Homolactic fermentation **5**: 3124
Homologation
  of alcohols **1**: 290
  of methanol, with synthesis gas **4**: 2059
Homopiperidine **1**: 466
Homopolymerization
  of aziridines **2**: 664
Honshu Chemical Industry *m*-cresol process **3**: 1666
Hopanes **7**: 4297
Horse fat **4**: 2471
  fatty acid content **4**: 2376
Hot acid olefin process **5**: 2884
Houben–Fischer reaction **6**: 3469
Houdry Catadiene butadiene process **2**: 908
Houdry Catofin propene process **7**: 4091
Houdry-Litol BTX process **2**: 702
HPC, *see hydroxypropyl cellulose*
HPMC, *see hydroxypropyl methyl cellulose*
HPN **1**: 329
HPO caprolactam process **2**: 1020
H-silanes **7**: 4329
HTC 1,2-dichloroethane process **3**: 1301
Hüls arc acetylene process **1**: 147
Hüls chloromethane process **3**: 1269
Hüls MTBE process **6**: 3340
Hüls nonylphenol process **6**: 3750
Hüls plasma arc acetylene process **1**: 153
HXODP
  physical properties **7**: 3886
Hydeal benzene process **2**: 709
Hydnocarpus oils **4**: 2467
Hydracrylic acid **5**: 2967
  physical properties **5**: 2958
Hydracrylonitrile **6**: 3464
Hydration
  of butenes **2**: 986
  of cyclohexane **3**: 1813
  of ethylene to ethanol, direct **4**: 2053
  of ethylene to ethanol, indirect **4**: 2056
  of olefins to aliphatic alcohols **1**: 290
  of pentenes **6**: 3617
  of propene **7**: 4067
Hydration degumming
  of fats and fatty oils **4**: 2413
Hydrazine
  for reduction of aromatic nitro compounds **1**: 516

Hydrazobenzene
  in benzidine production **2**: 794
  rearrangement **2**: 797
Hydrindene **5**: 2916
Hydrocarbon conversion **5**: 2862
Hydrocarbon cracking, *see hydrocarbon pyrolysis*
Hydrocarbon fractionation
  in ethylene production **4**: 2270
Hydrocarbon oxidation
  for aliphatic aldehyde production **1**: 359
  for maleic anhydride production **5**: 3165
  for production of aliphatic alcohols **1**: 288
  for propionic acid production **7**: 4108
Hydrocarbon pyrolysis
  acetylene as byproduct **1**: 167
  coking and decoking of furnaces **4**: 2254
  coproducts from ethylene production **4**: 2265
  cracking yield and recovery **4**: 2264
  cracking yields in ethylene production **4**: 2234
  energy consumption **4**: 2281
  ethylene production **4**: 2226
  furnace design **4**: 2230, 2239
  heat requirement **4**: 2233
  isolation of butadiene **2**: 908
  mechanism **4**: 2228
  propylene production **7**: 4084
  quenching **4**: 2256
  thermal efficiency of ethylene furnaces **4**: 2252
  toluene production **8**: 4732
  xylene production **8**: 4997
Hydrocarbons **5**: 2853
  alkylbenzenes **5**: 2893
  biphenyls and polyphenyls **5**: 2906
  chlorinated, *see chlorinated hydrocarbons*
  from coal tar **5**: 2913
  saturated, *see alkanes*
  toxicology **5**: 2924
  unsaturated, *see olefins*
  volatility **2**: 911
Hydrocarbonylation
  of methanol **4**: 2059
  of methyl acetate for acetaldehyde production **1**: 16
Hydrocarboxylation
  of olefins for fatty acid production **4**: 2522
Hydrocarboxymethylation
  for production of aliphatic alcohols **1**: 291
Hydrochlorination
  of acetylene **3**: 1334
  of ethylene **3**: 1291
  of methanol **3**: 1264
  of methanol, liquid-phase **3**: 1266
Hydrocracker residue pyrolysis
  cracking yield **4**: 2238
Hydrocracking **5**: 2863
Hydrocyanation **6**: 3459

5133

Hydrocyanic acid
    chlorination **3**: 1775
    for acrylonitrile production **1**: 254
Hydrodealkylation
    for benzene production **2**: 708
    of xylenols **3**: 1681
Hydrodechlorination
    of polychlorophenols **3**: 1610
Hydroformylation
    of allyl alcohol **1**: 421
    of butenes **6**: 3616
    of dienes, for production of dialdehydes **1**: 391
    of ethylene **7**: 4066
    of methanol, for acetaldehyde production **1**: 16
    of olefins for fatty acid production **4**: 2521
    of olefins, for fatty alcohol production **5**: 2544
    of propylene **2**: 926
    of styrenes, for production of araliphatic aldehydes **1**: 383
Hydrogen
    for reduction of aromatic nitro compounds **1**: 517
Hydrogenated-tallow amine **1**: 472
Hydrogenation
    of acetylene in ethylene production **4**: 2276
    of adiponitrile **5**: 2847
    of aldehydes **1**: 357
    of alkylphenols **6**: 3717
    of aromatic nitro compounds **1**: 516
    of benzaldehyde to benzyl alcohol **2**: 853
    of benzene **3**:
    of benzoic acid **2**: 1029
    of butyraldehydes **2**: 931
    of carbon monoxide for butanol production **2**: 958
    of C_3 and C_4 fractions in ethylene cracked gas **4**: 2279
    of coal, for production of saturated hydrocarbons **5**: 2861
    of cresols **3**: 1659
    of crotonaldehyde **2**: 927, 957
    of N,N'-dialkylaminomethylphenols **3**: 1680
    of esters **3**: 2019
    of fats and fatty oils **4**: 2397
    of fatty acids **4**: 2517
    of fatty oils for production of fatty alcohols **5**: 2538
    of halide-substituted nitro compounds **1**: 518
    of hydrocarbon **5**: 2862
    of maleic anhydride **2**: 1007; **8**: 4625
    of propanal **7**: 4066
    of pyrolysis gasoline **2**: 716
Hydrogenation processes
    for production of aliphatic alcohols **1**: 286
Hydrogen cyanide
    for production of nitriles **6**: 3459
Hydrogen cyanide malonate process **5**: 3180
Hydrogen fluoride addition **5**: 2571
Hydrogen peroxide
    acylation **6**: 3652
    for amine oxidation **1**: 447
    for epoxidation **4**: 1995
    for percarboxylic acid production **6**: 3644
    for propene epoxidation **7**: 4143
    nitrile activated **4**: 1995
    reaction with acetic anhydride **1**: 66
Hydrogen peroxide caprolactam processes **2**: 1032
Hydrogen phosphonates **6**: 3838
Hydrogen removal
    in styrene production **7**: 4427
Hydrogen replacement
    by bromine **2**: 872
Hydrogensilanes **7**: 4329
Hydrogen substitution
    fluorination **5**: 2569
Hydrogensulfite
    for arylamine production **1**: 514
Hydrolysis
    of acrolein **7**: 4056
    of aminonaphthalenesulfonic acids **6**: 3387
    of benzal chloride **2**: 677
    of benzyl chloride **2**: 852
    of cellulosic materials **4**: 2103
    of chlorobenzene **6**: 3701
    of chlorobenzenes **3**: 1610
    of 2-chlorophenol for catechol production **6**: 3760
    of dimethyl terephthalate **8**: 4583
    of esters **3**: 2016
    of fats, for fatty acid production **4**: 2498
    of formamide **5**: 2727
    of glycerides **4**: 2396
    of melamine **5**: 3201
    of methyl formate **5**: 2719
    of propylene oxide **7**: 4051
    of pyridine carbonitriles **7**: 4204
    of saccharides **2**: 1077
    of 1,2,4,5-tetrachlorobenzene **3**: 1621
    of trichloroethylene **3**: 1543
    of urea, in urea production **8**: 4799
    of wax esters **5**: 2537
Hydronaphthalenes **6**: 3351, 3358
Hydroperoxidation
    of p-diisopropylbenzene **5**: 2945
    of m-diisopropylbenzene, for resorcinol production **7**: 4265
Hydroperoxides
    esterification **6**: 3657
    for epoxidation **4**: 1996
    production **6**: 3635
    properties **6**: 3632
1-Hydroperoxy-1-azocyclohexane
    physical properties **6**: 3670
α-Hydroperoxy-α'-hydroxydicyclohexyl peroxide
    physical properties **6**: 3664
2-Hydroperoxy-2-isopropyl-1,3-dioxolane
    production **6**: 3666

Hydrophilization process
  for fatty acid separation **4**: 2513
Hydrophobic surfaces
  silanes **7**: 4354
Hydroquinol, *see hydroquinone*
Hydroquinone **5**: 2941
  by carbonylation of acetylone **1**: 127
  environmental protection and specifications **5**: 2948
  properties **5**: 2941
  toxicology **5**: 2950
  uses **5**: 2949
Hydroquinone derivatives
  uses **5**: 2949
Hydroquinone production **5**: 2945
Hydrotreating
  of deoiled slack wax **8**: 4926
2-Hydroxyacetanilide **1**: 612
4-Hydroxyacetanilide **1**: 619
Hydroxyacetic acid, *see glycolic acid*
Hydroxyacetonitrile **5**: 2681; **6**: 3463
3-Hydroxyacrolein, *see malondialdehyde*
Hydroxyaldehydes **1**: 374
  production **1**: 376
1-Hydroxyalkanesulfinates **7**: 4465
Hydroxyalkanesulfonic acids **8**: 4504
  production **8**: 4506
2-Hydroxyalkyl acrylates **1**: 239
α-Hydroxyalkylureas **8**: 4823
4-Hydroxy-1-aminonaphthalene **6**: 3425
5-Hydroxy-1-aminonaphthalene **6**: 3426
8-Hydroxy-2-aminonaphthalene **6**: 3426
2-Hydroxyaniline **1**: 602
3-Hydroxyaniline **1**: 603
4-Hydroxyaniline **1**: 603
2-Hydroxybenzaldehyde **2**: 686
3-Hydroxybenzaldehyde **2**: 687
4-Hydroxybenzaldehyde **2**: 687
Hydroxybenzene, *see phenol*
2-Hydroxybenzenecarboxylic acid, *see salicylic acid*
4-Hydroxy-1,3-benzenedisulfonic acid **2**: 776
4-Hydroxy-1,3-benzenedisulfonyl chloride **2**: 776
2-Hydroxybenzenesulfonic acid **2**: 774
3-Hydroxybenzenesulfonic acid **2**: 775
4-Hydroxybenzenesulfonic acid **2**: 775
2-Hydroxybenzoic acid, *see salicylic acid*
3-Hydroxybenzoic acid **5**: 2977
4-Hydroxybenzoic acid **5**: 2978
o-Hydroxybenzoic acid, *see salicylic acid*
4-Hydroxybenzoic acid benzyl ester **5**: 2978
4-Hydroxybenzoic acid *n*-butyl ester **5**: 2978
4-Hydroxybenzoic acid ethyl ester **5**: 2978
4-Hydroxybenzoic acid methyl ester **5**: 2978
4-Hydroxybenzoic acid *n*-propyl ester **5**: 2978
Hydroxybenzoic acids
  properties **5**: 2974
4-Hydroxybenzonitrile **6**: 3472

α-Hydroxybenzyl phenyl ketone **5**: 3113
2-Hydroxybiphenyl
  physical properties **6**: 3775
  uses and toxicology **6**: 3777
4-Hydroxybiphenyl
  uses and toxicology **6**: 3777
2-Hydroxybiphenyl production **6**: 3775
4-Hydroxybiphenyl production **6**: 3776
Hydroxybiphenyls
  properties **6**: 3774
  uses and toxicology **6**: 3777
4-Hydroxybutanal **1**: 375, 379
(*R,S*)-Hydroxybutanedioic acid, *see (R,S)-malic acid*
(*R,S*-)-3-Hydroxybutanoic Acid, *see (R,S-)-β-hydroxybutyric acid*
Hydroxybutyl methyl cellulose **2**: 1215
(*R,S*-)-α-Hydroxybutyric acid **5**: 2968, 2968, 2968
γ-Hydroxybutyric acid
  physical properties **5**: 2958
(*R,S*)-α-Hydroxybutyric acid
  physical properties **5**: 2958
(*R,S*)-β-Hydroxybutyric acid
  physical properties **5**: 2958
γ-Hydroxybutyric acid lactone, *see butyrolactone*
ε-Hydroxycaproic acid – ε-caprolactone process **2**: 1034
Hydroxycarboxylic acids, aliphatic **5**: 2955
  preparation **5**: 2960
  properties **5**: 2956
  toxicology **5**: 2970
Hydroxycarboxylic acids, aromatic **5**: 2973
  production **5**: 2975
  properties **5**: 2974
  toxicology **5**: 2983
Hydroxycitronellal **1**: 375, 380
β-Hydroxy compounds
  formation from epoxides **4**: 1988
4-Hydroxydimethylaniline **1**: 619
1-Hydroxy-1-dimethylphosphinoacetic acid **6**: 3834
3-Hydroxy-2,2-dimethylpropyl)-3-hydroxy-2,2-dimethylpropionate **1**: 329
2-Hydroxy-1,2-diphenylethanone **5**: 3113
3-Hydroxyethanal, *see acetaldol*
1-Hydroxyethane-1,1-diphosphonic acid **1**: 67
2-Hydroxyethanesulfonic acid **8**: 4505
2-Hydroxyethanoic acid, *see glycolic acid*
Hydroxyethylaniline *see also ethanolamines, N-alkylated*
  physical properties **4**: 2160
Hydroxyethyl cellulose **2**: 1224
  properties and uses **2**: 1225
Hydroxyethylethylenediaminetriacetic acid
  physical properties **4**: 2297
Hydroxyethyl hydroxypropyl cellulose **2**: 1229
3-(2-Hydroxyethyl)indole **5**: 3003
Hydroxyethyl methyl cellulose **2**: 1215
1-(2-Hydroxyethyl)-2-methyl-5-nitroimidazole
  physical properties **5**: 2988

4-(2-Hydroxyethyl)-morpholine *see also ethanolamines, N-alkylated*
   physical properties **4**: 2160
Hydroxyethyl phosphonomethyl cellulose **2**: 1235
1-(2-Hydroxyethyl)-piperazine *see also ethanolamines, N-alkylated*
   physical properties **4**: 2160
Hydroxyhydroquinone **6**: 3765
Hydroxylamine–phosphate oxime caprolactam process **2**: 1020
Hydroxylation
   of allyl alcohol **1**: 421
   of chlorobenzenes **3**: 1611
   of phenol for catechol production **6**: 3760
   of phenol, for production of hydroquinone **5**: 2946
   of toluene **3**: 1678
Hydroxylethyl starch **7**: 4407
Hydroxyl exchange
   nucleophilic aromatic **1**: 523
Hydroxylpropyl starch **7**: 4407
Hydroxyl value **1**: 307
   of fats and fatty oils **4**: 2437
4-Hydroxy-6-methylaminonaphthalene-2-sulfonic acid **6**: 3438
4-Hydroxy-7-methylaminonaphthalene-2-sulfonic acid **6**: 3435
4-Hydroxy-N-methylaniline **1**: 618
2-Hydroxy-1-methyl-3,5-dinitrobenzene **6**: 3566
4-Hydroxy-1-methyl-3,5-dinitrobenzene **6**: 3566
Hydroxymethylimidazoles **5**: 2991
2-Hydroxymethyl-2-methyl-1,3-propanediol **1**: 339
1-Hydroxy-3-methyl-4-nitrobenzene **6**: 3565
1-Hydroxy-4-methyl-2-nitrobenzene **6**: 3565
Hydroxymethyloxirane **4**: 2003
4-Hydroxy-4-methyl-2-pentanone **1**: 111
2-Hydroxy-2-methylpropanenitrile **1**: 109
2-(Hydroxymethyl)tetrahydrofuran **5**: 2762
1-Hydroxynaphthalene **6**: 3373
2-Hydroxynaphthalene **6**: 3377
1-Hydroxynaphthalene-3,6-disulfonic acid **6**: 3394
1-Hydroxynaphthalene-3,8-disulfonic acid **6**: 3394
1-Hydroxynaphthalene-4,7-disulfonic acid **6**: 3395
1-Hydroxynaphthalene-4,8-disulfonic acid **6**: 3395
2-Hydroxynaphthalene-3,6-disulfonic acid **6**: 3396
2-Hydroxynaphthalene-3,7-disulfonic acid **6**: 3397
2-Hydroxynaphthalene-5,7-disulfonic acid **6**: 3397
2-Hydroxynaphthalene-6,8-disulfonic acid **6**: 3397
3-Hydroxynaphthalene-2,6-disulfonic acid **6**: 3397
3-Hydroxynaphthalene-2,7-disulfonic acid **6**: 3396
4-Hydroxynaphthalene-1,5-disulfonic acid **6**: 3395
4-Hydroxynaphthalene-1,6-disulfonic acid **6**: 3395
4-Hydroxynaphthalene-2,7-disulfonic acid **6**: 3394
6-Hydroxynaphthalene-1,3-disulfonic acid **6**: 3397
7-Hydroxynaphthalene-1,3-disulfonic acid **6**: 3397
8-Hydroxynaphthalene-1,6-disulfonic acid **6**: 3394
1-Hydroxynaphthalene-2-sulfonic acid **6**: 3389
1-Hydroxynaphthalene-3-sulfonic acid **6**: 3389
1-Hydroxynaphthalene-4-sulfonic acid **6**: 3390
1-Hydroxynaphthalene-5-sulfonic acid **6**: 3390
1-Hydroxynaphthalene-7-sulfonic acid **6**: 3391
1-Hydroxynaphthalene-8-sulfonic acid **6**: 3391
2-Hydroxynaphthalene-1-sulfonic acid **6**: 3392
2-Hydroxynaphthalene-4-sulfonic acid **6**: 3392
2-Hydroxynaphthalene-6-sulfonic acid **6**: 3392
2-Hydroxynaphthalene-7-sulfonic acid **6**: 3393
2-Hydroxynaphthalene-8-sulfonic acid **6**: 3393
3-Hydroxynaphthalene-1-sulfonic acid **6**: 3392
4-Hydroxynaphthalene-1-sulfonic acid **6**: 3390
4-Hydroxynaphthalene-2-sulfonic acid **6**: 3389
5-Hydroxynaphthalene-1-sulfonic acid **6**: 3390
6-Hydroxynaphthalene-2-sulfonic acid **6**: 3392
7-Hydroxynaphthalene-1-sulfonic acid **6**: 3393
7-Hydroxynaphthalene-2-sulfonic acid **6**: 3393
8-Hydroxynaphthalene-1-sulfonic acid **6**: 3391
8-Hydroxynaphthalene-2-sulfonic acid **6**: 3391
Hydroxynaphthalenesulfonic acids
   production and properties **6**: 3384
1-Hydroxynaphthalene-3,6,8-trisulfonic acid **6**: 3398
2-Hydroxynaphthalene-3,6,8-trisulfonic acid **6**: 3398
7-Hydroxynaphthalene-1,3,6-trisulfonic acid **6**: 3398
8-Hydroxynaphthalene-1,3,6-trisulfonic acid **6**: 3398
5-Hydroxy-1-naphthoic acid **6**: 3382
1-Hydroxy-2-naphthoic acid **6**: 3383
2-Hydroxy-1-naphthoic acid **6**: 3382
3-Hydroxy-1-naphthoic acid **6**: 3382
3-Hydroxy-2-naphthoic acid **6**: 3383
4-Hydroxy-1-naphthoic acid **6**: 3382
4-Hydroxy-2-naphthoic acid **6**: 3384
5-Hydroxy-2-naphthoic acid **6**: 3384
6-Hydroxy-1-naphthoic acid **6**: 3382
6-Hydroxy-2-naphthoic acid **6**: 3384
7-Hydroxy-1-naphthoic acid **6**: 3382
7-Hydroxy-2-naphthoic acid **6**: 3384
8-Hydroxy-1-naphthoic acid **6**: 3383
8-Hydroxy-2-naphthoic acid **6**: 3384
6-Hydroxy-2-naphthoic acid nitrile **6**: 3473
2-Hydroxy-5-nitroaniline **1**: 612
3-Hydroxyphenol, *see resorcinol*
α-Hydroxyphenylacetic acid **5**: 2982
4-Hydroxyphenylaminoacetic acid **1**: 621
4-Hydroxy-7-phenylaminonaphthalene-2-sulfonic acid **6**: 3435
4-Hydroxy-m-phenylenediamine **1**: 613
N-(4-Hydroxyphenyl)glycine **1**: 621
Hydroxypivaldehyde **1**: 327, 375, 379
Hydroxypivalic acid neopentyl glycol ester **1**: 323, 329
3-Hydroxypropanal **1**: 375, 378
2-Hydroxypropanenitrile **6**: 3463
3-Hydroxypropanenitrile **6**: 3464
2-Hydroxypropane-1,2,3-tricarboxylic acid, *see citric acid*
3-Hydroxypropionic acid **5**: 2967
2-Hydroxypropionic acid, *see lactic acid*
β-Hydroxypropionic acid **5**: 2967

β-Hydroxy propionic acid
  toxicology   **5**: 2971
3-Hydroxypropionitrile   **6**: 3464
Hydroxypropyl cellulose   **2**: 1228
2-Hydroxypropylene-1,3-diurea   **8**: 4820
5-Hydroxypropyleneurea   **8**: 4820
Hydroxypropyl methyl cellulose   **2**: 1215
4-(2-Hydroxypropyl)-morpholine   **4**: 2171
1-(2-Hydroxypropyl)-piperazine   **4**: 2171
1,4-Bis(2-hydroxypropyl)-piperazine   **4**: 2171
6-Hydroxypurine, *see purine derivatives*
8-Hydroxyquinoline   **7**: 4255
p-Hydroxysalicylic acid   **5**: 2980
(*R,S*)-Hydroxysuccinic acid, *see (R,S)-malic acid*
2-Hydroxy-5-sulfobenzoic acid   **2**: 780
β-Hydroxytricarballylic acid, *see citric acid*
Hygroscopicity
  of urea   **8**: 4785
Hyperoxides
  in autooxidation of unsaturated fatty acids   **4**: 2399
Hypochlorination
  of allyl alcohol   **3**: 1593
  of ethylene   **3**: 1591
  of propene   **3**: 1592
Hypochlorous acid
  reaction with olefins   **3**: 1590
Hypoxanthine, *see purine derivatives*

ICI isopropanol process   **7**: 4070
ICI low-pressure methanol process   **6**: 3303
ICI phthalocyanine process   **7**: 3939
n-Icosane
  physical properties   **5**: 2855
1-Icosene
  physical properties   **5**: 2871
D-Idose   **2**: 1065
IDR urea process   **8**: 4808
IFP/CR & L MTBE process   **6**: 3340
IFP methanol process   **1**: 285
IFP toluene extraction process   **8**: 4732
Ignition
  of ethylene oxide   **4**: 2345
  of saturated hydrocarbons   **5**: 2858
Imferon   **3**: 1849
Imidazole   **5**: 2987
  functionalized   **5**: 2990
  physical properties   **5**: 2988
  properties   **5**: 2987
  specifications   **5**: 2991
  toxicology   **5**: 2992, 2993
Imidazole nitro derivatives   **6**: 3570
Imidazole production   **5**: 2989
Imidazole-2-thiol   **8**: 4668
2-Imidazolidinone   **8**: 4820
Δ^2-Imidazolines
  dehydrogenation   **5**: 2989
Imidodicarbonic diamide   **8**: 4818

7,7′-Iminobis(4-hydroxynaphthalene-2-sulfonic
  acid)   **6**: 3435
β-Iminobutyronitrile   **6**: 3465
2,2′-Iminodiethanol, *see diethanolamine*
β-Iminonitriles   **6**: 3465
Immobilized cells
  for fermentation   **4**: 2089
Impregnating oils   **8**: 4537
Incomplete conversion formaldehyde process   **5**: 2652
Indan   **5**: 2916
2-Indanol   **5**: 2917
4-Indanol   **5**: 2917
5-Indanol   **5**: 2917; **6**: 3757
Indene   **5**: 2915
Indian gum   **7**: 4016, 4017
Indole   **5**: 3001
  toxicology   **5**: 3004
  uses   **5**: 3003
Indole-3-aldehyde   **5**: 3002
Indoline   **5**: 3003
Insect waxes   **8**: 4904
Inst. Français du Pétrole, *see IFP*
Integrated vinyl chloride processes   **3**: 1341
Interesterification
  of fats and fatty oils   **4**: 2428
  of glycerides   **4**: 2396
Internal esterification
  of lactic acid   **5**: 3121
Inulin   **2**: 1068
Inventa caprolactam process   **2**: 1032
Iodine compounds   **5**: 3007
Iodine value   **1**: 307
  of fats and fatty oils   **4**: 2438
Iodobenzene   **5**: 3008
Iodofluoroalkanes   **5**: 2581
Iodoform   **5**: 3007
Iodogen   **3**: 1570
Ion-exchange membranes   **5**: 2604
Ion-exchange resins
  for production of higher alkyl phenols   **6**: 3749
Ion exchangers
  as catalysts for esterification of carboxylic
    acids   **3**: 2023
Iron
  for hydrazobenzene production   **2**: 795
Iron 2-ethylhexanoate   **6**: 3245
Iron(III) stearate   **6**: 3245
Iron oleate   **6**: 3245
Iron reduction
  of nitrophenols   **1**: 606
Iron soaps   **6**: 3244
Isoalcohols
  $C_{13}-C_{18}$ **1**: 302
Isoamyl alcohol   **6**: 3612
sec-Isoamyl alcohol   **6**: 3612
Isoamylene oxide   **4**: 2001
Isoamyl salicylate   **7**: 4277

Isobaric double-recycle urea process **8**: 4808
Isobenzofuran-1,3-dione, *see phthalic anhydride*
Isobutane
   uses **5**: 2867
Isobutanol **1**: 281, 284
Isobutene **2**: 987
   as feedstock for MTBE production **6**: 3339
   for isoprene synthesis **5**: 3037
   for methacrylic acid production **6**: 3256
   physical properties **5**: 2871
Isobutene chlorohydrins **3**: 1589
Isobutene dichlorohydrins **3**: 1589
Isobutene ozonide
   physical properties **6**: 3664
Isobutene separation
   from cracking mixtures **2**: 992
Isobutyl acetate
   physical properties **3**: 2014
   specifications **3**: 2037
   toxicology **3**: 2043
Isobutyl acrylate **1**: 226
Isobutylamine **1**: 454
2-Isobutylamino-5-sulfobenzoic acid **2**: 780
Isobutylbenzene
   physical properties **5**: 2902
Isobutyl carbamate **2**: 1047
Isobutylcyclohexane
   physical properties **3**: 1798
Isobutylene chlorination **3**: 1391
Isobutyl formate **5**: 2738
Isobutylidenediurea **8**: 4828
Isobutyl isobutyrate
   physical properties **3**: 2014
   specifications **3**: 2037
   toxicology **3**: 2043
2-(4-Isobutylphenyl)propanal **1**: 382
2-(4-Isobutylphenyl)propionaldehyde **1**: 387
Isobutyl vinyl ether
   physical properties **8**: 4867
   toxicology **8**: 4874
Isobutyraldehyde **1**: 352
   physical properties **2**: 925
Isobutyraldehyde dehydrogenation
   for methacrolein production **1**:
Isobutyric acid **2**: 1097, 1108
   for methacrylic acid production **6**: 3258
   production conditions **2**: 1101
   specification **2**: 1106
Isobutyronitrile **6**: 3461
Isobutyryl fluoride **2**: 1114
Isocaproic acid **2**: 1097
Isochloroprene
   properties **3**: 1384
Isocrotonic acid
   physical properties **3**: 1716
Isocrotyl chloride
   from isobutene chlorination **3**: 1393

Isocyanates, organic **5**: 3009
   economic aspects **5**: 3026
   environmental protection and specifications **5**: 3023
   from dithiocarbamic acids or acid esters **4**: 1966
   handling **5**: 3024
   production **5**: 3015
   properties **5**: 3010
   reaction with alcohols **2**: 1046
   toxicology **5**: 3027
   uses **5**: 3025
Isocyanic acid
   as side product in urea production **8**: 4799
Isocyanuric acid, *see cyanuric acid*
Isodecanoic acid **2**: 1110
   specification **2**: 1106
Isodecanol **1**: 294
Isodecanol mixtures **1**: 301
4-Isododecylphenol **6**: 3748
Isodurene **5**: 2895
   physical properties **5**: 2894
Isoheptanols **1**: 295
Isohexadecanols **1**: 303
Isohexanols **1**: 293
Isomar xylene isomerization process **8**: 5002
Isomeric C_{13} acid, *see isotridecanoic acid*
Isomeric C_8 acids, *see isooctanoic acid*
Isomerization
   of cresols **3**: 1676
   of fatty acids **4**: 2398, 2519, 2520
   of hydrocarbon **5**: 2863
   of thiocyanates **8**: 4632
   of xylenes **8**: 5002
Isononanal **1**: 364
Isononanoic acid **2**: 1110
   specification **2**: 1106
Isononanols **1**: 300
Isooctadeanol **1**: 303
Isooctalcohol mixture **1**: 294
Isooctanal **1**: 364
Isooctanoic acid **2**: 1097, 1109
   specification **2**: 1106
Isooctylphenol **6**: 3748
Isopentane
   uses **5**: 2868
Isopentane dehydrogenation **5**: 3040
Isopentanol **1**: 281
Isopentene dehydrogenation **5**: 3040
Isopentyl propionate
   properties **7**: 4119
Isophorone **5**: 3105
Isophorone demethanization **3**: 1695
Isophthalic acid **2**: 1121; **8**: 4573
   physical properties **8**: 4574
   uses **8**: 4588
Isophthalic acid production **8**: 4576

Isoprene 5: 3033
   economic aspects 5: 3050
   handling 5: 3047
   properties 5: 3034
   specifications 5: 3046
   toxicity 5: 3052
   uses 5: 3048
Isoprene production 5: 3036
Isoprene rubber 5: 3048
Isoprene separation
   from C_5 cracking fractions 5: 3041
Isopropanol
   from propene 7: 4097
   properties 7: 4062, 4062
   specifications and economic aspects 7: 4073, 4073
   toxicology 7: 4075, 4075
   uses 7: 4071, 4071
Isopropanolamines 4: 2163
   production 4: 2166
   properties 4: 2164
   specifications 4: 2168
   uses 4: 2169
Isopropanol oxidation
   for acetone production 1: 99
Isopropanol production 7: 4065, 4067
Isopropenyl acetate production
   from ketene 5: 3067
4-Isopropenylphenol 6: 3756
2-Isopropoxyisopropane, see diisopropyl ether
2-Isopropoxyphenol 6: 3781
Isopropyl acetate
   physical properties 3: 2014
   specifications 3: 2037
   toxicology 3: 2043
2-Isopropylacrolein 1: 373
Isopropyl alcohol, see isopropanol
Isopropylamine 1: 454
Isopropylbenzene, see cumene
Isopropyl carbamate 2: 1047
Isopropyl chloride 3: 1373
Isopropyl chloroformate 3: 1576
Isopropyl cyanide 6: 3461
Isopropylcyclohexane
   physical properties 3: 1798
Isopropyl ether, see diisopropyl ether
Isopropyl hydroperoxide
   physical properties 6: 3634
4,4'-Isopropylidenediphenol, see bisphenol A
4-Isopropylidene-1-methyl-1-cyclohexane 8: 4602
2-Isopropylimidazole
   physical properties 5: 2988
   toxicology 5: 2994
4-(Isopropyl)methylbenzene 8: 4603
4-Isopropyl-1-methyl-1,3-cyclohexadiene 8: 4602
1-Isopropyl-4-methylcyclohexane 8: 4601
3-Isopropyl-4-methylphenol 6: 3735
4-Isopropyl-2-methylphenol 6: 3735
5-Isopropyl-3-methylphenol 6: 3734
6-Isopropyl-2-methylphenol 6: 3735
6-Isopropyl-3-methylphenol 6: 3734
2-Isopropylnaphthalene 6: 3357
2-Isopropyl-4(5)-nitroimidazole
   physical properties 5: 2988
2-Isopropylphenol 6: 3732
3-Isopropylphenol 6: 3733
4-Isopropylphenol 6: 3732
Isopropyl phenyl ether 6: 3732
3-(4-Isopropylphenyl)-2-methyl-propanal, see cyclamenaldehyde
2-Isopropylpropenal 1: 368
2-Isopropyltoluene
   physical properties 5: 2902
3-Isopropyltoluene
   physical properties 5: 2902
4-Isopropyltoluene
   physical properties 5: 2902
Isoquinoline 7: 4253
   production and uses 7: 4257
   properties 7: 4256
   toxicology 7: 4257
Isostearyl alcohol 1: 303
Isothermal dehydrogenation
   of ethylbenzene 7: 4425
Isothiocyanates 8: 4631, 4635
   from dithiocarbamic acids or acid esters 4: 1966
   production 8: 4638
   properties 8: 4636
   specifications, handling and uses 8: 4641
   toxicology 8: 4643
Isothionic acid 8: 4505
Isothymol 6: 3734
Isotridecanals 1: 366
Isotridecanoic acid
   specification 2: 1106
Isotridecanol 1: 294
Isotridecyl alcohol 1: 302
Isovaleraldehyde 1: 352, 361
Isovaleric acid 2: 1097, 1108
   specification 2: 1106
Itaconic acid 3: 1922
Izod notched impact strength
   of cellulose ester thermoplastics 2: 1188

J acid 6: 3434
J acid imide 6: 3435
J acid urea 6: 3436
Japanese fish oil
   fatty acid content 4: 2376
Japan tallow wax 8: 4900
Jerusalem artichoke
   as raw material for ethanol production 4: 2100
Jojoba oil 8: 4898

K acid 6: 3441

Kadaya gum **7**: 4016
Kapok oil **4**: 2453
Karité butter **4**: 2451
Kawasaki naphthoquinone process **6**: 3449
Kellogg methanol process **6**: 3304
Kellogg millisecond furnace **4**: 2246
Kellogg millisecond primary quench exchanger **4**: 2260
Kemira–Leonard formic acid process **5**: 2723
Kennedy extractor
  for oilseeds **4**: 2408
Ketene
  properties **5**: 3062
  toxicology **5**: 3078
Ketene acetic anhydride process **1**: 72
Ketene acrylic acid process **1**: 230
Ketene insertion **5**: 3064
Ketene process
  for acetic anhydride recovery **2**: 1175
Ketene production **1**: 72
Ketenes **5**: 3061
  for production of esters **3**: 2028
  production **5**: 3065
  uses **5**: 3066
β-Ketobutyronitrile **6**: 3465
β-Ketoglutaric acid **6**: 3607
Ketoketenes **5**: 3076
Ketone-alcohol **1**: 266
Ketones **5**: 3083
  aromatic **5**: 3110
  cyclic **5**: 3099
  dialkyl **5**: 3095
  for amine production **1**: 450
  for production of pyridines **7**: 4183
  methyl alkyl, C_9–C_{18} **5**: 3094
  toxicology **5**: 3113
  unsaturated **5**: 3101
β-Ketonitriles **6**: 3465
α-Ketopropionic acid **6**: 3600
Ketoses **2**: 1063
γ-Ketovaleric acid **6**: 3606
K factor
  of naphtha **4**: 2225
Kinetics
  of fat hydrolysis **4**: 2498
  of methane chlorination **3**: 1261
  of methanol carbonylation **5**: 2721
  of methanol production from synthesis gas **6**: 3293
  of toluene side-chain chlorination **3**: 1453
Kinnear–Perren reaction
  for production of chlorophosphines **6**: 3826
Klett-Summerson value **2**: 1037
Knapsack dry generator **1**: 161
Knapsack rotary furnace calcium cyanamide process **3**: 1742
Knecht compound **2**: 1154

Knoevenagel condensation
  of aryl sulfones with aldehydes or ketones **7**: 4483
Knorr pyrrole synthesis **7**: 4237
Koch acid **6**: 3421
Koch carboxylic acid process **2**: 1102
Koch–Haaf method **4**: 2522
Kolbe–Schmitt reaction **2**: 1125
Kolbe–Schmitt synthesis
  of hydroxybenzoic acids **5**: 2975
  of salicylic acid **7**: 4273
Koppers pitch coking process **8**: 4530
Krupp Kontipress process **4**: 2410
Krupp–Koppers *p*-xylene crystallization process **8**: 4999
KTI cracking furnace **4**: 2248
Kureha, Chiyoda, Union Carbide acetylene process **1**: 165

LABS **5**: 2903
Lacceryl alcohol
  physical properties **5**: 2535
L acid **6**: 3412
Lacquers
  cellulose nitrate **2**: 1160
  from cellulose nitrate **2**: 1143
Lactamides **5**: 3130
Lactams **2**: 1115
Lactates **5**: 3129
Lactic acid **5**: 3119
  economic aspects and toxicology **5**: 3131
  physical properties **5**: 2958
  properties **5**: 3120
  purification **5**: 3125
  specifications **5**: 3127
  uses **5**: 3128
Lactic acid esters **5**: 3130
Lactic acid fermentation **5**: 3124
Lactic acid production **5**: 3123
Lactic acid salts **5**: 3129
Lactic acid synthesis **5**: 3125
Lactobionic acid **5**: 3146
Lactones **3**: 2033
  of gluconic acid **5**: 2774
Lactonitrile **6**: 3463
Lactonitriles **5**: 3130
Lactose **5**: 3135
  economic aspects **5**: 3140
  physiological properties **5**: 3141
  properties **5**: 3136
  specifications and uses **5**: 3139
Lactose production **5**: 3137
Lactose solutions **5**: 3137
Lactose syrup
  hydrolyzed **5**: 3146
Lactoyllactic acid **5**: 3121
Lactulose
  properties **5**: 3142

specifications, toxicology, physiology **5**: 3144
   uses **5**: 3145
Lactulose production **5**: 3143
Ladenburg rearrangement **7**: 4192
Lake Red C **2**: 785
Laminates
   melamine resins **5**: 3211
Lanza fatty acid separation process **4**: 2514
Lard **4**: 2468
   fatty acid content **4**: 2376
Latex **7**: 4191
Laurent's acid **6**: 3412
n-Lauric acid **2**: 1097
   physical properties **4**: 2485
Lauric acid oils **4**: 2445, 2449
Lauric aldehyde **1**: 353, 366
Laurolactam **3**: 1783, 1786
   toxicology **3**: 1787
Lauryl alcohol
   physical properties **5**: 2535
Lauryl bromide, see 1-bromo dodecane
Lawesson's reagent **8**: 4652
Lead cyanamide **3**: 1753
Lead 2-ethylhexanoate **6**: 3241
Lead naphthenate **6**: 3241
Lead oleate **6**: 3241
Lead soaps **6**: 3240
Lead stearate **6**: 3241
Leaf aldehyde **1**: 372
Lecithin **4**: 2380; **5**: 3149
   commercial grades **5**: 3152
   composition **5**: 3150
   properties **5**: 3154
   specifications **5**: 3157
   uses **5**: 3156
Lecithin production **5**: 3151
Lees **8**: 4563
Leguminous oils **4**: 2461
Lepidine **7**: 4255
Letter acids **6**: 3366
Levotartaric acid **8**: 4559
Levulinic acid **6**: 3606
Light oil **8**: 4544
Light solvent extraction process **1**: 233
Lignite extraction **8**: 4908
Lignite tars **8**: 4510
   processing **8**: 4550
   properties and composition **8**: 4513
Lignocellulose
   as raw material for ethanol production **4**: 2101
Lignoceric acid
   physical properties **4**: 2486
Lignoceryl alcohol
   physical properties **5**: 2535
Lilial **1**: 382, 388
Lime
   in propylene oxide production **7**: 4140

Linde propene process **7**: 4093
Lindepyrocrack furnace **4**: 2248
Linoleic acid
   physical properties **4**: 2487
Linolelaidic acid
   physical properties **4**: 2487
Linolenelaidic acid
   physical properties **4**: 2487
Linolenic acid
   physical properties **4**: 2487
Linolenyl alcohol
   physical properties **5**: 2549
Linoleyl alcohol
   physical properties **5**: 2549
Linseed oil **4**: 2458
   fatty acid content **4**: 2376
Linters
   as cellulose raw material **2**: 1146
   as raw materials for cellulose **2**: 1169
Lipofrac fatty acid separation process **4**: 2514
Liquefaction of coal
   for production of saturated hydrocarbons **5**: 2861
Liquefied petroleum gas
   as raw material for ethylene procuction **4**: 2224
Liquid–gas equilibrium
   in a ternary system with binary azeotrope **8**: 4793
Liquid–liquid extraction
   in benzene production **2**: 717
Liquid-phase hydrogenation
   of nitrobenzene **1**: 630
Liquid-phase oxidation
   for production of aromatic carboxylic
      acids **2**: 1128
   of acetaldehyde **1**: 75
   of butane **1**: 38
   of butane industrial operation **1**: 42
   of olefins, for allyl ester production **1**: 427
   of toluene **2**: 680
   of toluene with molecular oxygen **2**: 821
Lithium 12-hydroxystearate **6**: 3235
Lithium soaps **6**: 3235
Lithium stearate **6**: 3235
Litol benzene process **2**: 712
Litol BTX process **2**: 702
Liver oils
   manufacture and processing **4**: 2412
LM-pectin gelation **7**: 3985
Lobry de Bruyn-van Ekenstein rearrangement **2**: 1078
Locust bean
   production and properties **7**: 4025
Locust bean gum **7**: 4023
Longifolene **8**: 4612
Lonza aniline process **1**: 631
Loop hydrogenation reactor
   for fat hardening **4**: 2428
Low-methoxyl pectin lyase **7**: 3979

Low-temperature carbonization
  for benzene production **2:** 701
Low-temperature coal-tar pitch **8:** 4548
Low-temperature coking
  for benzene production **2:** 701
Low-temperature–disproportionation benzene process **2:** 713
Low-water-activity pectin gels **7:** 3985
LPO, *see liquid-phase oxidation*
LTC 1,2-dichloroethane process **3:** 1301
Lubricants
  from higher olefins **5:** 2886
  metallic soaps **6:** 3235
Lubricating oil distillates
  dewaxing **8:** 4920
Lubricats
  xanthate oils **8:** 4979
Lummus cracking furnace **4:** 2245
Lummus horizontal transfer-line exchanger **4:** 2262
Lummus propylene oxide process **7:** 4142
Lummus soft pitch process **8:** 4531
Lupanes **7:** 4297
Lupine oil **4:** 2463
Lurgi fatty acid hydrogenation process **5:** 2540
Lurgi low-pressure methanol process **6:** 3304
Lurgi Phenoraffin process **3:** 1661
Lurgi Phenosolvan process **8:** 4549
Lye
  for neutralization of fats and fatty oils **4:** 2414

MA, *see maleic anhydride*
Macrocrystalline waxes, *see paraffin waxes*
Magnesium behenate **6:** 3235
Magnesium oxide
  as catalyst for phenol methylation **3:** 1673
Magnesium soaps **6:** 3236
Magnesium stearate
  technical **6:** 3235
Maize oil **4:** 2454
MAK, *see methyl amyl ketone*
Maleic acid **5:** 3159
  physical properties **3:** 1905
  toxicology **5:** 3172
Maleic acids **5:** 3159
Maleic acid solutions
  dehydration **5:** 3164
Maleic anhydride
  properties **5:** 3161
  purification **5:** 3166
  specifications and handling **5:** 3168
  toxicology **5:** 3172
  uses **5:** 3169
Maleic anhydride hydrogenation **2:** 1007; **8:** 4625
Maleic anhydride production **5:** 3162
  byproducts and construction materials **5:** 3167
(R,S)-Malic acid **5:** 2969
  physical properties **5:** 2958

Malonates **5:** 3179
Malondialdehyde **1:** 390
Malonic acid **5:** 3177
  physical properties **3:** 1905
  properties **5:** 3178
  specifications and economic aspects **5:** 3192
  toxicology **5:** 3193, 3194
Malonic acid esters **5:** 3179
Malononitrile **5:** 3188
  physical properties **6:** 3457
  properties **5:** 3188
  uses **5:** 3191
Malononitrile dimer **6:** 3465
Malononitrile production **5:** 3189
(R,S)-Mandelic acid **5:** 2982, 2983
  physical properties **5:** 2958
D,L-Mandelonitrile **6:** 3470
Mandioc
  as raw material for ethanol production **4:** 2098
Manganese
  as catalyst in acetaldehyde oxidation **1:** 45
Manganese 2-ethylhexanoate **6:** 3245
Manganese(II) oleate **6:** 3245
Manganese naphthenate **6:** 3245
Manganese soaps **6:** 3244
Manganese stearate **6:** 3245
Manioc
  as raw material for ethanol production **4:** 2098
Mannitol **2:** 1074
D-Mannose **2:** 1065
α-L-Gulurono-β-D-mannuronans, *see alginates*
Marassé carboxylation **5:** 2975
Marassé reaction **6:** 3382
Margaric acid
  physical properties **4:** 2485
Margaryl alcohol **1:** 282
  physical properties **5:** 2535
Marine oils **4:** 2471
  manufacture and processing **4:** 2411
Masking
  of heavy-metal ions **6:** 3483
Matrin wheat starch process **7:** 4391
Mazzoni fatty acid fractionation **4:** 2511
MC, *see methyl cellulose*
McCormack reaction **6:** 3829
MCOPP **6:** 3786
MCPA **3:** 1620
MCPB **3:** 1620
MDI **1:** 634; **5:** 3012
MEA, *see monoethanolamine*
Mecoprop **3:** 1620
Meerwein-Ponndorf-Verley reaction **1:** 312
Melam **5:** 3200
Melamine **3:** 1747; **5:** 3197; **8:** 4818
  condensation **5:** 3199
  handling, uses and economic aspects **5:** 3211
  properties **5:** 3198

specifications **5:** 3210
toxicology **5:** 3212
Melamine Chemicals process **5:** 3207
Melamine hydrolysis **5:** 3201
Melamine ion **5:** 3201
Melamine production **5:** 3202
Melamine salts **5:** 3202
Melem **5:** 3199
Melissic acid
  physical properties **4:** 2486
Melissyl alcohol
  physical properties **5:** 2535
Melle Boinot fermentation process **4:** 2081
Mellitene **5:** 2896
  physical properties **5:** 2894
  toxicology **5:** 2925
Mellitic acid **2:** 1121
Mellitic trianhydride **2:** 1130
Mellophanic acid **2:** 1121
Mellophanic dianhydride **2:** 1130
Melone **5:** 3200
Melon seed oil **4:** 2454
Melt flow index
  of cellulose ester thermoplastics **2:** 1189
Melting point
  of fatty acids **4:** 2483
Membrane bioreactor
  for fermentation **4:** 2086
Menadione **6:** 3450
Menhaden fish oil
  fatty acid content **4:** 2376
p-Menthane **8:** 4601
1-*p*-Menthene-8-thiol production **8:** 4653
Mercaptans, *see thiols*
Mercaptoacetic acid **6:** 3221
  handling **6:** 3223
  properties **6:** 3221
  toxicology **6:** 3224
Mercaptoacetic acid production **6:** 3222
2-Mercaptobenzothiazole
  toxicology **8:** 4678
Mercaptoethanol production **8:** 4654
2-Mercapto-1-methylimidazole
  toxicology **8:** 4678
2-Mercaptonicotinic acid **8:** 4668
6-Mercaptopurine, *see purine derivatives*
2-Mercapto-4[3*H*]-quinazolinone **8:** 4668
Mercaptosilanes **7:** 4338
Mercury(II) chloride
  as catalyst in acetylene hydrochlorination **3:** 1334
Mesaconic acid **5:** 3169
Mesitylene **5:** 2893
  physical properties **5:** 2894
Mesityl oxide **5:** 3104
Mesophase pitches **8:** 4534
meso-Tartaric acid
  physical properties **5:** 2958

Metal acetylacetonates **5:** 3108
Metaldehyde **1:** 22
  toxicology **1:** 24
Metallic soaps **6:** 3227
  definition and properties **6:** 3227
  environmental protection **6:** 3234
  fire and explosion protection **6:** 3233
  production **6:** 3229
  specifications **6:** 3247
Metal treatment
  with oxalic acid **6:** 3591
Metanilic acid **2:** 771
Metatartaric acid **8:** 4567
Metathesis
  of olefins, for isoprene synthesis **5:** 3040
  of olefins, for production of dienes and polyenes **5:** 2892
  SHOP process **5:** 2880
Methacrolein **1:** 210, 368
  physical properties **1:** 210; **6:** 3252
  toxicology **1:** 217
  uses **1:** 213
Methacrylamide
  physical properties **6:** 3252
Methacrylic acid **6:** 3249
  properties **6:** 3251
  specifications and handling **6:** 3261
  toxicology **6:** 3263
  uses **6:** 3262
Methacrylic acid production **6:** 3254
Methacrylic anhydride
  physical properties **6:** 3252
Methacrylonitrile **6:** 3462
  physical properties **6:** 3252
Methacryloyl chloride
  physical properties **6:** 3252
Methallyl acetate **1:** 425, 426, 427
Methallyl alcohol **1:** 424
Methallylamine **1:** 433
Methallyl chloride **3:** 1390
  from isobutene chlorination **3:** 1393
  specifications and uses **3:** 1394
Methallylsulfonic acid **8:** 4506
Methallylsulfonic acid production **8:** 4507
Methanal, *see formaldehyde*
Methanamine, *see monomethylamine*
Methane **6:** 3269
  as raw material for chlorination **3:** 1270
  biological formation **6:** 3272
  biomethanation, environmental aspects **6:** 3281
  biomethanation process parameters **6:** 3278
  biotechnological formation **6:** 3273
  economic aspects **6:** 3284
  feedstocks for biomethanation **6:** 3277
  handling of biogas **6:** 3282
  in biogas **6:** 3271
  physical properties **5:** 2855

toxicology **5**: 2924
Methane chlorination **3**: 1261, 1268
Methane cracking reaction **1**: 131
p-Methane hydroperoxide
  toxicology **6**: 3676
Methanesulfinyl chloride
  synthesis **7**: 4459
Methanesulfonic acid
  physical properties **8**: 4504
Methanethiol
  toxicology **8**: 4677
Methanethiol production **8**: 4652
Methanilic acid **1**: 637
Methanogenesis
  biological **6**: 3272
  biological, process parameters **6**: 3279
Methanol **1**: 281; **6**: 3287
  addition to isobutene **6**: 3340
  as ethylene feedstock **4**: 2284
  chemical properties **6**: 3290
  economic aspects **6**: 3315
  environmental protection **6**: 3310
  handling and safety **6**: 3306
  physical properties **6**: 3288
  specifications and grades **6**: 3308
  specifications for formaldehyde production **5**: 2648
  toxicology **6**: 3317
  uses **6**: 3310
Methanol carbonylation **1**: 34; **5**: 2720
  for ethanol production **4**: 2059
Methanol distillation **6**: 3305
Methanol homologation
  with synthesis gas **4**: 2059
Methanol hydrochlorination **3**: 1264
Methanol oxidation
  in formaldehyde production **5**: 2647
Methanol partial oxidation
  for formaldehyde production **5**: 2652
Methanol production
  byproducts **6**: 3295
  catalysts **6**: 3296
  process technology **6**: 3299
  reactor design **6**: 3303
  thermodynamics and kinetics **6**: 3292
Methanol solutions **6**: 3290
2-Methoxyacetanilide **1**: 614
2-Methoxyaniline **1**: 612
4-Methoxyaniline **1**: 619
4-Methoxybenzenesulfonic acid **2**: 776
Methoxycarbonyl isothiocyanate
  physical properties **8**: 4636
Methoxycarbonyl isothiocyanate production **8**: 4640
2-Methoxy-5-chloroaniline **1**: 614
1-Methoxy-2,4-dinitrobenzene **6**: 3564
Methoxyethyl acetate
  physical properties **3**: 2014

specifications **3**: 2037
toxicology **3**: 2043
Methoxymethane, *see dimethyl ether*
2-Methoxy-5-methylaniline **1**: 614
Methoxymethyl carbamoyl chloride **2**: 1049
2-Methoxy-2-methylpropane, *see methyl tert-butyl ether*
2-Methoxy-4-nitroaniline **1**: 614
2-Methoxy-5-nitroaniline **1**: 615
2-Methoxynitrobenzene **6**: 3561
4-Methoxynitrobenzene **6**: 3563
2-Methoxyphenol **6**: 3780
3-Methoxypropylamine **1**: 455; **4**: 2173
  toxicology **4**: 2179
Methyl acetate **1**: 53
  carbonylation **1**: 76
  for ethanol production **4**: 2059
  for vinyl acetate production **8**: 4851
  physical properties **3**: 2014
  specifications **3**: 2037
  toxicology **3**: 2043
Methyl acetoacetate
  specifications **3**: 2037
Methylacetylene, *see propyne*
N-Methyl γ acid **6**: 3438
α-Methylacrolein, *see methacrolein*
Methyl acrylate **1**: 226
  physical properties **3**: 2014
  specifications **3**: 2037
Methyl acrylate production **3**: 2030
α-Methylacrylic acid, *see methacrylic acid*
Methylal **1**: 395
Methyl alcohol, *see methanol*
Methylamines **6**: 3325
  economic aspects **6**: 3333
  environmental protection and specifications **6**: 3330
  handling **6**: 3331
  production **6**: 3327
  properties **6**: 3325
  purification **6**: 3329
  solutions **6**: 3327
  uses **6**: 3332
Methyl-3-amino-4-hydroxybenzoate **1**: 616
Methyl-3-amino-4-methylthiophene 2-carboxylate **8**: 4703
2-(*N*-Methylamino)phenol **1**: 611
3-(*N*-Methylamino)phenol **1**: 617
4-(*N*-Methylamino)phenol **1**: 618
Methyl-3-aminothiophene 2-carboxylate **8**: 4703
Methylamyl alcohol **1**: 281
Methyl amyl ketone **1**: 113; **5**: 3091
N-Methylaniline **1**: 637
4-Methylanisole **6**: 3780
Methyl anthranilate
  physical properties **3**: 2015
Methylating agents
  monochloromethane **3**: 1258

Methylation
  of phenol for cresol production **3:**
  of toluene, oxidative **3:** 1679
  with dimethyl sulfate **3:** 1865
2-Methylaziridine, *see propylenimine*
Methylbenzaldehyde
  for cresol production **3:** 1677
2-Methylbenzaldehyde **2:** 688
3-Methylbenzaldehyde **2:** 688
4-Methylbenzaldehyde **2:** 688
Methylbenzene, *see toluene*
4-Methyl-1,3-benzenedisulfonamide **2:** 764
4-Methyl-1,3-benzenedisulfonic acid **2:** 764
Methylbenzenedisulfonic acids **2:** 761
4-Methyl-1,3-benzenedisulfonyl chloride **2:** 764
2-Methylbenzenesulfonamide **2:** 762
3-Methylbenzenesulfonamide **2:** 762
4-Methylbenzenesulfonamide **2:** 763
N-(Methyl)benzenesulfonamide **2:** 758
Methyl benzenesulfonate **2:** 758
2-Methylbenzenesulfonic acid **2:** 762
3-Methylbenzenesulfonic acid **2:** 762
4-Methylbenzenesulfonic acid **2:** 763
Methylbenzenesulfonic acids **2:** 761
2-Methylbenzenesulfonyl chloride **2:** 762
3-Methylbenzenesulfonyl chloride **2:** 762
4-Methylbenzenesulfonyl chloride **2:** 763
Methyl benzoate **2:** 826
  physical properties **3:** 2015
2-Methylbenzoic acid **2:** 831
3-Methylbenzoic acid **2:** 831
4-Methylbenzoic acid **2:** 831
Methylbenzoic acids
  oxidative decarboxylation, for cresol production **3:** 1677
2-Methylbenzoxazole **1:** 616
2-Methyl-4,4-bis(chloromethyl)-1-pentene
  from isobutene chlorination **3:** 1393
Methyl bromide, *see bromomethane*
2-Methyl-1,3-butadiene, *see isoprene*
3-Methylbutanal, *see isovaleraldehyde*
2-Methylbutane
  physical properties **5:** 2855
3-Methylbutanoic acid, *see isovaleric acid*
2-Methyl-1-butanol **1:** 281, 284; **6:** 3612
  uses **6:** 3619
2-Methyl-2-butanol **1:** 281, 284; **6:** 3612
3-Methyl-1-butanol **1:** 281, 284; **6:** 3612
  uses **6:** 3619
3-Methyl-2-butanol **1:** 281; **6:** 3612
3-Methyl-2-butanone **5:** 3085
2-Methyl-2-butenal, *see tiglic aldehyde*
2-Methyl-1-butene
  physical properties **5:** 2871, 2871
3-Methyl-1-butene
  physical properties **5:** 2871
2-Methyl-*cis*-2-butenedioic acid, *see citraconic acid*
2-Methyl-*trans*-2-butenedioic acid, *see mesaconic acid*
2-Methyl-2-butene oxide **4:** 2001
3-Methyl-3-buten-2-one **5:** 3103
1-Methylbutylamine **1:** 454
3-Methylbutylamine **1:** 454
N-Methylbutylamine **1:** 455
Methyl *tert*-butyl ether **6:** 3337
Methyl *n*-butyl ether
  physical properties **4:** 2187
Methyl *tert*-butyl ether
  from methanol economic aspects **6:** 3311
  physical properties **4:** 2187
  properties **6:** 3338
  specifications and handling **6:** 3343
  toxicology **6:** 3346
  uses **6:** 3344
Methyl *tert*-butyl ether production **6:** 3340
  from butenes **2:** 997
  from C$_4$ cracking mixtures **2:** 993
Methyl *sec*-butyl ketone **5:** 3089
Methylbutynol **1:** 310
2-Methyl-3-butyn-2-ol **1:** 309, 310
Methyl butyrate
  physical properties **3:** 2014
2-Methylbutyric acid **2:** 1097, 1108
  specification **2:** 1106
Methyl carbamate **2:** 1047
Methyl carbamoyl chloride **2:** 1049
Methyl cellulose **2:** 1214
  properties **2:** 1216
Methyl chloride, *see monochloromethane*
Methyl chloride methyl cellulose process **2:** 1215
Methyl chloroacetate **3:** 1548
Methyl chloroformate **3:** 1576
Methyl 4-(chloromethyl)benzenesulfonate **2:** 764
Methyl 2-chloropropionate **7:** 4119
α-Methylcinnamaldehyde **1:** 382, 389
Methyl crotonate
  physical properties **3:** 1716
Methyl cyanide **6:** 3460
Methyl cyanoacetate **5:** 3185
  toxicology **5:** 3193
Methylcyclohexane **3:** 1804
  physical properties **3:** 1797
Methylcyclohexanol **3:** 1817
Methylcyclohexanone **3:** 1818
N-Methylcyclohexylamine **1:** 461
Methylcyclopentane **3:** 1804
  physical properties **3:** 1797
2-Methyldecanal **1:** 353, 365
Methyl dichloroacetate **3:** 1552
N-Methyldicyclohexylamine **1:** 463
Methyldiethanolamine *see also ethanolamines, N-alkylated*
  physical properties **4:** 2160
Methyldiisopropanolamine **4:** 2171
Methyl *N,N*-dimethyl carbamate **2:** 1047

2-Methyl-4,6-dinitrophenol **6:** 3566
4-Methyl-2,6-dinitrophenol **6:** 3566
N-Methyldioctadecylamine **1:** 474
2-Methyl-1,3-dioxane **1:** 395
2-Methyl-1,3-dioxolane **1:** 395
N-Methyldipropylenetriamine **1:** 488
2,2'-Methylenebis(6-bromo-4-chlorophenol) **6:** 3790
2,2'-Methylene bis(6-*tert*-butyl-3-methylphenol)
   physical properties **6:** 3769
2,2'-Methylenebis(4-chlorophenol) **6:** 3789
4,4'-Methylenebis(2,6-dichlorophenol) **6:** 3789
4,4'-Methylene bis(phenyl isocyanate) **1:** 634
Methylenebis(thiocyanate)
   physical properties **8:** 4632
   toxicology **8:** 4635
   uses **8:** 4634
Methylenebis(thiocyanate) production **8:** 4633
2,2'-Methylenebis(3,4,6-trichlorophenol) **6:** 3790
Methylene bromide, *see dibromomethane*
Methylene chloride, *see dichloromethane*
Methylene chloride cellulose acetate process **2:** 1171
Methylene chlorobromide, *see bromochloromethane*
Methylenediphenyl diisocyanate **5:** 3012
Methylene diphenylene isocyanate **1:** 634
2-Methyleneglutaronitrile **6:** 3462
   physical properties **6:** 3457
Methylene glycol
   in formaldehyde solutions **5:** 2642
Methylene iodide **5:** 3007
2-Methylenepentanedinitrile **6:** 3462
Methyl N-ethyl carbamate **2:** 1047
4,4'-(1-Methylethylidene)bis-(2,6-di-bromophenol, *see
   tetrabromobisphenol A*
Methylethylketone
   physical properties **5:** 3077
Methyl ethyl ketone peroxide
   toxicology **6:** 3676
N-Methylformamide **5:** 2705
Methyl formate **5:** 2736
   physical properties **3:** 2014
   toxicology **3:** 2043
Methyl formate hydrolysis
   for formic acid production **5:** 2719
2-Methylfuran **5:** 2760
   physical properties **5:** 2747
   toxicology **5:** 2767
Methyl glycolates **5:** 2966
2-Methylheptanal **1:** 352
2-Methylheptane
   physical properties **5:** 2855
3-Methylheptane
   physical properties **5:** 2855
4-Methylheptane
   physical properties **5:** 2855
5-Methyl-3-heptanone **5:** 3098
2-Methylhexanal **1:** 352

2-Methylhexane
   physical properties **5:** 2855
3-Methylhexane
   physical properties **5:** 2855
5-Methyl-2-hexanone **5:** 3092
Methyl hexyl ketone **5:** 3093
Methyl hydroperoxide
   physical properties **6:** 3634
4(5)-Methyl-5(4)-hydroxymethylimidazole
   physical properties **5:** 2988
1-Methylimidazole
   physical properties **5:** 2988
   toxicology **5:** 2994
2-Methylimidazole
   physical properties **5:** 2988
   toxicology **5:** 2993, 2994
4-Methylimidazole
   toxicology **5:** 2993, 2994
4(5)-Methylimidazole
   physical properties **5:** 2988
3-Methylindole **5:** 3002
Methyl iodide **5:** 3008
Methyl isoamyl ketone **1:** 113; **5:** 3092
Methyl isobutyl ketone **1:** 106; **5:** 3087
Methyl isobutyrate
   physical properties **3:** 2014
Methyl isocyanate **5:** 3011
Methyl isopropenyl ketone **5:** 3103
1-Methyl-4-isopropylcyclohexane
   physical properties **3:** 1798
Methyl isopropyl ether
   physical properties **4:** 2187
Methyl isopropyl ketone **5:** 3085
1-Methyl-2-isopropyl-5-nitroimidazole
   physical properties **5:** 2988
Methyl isothiocyanate
   physical properties **8:** 4636
   toxicology **8:** 4643
   uses **8:** 4641
Methyl isothiocyanate production **8:** 4639
N-Methyl J acid **6:** 3435
Methylketene
   physical properties **5:** 3077
Methyl lactate **5:** 3130
Methyl methacrylate **6:** 3265
   economic aspects **6:** 3263
   physical properties **3:** 2014; **6:** 3252
   uses **6:** 3262
Methyl methacrylate production **3:** 2030; **6:** 3254
   from acetone **1:** 103
N-Methylmethanamine, *see dimethylamine*
Methyl methanesulfinate
   synthesis **7:** 4462
Methyl 4-methylbenzenesulfonate **2:** 763
Methyl N-methyl carbamate **2:** 1047
(+)-2-Methyl-6-(4-methyl-3-cyclohexenyl)-6-hepten-2-ol **8:** 4611

(−)-2-Methyl-6-(4-methyl-3-cyclohexenyl)-6-hepten-2-ol **8:** 4611
2-Methyl-6-(4-methyl-3-cyclohexenyl)-6-hepten-2-ol **8:** 4611
4-Methyl-2-(1-methylcyclohexyl)phenol
   physical properties **6:** 3754
2-Methyl-6-methylene-2,7-octadiene **8:** 4598
1-Methyl-4-(5-methyl-1-methylene-4-hexenyl)-1-cyclohexane **8:** 4610
N-Methyl-morpholine **1:** 468
1-Methylnaphthalene **6:** 3355
   toxicology **6:** 3359
2-Methylnaphthalene **6:** 3355
   toxicology **6:** 3359
2-Methyl-1,4-naphthalenedione **6:** 3450
2-Methyl-1,4-naphthoquinone **6:** 3450
2-Methyl-3-nitroaniline **6:** 3545
2-Methyl-4-nitroaniline **6:** 3545
2-Methyl-5-nitroaniline **6:** 3545
4-Methyl-2-nitroaniline **6:** 3545
1-Methyl-2-nitrobenzene, *see 2-Nitrotoluene*
1-Methyl-3-nitrobenzene, *see 3-Nitrotoluene*
1-Methyl-5-nitroimidazole
   physical properties **5:** 2988
2-Methyl-4(5)-nitroimidazole
   physical properties **5:** 2988
Methyloctylacetaldehyde **1:** 365
Methyloligosiloxanes **7:** 4324
Methyloxirane, *see propylene oxide*
2-Methylpentanal **1:** 352, 362
3-Methylpentanal **1:** 352, 362
4-Methylpentanal **1:** 352
2-Methylpentane
   physical properties **5:** 2855
3-Methylpentane
   physical properties **5:** 2855
2-Methylpentanoic acid, *see isocaproic acid*
   specification **2:** 1106
2-Methyl-1-pentanol **1:** 281, 293, 294
   physical properties **5:** 2551
4-Methyl-1-pentanol **1:** 281
4-Methyl-2-pentanol **1:** 281, 293
3-Methyl-2-pentanone **5:** 3089
4-Methyl-2-pentanone **5:** 3087
2-Methyl-2-pentenal **1:** 368, 373
4-Methyl-4-penten-2-one **5:** 3104
Methyl pentyl ketone **5:** 3091
3-Methyl-1-pentyn-3-ol **1:** 309
2-Methylphenol, *see o-cresol*
4-Methylphenylacetaldehyde **1:** 382, 385
Methylphenylketene
   physical properties **5:** 3077
Methyl phenyl ketone **5:** 3110
Methyl phosphine
   physical properties **6:** 3821
Methyl phosphinic acid **6:** 3837

1-Methylpiperidine
   physical properties **7:** 4197
2-Methylpiperidine
   physical properties **7:** 4197
3-Methylpiperidine
   physical properties **7:** 4197
4-Methylpiperidine
   physical properties **7:** 4197
1-Methyl-2-piperidylmethanol
   physical properties **7:** 4197
2-Methylpropanal, *see isobutyraldehyde*
2-Methylpropane
   physical properties **5:** 2855
2-Methyl-1,3-propanediol **1:** 323, 331
2-Methylpropanenitrile **6:** 3461
2-Methyl-2-propanethiol production **8:** 4652
2-Methylpropanoic acid, *see isobutyric acid*
2-Methyl-1-propanol **1:** 281; **2:** 956
   toxicology **2:** 966
   uses **2:** 962
2-Methyl-2-propanol, *see tert-butanol*
2-Methylpropenal, *see methacrolein*
2-Methyl-2-propenenitrile **6:** 3462
2-Methylpropene oxidation
   for methacrolein production **1:** 200
2-Methyl-1-propene-3-sulfonic acid **8:** 4506
2-methylpropenoic acid, *see methacrylic acid*
2-Methyl-2-propenoic acid 2-propenyl ester **1:** 426
2-Methyl-2-propen-1-ol **1:** 424
Methyl propionate
   physical properties **3:** 2014
   production **3:** 2029
   properties **7:** 4119
2-Methylpropionitrile **6:** 3461
1-Methylpropylamine **1:** 454
Methyl n-propyl ether
   physical properties **4:** 2187
2-Methyl-2-propyl-1,3-propanediol **1:** 323, 330
2-Methylpyridine
   physical properties **7:** 4178
   uses **7:** 4188
3-Methylpyridine
   physical properties **7:** 4178
   uses **7:** 4188
4-Methylpyridine
   physical properties **7:** 4178
   uses **7:** 4188
6-Methyl-2-pyridinecarbonitrile
   physical properties **7:** 4202
2-Methylpyridine N-oxide
   physical properties **7:** 4196
3-Methylpyridine N-oxide
   physical properties **7:** 4196
4-Methylpyridine N-oxide
   physical properties **7:** 4196
2-Methylpyridine production **7:** 4180
3-Methylpyridine production **7:** 4181

4-Methylpyridine production **7:** 4180
6-Methyl-2-pyridinol
   physical properties **7:** 4209
6-Methyl-3-pyridinol
   physical properties **7:** 4209
1-Methyl pyrrole, *see N-methylpyrrole*
1-Methyl-1*H*-pyrrole, *see N-methylpyrrole*
N-Methyl pyrrole **7:** 4238
   toxicology **7:** 4239
1-Methyl-2-pyrrolidone **2:** 721
N-Methylpyrrolidone **1:** 139
2-Methylquinoline **7:** 4255
4-Methylquinoline **7:** 4255
Methyl salicylate **7:** 4277
   physical properties **3:** 2015
Methyl stearate
   physical properties **3:** 2014
α-Methylstyrene **1:** 96
α-Methylstyrene **7:** 4436
p-Methylstyrene **7:** 4434
2-Methyltetrahydrofuran **5:** 2761
   physical properties **5:** 2747
1-Methyl-[1*H*]-tetrazole-5-thiol **8:** 4668
5-Methyl-1,3,4-thiadiazole-2-thiol **8:** 4668
Methyl thiocyanate
   physical properties **8:** 4632
   toxicology **8:** 4635
   uses **8:** 4634
Methyl thiocyanate production **8:** 4633
2-Methylthiophene **8:** 4701
   specifications **8:** 4696
   toxicology **8:** 4704
3-Methylthiophene **8:** 4701
   specifications **8:** 4696
   toxicology **8:** 4704
N-Methyl-*m*-toluidine **8:** 4749
N-Methyl-*o*-toluidine **8:** 4749
N-Methyl-*p*-toluidine **8:** 4749
Methyl trichloroacetate **3:** 1555
2-Methylundecanal **1:** 353
6-Methyluracil
   from diketene **5:** 3075
Methyl vinyl ether **8:** 4868
   physical properties **8:** 4867
   toxicology **8:** 4874
Methyl vinyl ketone **5:** 3101
3-Methylxanthine, *see purine derivatives*
MHC benzene process **2:** 709
MIAG basket conveyor extractor
   for oilseeds **4:** 2407
MIAK, *see methyl isoamyl ketone*
MIBK, *see methyl isobutyl ketone*
Michaelis–Arbusov reaction **6:** 3833
   for production of phosphonic acids **6:** 3843
Microbiocides
   halogenated phenols **6:** 3782
Microcrystalline waxes, *see microwaxes*

Microwaxes
   composition and properties **8:** 4931
   paraffin waxes, ecology **8:** 4940
   paraffin waxes, economic aspects **8:** 4941
   product classes **8:** 4932
   production **8:** 4934
   specifications **8:** 4937
   toxicology **8:** 4961
   uses **8:** 4938
Milk sugar, *see lactose*
Millisecond cracking furnace **4:** 2246
Mitsubishi cracking furnace **4:** 2249
Mitsubishi Gas Chemical methanol process **6:** 3304
Mitsubishi-Kasei THF process **8:** 4624
Mitsubishi quench exchanger **4:** 2262
Mitsui advanced cracker **4:** 2249
Mitsui quench exchanger **4:** 2263
Mixed acid
   for nitration of aromatics **6:** 3506
Mixed soaps **6:** 3228
MNA **6:** 3541
MNPT **6:** 3545
Mobil–Badger ethylbenzene process **4:** 2213
Mobil LTD benzene process **2:** 713
Mobil selective toluene disproportionation process **8:** 4733
Mofex toluene extraction process **8:** 4732
Molar substitution
   in cellulose ethers **2:** 1204
Molecular mass
   of pectin, determination **7:** 3992
Molecular-membrane lecithin process **5:** 3152
Molybdenum-vanadium catalysts
   for propene oxidation **1:** 232
Monobutylamine **1:** 458
Monochloroacetaldehyde, *see chloroacetaldehyde*
Monochloroacetic acid, *see chloroacetic acid*
Monochloroacetylene **1:** 130
Monochloroamine **3:** 1566
Monochlorobenzene
   chemical properties **3:** 1411
   physical properties **3:** 1407
   specifications **3:** 1421
   uses **3:** 1422
Monochlorobenzene production **3:** 1417
Monochloroethane
   production **3:** 1291
   properties **3:** 1290
   toxicology **3:** 1472
   uses and economic aspects **3:** 1294
Monochloroethylene, *see vinyl chloride*
Monochloromethane **3:** 1278
   see also chloromethanes
   chemical properties **3:** 1257
   physical properties **3:** 1254
   toxicology **3:** 1470
Monochloromethane production **3:** 1265

Monochloromethanes **3**: 1252
Monochloronaphthalene
    specifications **3**: 1443
Monochloro-*o*-phenylphenol **6**: 3786
Monochlorotoluenes
    production **3**: 1429
Monodesmosides **7**: 4284
Monoethanolamine **4**: 2152
    see also ethanolamines
    physical properties **4**: 2152
    toxicology **4**: 2177
Monoethylamine **1**: 458
Monoethylene glycol, *see ethylene glycol*
Monofluoroethylene **5**: 2587
Monoglycerides
    in fats and fatty oils **4**: 2374
Monoisopropanolamine **4**: 2164
    see also Isopropanolamines
    toxicology **4**: 2179
Monoisopropylamine **1**: 458
Monomethylamine **6**: 3326
    specifications and handling **6**: 3331
    toxicology **6**: 3334
    uses **6**: 3332
Monomethylamines **6**: 3325
    see also methylamines
Monomethylethanolamine *see also ethanolamines,*
        N-alkylated
    physical properties **4**: 2160
Monomethylisopropanolamine **4**: 2171
    toxicology **4**: 2179
Monomethylurea **8**: 4819
Monoolefins, *see alkenes*
Monoperoxymaleic acid *tert*-butyl ester
    physical properties **6**: 3655
Monoperoxyphthalic acid
    physical properties **6**: 3643
Monoperoxyphthalic acid *tert*-butyl ester
    physical properties **6**: 3655
Monoperoxyphthalic acid production **6**: 3647
Monopotassium tartrate **8**: 4567
Monosaccharide phosphates **2**: 1070
Monosaccharides **2**: 1063
Monoterpenes
    acyclic **8**: 4598
    basic structure **8**: 4593
    bicyclic **8**: 4604
    monocyclic **8**: 4601
Monothiophosphates **6**: 3852
Monovinylacetylene
    in chloroprene production **3**: 1387
Monsanto acetic acid process **1**: 36
Monsanto benzene sulfonic acid process **2**: 756
Monsanto ethyl benzene process **4**: 2211
Montanic acid
    physical properties **4**: 2486

Montan wax
    derivatization **8**: 4912
    formation occurence and extraction **8**: 4907
    properties and composition **8**: 4908
    refining **8**: 4910
    toxicology **8**: 4959
    uses and economic aspects **8**: 4913
Montanyl alcohol
    physical properties **5**: 2535
Montecatini acetylene process **1**: 141
Montecatini – Edison 5-ethyl-2-methyl-pyridine process **7**: 4188
Montedison chlorofluorination process **5**: 2578
Montedison melamine process **5**: 3208
Morpholine **1**: 467
    toxicology **1**: 493
Morpholinocarbamoyl chloride
    toxicology **2**: 1054
Morphylane process **2**: 725
Morphylane toluene extraction process **8**: 4732
Morphylex toluene extraction process **8**: 4732
Motor fuel ethanol **4**: 2118
2MT **8**: 4701
3MT **8**: 4701
MTB, *see methyl* tert-*butyl ether*
MTBE, *see methyl* tert-*butyl ether*
MTC conventional urea process **8**: 4801
Mustard seed oil **4**: 2465
Mutton tallow **4**: 2470
    fatty acid content **4**: 2376
Myrcene **8**: 4598
Myricyl alcohol
    physical properties **5**: 2535
Myristic acid
    physical properties **4**: 2485
Myristoleic acid
    physical properties **4**: 2487
Myristyl alcohol **1**: 282
    physical properties **5**: 2535
    toxicology **5**: 2559

N acid **6**: 3433
Naphtha
    as feedstock for xylenes **8**: 4994
    as raw material for ethylene production **4**: 2224
    as raw material in benzene production **2**: 698
    composition **2**: 701
    cyclohexane content **3**: 1802
Naphthalene **6**: 3351
    chlorinated, *see chlorinated naphthalenes*
    economic aspects **6**: 3358
    production **6**: 3352
    properties **6**: 3351
    toxicology **6**: 3359
    uses **6**: 3354
1-Naphthaleneamine **6**: 3401
Naphthalene anthraquinone processes **2**: 653

Naphthalene derivatives **6**: 3363, 3433
1,5-Naphthalenediamine **6**: 3404
1,8-Naphthalenediamine **6**: 3405
1,8-Naphthalenedicarboxylic acid, *see naphthalic acid*
2,6-Naphthalenedicarboxylic acid **2**: 1133
Naphthalene-1,4-dicarboxylic acid dinitrile **6**: 3473
Naphthalene-2,6-dicarboxylic acid dinitrile **6**: 3473
2,6-Naphthalenedicarboxyxlic acid **2**: 1121
Naphthalene 1,5-diisocyanate **5**: 3012
1,2-Naphthalenediol **6**: 3380
1,3-Naphthalenediol **6**: 3380
1,4-Naphthalenediol **6**: 3380
1,5-Naphthalenediol **6**: 3379
1,6-Naphthalenediol **6**: 3380
1,7-Naphthalenediol **6**: 3380
1,8-Naphthalenediol **6**: 3380
2,3-Naphthalenediol **6**: 3381
2,6-Naphthalenediol **6**: 3381
2,7-Naphthalenediol **6**: 3381
1,2-Naphthalenedione **6**: 3451
1,4-Naphthalenedione **6**: 3448
Naphthalene-1,3-disulfonic acid **6**: 3369
Naphthalene-1,5-disulfonic acid **6**: 3369
Naphthalene-1,6-disulfonic acid **6**: 3370
Naphthalene-1,7-disulfonic acid **6**: 3370
Naphthalene-2,6-disulfonic acid **6**: 3370
Naphthalene-2,7-disulfonic acid **6**: 3371
Naphthalene intermediates
   code-named **6**: 3366
2-Naphthaleneneamine **6**: 3403
Naphthalene nitration process **6**: 3522
Naphthalene oil **8**: 4539
Naphthalene oxidation
   for production of 1,4-naphthoquione **6**: 3448
   to phthalic anhydride **7**: 3869
Naphthalene oxidation process **7**: 3876
Naphthalene sulfonation **6**: 3365
Naphthalene-1-sulfonic acid **6**: 3367
Naphthalene-2-sulfonic acid **6**: 3368
Naphthalene-α-sulfonic acid **6**: 3367
Naphthalene-β-sulfonic acid **6**: 3368
Naphthalene sulfonic acids
   nitration–reduction **6**: 3407
   production and properties **6**: 3365
   salts **6**: 3368
1,4,5,8-Naphthalenetetracarboxylic acid **2**: 1121, 1133
1,8;4,5-Naphthalenetetracarboxylic dianhydride **2**: 1130, 1133
Naphthalene-1,3,5,7-tetrasulfonic acid **6**: 3372
1,4,5,8-Naphthalenetetrone **6**: 3452
Naphthalene-1,3,5-trisulfonic acid **6**: 3371
Naphthalene-1,3,6-trisulfonic acid **6**: 3371
Naphthalene-1,3,7-trisulfonic acid **6**: 3372
1-Naphthalenol **6**: 3373
2-Naphthalenol **6**: 3377
Naphthalic acid **2**: 1121, 1132
Naphthalic anhydride **2**: 1130, 1132

Naphtha oxidation
   for propionic acid production **7**: 4108
Naphtha pyrolysis
   cracking yield **4**: 2236
ortho-Naphthionic acid **6**: 3410, 3411
1,2,3,4-Naphthodiquinone **6**: 3452
1-Naphthoic acid nitrile **6**: 3473
2-Naphthoic acid nitrile **6**: 3473
1-Naphthol **6**: 3373
2-Naphthol **6**: 3377
α-Naphthol **6**: 3373
β-Naphthol **6**: 3377
Naphthol sulfonation **6**: 3385
α-Naphthonitrile **6**: 3473
β-Naphthonitrile **6**: 3473
Naphthonitriles **6**: 3472
Naphtho[1,8-c,d]pyran-1,3-dione, *see naphthalic anhydride*
1,4-Naphthoquinone **6**: 3448
1,5-Naphthoquinone **6**: 3451
1,7-Naphthoquinone **6**: 3452
2,3-Naphthoquinone **6**: 3452
2,6-Naphthoquinone **6**: 3451
α-Naphthoquinone **6**: 3448, 3451
1,4-Naphthoquinone derivatives **6**: 3450
Naphthoquinone diazides **3**: 1890
Naphthoquinones **6**: 3447, 3451
1-Naphthylamine **6**: 3401
2-Naphthylamine **6**: 3403
α-Naphthylamine **6**: 3401
β-Naphthylamine **6**: 3403
Naphthylamines
   sulfonation **6**: 3407
1-Naphthyl *N*-methyl carbamate **2**: 1047
NaTA **3**: 1555
Native starches **7**: 4388
   uses **7**: 4395
Natural gas
   as source of saturated hydrocarbons **5**: 2860
   for synthesis gas production **6**: 3300
Natural waxes, *see waxes, natural*
Naval stores **8**: 4762
NDI **5**: 3012
NDUP
   physical properties **7**: 3886
Needle cokes **8**: 4532
Neoalloocimen **8**: 4600
Neopentyl alcohol **1**: 281; **6**: 3612
Neopentyl glycol **1**: 323, 326
   uses **1**: 328
Neopentyl glycol production **1**: 327
Neral **1**: 368
Neutralization
   of fats and fatty oils **4**: 2414
Neutral soaps **6**: 3228
Nevile and Winther's acid **6**: 3390

Nickel catalysts
   for fatty acid hardening **4:** 2518
Nickel 2-ethylhexanoate **6:** 3245
Nickel formate **5:** 2739
Nickel naphthenate **6:** 3245
Nickel oleate **6:** 3245
Nickel soaps **6:** 3246
Nickel stearate **6:** 3245
Nicotinic acid N-oxide
   physical properties **7:** 4196
Niger seed oil **4:** 2459
Nippon Shokubai VGR phthalic anhydride process **7:** 3870
Nissan melamine process **5:** 3209
Nitrating acid
   in production of cellulose nitrate **2:** 1146
Nitrating agents **6:** 3506
Nitrating systems
   for production of cellulose nitrate **2:** 1154
Nitration
   of anthraquinone **2:** 652
   of aromatics **6:** 3505
   of aromatics reaction mechanisms **6:** 3507
   of arylamines **1:** 510
   of cellulose **2:** 1150
   of toluene **8:** 4747
Nitration – reduction
   of naphthalenesulfonic acids **6:** 3407
Nitric acid oxidation
   of cyclohexanol, for adipic acid production **1:** 265
Nitriles **6:** 3455
   alcoholysis to esters **3:** 2030
   aliphatic, individual **6:** 3460
   aliphatic, properties and production **6:** 3456
   aliphatic, toxicology **6:** 3466
   aromatic and araliphatic, individual **6:** 3469
   aromatic and araliphatic, production **6:** 3468
   aromatic and araliphatic, properties and production **6:** 3467
   aromatic and araliphatic, toxicology **6:** 3473
   for amine production **1:** 450
   for production of pyridines **7:** 4187
   reactions **6:** 3457
Nitrilotriacetic acid **4:** 2295; **6:** 3479
   economic aspects and toxicology **6:** 3484
   properties **6:** 3479
   uses **6:** 3483
Nitrilotriacetic acid chelates **6:** 3481
Nitrilotriacetic acid production **6:** 3481
2,2′,2′′-Nitrilotriethanol, see triethanolamine
2-Nitroacetophenone **6:** 3567
3-Nitroacetophenone **6:** 3568
4-Nitroacetophenone **6:** 3568
Nitroalkanes
   for amine production **1:** 452
Nitroalkylphenols **6:** 3565
Nitroamino aromatics **6:** 3538
2-Nitroaniline **6:** 3540
3-Nitroaniline **6:** 3541
4-Nitroaniline **6:** 3541
m-Nitroaniline **1:** 637; **6:** 3541
o-Nitroaniline **1:** 637; **6:** 3540
p-Nitroaniline **1:** 637; **6:** 3541
   uses **1:** 505
2-Nitroanisole **6:** 3561
4-Nitroanisole **6:** 3563
Nitro aromatics **6:** 3503
   nitration process **6:** 3505
Nitroaromatic sulfonic acids **6:** 3548
2-Nitrobenzaldehyde **2:** 685
3-Nitrobenzaldehyde **2:** 685
4-Nitrobenzaldehyde **2:** 685
Nitrobenzene **6:** 3509
   in aniline production **1:** 630
   uses and toxicology **6:** 3513
Nitrobenzene production **6:** 3510
Nitrobenzene reduction **1:** 607
   products **6:** 3511
2-Nitrobenzenesulfonic acid **6:** 3549
3-Nitrobenzenesulfonic acid **6:** 3550
4-Nitrobenzenesulfonic acid **6:** 3551
m-Nitrobenzenesulfonic acid **6:** 3550
o-Nitrobenzenesulfonic acid **6:** 3549
p-Nitrobenzenesulfonic acid **6:** 3551
2-Nitrobenzenesulfonyl chloride **6:** 3549
3-Nitrobenzenesulfonyl chloride **6:** 3550
4-Nitrobenzenesulfonyl chloride **6:** 3551
2-Nitrobenzoic acid **2:** 835
3-Nitrobenzoic acid **2:** 835
4-Nitrobenzoic acid **2:** 835
4-Nitrobenzyl chloride **6:** 3536
Nitrocellulose, see cellulose nitrate
2-Nitrochlorobenzene **6:** 3526
3-Nitrochlorobenzene **6:** 3529
4-Nitrochlorobenzene **6:** 3529
Nitrochlorobenzenes
   toxicology **6:** 3531
2-Nitro-4-chloro-1-methylbenzene **6:** 3535
2-Nitro-6-chloro-1-methylbenzene **6:** 3535
4-Nitro-2-chloro-1-methylbenzene **6:** 3535
Nitro compounds
   aromatic, see nitro aromatics
2-Nitro-p-cresol **6:** 3565
4-Nitro-m-cresol **6:** 3565
2-Nitrocyclohexanone **1:** 67
2-Nitrocyclohexanone caprolactam process **2:** 1034
2-Nitro-1,4-dichlorobenzene **6:** 3533
3-Nitro-1,2-dichlorobenzene **6:** 3533
5-Nitro-2-furaldehyde **6:** 3571
Nitrofurans **6:** 3571
Nitroglycerin **5:** 2804
Nitroguanidine **5:** 2832
Nitrohalo aromatics **6:** 3526
Nitroheterocycles **6:** 3568

2-Nitrohydroquinone diethyl ether **6**: 3564
2-Nitrohydroquinone dimethyl ether **6**: 3564
Nitroimidazole **6**: 3570
　physical properties **5**: 2988
Nitroimidazoles **5**: 2990
　toxicology **5**: 2995
Nitroketones **6**: 3567
4-Nitro-1-methylbenzene **6**: 3519
3-Nitro-4-methylbenzenesulfonic acid **6**: 3552
5-Nitro-2-methylbenzenesulfonic acid **6**: 3553
1-Nitronaphthalene **6**: 3523
2-Nitronaphthalene **6**: 3524
α-Nitronaphthalene **6**: 3523
β-Nitronaphthalene **6**: 3524
5-Nitro-1,4-naphthalenedione **6**: 3450
5-Nitro-1,4-naphthoquinone **6**: 3450
4-Nitroperoxy-benzoic acid
　physical properties **6**: 3643
2-Nitrophenetole **6**: 3561
4-Nitrophenetole **6**: 3563
2-Nitrophenol **6**: 3560
3-Nitrophenol **6**: 3562
4-Nitrophenol **6**: 3562
m-Nitrophenol **6**: 3562
o-Nitrophenol **6**: 3560
p-Nitrophenol **6**: 3562
Nitrophenols
　reduction with iron turnings **1**: 606
1-(3-Nitrophenyl)ethanone **6**: 3568
2-Nitrophenyl ethyl ether **6**: 3561
4-Nitrophenyl ethyl ether **6**: 3563
2-Nitrophenyl methyl ether **6**: 3561
4-Nitrophenyl methyl ether **6**: 3563
4-Nitrophenyl methyl ketone **6**: 3568
Nitropyridine **6**: 3569
2-Nitro-3-pyridinol
　physical properties **7**: 4209
Nitroquinoline **6**: 3570
Nitrosation
　of saturated hydrocarbons **5**: 2866
5-Nitro-2-thiazolamine **6**: 3572
Nitro thiazoles **6**: 3572
Nitrothiophene **6**: 3571
2-Nitrotoluene **6**: 3515
2-Nitrotoluene
　production **6**: 3517
3-Nitrotoluene **6**: 3518
4-Nitrotoluene **6**: 3519
m-Nitrotoluene **6**: 3518
o-Nitrotoluene, *see* 2-Nitrotoluene
p-Nitrotoluene **6**: 3519
Nitrotoluenes
　reduction **8**: 4747
　toxicity **6**: 3520
2-Nitrotoluene-4-sulfonic acid **6**: 3552
4-Nitrotoluene-2-sulfonic acid **6**: 3553
p-Nitrotoluene-o-sulfonic acid **6**: 3553

3-Nitro-p-toluidine **6**: 3545
4-Nitro-o-toluidine **6**: 3545
5-Nitro-o-toluidine **6**: 3545
6-Nitro-o-toluidine **6**: 3545
m-Nitro-p-toluidine **6**: 3545
1-Nitro-2,4,5-trichlorobenzene **6**: 3534
Nitrous acid
　for amine oxidation **1**: 448
Nitroxylenes **6**: 3521
　reduction **8**: 5021
NMP cyclopentene process **3**: 1835
NMR shift reagents
　complexes of fluorinated 1,3-diketones **5**: 2600
1-Nonacosanol
　physical properties **5**: 2535
n-Nonadecane
　physical properties **5**: 2855
1,19-Nonadecanediol
　physical properties **5**: 2553
1-Nonadecanol **1**: 282
　physical properties **5**: 2535
2-Nonadecyl-1-tricosanol
　physical properties **5**: 2551
Nonanal, *see pelargonaldehyde*
n-Nonane
　physical properties **5**: 2855
Nonanedioic acid, *see azelaic acid*
1,9-Nonanediol
　physical properties **5**: 2553
n-Nonanoic acid, *see* n-*pelargonic acid*
1-Nonanol, *see pelargonic alcohol*
1-Nonene
　physical properties **5**: 2871
2-Nonene
　physical properties **5**: 2871
3-Nonene
　physical properties **5**: 2871
4-Nonene
　physical properties **5**: 2871
Nonfossil waxes **8**: 4881
4-Nonylphenol **6**: 3748
Nonylphenol[c] **6**: 3748
Nonylphenol production **6**: 3750
2-Nonyl-1-tridecanol
　physical properties **5**: 2551
Notched impact strength
　of cellulose ester thermoplastics **2**: 1187
NTA, *see nitrilotriacetic acid*
Nuarimol **7**: 4233
Nucleophilic displacement
　of bromine **2**: 869
Nucleophilic substitution
　for arylamine production **1**: 521
　of pyridines **7**: 4179
　of pyrimidines **7**: 4228
Nutrition
　agar **7**: 4012

alginate **7**: 4002
carrageenan **7**: 4009
pectin **7**: 3992
starches **7**: 4395
NW acid **6**: 3390
Nylon **5**: 2845
Nylon fibers
  from adipic acid **1**: 272
Nylon-6 waste
  recovery of caprolactam **2**: 1031

Ocimene **8**: 4600
OCNB **6**: 3526
Octabromobiphenyl oxide
  toxicology **2**: 891
Octacosanoic acid, *see montanic acid*
1-Octacosanol, *see montanyl alcohol*
trans,trans-9,12-Octadecadienoic acid, *see linolelaidic acid*
(Z,Z)-9,12-Octadecadien-1-ol, *see linoleyl alcohol*
n-Octadecane
  physical properties **5**: 2855
1,18-Octadecanediol
  physical properties **5**: 2553
Octadecanoic acid, *see stearic acid*
1-Octadecanol, *see stearyl alcohol*
  toxicology **5**: 2559
cis,cis,cis-9,12,15-Octadecatrienoic acid, *see linolenic acid*
cis,trans,trans-9,11,13-Octadecatrienoic acid, *see α-eleostearic acid*
trans,trans,trans-9,11,13-Octadecatrienoic acid, *see β-eleostearic acid*
trans,trans,trans-9,12,15-Octadecatrienoic acid, *see linolenelaidic acid*
(Z,Z,Z)-9,12,15-Octadecatrien-1-ol, *see linolenyl alcohol*
1-Octadecene
  physical properties **5**: 2871
(Z)-9-Octadecene-1,12-diol
  physical properties **5**: 2553
cis-6-Octadecenoic acid, *see petroselinic acid*
cis-9-Octadecenoic acid, *see oleic acid*
trans-6-Octadecenoic acid, *see petroselaidic acid*
trans-9-Octadecenoic acid, *see elaidic acid*
9-Octadecen-1-ol **1**: 309
cis-9-octadecen-1-ol, *see oleyl alcohol*
(E)-9-Octadecen-1-ol, *see elaidyl alcohol*
Octadecylamine **1**: 473
2-Octadecyl-1-docosanol
  physical properties **5**: 2551
Octadecyl hydroperoxide
  physical properties **6**: 3634
Octadecyl vinyl ether
  physical properties **8**: 4867
Octafluorocyclobutane
  properties **5**: 2574
Octafluoropropane
  properties **5**: 2574

n-Octane
  physical properties **5**: 2855
  toxicology **5**: 2925
Octanedioic acid, *see suberic acid*
1,2-Octanediol **1**: 337
1,8-Octanediol
  physical properties **5**: 2553
Octane enhancer
  MTBE **6**: 3344
Octane number
  of saturated hydrocarbons **5**: 2858
  of toluene and motor fuel **8**: 4731
  of xylenes **8**: 4993
Octane ratings **2**: 698
tert-Octanethiol
  toxicology **8**: 4678
Octanoic acid, *see caprylic acid*
1-Octanol, *see caprylic alcohol*
  toxicology **5**: 2558, 2559
2-Octanol **1**: 298
2-Octanone **5**: 3093
1-Octene
  physical properties **5**: 2871
cis-2-Octene
  physical properties **5**: 2871
cis-3-Octene
  physical properties **5**: 2871
cis-4-Octene
  physical properties **5**: 2871
trans-2-Octene
  physical properties **5**: 2871
trans-3-Octene
  physical properties **5**: 2871
trans-4-Octene
  physical properties **5**: 2871
Octylamine **1**: 455, 458, 473
2-Octyl-1-dodecanol
  physical properties **5**: 2551
  toxicology **5**: 2559
4-*tert*-Octylphenol **6**: 3748
tert-Octylphenol **6**: 3748
5-*tert*-Octylphenol production **6**: 3749
Octylphenols
  production **6**: 3750
Odorants
  benzaldehyde **2**: 683
Off-gas
  from thermal methane chlorination **3**: 1271
Oil
  crude, composition **2**: 700
Oil of mirbane, *see nitrobenzene*
Oil of turpentine, *see turpentine*
Oil of wintergreen **7**: 4277
Oil quench
  prequench, in acetylene arc processes **1**: 149
Oil seeds
  processing and conditioning **4**: 2401

5153

Oil-shale tars
    processing **8**: 4551
    properties and composition **8**: 4513
Oiticica oil **4**: 2465
Oleananes **7**: 4295
Olefin hydration
    for butanol production **2**: 959
    for production of aliphatic alcohols **1**: 290
Olefin metathesis
    for isoprene synthesis **5**: 3040
Olefin oxidation
    for aliphatic aldehyde production **1**: 359
    for production of allyl esters **1**: 426
    for production of unsaturated aldehydes **1**: 370
Olefins **5**: 2870
    acylation **3**: 2031
    addition to phosphine **6**: 3823
    carbonylation **3**: 2029
    epoxidation with percarboxylic acids **4**: 1992
    for amine production **1**: 452
    hydroformylation for fatty alcohol production **5**: 2544
    lower, oligomerization **5**: 2876
    reactions **5**: 2872
    removal from pyrolysis gasolines **2**: 715
Oleflex propene process **7**: 4090
Oleic acid
    in fatty acid biosynthesis **4**: 2378
    ozonolysis **3**: 1908
    physical properties **4**: 2487
Oleic–linoleic acid oils **4**: 2455
Olex olefin extraction process **5**: 2875
Oleyl alcohol **1**: 309, 310
    physical properties **5**: 2549
    toxicology **5**: 2559
Oleyl amine **1**: 472
Oligoamines **1**: 487
Oligomerization
    of ethylene, SHOP process **5**: 2879
    of propene and butenes **5**: 2883
    production of dienes and polyenes **5**: 2891
Oligosaccharides **2**: 1066
Olive oil **4**: 2443
    fatty acid content **4**: 2376
ONA **6**: 3540
Once–through processes
    in urea production **8**: 4794
ONDP
    physical properties **7**: 3886
ONT, *see 2-Nitrotoluene*
Oppenauer oxidation **1**: 283
Optical isomerism
    of tartaric acid **8**: 4562
Organophosphorus compounds **6**: 3819
    economic aspects **6**: 3855
    toxicology **6**: 3856

Organosilicon compounds **7**: 4305
    economic aspects **7**: 4356
    nomenclature and classification **7**: 4307
    organofunctional **7**: 4330
    silicon-functional **7**: 4319
    structure and reactivity **7**: 4308
    synthesis **7**: 4310
    toxicology and environmental aspects **7**: 4355
    uses **7**: 4345
Orthanilic acid **1**: 637
Ortho-alkylation
    of phenols **6**: 3722
Orthocaine **1**: 616
Orthocarboxylic acid esters **3**: 2033
Orthoform New **1**: 616
Orthosilicic acid esters **7**: 4326
Ortic acid **7**: 4230
OTBG **5**: 2840
Otto engines
    methanol as fuel **6**: 3312
Ouricury wax **8**: 4895
7-Oxabicyclo[4.1.0]heptane-3-carboxylic acid-7′-oxabicyclo[4.1.0]hept-3′-ylmethyl ester **4**: 2004
Oxalates
    metal **6**: 3593
    organic **6**: 3595
Oxalic acid **6**: 3577
    anhydrous **6**: 3578
    anhydrous, production **6**: 3589
    economic aspects **6**: 3592
    handling **6**: 3593
    physical properties **3**: 1905
    properties **6**: 3578
    toxicology **6**: 3597
    uses **6**: 3590
Oxalic acid derivatives **6**: 3593
Oxalic acid dihydrate **6**: 3579
Oxalic acid esters **6**: 3595
Oxalic acid production **6**: 3581
Oxaloacetic acid **6**: 3608
Oxamide **6**: 3595
Oxetan-2-one, *see β-propiolactone*
Oxidation
    of acetaldehyde **1**: 6, 43, 75
    of aldehydes **1**: 356
    of alkylphenols **6**: 3717
    of amines **1**: 447
    of arylamines **1**: 511
    of butanols **2**: 955
    of butene, mechanism **1**: 39
    of butenes **2**: 989
    of butyraldehydes **2**: 931
    of chlorophenols **3**: 1608
    of cresols **3**: 1659
    of *p*-diisopropyl benzene **1**: 100
    of hydrocarbons, for aliphatic aldehyde production **1**: 359

of montan wax **8:** 4911
of olefins, for aliphatic aldehyde production **1:** 359
of olefins, for production of unsaturated aldehydes **1:** 370
of pentanols **6:** 3614
of phthalocyanines **7:** 3928
of primary alcohols to aldehydes **1:** 358
of 2-propanol **1:** 99
of propene, for acetone production **1:** 98
of toluene **2:** 679, 821, 855; **6:** 3698
phthalic anhydride **7:** 3869
with molecular oxygen, for production of aromatic carboxylic acids **2:** 1127
Oxidative cleavage
of cyclic compounds **3:** 1910
Oxidative dehydrogenation
of primary alcohol **1:** 358
N-Oxides
reaction with acetic anhydride S-Oxides **1:** 69
Oxidized petrolatum
uses **8:** 4938
Oxidized starch **7:** 4399
Oximation process
for ε-caprolactam production **2:** 1017
Oximinosilanes **7:** 4327
Oxine **7:** 4255
Oxirane, *see ethylene oxide*
Oxiranes, *see epoxides*
Oxoacetic acid, *see glyoxylic acid*
Oxo alcohols
from α-olefins **5:** 2884
3-Oxo-butanenitrile **6:** 3465
Oxocarboxylic acids **6:** 3599
4-Oxo-3,5-dialkoxymethyltetrahydro-1,3,5-oxadiazines **8:** 4829
Oxo-D process **2:** 909
3-Oxoglutaric acid **6:** 3607
α-Oxoglutaric acid **6:** 3608
2-Oxohexahydropyrimidine **8:** 4820
2-Oxohexamethylenimine, *see ε-caprolactam*
2-Oxo-4-hydroxy-5,5-dimethylhexahydropyrimidyl-*N,N'*-bisneopentals **8:** 4833
2-Oxo-4-hydroxy-5,5-dimethyl-6-isopropylhexahydropyrimidine **8:** 4831
2-Oxo-5-hydroxyhexahydropyrimidine **8:** 4820
Oxolane, *see tetrahydrofuran*
Oxole, *see furan*
Oxolene **6:** 3452
3-Oxopentane dicarboxylic acid **6:** 3607
4-Oxopentanoic acid **6:** 3606
Oxo process
for fatty alcohol production **5:** 2544
2-Oxopropanoic acid **6:** 3600
2-Oxosuccinic acid **6:** 3608
Oxo synthesis
for production of aliphatic alcohols C_3-C_{20} **1:** 285
for production of butanols **2:** 956

of aldehydes **1:** 358
2-Oxo-4-ureido-6-methylhexahydropyrimidine **8:** 4832
Oxyacetylene flames **1:** 182
Oxyalkylation
of amines **1:** 446
4-Oxyanilinoacetic acid **1:** 621
1,1'-Oxybis-(2-chloroethanol) **3:** 1522
1,1'-Oxybis(2,3,4,5,6-pentabromobenzene) **2:** 885
Oxy C acid **6:** 3442
Oxy Chicago acid **6:** 3395
Oxychlorination
allyl chloride production **1:** 416
of benzene **3:** 1416
of ethylene in the gas phase **3:** 1302
of ethylene in the liquit-phase **3:** 1308
of toluene **3:** 1428
of toluene, in cresol production **3:** 1676
tetrachloromethane production **3:** 1278
Oxydiethylene carbamoyl chloride **2:** 1049
Oxy-EDC process **3:** 1302, 1306
monochloroethane as a byproduct **3:** 1294
Oxygen
production of dialkyl peroxides **6:** 3639
production of hydroperoxides **6:** 3636
Oxy Koch acid **6:** 3398
Oxy L acid **6:** 3390
α-Oxyperoxides **6:** 3662
production **6:** 3666
uses **6:** 3669
Oxy Tobias acid **6:** 3392
Ozocerite **8:** 4881
Ozonides **6:** 3663
Ozonolysis
of oleic Acid **3:** 1908

Paal–Knorr pyrrole synthesis **7:** 4237
Pacol–Olex olefin process **5:** 2875
PAH, *see polycyclic aromatic hydrocarbons*
Paints
use of cellulose ethers **2:** 1238
Palladium complexes
as catalysts in ethylene oxidation to ethylene glycol **4:** 2314
Palmitic acid
physical properties **4:** 2485
Palmitic acid oils **4:** 2452
Palmitic–stearic acid oils **4:** 2449
Palmitoleic acid
physical properties **4:** 2487
Palm kernel oil **4:** 2447
fatty acid content **4:** 2376
Palm oil **4:** 2442
fatty acid content **4:** 2376
Palm seed oil **4:** 2448
Paper
use of cellulose ethers **2:** 1239

Paracetamol **1:** 619
Paraffin chlorination **3:** 1400
Paraffins
  chlorinated, *see chlorinated paraffins*
  for fatty alcohol production **5:** 2545
  for production of higher olefins **5:** 2873
Paraffin waxes
  composition and properties **8:** 4915
  handling **8:** 4927
  in crude petroleum **8:** 4920
  product classes **8:** 4916
  production **8:** 4920
  specifications **8:** 4928
  toxicology **8:** 4961
  uses **8:** 4930
Paraformaldehyde **5:** 2674
  specifications and uses **5:** 2676
Paraformaldehyde production **5:** 2675
Paraldehyde **1:** 20
  for production of pyridines **7:** 4185
  toxicology **1:** 24
Parex *p*-xylene process **8:** 5001
Partial combustion acetylene processes **1:** 132
Partial combustion carbide acetylene process **1:** 133, 144
Partial combustion cracking acetylene process **1:** 166
Partial oxidation
  of natural gas **6:** 3300
  of saturated hydrocarbons **5:** 2865
Payne hydrogen peroxide system **4:** 1995
PCB, *see chlorinated biphenyls*
PCMX **6:** 3786
PCNB **6:** 3529
Peanut oil **4:** 2462
  fatty acid content **4:** 2376
Peat tar **8:** 4510
  processing **8:** 4551
  properties and composition **8:** 4513
Pectase **7:** 3978
Pectate hydrolase **7:** 3978
Pectate lyase **7:** 3979
Pectic acid **2:** 1068
Pectic acid lyase **7:** 3979
Pectic acid transeliminase **7:** 3979
Pectin
  analysis **7:** 3990
  application in the food industry **7:** 3993
  derivatization **7:** 3989
  occurrence **7:** 3975
  properties **7:** 3982
  stability and chemical reactions **7:** 3988
  structure **7:** 3976
Pectinase **7:** 3978, 3979
Pectin demethoxylase **7:** 3978
Pectinesterase **7:** 3978
Pectin gel
  properties **7:** 3984

Pectin lyase **7:** 3979
Pectin methylesterase **7:** 3978
Pectin pectylhydrolase **7:** 3978
Pectin production **7:** 3979
Pectins **2:** 1071
Pectin – sugar – acid gels **7:** 3985
Pectin transeliminase **7:** 3979
Pectolytic enzymes **7:** 3978
Pelargonaldehyde **1:** 353, 364
n-Pelargonic acid **2:** 1097, 1110
  physical properties **4:** 2485
Pelargonic acid production **4:** 2522
Pelargonic alcohol
  physical properties **5:** 2535
Pentabromobiphenyl oxide **2:** 883
Pentabromoethylbenzene **2:** 883
2,3,4,5,6-Pentachlorobenzal chloride
  physical properties **3:** 1466
Pentachlorobenzene
  chemical properties **3:** 1412
  physical properties **3:** 1410
Pentachlorobenzene production **3:** 1420
2,3,4,5,6-Pentachlorobenzyl chloride
  physical properties **3:** 1466
Pentachloroethane
  properties **3:** 1325
  toxicology **3:** 1474
  uses and economic aspects **3:** 1327
Pentachloroethane production **3:** 1327
Pentachloronitrobenzene **6:** 3534
Pentachlorophenol **3:** 1606
  physical properties **3:** 1606
Pentachloropyridine
  physical properties **7:** 4199
Pentachlorothiophenol
  toxicology **8:** 4678
Pentachlorothiophenol production **8:** 4664
Pentachlorotoluene
  physical properties **3:** 1425
  production **3:** 1432
1,25-Pentacosanediol
  physical properties **5:** 2553
1-Pentacosanol
  physical properties **5:** 2535
n-Pentadecane
  physical properties **5:** 2855
1,15-Pentadecanediol
  physical properties **5:** 2553
Pentadecanoic acid
  physical properties **4:** 2485
1-Pentadecanol **1:** 282
  physical properties **5:** 2535
1-Pentadecene
  physical properties **5:** 2871
2-Pentadecyl-1-nonadecanol
  physical properties **5:** 2551
1,3-Pentadiene **5:** 2889

Pentaerythritol **1:** 324, 340
  toxicology **1:** 343
Pentaerythritol allyl ethers **1:** 431
Pentaerythritol esters **1:** 342
Pentaethylbenzene
  physical properties **5:** 2894
Pentafluoroethane
  properties **5:** 2574
Pentafluoroiodoethane
  properties **5:** 2582
1,2,3,4,5-Pentahydroxypentane-1-carboxylic acid, *see* D-gluconic acid
Pentamethylbenzene **5:** 2896
  physical properties **5:** 2894
Pentamethylenediamine **1:** 486
3,3-Pentamethyleneoxaziridine caprolactam process **2:** 1035
Pentanal, *see valeraldehyde*
Pentandial, *see glutardialdehyde*
n-Pentane
  physical properties **5:** 2855
  toxicology **5:** 2924
  uses **5:** 2868
Pentanedioic acid, *see glutaric acid*
1,2-Pentanediol **1:** 324, 337
1,5-Pentanediol **1:** 323, 326
2,3-Pentanedione **5:** 3109
2,4-Pentanedione **5:** 3106
Pentane-1-sulfonic acid
  physical properties **8:** 4504
1-Pentanol **1:** 281, 284; **6:** 3612
  uses **6:** 3618
2-Pentanol **1:** 281, 284; **6:** 3612
3-Pentanol **1:** 281, 284; **6:** 3612
Pentanols **6:** 3611
  production **6:** 3615
  specifications **6:** 3620
  toxicology **6:** 3621
  uses **6:** 3618
3-Pentanone **5:** 3095
Pentaoxacyclodecane **5:** 2681
Pentaoxane **5:** 2681
Pentaoxymethylene **5:** 2681
2-Pentenal **1:** 368
1-Pentene
  physical properties **5:** 2871
cis-2-Pentene
  physical properties **5:** 2871
trans-2-Pentene
  physical properties **5:** 2871
Pentenes
  hydration to pentanols **6:** 3617
Pentenic acid, *see diethylenetriamine pentaacetic acid*
Pentosan/furfural THF processes **8:** 4626
Pentoses **2:** 1063
  fermentation by bacteria **4:** 2080
  fermentation pathways in yeasts **4:** 2069

1,3,5,7,9-Pentoxecane **5:** 2681
Pentyl acetate
  physical properties **3:** 2014
Pentylamine **1:** 454
tert-Pentylbenzene
  physical properties **5:** 2902
n-Pentyl carbamate **2:** 1047
2-Pentyl-1-nonanol
  physical properties **5:** 2551
2-*tert*-Pentyl-phenol
  physical properties **6:** 3747
3-*tert*-Pentyl-phenol
  physical properties **6:** 3747
4-*tert*-Pentyl-phenol
  physical properties **6:** 3747
tert-Pentylphenols **6:** 3745
Per, *see tetrachloroethylene*
Peracetic acid
  buffered equilibrium **4:** 1994
  physical properties **6:** 3643
  toxicology **6:** 3676
Peracetic acid *tert*-butyl ester
  physical properties **6:** 3655
Perbenzoic acid **6:** 3643
Perbenzoic acid *tert*-butyl ester
  physical properties **6:** 3655
Perbenzoic acid cumyl ester
  physical properties **6:** 3655
Perc, *see tetrachloroethylene*
Percarboxylic acids
  for epoxidation **4:** 1992
Perchlorfluoroacetaldehydes **5:** 2598
Perchlorination
  for production of tetrachlormethane **3:** 1274
Perchlorobromo copper phthalocyanines **7:** 3945
Perchloro copper phthalocyanines **7:** 3945
Perchloroethylene, *see tetrachloroethylene*
Percolation process
  for refining deoiled slack waxes **8:** 4924
Perfluoroacetic acid **5:** 2602
Perfluoroadipic acid **5:** 2604
Perfluoroalcohols **5:** 2591
Perfluoroalkanedisulfonyl fluorides **5:** 2606
Perfluoroalkanes **5:** 2573
Perfluoroalkanesulfonic acids **5:** 2605; **8:** 4504
Perfluoroalkyl-*tert*-amines **5:** 2607
Perfluorobutyric acid **5:** 2602
Perfluorocapric acid **5:** 2602
Perfluorocaproic acid **5:** 2602
Perfluorocaprylic acid **5:** 2602
Perfluorocarboxylic acids **5:** 2601
Perfluorodicarboxylic acids **5:** 2604
Perfluorododecanoic acid **5:** 2602
Perfluoroepoxides **5:** 2592
Perfluoroethers **5:** 2591
Perfluoroethylene, *see tetrafluoroethylene*
Perfluoroglutaric acid **5:** 2604

Perfluoroheptanoic acid **5**: 2602
Perfluoro-1-iodo-*n*-butane
   properties **5**: 2582
Perfluoro-1-iodo-*n*-decane
   properties **5**: 2582
Perfluoro-1-iodo-*n*-heptane
   properties **5**: 2582
Perfluoro-1-iodo-*n*-hexane
   properties **5**: 2582
Perfluoro-1-iodo-*n*-pentane
   properties **5**: 2582
Perfluoro-1-iodopropane
   properties **5**: 2582
Perfluoro-2-iodopropane
   properties **5**: 2582
Perfluoromalonic acid **5**: 2604
Perfluorononanoic acid **5**: 2602
Perfluoropropionic acid **5**: 2602
Perfluorosuberic acid **5**: 2604
Perfluorosuccinic acid **5**: 2604
Perfluorotriamylamine **5**: 2608
Perfluorotributylamine **5**: 2608
Perfluorotriethylamine **5**: 2608
Perfluorotrihexylamine **5**: 2608
Perfluorotrimethylamine **5**: 2608
Perfluorotripropylamine **5**: 2608
Perfluoroundecanoic acid **5**: 2602
Perfluorovaleric acid **5**: 2602
Perfluorovinyl ether copolymers **5**: 2604, 2607
Perfluorovinyl ethers **5**: 2594
Performic acid
   physical properties **6**: 3643
Performic acid *tert*-butyl ester
   physical properties **6**: 3655
Peri acid **6**: 3413
Perilla oil **4**: 2459
Perkin reaction **1**: 70; **3**: 1641
Permanganate absorption number **2**: 1037
Permanganate number **2**: 1038
Peroxides, organic **6**: 3629
   classification **6**: 3630
   dialkyl, production **6**: 3639
   dialkyl, properties **6**: 3638
   dialkyl, uses **6**: 3641
   formation in dioxane **4**: 1938
   safety and handling **6**: 3672
   toxicology **6**: 3675
Peroxyacetic acid
   physical properties **6**: 3643
Peroxybenzoic acid
   physical properties **6**: 3643
Peroxybutanoic acid
   physical properties **6**: 3643
Peroxycarbamic acid *tert*-butyl ester
   physical properties **6**: 3655
Peroxycarboxylates **6**: 3653

Peroxycarboxylic acids
   acylation **6**: 3651
   for propene epoxidation **7**: 4144
   production **6**: 3644
   properties **6**: 3641
   uses **6**: 3649
Peroxycarboxylic esters **6**: 3653
   production **6**: 3657
Peroxy compounds
   organic, *see peroxides, organic*
1,1'-Peroxydicyclohexylamine
   physical properties **6**: 3670
1,1'-Peroxydicyclohexylamine caprolactam process **2**: 1033
1,1'-Peroxydicyclopentylamine
   physical properties **6**: 3670
2,2'-Peroxydiisobutylamine
   physical properties **6**: 3670
2,2'-Peroxydiisopropylamine
   physical properties **6**: 3670
Peroxydodecanoic acid
   physical properties **6**: 3643
Peroxydodecanoic acid production **6**: 3648
Peroxyesters **6**: 3653
   uses **6**: 3662
Peroxyformic acid
   physical properties **6**: 3643
Peroxyisobutyric acid *tert*-butyl ester
   physical properties **6**: 3655
Peroxynonanoic acid
   physical properties **6**: 3643
Peroxynonanoic acid *tert*-butyl ester,
   physical properties **6**: 3655
Peroxypropanoic acid
   physical properties **6**: 3643
Pervaporation
   ethanol purification **4**: 2122
Perylene dianhydride **2**: 1130, 1132
3,4,9,10-Perylenetetracarboxylic acid **2**: 1121, 1132
Perylo[3,4-c,d:9,10-c',d']dipyran-1,3,8,10-tetrone, *see perylene dianhydride*
Pesticides
   benzotrifluorides **5**: 2612
   dithiocarbamic acids **4**: 1971
   thioureas **8**: 4718
   xanthates **8**: 4979
n-Petanoic acid, *see n-valeric acid*
Petrolatum **8**: 4932
Petroleum
   as feedstock for xylenes **8**: 4994
   as raw material in benzene production **2**: 698
   as source of saturated hydrocarbons **5**: 2860
   for BTX production **2**: 704
   paraffin wax content **8**: 4920
Petroleum waxes
   classification **8**: 4914

Petroselaidic acid
   physical properties **4:** 2487
Petroselinic acid
   physical properties **4:** 2487
Pharmaceuticals
   based on guanidine **5:** 2830
   chlorophenols **3:** 1613
   from imidazoles **5:** 2991
   from indoles **5:** 3004
   from isothiocyanates **8:** 4642
   from 1-naphthol **6:** 3375
   from thiols **8:** 4656
   from xylidines **8:** 5024
   heterocyclic thiols and sulfides **8:** 4674
   lactulose **5:** 3145
   organosilicon compounds **7:** 4347
   purine derivatives **7:** 4170
   pyridines **7:** 4213
   pyrimidines **7:** 4232
   saponins **7:** 4287
   thioureas **8:** 4718
   xanthates **8:** 4979
Phase equilibria
   in urea production **8:** 4792
Phenacetin **1:** 620
Phenanthrene **5:** 2920
Phenanthrene-9-10-quinone **5:** 2922
β-Phenethyl bromide, *see 1-bromo-2-phenylethane*
o-Phenetidine **1:** 612, 636
p-Phenetidine **1:** 636
Phenetole **6:** 3779
Phenol **6:** 3689
   economic aspects **6:** 3706
   environmental protection **6:** 3704
   from tar fractions **8:** 4540
   properties **6:** 3690
   reaction with acrolein **1:** 204
   specifications and handling **6:** 3705
   toxicology **6:** 3709
Phenol alkylation **6:** 3721
Phenol amination
   for aniline production **1:** 632
Phenol chlorination **3:** 1609
Phenol derivatives **6:** 3713
Phenol distillation **8:** 4544
Phenol ethers **6:** 3778
Phenol hydrogenation **3:** 1609
Phenol hydroxylation
   for catechol production **6:** 3760
   for production of hydroquinone **5:** 2946
Phenol methylation
   for cresol production liquid-phase **3:** 1675
   for xylenol production **3:** 1934
Phenol production **6:** 3693
Phenol-water mixtures **6:** 3692
Phenoraffin process **8:** 4542
Phenoxyacetic acid **6:** 3780

m-Phenoxybenzyl bromide, *see 1-(Bromomethyl)-3-phenoxybenzene*
2-Phenoxyethanol **6:** 3780
Phenoxy herbicides, *see chlorophenoxyalkanoic acids*
Phenylacetaldehyde **1:** 384
2-Phenylacetaldehyde **1:** 382
Phenylacetic acid **1:** 57; **6:** 3805
Phenylacetone **5:** 3112
Phenylacetonitrile **6:** 3469
Phenylamine, *see aniline*
N-Phenylaminobenzene, *see diphenylamine*
Phenyl 2-aminobenzenesulfonate **2:** 770
3-(*N*-Phenylamino)phenol **1:** 617
Phenylated oxiranes
   for production of araliphatic aldehydes **1:** 383
Phenylbenzene, *see biphenyl*
Phenyl benzoate **2:** 827
Phenyl carbamate **2:** 1047
Phenyl carbamoyl chloride **2:** 1049
Phenyl chloroform, *see benzotrichloride*
Phenyl chloroformate **3:** 1576
N-Phenyl-diethylamine, *see N,N-diethylaniline*
1,2-Phenylenediacetonitrile **6:** 3470
m-Phenylenediamine
   properties **6:** 3810
   toxicology **6:** 3816
   uses **6:** 3815
o-Phenylenediamine
   properties **6:** 3810
   toxicology **6:** 3816
   uses **6:** 3814
p-Phenylenediamine
   properties **6:** 3810
   toxicology **6:** 3816
   uses **6:** 3815
m-Phenylenediamine production **6:** 3812
o-Phenylenediamine production **6:** 3811
p-Phenylenediamine production **6:** 3812
Phenylenediamines **6:** 3809
   properties **6:** 3810
   specifications and handling **6:** 3813
p-Phenylene diisocyanate **5:** 3012
N,N'-o-Phenyleneguanidine **5:** 2838
Phenylethane, *see ethylbenzene*
Phenylethylene, *see styrene*
Phenylethylene oxide **4:** 2003
Phenyl ethyl ether **6:** 3779
Phenylglycolic acid **5:** 2982
Phenylglyoxylonitrile **6:** 3470
2-Phenylimidazole
   physical properties **5:** 2988
Phenylimidazoles
   toxicology **5:** 2995
Phenyl isocyanate **5:** 3011
Phenyl isothiocyanate
   physical properties **8:** 4636
   toxicology **8:** 4643

N-Phenyl J acid **6:** 3435
Phenylketene
　physical properties **5:** 3077
Phenyl methyl ether **6:** 3779
Phenyloxirane **4:** 2003
2-Phenylphenol production **6:** 3775
Phenyl phosphine
　physical properties **6:** 3821
Phenylphosphinic acid **6:** 3837
2-Phenylpropanal **1:** 382, 385
3-Phenylpropanal **1:** 382, 386
2-Phenylpropane, *see cumene*
1-Phenyl-1-propanone **5:** 3111
1-Phenyl-2-propanone **5:** 3112
3-Phenyl-2-propenal, *see cinnamaldehyde*
3-Phenylpropenoic acid, *see cinnamic acid*
trans-3-Phenyl-propenoic acid, *see trans-cinnamic acid*
Phenyl salicylate **7:** 4277
　　physical properties **3:** 2015
Phenylsulfinyl chloride
　synthesis **7:** 4460
1-Phenylsulfonyl-aceto-4-nitrobenzene
　synthesis **7:** 4457
N-(Phenylsulfonyl)benzenesulfonamide **2:** 759
Phenylurea **8:** 4819
Phenyl vinyl sulfide
　oxidation to phenyl vinyl sulfone **7:** 4479
Phenyl vinyl sulfoxide
　synthesis **7:** 4489
Philipoff equation **2:** 1217
Phillips Petroleum butane dehydrogenation process **2:** 908
Phillips STAR propene process **7:** 4092
Phloroglucine **6:** 3766
Phloroglucinol **6:** 3766
Phosgenation
　of amine hydrochlorides **5:** 3019
　of amines **8:** 4821
　of carbamate salts **5:** 3020
　of free amines **5:** 3017
　of hydroxy compounds **2:** 1087
　of ureas **5:** 3020
Phosgene
　for carbamoyl chloride production **2:** 1050
　reaction with amines **1:** 447
　reaction with anhydros alcohols **3:** 1577
Phosphabicyclononanes **6:** 3822
Phosphate esters **6:** 3847
　　toxicology **6:** 3857
Phosphates
　of monosaccarides **2:** 1070
Phosphatidylcholine **4:** 2380
Phosphatidylethanolamine **4:** 2380
Phosphatidylinositol **4:** 2380
Phosphatidylserine **4:** 2380
Phosphinate esters **6:** 3837

Phosphinates
　addition to double bonds **6:** 3836
Phosphine oxides **6:** 3831
　uses **6:** 3833
Phosphines
　formation of quaternary compounds **6:** 3828
　production **6:** 3821
　properties **6:** 3820
　toxicology **6:** 3856
　uses **6:** 3824
Phosphine sulfides **6:** 3831
　uses **6:** 3833
Phosphinic acids **6:** 3835, 3837
Phosphites **6:** 3838
　toxicology **6:** 3859
Phospholipids
　in fats and fatty oils **4:** 2380
Phosphonate esters **6:** 3845
Phosphonates **6:** 3841
　addition to double bonds **6:** 3843
Phosphonic acid acylation **6:** 3844
Phosphonic acid esters **6:** 3845
Phosphonic acids **6:** 3842, 3844
Phosphonium salts **6:** 3827
　uses **6:** 3830
Phosphonocarboxylic acids **6:** 3844
Phosphonomethyl cellulose **2:** 1235
Phosphonous acid derivatives **6:** 3834
Phosphoric acid esters **6:** 3847
Phosphoric acid tri-2-propenyl ester **1:** 426
Phosphororganosilanes **7:** 4343
Phosphorus compounds
　organic, *see organophosphorus compounds*
Phosphorus oxychloride
　for cross-linking of starch **7:** 4402
Phosphorus trichloride
　reaction with alcohols or phenols **6:** 3839
Photochlorination
　of 1,1-dichloroethan **3:** 1312
Photocopying processes
　use of diazonium salts **3:** 1883
Photodehydrochlorination
　of 1,2-dichloroethane **3:** 1341
Photoglycine **1:** 621
Photography
　aminophenols **1:** 610
　dithiocarbamic acids and derivatives **4:** 1971
　xanthate derivatives **8:** 4979
Photolysis
　of dithiocarbamates **4:** 1967
Photonitrosation
　of cyclohexane **2:** 1026
　of saturated hydrocarbons **5:** 2866
Photooximation
　of cyclohexane **2:** 1026
Photooxygenation **6:** 3639

Phthalate esters
   metabolism and toxicokinetics   **7**: 3903
Phthalates
   economic aspects   **7**: 3893
   environmental protection   **7**: 3890
   production   **7**: 3888
   properties   **7**: 3886
   specifications and handling   **7**: 3892
   toxicology   **7**: 3896
   uses   **7**: 3892
Phthalic acid   **2**: 1121; **7**: 3865, 3866
   toxicology   **7**: 3894
Phthalic acid derivatives   **7**: 3865
Phthalic anhydride   **2**: 1130
   environmental protection   **7**: 3877
   for anthraquinone production   **2**: 656
   handling and uses   **7**: 3880
   in phthalimide production   **7**: 3881
   properties   **7**: 3866
   specifications and economic agents   **7**: 3878
   toxicology   **7**: 3894
Phthalic anhydride esterification   **7**: 3887
Phthalic anhydride production   **7**: 3869
Phthalic anhydride – urea CuPc process   **7**: 3936
Phthalimide
   properties   **7**: 3880
   toxicology   **7**: 3894
   uses   **7**: 3883
Phthalimide production   **7**: 3881
Phthalocyanine
   toxicology   **7**: 3957
Phthalocyanine analogues   **7**: 3949
Phthalocyanine complexes, *see phthalocyanines*
Phthalocyanine derivatives   **7**: 3944
Phthalocyanine dyes   **7**: 3955
Phthalocyanine oxidation   **7**: 3928
Phthalocyanine pigments
   finishing   **7**: 3939
   uses   **7**: 3954
Phthalocyanine polymers   **7**: 3921
Phthalocyanine production   **7**: 3934, 3939
Phthalocyanine reduction   **7**: 3931
Phthalocyanines   **7**: 3919
   as catalysts   **7**: 3932
   chemical properties   **7**: 3928
   color properties and particle size   **7**: 3924
   economic aspects   **7**: 3956
   environmental protection   **7**: 3952
   physical poperties   **7**: 3923
   polymeric   **7**: 3949
   production   **7**: 3933
   specifications and uses   **7**: 3953
   stabilization   **7**: 3943
Phthalocyanine sulfonic acids   **7**: 3947
Phthalocyanine sulfonyl chlorides   **7**: 3947
Phthalodinitrile CuPc process   **7**: 3937

Phthalonitrile   **7**: 3883
   toxicology   **7**: 3895
   uses   **7**: 3885
Phygon   **6**: 3450
Phytol   **8**: 4613
Picramic acid   **1**: 613
Picramide   **6**: 3544
Pigments
   phthalocyanines, finishing   **7**: 3939
   phthalocyanines, uses   **7**: 3953
Pimelic acid   **3**: 1915
   physical properties   **3**: 1905
Pinacol   **1**: 324, 337
Pinacolone   **5**: 3090
Pinane   **8**: 4605
2-Pinane hydroperoxide   **8**: 4606
2-Pinanol   **8**: 4607
α-Pinene oxide   **4**: 2002
Pine oil   **8**: 4762
   handling   **8**: 4777
   production and properties   **8**: 4776
   uses and economic aspects   **8**: 4778
Piperazine   **1**: 468
   toxicology   **1**: 492
Piperidine   **1**: 465
   physical properties   **7**: 4197
2-Piperidinecarboxylic acid
   physical properties   **7**: 4197
3-Piperidinecarboxylic acid
   physical properties   **7**: 4197
4-Piperidinecarboxylic acid
   physical properties   **7**: 4197
Piperidines   **7**: 4196
4-Piperidinol
   physical properties   **7**: 4197
2-(2-Piperidyl)ethanol
   physical properties   **7**: 4197
2-Piperidylmethanol
   physical properties   **7**: 4197
Piperylene   **5**: 2889
Pipe stills
   for tar distillation   **8**: 4521
Piria reaction   **1**: 514
Piribedil   **7**: 4232
Pitch, *see tar and pitch*
Pitch coke
   properties and production   **8**: 4529
Pitch – oil blends   **8**: 4535
Pitch paints   **8**: 4536
Pitch – polymer-combinations   **8**: 4533
Pivaldehyde   **1**: 352, 361
Pivalic acid   **2**: 1097, 1111
Plasma arc acetylene process   **1**: 153
Plasticizer content
   in plastics from cellulose esters   **2**: 1185
Plasticizers
   adipic esters   **1**: 272

alcohols **1**: 293, 294
alcohols, economic aspects **1**: 304
chlorinated paraffins **3**: 1405
esters **3**: 2039
2-ethylhexanol **4**: 2366
from higher olefins **5**: 2884
phthalates **7**: 3892
Plastic microwaxes **8**: 4933
uses **8**: 4938
Plastic molding compounds
from cellulose esters **2**: 1184
PMC **2**: 1235
PMDI **5**: 3012
PNA **6**: 3541
PNC **2**: 1026
PN salt **6**: 3555
PNT **6**: 3519
PO, *see propylene oxide*
Podands **3**: 1723
Point of complete fusion **4**: 2434
Point of incipient fusion **4**: 2434
Pokorny phenoxy acid process **3**: 1620
Polar polyolefin waxes **8**: 4956
Polyacetaldehyde **1**: 22
Polyacetylene **1**: 128
Polyacrolein **1**: 206
Polyamines **1**: 487
Polybutadiene oxide **4**: 2005
Polybutene
degradation **8**: 4953
from butenes **2**: 999
Poly(butylene terephthalate)
uses **8**: 4587
Polychlorinated biphenyls, *see chlorinated biphenyls*
Polychloroacetaldehydes **3**: 1528
Polycyclic aromatic hydrocarbons **5**: 2913
toxicology **5**: 2927
Polydichloroacetaldehydes **3**: 1531
Polydiisocyanate **5**: 3012
Polyenes **5**: 2889
production by oligomerization **5**: 2891
Polyepoxides **4**: 2005
Polyester production
terephthalic acids as feedstock **8**: 4586
Polyester resins
with propylene glycol **7**: 4053
Polyesters **3**: 2039
Polyether polyols
uses **7**: 4155
Polyethylene
cross-linking with silanes **7**: 4348
degradation **8**: 4953
product finishing **8**: 4947
Ziegler–Natta **8**: 4952
Poly(ethylene glycols) **4**: 2317
physical properties **4**: 2319
toxicology **4**: 2325

Poly(ethylene terephthalate)
uses **8**: 4587
Polyethylene waxes **8**: 4944
copolymeric **8**: 4949
properties **8**: 4948
toxicology **8**: 4961
Polyethylenimine **2**: 664
Polyethylen waxes
production **8**: 4945
Polygalacturonase **7**: 3978
Poly-α-1,4-D-galacturonide glycanohydrolase **7**: 3978
Poly-α-1,4-D-galacturonide lyase **7**: 3979
Polygasoline **7**: 4095
with butenes **2**: 997
Polyglycerols **5**: 2805
Polyhydric alcohols, *see alcohols, polyhydric*
Polyhydroxy acids
production **5**: 2962
Poly(*cis*-1,4-isoprene) **5**: 3048
Polymerization
methacrylic acid **6**: 3253
of acrolein **1**:
of aldehydes **1**: 356
of butadiene **2**: 900
of butenes **2**: 988
of ε-caprolactam **2**: 1015
of cyanamide **3**: 1747
of cyclopentadiene **3**: 1829
of ethylene, reaction conditions **8**: 4946
of ethylene, reaction mechanism **8**: 4945
of fatty acids **4**: 2399
of formaldehyde **5**: 2646
of furfuryl alcohol **5**: 2757
of ketene **5**: 3062
of styrene **7**: 4419
Polymer modification
with silanes **7**: 4348
Polymers
from isocyanates **5**: 3025
from isoprene **5**: 3048
phthalocyanines **7**: 3949
silane and siloxane **7**: 4350
silane coupling agents **7**: 4352
vinyl acetate **8**: 4854
vinyl ethers **8**: 4874
vinyl propionate **8**: 4857
Polymer stabilizers
silanes **7**: 4351
Poly-α-1,4-D-methoxygalacturonide lyase **7**: 3979
Poly(methyl methacrylate) **6**: 3262
Polymethylphenols **6**: 3727
Polymorphism
of glycerides **4**: 2385
of phthalocyanines **7**: 3925
Polynaphthoquinone **6**: 3452
Polyolefins
of allyl esters **1**: 428

Polyolefin waxes **8:** 4883
   economic importance **8:** 4959
   production and properties **8:** 4944
   toxicology **8:** 4961
   uses **8:** 4957
Polyols **1:** 342
Poly(oxymethylene) diacetates
   physical properties **5:** 2673
Poly(oxymethylene) dimethyl ethers
   physical properties **5:** 2673
Poly(oxymethylene) glycols
   physical properties **5:** 2673
Polyoxymethylene plastics **5:** 2680
Polyoxymethylenes
   cyclic **5:** 2677
   linear **5:** 2672
Polyphenylamines **1:** 635
Polyphenyls **5:** 2912
Polypropylene
   degradation **8:** 4953
   uses **7:** 4096
   Ziegler–Natta **8:** 4952
Polypropylene glycols
   uses **7:** 4155
Polypropylene waxes
   toxicology **8:** 4961
Polysaccharide esters **2:** 1071
Polysaccharides **2:** 1068; **7:** 3971
   applications **7:** 3974
   characterization **7:** 3973
   dextran, *see dextran*
   direct fermentation **4:** 2074
   sources **7:** 3973
Polysilanes **7:** 4345
Polystyrene
   uses **7:** 4432
Polysulfides
   aliphatic production and uses **8:** 4661
   aliphatic properties **8:** 4657
   aromatic production and uses **8:** 4666
   aromatic properties **8:** 4662
   toxicology **8:** 4677
Polytetrafluoroethylene **5:** 2586
Poly(tetramethylene oxide)
   uses **8:** 4627
Poly(vinylpyridines) **7:** 4191
Polzeniusz–Krauss channel furnace calcium cyanamide
   process **3:** 1741
Poppyseed oil **4:** 2460
Porous silica
   for acetylene storage **1:** 180
Porphyrins **7:** 3949
Potassium acetate **1:** 52
Potassium antimonyl tartrate **8:** 4567
   uses **8:** 4568
Potassium benzoate **2:** 826
Potassium bitartrate **8:** 4567

Potassium hydrogen oxalate **6:** 3593
Potassium oxalate **6:** 3593
Potassium sorbate **7:** 4366
Potato
   as raw material for ethanol production **4:** 2099
Potato starch process **7:** 4392
PPDI **5:** 3012
Pregelatinized starch **7:** 4396
Prehnitene **5:** 2895
   physical properties **5:** 2894
Prehnitic acid **2:** 1121
Preserving agents
   carnaúba wax **8:** 4891
Prilezhaev reaction **4:** 1992
Prilling
   of urea **8:** 4811
Prins reaction **1:** 397; **2:** 990; **5:** 3037
Promoters
   in 1,2-dichloroethane cracking **3:** 1337
Propadiene
   recovery from cracked gas in ethylene production **4:** 2277
Propadiene hydrogenation
   in cracked gas from ethylene production **4:** 2279
Propanal, *see propionaldehyde; see also aldehydes, aliphatic*
Propane
   physical properties **5:** 2855
   toxicology **5:** 2924
   uses **5:** 2867
Propane dehydrogenation **7:** 4089
Propanedial, *see malondialdehyde*
Propanedicarboxylic acid, *see malonic acid*
Propanedinitrile, *see malononitrile*
Propanedioic acid, *see malonic acid*
2,2-[Oxybis(methylene)]-bis(2-ethyl-1,3-propanediol **1:** 340, 342, 343
1,2-Propanediol, *see propylene glycol*
1,3-Propanediol, *see trimethylene glycol*
Propanediols **7:** 4047
Propane pyrolysis
   cracking yield **4:** 2235
1,3-Propanesulfone **8:** 4505
Propane-1-sulfonic acid
   physical properties **8:** 4504
n-Propanethiol
   toxicology **8:** 4677
1,2,3-Propanetriol, *see glycerol*
2-Propanol, *see isopropanol*
Propanolamines **4:** 2149
   N-alkylated **4:** 2170
   N-alkylated, production **4:** 2174
   N-alkylated, specifications, uses and economic aspects **4:** 2175
   toxicology **4:** 2179
Propanols **7:** 4061
   handling **7:** 4074

2-Propanone, *see acetone*
Propargyl alcohol   **1:** 309, 310
  by ethynylation   **1:** 126
Propellants
  guanidine nitrate   **5:** 2831
Propenal, *see acrolein*
2-Propenamide, *see acrylamide*
Propene, *see propylene*
2-Propene-1-sulfonic acid   **8:** 4506
2-Propenoic acid *see acrylic acid*
2-Propenoic acid 2-propenyl ester   **1:** 426
2-Propen-1-ol, *see allyl alcohol*
β-Propiolactone   **3:** 2033; **5:** 2967
  physical properties   **5:** 2958
Propiolic acid   **2:** 1097, 1111
  production   **2:** 1104
Propionaldehyde   **1:** 352; **7:** 4037
  chemical properties and uses   **7:** 4038
  for production of pyridines   **7:** 4182
  handling and toxicology   **7:** 4042
  hydrogenation   **7:** 4066
  production and specifications   **7:** 4041
Propionaldehyde diethyl acetal   **1:** 395
Propionaldehyde dimethyl acetal   **1:** 395
Propionaldehyde oxidation   **7:** 4107
Propionic acid   **7:** 4101
  chemical properties   **7:** 4104
  economic aspects   **7:** 4114
  for vinyl propionate production   **8:** 4856
  handling   **7:** 4110
  physical properties   **7:** 4102
  specifications and environmental protection   **7:** 4110
  toxicology   **7:** 4115
  uses   **7:** 4111
Propionic acid derivates   **7:** 4101
Propionic acid esters   **7:** 4118
Propionic acid production   **7:** 4105
Propionic anhydride   **7:** 4117
Propionitrile   **6:** 3461
  physical properties   **6:** 3457
Propionyl chloride   **7:** 4118
Propiophenone   **5:** 3111
Propyl acetate
  physical properties   **3:** 2014
  specifications   **3:** 2037
  toxicology   **3:** 2043
Propylamine   **1:** 454
  uses   **1:** 458
n-Propyl carbamate   **2:** 1047
n-Propyl carbamoyl chloride   **2:** 1049
n-Propyl chloroformate   **3:** 1576
Propyl cyanide   **6:** 3461
n-Propylcyclohexane
  physical properties   **3:** 1798
Propylene   **7:** 4081
  economic aspects   **7:** 4098

for glycerol synthesis   **5:** 2797
oligomerization   **5:** 2883
physical properties   **5:** 2871
properties   **7:** 4081
specifications and handling   **7:** 4093
uses   **7:** 4094
Propylene carbonate   **2:** 1086
Propylene chlorohydrin
  for propylene oxide production   **7:** 4135
Propylene chlorohydrination   **7:** 4135
Propylene chlorohydrin production   **3:** 1592
Propylene chlorohydrins   **3:** 1588
Propylene–chloronium complex   **7:** 4135
Propylene-1,2-diamine   **1:** 484
Propylene-1,3-diamine   **1:** 484
Propylene dichloride   **3:** 1374
1,3-Propylenediurea   **8:** 4820
Propylene epoxidation   **7:** 4143
Propylene fractionation
  in ethylene production   **4:** 2280
Propylene glycol
  properties   **7:** 4047
  specifications   **7:** 4052
  toxicology   **7:** 4057
  uses   **7:** 4053
Propylene glycol alginate   **7:** 3998
Propylene glycol dibenzoate   **2:** 827
Propylene glycol ethers
  uses   **7:** 4155
Propylene glycol production   **7:** 4051
Propylene homopolymer wax
  structure and properties   **8:** 4952
Propylene hydration   **7:** 4067
Propylene hydroformylation   **2:** 926
Propylene oligomers   **7:** 4097
Propylene oxidation
  for acetone production   **1:** 98
  for acetylene production   **1:**
  for acrylic acid production   **1:** 231
  oxalic acid production   **6:** 3585
Propylene oxide   **7:** 4129
  economic aspects and toxicology   **7:** 4156
  environmental protection   **7:** 4149
  for glycerol production   **5:** 2800
  from propene   **7:** 4097
  handling   **7:** 4152
  hydrolysis   **7:** 4051
  properties   **7:** 4130
  reactions   **7:** 4133
  reaction with ammonia   **4:** 2166
  specifications   **7:** 4151
  uses   **7:** 4154
Propylene oxide isomerization
  for allyl alcohol production   **1:** 422
Propylene oxide production   **7:** 4134
  routes   **7:** 4134
Propylene oxide–styrene process   **7:** 4428

Propylene oxide THF process **8:** 4624
Propylene oxide – water system **7:** 4132
Propylene production **7:** 4082
Propylene – propane separation **7:** 4086
Propyleneurea **8:** 4820
Propylenimine **2:** 663
  toxicology **2:** 670
2-Propyl-1-heptanol
  physical properties **5:** 2551
n-Propyl hydroperoxide
  physical properties **6:** 3634
1-Propylimidazole
  toxicology **5:** 2994
2-Propylpentanal **1:** 353
2-n-Propylpentanoic acid, *see di-n-propylacetic acid*
Propyl propionate
  properties **7:** 4119
Propyl vinyl ether
  physical properties **8:** 4867
Propyne **1:** 185
  recovery from cracked gas in ethylene production **4:** 2277
Propyne hydrogenation
  in cracked gas from ethylene production **4:** 2279
Propynoic acid, *see propiolic acid*
2-Propyn-1-ol, *see propargyl alcohol*
Protecting groups
  silanes **7:** 4345
Protein denaturants
  guanidine salts **5:** 2831
Protocatechuic acid **5:** 2980
Pseudocumene **5:** 2893
  physical properties **5:** 2894
Pseudohalosilanes **7:** 4321
Pseudoureas **3:** 1748
Pulping
  of slack waxes **8:** 4922
Pummerer rearrangement **1:** 69; **7:** 4492
Pumpkin seed oil **4:** 2453
  fatty acid content **4:** 2376
Purine **7:** 4165
Purine derivatives **7:** 4165
  occurrence and production **7:** 4167
  properties **7:** 4166
  specifications and uses **7:** 4170
6-Purinethiol hydrate **8:** 4668
Purpurin acid **6:** 3412
Purpurol **6:** 3426
Putrescine **1:** 486
Pyranoses **2:** 1063
Pyrene **5:** 2922
Pyridine **7:** 4175
  economic aspects **7:** 4189
  properties **7:** 4176
  specifications **7:** 4187
  toxicology **7:** 4213
  uses **7:** 4188

2-Pyridine aldoxime
  physical properties **7:** 4213
3-Pyridine aldoxime
  physical properties **7:** 4213
4-Pyridine aldoxime
  physical properties **7:** 4213
2-Pyridinecarbaldehyde
  physical properties **7:** 4213
3-Pyridinecarbaldehyde
  physical properties **7:** 4213
4-Pyridinecarbaldehyde
  physical properties **7:** 4213
Pyridinecarbaldehydes **7:** 4212
2-Pyridinecarbonitrile
  physical properties **7:** 4202
3-Pyridinecarbonitrile
  physical properties **7:** 4202
4-Pyridinecarbonitrile
  physical properties **7:** 4202
3-Pyridinecarbonitrile N-oxide
  physical properties **7:** 4196
Pyridinecarbonitriles **7:** 4202
  hydrolysis **7:** 4204
2-Pyridinecarboxamide
  physical properties **7:** 4202
3-Pyridinecarboxamide
  physical properties **7:** 4202
4-Pyridinecarboxamide
  physical properties **7:** 4202
Pyridinecarboxamides **7:** 4205
2-Pyridinecarboxylic acid
  physical properties **7:** 4202
3-Pyridinecarboxylic acid
  physical properties **7:** 4202
4-Pyridinecarboxylic acid
  physical properties **7:** 4202
Pyridinecarboxylic acids **7:** 4204
Pyridine derivatives **7:** 4175
  production **7:** 4179
  toxicology **7:** 4213
2,6-Pyridinedicarbonitrile
  physical properties **7:** 4202
2,3-Pyridinedicarboxylic acid
  physical properties **7:** 4202
2,4-Pyridinedicarboxylic acid
  physical properties **7:** 4202
2,6-Pyridinedicarboxylic acid
  physical properties **7:** 4202
Pyridine nitro derivatives **6:** 3569
Pyridine N-oxide
  physical properties **7:** 4196
Pyridine N-oxides **7:** 4195
Pyridine production **7:** 4179
Pyridines
  fluorinated **5:** 2619, 2620
  from tar fractions **8:** 4544
Pyridine-2-thiol **8:** 4668

Pyridinium salts
    quaternary **7**: 4192
2-Pyridinol
    physical properties **7**: 4209
3-Pyridinol
    physical properties **7**: 4209
4-Pyridinol
    physical properties **7**: 4209
Pyridinols **7**: 4209
Pyridyl alcohols **7**: 4211
2,6-Pyridyldimethanol
    physical properties **7**: 4211
2-(2-Pyridyl)ethanol
    physical properties **7**: 4211
2-(4-Pyridyl)ethanol
    physical properties **7**: 4211
Pyridylethanols **7**: 4212
2-Pyridylmethanol
    physical properties **7**: 4211
3-Pyridylmethanol
    physical properties **7**: 4211
4-Pyridylmethanol
    physical properties **7**: 4211
Pyridylmethanols **7**: 4211
Pyrimidines **7**: 4227
    chemical properties **7**: 4227
    production **7**: 4229
Pyrimidine-2-thiol **8**: 4668
Pyrocatechol, *see catechol*
Pyrogallol **6**: 3763
Pyrolysis
    of acetic acid **5**: 3065
    of chlorodifluoromethane **5**: 2585
    of diketene **5**: 3065
    of esters **3**: 2021
    of hydrocarbons for ethylene production **4**: 2226
    of hydrocarbons, isoprene isolation **5**: 3041
Pyrolysis gasoline
    for BTX production **2**: 706
    for styrene production **7**: 4430
    xylene content **8**: 4995
Pyrolysis residual oils
    processing **8**: 4551
    properties and composition **8**: 4513
Pyromellitic acid **2**: 1121, 1131
Pyromellitic dianhydride **2**: 1130, 1131
Pyropissite **8**: 4881
Pyroracemic acid **6**: 3600
Pyrotol benzene process **2**: 711
Pyrrole **7**: 4235
    toxicology **7**: 4238
1*H*-Pyrrole, *see pyrrole*
Pyrrole derivatives **7**: 4237
Pyrrolidine **1**: 464
    toxicity **1**: 492
Pyruvic acid **6**: 3600

Quenching
    of hot cracked gas in ethylene production **4**: 2256
Quick contact ethylene process **4**: 2284
Quinaldine **7**: 4255
Quinol, *see hydroquinone*
Quinoline **7**: 4253
    production and uses **7**: 4254
    properties **7**: 4253
    toxicology **7**: 4257
Quinoline nitro derivatives **6**: 3570
2-Quinolinethiol **8**: 4668
8-Quinolinol **7**: 4255
Quinone diazide **3**: 1890
Quinones, *see benzoquinone*

R acid **6**: 3396
2 R acid **6**: 3442
Radiant coils
    in cracking furnaces **4**: 2240
Radiative heat transfer
    in cracking furnaces **4**: 2253
Radical decomposition
    in hydrocarbon cracking **4**: 2229
Radziszewski reaction **5**: 2989
Raffinose **2**: 1067
Ramberg–Bäcklund reaction **7**: 4484
Rapeseed oil **4**: 2464
    fatty acid content **4**: 2376
Rapidogen dyes **2**: 779
Rare earth elements
    isolation with oxalic acid **6**: 3592
Raschig–Hooker phenol process **6**: 3702
Raw waxes **8**: 4917
Reaction mechanism
    of aromatic nitration **6**: 3507
    of ethylbenzene dehydrogenation **7**: 4422
    of high-pressure polymerization of ethylene **8**: 4945
    of hydrosilylation **7**: 4317
    of phthalocyanine formation **7**: 3934
    of *o*-xylene or naphthalene oxidation to phthalic anhydride **7**: 3871
    propylene chlorohydrination **7**: 4135
Reaction rate, *see kinetics*
Rearrangement
    of epoxides **4**: 1991
    of sulfinates **7**: 4482
Redox reactions
    of sulfinic acids **7**: 4458
Reduction
    of aromatic nitro compunds **1**: 513
    of arylamines **1**: 512
    of esters **3**: 2019
    of nitrobenzene to hydrazobenzene **2**: 794
    of phthalocyanine **7**: 3931
    of wax esters **5**: 2537

Refined waxes **8**: 4919
   uses **8**: 4930
Refinery cracking processes
   propylene production **7**: 4089
Refining
   of microwaxes **8**: 4936
Reformate
   as feedstock for xylenes **8**: 4994
   in benzene production **2**: 704
Reformate composition **2**: 706
Refractive index
   of oils, fats, and fatty acids **4**: 2393
Refrigerant numbering system **5**: 2573
   for bromofluoroalkanes **5**: 2580
Refrigerants
   chlorofluoroalkanes **5**: 2579
   fluoroalkanes **5**: 2574
Refrigeration
   in ethylene plants **4**: 2281
   with methanol **6**: 3315
Reimer-Tiemann reaction **2**: 686
Reppe acrylic acid and process **1**: 230
Reppe aliphatic alcohol process **1**: 291
Reppe butanol process **2**: 957
Reppe ester synthesis **3**: 2029
Reppe propionic acid synthesis **7**: 4106
Reppe THF process **8**: 4623
Resinification
   of furfuryl alcohol **5**: 2757
Resins
   from formaldehyde **5**: 2667
   from montan wax **8**: 4909
   guanamines **5**: 3216
Resorcinol **7**: 4261
   environmental protection and specifications **7**: 4267
   properties **7**: 4261
   toxicology **7**: 4268
   uses and handling **7**: 4268
Resorcinol production **7**: 4263
α-Resorcylic acid **5**: 2981
β-Resorcylic acid **5**: 2980
Retamo wax **8**: 4898
Retarded dissolution
   of methyl cellulose **2**: 1221
Retrogradation
   of starches **7**: 4388
Reverse-Bucherer reaction **6**: 3388
Reverse osmosis
   for purification of ethanol **4**: 2123
RG acid **6**: 3394
Rhamnogalacturonans, *see pectin*
L-Rhamnose
   in pectin **7**: 3976
Rhône-Poulenc phthalic anhydride process **7**: 3871
Rhodanine **4**: 1961

Rice bran oil **4**: 2455
   fatty acid content **4**: 2376
Rice starch production **7**: 4394
Ricinoleic acid
   alkaline cleavage **3**: 1909
   physical properties **5**: 2958
Ritter reaction **1**: 452, 476
RM acid **6**: 3434
Road tars **8**: 4535
Rochelle salt **8**: 4567
Rochow synthesis **3**: 1258
Rockwell hardness
   of cellulose ester thermoplastics **2**: 1188
Rohm & Haas acetone cyanohydrin process **1**: 110
Roofing tars **8**: 4536
Rosenmund reduction **1**: 360
Rotofermentor **4**: 2087
RR acid **6**: 3442
Rütgers continuous tar distillation process **8**: 4523
Rütgers pencil-pitch process **8**: 4525
Russell chain termination mechanism **1**: 41

Saccharides, *see carbohydrates*
S acid **6**: 3433
2 S acid **6**: 3439
Safflower oil **4**: 2459
   fatty acid content **4**: 2376
Sago starch production **7**: 4395
Salicylaldehyde **2**: 686
Salicylamide **7**: 4278
Salicylanilide **7**: 4278
Salicylic acid **2**: 1121; **7**: 4271
   properties **7**: 4272
   specifications and handling **7**: 4275
   toxicology **7**: 4279
   uses and economic aspects **7**: 4276
Salicylic acid derivatives **7**: 4277
Salicylic acid production **7**: 4273
S-Alkylation
   with dialkyl sulfates **3**: 1869
Salol **7**: 4277
Salt intolerance
   of hydroxyethyl methyl cellulose **2**: 1219
SAN **1**: 256
Sandmeyer reaction **1**: 510; **3**: 1897
Sapogenins **7**: 4289
Saponification number **1**: 307
Saponification value
   of fats and fatty oils **4**: 2437
Saponifier
   in propylene oxide production **7**: 4138
Saponins **7**: 4283
   glycoalkaloids **7**: 4292
   isolation **7**: 4286
   occurrence **7**: 4285
   properties **7**: 4284
   steroid **7**: 4288

toxicology **7**: 4287
triterpene **7**: 4295
Saponin structures **7**: 4286
SASOL slurry bed process
   for production of Fischer–Tropsch paraffins **8**: 4942
SBA acetylene process **1**: 141
Scale waxes **8**: 4917
   uses **8**: 4930
Scandinavian fish oil
   fatty acid content **4**: 2376
Schaeffer acid **6**: 3392
Scheele–Lowitz tartaric acid process **8**: 4563
Scheurer–Kestner tartaric acid process **8**: 4563
Schiff bases **1**: 445
Schmidt'sche transfer-line exchangers **4**: 2258
Schollkopf's acid **6**: 3395
Scientific Design–Bethlehem Steel formic acid process **5**: 2726
Screw extractor
   for oilseeds **4**: 2408
Screw press
   for oil seeds **4**: 2404
Scrubber acetic anhydride process **1**: 73
Seaweed extracts
   general classification **7**: 3994
Seaweeds
   alginate content **7**: 3996
Sebacic acid **3**: 1917
   electrochemical production **3**: 1912
   physical properties **3**: 1905
SEC **2**: 1235
Seed-kernel fats **4**: 2445
Seignette's salt **8**: 4567
   uses **8**: 4568
Selective oxidation
   of saturated hydrocarbons **5**: 2865
Self-accelerating decomposition temperature
   of peroxides **6**: 3674
Semicontinuous decolorizing process
   for refining deoiled slack waxes **8**: 4925
Senegal gum, *see gum arabic*
Sephadex gels **3**: 1850
Sesame oil **4**: 2456
   fatty acid content **4**: 2376
Sesquiterpenes
   acyclic **8**: 4609
   basic structures **8**: 4593
   bicyclic and tricyclic **8**: 4611
   monocyclic **8**: 4610
Sesterpenes
   basic structures **8**: 4593
SFP, *see submerged flame acetylene process*
Shaw synthesis **7**: 4230
Shea butter **4**: 2451
Shear modulus
   of cellulose ester thermoplastics **2**: 1189

Shellac wax **8**: 4905
Shell higher olefin process **5**: 2878
Shell monochloroethane process **3**: 1292
Shell oxo synthesis process **1**: 285
Shell vinyl ester process **8**: 4859
SHG transfer-line exchanger **4**: 2258
SHOP process **5**: 2878
Side-chain chlorination
   of aromatics **3**: 1450
   of methylbenzenes **2**: 1126
   of xylenes **3**: 1462
Silane coupling agents **7**: 4352
Silanes
   addition reactions, hydrosilylation **7**: 4316
   as protecting groups **7**: 4345
   chloromethyl, synthesis **7**: 4310
   for enzyme immobilization **7**: 4346
   for hydrophobic surfaces **7**: 4354
   for silicone and organic polymer modification **7**: 4348
   in catalysis **7**: 4351
   in liquid media **7**: 4352
   organofunctional **7**: 4330
   silicon-functional **7**: 4319
Silanols **7**: 4324
Silazanes **7**: 4328
Silicates
   ethyl **7**: 4327
Silicon
   as raw material for organosilicon compounds **7**: 4312
Silicon bond energies **7**: 4309
Silicon compounds
   organic, *see organosilicon compounds*
Siloxanes
   addition reactions, hydrosilylation **7**: 4316
   olgomers **7**: 4324
Silver catalyst formaldehyde processes **5**: 2648
Silver catalysts
   in ethylene oxidation **4**: 2336
Silver ketenide
   production from acetic anhydride **1**: 71
Silylating agents **7**: 4345
Simons fluorination process **5**: 2569
Single-column propene–propane separation process **7**: 4086
Sioplas process **7**: 4350
Skatole **5**: 3002
Skraup quinoline synthesis **7**: 4254
Skraup synthesis **1**: 204
Slack waxes **8**: 4916
   deoiling **8**: 4922
   uses **8**: 4930
Slack wax raffinates **8**: 4917
Slip point **4**: 2434
Smiles rearrangement **7**: 4454, 4485
Smoke point **4**: 2434

Snamprogetti ammonia-stripping urea process **8**: 4804
Snamprogetti MTBE process **6**: 3340
Snamprogetti self-stripping urea process **8**: 4804
Snia caprolactam process **2**: 1028
Soaps **6**: 3227
   see also metallic soaps
Société Belge de l'Azote, *see SBA*
Sodium acetate **1**: 52
Sodium acetate-cresol adduct **3**: 1684
Sodium alginate production **7**: 3997
Sodium amalgam
   for hydrazobenzene production **2**: 796
Sodium benzoate **2**: 825
Sodium carboxylates
   biodegradation **2**: 1104
Sodium chloroacetate **3**: 1547
Sodium *N*-chloro-*p*-toluenesulfonamide **3**: 1571
Sodium dichloroisocyanurate **3**: 1568
Sodium dithionite **1**: 515
Sodium ethanesulfinate
   synthesis **7**: 4452
Sodium formate
   for production of formic acid **5**: 2728
Sodium gluconate **5**: 2781
Sodium hydroxymethanesulfinate **7**: 4465
Sodium lactate **5**: 3130
Sodium 4-nitrobenzene-anti-diazotate
   preparation **3**: 1885
Sodium oxalate **6**: 3594
Sodium potassium tartrate **8**: 4567
Sodium propionate **7**: 4116
Sodium salicylate **7**: 4277
Sodium trichloroacetate **3**: 1555
Softeners
   fatty amines **1**: 478
Soft pitch **8**: 4531
Soft waxes
   uses **8**: 4930
Sohio acrylonitrile process **1**: 252; **6**: 3458
Solanidanes **7**: 4292
Solanocapsines **7**: 4292
Solubility
   of aminophenols **1**: 602
   of acetylene **1**: 121
   of acrylonitrile **1**: 251
   of adipic acid **1**: 264
   of alkali-metal xanthates **8**: 4972
   of benzal chloride **3**: 1457
   of benzotrichloride **3**: 1459
   of benzyl chloride **3**: 1449
   of bisphenol A **6**: 3769
   of butadiene in water **2**: 902
   of butanols **2**: 953
   of butyraldehydes **2**: 926
   of caprolactam **2**: 1015
   of catechol **6**: 3758
   of cellulose acetate **2**: 1177
   of cellulose ethers **2**: 1208
   of chlorine in benzene and chlorobenzenes **3**: 1415
   of chloroacetic acid **3**: 1541
   of citric acid **3**: 1647, 1648
   of cresols **3**: 1657
   of cyanamide **3**: 1747
   of dichloromethane **3**: 1255
   of diethyl ether **4**: 2189
   of diisopropyl ether **4**: 2193
   of dimethyl ether in dimethyl sulfate **3**: 1854
   of dimethyl terephthalate **8**: 4575
   of fats and fatty oils **4**: 2390
   of fatty acids **4**: 2494
   of gases in dimethylformamide **5**: 2699
   of gases in ethylene oxide **4**: 2332
   of gum arabic **7**: 4014
   of hydroquinone **5**: 2943
   of 4-hydroxybenzoic acid esters **5**: 2978
   of lactose **5**: 3138
   of lecithin and its components **5**: 3155
   of lower esters **3**: 2015
   of methyl cellulose **2**: 1219
   of monochloroacetaldehyde hemihydrate **3**: 1523
   of monochloromethane **3**: 1254
   of MTBE **6**: 3338
   of naphthalene sulfonic acid salts **6**: 3368
   of nitrilotriacetic acid **6**: 3479
   of oxalic acid **6**: 3580
   of pentanols **6**: 3612
   of phloroglucinol **6**: 3766
   of phthalic anhydride **7**: 3867
   of phthaloxyanines **7**: 3923
   of propyne **1**: 186
   of pyridines **7**: 4178
   of resorcinol **7**: 4261
   of salicylic acid **7**: 4272
   of sodium gluconate **5**: 2781
   of styrene **7**: 4420
   of sulfamic acid **7**: 4444
   of tartaric acid **8**: 4561
   of terephthalic acid and isophthalic acid **8**: 4575
   of thiourea **8**: 4708
   of urea **8**: 4786
Solvay chloromethane process **3**: 1270
Solvay crack gas vinyl chloride process **3**: 1336
Solvay oil **8**: 4539
Solvent deoiling
   of slack waxes **8**: 4922
Solvent extraction
   of ethanol **4**: 2121
   of oil seeds **4**: 2405
   use of phosphine oxides **6**: 3833
Solvents
   diisopropyl ether **4**: 2194
   N,N-dimethylformamide **5**: 2703
   dimethyl sulfoxide **7**: 4495

esters **3:** 2036
ethanol **4:** 2131
methyl isobutyl ketone **5:** 3089
pentanols **6:** 3618
propanols **7:** 4071
tetrahydrofuran **8:** 4622
turpentine **8:** 4772
Sorbates **7:** 4366
Sorbic acid **7:** 4365
   environmental protection **7:** 4369
   properties **7:** 4366
   specifications **7:** 4369
   toxicology **7:** 4373
   uses **7:** 4371
Sorbic acid esters **7:** 4367
Sorbic acid production **7:** 4368
   from ketene **5:** 3066
Sorbitol **2:** 1074
Sorghum starch production **7:** 4389
South American fish oil
   fatty acid content **4:** 2376
Soya amine **1:** 472
Soybean lecithin
   specifications **5:** 3153
Soybean oil **4:** 2461
   epoxidized **4:** 2004
   fatty acid content **4:** 2376
   for lecithin production **5:** 3151
Specific heat
   of formic acid **5:** 2715
   of oils and fats **4:** 2388
Spent refinery caustics
   recovery of cresols **3:** 1662
Spinning
   of cellulose acetate fibers **2:** 1183
Spirit of turpentine, *see turpentine*
Spirostanol saponins **7:** 4289
Spray deoiling
   of slack waxes **8:** 4923
Stability
   of cellulose ethers **2:** 1210
Stability constants
   of crown ether–metal ion complexes **3:** 1728
   of metal chelates of EDTA, HEEDTA, and
      DTPA **4:** 2298
Stabilizers
   for cellulose ester molding compounds **2:** 1194
Stachyose **2:** 1067
Stage
   fatty acid fractionation **4:** 2510
Stamicarbon CO_2-stripping urea process **8:** 4796, 4802
Stamicarbon melamine process **5:** 3206
Starch **7:** 4379
   as raw material for ethanol production **4:** 2095
   biosynthesis and structure **7:** 4381
   direct fermentation **4:** 2074

economic aspects **7:** 4411
fermentation by bacteria **4:** 2078
fermentation in yeasts **4:** 2066
modified **7:** 4396
native starches **7:** 4388
physicochemical properties **7:** 4382
rheological properties **7:** 4385
use of native starches **7:** 4395
water activity **7:** 4387
Starch acetates **7:** 4404
Starch anthranilates **7:** 4409
Starch crystals **7:** 4385
Starch esters **7:** 4404
Starch ethers **7:** 4407
Starch gelatinization **7:** 4384
Starch granules **7:** 4382
Starch gum confectionery **7:** 4399
Starch phosphate esters **7:** 4406
Starch production **7:** 4389
Staudinger equation **2:** 1217
Stauffer chloromethane process **3:** 1272
STCA **3:** 1555
Steam cracking, *see hydrocarbon pyrolysis*
Steam reforming
   of natural gas **6:** 3300
Stearic acid
   physical properties **4:** 2485
Stearyl alcohol **1:** 282
   physical properties **5:** 2535
Steels
   in formic acid production **5:** 2731
   in urea production **8:** 4799
Sterculia gum **7:** 4016
Stereoisomerism
   of monosaccharides **2:** 1063
Steroids
   esterification **3:** 2027
Steroid saponins **7:** 4288
Sterols
   in fats and fatty acids **4:** 2382
Stetter dicarboxylic acid synthesis **3:** 1913
Still distillation
   of fatty acids **4:** 2506
Stobbe condensation **3:** 1907
Stone & Webster cracking furnace **4:** 2246
Stone & Webster quench system **4:** 2260
Stovarsol **1:** 616
Streptose **2:** 1066
Stress–strain curves
   of cellulose ester thermoplastics **2:** 1190, 1191
Stripping processes
   in urea production **8:** 4796
Strong-acid indirect propene hydration process **7:** 4069
Styrene **7:** 4417
   from ethylbenzene **4:** 2209
   properties **7:** 4418

specification and handling **7**: 4431
　toxicology **7**: 4439
　uses and economic aspects **7**: 4432
Styrene–acrylonitrile resins **1**: 256
Styrene anthraquinone process **2**: 658
Styrene chlorohydrin **3**: 1589
　toxicology **3**: 1600
Styrene chlorohydrin production **3**: 1595
Styrene distillation **7**: 4426
Styrene–isoprene–styrene block copolymers **5**: 3048
Styrene oxide **4**: 2003
Styrene production **7**: 4422
Styrene–propylene oxide process **7**: 4428
Styrol, *see styrene*
Suberic acid **3**: 1916
　physical properties **3**: 1905
Submerged fermentation
　in citric acid production **3**: 1652
Submerged flame acetylene process **1**: 133, 143
Substituent analysis
　of cellulose ethers **2**: 1236
Substitution reactions
　of silicon-hydrogen bonds **7**: 4318
Succinaldehyde **1**: 392
Succinedialdehyde **1**: 390
Succinic acid **3**: 1914
　byproduct in adipic acid production **1**: 268
　physical properties **3**: 1905
Succinonitrile **6**: 3461
　physical properties **6**: 3457
Sucrose **2**: 1067
Sudan gum, *see gum arabic*
Sugar beet
　for fermentation **4**: 2094
Sugarcane
　for fermentation **4**: 2091
Sugarcane wax **8**: 4896
Sugars, *see carbohydrates*
Sulfamates **7**: 4446
Sulfamic acid **7**: 4443
　uses and toxicology **7**: 4446
Sulfanilic acid **1**: 637; **2**: 771
Sulfate turpentine **8**: 4762
　composition **8**: 4768
Sulfate turpentine production **8**: 4763
Sulfides **8**: 4647
　aliphatic, production **8**: 4659
　aliphatic, properties **8**: 4657
　aliphatic, uses **8**: 4661
　aromatic, production and uses **8**: 4666
　aromatic, properties **8**: 4662
　for arylamine production **1**: 514
　heterocyclic, production **8**: 4673
　heterocyclic, properties **8**: 4667
　heterocyclic, uses **8**: 4674
　oxidation to sulfones **7**: 4479
　oxidation to sulfoxides **7**: 4487

　toxicology **8**: 4677
Sulfinates
　for production of sulfoxides **7**: 4490
　rearrangement **7**: 4482
Sulfination
　with sulfur dioxide **7**: 4453
Sulfinic acid amides **7**: 4464
Sulfinic acid derivatives **7**: 4458
Sulfinic acid esters **7**: 4462
Sulfinic acids **7**: 4449
　addition reactions **7**: 4481
　alkylation to sulfones **7**: 4480
　preparation **7**: 4452
　properties **7**: 4450
　reactions **7**: 4457
　uses **7**: 4469
α-Sulfinyl carbanions
　reactions **7**: 4491
Sulfinyl chlorides **7**: 4458
　for sulfinic acid ester production **7**: 4462
Sulfite reaction
　in benzenesulfonic acid production **2**: 748
Sulfite turpentine **8**: 4762
Sulfite turpentine production **8**: 4764
3-Sulfobenzoic acid **2**: 836
Sulfoethyl cellulose **2**: 1235
Sulfoethylsilanols **7**: 4339
Sulfo group
　nucleophilic aromatic exchange **1**: 524
Sulfo J acid **6**: 3442
Sulfolane
　for aromatic extraction **2**: 719
Sulfolane aromatic separation process **2**: 719
Sulfolane toluene extraction process **8**: 4732
Sulfolenes
　cleavage **7**: 4456
Sulfonamides
　synthesis from amines **1**: 445
Sulfonation
　aromatic **2**: 744
　aromatic, reaction mechanism **2**: 745
　aromatic, technology **2**: 747
　of arylamines **1**: 511
　of benzene, for phenol production **6**: 3701
　of benzene, for resorcinol production **7**: 4265
　of naphthalene **6**: 3365
　of naphthols **6**: 3385
　of naphthylamines **6**: 3407
Sulfonation–desulfonation
　of chlorobenzenes **3**: 1610
Sulfones **7**: 4477
　nucleophilic cleavage to sulfinic acids **7**: 4454
　properties and synthesis **7**: 4477
　reactions **7**: 4484
　toxicology **7**: 4496
　uses **7**: 4485

Sulfonic acids
  aliphatic **8:** 4503
  aliphatic, production **8:** 4506
  nitroaromatic **6:** 3548
  reduction to sulfinic acids **7:** 4452
  toxicology **7:** 4470
Sulfonyl carbanions
  for sulfone preparation **7:** 4482
Sulfoquinovose **2:** 1066
Sulforganosilanes **7:** 4338
5-Sulfo-Tobias acid **6:** 3418
Sulfoxides **7:** 4477
  properties **7:** 4486
  reactions **7:** 4491
  synthesis **7:** 4487
  uses **7:** 4493
Sulfuranes
  for sulfinyl chloride production **7:** 4459
Sulfur dioxide
  for arylamine production **1:** 514
  for production of sulfones **7:** 4483
  for production of sulfoxides **7:** 4489
  for sulfination **7:** 4453
Sulfuric acid
  addition to olefins **3:** 1857
Sulfuric acid ethanol process **4:** 2056
Sulfuric acid ether process **4:** 2190
Sulfuric acid refining
  of deoiled slack waxes **8:** 4925
Sulfur trioxide
  for production of dimethyl sulfate **3:** 1857
Sulfuryl chloride
  for production of dialkyl sulfates **3:** 1859
Sulzer–MWB fatty acid separation process **4:** 2515
Sumitomo–Nippon Shokubai methacrylic acid process **6:** 3257
Sunflower oil **4:** 2456
  fatty acid content **4:** 2376
Surface coatings
  cellulose ester lacquers **2:** 1180
Surface fermentation
  citric acid production **3:** 1651
Surface properties
  of phthalocyanides **7:** 3927
Surface tension
  of fatty oils **4:** 2393
  of saturated fatty acids **4:** 2495
Surfactants
  esters **3:** 2039
  ethanolamines **4:** 2156
  fatty alcohol sulfates **5:** 2547
  isopropanolamines **4:** 2169
  nonionic, from alkylphenols **6:** 3751
  perfluoroalkanecarboxylic acids **5:** 2603
  saponins **7:** 4284
Suspension hydrogenation
  for production of fatty alcohols **5:** 2539

Sweat deoiling
  of slack waxes **8:** 4924
Sweet sorghum
  as raw material for ethanol production **4:** 2098
Switch Condenser
  in phthalic anhydride production **7:** 3874
Sylvan **5:** 2760
Symclosene TCC **3:** 1568
Syngas, *see synthesis gas*
Synthesis gas
  as ethylene feedstock **4:** 2284
  conversion to ethanol **4:** 2061
  conversion to methanol **6:** 3302
  in methanol production **6:** 3300
  reaction with methanol **4:** 2059
  thermodynamics of methanol production **6:** 3292

2,4,5-T **3:** 1620
T2A **8:** 4697
T2AA **8:** 4699
T acid **6:** 3421
Tail-end hydrogenation
  of acetylene in ethylene production **4:** 2276
Tallow amine **1:** 472
D-Talose **2:** 1065
Tannin **5:** 2974
Tapioca
  as raw material for ethanol production **4:** 2098
Tapioca starch production **7:** 4394
Tar
  anthracene content **1:** 644
  as feedstock for quinoline **7:** 4254
  as feedstock for pyridines **7:** 4179
Tar acids **3:** 1656
  high boiling **8:** 4550
Tara gum **7:** 4024
  production and properties **7:** 4025
Tar and pitch **8:** 4509
  ecotoxicology **8:** 4555
  industrial importance **8:** 4511
  origin and classification **8:** 4509
  processing of coke-oven coal tar **8:** 4515
  properties and composition **8:** 4512
  toxicology **8:** 4553
  uses and economic importance **8:** 4552
Tar–bitumen blends **8:** 4535
Tar distillates
  processing **8:** 4536
Tartar **8:** 4563
Tartar emetic **8:** 4567
Tartaric acid **8:** 4559, 4568
  chemical properties **8:** 4562
  chemical synthesis **8:** 4565
  configurations **8:** 4562
  economic aspects and toxicology **8:** 4569
  physical properties **8:** 4560
  specifications **8:** 4566

(R,S)-Tartaric acid
  physical properties **5**: 2958
Tartaric acid production **8**: 4563
Tartrate **8**: 4563
Tartronic acid
  physical properties **5**: 2958
  production **5**: 2962
Tassel-on-a-string model
  of amylopectin **7**: 4383
Tatoray benzene process **2**: 710
Tatoray toluene disproportionation process **8**: 4733
Tatoray xylene isomerization process **8**: 5002
TBBPA, *see tetrabromobisphenol A*
TBPA, *see tetrabromophthalic anhydride*
TBS **6**: 3793
TCA **3**: 1555
T2CA **8**: 4698
TCD dialdehyde **1**: 390, 393
TCNE **6**: 3462
TDI **5**: 3012
TEA, *see triethanolamine*
Teaseed oil **4**: 2459
TEC MTC conventional urea processes **8**: 4801
Telomerization
  of isoprene **5**: 3050
Tensile creep strength
  of cellulose ester thermoplastics **2**: 1190
Tensile modulus
  of cellulose ester thermoplastics **2**: 1187
Tensile strength
  of cellulose ester thermoplastics **2**: 1185
Terephthalic acid **2**: 1121; **8**: 4573
  economic aspects **8**: 4588
  physical properties **8**: 4574
  specifications **8**: 4586
  toxicology **8**: 4589
  uses **8**: 4587
Terephthalic acid esterification **8**: 4582
Terephthalic acid production **8**: 4576
Terpenes **8**: 4593
  as food additives **8**: 4616
  economic aspects **8**: 4597
  in turpentines **8**: 4765
  occurrence and biosynthesis **8**: 4596
  toxicology **8**: 4614
Terpene synthesis
  from isoprene **5**: 3049
Terpenoids
  in fats and fatty acids **4**: 2382
Terphenyls **5**: 2911
  toxicology **5**: 2927
α-Terpinene **8**: 4602
Terpinolene **8**: 4602
Tetra
  asymmetric, *see 1,1,1,2-tetrachloroethane*
  symmetric, *see 1,1,2,2-tetrachloroethane*

2,4,6,8-Tetraacetylazabicyclo [3.3.1] nonane-3,7-
  dione **1**: 67
N,N,N',N'-Tetraacetylethylenediamine **1**: 66
Tetraallyl pentaerythritol ether **1**: 431
3,3',4,4'-Tetraaminodiphenyl **2**: 805
Tetraazaporphyrin derivatives
  as phthalocyanine analogues **7**: 3951
Tetrabromobisphenol A **2**: 882
  toxicology **2**: 891
3,4,5,6-Tetrabromo-*o*-cresol **6**: 3788
Tetrabromoethene **2**: 869
4,5,6,7-Tetrabromo-1,3-isobenzofurandione, *see Tetrabromophthalic anhydride*
Tetrabromomethane **2**: 869
Tetrabromo-2-methylphenol **6**: 3788
Tetrabromophthalic anhydride **2**: 883, 885
  toxicology **2**: 892
3,5,3',5'-Tetra-*tert*-butyl-4,4'-dihydroxybiphenyl
  physical properties **6**: 3775
Tetra-*n*-butylphosphonium bromide
  physical properties **6**: 3829
Tetrachlorobenzene
  physical properties **3**: 1410
  specifications **3**: 1421
1,2,3,4-Tetrachlorobenzene production **3**: 1420
1,2,3,5-Tetrachlorobenzene production **3**: 1420
1,2,4,5-Tetrachlorobenzene production **3**: 1420
Tetrachlorobenzenes
  chemical properties **3**: 1411
2,2',5,5'-Tetrachlorobenzidine **2**: 808
2,3,5,6-Tetrachloro-1,4-benzoquinone **2**: 842
2,2',5,5'-Tetrachloro-4,4'-biphenyl-diyldiamine **2**: 808
Tetrachloro-1,1-difluoroethane
  properties **5**: 2576
Tetrachloro-1,2-difluoroethane
  properties **5**: 2576
1,3,4,6-Tetrachloro-3α,6α-diphenylglycoluril **3**: 1570
1,1,1,2-Tetrachloroethane **3**: 1320
1,1,2,2-Tetrachloroethane
  properties **3**: 1321
  toxicology **3**: 1474
  uses and economic aspects **3**: 1325
Tetrachloroethane dehydrochlorination **3**: 1358
1,1,2,2-Tetrachloroethane production **3**: 1323
Tetrachloroethylene
  properties **3**: 1361
  specifications **3**: 1370
  toxicology **3**: 1476
  uses and economic aspects **3**: 1368
Tetrachloroethylene production **3**: 1363
Tetrachlorometaldehyde **3**: 1528
Tetrachloromethane **3**: 1253, 1278
  see also chloromethanes
  chemical properties **3**: 1260
  physical properties **3**: 1257
  toxicology **3**: 1472
Tetrachloromethane production **3**: 1273

2,3,4,5-Tetrachlorophenol **3**: 1606
2,3,4,6-Tetrachlorophenol **3**: 1606
2,3,5,6-Tetrachlorophenol **3**: 1606
2,3,5,6-Tetrachloropyridine
   physical properties **7**: 4199
2,4,6,8-Tetrachloro-2,4,6,8-tetrazabicyclo[3.3.0]octane-3,7-dione **3**: 1570
2,3,4,5-Tetrachlorotoluene
   physical properties **3**: 1425
   production **3**: 1432
2,3,4,6-Tetrachlorotoluene
   physical properties **3**: 1425
   production **3**: 1432
2,3,5,6-Tetrachlorotoluene
   physical properties **3**: 1425
   production **3**: 1432
n-Tetracontane
   physical properties **5**: 2855
1,24-Tetracosanediol
   physical properties **5**: 2553
Tetracosanoic acid, see lignoceric acid
1-Tetracosanol, see lignoceryl alcohol
Tetracyanoethylene **6**: 3462
n-Tetradecane
   physical properties **5**: 2855
Tetradecanedioic acid
   physical properties **3**: 1905
1,14-Tetradecanediol
   physical properties **5**: 2553
Tetradecanoic acid, see myristic acid
1-Tetradecanol, see myristyl alcohol
Tetradecanonitrile
   physical properties **6**: 3457
1-Tetradecene
   physical properties **5**: 2871
cis-9-Tetradecenoic acid, see myristoleic acid
Tetradecylamine **1**: 473
2-Tetradecyl-1-octadecanol
   physical properties **5**: 2551
1,2,3,4-Tetraethylbenzene
   physical properties **5**: 2894
1,2,3,5-Tetraethylbenzene
   physical properties **5**: 2894
1,2,4,5-Tetraethylbenzene
   physical properties **5**: 2894
Tetraethylene glycol
   physical properties **4**: 2319
   toxicology **4**: 2325
Tetraethylenepentamine **1**: 488
Tetraethylthiuram disulfide **4**: 1971
Tetraethylthiuram monosulfide **4**: 1971
Tetrafluoroethylene **5**: 2584
   physical properties **5**: 2584
Tetrafluoroethylene copolymers **5**: 2604, 2607
Tetrafluoroethylene ethers **5**: 2604
Tetrafluoromethane
   properties **5**: 2574

1,1,3,3-Tetrafluoro-2-propanone **5**: 2597
Tetrahydrofuran **8**: 4621
   economic aspects and toxicology **8**: 4628
   handling and uses **8**: 4627
   physical properties **8**: 4622
   specifications and environmental protection **8**: 4626
Tetrahydrofuran production **8**: 4623
Tetrahydrofuranyl alcohol **5**: 2762
2-Tetrahydrofuranyl methanol **5**: 2762
Tetrahydrofurfuryl alcohol
   physical properties **5**: 2747
   toxicology **5**: 2768
Tetrahydroimidazo[4,5-d]-imidazole-2,5(1H,3H)dione **8**: 4830
1,2,3,4-Tetrahydronaphthalene, see tetralin
Tetrahydrothiophene **8**: 4701
2,4,6,8-Tetrakis(chloromethyl)-1,3,5,7-tetroxocane **3**: 1528
Tetrakis(hydroxymethyl)phosphonium chloride
   physical properties **6**: 3829
N,N,N',N'-Tetrakis-2-hydroxypropylethylene-diamine **4**: 2171
Tetralin **6**: 3358
   toxicology **6**: 3360
Tetramethoxymethylbenzoguanamine
   uses **5**: 3217
1,2,3,4-Tetramethylbenzene, see prehnitene
1,2,3,5-Tetramethylbenzene, see isodurene
1,2,4,5-Tetramethylbenzene, see durene
Tetramethylbenzenes **5**: 2895
   toxicology **5**: 2925
2,2,3,3-Tetramethylbutane
   physical properties **5**: 2855
4-(1,1,3,3-Tetramethylbutyl)phenol **6**: 3748
Tetramethylenediamine **1**: 486
Tetramethylene oxide, see tetrahydrofuran
Tetramethylethylene ozonide
   physical properties **6**: 3664
N,N,N',N'-Tetramethylguanidine **5**: 2836
(7R, 11R)-3,7,11,15-Tetramethyl-trans-2-hexadecen-1-ol **8**: 4613
1,1,3,3-Tetramethyl-5-indanol **6**: 3757
2,3,5,6-Tetramethylphenol **6**: 3730
Tetramethylphosphonium chloride
   physical properties **6**: 3829
Tetramethylsilane **7**: 4343
3,3,5,5-Tetramethyl-1,2,4-trioxolane
   physical properties **6**: 3664
Tetramethylurea **8**: 4821
1,3,6,8-Tetranitrocarbazole **2**: 1059
Tetraorganosilanes **7**: 4343
1,3,5,7-Tetraoxane **5**: 2680
2,5,8,10-Tetraoxa-9-oxotridec-12-enoic acid 2-propenyl ester **1**: 426
Tetraoxymethylene, see 1,3,5,7-tetraoxane

Tetraphenylphosphonium bromide
    physical properties   **6:** 3829
Tetraphenylurea   **8:** 4822
Tetraterpenes
    basic structures   **8:** 4593
1-Tetratriacontanol, *see geddyl alcohol*
1,3,5,7-Tetroxocane, *see 1,3,5,7-tetraoxane*
Texaco isopropanol process   **7:** 4069
Textile dyeing
    use of EDTA-type chelating agents   **4:** 2300
Textile fibers
    from acrylonitrile   **1:** 256
Textile manufacture
    acid-modified starches   **7:** 4399
    cellulose ethers   **2:** 1239
    oxalic acid   **6:** 3591
Textile yarns
    for ethanol purification   **4:** 2123
TFE, *see tetrafluoroethylene*
TFP   **5:** 2588
THD benzene process   **2:** 709
Theobromine, *see purine derivatives*
Theophylline monohydrate, *see purine derivatives*
Thermal conductivity
    of fatty acids   **4:** 2492
Thermal stability
    of phthalocyanines   **7:** 3924
Thermodynamic properties, of saturated hydrocarbons   **5:** 2856
Thermodynamics
    of methanol production from synthesis gas   **6:** 3292
    of silver-catalysed formaldehyde production   **5:** 2649
    of urea production   **8:** 4787
Thermoplastics
    from cellulose esters   **2:** 1184
THF, *see tetrahydrofuran*
THFA   **5:** 2762
Thiadiazinethiones   **4:** 1964
Thiamine   **7:** 4231
1,3-Thiazines   **4:** 1964
Thiazole nitro derivatives   **6:** 3572
2-Thiazoline-2-thiol   **8:** 4668
Thin-film distillation
    of glycerol   **5:** 2794
Thioacetals   **2:** 1075
2,2'-Thiobis(4-chlorophenol)   **6:** 3791
2,2'-Thiobis(4,6-dichlorophenol)   **6:** 3791
2,2'-Thiobis(3,4,6-trichlorophenol)   **6:** 3791
3,3'-Thiobis(2,4,6-trichlorophenol)   **6:** 3791
Thiocarbamoyl chlorides   **4:** 1957
Thiocarbamoylsulfenamides   **4:** 1956
p-Thiocresol production   **8:** 4663
Thiocyanates   **8:** 4631
    for production of heterocyclic thiols   **8:** 4673
    production   **8:** 4633
    properties   **8:** 4631
    toxicology   **8:** 4635
    use and commercial names   **8:** 4634
2-(Thiocyanatomethylthio)benzothiazole
    toxicology   **8:** 4635
    uses   **8:** 4634
3-Thiocyanatopropyltriethoxysilane
    physical properties   **8:** 4632
Thiocyanic acid esters, *see thiocyanates*
Thioesters
    for sulfinyl chloride production   **7:** 4460
Thioethers
    oxidation to sulfones   **7:** 4479
Thiofurfuran, *see thiophene*
Thioglycolic acid, *see mercaptoacetic acid*
Thioguanine, *see purine derivatives*
Thiolation
    of aromatics   **8:** 4664
Thiols   **8:** 4647
    aliphatic, handling   **8:** 4655
    aliphatic, production   **8:** 4651
    aliphatic, properties   **8:** 4648
    aliphatic, uses   **8:** 4656
    aromatic, properties   **8:** 4662, 4663
    aromatic, uses   **8:** 4665
    heterocyclic, production   **8:** 4670
    heterocyclic, properties   **8:** 4667
    heterocyclic, uses   **8:** 4674
    oxidation to sulfinic acids   **7:** 4454
    toxicology   **8:** 4677
Thionyl chloride
    for production of dialkyl sulfates   **3:** 1859
    for sulfinic acid production   **7:** 4456
Thioorganosilanes   **7:** 4338
Thiophene   **8:** 4691
    properties   **8:** 4691
    reactions   **8:** 4693
    specifications and handling   **8:** 4695
    toxicology   **8:** 4704
    uses   **8:** 4702
Thiophene 2-acetic acid   **8:** 4699
Thiophene 2-aldehyde   **8:** 4697
Thiophene 2-carboxylic acid   **8:** 4698
Thiophene dicarboxylic acids   **8:** 4700
Thiophene nitro derivatives   **6:** 3571
Thiophene production   **8:** 4694
Thiophenol
    toxicology   **8:** 4678
Thiophenol production   **8:** 4663
Thiophosphinates   **6:** 3838
Thiophosphoric acid esters   **6:** 3851
Thiosulfinic acid esters   **7:** 4464
Thiosulfonic acid esters
    cleavage   **7:** 4456
2-Thiouracyls   **4:** 1963
Thiourea   **3:** 1748; **8:** 4707
    economic aspects and toxicology   **8:** 4713

for production of heterocyclic thiols **8**: 4672
from dithiocarbamic acids or acid esters **4**: 1966
properties **8**: 4708
specifications **8**: 4711
uses **8**: 4712
Thiourea derivatives **8**: 4707
production **8**: 4716
properties **8**: 4716
toxicology **8**: 4719
uses **8**: 4718
Thiourea production **8**: 4710
Thiuram disulfides **4**: 1953
Thiuram monosulfides **4**: 1955
Thiuram tetrasulfides **4**: 1955
Thiuram trisulfides **4**: 1955
Thonzylamine hydrochloride **7**: 4232
Three-component oxidation
of saturated hydrocarbons **5**: 2866
THT **8**: 4701
Thymine **7**: 4231
Thymol **6**: 3734
Tiglic aldehyde **1**: 368, 372
Tiron **2**: 777
Tishchenko reaction **1**: 312; **3**: 2029
TLE, *see transfer-line exchangers*
TMG **5**: 2836
TMPD glycol **1**: 336
toxicology **1**: 343
Toa Gosei caprolactam process **2**: 1032
Tobias acid **6**: 3414
Tocopherols
in fats and fatty acids **4**: 2382
Tokuyama Soda chloromethane process **3**: 1272
Tokuyama Soda isopropanol process **7**: 4070
m-Tolidine **2**: 806
o-Tolidine **2**: 805
toxicology **2**: 814
o-Tolidine-6,6'-disulfonic acid **2**: 811
Toluene **2**: 695; **8**: 4727
chlorinated, *see chlorotoluenes*
economic aspects **2**: 731; **8**: 4735
for caprolactam production **2**: 1028
for styrene production **7**: 4430
handling and environmental aspects **8**: 4739
oxidation **2**: 679
properties **8**: 4728
specifications **8**: 4738
toxicology **8**: 4739
Toluene chlorination
liquid phase **3**: 1427
side-chain, kinetics **3**: 1453
Toluenediamine
toxicology **6**: 3816
2,3-Toluenediamine
properties **6**: 3810
2,4-Toluenediamine
properties **6**: 3810

2,5-Toluenediamine
properties **6**: 3810
2,6-Toluenediamine
properties **6**: 3810
3,4-Toluenediamine
properties **6**: 3810
3,5-Toluenediamine
properties **6**: 3810
Toluenediamines **6**: 3809
production **6**: 3812
properties **6**: 3810
specifications and handling **6**: 3813
uses **6**: 3815
Toluene diisocyanate **5**: 3012
Toluenedisulfonic acids **2**: 761
Toluene hydroxylation
for cresol production **3**: 1678
Toluene methylation
oxidative, for cresol production **3**: 1679
Toluene nitration process **6**: 3517
Toluene oxidation **2**: 821
for benzyl alcohol production **2**: 855
for phenol production **6**: 3697
Toluene oxidation process **6**: 3699
Toluene oxychlorination **3**: 1429
in cresol production **3**: 1676
Toluene processing **8**: 4733
Toluene production
integration into refineries **8**: 4734
Toluene separation **8**: 4731
m-Toluenesulfonic acid **2**: 762
o-Toluenesulfonic acid **2**: 762
Toluenesulfonic acids **2**: 761
p-Toluenesulfonyl isocyanate **5**: 3011
o-Toluic acid **2**: 831; **6**: 3805
Toluidine **8**: 4745
production **8**: 4747
properties **8**: 4745
specifications, handling and uses **8**: 4748
toxicology **8**: 4754
m-Toluidine **1**: 636
p-Toluidine **1**: 636
Toluidine derivatives **8**: 4749
p-Toluidine salt **2**: 766
Toluidinesulfonic acids **8**: 4751
o-Tolylbiguanide **5**: 2840
TOPO **6**: 3833
Toray PNC caprolactam process **2**: 1026
Tosyl chloride **2**: 763
Tower reactor
for fermentation **4**: 2085
TPP, *see triphenyl phosphine*
Tragacanthic acid
in gum tragacanth **7**: 4015
Transalkylation
of toluene **8**: 4733
Transcat vinyl chloride process **3**: 1346

Transesterification 3: 2017
 for production of esters 3: 2028
 of allyl esters 1: 427
 of glycerides 4: 2396
Transfer-line exchangers
 in hydrocarbon pyrolysis for ethylene production 4: 2256
Transmethylation
 of cresols 3: 1676
Transvinylation 8: 4858
α,α-Trehalose 2: 1067
Tri, see trichloroethylene
n-Triacontane
 physical properties 5: 2855
Triacontanoic acid, see melissic acid
1-Triacontanol, see myricyl alcohol
1-Triacontene
 physical properties 5: 2871
Trialkylaluminum
 for production of aliphatic alcohols 1: 287
Triallylamine 1: 432
Triallyl cyanurate 1: 426, 430
Triallyl phosphate 1: 426, 430
Triaminoguanidine 5: 2834
2,4,6-Triamino-1,3,5-triazine, see melamine
Triazenes
 preparation 3: 1888
 properties 3: 1887
 uses 3: 1889
Triazine
 fluorinated 5: 2621
Triazines 3: 1767
1,3,5-Triazine-2,4,6-triamine 8: 4818
1,3,5-Triazine-2,4,6(1H,3H,5H)-trione, see cyanuric acid
[1H]-1,2,4-Triazole-3-thiol 8: 4668
2,4,6-Tribromoaniline 2: 870
1,3,5-Tribromobenzene 2: 870
Tribromoethene 2: 869
Tribromofluoromethane
 properties 5: 2581
Tribromomethane 2: 869
2,4,6-Tribromophenol 2: 870
3,5,4'-Tribromosalicylanilide 6: 3793
Tributylamine 1: 454, 458
2,4,6-Tri-*tert*-butylphenol 6: 3737
Tributyl phosphate
 physical properties 6: 3849
 toxicology 6: 3858
Tri-*n*-butyl phosphine
 physical properties 6: 3822
Trichloroacetaldehyde, see chloral
Trichloroacetate
 toxicology 3: 1560
Trichloroacetic acid 3: 1552
 toxicology 3: 1560
Trichloroacetic acid esters 3: 1555
Trichloroacetonitrile 6: 3463

Trichloroacetyl chloride 3: 1554
2,4,5-Trichloroaniline 1: 636
Trichlorobenzene
 chemical properties 3: 1411
 physical properties 3: 1407
1,2,4-Trichlorobenzene
 specifications 3: 1421
 uses 3: 1423
1,2,3-Trichlorobenzene production 3: 1418
1,2,4-Trichlorobenzene production 3: 1418
1,3,5-Trichlorobenzene production 3: 1419
2,4,5-Trichlorobenzenesulfonic acid 2: 769
2,4,5-Trichlorobenzenesulfonyl chloride 2: 769
Trichlorobromomethane 2: 869
2,3,4-Trichloro-1-butene 3: 1382
1,1,2-Trichloro-2,2-difluoroethane
 properties 5: 2576
2,2,2-Trichloroethanal, see chloral
1,1,1-Trichloroethane
 production 3: 1311
 properties 3: 1309
 specifications 3: 1370
 stabilization with dioxane 4: 1942
 toxicology 3: 1473
 uses and economic aspects 3: 1316
1,1,2-Trichloroethane
 for production of 1,1,1-trichloroethane 3: 1314
 production 3: 1318
 properties 3: 1317
 toxicology 3: 1473
 uses and economic aspects 3: 1320
1,1,2-Trichloroethane dehydrochlorination 3: 1350
Trichloroethanenitrile 6: 3463
Trichloroethylene
 properties 3: 1356
 specifications 3: 1370
 toxicology 3: 1475
 uses and economic aspects 3: 1361
Trichloroethylene hydrolysis 3: 1543
Trichloroethylene production 3: 1357
Trichlorofluoromethane
 properties 5: 2576
Trichlorohydrin 3: 1376
2,4,4'-Trichloro-2'-hydroxydiphenylether 6: 3788
1,2,3-Trichloroisobutane
 from isobutene chlorination 3: 1393
Trichloroisocyanuric acid 3: 1566, 1568
N,N',N''-Trichloromelamine 3: 1571
Trichloromethane 3: 1253, 1278
 see also chloromethanes
 chemical properties 3: 1259
 physical properties 3: 1255
 toxicology 3: 1471
Trichloromethane production 3: 1268
Trichloromethyl benzene, see benzotrichloride
Trichloromethyl cyanide 6: 3463
1,2,4-Trichloro-5-nitrobenzene 6: 3534

Trichloroparaldehyde **3**: 1528
2,3,4-Trichlorophenol **3**: 1606
2,3,5-Trichlorophenol **3**: 1606
2,3,6-Trichlorophenol **3**: 1606
2,4,5-Trichlorophenol **3**: 1606
   production **3**: 1621
2,4,6-Trichlorophenol **3**: 1606
3,4,5-Trichlorophenol **3**: 1606
(2,4,5-Trichlorophenoxy)acetic acid **3**: 1620
2-(2,4,5-Trichlorophenoxy)propionic acid **3**: 1620
1,2,3-Trichloropropane **3**: 1376
4,3',4'-Trichlorosalicylanilide **6**: 3792
5,3',4'-Trichlorosalicylanilide **6**: 3792
2,3,4-Trichlorotoluene
   physical properties **3**: 1425
2,3,5-Trichlorotoluene
   physical properties **3**: 1425
2,3,6-Trichlorotoluene
   physical properties **3**: 1425
   uses **3**: 1435
2,4,5-Trichlorotoluene
   physical properties **3**: 1425
2,4,6-Trichlorotoluene
   physical properties **3**: 1425
3,4,5-Trichlorotoluene
   physical properties **3**: 1425
α,α,α-Trichlorotoluene, *see benzotrichloride*
2,3,4-Trichlorotoluene production **3**: 1431
2,3,6-Trichlorotoluene production **3**: 1431
2,4,5-Trichlorotoluene production **3**: 1431
2,4,6-Trichlorotoluene production **3**: 1432
2,4,6-Trichloro-1,3,5-triazine, *see cyanuric chloride*
1,3,5-Trichloro-1,3,5-triazine-
   2,4,6(1H,3H,5H)Trione **3**: 1568
1,1,1-Trichlorotrifluoroethane
   properties **5**: 2576
1,1,2-Trichlorotrifluoroethane
   properties **5**: 2576
Tricholine citrate **3**: 1635
Trickle-bed hydrogenation
   of fatty acid methyl esters **5**: 2541
Triclosan **6**: 3788
1,23-Tricosanediol
   physical properties **5**: 2553
1-Tricosanol
   physical properties **5**: 2535
Tricyclohexylidene triperoxide
   physical properties **6**: 3664
Tricyclohexyl phosphine
   physical properties **6**: 3822
Tridecanal **1**: 353, 366
n-Tridecane
   physical properties **5**: 2855
Tridecanedioic acid, *see brassylic acid*
1,13-Tridecanediol
   physical properties **5**: 2553

Tridecanoic acid
   physical properties **4**: 2485
1-Tridecanol **1**: 282
   physical properties **5**: 2535
1-Tridecene
   physical properties **5**: 2871
2-Tridecyl-1-heptadecanol
   physical properties **5**: 2551
Tridodecylamine **1**: 473
Triethanolamine **4**: 2152
   see also ethanolamines
   physical properties **4**: 2152
   toxicology **4**: 2177
Triethylamine **1**: 454, 458
Triethylbenzene
   toxicology **5**: 2926
1,2,3-Triethylbenzene
   physical properties **5**: 2894
1,2,4-Triethylbenzene
   physical properties **5**: 2894
1,3,5-Triethylbenzene
   physical properties **5**: 2894
Triethylbenzenes **5**: 2897
Triethylenediamine **1**: 470
Triethylene glycol
   physical properties **4**: 2319
   specifications **4**: 2318
   toxicology **4**: 2325
Triethylenetetramine **1**: 487
   toxicology **1**: 493
Triethylphosphate
   physical properties **6**: 3849
   toxicology **6**: 3857
Triethyl phosphine
   physical properties **6**: 3822
Triethyl phosphite
   toxicology **6**: 3859
Trifluoroacetaldehyde **5**: 2598
3-(Trifluoroacetyl)camphor **5**: 2602
Trifluorobenzenes **5**: 2617
Trifluorobromomethane **2**: 881
   toxicology **2**: 891
1,1,1-Trifluoro-5,5-dimethyl-2,4-phexane-
   dione **5**: 2602
1,1,1-Trifluoroethane
   properties **5**: 2574
2,2,2-Trifluoroethanol **5**: 2590
Trifluoroiodomethane
   properties **5**: 2582
Trifluoromethane
   properties **5**: 2574
Trifluoromethanesulfonic acid **5**: 2605
Trifluoromethylbenzene **5**: 2609
3-Trifluoromethylbenzoyl fluoride
   physical properties **5**: 2609
Trifluoromethylpyridines **5**: 2620
1,1,1-Trifluoro-2,4-pentanedione **5**: 2602

4,4,4-Trifluoro-1-phenyl-1,3-butanedione **5**: 2602
1,1,1-Trifluoro-2-propanone **5**: 2597
3,3,3-Trifluoropropene **5**: 2588
  physical properties **5**: 2584
4,4,4-Trifluoro-1-1(2-thienyl)-1,3-butanedione **5**: 2602
2,4,6-Trifluoro-1,3,5-triazine **5**: 2621
Trifluoro(trifluoromethyl)oxirane **5**: 2592
3,3,3-Trifluoro-1-(trifluoromethyl)propene **5**: 2589
2,2,2-Trifluoro-1-vinyloxyethane **5**: 2595
m-Trigallic acid **5**: 2974
Triglycerides
  in fats and fatty oils **4**: 2374
  melting points **4**: 2386
4,4,4-Trihaloacetoacetates production
  from ketene **5**: 3067
Trihydroxybenzene **6**: 3763
1,2,4-Trihydroxybenzene **6**: 3765
1,3,5-Trihydroxybenzene **6**: 3766
3,4,5-Trihydroxybenzoic acid **5**: 2981
2,6,8-Trihydroxypurine, *see purine derivatives*
Triiodomethane **5**: 3007
Triisobutylamine **1**: 454
Triisopropanolamine **4**: 2164
  see also isopropanolamines
  toxicology **4**: 2179
2,4,6-Triisopropylphenol **6**: 3733
Trimellitic acid **2**: 1121, 1129
Trimellitic anhydride **2**: 1129, 1130
2,4,4-Trimethyladicipic acid
  physical properties **3**: 1905
Trimethyladipic acid **3**: 1915
Trimethylamine **6**: 3326, 3325
  see also methylamines
  handling **6**: 3332
  specifications **6**: 3331
  toxicology **6**: 3334
  uses **6**: 3333
1,2,4-Trimethylbenzene, *see pseudocumene*
Trimethylbenzenes **5**: 2893
  toxicology **5**: 2925
2,6,6-Trimethylbicyclo[3.1.1]heptane **8**: 4605
2,6,6-Trimethylbicyclo[3.1.1]heptane-2-hydroperoxide **8**: 4606
2,6,6-Trimethylbicyclo[3.1.1]heptan-2-ol **8**: 4607
3,7,7-Trimethylbicyclo-[4.1.0]hept-3-ene **8**: 4604
2,2,3-Trimethylbutane
  physical properties **5**: 2855
Trimethylcyclohexane
  physical properties **3**: 1798
3,3,5-Trimethylcyclohexanol **3**: 1818
3,3,5-Trimethylcyclohexanone **3**: 1819
3,3,5-Trimethyl-2-cyclohexen-1-one **5**: 3105
2,6,10-Trimethyl-2,6,9,11-dodecatetraene **8**: 4609
Trimethylene chlorobromide, *see 1-bromo-3-chloropropane*
Trimethylenediamine **1**: 484

Trimethylene glycol
  from acrolein **1**:
  properties **7**: 4055
  toxicology **7**: 4059
  uses **7**: 4057
Trimethylene glycol production **7**: 4056
3,5,5-Trimethylhexanal **1**: 353
2,4,4-Trimethylhexanedioic acid, *see 2,4,4-trimethyladipic acid*
3,5,5-Trimethyl-1-hexanol **1**: 282
Trimethyl(2-hydroxyethyl)ammonium hydroxide **3**: 1633
6,10,10-Trimethyl-2-methylenebicyclo[7.2.0]-5-undecene **8**: 4611
3,3,7-Trimethyl-8-methylenetricyclo[5.4.0.9-Z9]undecane **8**: 4612
2,6,8-Trimethyl-4-nonanol **1**: 282
2,6,6-Trimethylnorpinane **8**: 4605
Trimethylolethane **1**: 324, 339
Trimethylolpropane **1**: 324, 338
  allyl ethers **1**: 431
  toxicology **1**: 343
2,7,7-Trimethyl-3-oxatricyclo[4.1.1.0^{2,4}]octane **4**: 2002
2,2,3-Trimethylpentane
  physical properties **5**: 2855
2,2,4-Trimethylpentane
  physical properties **5**: 2855
2,3,3-Trimethylpentane
  physical properties **5**: 2855
2,3,4-Trimethylpentane
  physical properties **5**: 2855
2,2,4-Trimethyl-1,3-pentanediol **1**: 323, 336
2,4,4-Trimethyl-2-pentanethiol production **8**: 4653
2,3,5-Trimethylphenol **6**: 3730
2,3,6-Trimethylphenol **6**: 3727
2,4,6-Trimethylphenol **6**: 3729
Trimethyl phosphate
  physical properties **6**: 3849
  toxicology **6**: 3857
Trimethyl phosphine
  physical properties **6**: 3822
Trimethyl phosphite
  toxicology **6**: 3859
1,7,7-Trimethyl-3-(trifluoroacetyl)-bicyclo[2,2,1]heptan-2-one **5**: 2602
Trimethyl trimellitate
  physical properties **3**: 2015
2,4,6-Trimethyl-1,3,5-trioxane, *see paraldehyde*
1,3,7-Trimethylxanthine, *see purine derivatives*
2,4,6-Trinitroaniline **1**: 637; **6**: 3544
1,3,5-Trinitrobenzene **6**: 3514, 3514
Trioctadecylamine **1**: 473
Trioctylamine **1**: 473
Trioctyl phosphine oxide **6**: 3833
  in liquid-liquid extraction of acetic acid **1**: 47

1,3,5-Trioxane **5**: 2677
  uses **5**: 2680
1,3,5-Trioxane production **5**: 2678
1,2,4-Trioxolane
  physical properties **6**: 3664
1,3,5-Trioxycyclohexane, see 1,3,5-trioxane
Trioxymethylene, see 1,3,5-trioxane
Tripentaerythritol **1**: 324, 343
Tripentylamine **1**: 454
Triphenylmethane dyes
  use of benzaldehyde **2**: 683
Triphenylmethanethiol production **8**: 4652
Triphenylmethylphosphonium iodide
  physical properties **6**: 3829
Triphenyl phosphine
  physical properties **6**: 3822
  production **6**: 3823
  uses **6**: 3824
Triphenyls
  toxicology **5**: 2926
Triphosphate esters **6**: 3851
Tripropylamine **1**: 454
Tripropylene glycol
  physical properties **7**: 4048
  toxicology **7**: 4058
  uses **7**: 4054
Trisanyl **6**: 3793
Tris(2-chloroethyl) phosphate
  toxicology **6**: 3858
2,4,6-Tris-(chloromethyl)-1,3,5-trioxane **3**: 1528
Tris(2-cyanoethyl) phosphine
  physical properties **6**: 3822
  production **6**: 3823
Tris(2,3-dibromopropyl) phosphate
  toxicology **6**: 3858
2,4,6-Tris(dichloromethyl)-1,3,5-trioxane **3**: 1531
Tris(3-methylbutyl)amine **1**: 454
2,4,6-Tris(2-propenyloxy)-1,3,5-triazine **1**: 426
Triterpenes
  acyclic **8**: 4613
  basic structures **8**: 4593
Triterpene saponins **7**: 4295
Trithianes
  for sufinyl chloride production **7**: 4459
1-Tritriacontanol
  physical properties **5**: 2535
Triuret **8**: 4818
(R,S)-Tropic acid
  physical properties **5**: 2958
Trostberg rotary furnace calcium cyanamide process **3**: 1741
Truce–Smiles rearrangement **7**: 4485
Tryptamine **5**: 3003
Tryptophol **5**: 3003
Tube metallurgy hydrocarbon cracking
  furnace **4**: 2250
Tung oil **4**: 2465

Turkey gum, see gum arabic
Turpentine **8**: 4759
  composition **8**: 4765
  definition and classification **8**: 4761
  destructively distilled, production **8**: 4765
  economic aspects **8**: 4773
  handling and ecological aspects **8**: 4771
  occurrence **8**: 4759
  properties **8**: 4767
  safety **8**: 4770
  toxicology **8**: 4774
  uses **8**: 4772
Turpentine production **8**: 4762
Turpentine substitute **8**: 4776

Ube oxalic acid process **6**: 3586
Udex aromatic separation process **2**: 718
Ultra-selective exchanger quench cooler **4**: 2260
Undecanal **1**: 353, 365, 368
n-Undecane
  physical properties **5**: 2855
Undecanedioic acid
  physical properties **3**: 1905
1,11-Undecanediol
  physical properties **5**: 2553
n-Undecanoic acid, see n-undecylic acid
1-Undecanol **1**: 282
  physical properties **5**: 2535
1-Undecene
  physical properties **5**: 2871, 2871
10-Undecenoic acid
  physical properties **4**: 2487
10-Undecen-1-ol **1**: 309
  physical properties **5**: 2549
n-Undecylic acid **2**: 1097
  physical properties **4**: 2485
2-Undecyl-1-pentadecanol
  physical properties **5**: 2551
Unidak naphthalene process **6**: 3354
UOP cumene process **5**: 2899
UOP Cymex process **3**: 1671
UOP-Dow aromatic separation process **2**: 718
UOP Pacol olefin process **5**: 2875
UOP technology complex **8**: 5003
Uracil **7**: 4230
Urea **8**: 4783
  chemical properties **8**: 4786
  economic aspects **8**: 4817
  for melamin production **5**: 3202
  form, grades and handling **8**: 4813
  for production of guanidine salts **5**: 2828
  for production of phthalimide **7**: 3882
  phosgenation **5**: 3020
  physical properties **8**: 4784
  product-shaping **8**: 4811
  reaction with acrolein **1**: 205
  specifications **8**: 4815

uses **8**: 4816
Urea–aldehyde condensation products
   cyclic **8**: 4828
Urea–ammonium nitrate production **8**: 4809
Urea cresol adducts **3**: 1683
Urea derivatives **8**: 4818
Urea–formaldehyde resins
   uses **8**: 4816
Urea production
   conditions **8**: 4787
   ecological aspects **8**: 4809
   materials **8**: 4797
   process design **8**: 4794
   processes **8**: 4800
   side reactions **8**: 4799
Urea yield
   as a function of $NH_3:CO_2$ ratio **8**: 4791
Uric acid, see purine derivatives
Urotropin, see hexamethylenetetramine
Ursanes **7**: 4297
7,7′-Urylenebis(4-hydroxynaphthalene-2-sulfonic acid) **6**: 3436
USSR formic acid process **5**: 2725
UTI heat recycle urea process **8**: 4800

Vacuum bioreactor
   for fermentation **4**: 2088
Valeraldehyde **1**: 352, 361
Valeraldehyde diethyl acetal **1**: 395
n-Valeric acid **2**: 1097, 1108
   specification **2**: 1106
Valeronitrile
   physical properties **6**: 3457
Vapor-phase hydrogenation
   of Nitrobenzene **1**: 631
Vapor-Phase oxidation
   of anthracene with air **2**: 654
Vapor pressure
   of acetylene **1**: 120
   of acrylic acid **1**: 224
   of benzal chloride **3**: 1456
   of benzaldehyde **2**: 674
   of benzoic acid **2**: 820
   of benzotrichloride **3**: 1459
   of benzyl alcohol **2**: 850
   of benzyl chloride **3**: 1449
   of biphenyl **5**: 2907
   of bis(2-chloroethyl) ether **4**: 2199
   of butadiene **2**: 901
   of 1,4-butanediol **2**: 942
   of butenediol **2**: 941
   of butenes **2**: 987
   of butyrolactone **2**: 1005
   of ε-caprolactam **2**: 1014
   of catechol **6**: 3759
   of chlorinated hydrocarbons **3**: 1288
   of chlorinated naphthalenes **3**: 1441
   of chloroacetic acid **3**: 1541
   of cresols **3**: 1658
   of cumene **5**: 2899
   of cyclohexanone oxime **2**: 1017
   of diamyl ether **4**: 2198
   of di-*n*-butyl ether **4**: 2196
   of 1,2-dichloropropane **3**: 1375
   of diethyl ether **4**: 2189
   of diisopropyl ether **4**: 2193
   of dimethyl ether **4**: 1932
   of dimethyl sulfate **3**: 1854
   of dimethyl terephthalate **8**: 4575
   of DMF **5**: 2698
   of ethylbenzene **8**: 4992
   of ethylene **4**: 2223
   of ethylene glycol–water mixtures **4**: 2307
   of ethylene oxide **4**: 2332
   of 2-ethylhexanol **4**: 2363
   of fatty acids, straight-chain **4**: 2490
   of formaldehyde, liquid **5**: 2641
   of formaldehyde released from paraformaldehyde **5**: 2675
   of formamide **5**: 2692
   of formic acid **5**: 2712
   of propylene glycols **7**: 4050
   of furfural **5**: 2750
   of furfuryl alcohol **5**: 2756
   of glycerol **5**: 2789
   of glycerol–water solutions **5**: 2789
   of isoprene **5**: 3034
   of ketene **5**: 3062
   of maleic anhydride **5**: 3161
   of methanol **6**: 3289
   of methylamine solutions **6**: 3327
   of MTBE **6**: 3338
   of pentanols **6**: 3612
   of phenol **6**: 3691
   of phthalimide **7**: 3880
   of propene **7**: 4083
   of propionic acid **7**: 4103
   of propylene oxide **7**: 4131
   of propyne and propadiene **1**: 187
   of resorcinol **7**: 4261
   of styrene **7**: 4419
   of thiophene **8**: 4692
   of toluene **8**: 4730
   of triglycerides **4**: 2388
   of 1,3,5-trioxane **5**: 2677
   of urea **8**: 4785
   of vinyl chloride **3**: 1331
   of xylene isomers **8**: 4992
   of xylenes **8**: 4993

VCM, see vinyl chloride

VDC, see 1,1-dichloroethylene

Veba-Chemie acetaldehyde process **1**: 9

Vegetable butters **4**: 2451

Vegetable oils **4**: 2441
   manufacture and processing **4**: 2401
Vegetable waxes **8**: 4887
Veratrole **6**: 3781
Verkleisterung **7**: 4385
Vernonia oil **4**: 2467
Versatic Acid **2**: 1111
Versatic acid vinyl esters **8**: 4860
Vicat softening temperature
   of cellulose ester thermoplastics **2**: 1188
Vicinal diols
   by hydroxylation of olefins with peracids **1**: 337
Vinegar, see acetic acid
Vinyl acetate
   by vinylation with acetylene **1**: 125
   physical properties **3**: 2014
   properties and production **8**: 4839
   specifications **3**: 2037
   specifications and handling **8**: 4852
   toxicology **8**: 4853
   uses and economic aspects **8**: 4854
Vinyl acetate–ethylene copolymers **8**: 4854
Vinyl acetate production **3**: 2031
Vinyl acetates **1**: 69
Vinylacetyl chloride **2**: 1114
Vinylation
   of acid amides **1**: 125
   of ammonia **1**: 125
   with acetylene **1**: 124
Vinylbenzene, see styrene
4-Vinylbenzenesulfonamide **2**: 766
4-Vinylbenzenesulfonic acid **2**: 766
Vinylbenzyl chloride **7**: 4438
N-Vinylcarbazole **2**: 1059
Vinyl chloride **3**: 1330
   by vinylation with acetylene **1**: 124
   for production with 1,1,2-trichloroethane **3**: 1320
   physical properties **3**: 1331
   toxicology **3**: 1474
   uses and economic aspects **3**: 1347
Vinyl chloride conversion
   for production of 1,1-dichloroethane **3**: 1296
Vinyl chloride production **3**: 1332
   chlorine-free **3**: 1341
   from acetylene **3**: 1334
   from 1,2-dichloroethane **3**: 1336
   from ethane **3**: 1345
   from ethylene **3**: 1343
   integrated balanced **3**: 1342
4-Vinylcyclohex-1-ene **5**: 2891
Vinyl esters **8**: 4839
   by vinylation with acetylene **1**: 125
   of higher carboxylic acids **8**: 4857
   of versatic 911, 10, and 1519 acids **8**: 4860
Vinyl ethers **8**: 4865
   by vinylation with acetylene **1**: 125
   environmental protection **8**: 4872

   production **8**: 4866
   production, process description **8**: 4870
   production, reaction conditions **8**: 4869
   properties **8**: 4865
   specifications and handling **8**: 4873
   uses and toxicology **8**: 4874
Vinyl fluoride **5**: 2587
Vinylidene chloride, see 1,1-dichloroethylene
1-Vinylimidazole
   physical properties **5**: 2988
   toxicology **5**: 2993
Vinyl isononanoate **8**: 4860
Vinyl isothiocyanate
   physical properties **8**: 4636
Vinyl isothiocyanate production **8**: 4640
Vinyl laurate **8**: 4860
4-Vinylphenol **6**: 3756
Vinyl phenyl ether
   by vinylation with acetylene **1**: 125
Vinyl propionate
   properties and production **8**: 4855
   specifications and uses **8**: 4857
Vinylpyridines **7**: 4189
Vinylsilanes **7**: 4330
Vinyl sulfides
   by vinylation with acetylene **1**: 125
Vinylsulfonic acid **8**: 4505
Vinylsulfonic acid production **8**: 4507
Vinyltoluene **7**: 4434
Violet acid **6**: 3394
Visbreaking
   propylene production **7**: 4089
Viscosity
   of alginate solutions **7**: 3999
   of aqueous propylene glycols **7**: 4049
   of carboxymethyl cellulose **2**: 1232
   of fatty acids **4**: 2492
   of fatty oils **4**: 2390
   of galactomannans **7**: 4026
   of glycerol–water solutions **5**: 2791
   of gum arabic solutions **7**: 4014
   of hydroxyethyl cellulose **2**: 1226
   of methyl cellulose **2**: 1216
   of pectin solutions **7**: 3983
   of propionic acid **7**: 4104
   of propylene oxide **7**: 4131
   of resorcinol **7**: 4261
   of starch gels **7**: 4386
   of styrene **7**: 4420
   of urea **8**: 4785
Viscosity adjustment
   of cellulose nitrate **2**: 1151
Visxosity
   of polysaccharide solutions **7**: 4000
Vitamin B_1 **7**: 4231
Vitamin K3 **6**: 3450

Volume expansion
 of benzene **2:** 698
V structure
 of starch **7:** 4385
VT **7:** 4434
Vulcanization
 diarylguanidines **5:** 2838
 dithiocarbamates **4:** 1967
 propionates **7:** 4114
 thiourea derivatives **8:** 4719
Vulcan Materials 1,1,1-trichloroethane process **3:** 1315

Wacker acetic anhydride process **1:** 73
Wacker–Hoechst acetaldehyde process **1:** 11
Wacker–Hoechst acetone process **1:** 98
Wacker phthalic anhydride process **7:** 3870
Wacker vinyl ester process **8:** 4859
Wash oil **8:** 4539
Wastes
 as feedstocks for biomethanation **6:** 3277
Waste sulfite liquors
 as raw material for ethanol production **4:** 2110
Water retention
 of methyl cellulose **2:** 1220
Water softening
 with nitrilotriacetic acid **6:** 3483
Waxes 8: 4877
 classification **8:** 4881
 definition **8:** 4880
 economic aspects **8:** 4885
 fully synthetic **8:** 4882
 history **8:** 4879
 in fats and fatty oils **4:** 2381
 microcrystalline, *see paraffin waxes*
 natural **8:** 4881
 natural, fossil **8:** 4907
 natural, nonfossil **8:** 4886
 natural, toxicology **8:** 4959
 paraffin, *see paraffin waxes*
 partially synthetic **8:** 4881
 properties and uses **8:** 4884
 thermal cracking **5:** 2873
 toxicology **8:** 4959
Wax esters **3:** 2022
 for fatty alcohol production **5:** 2537
Wax organizations **8:** 4886
Weak acid indirect propene hydration process **7:** 4068
Welding
 with acetylene flames **1:** 181
Wenker aziridine process **2:** 667
Wet generators
 for acetylene production from calcium carbide **1:** 159
Wetting agents
 for nitrobenzene reduction **1:** 607

Whale oil **4:** 2471
 fatty acid content **4:** 2376
 manufacture and processing **4:** 2411
Wheat germ oil **4:** 2455
 fatty acid content **4:** 2376
Wheat starch production **7:** 4391
Wheland intermediate **6:** 3508
Williamson synthesis **2:** 1072, 1207; **4:** 2188
Winterization
 of fats and fatty oils **4:** 2423
Witten dimethyl terephthalate process **8:** 4580
Wood pulp
 as raw material for cellulose **2:** 1169
 as raw material for cellulose **2:** 1146
 cellulose acetate **2:** 1182
 sulfite processed **2:** 1212
Wood tars **8:** 4510
 processing **8:** 4551
 properties and composition **8:** 4513
Wood turpentine **8:** 4761
 composition **8:** 4768
Wood turpentine production **8:** 4763
Wool wax **8:** 4906
Wulff thermal cracking acetylene process **1:** 163
Wurtz synthesis **3:** 1913

Xanthan gels **7:** 4021
Xanthan gum **7:** 4018
 application **7:** 4022
 structure and properties **7:** 4020
Xanthan gum production **7:** 4019
Xanthates 8: 4971
 chemical properties **8:** 4973
 economic aspects **8:** 4980
 physical properties **8:** 4972
 production **8:** 4976
 specifications and handling **8:** 4978
 toxicology **8:** 4981
 uses **8:** 4979
Xanthate solutions **8:** 4973
Xanthic acids
 chemical properties **8:** 4973
 physical properties **8:** 4971
Xanthine, *see purine derivatives*
Xanthogenates, *see xanthates*
XDI **5:** 3012
Xylan
 as resource for furfural production **5:** 2751
Xylene 8: 4987
 economic aspects **2:** 731; **8:** 5006
 environmental aspects and toxicology **8:** 5013
 handling **8:** 5012
 isomerization **8:** 5002
 occurrence and raw materials **8:** 4994
 properties **8:** 4990
 separation **8:** 4998
 side-chain chlorinated **3:** 1462

specifications **8**: 5010
m-Xylene **2**: 695
  physical properties **8**: 4991
o-Xylene **2**: 695
  for production of phthalimide **7**: 3882
  physical properties **8**: 4991
p-Xylene **2**: 695
  physical properties **8**: 4991
Xylene–air mixture **7**: 3873
o-Xylene ammonoxidation
  to phthalonitrite **7**: 3884
Xylene nitration process **6**: 3520
o-Xylene oxidation
  liquid-phase **7**: 3877
o-Xylene oxidation process **7**: 3873
Xylene production **8**: 4996
  integration into refinery and petrochemical complexes **8**: 5005
Xylenesulfonic acids **2**: 761
Xylenol **3**: 1655, 1691
  amination **8**: 5022
  environmental protection **3**: 1699
  handling and uses **3**: 1698
  hydrodealkylation **3**: 1681
  isolation **3**: 1691
  isomerization **3**: 1696
  production **3**: 1694
  separation **3**: 1693
  specifications **3**: 1697
  toxicology **3**: 1700
2,3-Xylenol
  physical properties **3**: 1692
2,4-Xylenol
  physical properties **3**: 1692
2,5-Xylenol
  physical properties **3**: 1692
2,6-Xylenol
  physical properties **3**: 1692
3,4-Xylenol
  physical properties **3**: 1692
3,5-Xylenol
  physical properties **3**: 1692

2,3-Xylidine **1**: 636
2,4-Xylidine **1**: 636, 639
2,5-Xylidine **1**: 636
2,6-Xylidine **1**: 636, 639
3,4-Xylidine **1**: 636
3,5-Xylidine **1**: 636
Xylidines **8**: 5019
  production **8**: 5021
  properties **8**: 5020
  specifications and handling **8**: 5022
  toxicology **8**: 5025
  uses **8**: 5023
D-Xylose
  fermentation in yeast **4**: 2069
m-Xylylene diisocyanate **5**: 3012

Yeast
  flocculating **4**: 2084
  in ethanol production **4**: 2062
Yeast nutrients **4**: 2063
Yuca
  as raw material for ethanol production **4**: 2098

Ziegler catalysts **8**: 4951
Ziegler–Natta polyolefin process **8**: 4950
Ziegler–Natta polyolefin waxes
  production **8**: 4950
  properties **8**: 4951
Ziegler process **1**: 287
Zinc arachidate **6**: 3243
Zinc dust
  for hydrazobenzene production **2**: 795
Zinc 2-ethylhexanoate **6**: 3243
Zinchydroxymethanesulfinate **7**: 4466
  uses **7**: 4470
Zinc 12-hydroxystearate **6**: 3243
Zinc laurate **6**: 3243
Zinc naphthenate **6**: 3243
Zinc oleate **6**: 3243
Zinc oxidomethanesulfinate **7**: 4466
Zinc soaps **6**: 3242